Phytomedicine

A Treasure of Pharmacologically
Active Products from Plants

Phytomedicine

A Treasure of Pharmacologically Active Products from Plants

Edited by

Rouf Ahmad Bhat
Cluster University Srinagar, Srinagar, India

Khalid Rehman Hakeem
Department of Biological Sciences, Faculty of Science, King Abdulaziz
University, Jeddah, Saudi Arabia

Moonisa Aslam Dervash
Department of Environmental Science, Cluster University Srinagar,
Srinagar, India

ELSEVIER

ACADEMIC PRESS

An imprint of Elsevier

Academic Press is an imprint of Elsevier
125 London Wall, London EC2Y 5AS, United Kingdom
525 B Street, Suite 1650, San Diego, CA 92101, United States
50 Hampshire Street, 5th Floor, Cambridge, MA 02139, United States
The Boulevard, Langford Lane, Kidlington, Oxford OX5 1GB, United Kingdom

Notices
Knowledge and best practice in this field are constantly changing. As new research and experience broaden our understanding, changes in research methods, professional practices, or medical treatment may become necessary.

Practitioners and researchers must always rely on their own experience and knowledge in evaluating and using any information, methods, compounds, or experiments described herein. In using such information or methods they should be mindful of their own safety and the safety of others, including parties for whom they have a professional responsibility.

To the fullest extent of the law, neither the Publisher nor the authors, contributors, or editors, assume any liability for any injury and/or damage to persons or property as a matter of products liability, negligence or otherwise, or from any use or operation of any methods, products, instructions, or ideas contained in the material herein.

Library of Congress Cataloging-in-Publication Data
A catalog record for this book is available from the Library of Congress

British Library Cataloguing-in-Publication Data
A catalogue record for this book is available from the British Library

ISBN 978-0-12-824109-7

For information on all Academic Press publications
visit our website at https://www.elsevier.com/books-and-journals

Publisher: Charlotte Cockle
Editorial Project Manager: Lena Sparks
Production Project Manager: Swapna Srinivasan
Cover Designer: Victoria Pearson

Typeset by SPi Global, India

Working together
to grow libraries in
developing countries

www.elsevier.com • www.bookaid.org

Contents

1. Phytomedicines: Diversity, extraction, and conservation strategies

Sumaira Rashid, Lone Rafiya Majeed, Bisma Nisar, Hina Nisar, Aftab Ahmad Bhat, and Bashir Ahmad Ganai

2. Biotechnological approaches for conservation of medicinal plants

Luis Jesús Castillo-Pérez, Angel Josabad Alonso-Castro, Javier Fortanelli-Martínez, and Candy Carranza-Álvarez

3. Plant biotechnologies for processing raw products in phytomedicines

Monica Butnariu and Alina Butu

4. Phytomedicine and phytonanocomposites—An expanding horizon

Mir Zahoor Gul, Mohd Yasin Bhat, Suresh Velpula,
Karuna Rupula, and Sashidhar Rao Beedu

5. Herbal medicine: Old practice and modern perspectives

Sami Ullah Qadir and Vaseem Raja

9.   Ethnobotanical perspectives in the treatment of communicable and noncommunicable diseases

Sateesh Suthari, Srinivas Kota, Omkar Kanneboyena, Mir Zahoor Gul, and Sadanandam Abbagani

10.   Health benefits of bioactive compounds from microalgae

Dig Vijay Singh, Atul Kumar Upadhyay, Ranjan Singh, and D.P. Singh

11. Some special diets used as neutraceuticals in Unani system of medicine with modern aspects

Huda Nafees, S. Nizamudeen, and Sana Nafees

12. Phytomedicines: Synergistic and antagonistic phytometabolites-drug interactions

Monica Butnariu and Marian Butu

17. Bioprospecting appraisal of Himalayan pindrow fir for pharmacological applications

Rezwana Assad, Zafar Ahmad Reshi, Showkat Hamid Mir,
Irfan Rashid, Yogesh Shouche, and Dhiraj Dhotre

18. Date palm (*Phoenix dactylifera* L.) secondary metabolites: Bioactivity and pharmaceutical potential

Heba I. Mohamed, Hossam S. El-Beltagi, S. Mohan Jain,
and Jameel M. Al-Khayri

19. The inhibitory role of the metabolites of *Moringa oleifera* seeds in cancer cells by apoptosis and cell cycle arrest activation

Ismail Abiola Adebayo, Hasni Arsad, and Mohammed Razip Samian

22. Hepatotoxicity: Its physiological pathways and control measures using phyto-polyphenols

Rajesh Kumar, Raksha Rani, Sanjay Kumar Narang, Seema Rai, and Younis Ahmad Hajam

23. Nanosized delivery systems for plant-derived therapeutic compounds and their synthetic derivative for cancer therapy

Henna Amin, Andleeb Khan, Hafiz A. Makeen, Hina Rashid, Insha Amin, Mubashir Hussain Masoodi, Rehan Khan, Azher Arafah, and Muneeb U. Rehman

24. Potential antioxidative response of bioactive products from *Ganoderma lucidum* and *Podophyllum hexandrum*

Saima Hamid

25. Phytomedicine and the COVID-19 pandemic

Muhammad Irfan Sohail, Ayesha Siddiqui, Natasha Erum, and Muhammad Kamran

26. Phytopharmaceutical marketing: A case study of USPs used for phytomedicine promotion

Sheikh Basharul Islam, Mushtaq Ahmad Darzi, and Suhail Ahmad Bhat

Contributors

Numbers in parenthesis indicate the pages on which the authors' contributions begin.

Sadanandam Abbagani (251), Department of Biotechnology, Kakatiya University, Warangal, Telangana, India

Ismail Abiola Adebayo (533,555), Microbiology and Immunology Department, Faculty of Biomedical Sciences, Kampala International University, Ishaka-Bushenyi, Uganda; Integrative Medicine Cluster, Advanced Medical and Dental Institute, Universiti Sains Malaysia, Penang, Malaysia

Tariq Oluwakunmi Agbabiaka (555), Department of Microbiology, Faculty of Life Sciences, University of Ilorin, Ilorin; Microbiology Unit, Department of Science Laboratory Technology, School of Science and Technology, Federal Polytechnic, Damaturu, Nigeria

Ajaz Ahmad (581), Department of Clinical Pharmacy, College of Pharmacy, King Saud University, Riyadh, Saudi Arabia

Jameel M. Al-Khayri (483), Agricultural Biotechnology Department, College of Agriculture and Food Sciences, King Faisal University, Al-Ahsa, Saudi Arabia

Angel Josabad Alonso-Castro (35,181), Department of Pharmacy, Division of Natural and Exact Sciences, University of Guanajuato, Guanajuato, Mexico

Saeed Alshahrani (581), Department of Pharmacology and Toxicology, College of Pharmacy, Jazan University, Jazan, Saudi Arabia

Bader Mohammed Alshehri (581), Medical Laboratories Department, College of Applied Medical Sciences, Majmaah University, Al Majma'ah, Kingdom of Saudi Arabia

Henna Amin (655), Department of Pharmaceutical Sciences, University of Kashmir, Srinagar, Jammu and Kashmir, India

Insha Amin (655), Division of Veterinary Biochemistry, Faculty of Veterinary Sciences and Animal Husbandry, Sheri Kashmir University of Agricultural Science and Technology (SKUAST-K), Srinagar, Jammu and Kashmir, India

Azher Arafah (581,655), Department of Clinical Pharmacy, College of Pharmacy, King Saud University, Riyadh, Saudi Arabia

Hasni Arsad (533), Integrative Medicine Cluster, Advanced Medical and Dental Institute, Universiti Sains Malaysia, Penang, Malaysia

Rezwana Assad (409,461), Department of Botany, University of Kashmir, Srinagar, Jammu and Kashmir, India

Lubna Azmi (225), Hygia Institute of Pharmaceutical Education & Research; Department of Chemistry, University of Lucknow, Lucknow, India

Sashidhar Rao Beedu (95), Department of Biochemistry, University College of Science, Osmania University, Hyderabad, Telangana, India

Aftab Ahmad Bhat (1), Government Degree College, Beerwah, Budgam, India

Mohd Yasin Bhat (95), Department of Plant Sciences, School of Life Sciences, University of Hyderabad, Hyderabad, Telangana, India

Suhail Ahmad Bhat (709), Department of Management Studies, University of Kashmir, Srinagar, Jammu and Kashmir, India

Monica Butnariu (59,343), Banat's University of Agricultural Sciences and Veterinary Medicine "King Michael I of Romania" from Timisoara, Timis, Romania

Alina Butu (59), National Institute of Research and Development for Biological Sciences, Bucharest, Romania

Marian Butu (343), National Institute of Research and Development for Biological Sciences, Bucharest, Romania

Candy Carranza-Alvarez (35,181), Faculty of Professional Studies, Huasteca Zone and Multidisciplinary Postgraduate Program in Environmental Sciences of the Autonomous University of San Luis Potosí, San Luis Potosí, Mexico

María Luisa Carrillo-Inungaray (181), Faculty of Professional Studies, Huasteca Zone and Multidisciplinary, Autonomous University of San Luis Potosí, San Luis Potosí, Mexico

Luis Jesús Castillo-Pérez (35), Programa Multidisciplinario de Posgrado en Ciencias Ambientales, Universidad Autónoma de San Luis Potosí, San Luis Potosí, Mexico

Mushtaq Ahmad Darzi (709), Department of Management Studies, University of Kashmir, Srinagar, Jammu and Kashmir, India

Dhiraj Dhotre (461), National Centre for Microbial Resource (NCMR), National Centre for Cell Science, Pune, Maharashtra, India

Rocio Del Carmen Díaz-Torres (181), Multidisciplinary Postgraduate Program in Environmental Sciences, Autonomous University of San Luis Potosi, San Luis Potosi, Mexico

Hossam S. El-Beltagi (483), Agricultural Biotechnology Department, College of Agriculture and Food Sciences, King Faisal University, Al-Ahsa, Saudi Arabia; Biochemistry Department, Faculty of Agriculture, Cairo University, Giza, Egypt

Natasha Erum (693), Department of Biochemistry, University of Agriculture, Faisalabad, Pakistan

Madiha Fatima (377), Department of Zoology, The University of Multan, Multan, Pakistan

Javier Fortanelli-Martínez (35), Instituto de Investigación de Zonas Desérticas, Universidad Autónoma de San Luis Potosí, San Luis Potosí, Mexico

Bashir Ahmad Ganai (1,427), Centre of Research for Development/Department of Environmental Science, University of Kashmir, Srinagar, Jammu and Kashmir, India

Mir Zahoor Gul (95,251), Department of Biochemistry, University College of Science, Osmania University, Hyderabad, Telangana, India

Younis Ahmad Hajam (621), Department of Biosciences, Division Zoology, School of Basic and Applied Sciences, Career Point University, Hamirpur, Himachal Pradesh, India

Saima Hamid (677), Centre of Research for Development/Department of Environmental Science, University of Kashmir, Srinagar, India

Takoua Ben Hlel (389), LIP-MB Laboratory LR11ES24, National Institute of Applied Sciences and Technology, University of Carthage, Tunis, Tunisia

Mubashir Hussain Masoodi (655), Department of Pharmaceutical Sciences, University of Kashmir, Srinagar, Jammu and Kashmir, India

Sheikh Basharul Islam (709), Department of Management Studies, University of Kashmir, Srinagar, Jammu and Kashmir, India

Sadaf Jahan (581), Medical Laboratories Department, College of Applied Medical Sciences, Majmaah University, Al Majma'ah, Kingdom of Saudi Arabia

S. Mohan Jain (483), Department of Agricultural Sciences, PL-27, University of Helsinki, Helsinki, Finland

Muhammad Kamran (693), Faculty of Agriculture, Gomal University, Dera Ismail Khan, Pakistan

Omkar Kanneboyena (251), Department of Botany, Kakatiya Government Degree College, Hanamkonda, Telangana, India

Andleeb Khan (581,655), Department of Pharmacology and Toxicology, College of Pharmacy, Jazan University, Jazan, Saudi Arabia

Rehan Khan (655), Institute of Nano Science and Technology, Mohali, Punjab, India

Srinivas Kota (251), Department of Biotechnology, Kakatiya University, Warangal, Telangana, India

Rajesh Kumar (621), Department of Biosciences, Division Zoology, School of Basic and Applied Sciences, Career Point University, Hamirpur, Himachal Pradesh, India

Hafiz A. Makeen (655), Department of Clinical Pharmacy, College of Pharmacy, Jazan University, Jazan, Saudi Arabia

Hina Ahmed Malik (377), Institute of Soil and Environmental Sciences, University of Agriculture, Faisalabad, Pakistan

M. Nejib Marzouki (389), LIP-MB Laboratory LR11ES24, National Institute of Applied Sciences and Technology, University of Carthage, Tunis, Tunisia

Showkat Hamid Mir (409,461), Department of Botany, University of Kashmir, Srinagar, Jammu and Kashmir, India

Juan José Maldonado Miranda (207), Faculty of Professional Studies, Huasteca Zone, Autonomous University of San Luis Potosí, San Luis Potosí, Mexico

Heba I. Mohamed (483), Biological and Geological Science Department, Faculty of Education, Ain Shams University, Cairo, Egypt

Huda Nafees (321), Department of Saidla, Faculty of Unani Medicine, Aligarh Muslim University, Aligarh, India

Sana Nafees (321), AIIMS, New Delhi, India

Sanjay Kumar Narang (621), S.V.G. College Ghumarwin, Bilaspur, Himachal Pradesh, India

Bisma Nisar (1), Department of Environmental Sciences, University of Kashmir, Srinagar, India

Hina Nisar (1), Department of Environmental Sciences, University of Kashmir, Srinagar, India

S. Nizamudeen (321), GUMC, Chennai, India

Sami Ullah Qadir (149), Department of Environmental Science and Engineering, Jamia Millia Islamia, New Delhi, India

Lone Rafiya Majeed (1), Vivekananda Global University, Jaipur, India

Seema Rai (621), Department of Zoology, Guru Ghasidas Vishwavidyalaya, Bilaspur, Chhattisgarh, India

Vaseem Raja (149), Department of Botany, University of Kashmir, Srinagar, Jammu and Kashmir, India

Raksha Rani (621), Department of Biosciences, Division Zoology, School of Basic and Applied Sciences, Career Point University, Hamirpur, Himachal Pradesh, India

Hina Rashid (655), Department of Pharmacology and Toxicology, College of Pharmacy, Jazan University, Jazan, Saudi Arabia

Irfan Rashid (409,461), Department of Botany, University of Kashmir, Srinagar, Jammu and Kashmir, India

Sumaira Rashid (1), Department of Environmental Sciences, University of Kashmir, Srinagar, India

Muneeb U. Rehman (581,655), Department of Clinical Pharmacy, College of Pharmacy, King Saud University, Riyadh, Saudi Arabia

Zafar Ahmad Reshi (409,461), Department of Botany, University of Kashmir, Srinagar, Jammu and Kashmir, India

Umair Riaz (377), Soil and Water Testing Laboratory for Research, Agriculture Department, Government of Punjab, Bahawalpur, Pakistan

Ascensión Rueda Robles (389), Institute of Nutrition and Food Technology "José Mataix", Center of Biomedical Research, University of Granada, Granada, Spain

Karuna Rupula (95), Department of Biochemistry, University College of Science, Osmania University, Hyderabad, Telangana, India

Aga Syed Sameer (581), Department of Basic Medical Sciences and Quality Unit, College of Medicine, King Saud Bin Abdulaziz University for Health Sciences, King Abdullah International Medical Research Centre (KAIMRC), Jeddah, Kingdom of Saudi Arabia

Mohammed Razip Samian (533), School of Biological Sciences, Universiti Sains Malaysia, Penang, Malaysia

Laila Shehzad (377), Sustainable Development Study Center, G.C. University, Lahore, Pakistan

Yogesh Shouche (461), National Centre for Microbial Resource (NCMR), National Centre for Cell Science, Pune, Maharashtra, India

Ila Shukla (225), Harsha Institute of Pharmacy, Itaunja, Lucknow, India

Ayesha Siddiqui (693), Department of Botany, University of Agriculture, Faisalabad, Pakistan

D.P. Singh (291), Department of Environmental Science, Babasahib Bhimrao Ambedkar University, Lucknow, India

Dig Vijay Singh (291), Department of Environmental Science, Babasahib Bhimrao Ambedkar University, Lucknow, India

Ranjan Singh (291), Department of Environmental Science, Babasahib Bhimrao Ambedkar University, Lucknow, India

Bhat Mohd Skinder (427), Centre of Research for Development/Department of Environmental Science, University of Kashmir, Srinagar, Jammu and Kashmir, India

Issam Smaali (389), LIP-MB Laboratory LR11ES24, National Institute of Applied Sciences and Technology, University of Carthage, Tunis, Tunisia

Muhammad Irfan Sohail (693), Institute of Soil and Environmental Sciences, University of Agriculture, Faisalabad, Pakistan

Sateesh Suthari (251), Department of Botany, Vaagdevi Degree & PG College, Hanamkonda, Telangana, India

Atul Kumar Upadhyay (291), Department of Environmental Science, Babasahib Bhimrao Ambedkar University, Lucknow, India

Suresh Velpula (95), Department of Biochemistry, University College of Science, Osmania University, Hyderabad, Telangana, India

Abdul Hamid Wani (427), Department of Botony, University of Kashmir, Srinagar, Jammu and Kashmir, India

Aadil Farooq War (409), Department of Botany, University of Kashmir, Srinagar, Jammu and Kashmir, India

About the editors

Dr. Rouf Ahmad Bhat (PhD) is working in Cluster University Srinagar, Jammu and Kashmir, India, in the capacity of assistant professor and has his specialization in limnology, toxicology, phytochemistry, and phytoremediation. Dr. Bhat has been teaching graduate and postgraduate students of environmental sciences for the past 3 years. He is an author of more than 50 research papers and 30 book chapters and has published more than 15 books with international publishers. He has presented and participated in numerous state, national, and international conferences, seminars, workshops, and symposium. Dr. Bhat has worked as an associate environmental expert in World Bank-funded Flood Recovery Project and also environmental support staff in Asian Development Bank (ADB)-funded development projects. He has received many awards, appreciations, and recognitions for his services to the science of water testing and air and noise analysis. He has served as editorial board member and reviewer of repute international journals. Dr. Bhat is still writing and experimenting with diverse capacities of plants for use in aquatic pollution.

Dr. Khalid Rehman Hakeem (PhD) is a professor at King Abdulaziz University, Jeddah, Saudi Arabia. After completing his doctorate (botany; specialization in plant ecophysiology and molecular biology) from Jamia Hamdard, New Delhi, India, in 2011, he worked as a lecturer at the University of Kashmir, Srinagar, for a short period. Later, he joined Universiti Putra Malaysia, Selangor, Malaysia, and worked there as a postdoctorate fellow in 2012 and fellow researcher (associate professor) from 2013 to 2016. Dr. Hakeem has more than 10 years of teaching and research experience in plant ecophysiology, biotechnology and molecular biology, medicinal plant research, and plant-microbe-soil interactions and in environmental studies. He is the recipient of several fellowships at both national and international levels; also, he has served as the visiting scientist at Jinan University, Guangzhou, China. Currently, he is involved with a number of international research projects with different government organizations. So far, Dr. Hakeem has authored and edited more than 50 books with international publishers, including *Springer Nature*, *Academic Press (Elsevier)*, and *CRC Press*. He also has to his credit more than 100 research publications in peer-reviewed international journals and 60 book chapters in edited volumes with international publishers. At present, Dr. Hakeem serves as an editorial board member and reviewer of several high-impact international scientific journals from *Elsevier*, *Springer Nature*, *Taylor & Francis*, *Cambridge*, and *John Wiley Publishers*. He is included in the advisory board of *Cambridge Scholars Publishing*, United Kingdom. He is also a fellow of Plantae group of the American Society of Plant Biologists; member of the World Academy of Sciences; member of the International Society for Development and Sustainability, Japan; and member of Asian Federation of Biotechnology, Korea. Dr. Hakeem has been listed in Marquis Who's Who in the World since 2014–19. Currently, Dr. Hakeem is engaged in studying the plant processes at ecophysiological and molecular levels.

Dr. Moonisa Aslam Dervash (PhD) is actively involved in teaching graduate and postgraduate students of environmental science for the past year at the Sri Pratap College Campus, Cluster University Srinagar, Jammu and Kashmir, India. She has received a number of awards and certificates of merit. Her specialization is in mesofauna and carbon sequestration. She has published scores of papers in international journals and has more than three books with national and international publishers. She is also the reviewer of various international journals.

Foreword

Prof. Dr. Münir Öztürk (M.Sc., Ph.D., D.Sc., FIAS),
Vice President of the Islamic World Academy of Sciences,
Professor (Emer.) of Ecology & Environmental Sciences,
Ex-Chairman Botany Department and Founder Director Centre for
Environmental Studies, Faculty of Science, Ege University, Izmir, Turkey;
Consultant Fellow, Faculty of Forestry, Universiti Putra Malaysia, Malaysia;
Distinguished Visiting Scientist, ICCBS, Karachi University, Pakistan
http://ege.academia.edu/MunirOzturk

Citations: http://scholar.google.com.pk/citations?user=ooL4g4wAAAAJ&hl=en
Books: http://www.amazon.com/-/e/B00JFW5DS8

Since time immemorial, phytomedicine has been used in healthcare systems at a global scale. A wealth of data is available on the studies undertaken to test the effectiveness of selected plants for the production of plant-based medicines. Today, this research has come to a stage where the value of products derived in this way has reached a value of 100 billion dollars a year in the global market. Chronic diseases have come to the forefront during the past century because of the dire modifications in our dietary lifestyle. The WHO has estimated that the use of conventional medicine is the main source for health needs of 80% population in Africa and Asia. The situation in the industrialized countries too has changed; nearly 40% of the Americans are using complementary and herbal medicines. In the past decade, major attempts have been made to extract bioactive drugs from plants. The understanding of the function of phytochemicals (e.g., polyphenols) has advanced to a very high level for a treatment of particular pathologies. During the early stages of certain diseases, such as cancer, different groups of phytochemicals (e.g., phytoestrogens) have been documented to work preventively against specific disorders.

The phytochemicals have been classified on the basis of their chemical composition and varying roles. Legumes, nuts, olive oil, vegetables, fruits, grape extracts, and tea are the sources of phenolic compounds and flavonoids. These show antioxidant properties with confirmed positive effects against diseases like thrombosis and tumorigenesis. Some epidemiologic studies have reported protective impacts between flavonoids or other phenolics and cardiovascular diseases and cancer, but some studies could not confirm these findings.

Various phytoestrogens are present not only in soy but also in flaxseed oil, whole grains, fruits, and vegetables with antioxidant properties. Several studies have demonstrated favorable effects on CVD risk factors, particularly in animal and cell culture models of cancer. However, as phytoestrogens act as both partial estrogen agonists and antagonists, their effects on cancer seem to be more complex. Hydroxytyrosol, one of many phenolics in olives and olive oil, is a potent antioxidant. Resveratrol, found in nuts and red wine, has antioxidant, antithrombotic, and antiinflammatory features and inhibits carcinogenesis. Lycopene, a potent antioxidant carotenoid in tomatoes and other fruits, is reported to act as a protective agent against prostate and other cancers and inhibits tumor cell growth in animals. Organosulfur compounds in garlic and onions, isothiocyanates in cruciferous vegetables, and monoterpenes in citrus fruits, cherries, and herbs show cardioprotective effects and anticarcinogenic actions as reported in experimental models.

There is a need for much more scientific research before we can start making scientific dietary recommendations. However, it has been sufficiently demonstrated that food sources rich in bioactive compounds are consumed. Throughout the manufacture of modern drugs, the use of medicinal plants as raw materials is that in terms of how they can counter the drug resistance problem in microorganisms. In both industrialized and developing countries, the demand for medicinal plants is rising as it is a leading research field. The problem of bioactivity-safety assessment and conservation of medicinal plants needs to be given greater attention. Conscious efforts need to be made in the design and implementation of such strategies to properly identify, recognize, and enlighten the position of medicinal plants.

Some in vitro screening experiments have been performed on these plants with the ultimate aim of including medicinal plants in the treatment systems of human and animal diseases. The applications are expected to enter the program orally or through some other way. However, these plants must be conserved so that the future source of these plants does not completely disappear. The growing trade in such plants has started producing serious consequences for the survival of such plant species, many of which are facing a threat of extinction.

This book includes twenty-six chapters by authors from around the world. The first chapter is from authors from India and is titled as "Phytomedicines: Diversity, Extraction, and Conservation Strategies," which stresses the fact that phytomedicines have a massive array of biological activities and therefore have been used worldwide since ancient times. The parts of plant used as phytomedicines include leaves, barks, tubers, roots, herbs, and the extracts. All such plant parts contain alkaloids, terpenes, phenolic compounds, basic metabolites, glycosides, and secondary metabolites. The important secondary metabolites possess some specific pharmacological activities for humans.

A researcher from Mexico has written "Biotechnological Approaches for Conservation of Medicinal Plants." The chapter presents evidence on the diversity and mass extraction of medicinal plants around the world, at the same time emphasizing that conservation strategies must be considered for sustainable use of such plants through biotechnological approaches. For example, plant tissue culture can be used for biodiversity conservation.

The third chapter has been prepared by a group of scientists from Romania, which presents information on the plant biotechnologies for processing raw products in phytomedicines. The authors have highlighted the application of such technologies as these offer new possibilities for developments in the treatment of human diseases. They have concluded that the inclusion of new characters contributes to increasing agricultural productivity and improving the quality of phytomedicines, which thus contribute toward the improvement of human health.

In Chapter 4, researchers from India have evaluated the topic of phytomedicine and phytonanocomposites—an expanding horizon in which a combination of phytobioactive compounds with the nanotechnology is presented as a practical and advanced approach to address limitations in the treatments. Drug delivery with lower toxicity and greater efficiency has been made possible with the introduction of nanotechnology. The chapter includes information on synthesizing of nanoparticles via biological entities especially plants that offer a clean, nontoxic, and environment-friendly method of synthesis. It also discusses safe therapeutic nanoparticles with a wide variety of sizes, shapes, and compositions.

Another group of authors has presented data on old practices and modern perspectives in herbal medicine in Chapter 5. They discuss that from prehistoric times until today nearly 53,000 taxa of plants have been exploited in the treatment of human diseases. As per these authors, a controlled accessibility of various agents needs to be considered if plant-based medicinal products are to be developed as chemotherapeutic medicines. For the judicious use of herbal medicines and their products, great care should be kept in mind, and their overexploitation must be controlled.

Chapter 6 focuses on bioactive compounds obtained from plants and their pharmacological applications and encapsulation. It discusses encapsulation methods along with the perspectives of phytomedicines with respect to useful strategies to preserve chemical composition and improve the pharmacological effects together with the bioavailability of bioactive compounds.

Researchers from Mexico have pooled up information on the medicinal plants and their traditional uses in Chapter 7. They stress on the fact that the large number of plant species with medicinal properties has led to the emergence of a branch called medicinal herbalism. This has allowed a wide development of traditional medicine in some countries. The traditional uses attributed to each medicinal plant depend on where the plant is consumed and the social groups using them.

An outline about the use of plants for diagnosis and treatment of ailments present in Chapter 8 is titled "Perspectives of phytotherapeutics: Diagnosis and cure" by the authors from India. Furthermore, they stress on phytotherapy as an alternative treatment when classical therapy is not giving satisfactory results.

Chapter 9 is titled "Ethnobotanical Perspectives in the Treatment of Communicable and Noncommunicable Diseases." The chapter presents perspectives of indigenous knowledge on botanicals in the treatment of infectious and noninfectious diseases. The authors stress on not only locally available plants but also modes of administration of plant drugs to cure various ailments for a healthy society. The chapter is intended to help policymakers appreciate the complexity of various diseases and disorders and assess the impact on health, while also allowing phytoanalysts, pharmacognosists, and other potential researchers to discover new drugs for a variety of diseases from natural products.

Chapter 10 deals with the health benefits of bioactive compounds from microalgae. It presents information about the huge biodiversity and variable composition of microalgae that can enhance the production of variable compounds and increase their availability commercially. Microalgae are the viable feedstock for production of bioactive compounds that can revolutionize the pharmaceutical, cosmetic, and food industries. The authors have focused on bioactive compounds produced by several microalgal species and their role in improving human health. The latter degrades with increasing living standards and exponential population growth.

Investigators have pooled up data on some special diets used as nutraceuticals in Unani medicine in Chapter 11. They have summarized data on nutraceuticals that show a growing interest in health benefits and provide options for preventing various lifestyle diseases, due to potential nutritional, safety, and therapeutic effects. They argued that there is an urge to develop nutraceuticals especially from herbs that are effective on difficult-to-cure disorders related to oxidative stress and to different diseases.

The authors present valuable information in Chapter 12 on synergistic and antagonistic interactions between phytometabolites and drugs. The authors have reviewed the data in two parts based on a study of the production mode and factors favoring the production of phytometabolite-drug interactions, comprising general aspects and the changes produced by medication. They stress the nutritional status and the effect phytometabolites and their compounds produce in drug metabolism, together with the interaction of grapefruit juice with certain classes of drugs to avoid accidents caused by active medication in an inappropriate dietary context, and the part that includes the mechanism of producing the interaction between grapefruit juice and diclofenac (DCF) based on the pharmacokinetic parameters determined experimentally.

In Chapter 13, the authors find information on the role of genetically modified plant repository in biopharmaceutical industries. The chapter emphasizes that biopharmaceuticals have conventionally been expressed in various transgenic systems, for example, cultures of mammalian, bacterial, and fungal cell lines. The authors discuss that in the near future, requirements for existing and new biotherapeutic products obtained via genomics will rise significantly.

Spanish scientists have pooled up information on the role of phytomedicines in metabolic disorders in Chapter 14. The chapter aims to give an overview of the ancient use of traditional medicine in the treatment of metabolic diseases and discusses the findings obtained from in vitro, preclinical, and clinical trials in an attempt to provide a realistic perspective on the contemporary medical role of medicinal plants and their phytochemicals in the field of metabolic disorders.

The aim of authors of Chapter 15 titled as "*Artemisia amygdalina* Decne: A Rich Repository of Pharmacologically Vital Phytoconstituents" has been to present the therapeutically active phytoconstituents of plant species.

The results related to the "bioprospecting of endophytic fungi for antibacterial and antifungal activities" are presented in the title of Chapter 16. The authors report that more than 35% endophytes are isolated from medicinal plants and more than 80% endophytic fungi produce biologically active compounds. They also report that secondary metabolites and medicinal plants contribute above 80% of natural drugs for pharmaceutical industries and markets.

The authors of Chapter 17 have evaluated the bioprospecting appraisal of Himalayan pindrow fir for pharmacological applications. They mention that the Himalayan pindrow fir possesses a multitude of therapeutic effects accredited to its valuable phytoconstituents.

Chapter 18, titled "Date Palm (*Phoenix dactylifera* L.) Secondary Metabolites: Bioactivity and Pharmaceutical Potential," discusses the medicinal and nutritional value of dates, with emphasis on secondary metabolites and various factors influencing their biosynthesis, pharmaceutical benefits, and potential use in functional food.

The seeds of *Moringa oleifera* are rich in phytometabolites, with an ability to activate apoptosis and induce arrest of cell cycle in cancer cells. Their activity results in the death of cancer cell as written by the authors from Uganda in Chapter 19. These seeds are mentioned as promising potent anticancer agents, which need to be further explored especially in clinical research.

In Chapter 20, researchers from Nigeria discuss the "medicinal properties of *Olax subscorpioidea*." They have highlighted, summarized, and identified gaps in studies reporting anticancer, antiinflammatory, antimicrobial, and antinociceptive activities and physiological modulatory effect. They have critically appraised the information on mechanism of action and safety. They say that there is a need to clearly elucidate the antimicrobial mechanism of action of different extracts of *Olax subscorpioidea.*

Chapter 21 is entitled "Phytotherapeutic Agents for Neurodegenerative Disorder: A Neuropharmacological Review." The authors report that the properties of phytochemicals can reduce the oxidative stress of nerve cells. They also report the antiapoptotic effects and restoration of neuronal cell death via agonists of neurotrophic factors inducing neuroprotection.

The title "Hepatotoxicity: Physiological Pathways and Control Measures Using Plant Metabolites" summarizes the data in Chapter 22. The chapter discusses the toxification of liver and hepatotoxins elucidating the hepatoprotective role of medicinal plants to modulate the detoxifying functions of the liver.

Various classes of anticancer drugs obtained naturally including paclitaxel, camptothecin, vincristine, and vinblastine are discussed in Chapter 23. In addition, other phytometabolites like curcumin, gingerol, resveratrol, and quercetin are reported to have well-established pharmacological effects as antineoplastic agents.

Information on nanosized delivery systems for plant-derived therapeutic compounds and their synthetic derivative for cancer therapy is given in Chapter 24. It is entitled "Potential Antioxidative Response of Bioactive Products from *Ganoderma lucidum* and *Podophyllum hexandrum.*" The chapter discusses the availability for such bioactive compounds, found in good proportion in the medicinal mushroom *G. lucidum* and the herb *P. hexandrum.* The synthesis of potential antioxidants such as extracellular polysaccharides from *G. lucidum* and podophyllotoxin from *P. hexandrum* is mentioned to be able to serve as the base for supplement industry to produce exogenous antioxidants for the consumption by humans.

Chapter 25, titled "Phytomedicine and the COVID-19 Pandemic," is an interesting chapter of this book. The authors stress that antiviral phytomedicines have already been used in the past two coronavirus outbreaks, that is, SARS-CoV and MERS-CoV. Some species such as *Lycoris radiata, Artemisia annua, Lindera aggregata, Isatis indigotica, Torreya nucifera,* and *Houttuynia cordata* have already proven their efficacies against such ailments.

Chapter 26 is entitled "Phytopharmaceutical Marketing: A Case Study of USPs Used for Phytomedicine Promotion." Researchers cover the belief of safety and efficacy among people, passed down through generations and strengthened by experience. Consequently, it has aided herbal pharmaceutical companies in securing conviction of people on herbal healthcare through misleading and persuasive marketing efforts, besides being merely informational. Propositions like "free from side effects," "efficacy," and "treatment of difficult-to-treat diseases" need to be verified through scientific studies.

This book is unique and would be an ideal source of scientific information to the research scholars, faculty, and scientists involved in human health systems, molecular biology, biochemistry, biotechnology, and food technology.

I greatly appreciate the efforts spent by the editors for successfully bringing together this impressive volume.

Preface

The domain of phytomedicine entwines the treasure of curative properties concentrated in plant metabolites (pharmacognosy) and their diversity, exploitation, conservation strategies, and practical applicability in the field of medical sciences (phytotherapy). Basically, phytomedicines have been used since time immemorial in the form of herbal magic potions, which were potent concoctions in curing the diseases much before the inception and advancement of technological revolution (medical diagnostic equipments). In spite of allopathic formulations in the present-day medical arena, phytomedicines are comparatively safer without malicious side effects. Phytometabolite research is on the forefront as far as the novel interventions in the medicinal sciences are concerned. In the past, phytomedicines were used and recommended on account of consumption experiences through generations. And such knowledge that was based on common experiences was transmitted to generations from ancestors and occupies a prominent position as far as traditional medical system is concerned. Due to various havocs of allopathic compositions, research and application of phytomedicine is being boosted at international level together with advent of novel technologies, such as biotechnological and nanotechnological interventions. Responding to increased demands of society, safety, toxicity, optimum dosage, phytometabolite-drug reactions (synergistic and antagonistic interactions), and quality control, phytotherapy is the prime need of the hour to have thorough insight in the concerned field.

The present book is a conglomerate of 26 comprehensive chapters based on exploration, research, and exploitation of phytometabolites on mass scale coupled with the efficacy, performance, and applicability on target organisms. The book is aimed to have in-depth information on phytopharmaceuticals extracted from plants and microalgae, endophytic fungi, and genetically modified plants, focused to deal with increased susceptibility of curable and fatal diseases. The readers will find a coherent package of phytotherapeutic information regarding inclusive assortment of research-based, scientific amplitude of metabolites from plant world encompassing various action plans such as special nutraceuticals used in Unani medicine with modern aspect, anticancerous properties, phytomedicines as hepatotonics, role of phytomedicines in metabolic disorders, active principle of some important medicinal plants (*Artemisia amygdalina*, *Abies pindrow*, *Phoenix dactylifera*, *Moringa oleifera*, *Olax subscorpioidea*, *Ganoderma lucidum*, and *Podophyllum hexandrum*) explored by the connoisseurs across the world. Information is presented sequentially regarding phytochemistry, biological activity, and the serviceable aspects of bioactive compounds. A nanotechnological intervention in the field of cancer drug delivery to target cells has been also discussed. The present volume of book also addresses various advancements and achievements of novel drugs from the phytoworld using molecular, enzymatic activities and various technological tools in an ecofriendly fashion vis-à-vis phytopharmaceutical marketing.

Academicians, researchers, and students shall find it a comprehensible volume about phytomedicines, and their diversity, exploration, exploitation, and management strategies thus will adequately suffice the requirements of training, teaching, and research.

We are extremely grateful to the authors who have contributed chapters in this book. We express our thanks to Elsevier for their cooperation and publication of this book.

Rouf Ahmad Bhat
Khalid Rehman Hakeem
Moonisa Aslam Dervash

About the book

This publication is a coherent conglomerate of diverse chapter titles significantly linked with the promising potential of phytomedicines for treatment of various diseases. The present book is conglomerate of 26 comprehensive chapters based on exploration, research, and exploitation of phytometabolites on mass scale coupled with the efficacy, performance, and applicability on target organisms. The book is aimed to have thorough information on phytopharmaceuticals extracted from plants and microalgae, endophytic fungi, and genetically modified plants, focused to deal with increased susceptibility of curable and fatal diseases. The readers will find a coherent package of phytotherapeutic information regarding inclusive assortment of research-based, scientific amplitude of metabolites from plant world vis-à-vis intervention of biotechnology and nanotechnology in exploitation of phytometabolites. Moreover, some important medicinal plants and their active principles have been vividly highlighted. Academicians, researchers, and students will find it as comprehensive bind about phytomedicines and phytometabolites for sustainable progression of mankind and sufficiently meet the prime necessities of training, teaching, and research.

Chapter 1

Phytomedicines: Diversity, extraction, and conservation strategies

Sumaira Rashid[a], Lone Rafiya Majeed[b], Bisma Nisar[a], Hina Nisar[a], Aftab Ahmad Bhat[c], and Bashir Ahmad Ganai[d]

[a]*Department of Environmental Sciences, University of Kashmir, Srinagar, India,* [b]*Vivekananda Global University, Jaipur, India,* [c]*Government Degree College, Beerwah, Budgam, India,* [d]*Centre of Research for Development/Department of Environmental Science, University of Kashmir, Srinagar, Jammu and Kashmir, India*

1. Introduction

Compounds that exhibit some therapeutic properties, phytopharmaceutical preparations, or herbal medicines are defined as medications obtained from the plants either whole plant or plant parts separately along with contrived in a rudimentary form or as a refined pharmacological preparation (Srivastava, Srivastava, Pandey, Khanna, & Pant, 2019). The occurrence of robust chemical defense system in these florae continuously recollects and regenerates the concentration of fervent scientists enthusiastic to discover novel medicines. The presence of nonnutrient but bioactive compounds in plants is referred as phytochemicals (from Greek word "Phyto" meaning "plant") or Phyto constituents being accountable in protection of plant from pathogenic contaminations or pest invasions (Doughari, Human, Bennade, & Ndakidemi, 2009). Medically active phytochemicals of plant parts are labeled as "secondary metabolites" by A. Kossel in 1891; he by then reported such bioactive components as parenthetically occurring chemicals in plants with no paramount importance to plant body itself. Majority of these compounds don't rightly contribute in development, progress, and reproduction of plants so named as "secondary metabolites" (Ahmed et al., 2017). The secondary metabolites possess an assortment of bioactive compounds chosen and propagated naturally to be utilized as a therapy in treating various health ailments and infections of humans (Wink, 2015). Plants are capable of producing a wide range of chemical compounds that possess various biological functions and uphold defense against pests, microbes, and herbivorous. So far, more than 12,000 such compounds have been isolated that account less than 10% of the total. Reports collected everywhere throughout the

Phytomedicine: A Treasure of Pharmacologically Active Products from Plants
https://doi.org/10.1016/B978-0-12-824109-7.00009-1

globe specify that approximately 35,000 plant species are presently being utilized in herbal treatments, whereas as per the available research data, just 20% of the total undergoes the phase of phytochemical investigation. At the same time, 10% make it to the biological screening stage. Utilizing today's modern technologies will help in exploring the remaining medicinal plants to some extent. Chemical compounds of both plant and synthetic origin arbitrate their impacts in human body to somewhat analogous ways. Thus phytomedicines do not vary from synthetic medications as far as their working is concerned, so phytomedicines are as viable as synthetic drugs in treatments while having equivalent probability of producing harmful adverse effects (Shah & Pandey, 2017). Phytomedicines have a massive range of biological activities thus utilized globally ever since the prehistoric times. However, various obstructions like chemical and biological barriers, insolvability, hydrophobicity, less take up rate, and soaring noxiousness hamper the applicability of these herbal drugs (Sajid, Cameotra, Khan, & Ahmad, 2019). For this purpose, drug-delivery skills encompass engrossed massive consideration. Innovative drug-delivery systems know how to proficiently carry herbal drugs with the capability of growing the therapeutic index and bioavailability of herbal drugs. Also, nanoparticle-based drug-delivery systems improve conveyance of phytomedicines to their particular targets (Khan & Ahmad, 2019). The parts of plant utilized for phytomedicines are leaves, barks, tubers, roots, herbs, and the plant extracts. These plant parts secrete substances in the form of secondary metabolites as terpenes, glycosides, phenolic compounds, alkaloids, and basic metabolites, and preparations created from herbal plants comprise elixirs, latex, balms, ointments, apozems, powders, and electroactives (Shah & Pandey, 2017). The future of plant-derived medicines is likely to have marvelous opportunity intended to discover several novel as well as innovative remedial strategies and products (Khan, 2015). Nowadays, two different types of herb-derived medicine are recognized. Modern herbalism as practiced in many Western countries is combined with conventional medicine. And the traditional systems include Chinese herbal medicine, which is based on the impressions of yin and yang and energy. In Japan plant–based medicine evolved into kampo, and in India plant-based medicine evolved as Ayurveda, which basically uses various herbal mixes. Phytomedicine is largely classified into a few basic systems as Ayurvedic herbalism, African herbalism, Chinese herbalism, and Western herbalism (Wickrama Singhe, 2006).

2. Prepreparation of plant samples for extraction of phytochemicals

Preparing plant samples for preserving biomolecules before extraction comprise the preliminary stage for contemplating therapeutic plants. Dried or fresh roots, flowers, and barks constitute the plant sample. As a rule, dried samples are favored at most times owing to the time required for designing experiments, so other processes like grinding and drying are also done to preserve the phytochemicals in the final extract (Azwanida, 2015). The time between in-gathering and experiment should be 3 hours at maximum to maintain the fresh-ness of fragile samples, which are predisposed to depreciate more rapidly than desiccated samples (Sulaiman, Sajak, Ooi, Supriatno, & Seow, 2011). Fresh and dried samples of *Moringa oleifera* leaf exhibited no disparity in phenolics comparatively; however, the content of flavonoids in dried samples showed up-per limits (Vongsak, Sithisarn, Mangmool, Thongpraditchote, & Wongkrajang, 2013). Proliferation of surface contact between plant sample and extraction menstruum is achieved by decreasing particle size. Grinding brings about coarse particle size; in the meantime, powdered sample results in increasingly homogenized sample with smaller particle size prompting enhanced surface contact with extraction menstruum. This significant prepreparation is important concerning effective extraction to happen; the menstruum must make contact with particular sample, and particle size lesser than 0.5 mm is considered nearly perfect for efficient extraction (Thermo Fisher Scientific, 2013).

Various methods employed for sample drying are air drying, lyophilization, freeze-drying, oven drying, and microwave drying.

2.1 Air drying

It can take from few days to several months for drying sample depending upon the plant material. Air drying at ambient temperature is ideal method as heat-liable compounds get preserved, but it takes long time when compared with microwave or oven drying combined with other drawbacks like susceptibility of contamination over fluctuating temperatures.

2.2 Microwave drying

Electromagnetic radiation used in microwave drying has both magnetic and electric fields. Through dipolar rotation, heating is caused by electric field simultaneously; electric field of the molecules in plant sample or menstruum having permanent or induced dipole moment gets aligned, and oscillation of molecules is produced by ionic induction (Kaufmann & Christen, 2002). The instability brings about minor collisions among molecules causing rapid heat-ing of samples at the same time. Although this process abbreviates drying time, squalling of phytochemicals may occur at occasions.

2.3 Oven drying

To eliminate moisture content out of plant samples, thermal energy is utilized in this method of preextraction. Being easiest and speedy heat induction pro-cess, it aims to preserve phytoconstituents. Oven drying for 4 hours at 44.5°C utilizing 80% methanol brought about highest antioxidant activity in *Cosmos caudatus*, and parallel results have been achieved while carrying out the same for 4.05 hours at 44.12°C (Mediani, Khatib, & Tan, 2013). Nonetheless, antioxidant activity of *Orthosiphon stamineus* after being oven dried revealed no significant effect, but sinesetin and rosmarinic acid phytoconstituent content was impacted signifying the compound sensitivity to temperature (Abdullah, Shaari, & Azimi, 2012).

2.4 Freeze-drying

Sublimation process governs this preextraction strategy. To solidify the moisture content present in plant extract, sample is frozen at −80°C to −20°C. After prolonged freezing, sample is instantaneously freeze-dried to avoid melting of frozen liquid. To prevent sample loss during process, mouth of container carrying sample is wrapped with parafilm. Freeze-drying produces higher phenolic levels paralleled to air drying aiming to preserve most phytochemicals. Meanwhile, it is a diverse and affluent process of drying in comparison with air and microwave drying. Heat-labile phytochemicals of high value are dried by this method only.

3. Extraction methods for phytochemicals

Standard procedures of separating the therapeutically active portions from medicinal plants via selective solvents are called as extraction with the purpose of separating soluble metabolites leaving behind cellular residue (Handa, Khanuja, Longo, & Rakesh, 2008). Crude extract used in this method contains composite blend of plant metabolites like glycosides, phenolics, alkaloids, flavonoids, and terpenoids. Several primarily acquired extracts may be all set for use as therapeutic agents in the form of infusions and fluid extracts; however, some require more dispensation. Some of the usually used extraction methods are discussed in the succeeding text.

3.1 Maceration

Maceration is winemaking technique and has been espoused in medicinal plant research. In maceration, whole or boorishly powdered plant material is kept in contact with the solvent in a stoppered container for a given period with repeated agitation until soluble matter is dissolved at room temperature for a period of 3 days (Handa et al., 2008). The process has the purpose to soften and break the plant's cell wall to release the soluble phytochemicals. The mixture then is strained, the marc (the damp solid material) is pressed, and the liquids are clarified by filtration or decantation after standing. This method suits best for thermolabile drugs (Pandey & Tripathi, 2014).

3.2 Decoction

Unlike infusion and maceration, compounds soluble in oil and water, heat-stable compounds, and compounds obtained from roots and barks (hard material) are produced by boiling crude drug with specified volume usually a ratio of 1:4 or 1:16 for about 15 minutes that is allowed to cool, strain, and pass enough cold water through the mixture to produce a concentrated extract, which is then filtered or used as such. For example, quath or Kawath, an Ayurvedic preparation, is produced by this process (Pandey & Tripathi, 2014).

3.3   Percolation

A narrow cone-shaped vessel with open endings at both sides called percolate is used in the process of percolation. Percolator is filled with ground dried sample to which boiling water/solvent is added to carry out maceration. A moderate speed of six drops for each minute is maintained until the active constituents get fully extracted and is usually used to extract biologically active constituents for preparation of fluid extracts and tinctures (Rathi, Bodhankar, & Baheti, 2006). The solid ingredients are soaked with a proper proportion of defined menstruum and permitted to run for roughly 4 hours in an all-around closed compartment after which the mass is stuffed and the upper ending of percolator is shut. Extra solvent is added to create a thin layer over the packed material, and the blend is permitted to macerate. The lower ending of the instrument is then opened, and fluid contained in that is permitted to dribble gradually. Extra solvent is poured in the percolator as needed, until the permeate quantifies around 75% of the necessary volume of the finished product. Blended fluid is clarified by filtration or by standing followed by emptying.

3.4   Infusion

Easily soluble constituents present in fine ground dried material when added with cold or boiling water for a limited time period result in softening and leaching of compounds from the material in menstruum producing a dilute solution commonly known as infusion (Bimakr, 2010).

3.5   Digestion

It is a sort of maceration where mild warmth enhances the process of extraction. This process is put into use when tolerably raised temperatures are acceptable and the dissolvable effectiveness of the menstruum gets enhanced by this process (Pandey & Tripathi, 2014).

3.6   Soxhlet extraction or hot continuous extraction

Finely ground plant material or crude drug is placed in a thimble chamber of Soxhlet assembly, which contains a porous cellulose or strong filter paper. Menstruum is poured in heating flask portion of assembly whereby it gets heated by a heating mantle, condensing the vapors in condenser, and dripping back into thimble filled with crude drug and coming in close contact with material.

After the solvent level in chamber rises atop siphon tube, the liquid contents or phytoconstituents get siphoned from siphon tube into heating flask. The process continues until solvent in siphon arm becomes colorless. Advantageously, smaller volumes of solvent are used to extract large amounts of drug. So it can be said that many benefits are reaped as far as time, energy, and finances are considered. It is utilized as a group procedure at small scale; however, it turns out to be substantially more conservative and practical when changed over into persistent extraction process on a medium or large scale (Sutar, Garai, Sharma, & Sharma, 2010).

3.7   Serial exhaustive extraction

It is another regular process of extraction involving progressive extraction with solvent of increasing polarity (hexane to methanol) to guarantee that a large range of compounds soluble in different polarity range of solvents could be obtained. Organic solvents are utilized by few researchers while using Soxhlet extraction of shade dried ground material. However, this process cannot be put into use for extraction of heat-sensitive compounds as protracted heating might cause compound degradation (Das, Tiwari, & Shrivastava, 2010).

3.8   Aqueous alcoholic extraction by fermentation

Asava and Arista are some Ayurvedic preparations where crude drug in powdered form (or decoction) is soaked for a limited time period wherein fermentation process sets up and alcohol generation takes place in situ prompting the extraction of active phytoconstituents from plant material. Although this Ayurvedic method yet lacks standardization, but with the progressive improvement in technology of fermentation, the day is not far when this process will be standardized for large-scale production of phytomedicine extracts or crude drugs. Meanwhile, alcohol generated in situ acts as a preservative itself. In small-scale manufacturing of crude drugs, earthen vessel is used for this process, but porcelain jars, metal vessels, wooden vats, etc. replace the earthen vessels in large-scale manufacturing units. Some well-known examples of such drug preparations include Dashmularista, Karpurasava, and Kanakasava.

3.9   Countercurrent extraction

Dampened raw material in the form of fine slurry minced by jagged disc disintegrators is moved one way inside a round and hollow extractor where it interacts with solvent of extraction. The extract becomes more concentrated as long as the raw material moves farther. Complete extraction is conceivable when the amounts of raw material and menstruum along with their flow rates are streamlined. The procedure is profoundly proficient, entailing brief period and representing no hazard from elevated temperatures. At last, adequately concentrated extract comes out toward one side of the extractor, while the marc drops out from the opposite end. Considerable benefits have been offered by this process of extraction.

- Unlike percolation, maceration, and decoction, this method extracts a unit quantity of raw material with very small quantities of menstruum.
- As room temperature is employed for carrying out this process, heat-sensitive compounds get relief from undue exposures to elevated temperatures.
- Heat produced throughout pulverization of crude drug gets nullified when carried out under wet conditions. This again provides a relief to heat-labile components from heat exposure.
- This method has been found to be more economical and efficient when compared with continuous hot extraction.

3.10 Microwave-assisted extraction (MAE)

It makes use of energy of microwaves to aid partition of analytes from the sample matrix into solvent. Microwave radiation interacts with dipoles of polar along with polarizable materials (e.g., solvents and sample) causes heating near surface of the materials and heat is transferred by conduction. Dipole rotation of molecules induced by microwave electromagnetic interrupts hydrogen bonding, enhancing the migration of dissolved ions while promoting solvent penetration into the matrix (Kaufmann & Christen, 2002; Patil & Shettigar, 2010). In non-polar solvents, poor heating occurs as the energy is transferred by dielectric absorption only (Handa et al., 2008). MAE can be considered as selective method that supports polar molecules and solvents with high dielectric constant.

Extraction solvent	Relative permittivity (20°C)
Hexane	1.89
Toluene	2.4
Dichloromethane	8.9
Acetone	20.7
Ethanol	24.3
Methanol	32.6
Water	78.5

3.11 Ultrasound-assisted extraction or sonication extraction

It entails application of ultrasound ranging from 20 to 2000 kHz (Handa et al., 2008). Acoustic cavitation from ultrasound produces mechanic effect that increases the surface contact between solvents and sample sand permeability of cell walls. Physical and chemical properties of the materials subjected to ultrasound are altered and disrupt the plant cell wall, facilitating release of compounds and enhancing mass transport of the solvents into the plant cells (Dhanani, Shah, Gajbhiye, & Kumar, 2013). The procedure is simple and relatively low-cost technology that can be used in both small and larger scale of phytochemical extraction. Although the process is useful in some cases, like extraction of Rauwolfia root, its large-scale application is limited due to the higher costs. One disadvantage of the procedure is the occasional but known deleterious effect of ultrasound energy (more than 20 kHz) on the active constituents of medicinal plants through formation of free radicals and consequently undesirable changes in the drug molecules (Azwanida, 2015; Pandey & Tripathi, 2014).

3.12 Accelerated solvent extraction (ASE)

It is a resourceful process of liquid solvent extraction compared with maceration and Soxhlet extraction as the method uses minimal amount of solvent. Sample is packed with inert material such as sand in the stainless steel extraction cell to prevent sample from aggregating and blocking the system tubing (Rahmalia, Fabre, & Mouloungui, 2015; Thermo Fisher Scientific, 2013). Packed ASE cell includes layers of sand-sample mixture in between cellulose filter paper and sand layers. This automated extraction technology is able to control temperature and pressure for each individual samples and requires less than an hour for extraction. Similar to other solvent technique, ASE also critically depend on the solvent types. Cyclohexane acetone solution at the ratio of 6:4 v/v with 5-minute heating (50°C) showed to yield highest bixin from *Bixa orellana* with 68.16% purity (Rahmalia et al., 2015). High recoveries (~94%) of flavonoids from *Rheum palmatum* were observed using 80% aqueous methanol by ASE, suggesting the suitability of this method for quality control evaluation (Azwanida, 2015; Tan, Jiang, & Hu, 2014).

3.13 Supercritical fluid extraction (SFE)

Also known by the name dense gas or supercritical fluid (SF) is a substance that shares the physical properties of both gas and liquid at its critical point (Tanase, Arca, & Muntean, 2019). Factors such as temperature and pressure are the determinants that push a substance into its critical region. SF behaves more like a gas but has the solvating characteristic of a liquid. An example of SF is CO_2 that become SF at above 31.1°C and 7380 kPa. Interestingly, supercritical CO_2 (SC-CO_2) extraction due to excellent solvent for nonpolar analytes and CO_2 is readily available at low cost and has low toxicity. Even though SC-CO_2 has poor solubility for polar compounds, modification such as adding small amount of ethanol and methanol enables it to extract polar compounds. SC-CO_2 also produces analytes at concentrated form as CO_2 vaporizes at ambient temperature. SC solvent strength can be easily altered by changing the temperature and pressure or by adding modifiers that lead to reduce extraction time. Optimization of SC-CO_2 on *Wedelia calendulacea* achieved its optimum yield at 25 MPa, 25°C temperature, 10% modifier concentration, and 90-minute extraction time (Patil, Sachin, Wakte, & Shinde, 2013; Patil & Shettigar, 2010). The various methods for extraction of plant metabolites are depicted in Table 1.

TABLE 1 Various methods for extraction of different phytometabolites.

Extraction method	Plant	Solvent	References
Classic water bath extraction	*Abies alba* Mill	Ethyl acetate	Benkovic et al. (2014)
	Acacia cornigera (L.) Willd.	Petroleum ether, chloroform, methanol	Maldini et al. (2009)
	Eucalyptus globulus Labill	Ethanol:water 80:20	Baptista, Pinto, Mota, Loureiro, and Rodrigues (2015)
	Guazuma ulmifolia Lam.	Petroleum ether, chloroform, methanol	Maldini et al. (2013)
	Liriodendron tulipifera L.	0.1-M oxalic acid	Um, Shin, and Lee (2017)
	Shorea roxburghii D.Don	Acetone:methanol	Subramanian, Raj, Manigandan, and Elangovan (2015)
	Punica granatum L.	Methanol	Tantray, Akbar, Khan, Tariq, and Shawl (2009)
Extraction by maceration	*Acanthopanax leucorrhizus* (Oliv.) Harms	90% ethanol	Hu et al. (2018)
	Anacardium occidentale L.	Water	Encarnaçao et al. (2016)
	Byrsonima crassifolia (L.) Kunth	Petroleum ether, chloroform:methanol	Maldini et al. (2009)
	Coutarea hexandra (Jacq.) K. Schum	95% ethanol	Nunes et al. (2012)
	Drypetes klainei Pierre ex Pax	Water	Brusotti et al. (2015)
	Lafoensia pacari A.St.-Hil	Absolute ethanol	Tamashiro Filho et al. (2012)
	Schinopsis brasiliensis Engl.	90% ethanol	Santos et al. (2012)

Continued

TABLE 1 Various methods for extraction of different phytometabolites—cont'd

Extraction method	Plant	Solvent	References
Soxhlet extraction	*Anogeissus leiocarpa* DC.	Ethanol	Salih et al. (2017)
	Caraipa densifolia Mart.	Hexane:methanol	Da Silveira et al. (2010)
	Eucalyptus grandis W.Hill ex. Maiden	Dichloroethane	Santos et al. (2012)
	Ficus talboti King.	Methanol	Arunachalam and Parimelazhagan (2014)
	Goniothalamus velutinus Airy Shaw	Absolute methanol	Iqbal, Salim, and Lim (2015)
	Rhus verniciflua (Stokes) F. Barkley	Ethanol	Bouras et al. (2015)
Ultrasound-assisted extraction	*Betula papyrifera* Marshall	Ethanol:water 80:20	Chen et al. (2009)
	Cassia auriculata (L.) Roxb.	Water	Sivakumar, Ilanhtiraiyan, Ilayaraja, Ashly, and Hariharan (2014)
	Chloroxylon swietenia DC.	Methanol	Enkhtaivan et al. (2015)
	Jatropha dioica Sesse	Ethanol	Paz, Márquez, Ávila, Cerda, and Aguilar (2015)
	Prunus domestica L.	7 ethanol and HCl 1%, 2,6-di-tert-butyl-4-methylphenol (BHT)	Usenik, Stampar, and Veberic (2009)
	Strychnos nux-vomica Dennst.	Methanol	Enkhtaivan et al. (2015)
	Turnera diffusa Willd	Ethanol	Paz et al. (2015)

Microwave-assisted extraction	Fagus sylvatica L.	Water, methanol:water, and ethanol:water	Hofmann, Nebehaj, Stefanovits-Banyai, and Albert (2015)
	Pinus pinaster Aiton	Ethanol:water 80: 20	Chupin et al. (2015)
	Quercus robur L.	Hydroalcoholic solution of methanol and ethanol	Bouras et al. (2015)
	Terminalia arjuna Wight and Arn	Distilled water	Yallappa et al. (2013)
	Ziziphus jujuba Mill.	Methanol	Dubey and Goel (2013)
Supercritical fluid extraction	Hymenaea courbaril L.	Carbon dioxide and water (9:1)	Veggi, Prado, Bataglion, Eberlin, and Meireles (2014)

4. Diversity of phytochemicals

Plants utilize sunlight to manufacture certain important biochemicals; they include a huge number of medicinally active compounds that are extracted from them. These compounds can be categorized into primary and secondary compounds. The normal plant growth and development is attributed to the primary compounds (Applezweig, 1980). The secondary compounds are derived biosynthetically from primary compounds. They help in the maintenance of plant and also are involved in the ecological interactions of plants with their surroundings. Primary compounds are most abundant in plants than secondary compounds. Generally speaking the natural compounds are derived from plants, and they form the basis for modern drug industry (Lahlou, 2007). The stereo structures with many chiral centers make the secondary compounds worth for the making of high cost synthetic and semisynthetic therapeutic compounds. These properties are lacking in the primary compounds; therefore they cannot be utilized for the synthesis of therapeutic agents (Atanasov et al., 2015; Moses, Pollier, Thevelein, & Goossens, 2013). The secondary compounds are synthesized at a particular developmental stage in a particular type of plant cells. The significant secondary metabolites that are found in higher plants are alkaloids, glycosides, carotenoids, flavonoids, tannins, phenolics, saponins, terpenes, anthraquinones, phytosterols, phytoestrogens, omega-3 fatty acids, and resveratrol (Table 1). They possess some explicit pharmacological properties for human body (Moses et al., 2013). Some of them are discussed in the succeeding text.

4.1 Alkaloids

Major group of secondary metabolites is constituted by alkaloids. It is mostly made of ammonia compounds having nitrogen bases produced from amino acid building blocks with many radicals replacing one or more hydrogen atoms in the peptide ring, most containing oxygen. The precursors for biosynthesis of alkaloid are amino acids (ornithine and lysine). Alkaloids are basic in nature. It is the position (1°, 2°, or 3° amines) of one or more nitrogen atoms in an alkaloid, which contributes to its basicity. Degree of basicity is determined by molecular structure, presence, and position of functional groups (Sarker & Nahar, 2007). Crystalline salts are formed on reaction of alkaloids with acids without producing water (Firn, 2010). Like atropine, most of the alkaloids exist in solid form and some as liquid possessing carbon, hydrogen, and nitrogen. Alkaloids are sparingly soluble in water and readily dissolve in alcohol. However, their salts are water soluble. Alkaloid solutions taste bitter; thereby, they defend plants against herbivory and pests. Owing to their powerful biological activities, they are used as narcotics, stimulants, pharmaceuticals, and poisons. The seeds and sand roots of plants possess higher percentage of alkaloids. Alkaloids are widely utilized in pharmacology as CNS stimulants and anesthetics (Madziga, Sanni, & Sandabe, 2010). In about 20% of plant species, more than 12,000 alkaloids are known to exist, but only few have been utilized for medicinal use. The suffix

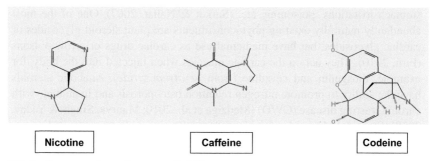

| Nicotine | Caffeine | Codeine |

FIG. 1 Structures of some important alkaloids.

"ine" is used for alkaloids, for example, tubocurarine and berberine. Some clinically useful alkaloids of plant origin are morphine and codeine (analgesics), (+)-tubocurarine (muscle relaxant), sanguinarine and berberine (antibiotics), vinblastine (anticancerous), ajmaline (antiarrhythmic), atropine (pupil dilator), and scopolamine (sedative). Some addictive stimulants include nicotine, caffeine, codeine, morphine, atropine, cocaine, ergotamine, and ephedrine (Fig. 1).

4.2 Glycosides

Simple sugars on condensation give glycosides, and one of the hydroxyl groups gets replaced in sugar molecule. They are crystalline and colorless containing carbon, hydrogen, and oxygen (some contain sulfur and nitrogen). Cell sap contains glycosides and is soluble in water. The chemical composition of glycosides is carbohydrates (glucose) and noncarbohydrates (aglycone, including phenol, glycerol, and alcohol) (Firn, 2010; Kar, 2007). They are neutral in reaction and easily hydrolyzed by mineral acids into its constituents. Glycosides are grouped into different categories based on sugar component, chemical nature of aglycone, or pharmacological action. The older naming system of glycosides includes its source and suffix "in," for example, digitoxin from *Digitalis* and salicin from *Salix*. However, the contemporary naming involves replacing the suffix "ose" of the parent sugar with "oside." Glycosides are usually utilized to improve appetite and digestion. The plant family Genitiaceae is the common source of glycosides. They are bitter in taste. The bitter property acts on gustatory nerves, which causes increased flow of saliva and gastric juices. Some of the bitter principles are either used to reduce thyroxin and metabolism or as astringents due to the presence of tannic acid, as antiprotozoan, for example, cardiac glycosides (act on the heart), anthracene glycosides (purgative and for curing skin disease), chalcone glycoside (anticancer), and amarogentin. Various medicinal preparations contain flavoring agents of cyanogenic glycosides (Sarker & Nahar, 2007). Amygdalin produces HCN in stomach that destroys malignant cells and thus used for cancer treatment and also as a cough suppressant. Excess of cyanogenic glycosides in food can be dangerous; they can cause stomach irritations, poisoning, etc. (Sarker & Nahar, 2007). One of the most abundantly naturally existing phytoconstituents are plant steroid glycosides or cardiac glycosides that have medicinal use as cardiac drugs or arrow poisons (Firn, 2010). They act on the cardiac muscles when injected into the body, for example, diosgenin and cevadine (from *Veratrum viride*). Anabolic steroids have the ability to promote nitrogen retention osteoporosis and in animals with chronic wasting disease (CWD) (Madziga et al., 2010; Maurya, Singh, & Yadav, 2008). Steroidal glycosides function when taken in small amounts, and excessive dose can be fatal.

4.3 Carotenoids

Red, orange, and yellow hues in many vegetables and fruits are due to a class of phytochemicals called carotenoids that neutralize free radicals generated during various metabolic processes due to their rich antioxidants. Many types of cancer, eye defects, and cardiovascular diseases are treated by carotenoids. The most common examples of carotenoids are carotene, lutein, zeaxanthin, and lycopene. Lutein is specifically utilized for uterine, prostate, breast, colorectal, lung, and gastrointestinal cancers (Gupta, 2015).

4.4 Phenolics

Phenolics are compounds that contain at least one benzene ring attached to one or more hydroxyl (–OH) groups. Chemically, phenols vary from simple phenols (like catechols and hydrobenzoic acid derivatives) to catechol melanins (C6)6 long-chain polymers having high molecular weight and condensed tannins (C6–C3–C6)n and lignins (C6–C3)n. Intermediate-molecular weight phenolic compounds include flavonoids (C6–C3–C6) and stilbenes (C6–C2–C6). Anthocyanins, flavonols (quercetin and myricetin), and isoflavones (daidzein and genistein) are some flavonoids that are formed from chalcone.

Due to the deterrent qualities of phenolic compounds, they protect plants from herbivores, pathogens, and other animals. The endogenous toxicity of phenolic compounds is reduced by the sugars attached to them. Phenolic compounds also protect plants from harmful UV radiations and cold stress (Ahmed et al., 2017). The color of fruits is imparted by phenolics, phenols, or polyphenolics (or polyphenol extracts); they are found in all plants as natural color pigments. Mostly, in plants, phenylalanine is the precursor of phenolics. It is synthesized by the action of phenylalanine ammonia lyase (PAL). They have an important role in plant defense against pathogens. The three classes of phenolics are (1) phenolic acids, (2) flavonoid phenolics (flavonones, flavones, xanthones, and catechins), and (3) nonflavonoid polyphenolics. The two most common phenolic compounds found in plants are caffic acid and chlorogenic acid; they are responsible for allergic dermatitis in humans (Kar, 2007). Phenolics are also rich in antioxidants and are abundant in apples, green tea, and red wine. They have nutraceutical applications. They have role in the treatment of cancer, heart ailment, and inflammation. Few more examples of phenolics include flavones, rutin, naringin, hesperidin, and chlorogenic.

4.5 Flavonoids

Flavans are the parent compounds forming an important class of polyphenols largely found in plants as flavonoids. Their structure consists of more than one benzene ring. They have antioxidant properties, and they take part in free radical mediated cellular signaling (Kar, 2007; Panche, Diwan, & Chandra, 2016). Like carotenoids, some flavonoids also impart color in higher plants. More than 4000 flavonoids are found in higher plants. In almost 70% of the plants, commonly found flavonoids are quercetin, kaempferol, and quercitrin. Other examples of flavonoids are flavones, dihydroflavones, flavans, flavonols, anthocyanidins, proanthocyanidins, chalcones, catechin, and leucoanthocyanidins.

Fruits and vegetables like berries, legumes, tea, grapes, olive oil, cocoa, walnuts, peanuts, spices, onions, and apples are good source of flavonoids. They are utilized for the treatment of bacterial diseases, viral infections, allergies, inflammation, tumors, and hepatotoxins. Flavonoids like quercetin and kaempferol are usually found in relatively low quantities of 12–28 mg/kg of fresh plant weight (Panche et al., 2016).

4.6 Saponins

The term saponin is actually derived from *Saponaria vaccaria* (*Quillaja saponaria*), a plant that contains large quantities of saponin and was once used as soap. Therefore saponins have "soap-like" behavior, that is, produce foam in water. Sapogenin (an aglycone) is formed when saponin is hydrolyzed. In saponins, mostly at C-3, sugar molecule is attached because of the presence of hydroxyl group at C-3. In *Quillaja saponaria* a toxic gylcoside quillajic acid and the sapogenin senegin is present. Senegin is also found in *Polygala senega*. Saponins are high-molecular weight compounds. They are mostly amorphous in nature. They are soluble in alcohol and water and insoluble in nonpolar organic solvents like benzene and *n*-hexane. They are very poisonous and are responsible for cattle poisoning and hemolysis of blood (Kar, 2007). They cause mucous membrane irritation and are bitter in taste. They also show therapeutic applications like hypolipidemic and anticancer. They also regulate the activity of cardiac glycosides. Steroidal and triterpenoidal sapogenins are the two types of sapogenin. The two major steroidal sapogenins are diosgenin and hecogenin. Steroidal sapogenins are used in the commercial production of sex hormones for clinical use. For instance, progesterone is derived from diosgenin, isolated from *Dioscorea* species, formerly supplied from Mexico and now from China. Hecogenin, isolated from *Sisal* leaves found extensively in East Africa, is used for the preparation of cortisone and hydrocortisone hormones (Sarker & Nahar, 2007).

4.7 Tannins

Tannins are widely distributed in plants. Chemically, they are phenolic compounds of high molecular weight. They are soluble in polar solvents like water and alcohol. They are present in the roots, bark, stem, and outer layers of plant tissue. They have peculiar feature to tan, that is, they give leathery texture to the things. Due to the presence of phenolics or carboxylic group, they are acidic in reaction (Kar, 2007). They make complexes with carbohydrates, proteins, gelatin, and alkaloids. There are two types of tannins: hydrolyzable and condensed. On hydrolysis, hydrolyzable tannins produce gallic acid and ellagic acid. On the basis of the kind of acid produced, the hydrolyzable tannins are known as gallotannins. On heating, they produce pyrogallic acid. The phenolic group in tannins makes them antiseptic. Aflavins (from tea), daidzein, genistein, and glycitein are some common examples of hydrolyzable tannins. Plants that are rich in tannins are utilized as healing agents in many diseases. They have been used for the treatment of diseases like leucorrhea, rhinorrhea, and diarrhea.

4.8 Terpenes

Terpenoids constitute the most widely distributed class of secondary metabolites. More than 40,000 different compounds are included in terpenoids, making it the largest group of plant metabolites. They are flammable unsaturated hydrocarbons, existing in liquid form commonly found in essential oils, resins, or oleoresins (Firn, 2010). It's very first member was isolated from turpentine oil that is why it was named as terpene or terpenoid. From isopentane skeleton (or isoprene units), all other terpenoids are derived. When terpenoids undergo thermal decomposition, they produce isoprene gas as a product. Terpenoids are often called as isoprenoids because isopentane (or isoprene units), under suitable chemical conditions, can be polymerized to produce variety of terpenoids. The in vivo terpenes are synthesized from acetyl CoA or from its intermediates. Terpenoids have many functions in plants. They function as photosynthetic pigments (phytol and carotenoids), as electron carriers (ubiquinone and plastoquinone), as hormones and sterols (gibberellins and abscisic acid), and also as structural components of cellular membranes. Carotenoids are precursors of abscisic acid that regulate developmental and stress responses. Besides, carotenoids are photooxidative protectants for other pigments. Xanthophylls, α-carotenes, β-carotenes, and lycopene are red-, orange-, and yellow-colored lipid-soluble pigments that are also carotenoids. Chlorophyll masks the carotenoids in most of the green leafy vegetables. The bright colors in tomatoes, carrots, pumpkins, and sweet potatoes are due to carotenoids. In citrus fruits the major source of terpenoids are limonoids. They have anticancer properties. D-Limonene, most common monocyclic monoterpene, found in orange peel oil is used to treat pancreatic cancer (Suntar, Khan, Patel, Celano, & Rastrelli, 2018).

Depending on the number of carbon atoms, terpenoids are classified as mono-, di-, tri-, and sesquiterpenoids. Terpinen-4-ol, thujone, camphor, eugenol, and menthol are common monoterpenes. Taxol (anticancer agent) is the common example of diterpenes (C20). The triterpenes (C30) include steroids, sterols, and cardiac glycosides that have antiinflammatory, sedative, insecticidal, or cytotoxic activity. Examples are amyrins, ursolic acid, and oleanic acid. Sesquiterpenes (C15) are major components of many essential oils (Martinez, Lazaro, del Olmo, & Benito, 2008). They have antimicrobial, antihelminthic, and neurotoxic actions (Table 2).

TABLE 2 Types of terpenoids according to the number of isopropene units.

Type of terpenoids	Example	Chemical structure
Monoterpene	Limonene $C_{10}H_{16}$	
Sesquiterpene	Artemisinin $C_{15}H_{22}O_5$	
Diterpene	Forskolin $C_{22}H_{34}O_7$	
Triterpene	α-Amyrin $C_{30}H_{50}$	
Tetraterpene	β-Carotene $C_{40}H_{56}$	

4.9 Anthraquinones

Anthraquinones are phenolic and glycosidic compounds, solely derived from anthracene. The two oxidized derivatives of anthraquinones are anthrones and anthranols (Firn, 2010; Maurya et al., 2008). Organic solvent is mixed with ground plant material and filtered, and a solution of sodium hydroxide or ammonium hydroxide (aqueous base) is added to it. Formation of a pink or violet color in the base layer indicates presence of anthraquinone in plant sample (Sarker & Nahar, 2007).

4.10 Omega-3 fatty acids

Omega-3 fatty acid exhibits many health benefits, so there is a huge interest in increasing consumption of it. Sardines, rainbow trout, salmon, and mackerel herring are the richest omega-3 fatty acid sources. Omega-3 fatty acids (18 carbon) and α-linolenic acid are present in flax oils, canola, and soybean. Osteoarthritis is treated by omega-3 fatty acid. Developing canola and soybean (oilseed crops) containing stearidonic acid provides an important strategy for increasing long-chain omega-3 fatty acid availability. Black currant seed oil and echium oil contain naturally omega-3 fatty acids (Bradberry & Hilleman, 2013).

4.11 Phytoestrogens

Nonsteroidal phytochemicals structurally and functionally much similar to gonadal estrogen hormone constitute phytoestrogens. They are the best alternative for hormone replacement therapy (HRT) and potential alternative to the synthetic selective estrogen receptor modulators, which are currently applied in HRT. Alleviation of menopausal symptoms and advantageous impacts on cardiovascular system are exhibited by phytoestrogens. Based on chemical structure, lignans, flavonoids, stilbenes, isoflavonoids, and coumestans constitute the classes of phytoestrogens. They are present either in plant or their seeds. Isoflavones are abundantly found in soybean, whereas coumestrol (the major coumestan) is found in soy sprout in good quantity. The richest source of coumestans is clover and soybean sprouts (Rietjens, Louisse, & Beekmann, 2017).

4.12 Resveratrol

Plants produce resveratrol, a natural phytoalexin during stress conditions and pathogen attack. They are produced after many physiological processes. It has been reported that resveratrol has antioxidant properties and exerts neuroprotective and cardioprotective effects at lower doses. The therapeutic applications of resveratrol include antithrombogenic, antiinflammatory, antiaging, and anticancer. Grapes and peanuts are the potential source of resveratrol (Keylor, Matsuura, & Stephenson, 2015).

4.13   Phytosterols

Plant sterols and stanols constitute the phytosterols. They are beneficial in lowering the cholesterol levels in blood by preventing absorption of cholesterol from the intestines. The natural sources of phytosterols are vegetables, fruits, nuts, and oils. The cardiovascular diseases are rising day by day due to sedentary lifestyle. Therefore market demand for products fortified with phytosterols is expected to pick up a pace in times to come. A new approach for reducing levels of cholesterol has been found as a functional food ingredient of phytosterol and presents a hope for great future concerning management of health in long term (Lin, Knol, & Trautwein, 2016).

5.   Regulation of herbal drugs

Most important apprehension of public health has arisen as far as safety of crude drugs is concerned due to progressive fame and growing market for herbal drugs across the globe. Adverse reactions caused due to poor-quality phytoproducts are attributed to the absence of laws or failure to implement herbal regulations and the online sales of phytoproducts. Inappropriate dosage, product tacit, potent pharmaceutical substances, replacement or wrong identification with toxic plant species, and interactions with conventional medicines turn out to be the frequent reasons of adulteration. Phytomedicines constitute an imperative alternative for modern allopathic medicine across the length and breadth of globe. Even though phytomedicines hold a promising place as far as its usage is concerned in the society, only limited uses of medicinal plants have been validated scientifically by running various clinical trials. Poor regulation of phytomedicines in majority of countries led to nonregistration by health authorities; hence, foremost fear of herbal drug safety always hovers over the heads. More than 50,000 adverse reactions produced by food supplement and botanical usages have been anticipated by the US Food and Drug Administration. Besides, efficacy and quality for majority of herbal medicines are unassured and not proved. Hence, the World Health Organization's strategy on traditional medicines from 2014 to 2023 centralizes endorsing efficiency, safety, and quality of herbal medicines by increasing the knowledge base while providing standards for quality assurance and regulations (Nooreen, Rai, & Yadav, 2018). Almost 119 member states of WHO had developed regulations for phytomedicines by 2012 (Bhatt, 2016), and this number saw a surge in 2018 when 124 member states put into practice the laws/regulations concerning herbal drugs while 98 member states made national policies on herbal medicines and 109 launched national laws or regulations on herbal medicines (WHO, 2019). The administrative ruling with respect to herbal drug preparation changes from nation to nation. Comprehensively a few assorted regulatory methodologies are in practice, for example, they utilize same regulatory prerequisites for all products, with specific kinds of proof (unusual for herbal drugs) and exclusion of phytomedicines for any regulatory necessity concerning enlistment or authorizing marketing. As far as India is concerned, following acts have been administered in the Indian System of Medicines (ISM) sector:

- Drugs and Cosmetics Act 1940 and Rules thereunder (amended in 2009)
- Drugs and Magic Remedies Act 1954. 1955 and Rules thereunder
- Central Council of Homoeopathy Act 1973
- Central Council of Indian Medicine Act 1973
- Medicinal and Toilet Preparation Acts and Rules 1995–1996

6. Global market potential of phytomedicines

Phytomedicines are progressively occupying one of the most significant aspects in expediently prospering worldwide commercial health venture. Herbal medicines are one of the most looked for after essential health care needs by about 3.5–4 billion individuals over the world as per the gauge of WHO with a significant portion of phytomedicines being decoctions and plant extract–derived medicines also named as modern herbal medicines (Pan et al., 2013) in alleviating various human ailments (Mohamed, Borhanuddin, & Fozi, 2012). Selling of these phytomedicines has extended extensively all throughout the world especially in the European nations of Spain, Germany, Italy, the United Kingdom, and France and all the more in the United States also. Of the total catalogued plant species of between 350,000 and 550,000 around the globe, Brazil continues as the largest center of biodiversity in the world by contributing about 35,000 recorded species, thereby performing an indispensable role in the field. A promising commitment of decreasing undue deaths, disabilities, and morbidities associated with malaria, diabetes, tuberculosis, AIDS, and sickle cell anemia has been demonstrated by phytomedicines (Elujoba, Odeleye, & Ogunyemi, 2005). In addition, it has diminished poverty by expanding monetary prosperity of communities and created healthcare frameworks by expanding people's access to medical services. The production, dispensation, and retailing of herbal products generate vast opportunities of employment for the producing nations. Phytomedicine trade in the United Kingdom alone accounts for more than 293 million euros for each year (IUCN, 2005). Around the globe, total market sales of herbals in some countries are given in the succeeding text (Table 3).

7. Causes for increased interest in phytomedicines

Even though synthetic drugs have taken a forefront from phytomedicines while treating ailments of people, the utilization of herbal drugs has expanded as of late globally as they are considered to be safer than modern drugs with not many or no adverse reactions at all. Herbal medicines allude to the utilization of plants and herbs with the end goal of fixing and moderating human diseases. Since antiquity, plants have been utilized by human race for therapeutic purposes.

TABLE 3 Total market sales of herbal medicines.

	2007	2008	2009
Bhutan (Bhutanese ngultrum)	10.53 million	5.54 million	12.94 million
Croatia	€3.1 million	€2.95 million	€4.01 million
Democratic Republic of the Congo	US $200,000	US $600,000	US $1,300,000
Estonia	€1.9 million	€1.8 million	€1.6 million
Mali	US $97,200	US $106,920	US $117,612
Mongolia	US $0.5 million	US $1 million	US $1.4 million
Pakistan	US $5.5 million	US $6.5 million	US $7 million
Serbia (Serbian dinar)	163.15 million	250.07 million	283.10 million
Sri Lanka	US $500 million	US $500 million	US $550 million
Yemen (Yemeni rial)	NA	NA	1,287,630,958

Different pharmacological investigations on phytochemicals have revealed the ability of phytomedicines for human healthcare systems. Archived accounts of medicinal plants can be traced back to 5000 years ago for Sumerians who portrayed entrenched therapeutic uses for plants like thyme, laurel, and caraway; archeological investigations have indicated that the act of phytomedicine usage dates back to 60,000 years in Iraq and 8000 years in China (Pan et al., 2014). With the approach of conventional medicine over the previous century, challenges have been posed for phytomedicines by mainstream medical practitioners in light of the absence of scientific proofs with regard to contemporary medication, notwithstanding its effective use from time immemorial. Fascinatingly, things change with time; lately, there has been a revival in the utilization of herbs because of the adverse reactions of synthetic drugs, absence of curative modern treatments for a few ceaseless maladies, microbial resistance, and at the same time remarkable interest in pharmaceutical research works. Since 1950 US Food and Drug Administration approved new drugs of around 1200 (Pan et al., 2014). Consequently the utilization of herbs and herbal products for well-being purposes has expanded in notoriety globally in the course of last 40 years in both developed and developing nations. Furthermore, worldwide pharmaceutical organizations acquainted with current scientific innovative thoughts have started to rediscover herbs as a potential source of novel drug development and reestablished their techniques for plant-derived drug developmental discoveries.

Herbal drugs are progressively picking up significance even among the literates in urban settlements most likely because of expanding inefficiency of synthetic drugs utilized for controlling numerous infections like tuberculosis, typhoid fever, and gonorrhea and resistance by a few microscopic organisms to different antimicrobials and the escalating cost of physicians' recommended drugs for the maintenance of individual's well-being (Smolinski, Hamburg, & Lederberg, 2003). Plants possess characteristic reserves for therapeutic agents practically free from adverse reactions ordinarily brought about by manufactured synthetic drugs (Fennell et al., 2004). WHO appraises that herbal drugs still hold a lion's share in the primary healthcare needs of developing nations owing to better cultural acceptability, easy access, and economic affordability of the phytomedicines. The overutilization of synthetic drugs bringing about higher occurrences of unfavorable medication responses has persuaded humanity to return to nature for more secure remedies. Because of different geographical locations where these plants grow combined with the issue of various vernacular names at different locations, WHO developed guidelines to ensure herbal safety while limiting their debasement and misuse (WHO, 1999). Based on various traditional uses of herbal products, a large number of drugs have been isolated from these plant sources where ethnomedicinal usage provides a hint for the kind of active ingredient present in a plant (Rizvi, Irshad, Hassadi, & Younis, 2009). It is realized that inhabitants who are exposed to interminable sicknesses will in general look for alternative treatments for their healthcare. Diseases that happen as often as possible used to have an extraordinary number of various local treatments available. An investigation by Waldstein demonstrated that Mexican vagrant ladies in the United States know both therapeutic plants and conventional medicine. But they preferred customary medicine since they accept that synthetic medication is risky to be taken while the utilization of plants is not considered to have any reactions. Womenfolk have been portrayed in numerous ethnobotanical studies accountable for family healthcare and thus have more acquaintance with medicinal herbs as compared with menfolk. This cozy connection between womenfolk and therapeutic plants might be the reason for their reluctance of accepting the modern or synthetic medicines (Nascimento, Medeiros, & Albuquerque, 2018).

8. Conservation strategies for phytomedicines

A profoundly conservationist approach estimates that the present loss of plant species is somewhere in the range of 100 and 1000 multiples higher than the normal common elimination rate as a result of which earth is losing one potential significant drug in every 2 years. As per the Worldwide Fund and International Union for Conservation of Nature, there are around 50,000–80,000 flowering species of plants utilized for therapeutic purposes around the world. Habitat destruction and excessive harvesting have resulted in undermining of around 15,000 species with extinction, and 20% of their wild assets have just been almost depleted owing to escalating human population growth with rocketing plant consumption. Despite the fact that this danger has been known for a considerable length of time, the quickened species loss along with habitat destruction around the world has amplified the threat of extermination of therapeutic plants particularly in Uganda, Tanzania, China, Nepal, India, and Kenya (Chen et al., 2016).

Almost one-tenth of plant species out of 50,000 are being utilized for drugs and healthcare products. Nevertheless, therapeutic plants are not uniformly allocated around the globe (Huang, 2011; Rafieian-Kopaei, 2013). For instance, highest number of 11,146 and 7500 therapeutic plant species are utilized by China and India, respectively, followed by Colombia, South Africa, the United States, and another 16 nations with therapeutic plant utilization percentage running from 7 in Malaysia to 44 in India versus their absolute number of plant species (Hamilton, 2003; Marcy, Balunasa, & Kinghornb, 2005; Srujana, Babu, & Rao, 2012). All-around considered opinions have been made numerous years prior to raising open mindfulness on the obliteration of seasonally dry monsoon forests and tropical rain forests. However, the arguments were to a great extent overlooked before today's dramatic endeavors are being made for conserving biodiversity. Inactions in terms of conservation have resulted in present crisis of habitat destruction, habitat fragmentation, environmental genocide, ecological deterioration, and genetic erosion. For whatever length of time forest destruction continues, therapeutic plants along with their natural habitats will stay under the danger of excessive collection and harvesting than at any other time. Therefore the definitive aim of conservation is to safeguard the natural habitats of vulnerable therapeutic plants and to accomplish their sustainable utilization in less vulnerable areas. Not every therapeutic plant is influenced similarly by over harvesting. Although factors like unrestricted deforestation, excessive utilization, habitat destruction, and indiscriminate collection methods have an impact on species rarity but are deficient to clarify the impact on individual species, resilience or susceptibility by harvesting pressures. Numerous biological factors of population size, habitat specificity, growth rate, distribution range, species diversity, reproductive system, etc. are found to be associated with risk of extinction (Liu, Yu, & Chen, 2011). Wild nurseries and natural reserves are distinctive instances of preserving the clinical adequacy of therapeutic plants in their natural habitats, while botanical gardens, gene banks, and seed banks are significant ideal models for ex situ conservation and future replanting.

(i) In situ conservation

Secondary metabolites responsible for medicinal properties in therapeutic plants respond to certain stimuli in particular natural habitats that may not be expressed under culture conditions hence contributing to the endemism of most medicinal plants (Figueiredo & Grelle, 2009). Maintenance of complex associational relationships of natural communities ensures preservation of indigenous plants in the entire plant communities by in situ conservation (Gepts, 2006). Moreover, in situ conservation enhances the kind of diversity that can be monitored for conservation and fortifies the connection of conserving resources while ensuring their sustainable utilization. In situ conservation endeavors around the world have concentrated on setting up protected areas network by adopting ecosystem-oriented approach rather than species oriented (Ma, Rong, & Cheng, 2012). Triumphant in situ conservation relies upon policies, rules, and prospective acquiescence of therapeutic plants within growth habitats (Volis & Blecher, 2010).

Natural reserves

Incessant degradation and decimation of natural habitats are a significant reason for the loss of therapeutic plant wealth. Natural reserves engross protected regions of significant untamed assets made to protect, safeguard, and reestablish biodiversity. Globally, 12,700 areas included in protected area network have been recognized representing 13.2 million square kilometer or 8.81% of Earth's land surface (Huang et al., 2002). Estimating functions of ecosystems in each separate habitat determines conserving efforts of protecting therapeutic plants by protecting key natural habitats.

Wild nurseries

Cost contemplations and contending land uses make it difficult to assign each natural wild habitat of plants as a protected area. A wild nursery establishment aims for species-oriented cultivation and domestication of endangered therapeutic plants in a natural habitat, protected area, or a region that is less distance away from the region where it naturally grows. Despite the fact that populaces of numerous wild species are under overwhelming pressure in view of excessive utilization, invasions, and habitat fragmentation, wild nurseries can give a successful advancement to deal with in situ conservation of therapeutic plants that are in-demand, endangered, and endemic (Liu et al., 2011).

(ii) Ex situ conservation

This includes establishment of ranches, upkeep of living assortments in ranch fields, home nurseries, botanical gardens, and arboreta in area outside the zone of their characteristic occurrence. Real aim of ex situ conservation involves swift expansion of substitute sources of therapeutic plant supply by cultivating huge enough plants at such a price that contends with costs acquired by gatherers of untamed therapeutic plant stocks. It will demotivate gatherers to go for wild collection and meanwhile create more jobs by fulfilling market demands for medicinal plants. On the off chance that this does not happen, naturally existing species in wild will vanish hence by decreasing the regional medicinal resource base. Ex situ conservation is not in every case stridently isolated from in situ conservation but a viable supplement to it for those over harvested and endangered therapeutic plants with sluggish growth, stumpy profusion, and high chances of developing diseases while replanting (Havens, Vitt, Maunder, Guerrant, & Dixon, 2006; Yu, Xie, Song, Zhou, & Chen, 2010). It intends to grow and naturalize threatened plant species to guarantee their survival with endurance and some of the time to deliver enormous planting material utilized in drug making, and it is over and again a prompt activity intended to sustain plant resources of medicinal value (Swarts & Dixon, 2009). Numerous kinds of formerly wild therapeutic plants when allowed to grow in gardens quite away from the places where they naturally grew not only maintain elevated potency but also can have their propagation material chosen and kept in seed banks for planting in future in case they become extinct in wild.

Field gene banks

To ensure protection of vulnerable therapeutic plant species against risk of extinction, gene banks are established as precautionary backup. Commercial cultivation of slow growing medicinal plants prompts maintaining gene banks where wild populations are imperiled (Cunningham, 1993). These gene banks help in reintroduction of extinct or endangered therapeutic plant species to their original habitats, stockpiling of hereditary material wherefrom researchers can design valuable plants in coming years, much after species in the wild have got extinct and a source of variants in the event of environmental disaster and disease when medicinal plant population has been wiped off entirely.

Seed banks

Seed banks provide an improved method of amassing genetic variety of numerous therapeutic plants ex situ, established to preserve genetic and biological diversity of wild therapeutic plants. Millennium Seed Bank Project at Royal Botanical Gardens in Britain is the most important seed bank. Seed banks permit moderately quick access to plant varieties for the assessment of their properties giving accommodating data for preserving the left out natural populaces. However, the difficult job in seed banking involves the reintroduction of plant species back into the wild and how to effectively aid in wild population restoration (Chen et al., 2016).

Botanical gardens

To improve continued existence of rare and endangered therapeutic plant species, botanical gardens exhibit a significant role in ex situ conservation by maintaining ecosystems. Living specimens comprise of few individuals of each species thus being of restricted usage in genetic conservation (Yuan et al., 2010) but botanical gardens have various distinctive features. They include a wide assortment of plant species grown collectively under common conditions and frequently include diverse flora in terms of taxonomy and ecology (Primack & Miller-Rushing, 2009). By developing protocols for cultivation and propagation besides taking responsibility of domesticating and cultivating different varieties, botanical gardens furthers in conserving therapeutic plant species.

(iii) Domestic cultivation practices

An extra possible alternative to the act of collecting plants present in their natural or wild habitats basically for nourishment or therapeutic purposes, that is, wild crafting would be farming. Yet around 66% of 50,000 therapeutic plant species in use around globe for medicinal purposes are still wild crafted (Canter, Thomas, & Ernst, 2005). To promote viable harvesting methods and reduce damage inflicted on environment due to wild crafting of medicinal plants, organizations like European Medicines Agency and World Health Organization figured out plans on good-quality agrarian and harvesting procedures. Beside this issue, protection and legitimate concern too have a considerable stress on plant availability as a premise of new drug discovery, especially guideline associated with plant access and dispersion of benefits and issues of patents with local governments in the nations of origin. The United Nation's Convention on Biological Diversity targets at saving the biodiversity, achievable utilization of its hereditary resources and dissemination of benefits achieved from their utilization in a reasonable and sensible manner (Atanasov et al., 2015). At numerous occasions, when a plant is promoted into phytocompound or when one of its components get acclaim as a remedial agent, the plant populace transform into endangered one on account of its far reaching unsustainable cultivation practices. The customary example of this was paclitaxel supply disaster. At the point when mixture ended up having phenomenal clinical importance in ovarian malignancy, suddenly demand for paclitaxel increased tremendously. Compound was only possible to be extracted from bark of *Taxus brevifolia* L. that gave a tough challenge as entire drug manufacturing involved dreary bark collection, which again laid heavy stresses on environment hence inflicting much damage (Cragg, Newman, & Rosenthal, 2012). Albeit wild reaped resources of therapeutic plants are broadly viewed as more effectual than those that are cultivated, but domestic cultivation is broadly utilized and for most part an acknowledged practice (Joshi & Joshi, 2014; Leung & Wong, 2010). To tackle issues experienced in therapeutic plant production like pesticide contamination, misidentification, low quantities of active ingredients, pesticide poisoning, all such issues get solved by utilizing new strategies of cultivation (Raina, Chand, & Sharma, 2011). Yield of target active compounds that are constantly secondary metabolites can be improved by cultivating medicinal plants under controlled conditions of providing optimum nutrient levels, water, temperature, humidity, and light (Liu et al., 2011; Wong et al., 2008). Besides, increasing domestic cultivation practices will lead to decrease of wild collection of therapeutic plants allowing their recovery in the wild and a significant drop of prices of medicinal plants to a sensible level (Larsen & Olsen, 2007; Schippmann, Leaman, Cunningham, & Walter, 2005).

(iv) Good agricultural practices

Regulation of production, quality assurance, and herbal drug standardization is possible by good agricultural practices (Chan et al., 2012). This approach includes quality aspects of pesticide detection, ecological environment of production sites, cultivation, macroscopic or microscopic authentication, germplasm collection, chemical identification of bioactive components, metal detection hence guaranteeing superior quality, and secure and pollution-free herbal drugs. Places in China where therapeutic plants are cultivated traditionally have promoted GAP approach for commonly used herbal medicines (Ma et al., 2012).

9. Conclusion

For anticipating and treating different health ailments like jaundice, gastric ulcers, diabetes, scabies, and diarrhea, phytomedicines have been used by almost every civilization from centuries together. This is basically of the general conviction that herbal medicines are safe, economical, and locally accessible. Accessibility to herbal medicines from any place of the earth has been tremendously increased owing to globalization. It is anticipated that sooner phytomedicines will be integrated with conventional system of medicine due to regulatory measures and quality control incorporation as the herbal age is expected to arrive. Successful drugs derived from medicinal plants still represent significant pool of novel therapeutic leads to be developed from rich source of medicinal plants. Plant production of secondary metabolites chemically diverse is optimized for their biological activities that are still much behind from being researched thoroughly. Hence the collaborative research needs to be carried out with benefit-sharing agreements between top institutes, drug-manufacturing companies of developed nations, and various organizations of developing nations where numerous therapeutic plants are still unexplored. This approach could help in coming up with more diverse range of phytocompounds for treating various human ailments. According to WHO, without consolidation of herbal medicines in primary healthcare needs the goal of "health for all" cannot be fructified. Accordingly, there is a dire need to improve drug discovery process from medicinal plants for many years to come as many new diseases (SARS Covid 19) are waiting to attack and bring the human race on the brink of destruction. Decentralization of selected aspects of research processes worldwide and utilization of local resources by pharmaceutical companies are the need of the hour. Continuous research by utilizing cutting-edge technologies can bless us with leads to develop novel drugs. Keeping in mind the prominence and adequacy of phytomedicines, the dream of a healthy world will materialize soon hopefully.

References

Abdullah, S., Shaari, A. R., & Azimi, A. (2012). Effect of drying methods on metabolites composition of misai kucing (Orthosiphon stamineus) leaves. *APCBEE Procedia, 2*, 178–182.

Ahmed, E., Arshad, M., Khan, M. Z., Amjad, M. S., Sadaf, H. M., Riaz, I., et al. (2017). Secondary metabolites and their multidimensional prospective in plant life. *Journal of Pharmacognosy and Phytochemistry, 6*(2), 205–214.

Applezweig, N. (1980). *Renewable resources: A systemic approach*. London, New York: Academic Press.

Arunachalam, K., & Parimelazhagan, T. (2014). Antidiabetic and enzymatic antioxidant properties from methanol extract of Ficus talboti bark on diabetic rats induced by streptozotocin. *Asian Pacific Journal of Reproduction, 3*, 97–105.

Atanasov, A. G., Waltenberger, B., Pferschy-Wenzig, E. M., Linderd, T., Wawroscha, C., & Uhrin, P. (2015). Discovery and resupply of pharmacologically active plant derived natural products: A review. *Biotechnology Advances, 33*(8), 1582–1614.

Azwanida, N. N. (2015). A review on the extraction methods use in medicinal plants, principle, strength and limitation. *Medicinal and Aromatic Plants, 4*, 196. https://doi.org/10.4172/2167-0412.1000196.

Baptista, E. A., Pinto, P. C., Mota, I. F., Loureiro, J. M., & Rodrigues, A. E. (2015). Ultrafiltration of ethanol/water extract of Eucalyptus globulus bark: Resistance and cake build up analysis. *Separation and Purification Technology, 144*, 256–266.

Benkovic, E. T., Grohar, T., Zigon, D., Svajger, U., Janes, D., Kreft, S., et al. (2014). Chemical composition of the silver fir (Abies alba) bark extract abigenol and its antioxidant activity. *Industrial Crops and Products, 52*, 23–28.

Bhatt, A. (2016). Phytopharmaceuticals: A new drug class regulated in India. *Perspectives in Clinical Research, 7*, 59–61.

Bimakr, M. (2010). Comparison of different extraction methods for the extraction of major bioactive flavonoid compounds from spearmint (Mentha spicata L.) leaves. *Food and Bioproducts Processing*, 1–6.

Bouras, M., Chadni, M., Barba, F. J., Grimi, N., Bals, O., & Vorobiev, E. (2015). Optimization of microwave-assisted extraction of polyphenols from Quercus bark. *Industrial Crops and Products, 77*, 590–601.

Bradberry, J. C., & Hilleman, D. E. (2013). Overview of omega-3 fatty acid therapies. *Pharmacy and Therapeutics, 38*, 681–691.

Brusotti, G., Andreola, F., Sferrazza, G., Grisoli, P., Merelli, A., della Cuna, F. R., et al. (2015). In vitro evaluation of the wound healing activity of Drypetes klainei stem bark extracts. *Journal of Ethnopharmacology, 175*, 412–421.

Canter, P. H., Thomas, H., & Ernst, E. (2005). Bringing medicinal plants into cultivation: Opportunities and challenges for biotechnology. *Trends in Biotechnology, 23*(180), 185.

Chan, K., Shaw, D., Simmonds, M. S., Leon, C. J., Xu, Q., Lu, A., et al. (2012). Good practice in reviewing and publishing studies on herbal medicine, with special emphasis on traditional Chinese medicine and Chinese materia medica. *Journal of Ethnopharmacology, 140*, 469–475.

Chen, Q., Fu, M., Liu, J., Zhang, H., He, G., & Ruan, H. (2009). Optimization of ultrasonic-assisted extraction (UAE) of betulin from white birch bark using response surface methodology. *Ultrasonics Sonochemistry, 16*, 599–604.

Chen, S. L., Yu, H., Luo, H. M., Wu, Q., Li, C. F., & Steinmetz, A. (2016). Conservation and sustainable use of medicinal plants: problems, progress, and prospects. *Chinese Medicine, 11*, 37. https://doi.org/10.1186/s13020-016-0108-7.

Chupin, L., Maunu, S., Reynaud, S., Pizzi, A., Charrier, B., & Bouhtoury, F. C.-E. (2015). Microwave assisted extraction of maritime pine (Pinus pinaster) bark: Impact of particle size and characterization. *Industrial Crops and Products, 65*, 142–149.

Cragg, G. M., Newman, D. J., & Rosenthal, J. (2012). The impact of the United Nations convention on biological diversity on natural products research. *Natural Product Reports, 29*, 1407–1423. https://doi.org/10.1039/c2np20091k.

Cunningham, A. B. (1993). *African medicinal plants: Setting priorities at the interface between conservation and primary healthcare. People and plants working paper I*. Paris: UNESCO. 92 p.

Da Silveira, C., Trevisan, M., Rios, J., Erben, G., Haubner, R., Pfundstein, B., et al. (2010). Secondary plant substances in various extracts of the leaves, fruits, stem and bark of Caraipa densifolia Mart. *Food and Chemical Toxicology, 48*, 1597–1606.

Das, K., Tiwari, R. K. S., & Shrivastava, D. K. (2010). Techniques for evaluation of medicinal plant products as antimicrobial agent: Current methods and future trends. *Journal of Medicinal Plant Research, 4*(2), 104–111.

Dhanani, T., Shah, S., Gajbhiye, N. A., & Kumar, S. (2013). Effect of extraction methods on yield, phytochemical constituents and antioxidant activity of *Withania somnifera*. *Arabian Journal of Chemistry, 10*, S1193–S1199.

Doughari, J. H., Human, I. S., Bennade, S., & Ndakidemi, P. A. (2009). Phytochemicals as chemotherapeutic agents and antioxidants: Possible solution to the control of antibiotic resistant verocytotoxin producing bacteria. *Journal of Medicinal Plant Research, 3*(11), 839–848.

Dubey, K. K., & Goel, N. (2013). Evaluation and optimization of downstream process parameters for extraction of betulinic acid from the bark of Ziziphus jujubae L. *Scientific World Journal, 2013*, 469674.

Elujoba, A. A., Odeleye, O. M., & Ogunyemi, C. M. (2005). Traditional medical development for medical and dental primary healthcare delivery system in Africa. *African Journal of Traditional, Complementary, and Alternative Medicines, 2*(1), 46–61.

Encarnaçao, S., de Mello-Sampayo, C., Graca, N. A., Catarino, L., da Silva, I. B. M., Lima, B. S., et al. (2016). Total phenolic content, antioxidant activity and pre-clinical safety evaluation of an Anacardium occidentale stem bark portuguese hypoglycemic traditional herbal preparation. *Industrial Crops and Products, 82*, 171–178.

Enkhtaivan, G., John, K. M., Ayyanar, M., Sekar, T., Jin, K.-J., & Kim, D. H. (2015). Anti-influenza (H1N1) potential of leaf and stem bark extracts of selected medicinal plants of South India. *Saudi Journal of Biological Sciences, 22*, 532–538.

Fennell, C. W., Lindsey, K. L., McGaw, L. J., Sparg, S. G., Stafford, G. I., Elgorashi, E. E., et al. (2004). Assessing African medicinal plants for efficacy and safety: Pharmacological screening and toxicology. *Journal of Ethnopharmacology, 94*, 205–217.

Figueiredo, M. S. L., & Grelle, C. E. V. (2009). Predicting global abundance of a threatened species from its occurrence: Implications for conservation planning. *Diversity and Distributions, 15*, 117–121.

Firn, R. (2010). *Nature's chemicals* (pp. 74–75). Oxford: Oxford University Press.

Gepts, P. (2006). Plant genetic resources conservation and utilization: The accomplishments and future of a societal insurance policy. *Crop Science, 46*, 2278–2292.

Gupta, C. (2015). Phytopharmaceuticals and their health benefits. *Dream, 20*, 47–57.

Hamilton, A. (2003). *Medicinal plant and conservation: Issues and approaches*. UK: WWF.

Handa, S. S., Khanuja, S. P. S., Longo, G., & Rakesh, D. D. (2008). *Extraction technologies for medicinal and aromatic plants* (1st ed., no. 66). Italy: United Nations Industrial Development Organization and the International Centre for Science and High Technology.

Havens, K., Vitt, P., Maunder, M., Guerrant, E. O., & Dixon, K. (2006). Ex situ plant conservation and beyond. *Bioscience, 56*, 525–531.

Hofmann, T., Nebehaj, E., Stefanovits-Banyai, E., & Albert, L. (2015). Antioxidant capacity and total phenol content of beech (Fagus sylvatica L.) bark extracts. *Industrial Crops and Products, 77*, 375–381.

Hu, H.-B., Liang, H.-P., Li, H.-M., Yuan, R.-N., Sun, J., Zhang, L.-L., et al. (2018). Isolation, purification, characterization and antioxidant activity of polysaccharides from the stem barks of Acanthopanax leucorrhizus. *Carbohydrate Polymers, 196*, 359–367.

Huang, H. (2011). Plant diversity and conservation in China: Planning a strategic bioresource for a sustainable future. *Botanical Journal of the Linnean Society, 166*, 282–300.

Huang, H., Han, X., Kang, L., Raven, P., Jackson, P. W., & Chen, Y. (2002). Conserving native plants in China. *Science, 297*, 935.

Iqbal, E., Salim, K. A., & Lim, L. B. (2015). Phytochemical screening, total phenolics and antioxidant activities of bark and leaf extracts of Goniothalamus velutinus (Airy Shaw) from Brunei Darussalam. *Journal of King Saud University-Science, 27*, 224–232.

IUCN. (2005). *Medicinal plants in North Africa: linking conservation and livelihoods. Malaga, 18 April, 2005 press release*. IUCN Center for Mediterranean Corporation.

Joshi, B. C., & Joshi, R. K. (2014). The role of medicinal plants in livelihood improvement in Uttarakhand. *International Journal of Herbal Medicine, 1*, 55–58.

Kar, A. (2007). *Pharmaocgnosy and pharmacobiotechnology* (revised-expanded 2nd ed., pp. 332–600). New Delhi: New Age International Limted Publishers.

Kaufmann, B., & Christen, P. (2002). Recent extraction techniques for natural products: Microwave-assisted extraction and pressurized solvent extraction. *Phytochemical Analysis, 13*, 105–113.

Keylor, M. H., Matsuura, B. S., & Stephenson, C. R. J. (2015). Chemistry and biology of resveratrol-derived natural products. *Chemical Reviews, 115*, 8976–9027.

Khan, H. (2015). Brilliant future of phytomedicines in the light of latest technological developments. *Journal of Phytopharmacology, 4*(1), 58–60.

Khan, M. S. A., & Ahmad, I. (2019). Herbal medicine: Current trends and future prospects. In K. MSA, I. Ahmad, & D. Chattopadhyay (Eds.), *New look to phytomedicine—Advancements in herbal products as novel drug leads* (pp. 3–11). Academic Press, ISBN:978-0-12-814619-4.

Lahlou, M. (2007). Screening of natural products for drug discovery. *Expert Opinion on Drug Discovery, 2*(7), 697–705.

Larsen, H. O., & Olsen, C. S. (2007). Unsustainable collection and unfair trade? Uncovering and assessing assumptions regarding Central Himalayan medicinal plant conservation. *Biodiversity and Conservation, 16*, 1679–1697.

Leung, K. W., & Wong, A. S. (2010). Pharmacology of ginsenosides: A literature review. *Chinese Medicine, 5*, 20.

Lin, Y., Knol, D., & Trautwein, E. A. (2016). Phytosterol oxidation products (POP) in foods with added phytosterols and estimation of their daily intake: A literature review. *European Journal of Lipid Science and Technology, 118*, 1423–1438.

Liu, C., Yu, H., & Chen, S. L. (2011). Framework for sustainable use of medicinal plants in China. *Zhi Wu Fen Lei Yu Zi Yuan Xue Bao, 33*, 65–68.

Ma, J., Rong, K., & Cheng, K. (2012). Research and practice on biodiversity in situ conservation in China: Progress and prospect. *Sheng Wu Duo Yang Xing, 20*, 551–558.

Madziga, H. A., Sanni, S., & Sandabe, U. K. (2010). Phytochemical and elemental analysis of Acalypha wilkesiana leaf. *Journal of American Science, 6*(11), 510–514.

Maldini, M., Di Micco, S., Montoro, P., Darra, E., Mariotto, S., Bifulco, G., et al. (2013). Flavanocoumarins from Guazuma ulmifolia bark and evaluation of their affinity for STAT1. *Phytochemistry, 86*, 64–71.

Maldini, M., Sosa, S., Montoro, P., Giangaspero, A., Balick, M. J., Pizza, C., et al. (2009). Screening of the topical anti-inflammatory activity of the bark of Acacia cornigera Willdenow, Byrsonima crassifolia Kunth, Sweetia panamensis Yakovlev and the leaves of Sphagneticola trilobata Hitchcock. *Journal of Ethnopharmacology, 122*, 430–433.

Marcy, J., Balunasa, A., & Kinghornb, D. (2005). Drug discovery from medicinal plants. *Life Sciences, 78*, 431–441.

Martinez, M. J. A., Lazaro, R. M., del Olmo, L. M. B., & Benito, P. B. (2008). Anti-infectious activity in the anthemideae tribe. In Atta-ur-Rahman (Ed.), *Vol. 35. Studies in natural products chemistry* (pp. 445–516). Elsevier.

Maurya, R., Singh, G., & Yadav, P. P. (2008). Antiosteoporotic agents from natural sources. In Atta-ur-Rahman (Ed.), *Vol. 35. Studies in natural products chemistry* (pp. 517–545). Elsevier.

Mediani, F. A., Khatib, A., & Tan, C. P. (2013). Cosmos caudatus as a potential source of polyphenolic compounds: Optimisation of oven drying conditions and characterisation of its functional properties. *Molecules, 18*, 10452–10464.

Mohamed, I. S. A., Borhanuddin, B., & Fozi, N. (2012). The application of phytomedicine in modern drug development. *The Internet Journal of Herbal and Plant Medicine, 1.*

Moses, T., Pollier, J., Thevelein, J. M., & Goossens, A. (2013). Bioengineering of plant (tri) terpenoids: From metabolic engineering of plants to synthetic biology in vivo and in vitro. *New Phytologist, 200*, 27–43.

Nascimento, A. L. B., Medeiros, P. M., & Albuquerque, U. P. (2018). Factors in hybridization of local medical systems: Simultaneous use of medicinal plants and modern medicine in Northeast Brazil. *PLoS One.* https://doi.org/10.1371/journal.pone.0206190.

Nooreen, Z., Rai, V. K., & Yadav, N. P. (2018). Phytopharmaceuticals: A new class of drug in India. *Annals of Phytomedicine, 7*(1), 27–37.

Nunes, L. G., Gontijo, D. C., Souza, C. J., Fietto, L. G., Carvalho, A. F., & Leite, J. P. V. (2012). The mutagenic, DNA-damaging and antioxidative properties of bark and leaf extracts from Coutarea hexandra (Jacq.) K. Schum. *Environmental Toxicology and Pharmacology, 33*, 297–303.

Pan, S. Y., Litscher, G., Gao, S. H., Zhou, S. F., Yu, Z. L., Chen, H. Q., et al. (2014). Historical perspective of traditional indigenous medical practices: The current renaissance and conservation of herbal resources. *Evidence-Based Complementary and Alternative Medicine*, 525340. 20 pages. Hindawi Publishing Corporation https://doi.org/10.1155/2014/525340.

Pan, S. Y., Zhou, S. F., Gao, S. H., Yu, Z. L., Zhang, S. F., & Tang, M. K. (2013). New perspectives on how to discover drugs from herbal medicines: CAM's outstanding contribution to modern therapeutics. *Evidence-Based Complementary and Alternative Medicine, 2013*, 627375.

Panche, A. N., Diwan, A. D., & Chandra, S. R. (2016). Flavonoids: An overview. *Journal of Nutritional Science, 5*, e47.

Pandey, A., & Tripathi, S. (2014). Concept of standardization, extraction and pre phytochemical screening strategies for herbal drug. *Journal of Pharmacognosy and Phytochemistry, 2*(5), 115–119.

Patil, A. A., Sachin, B. S., Wakte, P. S., & Shinde, D. B. (2013). Optimization of supercritical fluid extraction and HPLC identification of wedelolactone from Wedelia calendulacea by orthogonal array design. *Journal of Advanced Research, 5*, 629–635.

Patil, P. S., & Shettigar, R. (2010). An advancement of analytical techniques in herbal research. *Journal of Advances in Science and Research, 1*(1), 08–14.

Paz, J. E. W., Márquez, D. B. M., Ávila, G. C. M., Cerda, R. E. B., & Aguilar, C. N. (2015). Ultrasound-assisted extraction of polyphenols from native plants in the mexican desert. *Ultrasonics Sonochemistry, 22*, 474–481.

Primack, R. B., & Miller-Rushing, A. J. (2009). The role of botanical gardens in climate change research. *The New Phytologist, 182*, 303–313.

Rafieian-Kopaei, M. (2013). Medicinal plants and the human needs. *Journal of Herbmed Pharmacology, 1*, 1–2.

Rahmalia, W., Fabre, J. F., & Mouloungui, Z. (2015). Effects of cyclohexane/acetone ratio on bixin extraction yield by accelerated solvent extraction method. *Procedia Chemistry, 14*, 455–464.

Raina, R., Chand, R., & Sharma, Y. P. (2011). Conservation strategies of some important medicinal plants. *International Journal of Medicinal and Aromatic Plant, 1*, 342–347.

Rathi, B. S., Bodhankar, S. L., & Baheti, A. M. (2006). Evaluation of aqueous leaves extract of Moringa oleifera Linn for wound healing in albino rats. *Indian Journal of Experimental Biology, 44*, 898–901.

Rietjens, I. M. C. M., Louisse, J., & Beekmann, K. (2017). The potential health effects of dietary phytoestrogens. *British Journal of Pharmacology, 174*, 1263–1280.

Rizvi, M. M. A., Irshad, M., Hassadi, G. E., & Younis, S. B. (2009). Bioefficacies of Cassia fistula: An Indian labrum (review). *African Journal of Pharmacy and Pharmacology, 3*(6), 287–292.

Sajid, M., Cameotra, S. S., Khan, M. S. A., Ahmad, I., & Khan, M. S. A. (2019). Nanoparticle based delivery of phytomedicines. In I. Ahmad, & D. Chattopadhyay (Eds.), *New look to phytomedicine- advancements in herbal products as novel drug leads* Academic Press Elsevier, ISBN:978-0-12-814619-4.

Salih, E., Kanninen, M., Sipi, M., Luukkanen, O., Hiltunen, R., Vuorela, H., et al. (2017). Tannins, flavonoids and stilbenes in extracts of african savanna woodland trees Terminalia brownii, Terminalia laxiflora and Anogeissus leiocarpus showing promising antibacterial potential. *South African Journal of Botany, 108*, 370–386.

Santos, S. A., Villaverde, J. J., Freire, C. S., Domingues, M. R. M., Neto, C. P., & Silvestre, A. J. (2012). Phenolic composition and antioxidant activity of Eucalyptus grandis, E. urograndis (E. grandis_ E. urophylla) and E. maidenii bark extracts. *Industrial Crops and Products, 39*, 120–127.

Sarker, S. D., & Nahar, L. (2007). *Chemistry for pharmacy students general, organic and natural product chemistry* (pp. 283–359). England: John Wiley and Sons.

Schippmann, U., Leaman, D. J., Cunningham, A. B., & Walter, S. (2005). Impact of cultivation and collection on the conservation of medicinal plants: global trends and issues. In *III WOCMAP congress on medicinal and aromatic plants: Conservation, cultivation and sustainable use of medicinal and aromatic plants.*

Shah, V., & Pandey, S. (2017). Phytomedicine: An emerging opportunity of research and development in Vindhyan Ecoregion, India. *Journal of Medicinal Plants Studies, 5*(6), 11–14.

Sivakumar, V., Ilanhtiraiyan, S., Ilayaraja, K., Ashly, A., & Hariharan, S. (2014). Influence of ultrasound on avaram bark (Cassia auriculata) tannin extraction and tanning. *Chemical Engineering Research and Design, 92*, 1827–1833.

Smolinski, M. S., Hamburg, M. A., & Lederberg, J. (Eds.). (2003). *Microbial threats to health: Emergence, detection, and response* (pp. 203–210). Washington, DC: Institute of Medicine, National Academies Press.

Srivastava, A., Srivastava, P., Pandey, A., Khanna, V. K., & Pant, A. B. (2019). Phytomedine: A potential alternative medicine in controlling neurological disorders. In K. MSA, I. Ahmad, & D. Chattopadhyay (Eds.), *New look to phytomedicine-advancements in herbal products as novel drug leads* (pp. 3–11), ISBN:978-0-12-814619-4.

Srujana, S. T., Babu, K. R., & Rao, B. S. S. (2012). Phytochemical investigation and biological activity of leaves extract of plant Boswellia serrata. *Pharmaceutical Innovation, 1*, 22–46.

Subramanian, R., Raj, V., Manigandan, K., & Elangovan, N. (2015). Antioxidant activity of hopeaphenol isolated from Shorea roxburghii stem bark extract. *Journal of Taibah University for Science, 9*, 237–244.

Sulaiman, S. F., Sajak, A. A. B., Ooi, K. L., Supriatno, & Seow, E. M. (2011). Effect of solvents in extracting polyphenols and antioxidants of selected raw vegetables. *Journal of Food Composition and Analysis, 24*, 506–515.

Suntar, I., Khan, H., Patel, S., Celano, R., & Rastrelli, L. (2018). An overview on Citrus aurantium L.: Its functions as food ingredient and therapeutic agent. *Oxidative Medicine and Cellular Longevity, 78*, 642–669.

Sutar, N., Garai, R., Sharma, U. S., & Sharma, U. K. (2010). Anthelmintic activity of Platycladus orientalis leaves extract. *International Journal of Parasitology Research, 2*(2), 1–3.

Swarts, N. D., & Dixon, K. W. (2009). Terrestrial orchid conservation in the age of extinction. *Annals of Botany, 104*, 543–556.

Tamashiro Filho, P., Olaitan, B. S., de Almeida, D. A. T., da Silva Lima, J. C., Marson-Ascencio, P. G., Ascencio, S. D., et al. (2012). Evaluation of antiulcer activity and mechanism of action of methanol stem bark extract of Lafoensia pacari (Lytraceae) in experimental animals. *Journal of Ethnopharmacology, 144*, 497–505.

Tan, J. T., Jiang, G., & Hu, F. (2014). Simultaneous identification and quantification of five flavonoids in the seeds of Rheum palmatum L. by using accelerated solvent extraction and HPLC-PDA ESI/MSn. *Arabian Journal of Chemistry, 22.*

Tanase, C., Arca, S. C., & Muntean, D. L. (2019). A critical review of phenolic compounds extracted from the bark of Woody vascular plants and their potential biological activity. *Molecules, 24,* 1182.

Tantray, M. A., Akbar, S., Khan, R., Tariq, K. A., & Shawl, A. S. (2009). Humarain: A new dimeric gallic acid glycoside from Punica granatum L. Bark. *Fitoterapia, 80,* 223–225.

Thermo Fisher Scientific. (2013). *Methods optimization in accelerated solvent extraction in technical note 208* (pp. 1–4).

Um, M., Shin, G.-J., & Lee, J.-W. (2017). Extraction of total phenolic compounds from yellow poplar hydrolysate and evaluation of their antioxidant activities. *Industrial Crops and Products, 97,* 574–581.

Usenik, V., Stampar, F., & Veberic, R. (2009). Anthocyanins and fruit colour in plums (Prunus domestica L.) during ripening. *Food Chemistry, 114,* 529–534.

Veggi, P. C., Prado, J. M., Bataglion, G. A., Eberlin, M. N., & Meireles, M. A. A. (2014). Obtaining phenolic compounds from jatoba (Hymenaea courbaril L.) bark by supercritical fluid extraction. *Journal of Supercritical Fluids, 89,* 68–77.

Volis, S., & Blecher, M. (2010). Quasi in situ: A bridge between ex situ and in situ conservation of plants. *Biodiversity and Conservation, 19,* 2441–2454.

Vongsak, B., Sithisarn, P., Mangmool, S., Thongpraditchote, S., & Wongkrajang, Y. (2013). Maximizing total phenolics, total flavonoids contents and antioxidant activity of Moringa oleifera leaf extract by the appropriate extraction method. *Industrial Crops and Products, 44,* 566–571.

WHO. (1999). *WHO monographs on selected medicinal plants. Vol. 1* (pp. 1–295).

WHO. (2019). *WHO global report on traditional and complementary medicine.* World Health Organization 2019, ISBN:978-92-4-151543-6.

Wickrama Singhe, M. B. (2006). Quality control, screening, toxicity, and regulation of herbal drugs. In I. Ahmad, F. Aqil, & M. Owais (Eds.), *Modern phytomedicine. Turning medicinal plants into drugs.* Weinheim: WILEY-VCH Verlag GmbH & Co. KGaA, ISBN:3-527-31530-6. Copyright © 2006.

Wink, M. (2015). Modes of action of herbal medicines and plant secondary metabolites. *Medicine, 2*(3), 251–286.

Wong, K. L., Wong, R. N., Zhang, L., Liu, W. K., Ng, T. B., Shaw, P. C., et al. (2008). Recent approaches in herbal drug standardization. *International Journal of Intergrative Biology, 2*(3), 195–203.

Yallappa, S., Manjanna, J., Sindhe, M. A., Satyanarayan, N. D., Pramod, S. N., & Nagaraja, K. (2013). Microwave assisted rapid synthesis and biological evaluation of stable copper nanoparticles using T. arjuna bark extract. *Spectrochimica Acta. Part A, Molecular and Biomolecular Spectroscopy, 110,* 108–115.

Yu, H., Xie, C. X., Song, J. Y., Zhou, Y. Q., & Chen, S. L. (2010). TCMGIS-II based prediction of medicinal plant distribution for conservation planning: A case study of Rheum tanguticum. *Chinese Medicine, 5,* 31.

Yuan, Q. J., Zhang, Z. Y., Hu, J. A., Guo, L. P., Shao, A. J., & Huang, L. Q. (2010). Impacts of recent cultivation on genetic diversity pattern of a medicinal plant, Scutellaria baicalensis (Lamiaceae). *BMC Genetics, 11,* 52–59.

Chapter 2

Biotechnological approaches for conservation of medicinal plants

Luis Jesús Castillo-Pérez[a], Angel Josabad Alonso-Castro[b], Javier Fortanelli-Martínez[c], and Candy Carranza-Álvarez[d]

[a]*Programa Multidisciplinario de Posgrado en Ciencias Ambientales, Universidad Autónoma de San Luis Potosí, San Luis Potosí, Mexico,* [b]*Department of Pharmacy, Division of Natural and Exact Sciences, University of Guanajuato, Guanajuato, Mexico,* [c]*Instituto de Investigación de Zonas Desérticas, Universidad Autónoma de San Luis Potosí, San Luis Potosí, Mexico,* [d]*Facultad de Estudios Profesionales Zona Huasteca, Universidad Autónoma de San Luis Potosí, San Luis Potosí, México*

1. Introduction

Medicinal plants are an important global resource. Health systems in developed countries use folk medicine for primary health care. In some countries or regions, traditional medicine is called alternative or complementary medicine (Rosenbloom, Chaudhary, & Castro-Eschenbach, 2011; WHO, 2019). The World Health Organization (WHO) estimates that more than 80% of the world population use traditional medicine, mainly based on medicinal plants. In some countries, health professionals recommend the use of medicinal plants as an alternative treatment (Alonso-Castro et al., 2017). The International Union for Conservation of Nature and the World Wildlife Fund have recorded approximately 60,000 angiosperms used for medicinal purposes worldwide (Chen et al., 2016; Jamshidi-Kia, Lorigooini, & Amini-Khoei, 2018). A medicinal plant is defined as any plant species that contains secondary metabolites that can be used for therapeutic purposes or can be used as precursors for the synthesis of new drugs (Penso, 1980; Seca & Pinto, 2019; Yuan, Ma, Ye, & Piao, 2016).

However, it is estimated that approximately 15,000 wild populations of medicinal plants are threatened due to their extensive commercialization (Brower, 2008; Wang, Usmanovich, Luo, Xu, & Wu, 2019). Therefore it is necessary to establish useful strategies for the conservation of medicinal herbs.

This chapter contains evidence on the diversity and overexploitation of medicinal plants around the world. In addition, a special emphasis on the conservation of medicinal plants is given, addressing some strategies with biotechnological approaches such as the plant tissue culture (PTC), a technique

Phytomedicine: A Treasure of Pharmacologically Active Products from Plants
https://doi.org/10.1016/B978-0-12-824109-7.00002-9

that has managed to improve the yield of medicinal plants. The information will be useful to provide some strategies that promote a sustainable use of medicinal plants.

2. Knowledge about diversity of medicinal plants in the last decade

Plant and cultural diversity in the world offer new alternatives for the implementation of crops and products based on medicinal plants. However, a large number of plant species, distributed in many regions of the Earth, remain to be scientifically studied. A bibliographic search of scientific reports from the last decade was conducted on academic databases to analyze the diversity of medicinal plants in several regions of the planet and was carried out. The search of information was based on three pairs of keywords: medicinal plants, global diversity, and bioactive compounds.

The results show that the Asian continent, leading by India, presents the highest number of scientific reports (36%) (Fig. 1) (Table 1). The information contained in Table 1 does not duplicate medicinal plants. India has approximately 17,000 species of higher plants, due to its wide range of ecosystems, of which 45% are used for medicinal purposes and are enlisted in traditional medicinal systems such as Ayurveda (Singh, 2015).

Some reports cited in Table 1 indicate that ethnobotanical studies were performed for the first time in different countries, regions, and natural protected areas (i.e., Ullah et al., 2013). In addition, the ethnomedicinal uses of different herb species were documented for the first time (i.e., Eisenman, Zaurov, & Struwe, 2013; Shrestha, Shrestha, Koju, Shrestha, & Wang, 2016). This indicates that the biodiversity of many areas in the world remains to be studied. Many studies were carried out with indigenous populations, which preserve their ancient information. However, many studies pointed out the need to preserve folk knowledge (Eddouks, Ajebli, & Hebi, 2017; Ribeiro, Bieski, Balogun, & de Oliveira Martins, 2017). Ethnobotanical studies are necessary to preserve local ethnomedicinal information of medicinal plants and update their medicinal uses. Urbanization and loss of biodiversity are among the causes of the loss of traditional knowledge. In addition, many studies remark the need to carry out conservation programs in the studied areas (Li & Xing, 2016). Some plant species reported in Table 1 are classified as endangered (Rahmawati, Mustofa, & Haryanti, 2020). The information clearly indicates that medicinal plants continue to be used for primary health care in many regions of the world (Sarri, Mouyet, Benziane, & Cheriet, 2014). Furthermore, medicinal plants are now used to treat current emergent diseases such as obesity, AIDS, diabetes, and malaria (Alonso-Castro, Domínguez, Zapata-Morales, & Carranza-Álvarez, 2015; Ngarivhume, van'tKlooster, de Jong, & Van der Westhuizen, 2015). There is also a need to validate the ethnomedicinal uses by carrying out pharmacological and toxicological studies with plants mentioned in Table 1.

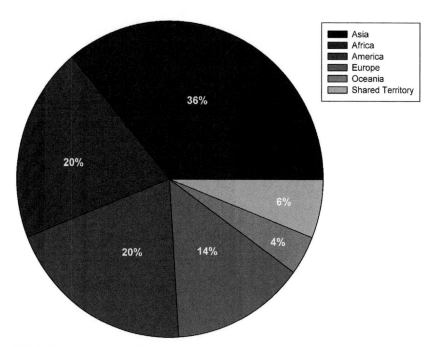

FIG. 1 Percentage of scientific research that studies the diversity of medicinal plants in the world ($n = 50$). *(Credit: Own elaboration.)*

The use of medicinal plants dates from centuries in countries like India, Japan, China, and Mexico (Jeyaprakash, Lego, Payum, Rathinavel, & Jayakumar, 2017; Leonti, Sticher, & Heinrich, 2003).

The use of medicinal plants is deeply rooted in the Asian continent. In contrast, Europe and Oceania showed less than 10 studies, which represent 18% of the diversity of medicinal plants in the last decade. Spain and Italy (in Europe) and three islands of Vanuatu and Northern New South Wales (in Oceania) are the countries where some research has been carried out on this topic (Table 1). In addition, developed countries such as the United States, the United Kingdom, Switzerland, Germany, and France do not have diverse studies on their medicinal plants during the last decade. In these countries, many pharmaceutical laboratories are situated as following: Pfizer, Merck, Johnson & Johnson, Abbott Laboratories, and Lilly in the United States of America; GlaxoSmithKline (GSK) and AstraZeneca, located in the United Kingdom; Novartis and Roche in Switzerland; and Sanofi-Aventis, in France (Hasegawa & Hambrecht, 2003; McCay, Lemer, & Wu, 2009). These countries are also considered as the world's largest producers and exporters of drugs. Germany, Switzerland, and the United States are the main exporters of medicines. In contrast the African continent, possessing a great diversity of plants, only exports 0.2% of all medicines in the world (Stevens & Huys, 2017). Countries like Russia do not make the information available for the international scientific community (Shikov et al., 2014).

TABLE 1 Studies that record the diversity of medicinal plants in several parts of the world in the last decade.

Region	Country	Continent	Diversity (number of species registered)	Relevant notes	Reference
Khulna City	Bangladesh	Asia	67	The use of medicinal plants has a great impact among allopathic physicians	Akber et al. (2011)
Chittagong Hill Tracts region	Bangladesh	Asia	73	Chakma population has an extensive use of medicinal plants	Rahmatullah et al. (2012)
Hainan Island	China	Asia	264	The loss of folk knowledge might occur due to degradation of the local environment	Li and Xing (2016)
Medog County, Tibet	China	Asia	37	The documentation of medicinal plants in this region had not been previously carried out	Yang, Chen, et al. (2020) and Yang, Yang, et al. (2020)
Almora District, Central Himalaya Region	India	Asia	188	Some medicinal plants require urgent actions for their protection and conservation	Kumari, Joshi, and Tewari (2011)
Odisha, East Coast of The Bay of Bengal	India	Asia	68	This research provides a list of medicinal plants, some of them without pharmacological studies, used to treat rheumatism	Panda et al. (2014)

Location	Country	Continent	Number	Notes	Reference
Indian Himalayan Region	India	Asia	90	Invasive medicinal plants can cause loss of local biodiversity. Monitorization of invasive medicinal plants should be carried out	Sekar (2012)
Ladakh, Kashmir (Himalaya Region)	India	Asia	111	More ethnobotanical studies are necessary to update the information in India	Dar, Shahnawaz, and Qazi (2017)
Daying Ering Memorial Wildlife Sanctuary	India	Asia	73	Most of the plants lack pharmacological studies that validate their ethnomedicinal uses	Jeyaprakash et al. (2017)
Manipur	India	Asia	145	Medicinal plants with magico-religious practices are mainly used by male members of the Zeliangrong community. Some side effects of excessive consumption of medicinal herbs are reported	Panmei, Gajurel, and Singh (2019)
Karo community, Sumatra Septentrional	Indonesia	Asia	344	There is a high demand to obtain medicinal herbs in the traditional Kabanjahe market. Women possess more knowledge of medicinal plants than men	Silalahi, Walujo, Supriatna, and Mangunwardoyo (2015)
Central Sulawesi Province	Indonesia	Asia	89	Some of the medicinal plants are classified as endangered species. Local conservation should be encouraged	Rahmawati et al. (2020)

Continued

TABLE 1 Studies that record the diversity of medicinal plants in several parts of the world in the last decade — cont'd

Region	Country	Continent	Diversity (number of species registered)	Relevant notes	Reference
Whole country	Kyrgyzstan	Asia	200	This is the first work reporting the ethnomedicinal use of plant species in this country	Eisenman et al. (2013)
Sankhuwasabha District	Nepal	Asia	48	Medicinal plants are mainly used to treat gastrointestinal disorders. The work reported the undocumented use of 10 medicinal plants	Shrestha et al. (2016)
Wana District South Waziristan Agency	Pakistan	Asia	50	This is the first report documenting the ethnomedicinal use of plant species in this region	Ullah et al. (2013)
Whole country	Thailand	Asia	2187	Most of plant species are used for the treatment of gastrointestinal disorders. The conservation of local herbs should be carried out	Phumthum et al. (2018)
Whole country	Uzbekistan	Asia	400	This is the first report documenting the ethnomedicinal use of plant species in this country	Eisenman et al. (2013)
Bac Huong Hoa nature reserve	Vietnam	Asia	111	This is the first report documenting the ethnomedicinal use of plant species in this nature reserve. The medicinal properties of 8 plants is reported for the first time	Lee et al. (2019)

M'Sila City	Algeria	Africa	58	Medicinal plants could be incorporated in the national health systems to reduce costs	Boudjelal et al. (2013)
M'Sila City	Algeria	Africa	36	Medicinal plants are used for primary health care problems, including gastrointestinal and respiratory acute diseases	Sarri et al. (2014)
Administrative departments of Tiaret, Saida, Naama, Djelfa and M'sila	Algeria	Africa	97	This is the first report documenting the ethnomedicinal use of plant species in this region. Many uses of herbs were documented for the first time	Miara, Bendif, Hammou, and Teixidor-Toneu (2018)
Ankober District, North Shewa Zone, Amhara Region	Ethiopia	Africa	158	Plants are used for primary health care needs in human and livestock. The local knowledge is deep-rooted	Lulekal (2014)
Whole country	Gabon	Africa	217	Twenty-seven tons of medicinal plant-based products are sold in Gabon's main markets, equivalent to $1.5 million per year	Towns, Quiroz, Guinee, de Boer, and van Andel (2014)
Whole country	Guinea-Bissau	Africa	218	The transmission of local knowledge about medicinal plants relies on original dialects among communities	Catarino, Havik, and Romeiras (2016)
Daraa-Tafilalet Region (Province of Errachidia)	Morocco	Africa	194	Preservation of traditional knowledge is encouraged. Local population have a great expertise on the use of medicinal plants	Eddouks et al. (2017)

Continued

TABLE 1 Studies that record the diversity of medicinal plants in several parts of the world in the last decade—cont'd

Region	Country	Continent	Diversity (number of species registered)	Relevant notes	Reference
Whole country	Nigeria	Africa	115	Many medicinal plants used for the empirical treatment of diabetes lacked preclinical studies	Ezuruike and Prieto (2014)
Arua, Dokolo, Mbale, Bushenyi, Iganga, Rakai, Luwero and Kaabong districts	Uganda	Africa	236	Opportunistic infections such AIDS, malaria, among others are treated with medicinal plants. The concomitant use of retroviral agents and medicinal plants is a common practice	Anywar et al. (2020)
Chipinge district in the Manicaland Province	Zimbabwe	Africa	28	The medicinal plants used for the treatment of malaria is documented	Ngarivhume et al. (2015)
Whole continent	Whole continent	Africa	More than 5400	This research mentions the largest amount of medicinal plants from Africa exported for their sale in markets around the world	Van-Wyk (2015)
Misiones province, in the subtropics of Argentina	Argentina	America	509	The updated information about medicinal uses of herbs is documented. Many plants need to be included in conservation programs	Kujawska, Hilgert, Keller, and Gil (2017)

Region	Country	Continent	No.	Description	Reference
Salta province (South American Gran Chaco Region)	Argentina	America	115	This is the first report documenting the ethnomedicinal use of native plant species in this region. The adaptation on knowledge about the use of medicinal plants from other indigenous people is documented	Suárez (2019)
Caatinga (semiarid region)	Brazil	America	108	The knowledge transmission about exotic medicinal plants from the region has occurred during centuries	Rangel, Ramos, de Amorim, and de Albuquerque (2010)
North Araguaia microregion	Brazil	America	309	This is the first report documenting the ethnomedicinal use of native plant species in this region. The study preserves local folk knowledge	Ribeiro et al. (2017)
Ecuadorian Amazon	Ecuador	America	101	This study proposes a new methodology for ethnobotanical studies based on phylogenetic biases	Arias, Cevallos, Gaoue, Fadiman, and Hindle (2020)
Huasteca Potosina region	Mexico	America	73	The information here reported updated local ethnomedicinal uses of medicinal plants	Alonso-Castro et al. (2012)
Xalpatláhuac, Guerrero	Mexico	America	67	The use of medicinal plants is an essential practice among inhabitants from one of the poorest municipalities in the country	Juárez-Vázquez et al. (2013)
Mexico, Central America, and the Caribbean	Several countries	America	139	This report indicates that medicinal plants with diuretic and antidiabetic effects are now used for weight loss	Alonso-Castro et al. (2015)

Continued

TABLE 1 Studies that record the diversity of medicinal plants in several parts of the world in the last decade—cont'd

Region	Country	Continent	Diversity (number of species registered)	Relevant notes	Reference
Several states	Mexico	America	77	Health professionals recommend and prescribe medicinal plants to treat some acute diseases	Alonso-Castro et al. (2017)
Several states	Mexico	America	343	The information of medicinal plants used for gastrointestinal illnesses is discussed	Sharma, del Carmen Flores-Vallejo, Cardoso-Taketa, and Villarreal (2017)
South-Eastern area of the Partenio Regional Park	Italy	Europe	87	Medicinal plants are reported for human and veterinary use. Preservation of local knowledge and biodiversity is encouraged	Menale and Muoio (2014)
Mainarde Mountains (central-southern Apennine)	Italy	Europe	106	This is the first work reporting the ethnomedicinal use of plant species in a natural protected area	Fortini, Di Marzio, Guarrera, and Iorizzi (2016)
Arribes del Duero	Spain	Europe	70	Magical and medicinal purposes of plant species are informed	González, García-Barriuso, and Amich (2010)
Balearic Islands, Mediterranean Sea	Spain	Europe	121	Provides the first report about ethnobotanical information of the Mallorca island	Carrió and Vallès (2012)
Navarra region	Spain	Europe	19	Medicinal plants used for the treatment of most common eye problems are indicated	Calvo and Cavero (2016)

South of Alava	Spain	Europe	36	Edible plants with medicinal uses are reported	Alarcón, Pardo-de-Santayana, Priestley, Morales, and Heinrich (2015)
Whole country	Switzerland	Europe	768	Neighboring countries have influenced the folk medicinal knowledge in Switzerland	Dal Cero, Saller, and Weckerle (2014)
Whole continent	Whole continent	Europe	400	Some threatened medicinal plants are reported in this study	Allen et al. (2014)
Northern New South Wales	Australia	Oceania	32	The indigenous community of Yaegl rely on the use of medicinal plants for primary health care	Packer et al. (2012)
Three islands of Vanuatu	Vanuatu	Oceania	133	This is the first work reporting the folk medicinal use of plant species in this region	Bradacs, Heilmann, and Weckerle (2011)
Central region of Abyan governorate	Yemen	Africa/Asia	195	This is the first work reporting the folk medicinal use of plant species in this region	Al-Fatimi (2019)
The Russian Federation	Russia	Europa/Asia	32	Information about the use of medicinal plants should be accessible to everyone	Shikov et al. (2014)
Ağrı Province	Turkey	Europa/Asia	118	Medicinal plants in this region have a potential to develop new pharmacological products	Dalar, Mukemre, Unal, and Ozgokce (2018)

Credit: Own elaboration.

3. The problems of subtracting medicinal herbs

Urban sprawl around the world has contributed to the destruction of several ecosystems, including forests, jungles, mangrove, and marine ecosystems. The loss of many plant species, as well as the imminent removal of some others, has been the consequence of this intrusion and other anthropogenic actions such as forest fires, loss of habitats, and illegal extraction of species for commercialization (Scanes, 2018). In this context, medicinal plants are a main target of many traffickers. In many cases medicinal plants are illegally extracted from their ecosystems and can be used locally or transported over long distances to big urbanized cities, which cause the loss of biodiversity (Rajeswara-Rao, 2016). The main threats from medicinal plants are habitat destruction, bioprospecting, and their overexploitation for illegal trafficking (Chi et al., 2017).

The methods for the conservation of medicinal plants are increasingly necessary, due the accelerated loss of species and natural habitats in the world. It is estimated that approximately 15,000 species of medicinal plants may be endangered worldwide. Therefore the discovery of new drugs could be missed (Chen et al., 2016; Van-Wyk & Wink, 2018).

The Global Convention on Biological Diversity, the Convention on International Trade in Endangered Species (CITES), and The National Center for the Preservation of Medicinal Herbs (NCPMH) have worked on the conservation of medicinal species. Nevertheless, more scientific and governmental efforts about the loss of medicinal plants should be carried out. On the other hand, some works have been published studying the commercialization of medicinal species products worldwide (Gänger, 2015; Street, Stirk, & Van Staden, 2008; Vasisht, Sharma, & Karan, 2016). For instance, many countries in Latin America illegally export medicinal plants. The trade of medicinal plants has been increased during the last decade because people are afraid of allopathic medicine because of its undesirable side effects (Gänger, 2015; Robbins, 2000).

Pharmaceutical industry that produces plant-based medicines is often accused to commit acts of biopiracy, which is defined as the practice of patenting and overexploiting traditional remedies, obtaining great profits, from which the indigenous communities will not obtain any economic benefit (Efferth et al., 2016; Zakrzewski, 2002). In 2001 a group of researchers found more than 5000 patent references from 90 medicinal plants (80% from India) (Robertson, 2008).

Finally, it is important to highlight that countries with high biodiversity on Earth face problems with the illegal trade of medicinal plants. Aboriginal population from Mexico has an ancient knowledge of traditional medicine. However, this population struggle to obtain economical resources, and many times, they need to commercialize medicinal plants to obtain an economic income (Barreda, 2001; Martínez-Moreno, Alvarado-Flores, Mendoza-Cruz, & Basurto-Peña, 2006).

4. Conservation strategies for medicinal plants

In the last two decades, the demand for wild populations of medicinal plants has been increased between 8% and 15% annually in Europe, North America, and Asia, causing the reproductive capacity of some species to be irreversibly reduced (Gupta, 2017; Majeed, 2017). Therefore some strategies for the conservation of medicinal plants, such as in situ to ex situ conservation methods, have been developed. The declaration of protected natural areas around the world and the establishment of wildlife nurseries are essential for the retention of plants in their natural habitats (in situ conservation). Worldwide, approximately 13,000 protected areas have been established, representing 13.2 million km^2 or 8.81% of the Earth's land surface (Huang et al., 2002). Furthermore the establishment of wildlife nurseries has contributed to species-oriented cultivation and the domestication of endangered medicinal plants in protected areas or natural habitats (Chen et al., 2016; Prins, 1996). In situ conservation is important because in several regions of the planet, a large percentage of medicinal plants are endemic, and their ecological interactions with other plants and animals, as well as the environmental conditions of each region, ease the production of their bioactive compounds. The production of bioactive compounds by medicinal plants can be diminished under cultivation conditions (Kusari, Hertweck, & Spiteller, 2012; Núñez-Pons & Avila, 2015).

Ex situ conservation processes are important methods for the conservation of medicinal plants threatened by illegal trafficking. Ex situ conservation is often applied in plants (for instance, Orchidaceae family) with slow growth rates or high environmental specifications to thrive (Merritt, Hay, Swarts, Sommerville, & Dixon, 2014; Yadav, 2016).

Ethnobiological gardens and germplasm and seeds banks are also important processes for ex situ conservation (Chen et al., 2016; Chen & Sun, 2018; Que et al., 2016; Singh, Ansari, Singh, Singh, & Pal, 2017). Ex situ conservation methods based on plant tissue culture have the ability to function as a large raw material factory by producing large quantities of plant material that can be used for the study and the production of bioactive compounds (Kasagana & Karumuri, 2011; Niazian, 2019).

5. Plant tissue culture: A strategy for conservation of medicinal plants

Biotechnological processes based on PTC offer an integrated approach for the rapid multiplication and production of secondary metabolites with applications in the pharmaceutical, agricultural, cosmetic, and food industries (Bhatia, Sharma, Dahiya, & Bera, 2015; Vanisree et al., 2004). PTC technologies have a great impact on ex situ conservation of plant genetic resources due the international exchange of germplasm (Streczynski et al., 2019).

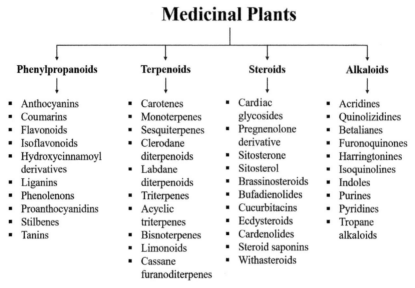

FIG. 2   Classification of plant-derived compounds. *(Credit: Own elaboration.)*

The establishment of in vitro cell cultures in the last decade has been an effective technique to produce specific medicinal compounds and finding new bioactive compounds. Fig. 2 shows different groups of bioactive compounds found in medicinal plants. With the use of plant biotechnology, many bioactive compounds are now synthesized by callus production.

Some advantages offered by the PTC are the following: (1) a higher production of plants in smaller facilities, (2) efficient intercontinental transfer of germplasm and vitroplants, (3) production of plants free from pathogenic microorganisms, (4) the production of the plants that is independent on the climate conditions, and (5) the production of the bioactive compounds that can be regulated (Cardoso, Oliveira, & Cardoso, 2019).

The cultivation of root sections of medicinal plants is a novel technique in which the genetic and biosynthetic stability of the plant is maintained, which provides accelerated growth and easy maintenance. From this methodology, many bioactive compounds have been synthesized (Kumar, Kumar, Nehra, Maan, & Kumar, 2017; Yang, Chen, et al., 2020; Yang, Yang, et al., 2020).

PTC has contributed to the production of many bioactive compounds (i.e., phenylpropanoids, alkaloids, quinones, terpenoids, steroids, essential oils, and amino acids) with pharmacological properties (Dias, Sousa, Alves, & Ferreira, 2016; Ochoa-Villarreal et al., 2016). In addition, research in several fields of experimental biology indicates that PTC will be one of the most recurrent

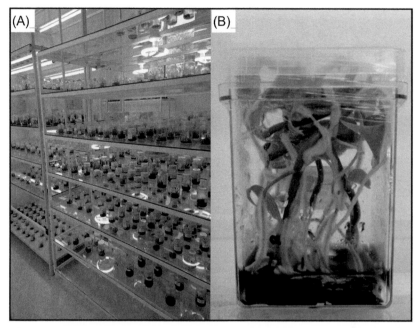

FIG. 3   (A) Cultivation room of the Environmental Sciences Research Laboratory, UASLP-Mexico and (B) Vitroplant of *Vanilla planifolia* (Orchidaceae). *(Photocredits: Castillo-Pérez, L.J.)*

techniques to produce bioactive compounds (Isah et al., 2018). The advantage of this methodology relies on providing a continuous and reliable source of natural products. The Research Laboratory of Environmental Sciences from the Universidad Autonoma de San Luis Potosi (UASLP) in San Luis Potosi, Mexico, has implemented a line of research focused on the study of bioactive compounds in species of the Orchidaceae family, especially with taxa that are distributed in the Huasteca Potosina region. This laboratory has also developed propagation protocols for more than 15 orchids (including vanilla) (Fig. 3). Research is carried out with the aim of conserving these orchid species and proposing sustainable and alternative methods for the study and production of bioactive compounds.

Table 2 shows a list of 20 representative studies about the obtention of bioactive compounds with medicinal or industrial potential using PTC during the last decade. This technique has been widely used since the beginning of the 20th century and has shown a significant demand. In vitro culture is an efficient strategy to produce plant species and their bioactive compounds (Table 2). In vitro culture also allows the alteration of biochemical pathways to ease the overproduction of specific bioactive compounds.

TABLE 2 Bioactive compounds isolated using plant tissue cultures in the last decade.

Plant name	Bioactive compound	Culture type	Reference
Aloe arborescens Mill	Iridoids, phenolics, flavonoids, and condensed tannins	Shoot	Amoo, Aremu, and Van-Staden (2012)
Arnebia hispidissima (Lehm). DC.	Alkannin	Callus and cell suspension	Shekhawat and Shekhawat (2011)
Baliospermum montanum (Willd.) Muell. Arg.	Steroids, triterpenoids, glycosides, saponins, alkaloids, flavanoids, phenolic compounds, and tannins	Callus	Johnson, Wesely, Hussain, and Selvan (2010)
Dendrobium fimbriatum Hook.	β-Sitosterol	Shoots	Paul, Joshi, Gurjar, Shailajan, and Kumaria (2017)
Dendrobium moniliforme (L.) Sw.	Polysaccharides, polyphenolics, and flavonoids	Suspension	Cui, Murthy, Moh, Cui, and Paek (2015)
Hypericum perforatum L.	Phenols and flavonoids	Roots	Cui, Chakrabarty, Lee, and Paek (2010)
Merwilla plumbea (Lindl.) Speta	Total phenolic, flavonoid, gallotannin and condensed tannin	Shoot	Baskaran, Ncube, and Van Staden (2012)
Moringa oleifera Lam	Kaempferol and quercetin	Suspension	Muhammad, Pauzi, Arulselvan, Abas, and Fakurazi (2013)
Opuntia robusta Wendl.	Phenolic acids and flavonoids	Callus	Astello-García, Robles-Martínez, Barba-de la Rosa, and del Socorro Santos-Díaz (2013)

Species	Target compounds	Culture type	Reference
Panax quinquefolius L.	Phenolic compounds	Suspension	Uchendu, Paliyath, Brown, and Saxena (2011)
Platanthera edgeworthii (Hook.f. ex Collett) R.K. Gupta	Phenols	Callus	Giri et al. (2012)
Passiflora spp.	Flavonoid C-glycosides	Callus and suspension	Ozarowski and Thiem (2013)
Prunella vulgaris L.	Alkaloid, saponins, phenolics, and tannins	Suspension	Rasool, Ganai, Akbar, Kamili, and Masood (2010)
Randia echinocarpa Moc. andSessé ex DC.	Melanins	Callus	Valenzuela-Atondo et al. (2020)
Ruta graveolens L.	Phenolic acids and furanocoumarins	Suspension	Szopa, Ekiert, Szewczyk, and Fugas (2012)
Saussurea involucrata (Kar. et Kir.)	Flavonoids (syringin and rutin)	Callus	Kuo et al. (2015)
Schisandra chinensis (Turcz.) Baill.	Dibenzocyclooctadiene lignans	Suspension	Szopa, Ekiert, and Ekiert (2017)
Sedum roseum (L.) Scop.	Salidroside, polysaccharides, phenolics, and flavonoids	Callus	Li et al. (2016)
Swertia chirayita H. Karst.	Lignans, alkaloids, flavonoids, terpenoids, iridoids, secoiridoids, and other compounds such as chiratin, ophelicacid, palmitic acid, oleic acid, and stearic acid	Suspension	Kumar and Van Staden (2016))
Zaleya decandra (L.) Burm.f.	Betalains	Callus	Radfar, Sudarshana, and Niranjan (2012)

Credit: Own elaboration.

6. Conclusions

The diversity of medicinal plants around the world is huge. However, some regions on the planet have not reported research on this topic. Asia, Africa, and America are the continents with most of the research on medicinal plants. Europe and Oceania have very few studies on this topic. There are some laws and recommendations for the conservation and sustainable use of medicinal plants. However, only a small portion of medicinal plants has accomplished sufficient protection through conventional conservation in nature reserves or botanical gardens. PTC has been proved to be an excellent biotechnological technique for the propagation, conservation, and production of medicinal plants.

References

Akber, M., Seraj, S., Islam, F., Ferdausi, D., Ahmed, R., Nasrin, D., et al. (2011). A survey of medicinal plants used by the traditional medicinal practitioners of Khulna City, Bangladesh. *American Eurasian Journal of Sustainable Agriculture*, *5*, 177–195.

Alarcón, R., Pardo-de-Santayana, M., Priestley, C., Morales, R., & Heinrich, M. (2015). Medicinal and local food plants in the south of Alava (Basque Country, Spain). *Journal of Ethnopharmacology*, *176*, 207–224.

Al-Fatimi, M. (2019). Ethnobotanical survey of medicinal plants in Central Abyan governorate, Yemen. *Journal of Ethnopharmacology*, *241*, 111973.

Allen, D., Bilz, M., Leaman, D. J., Miller, R. M., Timoshyna, A., & Window, J. (2014). *European red list of medicinal plants. Vol. 10* (p. 907382). Luxembourg: Publications Office of the European Union.

Alonso-Castro, A. J., Domínguez, F., Maldonado-Miranda, J. J., Castillo-Pérez, L. J., Carranza-Álvarez, C., Solano, E., et al. (2017). Use of medicinal plants by health professionals in Mexico. *Journal of Ethnopharmacology*, *198*, 81–86.

Alonso-Castro, A. J., Domínguez, F., Zapata-Morales, J. R., & Carranza-Álvarez, C. (2015). Plants used in the traditional medicine of Mesoamerica (Mexico and Central America) and the Caribbean for the treatment of obesity. *Journal of Ethnopharmacology*, *175*, 335–345.

Alonso-Castro, A. J., Maldonado-Miranda, J. J., Zarate-Martinez, A., del Rosario Jacobo-Salcedo, M., Fernández-Galicia, C., Figueroa-Zuñiga, L. A., et al. (2012). Medicinal plants used in the Huasteca Potosina, México. *Journal of Ethnopharmacology*, *143*, 292–298.

Amoo, S. O., Aremu, A. O., & Van-Staden, J. (2012). *In vitro* plant regeneration, secondary metabolite production and antioxidant activity of micropropagated *Aloe arborescens* mill. *Plant Cell, Tissue and Organ Culture*, *111*, 345–358.

Anywar, G., Kakudidi, E., Byamukama, R., Mukonzo, J., Schubert, A., & Oryem-Origa, H. (2020). Indigenous traditional knowledge of medicinal plants used by herbalists in treating opportunistic infections among people living with HIV/AIDS in Uganda. *Journal of Ethnopharmacology*, *246*, 112205.

Arias, D. M. R., Cevallos, D., Gaoue, O. G., Fadiman, M. G., & Hindle, T. (2020). Non-random medicinal plants selection in the Kichwa community of the Ecuadorian Amazon. *Journal of Ethnopharmacology*, *246*, 112220.

Astello-García, M. G., Robles-Martínez, M., Barba-de la Rosa, A. P., & del Socorro Santos-Díaz, M. (2013). Establishment of callus from *Opuntia robusta* Wendl., a wild and medicinal cactus, for phenolic compounds production. *African Journal of Biotechnology*, *12*, 3204–3207.

Barreda, A. (2001). Biopiratería y resistencia en México. *El cotidiano, 18*, 21–39.

Baskaran, P., Ncube, B., & Van Staden, J. (2012). *In vitro* propagation and secondary product production by *Merwillaplumbea* (Lindl.) Speta. *Plant Growth Regulation, 67*, 235–245.

Bhatia, S., Sharma, K., Dahiya, R., & Bera, T. (2015). *Modern applications of plant biotechnology in pharmaceutical sciences*. Academic Press.

Boudjelal, A., Henchiri, C., Sari, M., Sarri, D., Hendel, N., Benkhaled, A., et al. (2013). Herbalists and wild medicinal plants in M'Sila (North Algeria): An Ethnopharmacology survey. *Journal of Ethnopharmacology, 148*, 395–402.

Bradacs, G., Heilmann, J., & Weckerle, C. S. (2011). Medicinal plant use in Vanuatu: A comparative ethnobotanical study of three islands. *Journal of Ethnopharmacology, 137*, 434–448.

Brower, V. (2008). Back to nature: Extinction of medicinal plants threatens drug discovery. *Journal of the National Cancer Institute, 100*, 838–839.

Calvo, M. I., & Cavero, R. Y. (2016). Medicinal plants used for ophthalmological problems in Navarra (Spain). *Journal of Ethnopharmacology, 190*, 212–218.

Cardoso, J. C., Oliveira, M. E., & Cardoso, F. D. C. (2019). Advances and challenges on the *in vitro* production of secondary metabolites from medicinal plants. *Horticultura Brasileira, 37*, 124–132.

Carrió, E., & Vallès, J. (2012). Ethnobotany of medicinal plants used in eastern Mallorca (Balearic Islands, Mediterranean Sea). *Journal of Ethnopharmacology, 141*, 1021–1040.

Catarino, L., Havik, P. J., & Romeiras, M. M. (2016). Medicinal plants of Guinea-Bissau: Therapeutic applications, ethnic diversity and knowledge transfer. *Journal of Ethnopharmacology, 183*, 71–94.

Chandra, S. (2012). Invasive alien plants of Indian Himalayan region—Diversity and implication. *American Journal of Plant Sciences, 3*, 177–184.

Chen, G., & Sun, W. (2018). The role of botanical gardens in scientific research, conservation, and citizen science. *Plant Diversity, 40*, 181–188.

Chen, S. L., Yu, H., Luo, H. M., Wu, Q., Li, C. F., & Steinmetz, A. (2016). Conservation and sustainable use of medicinal plants: Problems, progress, and prospects. *Chinese Medicine, 11*, 37.

Chi, X., Zhang, Z., Xu, X., Zhang, X., Zhao, Z., Liu, Y., et al. (2017). Threatened medicinal plants in China: Distributions and conservation priorities. *Biological Conservation, 210*, 89–95.

Cui, X. H., Chakrabarty, D., Lee, E. J., & Paek, K. Y. (2010). Production of adventitious roots and secondary metabolites by *Hypericum perforatum* L. in a bioreactor. *Bioresource Technology, 101*, 4708–4716.

Cui, H. Y., Murthy, H. N., Moh, S. H., Cui, Y. Y., & Paek, K. Y. (2015). Establishment of protocorm suspension cultures of *Dendrobium candidum* for the production of bioactive compounds. *Horticulture, Environment, and Biotechnology, 56*, 114–122.

Dal Cero, M., Saller, R., & Weckerle, C. S. (2014). The use of the local flora in Switzerland: A comparison of past and recent medicinal plant knowledge. *Journal of Ethnopharmacology, 151*, 253–264.

Dalar, A., Mukemre, M., Unal, M., & Ozgokce, F. (2018). Traditional medicinal plants of Ağrı Province, Turkey. *Journal of Ethnopharmacology, 226*, 56–72.

Dar, R. A., Shahnawaz, M., & Qazi, P. H. (2017). General overview of medicinal plants: A review. *Journal of Phytopharmacology, 6*, 349–351.

Dias, M. I., Sousa, M. J., Alves, R. C., & Ferreira, I. C. (2016). Exploring plant tissue culture to improve the production of phenolic compounds: A review. *Industrial Crops and Products, 82*, 9–22.

Eddouks, M., Ajebli, M., & Hebi, M. (2017). Ethnopharmacological survey of medicinal plants used in Daraa-Tafilalet region (province of Errachidia), Morocco. *Journal of Ethnopharmacology, 198*, 516–530.

Efferth, T., Banerjee, M., Paul, N. W., Abdelfatah, S., Arend, J., Elhassan, G., et al. (2016). Biopiracy of natural products and good bioprospecting practice. *Phytomedicine, 23*, 166–173.

Eisenman, S. W., Zaurov, D. E., & Struwe, L. (2013). *Medicinal plants of Central Asia: Uzbekistan and Kyrgyzstan.* New York: Springer-Verlag.

Ezuruike, U. F., & Prieto, J. M. (2014). The use of plants in the traditional management of diabetes in Nigeria: Pharmacological and toxicological considerations. *Journal of Ethnopharmacology, 155*, 857–924.

Fortini, P., Di Marzio, P., Guarrera, P. M., & Iorizzi, M. (2016). Ethnobotanical study on the medicinal plants in the Mainarde Mountains (central-southern Apennine, Italy). *Journal of Ethnopharmacology, 184*, 208–218.

Gänger, S. (2015). World trade in medicinal plants from Spanish America, 1717–1815. *Medical History, 59*, 44–62.

Giri, L., Dhyani, P., Rawat, S., Bhatt, I. D., Nandi, S. K., Rawal, R. S., et al. (2012). *In vitro* production of phenolic compounds and antioxidant activity in callus suspension cultures of *Habenariaedgeworthii*: A rare Himalayan medicinal orchid. *Industrial Crops and Products, 39*, 1–6.

González, J. A., García-Barriuso, M., & Amich, F. (2010). Ethnobotanical study of medicinal plants traditionally used in the Arribes del Duero, western Spain. *Journal of Ethnopharmacology, 131*, 343–355.

Gupta, R. (2017). Agrotechnology of medicinal plants. In *The medicinal plant industry* Routledge.

Hasegawa, G. R., & Hambrecht, F. T. (2003). The confederate medical laboratories. *Southern Medical Journal, 96*, 1221–1231.

Huang, H., Han, X., Kang, L., Raven, P., Jackson, P. W., & Chen, Y. (2002). Conserving native plants in China. *Science, 297*, 935–936.

Isah, T., Umar, S., Mujib, A., Sharma, M. P., Rajasekharan, P. E., Zafar, N., et al. (2018). Secondary metabolism of pharmaceuticals in the plant *in vitro* cultures: Strategies, approaches, and limitations to achieving higher yield. *Plant Cell, Tissue and Organ Culture, 132*, 239–265.

Jamshidi-Kia, F., Lorigooini, Z., & Amini-Khoei, H. (2018). Medicinal plants: Past history and future perspective. *Journal of Herbmed Pharmacology, 7*, 1–7.

Jeyaprakash, K., Lego, Y. J., Payum, T., Rathinavel, S., & Jayakumar, K. (2017). Diversity of medicinal plants used by adi community in and around area of D'Ering wildlife sanctuary, Arunachal Pradesh, India. *World Scientific News, 65*, 135–159.

Johnson, M., Wesely, E. G., Hussain, M. Z., & Selvan, N. (2010). *In vivo* and *in vitro* phytochemical and antibacterial efficacy of *Baliospermummontanum* (Willd.) Muell. Arg. *Asian Pacific Journalof Tropical Medicine, 3*, 894–897.

Juárez-Vázquez, C. M., Carranza-Álvarez, C., Alonso-Castro, A. J., González-Alcaraz, V. F., Bravo-Acevedo, E., Chamarro-Tinajero, F. J., et al. (2013). Ethnobotany of medicinal plants used in Xalpatlahuac, Guerrero, México. *Journal of Ethnopharmacology, 148*, 521–527.

Kasagana, V. N., & Karumuri, S. S. (2011). Conservation of medicinal plants (past, present & future trends). *Journal of Pharmaceutical Sciences and Research, 3*, 1378.

Kujawska, M., Hilgert, N. I., Keller, H. A., & Gil, G. (2017). Medicinal plant diversity and intercultural interactions between indigenous guarani, criollos and polish migrants in the subtropics of Argentina. *PLoS One, 12*, e0169373.

Kumar, U., Kumar, A., Nehra, A., Maan, P., & Kumar, A. (2017). Biotechnological approaches for enhanced secondary metabolite production using hairy root cultures. *International Journal of Biological Sciences, 7*, 82–107.

Kumar, V., & Van Staden, J. (2016). A review of *Swertiachirayita* (Gentianaceae) as a traditional medicinal plant. *Frontiers in Pharmacology, 6*, 308.

Kumari, P., Joshi, G. C., & Tewari, L. M. (2011). Diversity and status of ethno-medicinal plants of Almora district in Uttarakhand, India. *International Journal of Biodiversity and Conservation, 3*, 298–326.

Kuo, C. L., Agrawal, D. C., Chang, H. C., Chiu, Y. T., Huang, C. P., Chen, Y. L., et al. (2015). *In vitro* culture and production of syringin and rutin in *Saussureainvolucrata* (Kar. et Kir.)–an endangered medicinal plant. *Botanical Studies, 56*, 12.

Kusari, S., Hertweck, C., & Spiteller, M. (2012). Chemical ecology of endophytic fungi: Origins of secondary metabolites. *Chemistry & Biology, 19*, 792–798.

Lee, C., Kim, S. Y., Eum, S., Paik, J. H., Bach, T. T., Darshetkar, A. M., et al. (2019). Ethnobotanical study on medicinal plants used by local Van Kieu ethnic people of Bac Huong Hoa nature reserve, Vietnam. *Journal of Ethnopharmacology, 231*, 283–294.

Leonti, M., Sticher, O., & Heinrich, M. (2003). Antiquity of medicinal plant usage in two macro-Mayan ethnic groups (Mexico). *Journal of Ethnopharmacology, 88*, 119–124.

Li, H., Piao, X. C., Gao, R., Jin, M., Jiang, J., & Lian, M. L. (2016). Effect of several physicochemical factors on callus biomass and bioactive compound accumulation of *R. sachalinensis* bioreactor culture. *In Vitro Cellular & Developmental Biology-Plant, 52*, 241–250.

Li, D. L., & Xing, F. W. (2016). Ethnobotanical study on medicinal plants used by local Hoklos people on Hainan Island, China. *Journal of Ethnopharmacology, 194*, 358–368.

Lulekal, E. (2014). *Plant diversity and ethnobotanical study of medicinal plants in Ankober District, north Shewa zone of Amhara Region, Ethiopia* (Doctoral dissertation). Addis Ababa University.

Majeed, M. (2017). Evidence-based medicinal plant products for the health care of world population. *Annals of Phytomedicine, 6*, 1–4.

Martínez-Moreno, D., Alvarado-Flores, R., Mendoza-Cruz, M., & Basurto-Peña, F. (2006). Plantas medicinales de cuatro mercados del estado de Puebla, México. *Botanical Sciences, 79*, 79–87.

McCay, L., Lemer, C., & Wu, A. W. (2009). Laboratory safety and the WHO world alliance for patient safety. *Clinica Chimica Acta, 404*, 6–11.

Menale, B., & Muoio, R. (2014). Use of medicinal plants in the south-eastern area of the partenio regional park (Campania, southern Italy). *Journal of Ethnopharmacology, 153*, 297–307.

Merritt, D. J., Hay, F. R., Swarts, N. D., Sommerville, K. D., & Dixon, K. W. (2014). *Ex situ* conservation and cryopreservation of orchid germplasm. *International Journal of Plant Sciences, 175*, 46–58.

Miara, M. D., Bendif, H., Hammou, M. A., & Teixidor-Toneu, I. (2018). Ethnobotanical survey of medicinal plants used by nomadic peoples in the Algerian steppe. *Journal of Ethnopharmacology, 219*, 248–256.

Muhammad, A. A., Pauzi, N. A. S., Arulselvan, P., Abas, F., & Fakurazi, S. (2013). *In vitro* wound healing potential and identification of bioactive compounds from *Moringa oleifera* Lam. *BioMed Research International, 2013*, 1–10.

Ngarivhume, T., van't Klooster, C. I., de Jong, J. T., & Van der Westhuizen, J. H. (2015). Medicinal plants used by traditional healers for the treatment of malaria in the Chipinge district in Zimbabwe. *Journal of Ethnopharmacology, 159*, 224–237.

Niazian, M. (2019). Application of genetics and biotechnology for improving medicinal plants. *Planta, 249*, 953–973.

Núñez-Pons, L., & Avila, C. (2015). Natural products mediating ecological interactions in Antarctic benthic communities: A mini-review of the known molecules. *Natural Product Reports, 32*, 1114–1130.

Ochoa-Villarreal, M., Howat, S., Hong, S., Jang, M. O., Jin, Y. W., Lee, E. K., et al. (2016). Plant cell culture strategies for the production of natural products. *BMB Reports, 49*, 149–158.

Ozarowski, M., & Thiem, B. (2013). Progress in micropropagation of *Passiflora spp.* to produce medicinal plants: A mini-review. *Revista Brasileira de Farmacognosia, 23*, 937–947.

Packer, J., Brouwer, N., Harrington, D., Gaikwad, J., Heron, R., Elders, Y. C., et al. (2012). An ethnobotanical study of medicinal plants used by the Yaegl Aboriginal community in northern New South Wales, Australia. *Journal of Ethnopharmacology, 139*, 244–255.

Panda, S. P., Sahoo, H. K., Subudhi, H. N., Sahu, A. K., Spmu, I. O., & Nagar, S. (2014). Potential medicinal plants of Odisha used in rheumatism and conservation. *American Journal of Ethnomedicine, 1*, 260–265.

Panmei, R., Gajurel, P. R., & Singh, B. (2019). Ethnobotany of medicinal plants used by the Zeliangrong ethnic group of Manipur, Northeast India. *Journal of Ethnopharmacology, 235*, 164–182.

Paul, P., Joshi, M., Gurjar, D., Shailajan, S., & Kumaria, S. (2017). *In vitro* organogenesis and estimation of β-sitosterol in *Dendrobium fimbriatum* Hook.: An orchid of biopharmaceutical importance. *South African Journal of Botany, 113*, 248–252.

Penso, G. (1980). The role of WHO in the selection and characterization of medicinal plants (vegetable drugs). *Journal of Ethnopharmacology, 2*, 183–188.

Phumthum, M., Srithi, K., Inta, A., Junsongduang, A., Tangjitman, K., Pongamornkul, W., et al. (2018). Ethnomedicinal plant diversity in Thailand. *Journal of Ethnopharmacology, 214*, 90–98.

Prins, E. E. (1996). Prohibitions and pollution at a medicinal plant nursery: Customary implications associated with ethnobotanical reserves in conservative areas of Kwa Zulu-Natal. *Southern African Humanities, 8*, 81–93.

Que, L., Yang, G., Miao, J. H., Wang, H. Y., Chen, M., & Zang, C. X. (2016). Current status and prospects of traditional Chinese medicine resource *ex situ* conservation. *Journal of Chinese Materia Medica, 41*, 3703–3708.

Radfar, M., Sudarshana, M. S., & Niranjan, M. H. (2012). Betalains from stem callus cultures of *Zaleyadecandra* LN Burm. f. A medicinal herb. *Journal of Medicinal Plants Research, 6*, 2443–2447.

Rahmatullah, M., Chowdhury, A. R., Esha, R. T., Chowdhury, M. R., Adhikary, S., Haque, K. M. A., et al. (2012). Ayurvedic influence on use of medicinal plants in Chakma traditional medicine. *American Eurasian Journal of Sustainable Agriculture, 6*, 107–112.

Rahmawati, N., Mustofa, F. I., & Haryanti, S. (2020). Diversity of medicinal plants utilized by to Manui ethnic of Central Sulawesi, Indonesia. *Biodiversitas Journal of Biological Diversity, 21*, 375–392.

Rajeswara-Rao, B. R. (2016). Genetic diversity, genetic erosion, conservation of genetic resources, and cultivation of medicinal plants. In M. Ahuja, & S. Jain (Eds.), *Genetic diversity and erosion in plants. Sustainable Development and Biodiversity*. Switzerland: Springer.

Rangel, C. D. F. C. B., Ramos, M. A., de Amorim, E. L. C., & de Albuquerque, U. P. (2010). A comparison of knowledge about medicinal plants for three rural communities in the semi-arid region of northeast of Brazil. *Journal of Ethnopharmacology, 127*, 674–684.

Rasool, R., Ganai, B. A., Akbar, S., Kamili, A. N., & Masood, A. (2010). Phytochemical screening of *Prunella vulgaris* L An important medicinal plant of Kashmir. *Pakistan Journal of Pharmaceutical Sciences, 23*, 399–402.

Ribeiro, R. V., Bieski, I. G. C., Balogun, S. O., & de Oliveira Martins, D. T. (2017). Ethnobotanical study of medicinal plants used by Ribeirinhos in the north Araguaia microregion, Mato Grosso, Brazil. *Journal of Ethnopharmacology, 205*, 69–102.

Robbins, C. S. (2000). Comparative analysis of management regimes and medicinal plant trade monitoring mechanisms for American ginseng and goldenseal. *Conservation Biology, 14*, 1422–1434.

Robertson, E. (2008). *Medicinal plants at risk. Nature's pharmacy, our treasure chest: Why we must conserve our natural heritage*. Tucson, AZ: Center for Biological Diversity.

Rosenbloom, R. A., Chaudhary, J., & Castro-Eschenbach, D. (2011). Traditional botanical medicine: An introduction. *American Journal of Therapeutics, 18*, 158–161.

Sarri, M., Mouyet, F. Z., Benziane, M., & Cheriet, A. (2014). Traditional use of medicinal plants in a city at steppic character (M'sila, Algeria). *Journal of Pharmacy & Pharmacognosy Research, 2*, 31–35.

Scanes, C. G. (2018). Human activity and habitat loss: Destruction, fragmentation, and degradation. In C. G. Scanes, & S. R. Toukhsati (Eds.), *Animals and human society* Academic Press.

Seca, A. M., & Pinto, D. C. (2019). Biological potential and medical use of secondary metabolites. *Medicines, 6*, 66.

Sharma, A., del Carmen Flores-Vallejo, R., Cardoso-Taketa, A., & Villarreal, M. L. (2017). Antibacterial activities of medicinal plants used in Mexican traditional medicine. *Journal of Ethnopharmacology, 208*, 264–329.

Shekhawat, M. S., & Shekhawat, N. S. (2011). Micropropagation of *Arnebiahispidissima* (Lehm). DC. And production of alkannin from callus and cell suspension culture. *Acta Physiologiae Plantarum, 33*, 1445–1450.

Shikov, A. N., Pozharitskaya, O. N., Makarov, V. G., Wagner, H., Verpoorte, R., & Heinrich, M. (2014). Medicinal plants of the Russian pharmacopoeia; their history and applications. *Journal of Ethnopharmacology, 154*, 481–536.

Shrestha, N., Shrestha, S., Koju, L., Shrestha, K. K., & Wang, Z. (2016). Medicinal plant diversity and traditional healing practices in eastern Nepal. *Journal of Ethnopharmacology, 192*, 292–301.

Silalahi, M., Walujo, E. B., Supriatna, J., & Mangunwardoyo, W. (2015). The local knowledge of medicinal plants trader and diversity of medicinal plants in the Kabanjahe traditional market, North Sumatra, Indonesia. *Journal of Ethnopharmacology, 175*, 432–443.

Singh, R. (2015). Medicinal plants: A review. *Journal of Plant Sciences, 3*, 50.

Singh, R. S., Ansari, I., Singh, R. K., Singh, S. K., & Pal, D. (2017). *Ex-situ* conservation of medicinal plants and its therapeutic in mine impacted lands in dry tropical forests of Jharkhand, India. *Eurasian Journal of Forest Science, 5*, 44–69.

Stevens, H., & Huys, I. (2017). Innovative approaches to increase access to medicines in developing countries. *Frontiers in Medicine, 4*, 218.

Streczynski, R., Clark, H., Whelehan, L. M., Ang, S. T., Hardstaff, L. K., Funnekotter, B., et al. (2019). Current issues in plant cryopreservation and importance for *ex situ* conservation of threatened Australian native species. *Australian Journal of Botany, 67*, 1–15.

Street, R. A., Stirk, W. A., & Van Staden, J. (2008). South African traditional medicinal plant trade-challenges in regulating quality, safety and efficacy. *Journal of Ethnopharmacology, 119*, 705–710.

Suárez, M. E. (2019). Medicines in the forest: Ethnobotany of wild medicinal plants in the pharmacopeia of the Wichí people of Salta province (Argentina). *Journal of Ethnopharmacology, 231*, 525–544.

Szopa, A., Ekiert, R., & Ekiert, H. (2017). Current knowledge of *Schisandra chinensis* (Turcz.) Baill. (Chinese magnolia vine) as a medicinal plant species: A review on the bioactive components, pharmacological properties, analytical and biotechnological studies. *Phytochemistry Reviews, 16*, 195–218.

Szopa, A., Ekiert, H., Szewczyk, A., & Fugas, E. (2012). Production of bioactive phenolic acids and furanocoumarins in *in vitro* cultures of *Rutagraveolens* L. and *Rutagraveolens* ssp. *divaricata* (Tenore) gams. Under different light conditions. *Plant Cell Tissue and Organ Culture, 110*, 329–336.

Towns, A. M., Quiroz, D., Guinee, L., de Boer, H., & van Andel, T. (2014). Volume, value and floristic diversity of Gabon's medicinal plant markets. *Journal of Ethnopharmacology, 155,* 1184–1193.

Uchendu, E. E., Paliyath, G., Brown, D. C., & Saxena, P. K. (2011). *In vitro* propagation of north American ginseng (*Panax quinquefolius* L.). *In Vitro Cellular & Developmental Biology-Plant, 47,* 710–718.

Ullah, M., Khan, M. U., Mahmood, A., Malik, R. N., Hussain, M., Wazir, S. M., et al. (2013). An ethnobotanical survey of indigenous medicinal plants in Wana district South Waziristan agency, Pakistan. *Journal of Ethnopharmacology, 150,* 918–924.

Valenzuela-Atondo, D. A., Delgado-Vargas, F., López-Angulo, G., Calderón-Vázquez, C. L., Orozco-Cárdenas, M. L., & Cruz-Mendívil, A. (2020). Antioxidant activity of *in vitro* plantlets and callus cultures of *Randiaechinocarpa*, a medicinal plant from northwestern Mexico. *In Vitro Cellular & Developmental Biology-Plant,* 1–7.

Vanisree, M., Lee, C. Y., Lo, S. F., Nalawade, S. M., Lin, C. Y., & Tsay, H. S. (2004). Studies on the production of some important secondary metabolites from medicinal plants by plant tissue cultures. *Botanical bulletin of Academia Sinica, 45,* 1–22.

Van-Wyk, B. E. (2015). A review of commercially important African medicinal plants. *Journal of Ethnopharmacology, 176,* 118–134.

Van-Wyk, B. E., & Wink, M. (2018). *Medicinal plants of the world.* Pretoria: Briza Publications.

Vasisht, K., Sharma, N., & Karan, M. (2016). Current perspective in the international trade of medicinal plants material: An update. *Current Pharmaceutical Design, 22,* 4288–4336.

Wang, X., Usmanovich, K. R., Luo, L., Xu, W. J., & Wu, J. H. (2019). Salvation of rare and endangered medicinal plants. In L. Huang (Ed.), *Molecular pharmacognosy.* Singapore: Springer.

WHO, World Health Organization. (2019). *WHO global report on traditional and complementary medicine 2019.* Geneva: World Health Organization.

Yadav, N. (2016). Conservation of some endangered and economically important medicinal plants of India–A review. *Journal of Integrated Science and Technology, 4,* 59–62.

Yang, J., Chen, W. Y., Fu, Y., Yang, T., Luo, X. D., Wang, Y. H., et al. (2020). Medicinal and edible plants used by the Lhoba people in Medog County, Tibet, China. *Journal of Ethnopharmacology, 249,* 112430.

Yang, J., Yang, X., Li, B., Lu, X., Kang, J., & Cao, X. (2020). Establishment of *in vitro* culture system for *Codonopsispilosula* transgenic hairy roots. *3 Biotech, 10,* 1–8.

Yuan, H., Ma, Q., Ye, L., & Piao, G. (2016). The traditional medicine and modern medicine from natural products. *Molecules, 21,* 559.

Zakrzewski, P. A. (2002). Bioprospecting or biopiracy? The pharmaceutical industry's use of indigenous medicinal plants as a source of potential drug candidates. *University of Toronto Medical Journal, 79,* 252–254.

Chapter 3

Plant biotechnologies for processing raw products in phytomedicines

Monica Butnariu[a] and Alina Butu[b]

[a]*Banat's University of Agricultural Sciences and Veterinary Medicine "King Michael I of Romania" from Timisoara, Timis, Romania*, [b]*National Institute of Research and Development for Biological Sciences, Bucharest, Romania*

1. Introduction

According to the definition of the Codex Alimentarius Commission, adapted to the Cartagena Protocol, modern biotechnology is defined as the in vitro application of nucleic acid technologies, including recombinant DNA and direct injection, which go beyond the natural barriers of reproductive physiology or the barriers of natural recombination, and they are not technologies used in reproduction and natural selection. The application of modern biotechnologies in phytomedicines offers new possibilities for human development and health (Tsuda, Watanabe, & Ohsawa, 2019).

The inclusion of the new characters contributes to the increase of the productivity of the plants or to the improvement of the quality of the phytomedicines, which contributes to the improvement of the health of the population.

Regarding the effects on public health, there may be indirect effects, such as reducing the use of chemicals and increasing revenues and increasing the viability of cereals and the security of PRPs, especially in developing countries. Genetic engineering is currently aimed at obtaining hypoallergenic products. The researchers created an experimental variety of low-allergen rice. The RNA-antisense strategy was applied to suppress the expression of allergenic genes (Whelan & Lema, 2015).

Although this rice is hypoallergenic, it has not yet been accepted for consumption by allergic people. New features of genetically modified organisms (GMOs) can also bring potential risks. Many of the genes and characters used in agricultural biotechnologies are new and not known as harmless foods. There is also the risk of indirect negative effects on human health through the negative influence on the environment and economy, including social and ethnic impact.

Phytomedicine: A Treasure of Pharmacologically Active Products from Plants
https://doi.org/10.1016/B978-0-12-824109-7.00023-6

Genetic engineering and its products have only appeared in the last decades. It is almost impossible to evaluate the potential impact of transgenic species (TSs) on the environment (Grantina-Ievina et al., 2019).

Cultivation plants, resulting from genetic engineering experiences, represent more than the next generation of plant varieties resulting from modern technology "high-tech."

They have two specific characteristics that can turn them into a particular danger to human health or the environment:

- First, genetically modified (GM) plants contain genes and characters that are completely new to the species—target, their environment, and their genetic background.
- Secondly the genetic engineering process is neither precise nor directed but rather a brutal intervention. The newly introduced gene can be integrated anywhere in the genome of the recipient plant (Babar et al., 2020).

The undesirable effects caused by GMOs to the human body may be due to the actual inclusion of the foreign DNA into the human one, which can lead to the change of the activity of certain genes of the recipient organism, or to the horizontal transmission of TSs to another organism. Synthesis of new transgenic proteins can be toxic and/or allergic to the recipient organism.

The main concerns regarding the safety of foods containing GM derivatives focus on the following aspects: Existing analytical tests and databases containing natural toxic substances or nutrients present in conventional foods are not adequate to test unintended changes in GM derivatives; genetic engineering can greatly affect toxins, allergens, or nutrients in food; food allergies can be exacerbated by genetic engineering; and the use of marker genes, which confer antibiotic resistance on some GM foods, raises health problems. Two reports highlighted the growing concern among professional doctors about the safety and regulation of GMOs (Kramkowska, Grzelak, & Czyżewska, 2013).

A report was prepared by the Environmental Association of Irish Physicians in response to a report requested by the government. This report questioned all three basic arguments, on the basis of which the report requested by the government based its conclusion on the fact that GM foods pose no danger to human health.

The Irish Physicians Association rejected the report's claims: "The scientific data on the safety of current phytomedicines GM are supported by the lack of any reports of adverse effects resulting from their consumption."

Lately, there has been a steady increase in the number of soy allergies among Irish children, but there has been no possibility to say whether it is related to foods containing GM soy, as there is no obligation to label these products under EU law (Seth, Poowutikul, Pansare, & Kamat, 2020).

The report of the Irish doctors' group was an echo of the concerns expressed by the British Doctors' Association (BMA), which in a report said: "Any conclusions regarding the safety of the introduction of GM material in England are premature, as there is currently insufficient data to inform decision makers." The BMA urged the application of the precautionary principle in the development of GM plants and foods, and stated that GMOs should not be released until the level of scientific certainty is sufficient to ensure safety for health and the environment (Goodman, Panda, & Ariyarathna, 2013).

The BMA has called for a moratorium on commercial planting of GM plants in England, establishing appropriate systems for maintaining and tracking the segregation and identity of GM plants and banning the use of marker genes that confer antibiotic resistance in GM foods.

In addition, they stressed the need for further research to determine the full potential impact of GMOs on the environment and health (Bawa & Anilakumar, 2013).

Thus the following are claimed: The risks and benefits of GMOs are not completely clear or universal; the ability to predict the ecological impact of new species (GMOs or traditional) is low, due to the uncertainty of available data; there may be certain benefits and risks that have not yet been identified or presented in the literature; the quantity and quality of GMOs, which can be developed deserve particular attention in terms of risk estimation, and the careful evaluation of potential benefits can help specialists in determining the balance between benefits and risks; and developing measures to prevent gene transfer to wild plants can reduce the impact on the environment and contribute to benefits.

2. Impact of GMOs on public health

Regarding the assessment of the safety of GM plants, it is considered that the food is completely digested and the consumed PRP genes cannot circulate to the recipient cells. However, this assumption has not been verified until the present time, when some studies have shown that gene fragments can circulate and sometimes integrate into the chromosomes of the "host" organism. It was then shown that DNA is not completely disintegrated in the gastrointestinal tract (Tulinská et al., 2018).

The ingested DNA segments may have the necessary size to contain entire genes that remain intact and can advance to tissues and may enter the bloodstream. DNAs contain genetic elements called "CpG motif," which can provoke the immune system to an inflammatory reaction. Once released into the environment, GM plants can no longer be restricted. Like all living organisms, GM plants reproduce, and this is an opportunity for genes to escape outside the growth zone originally intended.

The major way to escape the newly introduced genes in the wild, natural environment is through the transfer of pollen. Seeds can be picked up by birds and dropped elsewhere, larger mammals can carry potato tubers, or breeding organs can even be dislocated by the wind (Takaishi et al., 2019).

Most GM crop plants currently on the market contain marker genes that confer antibiotic resistance, along with desired characteristics such as resistance to insects or herbicides. There is a risk that these genes will be transferred from plants to disease-producing microorganisms if TSs are used as animal feed or food for humans.

These bacteria will therefore become immune to antibiotic treatment. Research on the existence of gene transfer and its size has only begun recently, which is why the existing scientific data are incomplete (Wang et al., 2019).

In 2002 English researchers demonstrated for the first time that genetically modified material (DNA) in food finds its way to bacteria from the human intestine, raising potential serious health problems. Many GM plants have inserted antibiotic resistance marker genes already at an early stage of development. If the genetic material of these marker genes finds a way to reach the stomach, which the researchers have suggested is possible, then the effectiveness of widely used antibiotics could be compromised. In one study the possibility of antibiotic resistance being transferred from TSs into the environment and thus reaching the digestive tract was emphasized.

The study draws attention to the fact that the widespread cultivation of TSs exacerbates the already difficult problem of resistant bacteria (Lemon et al., 2019).

There is sufficient scientific data to argue that genes can be relatively stable in the intestine, bacteria can in principle take up genes in the gut of mammals, horizontal transfer of genes from genetically modified microorganisms to bacteria has been observed in the intestine of insects (i.e., colembol), and soil bacteria can take up genes from the soil.

Because these genes are useless and dangerous to human health, many European regulatory authorities oppose the use of marker genes that confer antibiotic resistance (Duan et al., 2018).

3. GMOs: Allergy triggers

Allergic reactions are responses of adaptation and defense of the body to the toxic protein. In sensitive people the allergic reaction can be caused even by a very low dose of allergen. It is very difficult to set the minimum trigger dose individually. However, the consequences of the allergic reaction depend on the amount of allergen ingested, and thus the severity depends on the dose. Allergies are very common in all developed countries and affect on average 3%–4% of the adult population and up to 8% children (Ishii, 2017).

Allergies are based on immunological reactions, as opposed to food intolerance, defined as a nonimmunological reaction. Most allergens are water-soluble glycoproteins with molecular mass between 18 and 70 kDa. The molecular mass of 10 kDa represents the lower limit for allergic responses (Jyoti et al., 2019).

Following ingestion or inhalation, food allergens induce the production of specific antibodies (IgE), with which they then interact. About 10 types of products are known, which causes 90% of the severe allergies encountered in the western hemisphere according to Fig. 1.

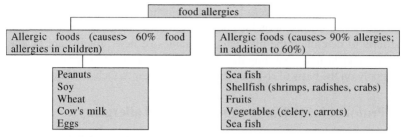

FIG. 1   Foods with allergic potential.

Allergic reactions to food known according to Fig. 2. Most food allergens are of protein structure and are contained in eggs, fish, seafood (including crustaceans), soybeans, nuts (almonds, peanuts, Brazil nuts, forest nuts, mahogany, pistachios, and nuts), and wheat.

Due to the fact that these allergens are known, it was possible to develop and use more advanced testing methods. However, traditional foods are usually not checked for allergen content before being delivered to the market. Since any food containing protein may be a potential allergen, the introduction of a new protein or any change in the structure of the existing ones or the quantitative modification may also affect the allergenic capacity of the derived products (Avsar, Sadeghi, Turkec, & Lucas, 2020).

The currently available biotechnological methods can be used for several purposes:

- inhibition of gene activity (reduction of protein production),
- activation of the existing gene (initiation of protein production),
- explication of the existing gene (increased production of specific proteins),
- changing existing genes (changing composition or eliminating protein),
- introduction of new genes from other species.

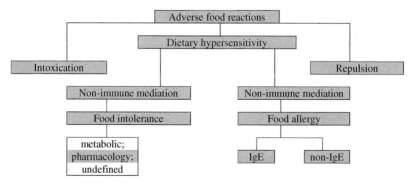

FIG. 2   Schematic presentation of allergic reactions to food.

The introduction of modern biotechnologies in agriculture could influence food safety if new food proteins could be the cause of allergic reactions, and true tests to determine the allergenic capacities of new proteins are lacking. Thus the conclusions regarding the allergenic properties of the proteins can be made only on the basis of risk factors (Loyola-Vargas & Ochoa-Alejo, 2018).

4. Protocol for risk assessment for food allergies

Evaluation of allergic capacity (if foods or certain ingredients have an allergic potential) includes quantitative assessment of the allergen (if the concentration of allergen is harmless), exposure assessment (which is the possible way to contact the allergen), and highlighting under the sensitive population (which would be the allergic response to new foods) (Dunn, Vicini, Glenn, Fleischer, & Greenhawt, 2017).

FAO/WHO experts have established a protocol for evaluating the allergenic potential of GM foods. The protocol is applicable to foods that contain genes derived from both known sources, allergic and nonallergic sources.

The expert group expressed its disagreement on the transfer of genes from food allergens, except only in cases where the lack of allergenic capacities of the proteins resulting from the transfer of genes was demonstrated. These principles have been applied by many GM food safety agencies and have served as a basis for the Codex Alimentarius Commission's Guide to Estimating Food Safety from Biotechnology (Bruetschy, 2019).

Over the past few years, researchers have become increasingly concerned about the increased incidence of allergies caused by GM plants.

GM foods contain proteins from other organisms whose genes have been transferred and allergic reactions caused by them are considered to be the main danger to human health. Thus genetic changes can create changes in plants, which result in the expression of a new allergen or increase the expression of endogenous allergens. As example, it can serve the allergen from Brazil nuts in GM soy (Rostoks et al., 2019).

Soy is an important source of protein in the diet of humans and animals but low in methionine. To improve its nutritional qualities, the researchers created GM soy, which produces a protein rich in methionine, whose gene was taken from Brazil nuts (Brazil nuts contain 18% methionine). Due to the fact that Brazil nuts are known as allergens, investigations have been made into the allergenic properties of GM soy.

It was determined that people allergic to Brazil nuts develop allergic reactions also to soy GM, from which the production was stopped. In 2004 information about the sickness of farmers in the Philippines appeared. They lived near fields sown with the transgenic variety of Bt-maize. Symptom manifestation coincided with the flowering and spreading period of pollen. Corn has been genetically modified to produce the insecticide Bt-toxin Cry1Ab toxic to corn worms (Szymczyk et al., 2018).

The symptoms presented referred to the respiratory system (sneezing, allergic cough, asthma, etc.), digestive tract, and teguments; also fever was present.

These symptoms disappeared with the departure to other districts of the country and reappeared on their return. It was supposed to be an infectious disease, but after different researches, this assumption was rejected.

Other examples of genetic modification are the introduction of Bt-toxin Cry1Ac into cotton, rice, tomatoes, and other cultures to protect them against lepidopterans. There were three studies in mice on the Bt-toxin Cry1Ac (Fernandes, Costa, Dutra, & Raizer, 2019).

Two of these studies show that Bt-toxin initiates an antibody-mediated response in the blood and mucous membranes of mice. The third study shows that Cry1Ac enhances the immune response to other antigens. In one study, it is mentioned that mice that had a diet based on GM cultures, which contained Cry1AC, presented structural and growth disorders at histological examination of the ileum (lower portion of the small intestine).

2BT—*Bacillus thuringiensis* produces insecticidal proteins, in the form of parasporal crystals (Cry protoxins), in the sporulation phase. These studies have shown that the Cry1Ac protoxin is an active immunogen in mammals, the immune response manifesting itself both systemically and at the mucous level, and therefore may be a threat to the human body (Botha, Erasmus, du Plessis, & Van den Berg, 2019).

Another variety of Bt maize—Strlink was produced by the company Aventis Crop Science, licensed in the United States (1998), for the animal feed and ethanol industry. Pathways of allergic properties in GM plants include insertion of a gene that synthesizes a protein already known as an allergen; transgenesis, which in addition to the desired character induces the synthesis of allergic proteins; and accidental disturbance at the genome level with modification of the resulting protein, which possesses allergic properties.

Following the tests performed, the cause was found to be the production of Cry9C protein (more than 60 such proteins are known. Three proteins are used in Bt-maize engineering and are expressed by the genes Cry1Ab, Cry1Ac, and Cry9c)—one of the crystalline protein family that gives the maize an insecticidal character (Zhao et al., 2001).

Table 1 presents some examples of food allergens with known functions in plant defense reactions.

Unlike other Cry proteins, Cry9C is thermostable and resistant to the action of gastric juice—these are the main indicators of allergic potential. Another aspect is that in this variety of corn, the expression of the gene shows that encoding Cry9C occurs not only in pollen and leaves but also in its seeds, which increases its health risk.

Thus the Cry9C concentration in seeds is 10–400 times higher than the production of Cry protein in other maize varieties. More than half of the transgenic proteins, which ensure plant resistance to insects and diseases caused by fungi and bacteria, are toxic and allergic (Latham, Love, & Hilbeck, 2017).

TABLE 1 Food allergens with known functions in plant defense reactions.

Allergen	Plant	Plant defense function
Lectin	Peanuts (*Arachis hypogaea*)	Delay in development *Callosobruchus maculatus*
The trypsin inhibitor	Soy (*Glycine max*)	Delay in development of larval *Spodoptera litura*
WGA wheat germ agglutinin with isolectin: *Triticum aestivum*: WGA A *Triticum aestivum*: WGA D *Triticum durum*: WGA B	Wheat (*Triticum* spp.)	Insecticidal activity for *Callosobruchus maculatus* Insecticidal activity for *Ostrinia nubilalis* Growth inhibition of larval *Diabrotica undecimpunctata*
α-Amylase/trypsin inhibitor	Barley (*Hordeum vulgare*)	Inhibition of α-amylase at *Tenebrio molitor*

There are groups of proteins, known as allergens, that participate in natural plant defense mechanisms. Some of these, α-amylase and trypsin inhibitors, lectins, and pathogenesis-related proteins, are used in genetic engineering to increase the resistance of agricultural crops. It is possible that a nonallergic protein in one organism becomes an allergen in another. If a bacterial protein is passed into the plant, it can cause it posttranslational changes, inducing allergic properties. Even the most frequent modification, glycosylation (endowment of the protein with polysaccharide chains), changes it substantially (Takaishi et al., 2019).

5. α-Amylase inhibitors

A number of allergic pathologies are caused by α-amylase inhibitors. For example, Baker's asthma is a type I allergic reaction, which most often is triggered by inhalation of cereal flour. The DNA responsible for the production of the α-amylase inhibitor in insects is found in the wheat seed endosperm—*Triticum aestivum*.

Similar allergens that cause IgE-mediated reactions have been detected in rice (*Oryza sativa*) and barley (*Hordeum vulgare*) (Osorio et al., 2019).

The DNA sequence for the allergenic inhibitors of wheat α-amylase denotes similarity to the DNA of the rice and barley inhibitors, 40% and 20%, respectively. Rice allergic protein is thermostable and resistant to proteolytic agents, which causes great problems for allergic patients (Zhou, Wang, Zhou, Ouyang, & Yao, 2017).

Attempts to create hypoallergenic rice have already yielded some results, but the results still do not confer full certainty on the tolerance of this rice by patients. Studies with the yellow mustard seeds (*Sinapis alba*) have found a similarity of the allergens of this plant with the α-amylase inhibitors from cereals, beans, peas, etc. Thus the results of the studies carried out suggest that the proteins in the family of α-amylase inhibitors are potential allergens, at least in cereals and vegetables (Malehorn, Borgmeyer, Smith, & Shah, 1994).

6. Trypsin inhibitors

Another example is the trypsin inhibitor in *Vigna unguiculata*, which protects a tobacco plant from *Heliothis virescens*. Trypsin inhibitors cause serious disturbances in the body of insects, the most affected being lepidopterans (especially their larvae). They consist of digestive disorders followed by weight loss until their death. These proteins become an insecticidal remedy used in genetic engineering (Savić et al., 2019).

An example could be the taking of the gene that encodes the trypsin inhibitor from the white mustard and transfers it to the GM tobacco.

But to reach the expected insecticidal effect, the expression level of this protein in GM plants must be not less than 1% of the total amount of soluble proteins, which can cause major undesirable effects in humans (O'Callaghan et al., 2007).

In addition to the allergic effects it causes, trypsin inhibitors can cause serious digestive disorders, reduce protein digestion, and cause amino acid absorption deficiency. In the tests performed on animals with a diet rich in trypsin inhibitors, it was observed favoring the pathological processes of the pancreas.

Considering that some proteins from cereals and vegetables from the α-amylase family and trypsin inhibitors are allergic, any transgenic PRP culture containing such inhibitors requires careful testing to ensure their harmlessness to allergic persons (Zhou et al., 2017).

7. Lectins

The term "lectin" (agglutinin) is defined as a class of proteins or glycoproteins, with binding properties to other cells, which can reversibly bind carbohydrates without altering their structure. By attaching to the cell membrane in various organs (most often the intestine), they cause agglutination or may initiate a cascade of immune or autoimmune responses, which may eventually cause cell death (Nielsen et al., 2018).

Different lectins target different organs and systems of the body. Once the intact lectin is located at a certain level in the body, it actually has a "magnetic" effect on the cells in that region. It clusters the cells, and they are targeted for destruction, because they are considered by their own body as invading foreign elements. This agglomeration may cause irritation syndrome in the intestine or liver cirrhosis, but it may also cause blockage of renal blood flow (Gürel, Uçarlı, Tufan, & Kalaskar, 2015).

The effects of the lectin-intestinal interaction include attachment to cells from the intestinal mucosa, destruction of the microvillus membrane, and "shedding" cells, reducing the absorption capacities of the small intestine.

Other effects of lectin-intestinal interaction can be increased endocytosis; induction of hyperplastic processes of the small intestine; increased turnover of epithelial cells. Interference with the immune system; hypersensitivity reactions. Interference with intestinal flora; exaggerated selective multiplication (Wang et al., 2016).

Direct and indirect effects on metabolism. Most lectins found in food are not dangerous, although they can cause a lot of problems. A study was conducted that found 30% of fresh and processed foods contain active lectins.

Lectins from salad leaves, fruits, spices, seeds, cereals, and nuts (even after frying) have shown a high level of toxicity. The same can be said about oils from plants containing lectins (Bhagyawant et al., 2019).

Unrefined soybean oil contains 858–2983 µg/kg lectins and refined oil 24–55 µg/kg. In most cases the immune system protects the organism against lectins. Ninety-five percent of the lectins we absorb from our typical diets are removed from the body, but 5% of them are filtered into the bloodstream, where they react and destroy red blood cells and leukocytes. The actions of lectins in the digestive tract may be even stronger.

At this level, they often cause a strong inflammation of the intestinal mucosa, and this binding action can mimic food allergies. Lectins are commonly produced in plants, contained in large quantities in cereals and in some vegetables, and can reach up to 10% of the total protein in seeds (Leal et al., 2018).

They are involved in the system of plant protection against pests. They are known for their antinutrient qualities and can be toxic even in the naturally produced concentration. The most cytotoxic lectins most often cause food allergies. As food allergens, four lectins were identified: two lectins from *Triticum aestivum* wheat, one from *Triticum durum* wheat, and one from peanuts *Arachis hypogaea*.

Each of these four allergens causes significant retardation in the development of insects such as peas ladybug (*Callosobruchus maculatus*). *N*-Acetylglucosamine agglutinin (lectin WGA) of wheat germ is next to all lethal for the larvae the European borer of *Ostrinia nubilalis* (European corn borer); even at a relatively low concentration, it inhibits the growth of the southern larva of the roots, which attacks the root of southern corn (*Diabrotica undecimpunctata*). It is obvious that a certain lectin with defined specificity for a certain carbohydrate cannot protect the plant against all pests, because the glycosyl structures of the external surfaces of different bacteria, fungi, and insects are different (He et al., 2018).

However, with regard to the high efficiency of lectins in pest control, one of the tasks of genetic engineering would become meticulous research and then the introduction of lectins with specific target in certain plants, since the pests have affinity toward certain plants anyway. But the basic purpose remains the detection of those lectins that are both toxic to pests and harmless to humans and animals and beneficial to insects (Sestito et al., 2017).

8. Proteins related to pathogenesis

Due to the modification of their physiological status, the plants have the capacity to protect themselves in the case of different stresses such as unfavorable conditions, the action of ultraviolet rays, heavy metals, pollutants in the atmosphere, the application of different chemicals such as phytohormones, injuries, and attacks by diseases and pests.

The protective reactions that the plants develop in these cases are called "protective responses," and the proteins actively synthesized in accordance with these reactions are called "defense-related proteins" (Lyu et al., 2020).

The latter are synthesized not only in the affected place but also throughout the plant. Defense proteins, especially those specifically induced in pathological states, are of particular interest for the future of agriculture and are called pathogenesis-related (PR) proteins. These proteins have been identified in many species of higher plants. Their synthesis and accumulation in the plant have been preserved throughout the evolution. In genetic engineering, they make a considerable contribution due to the fact that they can be synthesized by cells directly through the translation of a gene (Yang et al., 2020).

Thus PR proteins found in many plant species were classified into 17 families. At the base of the classification, sequential, serological, and immunological similarity and enzymatic properties were taken. Among the plant allergens included in the Official Allergen Database of the Union of Immunological Societies, approximately 25% refer to the group of proteins related to pathogenesis. They are resistant to proteolysis and low pH. These characteristics and expression level place PR in the group of substances that provoke human immune response upon contact with mucous membranes. PR proteins are classified into 17 groups. Seven of these groups contain proteins with allergic properties (2, 3, 4, 5, 8, 10, 14), and the other contain food allergens (Schwartz et al., 2019).

Similarity in amino acid sequence causes PR proteins to cause cross-reactions for allergens from different plants. The expression induced, by the environmental factors, of the characteristic allergens PR homologous, can explain the diversity of the allergies (Purohit et al., 2008). Anthropogenic pollutants also influence the expression of PR protein allergens as shown in Fig. 3.

For example, proteins that have common epitopes with Bet v.1 (major pollen allergen) are found in the pollen of several species of apples, celery, carrots, nuts, and soy. Patients allergic to pollen often have allergies to ingestion of certain fruits and vegetables, as well as products derived from them (Purohit et al., 2008).

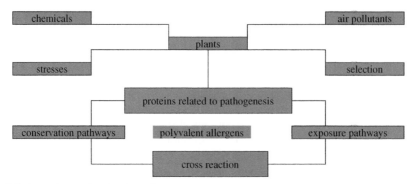

FIG. 3 Pathogenesis-related proteins.

Atmospheric pollen is the atopic allergen spread throughout the temperate zone and is responsible, depending on the country and localities, for a relatively large number of intermittent rhinitis, rhinoconjunctivitis, and bronchial asthma.

Plants with allergen pollen are divided into graminee, grasses, and trees. The graminee comprise about 9000 species (You et al., 2019). There are wide variations in the world regarding their pollination. *Cynodon dactylon*, *Lolium perenne*, *Sorghum halepense*, *Bromus inermis*, *Holcus lanatus*, *Phleum pratense*, *Triticum aestivum*, and *Festuca elatior* are the most important allergenic pollen plants (McInnes et al., 2017).

Allergens from graminee can be classified into 13 distinct groups of proteins with different biochemical properties. Thus groups 1 (e.g., Lol p1 and Phl p1), 2, and 3 (e.g., Lol p2 and Lol p3) are β-expansins present in most graminee species. Groups 4 and 13 (e.g., Lol p4, Phl p4, and Phl p13) are enzymes involved in pectin degradation, whereas group 5 (Lol p5, Phl p5, and Poa p5) and group 6 are ribonucleases. Minor allergens are included in group 7 of calcium-binding proteins, group 10 of cytochrome C proteins, and group 12 of profilins. The function of the allergens in group 11 is unknown (Masoumi, Fathi, Hajmoradi, & Baghaeifar, 2019).

Herb pollen is second in terms of pollen sensitivity to us in the country and comes from species of *Asteraceae*, *Urticaceae*, *Chenopodiaceae*, and *Brassicaceae*. Among the *Asteraceae*, of clinical importance are *Ambrosia*, *Artemisia vulgaris* and *Helianthus annuus* (sunflower); of *Urticaceae*, *Urtica dioica* (nettle), *Parietaria officinalis*, and *Judaica* are noted and among *Brassicaceae*, *Brassica napus* (turnip) (Jones et al., 2019).

Numerous and severe sensitization occurs from *Ambrosia* pollen (*Ambrosia artemisiifolia* L., *Ambrosia psilostachya*, and *Ambrosia trifida*). The pollen of *Artemisia absinthium*, *Artemisia vulgaris* (black pelin), and *Chrysanthemum* are of some importance. Major allergens from *Asteraceae* include pectin ligands (e.g., Amb a1,2), electron transfer proteins (e.g., Amb a3), and β-extensins; minor allergens are mainly represented by profiles (Amb a8 and Art v4) and lipid transfer proteins (e.g., Amb a6 and Art v3) (Denisow-Pietrzyk, Pietrzyk, & Denisow, 2019).

The trees, from an allergic point of view, reunite the *Fagale* family, with its subdivisions (e.g., *Betulaceae, Fagaceae, Ulmaceae, Platunacee,* and *Oleacee*). Although tree pollen is slightly allergenic, in some areas, sensitization may be important (Reitsma et al., 2018).

The major allergens from *Fagale* are pathogenesis-related proteins (PR-10) that possess ribonuclease activity (e.g., Bet v1), pectin-methyl esterases, globulin-like proteins, and lipid transfer proteins (LTP).

Minor allergens include profilins (e.g., Bet v2), isoflavone reductase (e.g., Bet v5), peptidyl-prolyl isomerase (e.g., Bet v 7), and calcium-binding proteins (e.g., Bet v 3 and 4). Homology within the limits of the registered inhalation allergens is 80%–90%. Other registered PR proteins belong to the PR-10 family (Picornell et al., 2019).

Homology within the lines of PR proteins is from 37% to 64%. The sequence identity between Mal dI and Bet vI is 56% (63% with the BV1SC AL Bet vI isoform) at the protein level. The sequential identity between Bet vI and p I49 is 55% (70% including conservative changes) at the protein level. The sequential identity between Bet vI and PR PvI is 44% at the protein level. Patients allergic to pollen often have allergies to ingestion of certain fruits and vegetables and their products. Most of the reactions are due to the four cross-reaction structures present in birch pollen (*Betula verrucosa*) (Larsen, Casals, From, Strøman, & Ipsen, 1993).

The allergens similar in structure to the allergens from the birch pollen are thermosensitive and are practically destroyed during the preparation of the products or during the digestion process unlike the classic food allergens, but there are cases of allergic reactions in the sensitive people. Food allergies are one of the most serious problems at the moment due to the health impact (Ahlstedt, Belin, Eriksson, & Hanson, 1975).

Food allergies are caused by the interaction of food components with the immune system. Factors that determine the severity of allergic reactions are type of allergen, condition and constitution of the affected person, and circumstances of consumption of a particular product. Any new protein introduced into the food chain is a potential new allergen. Most food allergens have common characteristics: molecular mass between 10 and 70 kDa, thermostability, resistance to digestion processes, proteolysis and hydrolysis, and water solubility (Worm et al., 2019).

The possibility of genetic engineering to produce various modifications of the products or of the PRPs determines the modification of the allergic potential; thus the risk assessment and the study of the allergic properties are vitally important.

9. Modification of plant toxicity

Plants have been and are an important source of food for humans. Through trial and error, people have learned that only certain plants can be consumed without danger to health and life. During evolution, they realized that some plants can only be consumed after preparation.

Food safety knowledge has been passed down from generation to generation, so that over the last century; over the past 50 years, with the help of chemical analyzes and toxicity tests, mankind has learned the causes and constituents of many plant toxicities. Therefore, in addition to the nutrients, the plants also contain toxic substances (Takaiwa, Ogo, & Wakasa, 2018). These substances (Table 2) are synthesized for protection against pests and microbes.

Here, we can also mention the dependence of dose toxicity, so that a small dose can have beneficial effects on the body, and if it exceeds a certain level, it becomes poisonous.

For this reason, it is recommended to respect the quantity of certain plants used for food consumption. Most of these toxins are present in ordinary foods, but their amount is insignificant, so they do not cause poisoning (Olcese et al., 2019).

Glucosinolates are substances that give spicy taste to mustard and horseradish. Toxic action on humans is manifested by antithyroid activity and disorders of the digestive tract, including liver damage (Liu et al., 2014).

Saponins' toxic action is manifested by irritation of the lining of the digestive tract (gastroenteritis with anorexia, diarrhea, and weight loss) (Zhang, Han, Liu, Guo, & He, 2014).

Cyanogenic glycosides include durin, linamarine, and lotaustraline. Acute intoxication has the following signs: respiration and accelerated pulse, headache, gastralgia, vomiting, diarrhea, confusion, and convulsions.

TABLE 2 Food allergens and inhalers with sequential homology with PR proteins.

Allergen	Protein. PR6 with significant sequential homology 7 with the registered allergens
Food allergen Mal d l, *Malus domestica* (apple)	PI49/p 1176 *Pisum sativum* (green peas) PvPR½ *Phaseolus vulgaris* (beans) PcPR 1–1 *Petroselinum crispum* (parsley)
The inhaler allergen Bet v l, *Betula verrucosa* (birch)	STH2 *Solanum tuberosum* (potato)
The inhaler allergen Aln g l, *Alnus glutinosa* (black alder)	
The inhaler allergen Cor a l, *Corylus avellana* (hazelnut tree)	

Upon ingestion of a higher dose, death may occur. In the case of prolonged consumption of plants containing cyanogenic glycosides, without their processing, chronic intoxication may occur, followed by chronic pathologies of the central nervous system (Kerre et al., 2018).

In some plants, toxic substances are always present; in others, they can be synthesized at certain stages of their life, on certain metabolic pathways, under the influence of different factors.

The probability of activating the mechanisms until then passive or increasing the activity of an active one is considered low in the ordinary plants, but it can become high due to the genetic changes (Van Hoang et al., 2018).

Thus, for example, in cassava, the level of the toxic substances is high, presenting a danger to humans (Table 3).

These mechanisms can be activated by mutations, chromosomal rearrangements, or new regulatory regions (portions) introduced in the process of growth (propagation), and toxins that were not specific to the species until now can be produced. Similarly, toxins produced in small quantities up to genetic modification may occur in large quantities as a result of this modification (Kurokawa et al., 2013).

Regarding the genetic modification of plants, there are aspects of toxicity caused by the introduction of new DNA into plants, namely:

- Direct changes through (a) the production of the new toxic substance from the moment of the genetic changes and (b) the change of the quantitative expression of the already existing toxic substance (increase or decrease).
- Indirect changes, which lead to the production or accumulation of toxic constituents. These adverse effects may occur as a result of any insertion, pleiotropies (unwanted and unexpected change caused by transgene expression), or soma clonal variations.

To determine the adverse effects of the genetic modification and to predict their impact on human health, it is important to determine the substantial equivalence-comparative appreciation of the GM plant with the same unmodified plant.

Estimates are made regarding not only the presence of certain toxicologically relevant substances but also from a nutritional point of view.

Even the smallest difference must be viewed as an indication of the unwanted effects that have occurred as a result of the change (van Wijk et al., 2005).

Unfortunately, even this method cannot guarantee us 100% the lack of undesirable toxic substances, and predicting the undesirable effects is impossible.

The most obvious example of toxicity became the case of the company, which marketed the food supplement GM tryptophan. The increased level of L-tryptophan was due to the GM bacterium *Bacillus amyloliquefaciens*.

The dietary supplement was marketed without being tested, as it is considered that the route of production is not important.

TABLE 3 Toxic natural substances of crop plants.

No.	Toxin	Toxin localization	Name of the plant
1	Durin	The young vegetative parts	Wheat
2	Lectins		
3	Protease inhibitors	Grains	
4	A trypsin inhibitor	Grains	Rice
5	Cyanogenic glycosides	The young vegetative parts	Corn
6	A trypsin inhibitor	Grains, cobs	
7	Epiheterodendrin	The young vegetative parts	Barley
8	A trypsin inhibitor	Grains	
9	Saponins	Seeds	Soy
10	Lectins		
11	Coumestrol		
12	Daidzein		
13	Genistein		
14	Protease inhibitors		
15	Durin	Young plants	Sorghum
16	Linamarin	Tubers	Cassava
17	Lotaustraline		
18	Protease inhibitors		
19	Alpha-chaconine	Tubers	Potato
20	Alpha-solanine		
21	Alpha-tomatine	Fruits	Tomatoes
22	Nicotine		
23	Lectins		
24	Glucosinolates	The vegetative parts	Rape
25	Erucic acid		
26	Saponins		
27	S-Methyl-L-cysteine sulfoxide toxic amino acid		
28	Lectins	Leaves and roots	Chicory
29	Lactucin		
30	8-Deoxylactucin		

Thus GM tryptophan was considered identical to the one previously marketed. As a result, in the United States, there was an outbreak of an unknown pathology until then, which was later called "eosinophilia-myalgia syndrome" (Chen, Chen, Ma, & Zeng, 2018).

Patients experienced acute pain in the muscles and joints and subsequently swelling of the joints. This disease was similar to scleroderma and practically did not respond to treatment. The death of 37 patients occurred and the invalidation of about 1500.

Research has shown that as a result of the transformation, the L-tryptophan-producing GM bacteria have acquired the property of producing in small quantities (less than 0.1%) of ethylenebistriptophan, an extremely toxic substance and a substance that was the cause of this syndrome. It is not clear at this time whether the toxicity of the product is due to the use of GM bacteria or insufficient purification, especially since the genetically modified bacterial stock was not available for investigation (Yin, Park, Gang, & Hulbert, 2014).

Some substances used to fight pests can block the digestive tract carriers in both insects and humans. A number of TSs of corn, tobacco, and tomatoes, resistant to insects, synthesize lignin, a substance that prevents plant damage.

This can be broken down into methanol and phenols, which in turn have a toxic and mutagenic effect on humans (Pesquet et al., 2013). GM tomatoes are obtained by the "antisense" method; they are the first GM plants marketed. Acute toxicity tests were performed for these tomatoes on laboratory mice. According to histological data, 7 out of 20 females had moderate erosive-necrotic lesions of the intestine and stomach.

Thus, in humans, there is a risk of developing digestive bleeding, particularly in the elderly and people who use aspirin as a prophylaxis of thrombosis. Seven out of 40 mice died within 2 weeks of unknown causes. The cause of the toxicity remained unpublished, and the marketing of these tomatoes was stopped due to the exaggerated costs of production and the unsatisfactory taste qualities (Blando et al., 2019).

Herbicide-resistant GM maize with the gene encoding the phosphinothricin acetyltransferase enzyme differs considerably from non-GM maize in its fat and carbohydrate content.

Toxicity tests were performed only on these proteins. Experiments in mice have shown that digestion capacity decreased after ingestion of GM maize (Zhang, Kang, et al., 2018; Zhang, Yook, et al., 2018).

After years of obtaining GM organisms for the benefit of farmers and PRP manufacturers (related to the first generation of GM foods), genetic engineering techniques have begun to be used to change the nutritional qualities of foods, leading to the emergence of the second generation of GM foods, foods designed for the benefit of consumers. Most GM organisms, used as sources of PRPs, belong to the plant world (Zhang, Kang, et al., 2018; Zhang, Yook, et al., 2018).

Table 4 shows the plants that, besides the agricultural changes, present genetic changes made for nutritional purposes.

TABLE 4 Characteristics of genetically modified crop plants.

No.	Nutritional and processing characteristics	Cultivation characteristics	Plant
1	Low cyanogenic glycoside content	Resistance to diseases Insect resistance	Cassava
2	Higher starch content	Resistance to mosaic virus Resistance to deposit pests	Corn
3	Modified starch		
4	High lysine and tryptophan content		
5	Improved protein content		
6	Increased oil content and change in the proportion of fatty acids		
7	Higher provitamin A content	Resistance to bacteria Insect resistance	Rice
8	Higher iron content	Resistance to pests of deposit Resistance to disease fungal nature Herbicide tolerance High photosynthesis capacity	
9	Improved oil composition	Insect protection Resistance to viruses	Soy
10	Increased content of cattle		
11	Low level of flatulent compounds		
12	Improved oil composition	Insect protection	Rape
		Increased resistance to disease	
13	Improved oil composition		Palm
14	High starch content	Resistance to viruses Insect resistance Resistance to pests of deposit Resistance to blows Resistance to fungal diseases	Potato

TABLE 4 Characteristics of genetically modified crop plants—cont'd

No.	Nutritional and processing characteristics	Cultivation characteristics	Plant
15	Increased lycopene content	Resistance to viruses Insect resistance Resistance to diseases	Tomatoes
16	Improved processing properties		
17	Delayed baking	Resistance to fungal diseases	Bananas

Thus genetic engineering techniques can contribute to the following:

– changing the proportion of macro- or micronutrients,
– elimination or reduction of the content of compounds with adverse effect on health (antinutrient compounds, toxins, and allergens).

One of the first genetic changes made for nutritional purpose was to change the composition of oils, given the efficiency of diets rich in unsaturated fatty acids, especially linolenic acid, for normalizing dyslipidemias. Soybean, palm, and rapeseed oil present, as a result of the genetic modifications suffered by the plants from which it is extracted, an improved ratio between saturated and unsaturated fatty acids (Xie et al., 2018).

Under these conditions, nutritionists may recommend the use of such oils to prevent the occurrence of cardiovascular disease and reduce the number of people affected by atherosclerosis. Another genetic modification, which has both nutritional and manufacturing implications, concerns the modification of starch (Menzel et al., 2015).

Changing the proportion normally existing between the two polymers contained in starch, amylose and amylopectin, leads to the obtaining of starch for well-defined purposes, such as improving the properties of gelling or thickening and reduced absorption of the oil during the frying process.

Modified starches from genetically modified plants reduce the postextraction chemical treatments they have to undergo to change their structure in accordance with the field of use. Also, for nutritional reasons, genetic engineering has been used to obtain wheat that allows the manufacture of bread with a high content of resistant starch (a form of starch that has beneficial effects on health, similar to those produced by dietary fibers) (Dong et al., 2019).

Genetic engineering used for improvements needs beta-carotene to turn it into active retinol. The bioavailability of retinol depends on the lipid composition of foods. Efficient use of beta-carotene is associated with that of lipids. Beta-carotene is liposoluble, that is, the solubilized form allows it to enter in enterocytes, the place of its conversion to retinol (Wu et al., 2017).

Bioconversion of carotene into retinol occurs in the intestinal cells and the liver. Vitamin A circulates in the blood in the form of retinol, being associated with a transport protein.

Among the classical methods of correction of vitamin A deficiency (drug supplementation and enrichment of food with vitamin A), methods of genetic improvement (carrots, potatoes with high beta-carotene content, or the production of golden rice) can be used. Rice is a product consumed by more than half of the world's population (Schmidt & Pendarvis, 2017).

It is the staple food in 33 developing countries, presenting 27% of the energy supply, 20% protein, and 3% lipid provided by the diet. It contains thiamine, riboflavin, niacin (group B vitamin), and zinc, in sufficient quantities, but lacks in vitamins C and D and beta-carotene (including due to peeling). Whole brown rice contains insignificant amounts of provitamin A (0.1-µg beta-carotene equivalent per gram), while husked rice does not contain it at all. Taking into account the rice husking habit, the geneticists decided that the expression of provitamin A should be targeted in the endocarp (Che et al., 2016).

The process is based on the use of rice embryos and *Agrobacterium* to introduce, through a single transgenic operation, specific genes. The vector pB19hpc includes the basic food composition of poor countries in Africa, Asia, and Latin America, where the choice of food is limited, which is why deficiencies of macro- and micronutrients (vitamins and minerals) frequently occur. Genetically modified rice with high provitamin A content was created (Ralley, Schuch, Fraser, & Bramley, 2016).

This rice, which has yellow grains and became known as "golden rice," has been specially designed for the approximately 700 million people who are deficient in vitamin A and who consume rice as a staple food. Conventional techniques for making such rice are not applicable, as there is no variety of rice producing provitamin A in the endosperm.

Using genetic engineering techniques the entire biosynthesis pathway of provitamin A was introduced to rice, using two genes: one from daffodils and the other from a bacterium. Animal products are the source of vitamin A (retinol), and those of plant origin contain provitamin A (beta-carotene) (Moghissi, Pei, & Liu, 2016).

The organism, in fact, the phytoene-synthase sequences (from the *Narcissus pseudonarcissus* plant) and the phytoene-desaturase sequences (from the *Erwinia uredovora* bacterium), allows generation of lycopene in the endosperm, where geranylgeranyl diphosphate is subsequently synthesized, which is subsequently transformed into phosphate (colorless carotene).

The amount of beta-carotene produced is 1.6–2.0 µg/g. Considering the lowest conversion rate (12-µg beta-carotene to obtain 1-µg retinol) and the recommended daily intake (400–500-µg retinol for children), a daily intake of 2.5 kg of golden rice would result (which is currently produced) (Swamy et al., 2019).

Based on the vitamin A content, estimated between 0.16 and 0.20 mg/100 g golden rice, bioavailability, and the conversion factors, the recommended intake for children is 0.3 mg/day, considering the level that would be sufficient for reduce mortality, morbidity, and vision disorders. But the research into increasing the content of provitamin A in golden rice does not stop there.

Thus the emergence of a new variety is expected, which will allow the eradication of the deficiencies and their consequences after consumption of approximately 30–224 g/day. In addition, the golden rice has also been modified to solve another major nutritional problem: iron deficiency, affecting 2.15 billion people (Bollinedi et al., 2019).

In addition to the genes for β-carotene synthesis, golden rice contains one gene that increases the absorption of iron in the soil and another that increases its absorption in the human body.

Although the most effective way of satisfying the needs of vitamin A and iron of an organism is the consumption of a wide range of foods, such as lean red meat and cereals, for iron intake, and liver, eggs, fatty fish, fruits, yellow and orange vegetables, dark-green leafy vegetables for β-carotene, many people in developing countries do not have access to them (Zhu et al., 2018).

Because rice is a staple food in these countries, it can successfully contribute to reducing vitamin A and iron deficiency in these parts of the world, where other strategies have proven ineffective. Gold rice is still undergoing changes, researchers at the International Rice Research Institute wanting to increase its β-carotene content.

Another genetically modified food, with potential benefits for the health of the inhabitants of the developing countries, is rice that has a yield of 30% higher than the normal rice. Like golden rice, it is still in the research stage (Napier, Haslam, Tsalavouta, & Sayanova, 2019).

But not always the nutritional improvements obtained using the techniques of genetic engineering are considered to be a success. For example, trying to improve the amino acid content of soy was unsuccessful on allergic considerations.

Soybeans have been genetically modified using genes from Brazil nuts, which contain high concentrations of amino acids with sulfur, and the transfer made included genes of an allergenic substance. Soybean derivatives are used as food ingredients; soybeans with improved sulfur amino acids have been designated unsuitable for use in nutrition (Chinnadurai et al., 2018).

Some sources of food used in human nutrition contain antinutrient factors (hemagglutinins, saponins, trypsin inhibitors, cyanogenic glycosides, thioglycosides, coumarin, gossypol, and flatulent carbohydrates), which affect the health of consumers. The importance of these substances varies worldwide depending on the frequency of their presence in the diet, the amount ingested, and the sensitivity of the population to a particular compound. Genetic engineering can intervene to reduce the content or eliminate the presence of antinutrient factors in food (Tang et al., 2013).

Some examples of the use of genetic engineering in this regard are presented in the succeeding text. An existing globulin in rice, resistant to the action of proteases in the intestine, produces atypical dermatitis that especially affects Japanese children.

This globulin is also thermostable, so the heat treatment of the rice does not ensure its distortion. Because the enzymatic hydrolysis of globulin to produce hypoallergenic rice is very expensive, and by chemical mutagenesis, sterile rice was obtained, whose grains still contained traces of globulin; genetic engineering techniques were used (Van Giap, Jung, & Kim, 2019).

Genetic engineering has achieved a significant reduction in allergenic globulin, without affecting the agrotechnical characteristics of rice. Similar approaches can work to remove allergenic proteins from peanuts and soybeans.

Some lectins pose health problems, so varieties of plants with low content or lacking these substances have been created. For French soybeans and beans, in the cultivator's plant varieties were found, varieties that do not have the gene for the formation of lectins, so for them it will be possible to create commercial varieties without lectins. For other bean species, which have multiple copies of lectin genes, it is not possible to apply the classical procedure, which is why inactivation of lectin genes can be achieved by genetic engineering techniques. Referring to the components of phytomedicines, it should be mentioned that every day we ingest a large number of live bacteria (Han, Baek, Jo, Yun, & Choi, 2019).

Sometimes these are parts of the saprophytic flora of foods (vegetables, fresh fruits, and some brands of bottled water). The number of opportunistic germs is low in quality foods. For other products, however, bacteria are introduced by humans and are involved in their transformation. As an example, they serve sausages, cheeses, and other fermented dairy products. The highest consumption of live bacteria refers to yogurts and fermented dairy products, because the number of bacteria is 108 bacteria/g and the consumption of these products exceeds 200 g/day (Jonsdottir et al., 2018).

Currently, there are four major patented sectors of PRPs, in which live genetically modified microorganisms (GMMs) (yeasts as fermentation agents) are used: the bakery, oenology, brewery, and ethanol production industry. Also, live GMMs are used in the production of enzymes, flavorings, and technological aids for phytomedicines. GMMs are still used as metabolic microwaves, and the obtained substances are purified and marketed.

For example, chymosin, an enzyme used in the cheese industry, is produced by the GM strains of *Escherichia coli*, *Aspergillus niger*, and *Kluyveromyces lactis*.

The live GMMs have the role of improving the production processes and the organoleptic qualities of the products. The microbial bodies are separated from the final products, except the bakery. But given the bread baking temperatures, it is assumed that most of the microorganisms are destroyed (Wakasa, Kawakatsu, Harada, & Takaiwa, 2018).

Some substances contained in foods are not essential nutrients, but they do contribute to promoting health. These beneficial effect substances, known as bioactive plant compounds, are represented by different classes of plant substances: flavonoids, phytoestrogens, and glucosinolates. Efforts are made both through conventional techniques and genetic engineering techniques to increase the content of bioactive plant compounds (e.g., increasing the content of glucosinolates in cabbage species), although increasing their content may not improve the health (Nugroho, Han, Pervitasari, Kim, & Kim, 2020).

Other PRPs, such as fibers, although not nutritionally valuable because they are not hydrolyzed by digestive enzymes, have beneficial effects on gut function. The use of genetic engineering techniques aims to increase the proportion of fiber in food, although, like other substances, it is harmful if consumed in excess.

The imbalance between the intake and the need for biologically active substances, which influence the pathology of contemporary humans, may explain the implication of genetic engineering in modifying food composition. Such research can encourage the consumption of health-promoting foods (Cuong et al., 2019).

Beginning with the 1980s cultivation of GMMs and their use to obtain molecules of therapeutic interest began. For example: insulin, growth hormone, vaccines (the hepatitis B vaccine obtained from the yeast strain *Saccharomyces cerevisiae*).

During the last 20 years, new species of bacteria have been introduced into these products. Their role is to modify the taste and textural qualities, but above all, they are used to induce beneficial effects for human health.

They are called "probiotics," and their main positive effects are regulation of intestinal transit and stimulation of the immune system. Thus genetically modified microorganisms (GMMs) that have acquired new properties can be used for medical purposes, to give the products the desired organoleptic properties or to introduce changes in the industrial production process (Gosálbez & Ramón, 2015).

According to the studies carried out, GMMs could be used in the future not only indirectly but also directly for the achievement of certain therapeutic and prophylactic purposes. This refers to GMMs that could replace a certain active principle (drug and enzyme) and could be consumed by humans via PRPs. It is also worth mentioning the ability of GMMs to transport a drug in the lower portions of the digestive tract (to prevent the degradation of the drug under the influence of gastric emptying) without resorting to sophisticated drug forms (Petrenko, Yamakura, Kohno, Sakimura, & Baba, 2010).

The classic vaccines are microbial cultures destroyed or with attenuated virulence or fragments of antigens, used in the prophylaxis of infectious diseases. The benefits of edible vaccines (expressed in plants) would be the elimination of the need for purification and the elimination of the injection route of administration.

Experiences have shown that the immune response is induced following vaccine administration. In addition, these antigens form an immune response appropriate to the host organism (Takaiwa et al., 2019).

The studies were carried out on corn, tobacco, tomatoes, soy, potato, salad, spinach, and peas. The first study to show that plants can express antigens occurred in 1990.

In this study, tobacco was genetically modified to express the surface antigen of the dental bacterium, *Streptococcus mutans*.

The experiment was performed on mice, the result being positive, that is, antibodies were formed against this bacterium. For the first time the efficacy of edible vaccines was reported in 1995, and the first report on human edible vaccines was made in 1998. Thus it was concluded that edible vaccines survive after digestion and effectively stimulate the immune response (Kauffmann et al., 2019).

Since then, many other vaccines have been expressed in different plants, for example, vaccines against rabies, cholera, traveler's diarrhea, dental caries, colon cancer, lung infections, hepatitis B, *Helicobacter pylori* (cause of ulcer), cytomegalovirus, and Norwalk virus (Stefanetti, Okan, Fink, Gardner, & Kasper, 2019).

In development is the edible vaccine against hepatitis B grown in banana. Its efficacy has already been demonstrated in mice. Vaccines obtained through genetic engineering may be cheaper than those produced in the laboratory.

Secondly, thanks to this method, the security technique is improved, because the classical production implies a high risk of infection of the personnel involved. Another advantage is that they can be stored at room temperature in the natural packaging—fruit (edible vaccines) (Green et al., 2019).

As a variety of the concept of using edible vaccines, involved in the development of immunity, researchers are experimenting with the use of GM foods to suppress immunity (in the case of autoimmune diseases, when the body produces antibodies against its own tissues: psoriasis, lupus erythematosus, rheumatoid arthritis, etc.).

Currently, it is working on obtaining the edible vaccine against diabetes, the plant involved being the potato containing glutamic acid decarboxylase—the enzyme involved in the autoimmune processes. In addition to vaccine production, researchers are also working on creating antibody-producing plants. For now, all the findings remain in the laboratory phase.

It is believed that these antibodies will be used more widely for the diagnosis of different diseases in humans and animals. Success in preventing disease through antibody-producing plants was first reported in 1998. The study described the first human clinical trials, in which antibodies obtained from the plant were applied to the teeth, rather than being consumed as edible vaccine (Alzan, Cooke, & Suarez, 2019).

This treatment prevents colonization with *Streptococcus mutans*, thus demonstrating the possibility of preventing dental pathologies. Also, in the research phase, there are studies related to cancer treatment. In 2000 the expression of the antibody against carcinoembryonic antigen in wheat and rice was announced.

About the creation of GM plants that produce antibodies for diagnosis was reported in Japan in 1998.

Thus a variety of GM tobacco was obtained expressing the HB core antigen of hepatitis B, which is used to determine hepatitis B virus contamination. Estimates show that the content of a GM tobacco leaf can be used to test 64,000–102,000 individuals (depending on the antigen concentration in the leaves) (Burton et al., 2019).

The list of human proteins that can be obtained with the help of GM plants is constantly increasing, due to the high demand for therapeutic and diagnostic use (Lu & Thum, 2019).

For example, enkephalins, ∂-interferon, albumin, hemoglobin, erythropoietin, the conversion enzyme of angioenzin-1.

- Collagen is the main protein of human skin. It is applied as a healing, covering the implants, traumas wound dressing, treating scars and wrinkles, etc. The collagen currently used is taken from cattle. This situation creates demand for human collagen synthesized by TSs, as this would eliminate immune reactions and the possibility of infection with various animal-specific pathogens. Two teams of French researchers succeeded in producing human collagen in GM tobacco. Research continues (Sipilä et al., 2018).
- Hemoglobin is a key molecule of respiration, as it ensures the transport of oxygen and carbon dioxide through blood. For several decades, scientists have been searching for a blood substitute, which could be easily stored and transported and which would not raise the issue of compatibility by blood groups and infectious risk. This substituent could be used in cases of emergency in cases of massive bleeding. Currently the possibility of producing transgenic tobacco is being evaluated, which will produce the human hemoglobin molecule. Hemoglobin is synthesized due to the activity of two genes. These two genes were transferred to tobacco, which allowed its synthesis in the cells of the plant. Further clinical studies are carried out, which would allow further application in practice (Monien et al., 2018).
- Gastric lipase is a protein used in the treatment of insufficient pancreatic exocrine function. In the absence of gastric lipase, lipid digestion becomes impossible. This problem particularly affects patients with mucoviscidosis or with pathologies of the pancreas. Mucoviscidosis is a common hereditary disease in Europe, constituting 1/2500 newborns. Gastric lipase is indispensable for these patients as a supportive treatment and to slow the evolution of the pathology. The current treatment is based on the administration of the pancreatic extract of pig, however, remaining inefficient for 15% of patients. In a laboratory in France, there is research into a gastric lipase produced from GM maize. Thus the gene that encodes this lipase is first transferred to tobacco (the test plant) and then to rapeseed, which are more efficient in the production of pharmaceutical molecules. Gastric lipase obtained from tobacco represents

0.5%–1% of leaf dry matter, 1 kg/ha, which allows to treat several dozen patients per year. In 2005 20 ha of GM maize producing gastric lipase were grown in France. The company "Meristem Therapeutics" has manufactured a variety of transgenic maize, which produces a gastric lipase of dogs intended to improve the condition of the digestive tract of sick children of mucoviscidosis. This has become an effective alternative to producing this protein. Until then, gastric lipase was produced by the use of transgenic yeasts; this process takes several weeks. The advantage of this method is the following: all proteins obtained by microorganisms require different chemical modifications to become active. For example, glycosylation is performed in the laboratory and involves both cost and quality losses. While the plants are able to do this (as well as other post-translational modifications) so that the active substance is ready to be obtained (Chahinian et al., 2006).

The production of a pharmaceutical substance in the open field is equivalent to opening a pharmacy "by nature." Thus genetic pollution takes place both horizontally (direct transfer of genetic material between two organisms without crossing, e.g., between plants and soil microorganisms) and vertically (pollination and cross-breeding between varieties) (Jänsch et al., 2018).

10. Trends in biotechnologies

The researchers working in this field estimate that in the next decade will be created varieties resistant to insects, which can better withstand droughts and salt and which will be enriched with omega-3, vitamin A, and other beneficial substances (Kraic et al., 2018).

One company, in the laboratory, made *an apple that does not oxidize after being cut* or if it is hit. The company managed to add in the DNA of the apple a natural gene, also of apple, which causes the apple not to blacken due to oxidation for 16–21 days.

This apple variety that has been genetically modified, having the property of not oxidizing (does not appear brown as soon as it was sliced), is about to receive approval from the US Department of Agriculture.

Practically the specialists have deactivated the genes responsible for producing the enzyme that induces this shade of apple called Arctic (Jia, Jiang, van Nocker, Gong, & Ma, 2019).

But first of all, officials will have to deal with public protests opposing genetically modified organisms. According to company representatives, their realization has many benefits. These include higher profits for apple producers (by selling sliced apples, more appealing to buyers) and higher consumer consumption (consuming 1.5 kg less apples per year than 20 years ago).

The Arctic apple will completely oxidize in about 3 weeks, because the genes responsible for oxidation have been blocked. This means that apples can be sold cheaper than the basic varieties undergoing antioxidant treatment (Fang et al., 2019).

The disadvantages listed by critics of this genetically modified apple variety outweigh the advantages that the company emphasizes.

First, representatives of US apple producers are concerned that such an apple would raise questions about the nutritional quality of an apple. They wonder how good it is to consume an apple after 3 weeks of collection, given that a natural one is already rotted. Consumers would buy apples that look good but no longer have the necessary nutrients (Maxmen, 2017).

Other voices in the apple industry believe that there was no need for such a genetic mutation, because preventing the oxidation of fresh foods is made simple, with a few splashes of lemon juice—vitamin C is a very good antioxidant. Other critics say that apple is the first step for fruit to become an industrial product.

Although Americans have been consuming genetically modified foods since the early 1990s, most of these products have been processed, not fresh. Now, there is a risk that apples will be packaged and no longer sold at fresh produce stands. So, consumers are not yet decided, if they want natural or genetically modified apples (An et al., 2018).

The researchers also plan to treat *zinc deficiency*, a nutritional deficiency that significantly affects the immune system and contributes to the death of 800,000 people annually. The outer layer of cereals contains zinc, but inside the grains, this element is found in very small quantities. In societies where grains are milled (e.g., to produce white rice instead of brown rice), zinc is not consumed in sufficient quantities.

The solution identified by the specialists is the addition in the varieties of cereals of a gene that would redistribute the zinc inside the plants, causing that the interior of the grain of cereals contains a greater quantity.

Scientists have made some successful steps in barley experiments, but more efforts are needed to create a variety that can be launched on the market (Khatun et al., 2018).

Researchers are working on another bold project: using genetic modifications to make *vegetable oils as healthy as fish oil*. One way to reduce the risk of cardiovascular disease is to consume fish oil. However, overfishing is a problem that has led to the massive reduction of fish stocks. Therefore it is impossible for every person on Earth to have access to a constant source of fish oil.

The researchers found that the fatty acids that underlie the beneficial effect of fish oil on human health come from the algae that fish consume. Scientists have succeeded in extracting from the algae the genes that produce these fatty acids and have inserted them into the *Arabidopsis* plant and aim to repeat the success on the oil producing plants, their first target being the rape.

Experts hope that when the healthy crops they create will be ready for marketing, a goal planned over a few years, the European public will be more receptive to genetically modified food than it is today (Lin et al., 2019).

Another innovation that promises to change the lives of millions of people is *golden rice*. Rice is the main food for more than half of the world's population, especially for the Asian population. Although rice has many calories, it does not contain vitamin A, a nutrient that plays a vital role in maintaining health.

Millions of people around the world do not have a varied diet, which means that this deficiency is extremely widespread.

Golden rice is a solution to this problem. The rice is modified to contain beta-carotene, a substance that is transformed by the body into vitamin A. Specialists expect gold rice that debut in countries such as the Philippines, Bangladesh, Indonesia, and Vietnam, to improve the lives of millions of children and children grown-ups (Gayen, Ali, Sarkar, Datta, & Datta, 2015).

Another ongoing project consists of *enriching the cassava with three essential elements: vitamin A, iron, and protein.* Cassava is the main food for 250 million people in Africa, and its enrichment with the three elements would eliminate the deficiencies in the body of several million children and adults on the continent.

The importance of this research is underlined by the studies conducted in the preschool children in Nigeria: 83% of those who ate peanut deficiency had vitamin A deficiency and 43% iron deficiency, while in Kenya 41% had vitamin A deficiency and 78% deficiency iron (Kobayashi et al., 2019).

11. Conclusions and recommendations

Genetically modified plants have been one of the most controversial fields of science in the last two decades. Introduced in 1995 in the United States and Canada, GM crops are used today by over 15 million farmers in 25 countries. The agricultural surface on which genetically modified plants are grown exceeded 1 billion hectares for the first time in 2010. The development that biotechnology has undergone in recent years has allowed to study these plants and their effects, allowing a better knowledge of the benefits of the crops. A study published in 2010 in the journal Nature Biotechnology, which provided an overview of 168 surveys conducted by farmers in 12 countries, showed that genetically modified plants had a largely positive effect. Of the 168 surveys comparing genetically modified and conventional crops, 124 showed positive results for farmers who opted for genetically modified crops, 32 did not indicate any difference, and 13 indicated favorable results for conventional agriculture. Initial fears about the effect that genetically modified plants may have had on the environment have not been confirmed. Crops that were genetically modified to be more resistant to pests required a much smaller amount of insecticide, and increasing production per hectare reduced the amount of cultivated land. A report prepared by the European Commission stated that "the main conclusion we can draw from the efforts of more than 130 research projects that have been spread over a period of 25 years and involving more than 500 independent research groups is that biotechnology, and especially genetically modified organisms, are no more risky than conventional plant growth technologies." In different countries the most important scientific forums that have studied the genetically modified organisms have announced their position toward this application of biotechnology, thus arguing based on their own scientific research, the use of agricultural biotechnologies as an integral part of EU agricultural policy, adding that reject any unscientific manifestation, any opinion of nonperformers in biotechnological scientific research, any tendency to manipulate public opinion, and any action by which the EU remains a massive and perpetual importer of food, including from countries with excellent agricultural biotechnologies.

Research shows that genetically modified plants will offer other benefits over the next few years in addition to reducing the amount of pesticides used. Scientists recommend the consumption of fruits and vegetables, foods whose effect on human health is well known.

References

Ahlstedt, S., Belin, L., Eriksson, N. E., & Hanson, L. A. (1975). Quantity and avidity of antibodies against birch pollen in atopic patients during hyposensitization. A preliminary study. *International Archives of Allergy and Applied Immunology, 48*(5), 632–641.

Alzan, H. F., Cooke, B. M., & Suarez, C. E. (2019). Transgenic *Babesia bovis* lacking 6-Cys sexual-stage genes as the foundation for non-transmissible live vaccines against bovine babesiosis. *Ticks and Tick-Borne Diseases, 10*(3), 722–728.

An, J. P., Yao, J. F., Xu, R. R., You, C. X., Wang, X. F., & Hao, Y. J. (2018). Apple bZIP transcription factor MdbZIP44 regulates abscisic acid-promoted anthocyanin accumulation. *Plant, Cell & Environment, 41*(11), 2678–2692.

Avsar, B., Sadeghi, S., Turkec, A., & Lucas, S. J. (2020). Identification and quantitation of genetically modified (GM) ingredients in maize, rice, soybean and wheat-containing retail foods and feeds in Turkey. *Journal of Food Science and Technology, 57*(2), 787–793.

Babar, U., Nawaz, M. A., Arshad, U., Azhar, M. T., Atif, R. M., Golokhvast, K. S., et al. (2020). Transgenic crops for the agricultural improvement in Pakistan: A perspective of environmental stresses and the current status of genetically modified crops. *GM Crops Food, 11*(1), 1–29.

Bawa, A. S., & Anilakumar, K. R. (2013). Genetically modified foods: Safety, risks and public concerns-a review. *Journal of Food Science and Technology, 50*(6), 1035–1046.

Bhagyawant, S. S., Narvekar, D. T., Gupta, N., Bhadkaria, A., Gautam, A. K., & Srivastava, N. (2019). Chickpea (*Cicer arietinum* L.) lectin exhibit inhibition of ACE-I., α-amylase and α-glucosidase activity. *Protein and Peptide Letters, 26*(7), 494–501.

Blando, F., Berland, H., Maiorano, G., Durante, M., Mazzucato, A., Picarella, M. E., et al. (2019). Nutraceutical characterization of anthocyanin-rich fruits produced by "Sun Black" tomato line. *Frontiers in Nutrition, 6*, 133.

Bollinedi, H., Dhakane-Lad, J., Gopala Krishnan, S., Bhowmick, P. K., Prabhu, K. V., Singh, N. K., et al. (2019). Kinetics of β-carotene degradation under different storage conditions in transgenic Golden Rice® lines. *Food Chemistry, 278*, 773–779.

Botha, A. S., Erasmus, A., du Plessis, H., & Van den Berg, J. (2019). Efficacy of Bt maize for control of *Spodoptera frugiperda* (Lepidoptera: Noctuidae) in South Africa. *Journal of Economic Entomology, 112*(3), 1260–1266.

Bruetschy, C. (2019). The EU regulatory framework on genetically modified organisms (GMOs). *Transgenic Research, 28*(Suppl. 2), 169–174.

Burton, S., Spicer, L. M., Charles, T. P., Gangadhara, S., Reddy, P. B. J., Styles, T. M., et al. (2019). HIV-1 envelope vaccination regimens differ in their ability to elicit antibodies with moderate neutralization breadth against genetically diverse tier 2 HIV-1 envelope variants. *Journal of Virology, 93*(7). pii: e01846-18.

Chahinian, H., Snabe, T., Attias, C., Fojan, P., Petersen, S. B., & Carrière, F. (2006). How gastric lipase., an interfacial enzyme with a Ser-His-Asp catalytic triad, acts optimally at acidic pH. *Biochemistry, 45*(3), 993–1001.

Che, P., Zhao, Z. Y., Glassman, K., Dolde, D., Hu, T. X., Jones, T. J., et al. (2016). Elevated vitamin E content improves all-trans β-carotene accumulation and stability in biofortified sorghum. *Proceedings of the National Academy of Sciences of the United States of America, 113*(39), 11040–11045.

Chen, L., Chen, M., Ma, C., & Zeng, A. P. (2018). Discovery of feed-forward regulation in L-tryptophan biosynthesis and its use in metabolic engineering of E. coli for efficient tryptophan bioproduction. *Metabolic Engineering, 47*, 434–444.

Chinnadurai, P., Stojšin, D., Liu, K., Frierdich, G. E., Glenn, K. C., Geng, T., et al. (2018). Variability of CP4 EPSPS expression in genetically engineered soybean (*Glycine max* L. Merrill). *Transgenic Research, 27*(6), 511–524.

Cuong, D. M., Park, C. H., Bong, S. J., Kim, N. S., Kim, J. K., & Park, S. U. (2019). Enhancement of glucosinolate production in watercress (*Nasturtium officinale*) hairy roots by overexpressing cabbage transcription factors. *Journal of Agricultural and Food Chemistry, 67*(17), 4860–4867.

Denisow-Pietrzyk, M., Pietrzyk, Ł., & Denisow, B. (2019). Asteraceae species as potential environmental factors of allergy. *Environmental Science and Pollution Research International, 26*(7), 6290–6300.

Dong, Q., Xu, Q., Kong, J., Peng, X., Zhou, W., Chen, L., et al. (2019). Overexpression of Zm-bZIP22 gene alters endosperm starch content and composition in maize and rice. *Plant Science, 283*, 407–415.

Duan, M., Gu, J., Wang, X., Li, Y., Zhang, S., Yin, Y., et al. (2018). Effects of genetically modified cotton stalks on antibiotic resistance genes, intI1, and intI2 during pig manure composting. *Ecotoxicology and Environmental Safety, 147*, 637–642.

Dunn, S. E., Vicini, J. L., Glenn, K. C., Fleischer, D. M., & Greenhawt, M. J. (2017). The allergenicity of genetically modified foods from genetically engineered crops: A narrative and systematic review. *Annals of Allergy, Asthma & Immunology, 119*(3), 214–222.e3.

Fang, H., Dong, Y., Yue, X., Chen, X., He, N., Hu, J., et al. (2019). MdCOL4 interaction mediates crosstalk between UV-B and high temperature to control fruit coloration in apple. *Plant and Cell Physiology, 60*(5), 1055–1066.

Fernandes, M. G., Costa, E. N., Dutra, C. C., & Raizer, J. (2019). Species richness and community composition of ants and beetles in Bt and non-Bt maize fields. *Environmental Entomology, 48*(5), 1095–1103.

Gayen, D., Ali, N., Sarkar, S. N., Datta, S. K., & Datta, K. (2015). Down-regulation of lipoxygenase gene reduces degradation of carotenoids of golden rice during storage. *Planta, 242*(1), 353–363.

Goodman, R. E., Panda, R., & Ariyarathna, H. (2013). Evaluation of endogenous allergens for the safety evaluation of genetically engineered food crops: Review of potential risks, test methods, examples and relevance. *Journal of Agricultural and Food Chemistry, 61*(35), 8317–8332.

Gosálbez, L., & Ramón, D. (2015). Probiotics in transition: Novel strategies. *Trends in Biotechnology, 33*(4), 195–196.

Grantina-Ievina, L., Ievina, B., Evelone, V., Berga, S., Kovalcuka, L., Bergspica, I., et al. (2019). Potential risk evaluation for unintended entry of genetically modified plant propagating material in Europe through import of seeds and animal feed-the experience of Latvia. *GM Crops Food*, *10*(3), 159–169.

Green, C. A., Sande, C. J., Scarselli, E., Capone, S., Vitelli, A., Nicosia, A., et al. (2019). Novel genetically-modified chimpanzee adenovirus and MVA-vectored respiratory syncytial virus vaccine safely boosts humoral and cellular immunity in healthy older adults. *The Journal of Infection*, *78*(5), 382–392.

Gürel, F., Uçarlı, C., Tufan, F., & Kalaskar, D. M. (2015). Enhancing T-DNA transfer efficiency in barley (*Hordeum vulgare* L.) cells using extracellular cellulose and lectin. *Applied Biochemistry and Biotechnology*, *176*(4), 1203–1216.

Han, J. Y., Baek, S. H., Jo, H. J., Yun, D. W., & Choi, Y. E. (2019). Genetically modified rice produces ginsenoside aglycone (protopanaxadiol). *Planta*, *250*(4), 1103–1110.

He, S., Simpson, B. K., Sun, H., Ngadi, M. O., Ma, Y., & Huang, T. (2018). *Phaseolus vulgaris* lectins: A systematic review of characteristics and health implications. *Critical Reviews in Food Science and Nutrition*, *58*(1), 70–83.

Ishii, T. (2017). The ethics of creating genetically modified children using genome editing. *Current Opinion in Endocrinology, Diabetes, and Obesity*, *24*(6), 418–423.

Jänsch, S., Bauer, J., Leube, D., Otto, M., Römbke, J., Teichmann, H., et al. (2018). A new ecotoxicological test method for genetically modified plants and other stressors in soil with the black fungus gnat *Bradysia impatiens* (Diptera): Current status of test development and dietary effects of azadirachtin on larval development and emergence rate. *Environmental Sciences Europe*, *30*(1), 38.

Jia, D., Jiang, Q., van Nocker, S., Gong, X., & Ma, F. (2019). An apple (*Malus domestica*) NAC transcription factor enhances drought tolerance in transgenic apple plants. *Plant Physiology and Biochemistry*, *139*, 504–512.

Jones, N. R., Agnew, M., Banic, I., Grossi, C. M., Colón-González, F. J., Plavec, D., et al. (2019). Ragweed pollen and allergic symptoms in children: Results from a three-year longitudinal study. *The Science of the Total Environment*, *683*, 240–248.

Jonsdottir, S., Stefansdottir, S. B., Kristinarson, S. B., Svansson, V., Bjornsson, J. M., Runarsdottir, A., et al. (2018). Barley produced *Culicoides allergens* are suitable for monitoring the immune response of horses immunized with E. coli expressed allergens. *Veterinary Immunology and Immunopathology*, *201*, 32–37.

Jyoti, A., Kaushik, S., Srivastava, V. K., Datta, M., Kumar, S., Yugandhar, P., et al. (2019). The potential application of genome editing by using CRISPR/Cas9, and its engineered and ortholog variants for studying the transcription factors involved in the maintenance of phosphate homeostasis in model plants. *Seminars in Cell & Developmental Biology*, *96*, 77–90.

Kauffmann, F., Van Damme, P., Leroux-Roels, G., Vandermeulen, C., Berthels, N., Beuneu, C., et al. (2019). Clinical trials with GMO-containing vaccines in Europe: Status and regulatory framework. *Vaccine*, *37*(42), 6144–6153.

Kerre, S., Strobbe, T., Naessens, T., Theunis, M., Foubert, K., & Aerts, O. (2018). Alkyl glucosides: Newly identified allergens in foam wound dressings. *Contact Dermatitis*, *79*(3), 191–193.

Khatun, M. A., Hossain, M. M., Bari, M. A., Abdullahil, K. M., Parvez, M. S., Alam, M. F., et al. (2018). Zinc deficiency tolerance in maize is associated with the up-regulation of Zn transporter genes and antioxidant activities. *Plant Biology (Stuttgart, Germany)*, *20*(4), 765–770.

Kobayashi, T., Ozu, A., Kobayashi, S., An, G., Jeon, J. S., & Nishizawa, N. K. (2019). OsbHLH058 and OsbHLH059 transcription factors positively regulate iron deficiency responses in rice. *Plant Molecular Biology*, *101*(4–5), 471–486.

Kraic, J., Mihálik, D., Klčová, L., Gubišová, M., Klempová, T., Hudcovicová, M., et al. (2018). Progress in the genetic engineering of cereals to produce essential polyunsaturated fatty acids. *Journal of Biotechnology*, *284*, 115–122.

Kramkowska, M., Grzelak, T., & Czyżewska, K. (2013). Benefits and risks associated with genetically modified food products. *Annals of Agricultural and Environmental Medicine, 20*(3), 413–419.

Kurokawa, S., Nakamura, R., Mejima, M., Kozuka-Hata, H., Kuroda, M., Takeyama, N., et al. (2013). MucoRice-cholera toxin B-subunit, a rice-based oral cholera vaccine, down-regulates the expression of α-amylase/trypsin inhibitor-like protein family as major rice allergens. *Journal of Proteome Research, 12*(7), 3372–3382.

Larsen, J. N., Casals, A. B., From, N. B., Strøman, P., & Ipsen, H. (1993). Characterization of recombinant bet vI, the major pollen allergen of *Betula verrucosa* (white birch), produced by fed-batch fermentation. *International Archives of Allergy and Immunology, 102*(3), 249–258.

Latham, J. R., Love, M., & Hilbeck, A. (2017). The distinct properties of natural and GM cry insecticidal proteins. *Biotechnology & Genetic Engineering Reviews, 33*(1), 62–96.

Leal, R. B., Pinto-Junior, V. R., Osterne, V. J. S., Wolin, I. A. V., Nascimento, A. P. M., Neco, A. H. B., et al. (2018). Crystal structure of Dly L, a mannose-specific lectin from *Dioclealasiophylla* Mart. Ex Benth seeds that display cytotoxic effects against C6 glioma cells. *International Journal of Biological Macromolecules, 114*, 64–76.

Lemon, D. J., Kay, M. K., Titus, J. K., Ford, A. A., Chen, W., Hamlin, N. J., et al. (2019). Construction of a genetically modified T7Select phage system to express the antimicrobial peptide 1018. *Journal of Microbiology, 57*(6), 532–538.

Lin, Y., Chen, G., Mietkiewska, E., Song, Z., Caldo, K. M. P., Singer, S. D., et al. (2019). Castor patatin-like phospholipase A IIIβ facilitates removal of hydroxy fatty acids from phosphatidylcholine in transgenic Arabidopsis seeds. *Plant Molecular Biology, 101*(6), 521–536.

Liu, M. S., Ko, M. H., Li, H. C., Tsai, S. J., Lai, Y. M., Chang, Y. M., et al. (2014). Compositional and proteomic analyses of genetically modified broccoli (*Brassica oleracea* var. italica) harboring an agrobacterial gene. *International Journal of Molecular Sciences, 15*(9), 15188–15209.

Loyola-Vargas, V. M., & Ochoa-Alejo, N. (2018). An introduction to plant tissue culture: Advances and perspectives. *Methods in Molecular Biology, 1815*, 3–13.

Lu, D., & Thum, T. (2019). RNA-based diagnostic and therapeutic strategies for cardiovascular disease. *Nature Reviews. Cardiology, 16*(11), 661–674.

Lyu, T., Liu, W., Hu, Z., Xiang, X., Liu, T., Xiong, X., et al. (2020). Molecular characterization and expression analysis reveal the roles of Cys2/His2 zinc-finger transcription factors during flower development of *Brassica rapa* subsp. chinensis. *Plant Molecular Biology, 102*(1–2), 123–141.

Malehorn, D. E., Borgmeyer, J. R., Smith, C. E., & Shah, D. M. (1994). Characterization and expression of an antifungal zeamatin-like protein (Zlp) gene from *Zea mays. Plant Physiology, 106*(4), 1471–1481.

Masoumi, S. M., Fathi, N., Hajmoradi, F., & Baghaeifar, Z. (2019). The micromorphological investigation of pollen grains of some important allergenic plants in Kermanshah (west of Iran). *Cellular and Molecular Biology (Noisy-le-Grand, France), 65*(8), 32–38.

Maxmen, A. (2017). Genetically modified apple reaches US stores, but will consumers bite? *Nature, 551*(7679), 149–150.

McInnes, R. N., Hemming, D., Burgess, P., Lyndsay, D., Osborne, N. J., Skjøth, C. A., et al. (2017). Mapping allergenic pollen vegetation in UK to study environmental exposure and human health. *The Science of the Total Environment, 599-600*, 483–499.

Menzel, C., Andersson, M., Andersson, R., Vázquez-Gutiérrez, J. L., Daniel, G., Langton, M., et al. (2015). Improved material properties of solution-cast starch films: Effect of varying amylopectin structure and amylose content of starch from genetically modified potatoes. *Carbohydrate Polymers, 130*, 388–397.

Moghissi, A. A., Pei, S., & Liu, Y. (2016). Golden rice: Scientific, regulatory and public information processes of a genetically modified organism. *Critical Reviews in Biotechnology, 36*(3), 535–541.

Monien, B. H., Sachse, B., Meinl, W., Abraham, K., Lampen, A., & Glatt, H. (2018). Hemoglobin adducts of furfuryl alcohol in genetically modified mouse models: Role of endogenous sulfotransferases 1a1 and 1d1 and transgenic human sulfotransferases 1A1/1A2. *Toxicology Letters, 295*, 173–178.

Napier, J. A., Haslam, R. P., Tsalavouta, M., & Sayanova, O. (2019). The challenges of delivering genetically modified crops with nutritional enhancement traits. *Nature Plants, 5*(6), 563–567.

Nielsen, M. I., Stegmayr, J., Grant, O. C., Yang, Z., Nilsson, U. J., Boos, I., et al. (2018). Galectin binding to cells and glycoproteins with genetically modified glycosylation reveals galectin-glycan specificities in a natural context. *The Journal of Biological Chemistry, 293*(52), 20249–20262.

Nugroho, A. B. D., Han, N., Pervitasari, A. N., Kim, D. H., & Kim, J. (2020). Differential expression of major genes involved in the biosynthesis of aliphatic glucosinolates in intergeneric Baemoochae (Brassicaceae) and its parents during development. *Plant Molecular Biology, 102*(1–2), 171–184.

O'Callaghan, M., Brownbridge, M., Stilwell, W. B., Gerard, E. M., Burgess, E. P., Barraclough, E. I., et al. (2007). Effects of tobacco genetically modified to express protease inhibitor bovine spleen trypsin inhibitor on non-target soil organisms. *Environmental Biosafety Research, 6*(3), 183–195.

Olcese, R., Silvestri, M., Del Barba, P., Brolatti, N., Barberi, S., Tosca, M. A., et al. (2019). Mal d 1 and Bet v 1 sensitization pattern in children with pollen food syndrome. *Allergology International, 68*(1), 122–124.

Osorio, C. E., Wen, N., Mejias, J. H., Liu, B., Reinbothe, S., von Wettstein, D., et al. (2019). Development of wheat genotypes expressing a glutamine-specific endoprotease from barley and a prolyl endopeptidase from *Flavobacterium meningosepticum* or *Pyrococcusfuriosus* as a potential remedy to celiac disease. *Functional & Integrative Genomics, 19*(1), 123–136.

Pesquet, E., Zhang, B., Gorzsás, A., Puhakainen, T., Serk, H., Escamez, S., et al. (2013). Non-cell-autonomous postmortem lignification of tracheary elements in *Zinnia elegans*. *The Plant Cell, 25*(4), 1314–1328.

Petrenko, A. B., Yamakura, T., Kohno, T., Sakimura, K., & Baba, H. (2010). Reduced immobilizing properties of isoflurane and nitrous oxide in mutant mice lacking the N-methyl-D-aspartate receptor GluR(epsilon)1 subunit are caused by the secondary effects of gene knockout. *Anesthesia and Analgesia, 110*(2), 461–465.

Picornell, A., Buters, J., Rojo, J., Traidl-Hoffmann, C., Damialis, A., Menzel, A., et al. (2019). Predicting the start, peak and end of the *Betula* pollen season in Bavaria, Germany. *The Science of the Total Environment, 690*, 1299–1309.

Purohit, A., Niederberger, V., Kronqvist, M., Horak, F., Grönneberg, R., Suck, R., et al. (2008). Clinical effects of immunotherapy with genetically modified recombinant birch pollen Bet v 1 derivatives. *Clinical and Experimental Allergy, 38*(9), 1514–1525.

Ralley, L., Schuch, W., Fraser, P. D., & Bramley, P. M. (2016). Genetic modification of tomato with the tobacco lycopene β-cyclase gene produces high β-carotene and lycopene fruit. *Zeitschrift fuer Naturforschung, C: Journal of Biosciences, 71*(9–10), 295–301.

Reitsma, M., Bastiaan-Net, S., Sijbrandij, L., de Weert, E., Sforza, S., Gerth van Wijk, R., et al. (2018). Origin and processing methods slightly affect allergenic characteristics of cashew nuts (*Anacardium occidentale*). *Journal of Food Science, 83*(4), 1153–1164.

Rostoks, N., Grantiņa-Ieviņa, L., Ieviņa, B., Evelone, V., Valciņa, O., & Aleksejeva, I. (2019). Genetically modified seeds and plant propagating material in Europe: Potential routes of entrance and current status. *Heliyon, 5*(2). https://doi.org/10.1016/j.heliyon.2019.e01242.

Savić, J., Nikolić, R., Banjac, N., Zdravković-Korać, S., Stupar, S., Cingel, A., et al. (2019). Beneficial implications of sugar beet proteinase inhibitor BvSTI on plant architecture and salt stress tolerance in *Lotus corniculatus* L. *Journal of Plant Physiology, 243*, 153055.

Schmidt, M. A., & Pendarvis, K. (2017). Proteome rebalancing in transgenic *Camelina occurs* within the enlarged proteome induced by β-carotene accumulation and storage protein suppression. *Transgenic Research, 26*(2), 171–186.

Schwartz, D. M., Farley, T. K., Richoz, N., Yao, C., Shih, H. Y., Petermann, F., et al. (2019). Retinoic acid receptor alpha represses a Th9 transcriptional and epigenomic program to reduce allergic pathology. *Immunity, 50*(1), 106–120.e10.

Sestito, S. E., Facchini, F. A., Morbioli, I., Billod, J. M., Martin-Santamaria, S., Casnati, A., et al. (2017). Amphiphilic guanidinocalixarenes inhibit lipopolysaccharide (LPS)- and lectin-stimulated toll-like receptor 4 (TLR4) signaling. *Journal of Medicinal Chemistry, 60*(12), 4882–4892.

Seth, D., Poowutikul, P., Pansare, M., & Kamat, D. (2020). Food allergy: A review. *Pediatric Annals, 49*(1), e50–e58.

Sipilä, K. H., Drushinin, K., Rappu, P., Jokinen, J., Salminen, T. A., Salo, A. M., et al. (2018). Proline hydroxylation in collagen supports integrin binding by two distinct mechanisms. *The Journal of Biological Chemistry, 293*(20), 7645–7658.

Stefanetti, G., Okan, N., Fink, A., Gardner, E., & Kasper, D. L. (2019). Glycoconjugate vaccine using a genetically modified O antigen induces protective antibodies to *Francisella tularensis*. *Proceedings of the National Academy of Sciences of the United States of America, 116*(14), 7062–7070.

Swamy, B. P. M., Samia, M., Boncodin, R., Marundan, S., Rebong, D. B., Ordonio, R. L., et al. (2019). Compositional analysis of genetically engineered GR2E "Golden Rice" in comparison to that of conventional rice. *Journal of Agricultural and Food Chemistry, 67*(28), 7986–7994.

Szymczyk, B., Szczurek, W., Świątkiewicz, S., Kwiatek, K., Sieradzki, Z., Mazur, M., et al. (2018). Results of a 16-week safety assurance study with rats fed genetically modified Bt maize: Effect on growth and health parameters. *Journal of Veterinary Research, 62*(4), 555–561.

Takaishi, S., Saito, S., Endo, T., Asaka, D., Wakasa, Y., Takagi, H., et al. (2019). T-cell activation by transgenic rice seeds expressing the genetically modified Japanese cedar pollen allergens. *Immunology, 158*(2), 94–103.

Takaiwa, F., Ogo, Y., & Wakasa, Y. (2018). Specific region affects the difference in accumulation levels between apple food allergen Mal d 1 and birch pollen allergen Bet v 1 which are expressed in vegetative tissues of transgenic rice. *Plant Molecular Biology, 98*(4–5), 439–454.

Takaiwa, F., Yang, L., Takagi, H., Maruyama, N., Wakasa, Y., Ozawa, K., et al. (2019). Development of Rice-seed-based oral allergy vaccines containing hypoallergenic Japanese cedar pollen allergen derivatives for immunotherapy. *Journal of Agricultural and Food Chemistry, 67*(47), 13127–13138.

Tang, M., He, X., Luo, Y., Ma, L., Tang, X., & Huang, K. (2013). Nutritional assessment of transgenic lysine-rich maize compared with conventional quality protein maize. *Journal of the Science of Food and Agriculture, 93*(5), 1049–1054.

Tsuda, M., Watanabe, K. N., & Ohsawa, R. (2019). Regulatory status of genome-edited organisms under the Japanese Cartagena Act. *Frontiers in Bioengineering and Biotechnology, 7*, 387.

Tulinská, J., Adel-Patient, K., Bernard, H., Líšková, A., Kuricová, M., Ilavská, S., et al. (2018). Humoral and cellular immune response in Wistar Han RCC rats fed two genetically modified maize MON810 varieties for 90 days (EU 7th Framework Programme project GRACE). *Archives of Toxicology*, *92*(7), 2385–2399.

Van Giap, D., Jung, J. W., & Kim, N. S. (2019). Production of functional recombinant cyclic citrullinated peptide monoclonal antibody in transgenic rice cell suspension culture. *Transgenic Research*, *28*(2), 177–188.

Van Hoang, V., Ochi, T., Kurata, K., Arita, Y., Ogasahara, Y., & Enomoto, K. (2018). Nisin-induced expression of recombinant T cell epitopes of major Japanese cedar pollen allergens in *Lactococcus lactis*. *Applied Microbiology and Biotechnology*, *102*(1), 261–268.

van Wijk, F., Nierkens, S., Hassing, I., Feijen, M., Koppelman, S. J., de Jong, G. A., et al. (2005). The effect of the food matrix on in vivo immune responses to purified peanut allergens. *Toxicological Sciences*, *86*(2), 333–341.

Wakasa, Y., Kawakatsu, T., Harada, T., & Takaiwa, F. (2018). Transgene-independent heredity of RdDM-mediated transcriptional gene silencing of endogenous genes in rice. *Plant Biotechnology Journal*, *16*(12), 2007–2015.

Wang, L., Fouts, D. E., Stärkel, P., Hartmann, P., Chen, P., Llorente, C., et al. (2016). Intestinal REG3 lectins protect against alcoholic steatohepatitis by reducing mucosa-associated microbiota and preventing bacterial translocation. *Cell Host & Microbe*, *19*(2), 227–239.

Wang, Y., Li, Y., He, S. P., Gao, Y., Wang, N. N., Lu, R., et al. (2019). A cotton (*Gossypium hirsutum*) WRKY transcription factor (GhWRKY22) participates in regulating anther/pollen development. *Plant Physiology and Biochemistry*, *141*, 231–239.

Whelan, A. I., & Lema, M. A. (2015). Regulatory framework for gene editing and other new breeding techniques (NBTs) in Argentina. *GM Crops Food*, *6*(4), 253–265.

Worm, M., Rak, S., Samoliński, B., Antila, J., Höiby, A. S., Kruse, B., et al. (2019). Efficacy and safety of birch pollen allergoid subcutaneous immunotherapy: A 2-year double-blind, placebo-controlled, randomized trial plus 1-year open-label extension. *Clinical and Experimental Allergy*, *49*(4), 516–525.

Wu, Y., Xu, Y., Du, Y., Zhao, X., Hu, R., Fan, X., et al. (2017). Dietary safety assessment of genetically modified rice EH rich in β-carotene. *Regulatory Toxicology and Pharmacology*, *88*, 66–71.

Xie, Z., Zou, S., Xu, W., Liu, X., Huang, K., & He, X. (2018). No subchronic toxicity of multiple herbicide-resistant soybean FG72 in Sprague-Dawley rats by 90-days feeding study. *Regulatory Toxicology and Pharmacology*, *94*, 299–305.

Yang, Y., Hu, Y., Yue, Y., Pu, Y., Yin, X., Duan, Y., et al. (2020). Expression profiles of glucosinolate biosynthetic genes in turnip (*Brassica rapa* var. rapa) at different developmental stages and effect of transformed flavin-containing monooxygenase genes on hairy root glucosinolate content. *Journal of the Science of Food and Agriculture*, *100*(3), 1064–1071.

Yin, C., Park, J. J., Gang, D. R., & Hulbert, S. H. (2014). Characterization of a tryptophan 2-monooxygenase gene from *Puccinia graminis* f. sp. tritici involved in auxin biosynthesis and rust pathogenicity. *Molecular Plant-Microbe Interactions*, *27*(3), 227–235.

You, R. I., Lee, Y. P., Su, T. Y., Lin, C. C., Chen, C. S., & Chu, C. L. (2019). A benzenoid 4, 7-dimethoxy-5-methyl-L, 3-benzodioxole from *Antrodiacinnamomea* attenuates dendritic cell-mediated Th2 allergic responses. *The American Journal of Chinese Medicine*, *47*(6), 1271–1287.

Zhang, T., Han, S., Liu, Q., Guo, Y., & He, L. (2014). Analysis of allergens in tubeimu saponin extracts by using rat basophilic leukemia 2H3 cell-based affinity chromatography coupled to liquid chromatography and mass spectrometry. *Journal of Separation Science*, *37*(22), 3384–3391.

Zhang, J., Kang, Y., Valverde, B. E., Dai, W., Song, X., & Qiang, S. (2018). Feral rice from introgression of weedy rice genes into transgenic herbicide-resistant hybrid-rice progeny. *Journal of Experimental Botany*, *69*(16), 3855–3865.

Zhang, C. J., Yook, M. J., Park, H. R., Lim, S. H., Kim, J. W., Song, J. S., et al. (2018). Evaluation of maximum potential gene flow from herbicide resistant *Brassica napus* to its male sterile relatives under open and wind pollination conditions. *The Science of the Total Environment*, *634*, 821–830.

Zhao, J. Z., Li, Y. X., Collins, H. L., Cao, J., Earle, E. D., & Shelton, A. M. (2001). Different cross-resistance patterns in the diamondback moth (Lepidoptera: Plutellidae) resistant to *Bacillus thuringiensis* toxin Cry1C. *Journal of Economic Entomology*, *94*(6), 1547–1552.

Zhou, W., Wang, X., Zhou, D., Ouyang, Y., & Yao, J. (2017). Overexpression of the 16-kDa α-amylase/trypsin inhibitor RAG2 improves grain yield and quality of rice. *Plant Biotechnology Journal*, *15*(5), 568–580.

Zhu, Q., Zeng, D., Yu, S., Cui, C., Li, J., Li, H., et al. (2018). From golden rice to aSTARice: Bioengineering astaxanthin biosynthesis in rice endosperm. *Molecular Plant*, *11*(12), 1440–1448.

Chapter 4

Phytomedicine and phytonanocomposites—An expanding horizon

Mir Zahoor Gul[a], Mohd Yasin Bhat[b], Suresh Velpula[a], Karuna Rupula[a], and Sashidhar Rao Beedu[a]

[a]*Department of Biochemistry, University College of Science, Osmania University, Hyderabad, Telangana, India,* [b]*Department of Plant Sciences, School of Life Sciences, University of Hyderabad, Hyderabad, Telangana, India*

1. Introduction

Phytomedicine is a medicinal system that encompasses plant materials in preventive, healing, and therapeutic procedures. It exists in the world right from the dawn of different human civilizations. These civilizations such as Chinese, African, and Indian had their systems of phytomedicine based on the availability of plants in or around the local habitations and experiences from random trials passed on from generation to generation. In the Indian context, it is a well-known fact that the system of medicine is there in place for the healthy living and well-being of people from antiquity. The system of medicine has a unique cultural history in the form of scriptures/sages and amalgamation and best influences from other civilizations like Greece (Unani), which together gave rise to the science of Ayurveda, Siddha, Yoga, and Naturopathy. This traditional system of medicine has changed and progressed over the centuries. Ayurveda that is a classical and traditional but currently well-recognized system of preventive and curative healthcare with origin in Vedas (5000 years back) largely uses plants or their products as raw materials for the manufacturing drugs. The Siddha and Unani systems also involve, to a certain extent, the standardization and use of the herbal drug that also covers safety issues, quality control, and adverse drug response monitoring (Samal, 2015). In India, several establishments are constantly working toward the goal of education, research, and development of new drugs such as Central Council for Research in Ayurvedic Sciences (CCRAS); Ministry of Ayurveda, Yoga, Naturopathy, Unani, Siddha, and Homoeopathy

Phytomedicine: A Treasure of Pharmacologically Active Products from Plants
https://doi.org/10.1016/B978-0-12-824109-7.00004-2

(AYUSH); Indian Institute of Integrative Medicine (IIIM); and Indian Pharmacopoeia Commission (IPC) according to their proficiency and facilities existing in their work domain. Indian Pharmacopoeia Commission is very actively engaged in setting the standards of modern medicine and compilation of herbs and herbal products (Prakash et al., 2017).

Despite the boom in production of synthetic drugs postindustrial revolution, it has been affirmed by the World Health Organization (WHO) that more than 70% of the human populace in under-developed nations still resort to alternative or traditional phytomedicinal system when it comes to primary medication and remedy of their various ailments (Karunamoorthi, Jegajeevanram, Vijayalakshmi, & Mengistie, 2013). Globally the herbal medicine market is divided into certain categories—functional foods, pharmaceuticals, beauty products, and dietary supplements. In 2017, the herbal pharmaceutical category commanded the global herbal medicine market with $50.974 billion. Going by the estimates, this category is anticipated to register a compound annual growth rate of 5.88% to touch $129.6893 billion by 2023 (Anonymous, 2019). Several plant or botanical sources are acknowledged to deliver various modules or chemical entities for therapeutic or pharmacological requirement, for example, certain bioactive molecules directly consumed as medication (drug), some pharmacologically active compounds with special structures (novel chemophore) acting as lead molecules, and herbal or botanical composite extracts as botanical drugs (Katiyar, Kanjilal, Gupta, & Katiyar, 2012). Apart from these, plant-derived nutraceuticals/functional foods, proteins, and gum resins have received significant attention and a consideration because of their recognized safety, prospective nutritional, and therapeutic benefits (Nasri, Baradaran, Shirzad, & Kopaei, 2014).

With the advancement of technology, numerous potent drugs to treat several diseases have been produced over the years, but most of these drugs are associated with the limitations of delivering them in biological systems. Their beneficial efficiency is significantly marred due to their incompatibilities and complexity of chemical structures. The application of conventional drugs is characterized by very poor bioavailability, limited effectiveness, undesirable toxicological effects, and lack of selectivity (Beutler, 2009). In the last few decades, the evolution of nanotechnology has opened some new vistas in advanced biology and medical research. Nanotechnology and nanoscience have appeared as new emergent areas of applied science and are receiving global attention due to their widespread applications. This has permitted novel research approaches to flourish in the area of drug delivery with lower toxicities and greater efficiencies. Size reduction methodologies and technologies have resulted in diverse nanostructures that show exceptional physicochemical and biological properties (Cho, Wang, Nie, Chen, & Shin, 2008; Sapsford et al., 2013; Wang et al., 2014). The greatest application of drug delivery technology is the design and utilization of nanoparticulate drug delivery systems (NPDDS). It has been possible because of the current developments in polymer and surface conjugation techniques and microfabrication procedures (Patra et al., 2018). Owing to the complexity that they possess ranging from a very meek metal-ceramic core structure to a complex lipid-polymer matrix, these submicron preparations are suited to act as therapeutic carriers for several conditions (Patra et al., 2018). The synthesis of stable nanoparticles (NPs) has received substantial attention by the introduction and recent progress of nano-revolution that has unfolded the essential role of plants in biosynthesis and green synthesis. There has been an escalation in applications of biologically synthesized NPs in various domains of biomedical science. Even though NPs can be synthesized by many conventional approaches, the biological route is more competent than the physical and chemical techniques (Baker et al., 2013; Baranwal, Mahato, Srivastava, Maurya, & Chandra, 2016; Makarov et al., 2014). Therefore, the synthesis of nanocomposites by utilizing plants associated with the benefits of nontoxicity, reproducibility, ease in scaling up, and well-defined structure has become a new trend in nanocomposite construction.

2. Approved or established plant-based drugs

Although the usage of bioactive herbal preparations has started from antiquity, contemporary drug discovery in the form of isolated and well-characterized chemical entities has been in place with the advent of the 19th century. The natural products from plants offer inestimable molecular skeletons and pharmacophores or novel scaffolds for conversion into proficient drug molecules for contending with several human diseases. The discovery and isolation of alkaloid morphine from poppy (*Papaver somniferum*) at the very beginning of the 19th century by Friedrich Serturner triggered the isolation and chemical characterization of molecules from plants with therapeutic claims in traditional systems with no looking back (Kong, Wang, & Du, 2018). The importance of phytomedicine throughout the world in contemporary times can be gauged from the statistic that almost a quarter of the drugs that are prescribed to the patients with various kinds of ailments have plants as their basic source. Even at the advent of 21st century, out of more than 250 drugs deliberated as basic, indispensable, and essential for basic healthcare system by the World Health Organization (WHO), 11% are completely plant based, and a considerable proportion of such synthetic drugs are attained from natural precursors with plants as their basic source (Rates, 2001). Recently, from a study conducted during the period 1981–2014, it was inferred that the US Food and Drug Administration (FDA) approved 1562 drugs, out of which 64 were intact natural products, 141 were herbal mixtures, and 320 were natural product derivatives (Calixto, 2019). An overview of certain important classes of drugs that are mostly approved by recognized agencies or are widely used by a large section of the global population with proven pharmacological efficacy and safety in the indigenous systems of medicine is represented in Section 2.1.

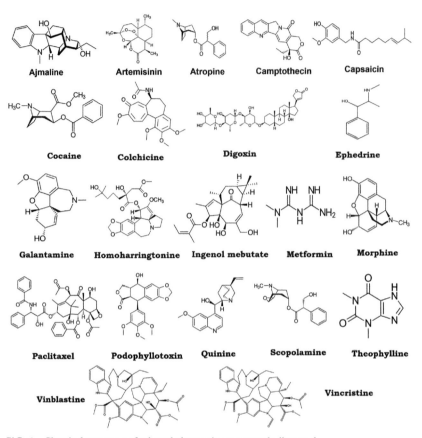

FIG. 1 Chemical structures of selected phytoactive compounds discussed.

2.1 Small molecules

A large number of plant-derived or plant-based small-molecule drugs are used to counter a multitude of disease indications and are an important domain of health and pharmaceutical sectors (Fig. 1). Some of the important ones are presented below:

(i) Ajmaline

(*Mol. weight* = 326.43 g/mol.; *Mol. formula*: $C_{20}H_{26}N_2O_2$)

Class: Alkaloid

Source: *Rauwolfia* spp.

History: Ajmaline was first isolated from the fresh undried roots of *Rauwolfia serpentina* by Salimuzzaman Siddiqui, an eminent organic chemist from Pakistan in 1931. Subsequently, he named this compound as Ajmaline after his mentor Hakim Ajmal Khan who was one of the most recognized practitioners of the Unani medicinal system in South Asia (Roy, 2018).

Therapeutic use: Ajmaline has a powerful antiarrhythmic effect for the management and cure of atrial and ventricular arrhythmias and exhibits

its action by altering the form and threshold of cardiac action potentials. It has also been shown to be useful for the treatment of Wolff-Parkinson-White (WPW) syndrome. Maximal cumulative dose is 1 mg/kg body weight (DrugBank).

Approvals: The Ajmaline is documented in the Japanese Pharmacopoeia (17th ed.). It is an ingredient of medicines in several countries.

Bioavailability: Oral absorption of this molecule is fast, but the bioavailability is very low. Pastoral administration, it usually takes about 20 min before it starts to show its efficacy. It takes 40 min to an hour for reaching the highest efficacy (Yang, Wang, & Du, 2018).

(ii) Artemisinin

(*Mol. weight* = 165.23 g/mol.; *Mol. formula*: $C_{15}H_{22}O_5$)

Class: Sesquiterpene lactone

Source: *Artemisia annua*

History: It was discovered in 1972 in an extract of *Artemisia annua* by Professor Tu Youyou who shared the Nobel Prize in Medicine/Physiology for its discovery in the year 2015. The discovery was made as a part of a program in the 523 Project launched by China to develop new antimalarial drugs in 1967 (Su & Miller, 2015).

Therapeutic use: Artemisinin along with its derivatives has been established to have potent efficacy against several forms of malarial causal organisms, including chloroquine-resistant *Plasmodium falciparum.* Their potency also encompasses phylogenetically disparate parasitic infectious diseases such as schistosomiasis. Artemisinin and one of its metabolites, dihydroartemisinin, also may be beneficial as anticancer agents since they have been revealed to impede the growth of different types of cancerous cells in laboratory research. The growth inhibition has been related to their antiangiogenic and apoptotic ability secondary to inherent endoperoxidase activity. Artemisinin has been used in trials studying the treatment of schizophrenia. It has been shown to exhibit antiinflammatory effects against autoimmune disorders and allergic asthma in a variety of disease models. A remarkable antiviral potency of artemisinin has also been reported against certain viruses such as herpes and hepatitis B and C viruses. Similarly, there have been studies that indicate the noteworthy role and potential benefit of artemisinin in diabetes via eliciting trans-differentiation of pancreatic α-cells to produce β-cells (Krishna, Bustamante, Haynes, & Staines, 2008; Wang et al., 2019).

Approval: Due to their safety and efficacy, artemisinin derivatives are now FDA approved for the treatment of malaria. The usual dosage schedule for oral administered artemisinin on the first day is 500–1000 mg (10–20 mg/kg) subsequently followed by 500 mg/day for 4 days (Maude, Woodrow, & White, 2010). The World Health Organization (WHO) endorses artemisinin-based combination therapies (ACTs) as first-line therapy for the treatment of uncomplicated malaria (Ikram & Simonsen, 2017).

Bioavailability: There are various administration courses for artemisinin, including intravenous, intramuscular, oral, and rectal. There is variability about the possibility of administration routes concerning artesunate and artemether. All three preparations can, however, be administrated through the oral route. Oral bioavailability is almost 30% because of the high first metabolism. While artesunate takes few minutes to reach peak levels, artemether peaking takes about 2–6 h. The plasma protein binding ranges between 43% and 81.5%. The conversion of both to dihydroartemisinin possesses considerable antimalarial activity (Medhi, Patyar, Rao, Byrav Ds, & Prakash, 2009).

(iii) **Camptothecin**

(*Mol. weight* = 348.34 g/mol.; *Mol. formula*: $C_{20}H_{16}N_2O_4$)

Class: Alkaloid

Source: *Camptotheca acuminata*

History: In the 1950s camptothecin was initially isolated from the stem of *Camptotheca acuminate* by researchers working in the US Department of Agricultural (USDA)'s Plant Introduction Division. The structure of camptothecin was well elucidated by Monroe Eliot Wall and Mansukh C. Wani in the year 1966.

Therapeutic use: The therapeutic importance of camptothecin mainly from the beginning has been its antitumor activity. Camptothecin and its semisynthetic analogs topotecan and irinotecan have been known to specifically inhibit the nuclear enzyme DNA topoisomerase-I and exercise anticancer activity by arresting the process of DNA synthesis (Li, Jiang, Li, & Ling, 2017).

Approval: Topotecan and irinotecan are widely used in cancer chemotherapy, while many more derivatives are still in different stages of clinic trials. Topotecan hydrochloride and irinotecan, two water-soluble camptothecin derivatives, were approved by the FDA in the year 1996 for the treatment of various cancers (ovarian, lung, cervical, colorectal, leukemia, etc.) (Sun et al., 2017).

Bioavailability: As per FDA label, after injection of topotecan hydrochloride at the dose level of 0.5–1.5 mg/m^2 administered as a 30-min infusion process, an area under the curve (AUC) upsurges proportionally to dose. The protein binding of topotecan is almost round about 35%. Following intravenous injection the terminal half-life of topotecan is 2–3 h.

(iv) **Capsaicin**

(*Mol. weight* = 305.42 g/mol.; *Mol. formula*: $C_{18}H_{27}NO_3$)

Class: Alkaloid

Source: *Capsicum* spp. (chili pepper)

History: In 1816 Christian Friedrich Bucholz first carried out the extraction from capsicum and obtained semipurified material and named it as capsaicin. In the year 1876 John C. Thresh, an English chemist, further carried out the purification process and obtained it in pure form and named it capsaicin. In the year 1919 E.K. Nelson elucidated its chemical structure.

Therapeutic use: This molecule is a neuropeptide liberating agent specific for primary sensory peripheral neurons. Capsaicin helps in alleviating peripheral nerve pain and aches after topical application. This drug molecule has been useful in controlling mucositis caused during chemotherapy and radiotherapy. It has been found to exhibit a protective influence on the cardiovascular system.

Approval: Capsaicin is recorded medication in the official pharmacopeia of several regions such as the United States, Britain, and Europe. In 2009 the FDA approved the capsaicin formulation patch dermal delivery system useful for neuropathic pain concomitant with postherpetic neuralgia (Yang & Du, 2018a).

Bioavailability: During oral absorption, capsaicin may be absorbed from the stomach and intestine with a range of absorption between 50% and 90% in animal models. The peak blood concentration is usually achieved within 1 h after administration. Following subcutaneous or intravenous administration, the concentrations in the brain, spinal cord, and liver were several folds greater than that in the blood. In human, topical capsaicin is quickly and well absorbed via the skin (DrugBank).

(v) **Cocaine**

(*Mol. weight* = 303.35 g/mol.; *Mol. formula*: $C_{17}H_{21}NO_4$)

Class: Alkaloid

Source: *Erythroxylum coca*

History: A German chemist, Albert Niemann, obtained highly pure material named cocaine in 1859 after it was extracted from the leaves of herbal ingredients in 1855. Thereafter, it was used as a medication as a local anesthetic. Because of its sexual stimulatory and antidepressant properties, it was also referred to as "a magical substance" by a renowned psychologist Sigmund Freud (Zhao, Wang, Yang, & Du, 2018).

Therapeutic use: Its use as pharmacological molecules is a local anesthetic and vasoconstrictor and is particularly used with that purpose in the eye, ear, nose, and throat. Considered as a drug of abuse, it possesses a potent central nervous system effect comparable with that of amphetamine. Cocaine acts by multiple mechanisms on neurons of the brain. The mode of action of its reinforcing effects is known to involve inhibition of the uptake of dopamine (DrugBank).

Approved system of medicine: Apart from its recorded use in the official pharmacopeia of various countries, the FDA approved cocaine hydrochloride nasal solution under certain regulations for indicated and limited usage as an anesthetic in 2017.

Bioavailability: This drug is absorbed easily from all places of application, which includes mucous membranes and gastrointestinal mucosa. Oral bioavailability data are very inadequate because the pharmacokinetic profiling of oral cocaine has never been complete. After oral dosing, cocaine peak in the systemic circulation shows around after 45 min with average systemic bioavailability as 33% (Fattinger, Benowitz, Jones, & Verotta, 2000).

(vi) Colchicine

(*Mol. weight* = 399.44 g/mol.; *Mol. formula*: $C_{22}H_{25}NO_6$)

Class: Alkaloid

Source: *Colchicum autumnale* (autumn crocus)

History: The autumn crocus was described for the treatment of rheumatism and inflammation in the Papyrus Ebers, an Egyptian medical papyrus of herbal knowledge in 1500 BC. Colchicine was isolated from *Colchicum autumnale* during 1820 by French chemists, P.S. Pelletier and J.B. Caventou. In 1833 P.L. Geiger further purified colchicine as an active constituent. After structural elucidation in 1959, it quickly assumed popularity as a remedy for gout (Zhang, Zhou, & Du, 2018).

Therapeutic use: Colchicine is commonly used in the management of gout. The antiinflammatory and antigout potentials were achieved by impeding the movement and cytoskeleton distortion of neutrophils, which is achieved by hindering tubulin polymerization. Apart from its well-known use in gout, colchicine is also utilized for the remedy for managing complications associated with a hereditary autoinflammatory condition such as familial Mediterranean fever. It has also been used in the medication of other disorders like coronary artery disease, Behcet's disease (BD), pericarditis, and other inflammatory conditions. Colchicine is known to interrupt microtubule dynamics by binding to tubulin hindering its polymerization ceasing mitosis. However, severe cytotoxicity to normal tissues at higher concentrations restricts its usage in cancer therapy (Leung, Yao Hui, & Kraus, 2015).

Approval: Colchicine is recorded medication in the official pharmacopeia of several regions such as the United States, Britain, China, and Europe (Zhang et al., 2018). In 2009 colchicine was approved by the FDA as a sole-standing medication for the remedy of an acute attack of gout and familial Mediterranean fever (Leung et al., 2015).

Bioavailability: Colchicine undergoes rapid absorption after oral dosage in the gastrointestinal tract. According to the FDA label, the bioavailability of colchicine is about 45%. However, another reference specifies that the bioavailability shows variability ranging from 24% to 88% (DrugBank).

(vii) Digoxin

(*Mol. weight* = 780.94 g/mol.; *Mol. formula*: $C_{41}H_{64}O_{14}$)

Class: Glycoside

Source: *Digitalis* spp. (foxglove plant)

History: Digoxin was first isolated by Sydney Smith in 1930 from *Digitalis lanata* (Hollman, 1996).

Therapeutic use: Digoxin is involved usually in the treatment of chronic congestive heart failure (CHF). It is considered as a lifeline for people with heart ailments. It is also be used to treat some arrhythmia and management of atrial fibrillation. Its recommended dose for intravenous is 8–12 µg/kg, and oral solution is 10–15 µg/kg.

Approvals: Being classified as a potent cardiac glycoside, it was initially approved for use by the FDA way back in 1954. Digoxin falls under the World Health Organization's List of Essential Medicines, the most effective, important, and safest medications required in healthcare systems throughout the world (WHO, 2019). It is also a recorded medication in the official pharmacopeia of several regions such as China, Britain, and Europe (Chen, Sun, & Du, 2018).

Bioavailability: The oral bioavailability of digoxin remains generally high (70%–80%), which can be attributed to the particle size and dissolution. Substantial metabolic degradation of digoxin within the gastrointestinal tract takes place in patients by hydrolysis in the acidic milieu of the stomach or due to digestive breakdown by gut bacteria. Therefore the bioavailability of digoxin may swiftly take an upsurge in such patients when they consume any broad-spectrum antibiotic (DrugBank).

(viii) **Ephedrine**

(*Mol. weight* = 165.23 g/mol.; *Mol. formula*: $C_{10}H_{15}NO$)

Class: Alkaloid

Source: *Ephedra* spp.

History: Ephedrine was isolated from *Ephedra sinica* by Changi Changyi, a Japanese organic chemist, in the year 1885 and subsequently elucidated its chemical structure in 1926. Thereafter, it became very famous owing to its pharmacological benefits (Gazaliev, Zhurinov, Fazylov, & Balitskii, 1989).

Therapeutic use: Ephedrine is an alpha- and beta-adrenergic agonist that can also augment the release and action of norepinephrine. Pure ephedrine can be used for the treatment of chest tightness, cold, and bronchial asthma, acting chiefly as a bronchodilator and decongestant. It can also be used in the treatment of several other disorders such as heart failure, rhinitis, and urinary problems. It exhibits its stimulatory effects on the central nervous system (CNS). It can also be used in dealing with narcolepsy and depression. The recommended ephedrine sulfate subcutaneous or intramuscular dose is 0.5 mg/kg of body weight every 4–6 h (in children) and higher for adults (Limberger, Jacques, Schmitt, & Arbo, 2013) (DrugBank).

Approved system of medicine: It is a recorded medication in the official pharmacopeia of several regions such as India, China, Britain, and Europe. Ephedrine has been approved for clinical practice and treatment by the FDA (Wang, Yang, & Du, 2018). This drug molecule is also on the World Health Organization's List of Essential Medicines, the most effective, important, and safest medications required in a healthcare system throughout the world (WHO, 2019).

Bioavailability: Postintravenous injection, ephedrine is entirely biologically available. However, after oral administration, the bioavailability of ephedrine has been described to be above 90% (Electronic Medicines Compendium).

(ix) Galantamine
(*Mol. weight* = 287.36 g/mol.; *Mol. formula*: $C_{17}H_{21}NO_3$)
Class: Alkaloid
Source: Amaryllidaceae spp. (*Galanthus*, *Narcissus*, etc.)
History: Galantamine was first isolated from the bulbs of *Galanthus nivalis* (the snowdrop) by a team of Bulgarian chemists headed by Dimitar Paskov in 1956. This isolated active ingredient was identified and studied particularly about its acetylcholinesterase (AChE)-inhibiting properties (Tewari et al., 2018).
Therapeutic use: This drug molecule is used for the medication of cognitive deterioration in Alzheimer's disease (AD) and several other memory impairments related to the central nervous system (CNS). The optimal dose is 16–24 mg/day for patients suffering from minor or moderate AD. It is an acetyl cholinesterase inhibitor that reverses the muscular effects of gallamine triethiodide and tubocurarine. Galantamine works in dual mode increasing the quantity of acetylcholine present in the brain by stimulation of the nicotinic receptors in the brain to release and enhancing the availability of more acetylcholine and inhibition of the acetyl cholinesterase enzyme (Lilienfeld, 2002) (DrugBank).
Approval: Galantamine hydrobromide is documented as a medication in the official pharmacopeia of quite a few regions such as the United States, China, England, and Europe (Lei, Wang, & Du, 2018). It is designated and approved by the FDA as a safe and effective medication for the treatment of various forms of dementia and Alzheimer's disease.
Bioavailability: Absorption of galantamine is quick and complete demonstrating linear pharmacokinetics over the range of 8–32 mg/day. Oral bioavailability is almost 90%. The terminal elimination half-life of this drug molecule is approximately about 7 h (Farlow, 2003).

(x) Homoharringtonine (HHT)/omacetaxine mepesuccinate
(*Mol. weight* = 545.62 g/mol.; *Mol. formula*: $C_{29}H_{39}NO_9$)
Class: Alkaloid ester
Source: *Cephalotaxus* spp.
History: This molecule was first isolated in 1963 from *Cephalotaxus* species. Its structure was properly elucidated, and anticancer effects were investigated 1970s onward using various leukemic cell lines from time to time by different researchers (Powell, Weisleder, & Smith, 1972).
Therapeutic use: Homoharringtonine is used for the treatment of those patients who are intolerant and/or resistant to tyrosine kinase inhibitors utilized for accelerated or chronic myeloid leukemia (CML). This drug is also used for the treatment of acute myeloid leukemia (AML) and myelodysplastic syndrome (MDS). The mechanism of action involves its unique property of binding to the 80S ribosomes in eukaryotic cells and impeding the protein synthesis by interference with the chain elongation (Lü & Wang, 2014; Wang, Li, & Du, 2018).

Approval: The FDA approved it as a medication to patients suffering from accelerated or chronic phase CML in October 2012. The induction dose is $1.25 \, mg/m^2$ injected subcutaneous two times daily for 14 consecutive days (DrugBank). It also is reported in the official pharmacopeia of the People's Republic of China, 2015 (Wang, Li, et al., 2018).

Bioavailability: Though there is an absence of any formal endeavors to evaluate the bioavailability of omacetaxine in patients, a cross-study comparative analysis of systemic exposure after subcutaneous and intravenous administration infers that bioavailability of this drug molecule is approximately 70%–90% (Nemunaitis et al., 2013).

(xi) **Ingenol mebutate (ingenol-3-angelate)**

(*Mol. weight* = 430.54 g/mol.; *Mol. formula*: $C_{25}H_{34}O_6$)

Class: Diterpene ester

Source: *Euphorbia peplus*

History: This molecule was initially derived from the latex of *Euphorbia peplus* L. in 1985 and was found to be having potent anticancer activity in subsequent studies. Some other *Euphorbia* species, such as *Euphorbia paralias*, *Euphorbia milii*, *Euphorbia palustris*, *Euphorbia marginata*, and *Euphorbia helioscopia*, also represent the sources of this drug molecule. Among the different species of *Euphorbia*, *Euphorbia myrsinites* provide the highest ingenol concentration of approximately 547 mg/kg of dry weight (Seca & Pinto, 2018).

Therapeutic use: This molecule is a selective and small-molecule activator of protein kinase C (PKC) with prospective antineoplastic activity. It activates several protein PKC isoforms, thereby eliciting apoptosis signaling pathway in some malignant cells such as myeloid leukemia, melanoma, and carcinoma cells. Ingenol mebutate exhibits a high degree of antiproliferative effects against melanoma and myeloid leukemia cell lines even at nanomolar concentrations (Hampson et al., 2005) (DrugBank).

Approved system of medicine: A gel preparation of this drug was approved in 2012 by the FDA and European Medicines Agency (EMA) for the topical management and treatment of actinic keratosis, a precancerous zone of thick, scaly, and crunchy skin (Kircik, Sung, Stein-Gold, & Goldenberg, 2017).

Bioavailability: The systemic absorption is negligible (< 0.1 ng/mL) as it is used in the form of topical treatment. Ingenol mebutate (0.05% gel) shows a decent safety profile for skin areas up to $100 \, cm^2$ with good tolerability and manageable local skin responses (Anderson et al., 2014).

(xii) **Metformin**

(*Mol. weight* = 129.16 g/mol.; *Mol. formula*: $C_4H_{11}N_5$)

Class: Biguanide

Source: *Galega officinalis*

History: Metformin was initially developed from natural molecules isolated from *Galega officinalis* (Goat's rue). Synthetic biguanides like buformin and phenformin were developed in the 1920s in Germany and

Britain. However, their use was restricted due to certain side effects. However, a French physician Jean Sterne during the 1940s synthesized a new biguanide called metformin (dimethyl biguanide). Although it was first studied for the cure of influenza, he recognized its hypoglycemic abilities, and it began to be used as a medication for diabetes after 1950. **Therapeutic use:** Metformin is an antidiabetic drug that lowers blood sugar levels in type 2 diabetes. Being an insulin sensitizer, it leads to a diminution in insulin resistance. In addition to reducing blood sugar/lipid levels, it plays a role in the regulation of cell growth and possesses antiinflammatory and antiaging properties. The main pharmacological mode of function of metformin involves the stoppage of hepatic gluconeogenesis, activation of 5′ AMP-activated protein kinase (AMPK), and regulation of the mitochondrial function (Yang & Du, 2018b).

Approval: Presently prescribed to at least 120 million people worldwide, metformin is the forerunner drug of choice for managing type II diabetes. Being available in regular and extended release forms, it was first approved as a medication in Canada in 1972 followed by the United States in 1995 (DrugBank). Metformin is on the World Health Organization's List of Essential Medicines, the most effective, important, and safest medications required in healthcare systems throughout the world (WHO, 2019).

Bioavailability: In the fasting conditions the absolute bioavailability of administered metformin is about 50%–60%. From single-dose studies, it has been inferred that with an increase in metformin dose, there is an absence of dose proportionality, which is due to decreased absorption rather than any changes in process of elimination (DrugBank).

(xiii) **Morphine**

(*Mol. weight* = 285.34 g/mol.; *Mol. formula*: $C_{17}H_{19}NO_3$)

Class: Alkaloid

Source: *Papaver somniferum* (opium poppy)

History: Morphine (morphium), the principal alkaloid of opium, was obtained from poppy seeds by a German pharmacist, Friedrich Sertürner in 1805. The discovery of this molecule brought human analgesic drug research at the forefront (Krishnamurti & Rao, 2016).

Therapeutic use: Morphine, an agonist of opioid receptors, is a potent analgesic and sedative as it acts on the central nervous system (CNS) and peripheral nervous system. It is used to relieve severe pain triggered by myocardial infarction and during labor. It is also used to treat patients with acute pulmonary edema due to its anxiolytic and vasodilation properties since it can promote the release of endogenous histamine. Morphine provides relief to patients with pulmonary edema symptoms (Ellingsrud & Agewall, 2016; Kong et al., 2018). The recommended direct intravenous injection of morphine sulfate to manage pain is 0.1–0.2 mg/kg body weight at the interval of 4 h, whereas the intramuscular injection is 10 mg/70 kg/adult at the interval of 4 h.

Approval: In 1941 morphine received FDA approval as a therapeutic drug molecule (DrugBank). Morphine is recorded medication in the official pharmacopeia of several regions such as the United States, Britain, China, Japan, and Europe (Kong et al., 2018). Additionally, it is on the World Health Organization's List of Essential Medicines, the most effective, important, and safest medications required in a healthcare system throughout the world (WHO, 2019).

Bioavailability: The absolute bioavailability of morphine was found to be 40%–50% of the dose after it is administrated orally (Kong et al., 2018). It is usually absorbed in the alkaline milieu of the upper part of the intestine and rectum. The bioavailability of morphine is 80%–100%. For dose compensation purposes, oral doses are usually sixfold higher than parenteral doses to realize the equivalent effect owing to a substantial amount of the first-pass metabolism. After 24–48 h, morphine attains generally a stable steady-state concentration (DrugBank).

(xiv) Paclitaxel

(*Mol. weight* = 853.91 g/mol.; *Mol. formula*: $C_{47}H_{51}NO_{14}$)

Class: Diterpene alkaloid

Source: *Taxus brevifolia* (Pacific yew tree)

History: This molecule was isolated and identified by Monroe E. Wall and Mansukh C. Wani from the bark of *Taxus brevifolia* tree in the year 1971 (Wani, Taylor, Wall, Coggon, & Mcphail, 1971).

Therapeutic use: Paclitaxel as an antineoplastic agent impedes the disassembly of microtubules by binding to tubulin and leading to inhibition of mitosis. It also sets cells on the path of apoptosis by binding to and obstructing the function of Bcl-2, the apoptosis inhibitor protein. Paclitaxel therapy is known to provide promising outcomes in patients suffering from various cancers such as cervical cancer, lung cancer, ovarian cancer, advanced squamous cell carcinoma of head and neck, melanoma, urothelial cancer, esophageal cancer, non-Hodgkin's lymphoma or multiple myeloma, breast cancer, Kaposi sarcoma, and pancreatic cancer (Khanna, Rosenberg, & Vail, 2015; Wang & Du, 2018).

Approval: Paclitaxel is on the World Health Organization's List of Essential Medicines, the most effective, important, and safest medications required in a healthcare system throughout the world (WHO, 2019). In 1992 and the following years, paclitaxel was approved as therapeutic medication by the FDA as a Taxol for the treatment of different cancers (e.g., ovarian, breast, lung, and Kaposi's sarcoma) (Khanna et al., 2015; Sun et al., 2017). The dosage is variable for different types of cancer as 100–175 mg/m^2 intravenous over different periods.

Bioavailability: The oral bioavailability is less than 10% of the drug and thus not so much feasible due to efficient transport by intestinal drug efflux pump P-glycoprotein (P-gp) (Malingré et al., 2001). Its most of the metabolism occurs through the liver making entry into the intestine with bile and finally its 90% removal from the body via fecal matter (Wang & Du, 2018).

(xv) Podophyllotoxin

(*Mol. weight* = 414.40 g/mol.; *Mol. formula*: $C_{22}H_{22}O_8$)

Class: Nonalkaloid lignan

Source: *Podophyllum peltatum* and *Podophyllum hexandrum*

History: Podophyllotoxin was first isolated from the *Podophyllum peltatum* L. in 1880. In 1942 it was found to be useful for the treatment of venereal warts. Later, podophyllotoxin was reported to inhibit the growth and progression of tumors. The chemical structure of this therapeutic molecule was elucidated in 1951. In the 1960s two of its important and well-known derivatives, etoposide and teniposide, were synthesized.

Therapeutic use: Podophyllotoxin has a significant inhibitory ability on the proliferation and division of cancerous cells by binding and disrupting the cell cytoskeleton via polymerization of tubulin-causing suppression of the formation of the mitotic-spindle microtubules leading to the cell cycle arrest and apoptosis. The antiproliferative efficacy of podophyllotoxin analogs (etoposide and teniposide) is related to a distinct mechanism causing the inhibition of DNA topoisomerase II activity and the establishment of stable nucleic acid drug-enzyme complex. Subsequently, this results in the induction of DNA double or single-strand break or DNA damage and eventually culminating in cell death. These derivatives were reported to possess antiinflammatory and immunosuppressive properties (Pang, Zhang, & Du, 2018). Podophyllotoxin and its derivatives were also reported to exhibit antiviral activity (Sudo, Konno, Shigeta, & Yokota, 1998).

Approval: Etoposide was approved by the FDA for several cancers such as lung cancer, choriocarcinoma, testicular and familial ovarian cancer, lymphoma, and acute myeloid leukemia in 1983. Furthermore, teniposide was also approved as a medication in the treatment of certain brain tumors, Hodgkin's lymphoma, acute lymphocytic leukemia, and other types of cancer in 1992 (Sun et al., 2017). The dosage varies for different types of cancer between 35 and 100 mg/m^2 intravenously over different periods.

Bioavailability: Clinical studies have shown that etoposide is a highly schedule-dependent drug. The oral bioavailability of etoposide is found to be 50%, although its absorption graph does not show a linear elevation to the dosage. Approximately 30%–70% of a dose of etoposide gets eliminated by the process of excretion, while it appears to be 5%–20% for teniposide (Clark & Slevin, 1987).

(xvi) Quinine

(*Mol. weight* = 324.42 g/mol.; *Mol. formula*: $C_{20}H_{24}N_2O_2$)

Class: Alkaloid

Source: *Cinchona* spp.

History: Quinine, an alkaloid, is used to treat malaria and babesiosis. It remains an important antimalarial drug since 17th century. In 1817 Joseph B. Caventou (French pharmacist) in collaboration with Pierre J. Pelletier isolated and purified quinine from the bark of cinchona tree (*Cinchona pubescens*), which replaced bark as the regular treatment (Achan et al., 2011).

Therapeutic use: Quinine is an efficient antimalarial drug that displays specific and high toxicity potential against *Plasmodium*. Although many derivatives such as chloroquine have also been used in the medication for malaria, quinine is still extensively used because it is the sole compound to which *Plasmodium* has no resistance (Achan et al., 2011; Izawa, Amino, Kohmura, Ueda, & Kuroda, 2010). It is also an ingredient of common cold preparations due to its antipyretic and analgesic properties. Quinine also finds a role in the treatment of certain disorders involving muscular tissue, such as nocturnal leg cramps and myotonia congenita owing to its uninterrupted effects on muscle membrane and sodium channels (DrugBank). In the ongoing pandemic COVID-19, chloroquine is being seriously viewed as a prospect for combating the virus. Although there are reports of its inhibition of SARS-CoV in vitro, whether it will be beneficial against the current SARS-CoV-2 needs deep study and thorough clinical evaluations (Wong, Yang, & He, 2020).

Approval: The sulfate product of quinine is recorded medication in the official pharmacopeia of several regions such as India, China, Britain, and Europe. The recommended dose is 10-mg sulfate/kg oral administration thrice a day for 3 or 7 days (Wen, Yuan, Kong, & Chen, 2018). Quinine is on the World Health Organization's List of Essential Medicines, the most effective, important, and safest medications required in a healthcare system throughout the world (WHO, 2019). Quinine is a prescription medication approved by the FDA in August 2005. Presently, Quinine is exclusively used in the United States for the management and treatment of uncomplicated malaria caused by *Plasmodium falciparum* parasite (Abed, Baniya, & Bachuwa, 2016).

Bioavailability: Quinine in its salt form is injected via the parenteral or oral route. Pastoral administration more than 70% of the drug gets absorbed. The maximum systemic concentration peak is achieved between 1 and 3 h, and almost 80% of the administered quantity is eliminated via hepatic route (Adehin et al., 2019).

(xvii) **Salicylic acid**

(*Mol. weight* = 138.12 g/mol.; *Mol. formula*: $C_7H_6O_3$)

Class: Phenolic acid

Source: *Salix* spp.

History: Father of modern medicine, Hippocrates, (460–370 BC), recommended the chewing of bitter powder from the willow tree bark to relieve fever and pain. In 1828 Johann Büchner isolated a yellow substance from the tannins of willow trees and named it salicin, the Latin word for willow. The process was further refined, and a pure crystalline form of salicin was isolated in 1829 by Henri Le Roux, a French pharmacist. These crystals were then used for the treatment of rheumatism. Subsequently an Italian chemist, Raffaele Piria, converted silicin into an acidic aromatic compound, which he named as salicylic acid. In 1853 the French chemist, Charles Gerhardt, modified salicylic acid into aspirin (acetylsalicylic acid) with the introduction of an acetyl group in place of a hydroxyl group and soon became a drug of choice because of its therapeutic properties (Desborough & Keeling, 2017).

Therapeutic use: Aspirin also known as salicylate is a nonsteroidal antiinflammatory drug (NSAID) and is known to be a nonselective cyclooxygenase (COX) enzyme inhibitor. Aspirin is widely used to relieve mild to moderate pain and reduce fevers. Aspirin also stops the aggregation of platelets and thus used in the prevention of blood clots, stroke, and myocardial infarction (Chen, Qiang, & Du, 2018) (DrugBank).

Approved system of medicine: Both salicylic acid and aspirin have recorded medication in the official pharmacopeia of several regions such as the United States, Britain, China, Japan, India, and Europe (Chen, Qiang, et al., 2018). It is widely consumed throughout the world and is on the World Health Organization's List of Essential Medicines, the most effective, important, and safest medications required in a healthcare system throughout the world (WHO, 2019).

Bioavailability: After oral administration of aspirin, 60%–80% will be absorbed in the stomach and small intestine. Rectal absorption of aspirin indicated that it is slower as compared with orally administered aspirin (Dalvi, Gupta, Pohujani, Vaidya, & Satoskar, 1985).

(xviii) **Scopolamine/hyoscine**
  (*Mol. weight* = 303.35 g/mol.; *Mol. formula*: $C_{17}H_{21}NO_4$)
  Class: Alkaloid
  Source: Solanaceae (nightshade family)
  History: Scopolamine, also known as hyoscine, was isolated first by E. Schmidt from *Scopolia japonica* in 1892 (Xu, Liu, & Du, 2018).
  Therapeutic use: Like atropine, scopolamine and its quaternary derivatives exhibit their power as antimuscarinics and have more effects apart from influencing the central nervous system. It is commonly used to treat certain stomach and intestinal problems, muscle spasms, and Parkinson-like conditions (DrugBank).
  Approval: Scopolamine butylbromide was patented in 1950 and approved for the medication in 1951. It is a recorded medication in the official pharmacopeia of several regions such as the United States, Britain, China, Japan, and Europe (Xu et al., 2018). It is on the World Health Organization's List of Essential Medicines, the most effective,

important, and safest medications required in a healthcare system throughout the world (WHO, 2019).

Bioavailability: Scopolamine is quickly and completely absorbed after the application of a transdermal patch. The oral bioavailability of is 27%. After the ocular injection of an ophthalmic solution (0.25%), scopolamine hydrobromide is very rapidly absorbed (Lemaire-Hurtel & Alvarez, 2014).

(xix) Theophylline

(Mol. weight= 180.16 g/mol.; *Mol. formula*: $C_7H_8N_4O_2$)

Class: Xanthine alkaloid

Source: *Camellia sinensis* (tea) and *Theobroma cacao* (cocoa)

History: Theophylline was isolated from the leaves of *Camellia sinensis* by the German biochemist, Albrecht Kossel, and determined its structure in 1888. Later in 1900 theophylline was chemically synthesized by another German chemist.

Therapeutic use: Theophylline is used in therapy for respiratory diseases such as bronchial asthma, chronic obstructive disease of the airways, and chronic obstructive pulmonary disease (COPD). It impedes the phosphodiesterase (PDE) enzyme, preventing the intracellular cyclic AMP (cAMP) breakdown. It leads to an increase in the upsurge of intracellular cAMP and reduction of the smooth muscle tone culminating in the dilation and relaxation of the air passages and pulmonary blood vessels.

Approval: Theophylline is recorded medication in the official pharmacopeia of several regions such as the United States, Britain, Japan, China, and Europe. Theophylline medication is considered safe and potent for the approved complications in pediatric patients.

Bioavailability: The absorption of theophylline is rapid and almost complete after oral administration with peak concentrations taking place within 1–2 h of a dose. The bioavailability of theophylline is around 80% in newborn, while it is higher, that is, 96%–100%, in adult patients. However, during intravenous administration of this drug, bioavailability is 100% (Gong, Da Du, & Du, 2018; Griffin, Posner, & Barker, 2013).

(xx) Vincristine

(Mol. weight= 824.96 g/mol.; *Mol. formula*: $C_{46}H_{56}N_4O_{10}$)

Vinblastine

(Mol. weight= 810.97 g/mol.; *Mol. formula:* $C_{46}H_{58}N_4O_9$)

Class: Alkaloids

Source: *Catharanthus roseus*

History: The antidiabetic properties of the extracts from leaves of the *Catharanthus roseus* in the folklore system led to the discovery and isolation of two complex vinca alkaloids, vinblastine and vincristine. These molecules were isolated in the 1950s by Robert Nobel and Charles T. Beer (Canadian scientists) from the Madagascar periwinkle plant, *Catharanthus roseus.* They were introduced in cancer chemotherapy in the late 1960s (Duffin, 2000).

Therapeutic use: Vinblastine and vincristine are well-known clinical cytotoxic drugs and have a strong inhibitory effect on cancer cell proliferation. Vinblastine works best on Hodgkin lymphoma and advanced testicular or breast cancers, whereas vincristine is used in the medication of acute leukemia and other lymphomas. These alkaloid molecules and their derivatives check cancerous cells from undergoing mitotic divisions. The mechanism of action of these cell cycle–dependent agents is the inhibition of tubulin polymerization into microtubules (DrugBank).

Approval: The FDA approved vinblastine and vincristine in 1961 and 1963, respectively as a pharmaceutical strategy against different cancer types (i.e., leukemia, osteosarcoma, lung cancer, breast cancer, and Hodgkin's lymphoma) (Sun et al., 2017). Vinblastine and vincristine are on the World Health Organization's List of Essential Medicines, the most effective, important, and safest medications required in a healthcare system throughout the world (WHO, 2019). The recommended dose of vincristine sulfate is 1.4–1.5 mg/m^2 up to an extreme highest weekly dose of 2 mg.

Bioavailability: After 15–30 min postinjection of vincristine sulfate, over 90% of the drug is dispersed into tissue from the blood, where it remains strong but not irreversibly bound. The total average excretion of vincristine and vinblastine over 4 days in urine and feces is 36.7% and 18.2%, respectively, of the injected dose (DrugBank).

2.2 Potential small drug candidate molecules

In addition to these approved molecules, there are certain other compounds of plant origin, which are currently used by people all over the world. However, due to some limitations, they remain unapproved. They are categorized as follows:

(i) **Molecules with poor bioavailability**

Many herbal drug molecules and even extracts, for example, andrographolide, puerarin, breviscapine, ginkgolide B, pilocarpine, baicalin, oleanolic acid, curcumin, and shikonin, despite their remarkable in vitro results exhibit very low or insignificant in vivo potency because of their poor lipid solubility or inappropriate molecular size leading to their poor systemic absorption and bioavailability (Kesarwani & Gupta, 2013).

(ii) **Molecules with adverse effects and toxicity**

The broad perception that botanical medications or drugs are very safe lacking any adverse effects on the human body is not only incorrect but also deceptive at times. They have been shown to generate a wide range of detrimental or hostile response reactions some of which can cause severe injuries and even fatal at times. Such problems are associated with certain botanical pharmacologically potent molecules and remain unapproved such as quinidine, reserpine, yohimbine, caffeine, camphor, cyclandelate, eserine, picrotoxin, strychnine, aconitine, brucine, gossypol, and areolae (Du, 2018; Ekor, 2014).

(iii) Molecules with chemical instability

Certain plant-based products are often vulnerable to deterioration reactions particularly during storage, leading to loss of active constituents, production of metabolites with no or minimal activity, and, in certain extreme cases, generation of toxic metabolite molecules. It is also difficult to carry out mechanistic research at the molecular level on these molecules, for example, allicin and eserine (Thakur, Ghodasra, Patel, & Dabhi, 2011).

(iv) Molecules with undefined mechanism of action

For any herbal drug to be approved for general use on a large scale, the mechanism of action should be well understood and comprehensible. Certain plant-derived molecules for example, rhomotoxin, chelidonine, berbamine, palmatine, and curcumin, lack such large-scale mechanistic clinical research to specify the indications and dosages with which they can be brought into mass use.

(v) Molecules with lack of clinical studies

Some drug molecules have wide ranging pharmacological effects or therapeutic potency, but the related research studies are still based on animal models or in vitro experiments, for example, daphnetin, luteolin, honokiol, magnolol, and quercetin (Du, 2018). These molecules are under various stages of development and improvement, and it becomes imperative to understand and improve their efficacy, limit their toxicity, and enhance bioavailability. One of the methods by which it can be achieved is to bring slight changes in their molecular structure. Intensive research-related pharmacodynamics, pharmacokinetics, and clinical studies are need of the hour for bringing these molecules into the therapeutic domain.

2.3 Botanical drugs/extracts

(i) Sativex

Sativex (Nabiximols) is *Cannabis sativa*-based product containing tetranabinex and nabidiolex (cannabidiol, CBD) as its principal components. It is indicated for the relief multiple sclerosis (MS) symptoms and the treatment of several neuropathic-related cancer pain. Sativex received regulatory approvals for the treatment for the alleviation of neuropathic pain, spasticity, and pain associated with multiple sclerosis in Canada in the year 2005. It was further approved in 2010 as a botanical drug in the United Kingdom. Its components have been known to have neuroprotective, anxiolytic, analgesic, anticonvulsant, muscle relaxant, antioxidant, and antipsychotic properties (Feliú et al., 2015).

(ii) Sinecatechins

Sinecatechins are natural substances found in the water extract of certain green tea (*Camellia sinensis*) leaves. Sinecatechins (Veregen) 15% topical formulation is the first botanical drug product approved by the FDA in 2006 for the treatment of external genital and perianal warts caused by human papillomavirus (HPV). Sinecatechins are mostly catechins

(85%–95%) of which epigallocatechin gallate is the most abundant component ($>55\%$) along with other derivatives such as epicatechin and epigallocatechin. Apart from these well-known catechin components, it also comprises gallic acid, caffeine, and theobromine, which collectively constitute about 2.5% of the drug. The residual amount comprises of undefined and unknown constituents from green tea leaves.

(iii) **Crofelemer (Mytesi)**

Crofelemer is a novel proanthocyanidin purified from the bark latex of *Croton lechleri*. It was approved in 2012 by the FDA for the treatment of diarrhea associated with anti-HIV drugs such as nucleoside analog reverse transcriptase inhibitors and protease inhibitors (DrugBank).

2.4 Health supplements (nutraceuticals)

Nutraceuticals refer to a food supplement that plays a fundamental role in maintaining a healthy body and provides basic nutrients that are needed for various metabolic processes to regulate functions of the body and thus help in the prevention and treatment of diseases. There is a plethora of medicinal herbs and food supplements that nourish, support, and stimulate our body system. These nutraceuticals are the kind of products that are more than foods but can be regarded as lesser than pharmaceuticals in terms of health benefits. Herbal nutraceuticals have received significant attention owing to the presumed safety and prospective therapeutic benefits. Herbal dietary supplements comprise herbal extracts, that is, complex mixtures of phytochemicals of which the pharmacologically active principle(s) often constitute only a small part. The minor constituents of these herbal extracts may augment the pharmacological potential of the main active principle compounds synergistically. The main herbal, pharmacologically active, constituents include polyphenols, lipids, carbohydrates, terpenes, steroids/sterols, and alkaloids. Most of the herbal medicines have been used in many diverse traditional systems. In the Indian context, not only there are well-documented health usage of medicinal plants in traditional or folk system, but also there is a well-organized system or department of Ayurveda, Yoga, and Naturopathy, Unani, Siddha, and Homoeopathy (AYUSH) that is actively involved in the generation pharmacopeia of standards for Indian systems of medicine of which these medicinal plants form an important part (Asif & Mohd, 2019; Pandey, Meena, Rai, & Pandey-Rai, 2011). The medicinal herbs not only are capable of normalizing the body functions but also have preventive and nutritive functions, thereby enhancing the immune system.

Triphala, veered tonic in Ayurveda, is an ancient powerful herbal remedy with antioxidant, antiinflammatory, and antibacterial properties. It is a polyherbal medicine, mainly a combination of three important herbs (*Terminalia bellerica*, *Terminalia chebula*, and *Emblica officinalis*). It is supposed that combining synergistic herbs results in extra therapeutic effectiveness. The well-known therapeutic uses of Triphala are immunomodulating, antibacterial, antimutagenic, etc. that are already well established (Peterson, Denniston, & Chopra, 2017).

Turmeric is a golden-colored spice that is used in many disciplines of medicine. Turmeric has more than 4000 years of history, and it has its significance in Ayurveda, Siddha, Unani, and traditional Chinese medicine. It is well documented for the treatment of inflammatory conditions, skin diseases, wounds, digestive ailments, and liver conditions. The FDA published a monograph on turmeric and rated its active component, curcumin, as generally regarded as safe (GRAS) (Prasad & Aggarwal, 2011).

According to Ayurveda, ginger and garlic are the foundation of all healing food recipes. Besides their significant roles in nutrition, they have been reported to possess myriad therapeutic properties such as antioxidant, cardioprotective, antiinflammatory, analgesic, antimicrobial, antidiabetic, rheumatologic, and cholesterol regulation. Besides, they regulate blood pressure, enhance immunity, and proved to be beneficial against cardiovascular disease and cancers (Gupta, 2010).

Fenugreek (*Trigonella foenum-graecum*) is one of the most promising medicinal herbs and has nutritional value found in the continents of Asia, Europe, Africa, and Australia. Fenugreek has a broad range of therapeutic effects, including pain relief, antidiabetes, antiatherosclerosis, antiinflammation, antispasmodic, anticancer, astringent, and heart tonic (Wani & Kumar, 2018).

Carotenoids (β-carotene, zeaxanthin, lycopene, and lutein) that are present mostly in fruits and vegetables have bioactive properties, such as antioxidant, antiinflammatory, and autophagy modulatory activities. Furthermore the consumption of carotenoid-rich diets has been linked with the treatment and prevention of various diseases, chiefly, including those neurodegenerative ones (Stahl & Sies, 2005).

Legumes like chickpeas and soybeans, grains, and palm oil contain noncarotenoids, reduce cholesterol levels, and are also found to be anticarcinogenic.

Omega-3 essential fatty acid (α-linolenic acid) is found in plant sources such as nuts and seeds. They play a critical role in the metabolism and cellular function of the body. There are reports of a substantial lessening in a radiation-induced inflammatory response attributed to the prospective decrease in prostaglandin, leukotrienes, TNF-α, etc. when these fatty acid supplements are consumed.

Green and black tea are a rich source of plant polyphenols, the most abundant being epigallocatechin-3-gallate and the aflavins. These polyphenols have shown to hinder peroxidation, restrict DNA impairment, and diminish ROS. They also subdue inflammatory promoters (i.e., cyclooxygenase-2) and stimulate cell cycle halt and apoptosis (Souyoul, Saussy, & Lupo, 2018). Furthermore, polyphenolic extracts obtained from Annurca apples have been designated as safe and with cholesterol-lowering ability.

A highlight of the herbal/plant-based nutraceuticals effective on ailments linked essentially to oxidative stress like cancer, cardiovascular allergy, neurodegenerative ailments, diabetes, immune disorders, inflammatory disorders, and Parkinson's diseases is tabulated (Table 1).

TABLE 1 Medicinal plants commonly found and used with documented therapeutic use as herbal nutraceuticals.

S. No.	Plant species/common name	Therapeutic potential	Major bioactive compounds	Form in which consumed
1	*Allium sativum* (garlic/lahsun)	Antibacterial, antifungal, cardiovascular, antioxidant, anticancer, antiinflammatory, etc.	Allicin, diallyl disulfide, *S*-allylcysteine, diallyl trisulfide, etc.	Fresh/dried cloves, capsules, tablets, tinctures, extracts
2	*Aloe vera* (ghritkumari)	Antioxidant, antiaging, antibacterial, wound healing, constipation, lower glucose level, ulcers, laxative, etc.	Anthraquinones (aloin and emodin)	Juice, gel, creams, extracts, sap
3	*Andrographis paniculata* (green chireta)	Antiinflammation, skin diseases, anticancer, antiobesity, antidiabetic, respiratory infections, etc.	Andrographolide, panicoline, paniculide	Powder
4	*Azadirachta indica* (neem tree)	Antihelminthic, antifungal, antidiabetic, sedative, skin diseases	Isomeldenin, nimbin, nimbinene, 6-desacetylnimbinene, nimbandiol, immobile, nimocinol	Vegetable, cream, toothbrush, neem oil (Ayurvedic mixture)
5	*Bacopa monnieri* (Indian pennywort)	Memory enhancement, antidepressant, anticonvulsant, antioxidant, gastrointestinal, antimicrobial, antiinflammatory, cardiovascular, muscle relaxant	Bacosides, bacopasides, brahmine, nicotine, herpestine	Leaves, powder, extract, Bacopa Plus (Ayurvedic mixture)
6	*Borago officinalis* (starflower)	Respiratory and urinary infections, cardiovascular, etc.	Gamma-linolenic acid	Vegetable, soups, beverage

7	Centella asiatica (Indian pennywort)	Wound healing, sedative, anxiolytic, antidepressant, ulcers, etc.	Asiaticoside, brahmoside, Asiatic acid, centelloside, etc.	Herb, powder, capsules, tea, tablets, tinctures
8	Curcuma longa (turmeric)	Antifungal, anticancer, antiinflammatory, indigestion, etc.	Curcuminoids	Dried roots, culinary powder, CoCurcumin (Ayurvedic mixture)
9	Cymbopogon citratus (lemon grass)	Analgesic, antipyretic, carminative, stimulant, antifungal, etc.	Citral, myrcene, citronella, citronellol	Herbal supplement, tea, culinary
10	Foeniculum vulgare (fennel)	Carminative, digestive, lactogogue, diuretic, respiratory and gastrointestinal disorders	Trans-anethole, estragole, fenchone, limonene	Seeds, capsules, tinctures, culinary
11	Glycyrrhiza glabra (licorice)	Expectorant, antimicrobial, antiulcer, anticancer, antiinflammatory, antidiabetic, hepatoprotective	Glycyrrhizin, glycyrrhetic acid, isoliquiritin, isoflavones	Powder, extract, tincture, table
12	Hypericum perforatum (St John's-wort)	Epilepsy, depression	Hyperforin and hypericin	Dried herb, tea, oil, tablets
13	Mentha arvensis (mint)	Analgesic, hypertensive, antidiarrhea, stomach problems, asthma, arthritis	Menthol, menthone, limonene, piperitone, β-caryophyllene, α- and β-pinene	Extract, powder, culinary, drinks
14	Moringa oleifera (drumstick tree)	Antioxidant, anticancer, antiinflammatory, antidiabetic, antimicrobial	Polyphenols, alkaloids, glucosinolates, and isothiocyonates	Fruit pods, leaves, oil, culinary
15	Ocimum sanctum (basil/tulsi)	Asthma, antimalaria, dysentery, skin diseases, arthritis, antipyretic, antifertility, anticancer, antidiabetic, antimicrobial, hepatoprotective, cardioprotective, antiemetic, analgesic	Eugenol, β-elemene, oleanolic acid, ursolic acid, rosmarinic acid, linalool	Powder, extract, beverages, basil oil (Ayurvedic)

Continued

S. No.	Plant species/common name	Therapeutic potential	Major bioactive compounds	Form in which consumed
16	*Phyllanthus emblica* (amla)	Antimicrobial, rejuvenator, antioxidant, antiinflammatory, analgesic, antipyretic, hepatoprotective, anticancer	Ellagitannins (emblicanin A, emblicanin B, punigluconin), punicafolin, phyllanemblinin, chebulagic acid	Fruit, oil, culinary, Triphala (Ayurvedic mixture)
17	*Trigonella foenum-graecum* (fenugreek)	Carminative, gastric stimulant, antidiabetic, galactogogue, hepatoprotective, antiinflammatory, antibacterial, antiulcer	Coumarin, fenugreekine, nicotinic acid, phytic acid, scopoletin, trigonelline	Herb, spice, vegetable, powder, tincture
18	*Terminalia chebula*	Antioxidant, antidiabetic, renal protective, hepatoprotective, immunomodulatory	Chebulin, ellagic acid, 2,4-chebulyl-D-glucopyranose, arjunglucoside I, arjungenin, chebulinic acid, punicalagin, terflavin A, terchebin	Fruits, Triphala (Ayurvedic mixture)
19	*Withania somnifera* (ashwagandha/Indian ginseng)	Anticancer, antioxidant, tonic, anxiolytic, antiinflammatory, antiaging, immunomodulatory, antiarthritic	Lactones, withaferin A, withanolides, withasomniferin A, withasomnidienone, withasomnierose A–C, withanone	Powder, ashwagandha Ayurvedic (rasayana), tinctures,
20	*Zingiber officinale*	Antiarthritic, nausea, sore throats, constipation, antihypertensive, dementia, anticancer, cardioprotective, asthma, antidiabetic	Gingerols, paradols, shogaols, zingiberene	Rhizome, spice, beverages, powder

2.5 Resins/gums

(i) Myrrh

It is a resin that comes out from the bark of trees that are members of the *Commiphora* species.

The chemical constituents apart from the gum are essential oils, which include terpenes, steroids, and lignans. Myrrh has long been used in traditional Chinese medicine and Ayurveda, which attribute tonic and rejuvenating properties to the resin. Myrrh has been widely used as an analgesic for toothaches and a liniment for bruises, aches, and sprains. Additionally, myrrh is claimed to act as a therapeutic remedy against ulcers, indigestion, colds, skin ailments, arthritis cough, asthma, lung congestion, arthritis pain, and cancer. Owing to its pharmacological benefits, myrrh resin is also suggested to be used for circulatory ailments, inflammation linked to rheumatoid arthritis, anesthetic, and uterine tumors (Cao et al., 2019; Nomicos, 2007).

(ii) Guggul

Guggul is an aromatic or fragrant oleo gum resin that exudes out from the bark of *Commiphora mukul* (Mukul myrrh tree) found in the northern parts of India, Central Asia, and North Africa. Guggul has been used thousands of years in the Indian traditional medicine system, Ayurveda, for a wide variety of applications such as arthritis, atherosclerosis, gout, blood, diuretic, thyroid stimulant, liver tonic, expectorant, appetite stimulant, acne, obesity, and urinary tract infections. This oleo gum resin is also known for its use in indolent ulcers and a gargle for the ulcerated throat. Additionally, guggul is extensively used in the Indian subcontinent to fight high cholesterol levels and to boost metabolism. Guggul is a complex mixture of many chemically active constituents like terpenoids, diterpenoids, steroids, and flavonoids apart from many inorganic compounds. Guggulsterones, myrrhanol, eugenol, etc. isolated from Guggul have been identified as the potent bioactive constituent responsible for many of its therapeutic benefits (Sarup, Bala, & Kamboj, 2015).

(iii) Frankincense/olibanum

Frankincense is a hard, gelatinous aromatic resin exuded from the trunk notches of trees of the genus *Boswellia*, particularly *Boswellia sacra*, *Boswellia carteri*, *Boswellia thurifera*, *Boswellia frereana*, and *Boswellia bhaw-dajiana* of the family Burseraceae mainly from Ethiopia, Somalia, and India. In traditional Chinese medicine and Ayurveda, frankincense is used as natural treatment for many chronic diseases. Medicinal preparations containing frankincense have been found to have definite curative properties on several diseases and are clinically proven. It is used for traditional remedies for promoting blood circulation, swelling, relieve menstrual pain, gum, mouth, and throat complaints and as a rejuvenating medicine. Chemical profiling of essential oil of frankincense showed the presence of diverse class of compounds like 11-keto-β-boswellic acid, 3-oxo-tirucallic acid, cembrene, elemolic acid, limonene, lupeol, α-amyrin, α-thujene, β-boswellic acid, β-pinene, and serratol with a wide range of pharmacological activities including antiinflammatory and anticancer (Cao et al., 2019).

(iv) Sweetgum (ancient chewing gum)

It is a resinous aromatic fragrant transparent yellowish sap (storax) secreted from the outer bark of *Liquidambar styraciflua*. *Liquidambar styraciflua* is a deciduous plant, native to warm temperate areas such as North America and Central America. Sweetgum has been traditionally used as antiseptic, diuretic, sedative, vulnerary, antidiarrhea, antifungal, antiinflammatory, antimicrobial, etc. The resin is chewed to treat sore throats, coughs, asthma, cystitis, dysentery, etc. Externally, it is applied to sores, wounds, piles, ringworm, and scabies. Some of the main constituents of sweetgum include cinnamyl cinnamate, styrene, cinnamyl alcohol, 2-phenylpropyl alcohol, cinnamic acid 3-phenylpropyl cinnamate, gallic acid, chlorogenic, caffeic acid, kaempferol, ferulic acid, and vanillin (Lingbeck, O'Bryan, Martin, Adams, & Crandall, 2015).

2.6 Proteins and peptides

Plants offer unlimited prospects as photosynthetic factories to produce pharmaceutically significant, commercially valued, and approved biomedicines for human use with minimal cost. In 2012, the FDA has approved the drug recombinant enzyme Elelyso (*Taliglucerase alfa*) formed by carrot cells for the treatment of Gaucher's disease. The successful prospects of biomedicine for Ebola patients by molecular farming have intensely enhanced due to the achievements made by an experimental drug called ZMapp produced in tobacco leaves. Pegunigalsidase alfa (PRX-102) is an innovative enzyme expressed in a BY2 tobacco cell culture beneficial for the management and treatment of Fabry disease. Three more biomedicines produced in tobacco plants, include vaccine Pfs25 VLP for malarial treatment, vaccine recombinant protective antigen for anthrax, and HIV P2G12 antibody that are either in process or have completed phase 1 clinical trials (Yao, Weng, Dickey, & Wang, 2015).

There are good number of examples for plant peptides linked with a beneficial and positive effects on human health. Peptides perform an abundant role in plant defense scheme as antimicrobial molecules. Several bioactive peptides from the food sources have been merged in fortified foods or dietary supplements and are subsequently commercialized to minimize the risk of chronic ailments like hypertension, hypercholesterolemia, and obesity. Since the upsurge in drug-resistant infections in humans, microbicides have been engaged to counter various pathogenic bacteria, viruses, fungi, and/or parasitic activities. Plant bioactive peptides have been isolated from roots, seeds, flowers, fruits, and leaves of many varieties of species and possess activities toward phytopathogens, as well as against bacteria pathogenic to humans. Some examples of plant antimicrobial peptides (AMPs) are thionein (*Triticum aestivum*), cyclotides (*Oldenlandia affinis*), knottin peptides (*Phytolacca americana*), snakins (*Solanum tuberosum*), vulgarinin (*Phaseolus vulgaris*), heveins (*Hevea brasiliensis*), arietin (*Cicer arietinum*), etc. These bioactive peptides are emerging as a promising antibiotic compound and might open new promising prospects to discover leads for the management of various diseases in the years to come. Apart from their microbicide activities, these AMPs display essential immunomodulatory and wound healing promoting activities such as induction of cell migration and proliferation and angiogenesis (López-Meza, Ochoa-Zarzosa, Barboza-Corona, & Bideshi, 2015). The current trend in this direction is that proteins from plant sources are enzymatically hydrolyzed to form biologically active peptides that are examined for the treatment of diseases such as cancer and diabetes. It has opened up a new therapeutic window for the development of novel therapeutic agents (Guzmán-Rodríguez, Ochoa-Zarzosa, López-Gómez, & López-Meza, 2015; Patil, Goswami, Kalia, & Kate, 2019) (Fig. 2).

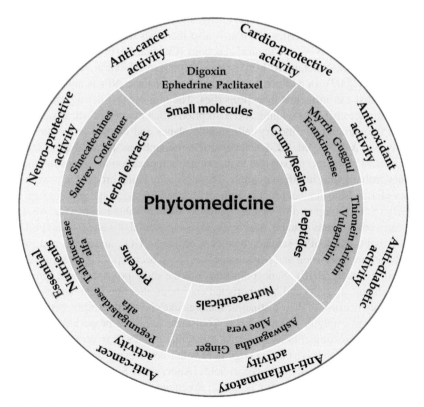

FIG. 2 Phytomedicine: various domains and applications in therapeutics.

3. Phytonanocomposites

Nanotechnology, an application-based technology, is a creation and careful maneuver of materials on an atomic level and therefore use of these materials at the nanoscale for numerous purposes (Jain, Huang, El-Sayed, & El-Sayed, 2008). This technology is advancing with every passing day, marking its footprints in all aspects of human life and generating a new excitement in biomedical research and biotechnology.

The advancement of research in nanotechnology has opened new windows of opportunities for certain fields of practical applicability and utility such as nanomedicine, nanobiotechnology, and theranostics. Nanoparticles (NPs) are a varied class of materials that comprise particulate substances, which have one or more dimensions of the order of 100 nm or less (Auffan et al., 2009). The NPs have attracted great interest due to their unusual and fascinating optoelectronic and physicochemical properties based on size, their distribution and morphology, and valuable applications over their bulk counterparts, thereby unlocking many new pathways in nanotechnology. Due to their very small size, nanocomposites possess the extraordinary high surface area-to-a volume ratio that changes their functional properties. As this ratio increases further, the behavior of atoms present on the outer boundary of these NPs also changes. This is a major contributory factor behind the unique properties of NPs. It also increases the possibility of a substrate to bind with NPS in chemical reactions (Cuenya, 2010; Prathna, Mathew, Chandrasekaran, Raichur, & Mukherjee, 2010; Rao, Kulkarni, John Thomas, & Edwards, 2002). In addition to size, the shape of these NPs is another major property that can influence their properties. The number and density of surface atoms, surface charges, bending, and adsorption efficiency can drastically change due to the variations in the crystallographic planes and thereby shape the NPs (Narayanan & El-Sayed, 2004). Nanoparticle research has attracted great scientific interest in the recent past due to its extensive application in the development of new technologies in different areas. These NPs are being brought into work at every phase of the clinical process with an outstanding speed and success rate. A variety of NPs composed of varying materials such as noble metals (Au, Ag, Pt, Pd, etc.), other metals (Cu, Fe, Al, Pb, Zn, etc.) and their alloys and oxides (Fe_2O_3, Fe_3O_4, FePt, and CoPt), semiconductors (CdS, ZnS, TiO_2, and PbS), and their combinations (core-shell and nanocomposites) have been reported in recent years (Thanh & Green, 2010). Nanocomposites are the most promising and remarkable biomedical agents and have attracted enormous attention from the scientific community because of their distinctive features like chemical stability, better conductivity, and catalytic abilities. Apart from these excellent properties, the ease of synthesis, biocompatibility, and customizability of NPs, it is very evident that nanoparticle research is a cutting-edge field with roles in every domain of science and technology especially the medical fields (Anderson & Shive, 2012; Jain et al., 2008; Li, Wang, Fan, Feng, & Cui, 2012; Sreeprasad & Pradeep, 2013). Owing to their design uniqueness and property, a combination that is lacking in conventional composites, they are described to be the materials of the 21st century.

In the recent past, nanocomposite materials have been explored for diverse biomedical applications and in the prospective interdisciplinary areas of nanobiotechnology (Abbasi et al., 2016; Jeong, Choi, Ellis, & Lee, 2019; Pereira et al., 2015; Rai, Ingle, Birla, Yadav, & Santos, 2016; Rao & Gan, 2014). Gold and silver nanocomposites are routinely used for biomedical practices and in the flourishing multifaceted fields of nanobiotechnology. Drug molecules such as paclitaxel, methotrexate, and doxorubicin are being specifically delivered through gold NPs. Additionally, Au nanocomposites have been employed for the detection of the tumor, angiogenesis, genetic abnormalities and diagnosis, photoimaging, and photothermal therapy (Baranwal et al., 2016; Jeong et al., 2019). Silver nanocomposites have been used for antimicrobial and anticancer therapeutics, diagnostics, and other clinical/pharmaceutical applications (Baranwal et al., 2016; Lee & Jun, 2019). There are surfeit applications of iron oxide nanocomposites such as cancer therapy, tissue restoration, cell tagging, targeting and immunological assays, magnetic resonance imaging, detoxification of biological fluids, and magnetically responsive drug delivery process (Bárcena, Sra, & Gao, 2009; Dadfar et al., 2019). Because of their biocompatible, less toxic, self-cleansing, skin-compatible behaviors, Zn and Ti nanocomposites have attracted tremendous interest in various fields of medicine, including anticancer, antibacterial, antioxidant, antidiabetic, antiinflammatory, cosmetic, ultraviolet (UV)-blocking agents, and drug delivery and bioimaging applications (Baranwal et al., 2016; Jiang, Pi, & Cai, 2018; Ziental et al., 2020). Copper nanocomposites have great desirable properties in the areas of the biomedical research domain such as detection of biomarkers associated with diseases, for example, diabetes, cancers, neurological disorders, and cardiac syndromes, apart from an effective antimicrobial activity against broad-spectrum pathogens and also anticancer properties (Baranwal et al., 2016; Rubilar et al., 2013; Verma & Kumar, 2019). Due to their intrinsic attributes, such as vivid catalytic, physical, mechanical, and optical properties, besides heterogeneity in shapes and sizes, palladium nanocomposites have displayed a wide array of applications such as photoacoustic agents, gene/drug carriers, prodrug activators, anticancer agents, and antimicrobial agents (Baranwal et al., 2016; Phan, Huynh, Manivasagan, Mondal, & Oh, 2020). Furthermore, metal nanocomposites have been very handy in the spatial examination of several important biomolecules like peptides, nucleic acids, lipids, fatty acids, glycosphingolipids, and drug molecules, to envisage these molecules with higher sensitivity and resolution (Chu, Unnikrishnan, Anand, Mao, & Huang, 2018).

3.1   Green synthesis of nanocomposites

Technological advancement over recent decades has facilitated the manipulation of nanostructured material synthesis with shaped-up properties to suit the requirements of varied applications. During the last two decades, due to the unique physicochemical properties and several applications of NPs, the research community has devoted considerable time toward developing a variety of new and promising approaches for the synthesis of NPs from a diverse range of materials (Dhand et al., 2015). There are plentiful physical, chemical, and conglomerate methods available to synthesize different types of NPs with desired characteristics. Unfortunately, many of these methods are technically laborious and economically expensive and involve the use of toxic, perilous chemicals as reducing and stabilizing agents and potentially hazardous to the environment that greatly limits their clinical applications (Bundschuh et al., 2018; Hussain, Singh, Singh, Singh, & Singh, 2016). To overcome such drawbacks, currently, researchers have diverted their attention toward **green synthesis approaches** for research and development on materials science and technology. Green nanotechnology, the amalgamation nanotechnology advancements with green chemistry principles, is a sensible approach for the development of a society with ecological sustainability (Fig. 3). Green nanotechnology is an interdisciplinary field with the ultimate objective to synthesize nanostructures in a responsible, effective, and sustainable manner and holds great potential in the area of medical science with a special emphasis on environmental aspects, health, and safety (EHS) (Duan, Wang, & Li, 2015). The efficient green synthesis methods including natural reducing, capping, and stabilizing materials without involving expensive, hazardous, and high energy consumption agents have diverted the attention of researchers toward biological approaches. Greener nanotechnology practices frequently involve the utilization of natural resources comparatively safer solvents and biodegradable and biocompatible materials via energy-reliant procedures in the nanoparticle formulation (Duan et al., 2015).

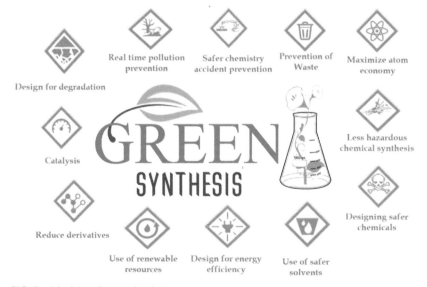

FIG. 3 Principles of green chemistry.

Biological synthesis has been established as a creditable alternative to conventional synthesis procedures for constructing nanocomposites in recent times. It involves the utilization of proenvironment green chemistry-based methods that engage unicellular and multicellular biological organisms like bacteria, fungi, viruses, and plants (Pooja, Joginder, & Suresh, 2014). The involvement of these biological entities especially plants offers a clean and nontoxic (safe) prospects of production of nanocomposites of a wide array of shapes, sizes and composition, and physicochemical properties and is safe for human therapeutic use (Mohanpuria, Rana, & Yadav, 2008). Many biological entities have the inherent ability to act as templates for the synthesis, assemblage, and organization of nanometer-scale materials, which help to engineer well-defined micro- and macroscale structures (Dhand et al., 2015). The additional advantage of the biosynthesis route is that it can reduce the requisite steps, including the biological activation by attachment of functional groups to the nanoparticle surface, which otherwise would be indispensable during physiochemical synthesis (Baker et al., 2013). As compared with the earlier described biological identities and their prospects of becoming proficient biological factories, nanocomposite synthesis by the plant (products) is a relatively straightforward and beneficial methodology. Since this synthesis tends to be more swift than other biological entities, it is an economical relatively easy approach for scaling up the production of NPs in large amounts (Baker et al., 2013; Pooja et al., 2014).

3.2 Phytofabrication of nanocomposites

Herbal medicine has been a fundamental component in the therapeutic approaches of humans from antiquity. Therefore nature bioactive molecules from plants are widely employed for the development of new drug entities. Because of their multifarious therapeutic functionality, research on medicinal plants remains a very capitative domain of investigation among researchers. To overcome their plethora of limitations like low bioavailability, water insolubility, in vivo stability, intestinal absorption, and undefined site of action, the introduction of novel drug delivery system (NDDS) for herbal medicines is of tremendous significance. The merger of nanotechnology as an NDDS in the conventional system of medicine augments the prospects of herbal drugs for being used to cure various chronic ailments. Hence, using nanotechnology to synthesize nanocomposites is such a propitious approach to form phytonanocomposites whose synergistic properties can be effective against several diseases. With the advancement in technology, it will be possible to determine the toxicity profiles of NPs due to their physicochemical properties.

Green mode of synthesized metal nanocomposites has been considered to have better biocompatibility than the chemically or physically synthesized nanostructures. Through various biological entities that have been exploited for the synthesis of nanocomposites, the use of plants for the simplistic and robust synthesis of nanocomposites is tremendous. The phytonanotechnology came up with new prospects for the synthesis of nanocomposites. Plant and plant products are better synthesizers due to the rich resources in comparison with other forms of biological resources. Due to the ease of handling and flexible reaction conditions, plant extract–mediated synthesis of nanocomposites is considered as more appropriate for extensive production as compared with another biological synthesis of metal nanocomposites. On account of varied functional and structural properties of tree gum polysaccharides/hydrocolloids, they are nature's versatile green materials to agglomerate diverse nanostructures. During the recent past, several new gum exudates from gum kondagogu, gum karaya, gum ghatti, gum arabic, and gum tragacanth have been used as natural biosorbent designed for sorption of noxious metals and as renewable and low-cost reducing and stabilizing agents for the synthesizing of nanocomposites (Padil, Wacławek, Černík, & Varma, 2018; Sashidhar, Raju, & Karuna, 2014). It has been proposed that wide array of molecules, ranging from proteins to varied low–molecular weight compounds (flavonoids, alkaloids, polyphenols, terpenoids, heterocyclic compounds, polysaccharides, or hydrocolloids), that are present in plants/plant extracts plays a vital role in the bioreduction of metal ions and act as capping and stabilizing agents for synthesized nanocomposite materials (Baker et al., 2013; Baranwal et al., 2016; Makarov et al., 2014). Hence the cornucopia of plants that have been effectively used for the biological synthesis of nanocomposites prompts the deeper insights of biological nanofactories to meet up the demands for nanoscale products in different fields. To date, numerous plants have been exploited to synthesize copious nanocomposites such as Ag, Au, Cu, Ti, Pt, Pd, Zn, and Fe. Various parts of plants (seeds, roots, stem, leaves, flowers, fruits, and peel) are being used to synthesize nanocomposites of different forms and sizes using biochemical approaches (Table 2; Fig. 4). The disparity in the composition and concentration of active biomolecules in plants and their consequent interactive modes with different metal ions is assumed to be the main cause of the diversity nanoparticle sizes and shapes generated (Marslin et al., 2018). Also, many parameters such as the method used for synthesis, metal salt concentration, pH, temperature, pressure, time, environment, and proximity have a significant impact on the production rate, characterization, and application of nanocomposites (Baranwal et al., 2016).

3.3 Phytonanocomposites: Modern nanomedicine

Plant-derived active constituents have been studied extensively for health benefits and for the prevention and treatment of some human diseases. Epigallocatechin gallate (EGCG), curcumin, quercetin, resveratrol, apigenin, anthocyanidins, etc. are among the major phytochemical active constituents used for multiple health therapies. These biologically active molecules have remarkable attributes such as exceptional chemical complexity and biological activities that make them conducive to the development of new drugs (Wang et al., 2014). However, there are some concerns associated with the biocompatibility, solubility, inefficacy,

TABLE 2 Plant-mediated synthesis of nanocomposites.

S. No	Plant	Plant part used for synthesis	Nature of NPs	Size (nm)	Shape	Applications/activity
1	*Bacopa monnieri*	Leaves	Gold (Au)	3–45	Spherical	–
2	*Cassia auriculata*	Leaves		15–25	Spherical	Antidiabetic
3	*Chloroxylon swietenia*	Leaves		18–37	Spherical	Larvicidal
4	*Cinnamomum zeylanicum*	Leaves		5–60	Spherical, triangle	Antimicrobial
5	*Citrus maxima*	Fruits		25.7 ± 10	Rod, spherical	Catalytic
6	*Coriandrum sativum*	Leaves		5–60	Spherical, triangular, decahedral	Drug delivery, tissue/tumor imaging, photothermal therapy
7	*Couroupita guianensis*	Flowers		29.2–43.8	Cubic	Cytotoxic
8	*Cymbopogon citratus*	Leaves		20–50	Spherical, rod	Antimalarial
9	*Dysosma pleiantha*	Rhizomes		50–130	Spherical	Antimetastatic
10	*Eclipta alba*	Leaves		–	–	Cytotoxic
11	*Ficus racemosa*	Latex		20–120		–
12	*Mangifera indica*	Leaves		~20	Spherical	–
13	*Mimosa pudica*	Seeds		–		Drug delivery
14	*Mirabilis jalapa*	Flowers		100	Spherical	Antimicrobial
15	*Nyctanthes arbor-tristis*	Flower		19.8 ± 5.0	Spherical	–

Continued

TABLE 2 Plant-mediated synthesis of nanocomposites — cont'd

S. No	Plant	Plant part used for synthesis	Nature of NPs	Size (nm)	Shape	Applications/activity
16	Olea europaea	Fruit		50–100	Triangular, hexagonal, spherical	Cytotoxic, antibacterial
17	Punica granutum	Fruit peel		–	–	Drug delivery
18	Salix alba	Leaves		50–80	Nonspherical	Antifungal
19	Sterculia acuminate	Fruit extract		10–40	Spherical	Cytotoxic, catalytic
20	Stevia rebaudiana	Leaves		5–20	Cubic	–
21	Syzygium aromaticum			5–100	Irregular	Detection and destruction of cancer cells
22	Terminalia arjuna	Leaves		20–50	Cubic, spherical	Induce mitotic cell division and pollen germination
23	Terminalia catappa	Leaves		10–35	Spherical	Antibacterial
24	Trigonella foenum-graecum	Seeds		15–25	Spherical	Antianemic, catalytic
25	Vitis vinifera	Peel		~20–40	Spherical	Cytotoxic
26	Ziziphus mauritiana	Leaves		20–40	Spherical	Antibacterial
27	Abelmoschus esculentus	Pulp	Silver (Ag)	3–11	Spherical	Cytotoxicity
28	Abutilon indicum	Leaves		5–25	Spherical	Antibacterial, cytotoxicity

#	Species	Part	Size	Shape	Activity
29	*Acacia nilotica*	Pod	20–30	Distorted spherical	Antibacterial
30	*Achillea biebersteinii*	Flowers	10–40	Spherical, pentagonal	Cytotoxicity
31	*Allium sativum*	Rhizome	4–6	Spherical	Antibacterial, antioxidant
32	*Alpinia calcarata*	Root	27	Spherical	Antibacterial, antioxidant
33	*Alternanthera sessilis*	Whole plant	40	Spherical	Antibacterial, antioxidant
34	*Andrographis paniculata*	Leaves	13–27	Spherical	Hepatocurative
35	*Anogeissus latifolia*	Gum powder	5.5–5.9	Spherical	Antibacterial
36	*Arachis hypogaea*	Shell extract	10–50	Spherical	Antifungal
37	*Artemisia nilagirica*	Leaves	70–90	Spherical	Antimicrobial
38	*Boswellia serrata*	Gum	7–10	Spherical	Antibacterial
39	*Butea monosperma*	Bark	~35	Spherical	–
40	*Calotropis gigantea*	Latex	5–30	Spherical	Antibacterial, cytotoxicity
41	*Carica papaya*	Leaves/latex	2–20	Spherical	Cytotoxicity, antibacterial
42	*Cassia auriculata*	Leaf	41	Spherical	Antimicrobial
43	*Chelidonium majus*	Whole plant	90	Spherical	Antioxidant, antimicrobial
44	*Cinnamomum zeylanicum*	Leaves	45	Spherical	Antibacterial
45	*Citrus sinensis*	Peel	35 ± 2	Spherical	Antibacterial
46	*Coleus aromaticus*	Leaves	44	Spherical	Antibacterial
47	*Coriandrum sativum*	Leaves	13	–	Antibacterial

Continued

TABLE 2 Plant-mediated synthesis of nanocomposites—cont'd

S. No	Plant	Plant part used for synthesis	Nature of NPs	Size (nm)	Shape	Applications/activity
48	Curcuma longa			10–15	Spherical	Cytotoxicity
49	Dalbergia spinosa	Leaves		18 ± 4	Spherical	Antiinflammatory, antibacterial
50	Dioscorea bulbifera	Tuber		8–20	Rod, triangular	Antimicrobial
51	Gelsemium sempervirens	Whole plant		112	Spherical	Cytotoxicity
52	Glycyrrhiza glabra	Root		7–45	Spherical	Antimicrobial
53	Moringa oleifera	Fruit/leaves		10–57	Irregular, crystalline, spherical	Cytotoxicity, antimicrobial
54	Morus (mulberry)	Leaves		15–20	Spherical	Antimicrobial
55	Nigella sativa	Leaves		15	Spherical	Cytotoxicity
56	Nyctanthes arbor-tristis	Flower/seeds		50–80	Spherical	Antibacterial, cytotoxicity
57	Ocimum sanctum	Leaves/roots		3–20	Spherical	Antimicrobial, catalytic
58	Phyllanthus niruri	Leaves		30–60	Spherical	Antibacterial
59	Picrasma quassioides	Bark		17.5–66.5	Spherical	Catalytic
60	Pinus densiflora	Cones		30–80	Oval, triangular	Antimicrobial
61	Piper longum	Fruits		15–200	Spherical	–
62	Pistacia atlantica	Seeds		27	Spherical	Antibacterial
63	Sambucus nigra	Fruits		20–80	Spherical	Antiinflammatory

64	*Solanum tuberosum*	Starch		20–50	Spherical	Antimicrobial, biomedicine, sensors
65	*Vitex negundo*	Leaves		5–47	Spherical	Antibacterial
66	*Withania somnifera*	Leaves		5–40	Irregular, spherical	Antimicrobial
67	*Zingiber officinale*	Rhizome		10–24	Spherical	Antibacterial, catalytic
68	*Acalypha indica*	Leaves	Ag and Au	20–30	Spherical	Antibacterial, cytotoxicity
69	*Aloe vera*	Leaves		50–350	Rectangular, triangular, spherical	Antimicrobial, cytotoxicity
70	*Artemisia indica*	Leaves		15–35	Spherical, triangular	Antibacterial, tyrosinase inhibitory
71	*Azadirachta indica*	Gum		5–100	Hexagonal spherical triangular,	Biolarvicidal, remediation of toxic metals
72	*Bauhinia purpurea*	Leaves		20–50	Cubic	Cytotoxicity, antimicrobial, antioxidant, catalytic
73	*Camellia sinensis*	Leaves		30–40	Spherical, triangular, irregular	Catalytic, sensors
74	*Cochlospermum gossypium*	Gum		2.5–12.73	Crystalline	Antibacterial, cytotoxic
75	*Coleus forskohlii*	Root		5–8	Rod shaped	Antioxidant, antibacterial, cytotoxic
76	*Eucalyptus macrocarpa*	Leaves		20–100	Spherical, triangular, hexagonal	Antibacterial

Continued

TABLE 2 Plant-mediated synthesis of nanocomposites—cont'd

S. No	Plant	Plant part used for synthesis	Nature of NPs	Size (nm)	Shape	Applications/activity
77	*Ocimum sanctum*	Leaves		10–30	Crystalline, hexagonal, triangular, spherical	Biolabeling, biosensor
78	*Panax ginseng*	Root		10–40	Spherical	Antibacterial
79	*Anacardium occidentale*	Leaves	Pt	100	Irregular rod	Catalytic, thermal
80	*Antigonon leptopus*	Whole plant		5–190	Spherical	–
81	*Azadirachta indica*	Leaf		5–50	Spherical	Cytotoxic
82	*Bacopa monnieri*	Leaves		5–20	Spherical	Antioxidant
83	*Camellia sinensis*	Leaves		30–60	Flower	Cytotoxic
84	*Cochlospermum gossypium*	Gum		2.4±0.7	Crystalline	
85	*Fumariae herba*	Whole		30	Hexagonal, pentagonal	–
86	*Prunus x yedoensis*	Gum		10–20	Circular	Antimicrobial
87	*Punica granatum*	Peel		16–23	Spherical	Cytotoxic
88	*Camellia sinensis*	Leaves	Pd	6–18	Crystalline	Antioxidant, antibacterial, cytotoxic
89	*Catharanthus roseus*	Leaves		38	Spherical	Antibacterial, catalytic remediation

90	Cinnamomum camphora	Leaves		3.2–6.0	Crystalline	Catalytic
91	Cochlospermum gossypium	Gum		6–10	Cubic, spherical	Catalytic
92	Euphorbia thymifolia	Leaves		30–35	Spherical	Catalytic
93	Garcinia pedunculata	Leaves		2–4	Spherical	Catalytic, antibacterial
94	Allium cepa	Root	ZnO	100	Spherical	Clastogenic, cytotoxic
95	Aloe vera	Leaves		50–350	Rectangular, triangular, spherical	Antimicrobial
96	Artocarpus gomezianus	Fruits		>20	Spherical	Luminescence, catalytic, antioxidant
97	Brassica oleracea	Leaves		79	Spherical and sheet	Antibacterial
98	Ricinus communis	Seeds		10–30	Crystalline	Antioxidant, antifungal, cytotoxic
99	Tecoma castanifolia	Leaves		70–75	Spherical	Antibacterial, antioxidant, cytotoxic
100	Cassia auriculata	Leaves	CuO	30–35	Spherical	Antibacterial
101	Citrus medica	Fruits		20	Spherical	Antimicrobial
102	Ginkgo biloba	Leaves		15–20	Spherical	Catalytic
103	Gloriosa superba	Leaves		5–10	Spherical	Antibacterial
104	Malva sylvestris	Leaves		5–30	Spherical	Antibacterial
105	Phyllanthus amarus	Leaves		20	Spherical	Antibacterial
106	Sterculia urens	Bark (gum karaya)		5–10	Needle	Antibacterial

Continued

TABLE 2 Plant-mediated synthesis of nanocomposites—cont'd

S. No	Plant	Plant part used for synthesis	Nature of NPs	Size (nm)	Shape	Applications/activity
107	*Lawsonia inermis*	Leaves	Fe	21	Hexagonal	Antibacterial
108	*Medicago sativa (alfalfa)*	Leaves		2–10	Crystalline	Cancer hyperthermia, drug delivery
109	*Green tea/eucalyptus*	Leaves		20–80	Spherical	Catalytic
110	*Gardenia jasminoides*	Leaves		32	Rocklike appearance	Antibacterial
111	*Sorghum*	Seeds (bran)		50	Spherical	Catalytic

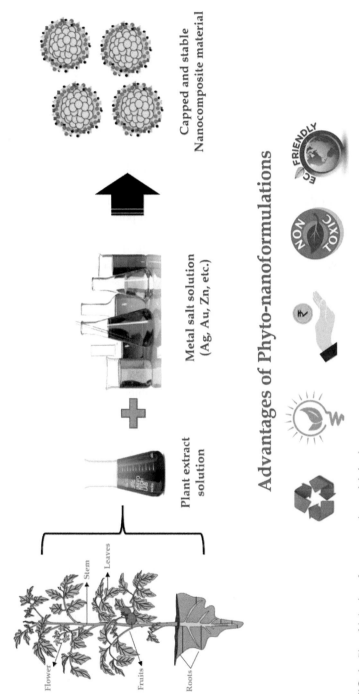

FIG. 4 Phytofabrication of nanocomposites and their advantages.

toxicity, and low tissue distribution that present a huge challenge to use them as therapeutics (Beutler, 2009). It is therefore imperative that innovative and efficient delivery mechanisms be built that can resolve these critical issues. New opportunities have begun to take shape to develop nanocomposites for versatile therapeutic applications with increased comprehension of the significance of nanocomposite topology (size, shape, and surface properties) for biological interactions at the molecular level.

Nanotechnology-based delivery systems such as NPs, nanocarriers, liposomes, nanosomes, and dendrites are being used to enhance the delivery of a variety of therapeutic drugs including phytochemicals to the target tissues (Aqil, Munagala, Jeyabalan, & Vadhanam, 2013). Innovative nanotechnology-based techniques have been developed in contemporary times to retain the bioactivity of these phytoconstituents. Driven by the immense development of novel bioconjugation and polymerization procedures, tremendous efforts in the direction of designing biomolecule-composite conjugates with extraordinary properties are being done. Nanoparticle-based biodegradable and biocompatible polymers are a promising tool to deal with frequent issues concerned with poor stability, low solubility, and less bioavailability faced in medical applications by these phytobioactive compounds (Wang et al., 2014). Poly(ethylene glycol) (PEG), poly(lactic acid) (PLA), polyglycolides (PGA), poly(lactide-*co*-glycolides) (PLGA), polycaprolactone (PCL), and poly(hydroxy butyrate) (PHB) are regularly used as phytobioactive compound carriers. Poly(lactic-*co*-glycolic acid) (PLGA) and poly(D,L-lactic acid) (PLA) is the FDA-approved and European Medicine Agency (EMA)–approved biodegradable polymers (Danhier et al., 2012). Apart from being biocompatible and biodegradable, these polymers have certain beneficial properties like good mechanical strength and stimulus responsiveness, which make them good material for the construction of several types of biomolecule-polymer nanocomposites (Wang & Wang, 2014). The application of these nanocarriers is one of the novel strategies that can optimize some unfavorable physicochemical characteristics of these phytobioactive compounds. A significant improvement of therapeutic response is attained by encapsulation of phytochemicals with the NPs of the biodegradable polymer with better internalization and prolonged retention time. It reduces the dose requisition of drugs, thereby reducing the chances of toxicity to nontarget normal tissues. Therefore the nanoformulation of phytobioactive compounds is an effective approach to enhance the bioavailability of therapeutic phytochemicals (Cho et al., 2008; Sapsford et al., 2013; Wang et al., 2014).

Numerous nanoformulation have been developed by encapsulating plant-derived bioactive compounds to ameliorate the solubility and stability of these constituents, thus providing a chance to reconsider those potential drugs, which were of limited use earlier due to their poor pharmacokinetics. Few examples of nanoconjugated/encapsulated phytobioactive compounds that have greater prospects of use in therapeutic managements are listed in the succeeding text.

(a) **Curcumin (diferuloylmethane)**, a principal curcuminoid of turmeric (*Curcuma longa*) major biologically active polyphenolic compound has been used for therapeutic applications against various diseases for centuries in Asian countries including India and China. Because of its tremendous antioxidant and antiinflammatory properties, it plays a multipronged and multitargeted therapeutic role at the molecular level in the pathological conditions like cancer, cardiovascular, inflammatory, and neurological disorders. However, because of its poor bioavailability, which results from low solubility, rapid metabolism, removal from the body, and weak permeability across the blood-brain barrier (BBB), the clinical implications of native curcumin are hindered. To overcome these limitations, in recent years, an assortment of nanomedicine-based drug delivery system has been considered to enhance the bioavailability of curcumin toward several therapies. Multifarious nanotechnology-based delivery approaches like solid lipid nanoparticles (SLNs), curcumin conjugates, and polymeric NPs have been employed to increase the oral bioavailability, biological activity, or tissue-targeting ability of curcumin (Wang et al., 2014; Yallapu, Nagesh, Jaggi, & Chauhan, 2015).

(b) **Epigallocatechin gallate (EGCG)** is a unique plant compound and has drawn the attention of the scientific community because of its multiple therapeutic properties, but its instability in biological milieu compromises the potential effectiveness. Experimental studies in recent past along these lines have found that nanoencapsulation of EGCG significantly increases its stability, pharmacokinetics, and target specificity, thus enhancing its therapeutic activities (Li et al., 2018).

(c) **Quercetin (3,3′,4′,5′-7-pentahydroxy flavone)** is a plant-derived flavonol widely present in fruits and vegetables that has unique therapeutic value. But its use as a drug molecule is limited due to low oral bioavailability and reduced solubility. Moreover, instant gastrointestinal digestion of quercetin impedes its clinical transformation. Production of nanoformulations for overcoming this drawback has been seriously considered and done in recent times. Many scientific studies suggested that nanoencapsulation increases the solubility of quercetin, improves its sustained release, and prevents modification and metabolism, thereby enhancing its bioactivity with less toxicity (Wang et al., 2014).

(d) **Resveratrol (3,5,4′-trihydroxy-*trans*-stilbene)** is a stilbenoid polyphenol present in strongly pigmented fruits and vegetables. Scientific studies have determined the great therapeutic potential of resveratrol such as antioxidant, antiinflammatory, cardio- and neuroprotective, antidiabetic, and anticancer. However, it has been seen that resveratrol is extensively metabolized and swiftly excreted; therefore it exhibits poor bioavailability, despite its lipophilic nature. To overcome these constraints, different strategies have been applied from time to time. In recent times, multiple studies have focused on novel approaches to stabilize resveratrol from degradation and optimize its solubility to boost bioavailability, achieve a sustained release, and eventually transport it to the specific targets. These operational methods include nanoencapsulation of resveratrol in lipid nanocarriers or liposomes, micelles, nanoemulsions, inclusion into polymeric particles, solid dispersions, and nanocrystals (Chimento et al., 2019; Wang et al., 2014).

(e) **Genistein (5,7-dihydroxy-3-(4-hydroxyphenyl)chromen-4-one)** is an abundant isoflavones present in a large variety of plant-derived foods, especially in soybean and soy foods. Genistein is gaining considerable attention due to its numerous therapeutic benefits. It serves as a phytoestrogen, antioxidant, and anticancer agent and can aid people with metabolic syndromes. Earlier studies have elucidated that genistein can suppress the growth of hormone-dependent cancers, such as different lineages of colon cancer cells, by influencing the expression of estrogen receptors and other tumor suppressor genes and by altering the function of antioxidant enzymes and by diminishing reactive oxygen species (ROS) generation. The antioxidant ability and antiproliferative properties of genistein were substantially improved by its encapsulation leading to enhanced water solubility and higher and sustained levels inside HT29 cells (Pool et al., 2018).

(f) **Gingerol**, one of the main bioactive compounds derived from the rhizomes of the ginger plant (*Zingiber officinale*), has attracted several researchers owing to its wide array of biological activities including anticancer and antiinflammatory. However, its applicability in the clinical perspectives has not been so much promising due to poor aqueous solubility, sensitivity to temperature, pH, oxygen, and photoinstability. Nanodrug delivery system is a propitious approach to deliver gingerol in a targeted based controlled pattern and attain better clinical efficacy. Encapsulation of gingerol in different nanosized carriers has shown improved water solubility and the oral bioavailability of gingerol. These nanocomposites are known to possess an appropriate size distribution, drug encapsulation proficiency, and pharmacokinetics. Further 6G-loaded PEGylated nanoniosomes also displayed more stability and slower release of the 6-gingerol drug at the target sites (Manatunga et al., 2018; Wei et al., 2018).

4. Concluding remarks and perspectives

Plants therapeutics have a long history of use in alleviating human diseases and improving people's health across diverse cultures. A plethora of drugs from plants have already been approved to counter several ailments. Botanical drugs as extracts have also been used as prospective therapeutic agents because of their low toxicity, greater affordability, and easy availability. However, poor aqueous solubility, degradation during gastric emptying, and extensive metabolism are a few limitations, which restrict the utility of these bioactive compounds into clinical translation. In conjunction with nanotechnology the nanoformulations of phytobioactive compounds amplify their therapeutic impact and offer a novel strategy that has a contemporary significance. The intervention of nanotechnology minimizes the dosage while enhancing bioavailability and bioactivity. In addition, the phytonanocomposites enable the targeted and sustained delivery that enhances the profile of pharmacokinetics and the diffusion of drugs into the biological systems. Thus plant-derived nanocomposites have the potential for biomedical applications in the future. Moreover, future research should focus on identifying novel phytoactive constituents and in designing and developing multifunctional smart nanocomposites as effective and novel therapeutics. Though it is challenging, it is the need of the hour to come out with potential, biocompatible therapeutic plant-derived drugs involving the major research institutes world over.

Acknowledgments

The author, Mir Zahoor Gul, is grateful to the Council for Science and Industrial Research (CSIR), New Delhi for Research Associate Fellowship vide award letter No. [09/132(0883)/2019-EMR-1].

References

Abbasi, E., Milani, M., Aval, S. F., Kouhi, M., Akbarzadeh, A., Nasrabadi, H. T., et al. (2016). Silver nanoparticles: Synthesis methods, bio-applications and properties. *Critical Reviews in Microbiology*. https://doi.org/10.3109/1040841X.2014.912200.

Abed, F., Baniya, R., & Bachuwa, G. (2016). Quinine-induced disseminated intravascular coagulation. *Case Reports in Medicine*, *2016*. https://doi.org/10.1155/2016/9136825.

Achan, J., Talisuna, A. O., Erhart, A., Yeka, A., Tibenderana, J. K., Baliraine, F. N., et al. (2011). Quinine, an old anti-malarial drug in a modern world: Role in the treatment of malaria. *Malaria Journal*. https://doi.org/10.1186/1475-2875-10-144.

Adehin, A., Igbinoba, S. I., Soyinka, J. O., Onyeji, C. O., Babalola, C. P., & Bolaji, O. O. (2019). Pharmacokinetic parameters of quinine in healthy subjects and in patients with uncomplicated malaria in Nigeria: Analysis of data using a population approach. *Current Therapeutic Research, Clinical and Experimental*, *91*, 33–38. https://doi.org/10.1016/j.curtheres.2019.100567.

Anderson, L., Jarratt, M., Schmieder, G., Shumack, S., Katsamas, J., & Welburn, P. (2014). Tolerability and pharmacokinetics of ingenol mebutate 0.05% gel applied to treatment areas up to 100cm2 on the forearm (s) of patients with actinic keratosis. *The Journal of Clinical and Aesthetic Dermatology*, *7*, 19–29.

Anderson, J. M., & Shive, M. S. (2012). Biodegradation and biocompatibility of PLA and PLGA microspheres. *Advanced Drug Delivery Reviews*. https://doi.org/10.1016/j.addr.2012.09.004.

Anonymous. (2019). *Herbal medicine market research report—Global forecast till 2023*. Global. https://www.marketresearchfuture.com/videos/herbal-medicine-market.

Aqil, F., Munagala, R., Jeyabalan, J., & Vadhanam, M. V. (2013). Bioavailability of phytochemicals and its enhancement by drug delivery systems. *Cancer Letters*. https://doi.org/10.1016/j.canlet.2013.02.032.

Asif, M., & Mohd, I. (2019). Prospects of medicinal plants derived nutraceuticals: A re-emerging new era of medicine and health aid. *Progress Chemical Biochemical Research*, 150–169. https://doi.org/10.33945/sami/pcbr.2019.4.1.

Auffan, M., Rose, J., Bottero, J. Y., Lowry, G. V., Jolivet, J. P., & Wiesner, M. R. (2009). Towards a definition of inorganic nanoparticles from an environmental, health and safety perspective. *Nature Nanotechnology*. https://doi.org/10.1038/nnano.2009.242.

Baker, S., Rakshith, D., Kavitha, K. S., Santosh, P., Kavitha, H. U., Rao, Y., et al. (2013). Plants: Emerging as nanofactories towards facile route in synthesis of nanoparticles. *BioImpacts*. https://doi.org/10.5681/bi.2013.012.

Baranwal, A., Mahato, K., Srivastava, A., Maurya, P. K., & Chandra, P. (2016). Phytofabricated metallic nanoparticles and their clinical applications. *RSC Advances*. https://doi.org/10.1039/c6ra23411a.

Bárcena, C., Sra, A. K., & Gao, J. (2009). Applications of magnetic nanoparticles in biomedicine. In *Nanoscale magnetic materials and applications* (pp. 591–626). https://doi.org/10.1007/978-0-387-85600-1_20.

Beutler, J. A. (2009). Natural products as a foundation for drug discovery. *Current Protocols in Pharmacology*. https://doi.org/10.1002/0471141755.ph0911s46.

Bundschuh, M., Filser, J., Lüderwald, S., McKee, M. S., Metreveli, G., Schaumann, G. E., et al. (2018). Nanoparticles in the environment: Where do we come from, where do we go to? *Environmental Sciences Europe*. https://doi.org/10.1186/s12302-018-0132-6.

Calixto, J. B. (2019). The role of natural products in modern drug discovery. *Anais da Academia Brasileira de Ciências*, *91*. https://doi.org/10.1590/0001-3765201920190105.

Cao, B., Wei, X. C., Xu, X. R., Zhang, H. Z., Luo, C. H., Feng, B., et al. (2019). Seeing the unseen of the combination of two natural resins, frankincense and myrrh: Changes in chemical constituents and pharmacological activities. *Molecules*. https://doi.org/10.3390/molecules24173076.

Chen, Y. C., Qiang, G. F., & Du, G. H. (2018). Salicylic acid. In *Natural small molecule drugs from plants* (pp. 455–460). https://doi.org/10.1007/978-981-10-8022-7_76.

Chen, Y., Sun, L., & Du, G. H. (2018). Digoxin. In *Natural small molecule drugs from plants* (pp. 49–58). https://doi.org/10.1007/978-981-10-8022-7_8.

Chimento, A., De Amicis, F., Sirianni, R., Sinicropi, M. S., Puoci, F., Casaburi, I., et al. (2019). Progress to improve oral bioavailability and beneficial effects of resveratrol. *International Journal of Molecular Sciences*. https://doi.org/10.3390/ijms20061381.

Cho, K., Wang, X., Nie, S., Chen, Z., & Shin, D. M. (2008). Therapeutic nanoparticles for drug delivery in cancer. *Clinical Cancer Research*. https://doi.org/10.1158/1078-0432.CCR-07-1441.

Chu, H. W., Unnikrishnan, B., Anand, A., Mao, J. Y., & Huang, C. C. (2018). Nanoparticle-based laser desorption/ionization mass spectrometric analysis of drugs and metabolites. *Journal of Food and Drug Analysis*. https://doi.org/10.1016/j.jfda.2018.07.001.

Clark, P. I., & Slevin, M. L. (1987). The clinical pharmacology of etoposide and teniposide. *Clinical Pharmacokinetics*. https://doi.org/10.2165/00003088-198712040-00001.

Cuenya, B. R. (2010). Synthesis and catalytic properties of metal nanoparticles: Size, shape, support, composition, and oxidation state effects. *Thin Solid Films*, *518*, 3127–3150. https://doi.org/10.1016/j.tsf.2010.01.018.

Dadfar, S. M., Roemhild, K., Drude, N. I., von Stillfried, S., Knüchel, R., Kiessling, F., et al. (2019). Iron oxide nanoparticles: Diagnostic, therapeutic and theranostic applications. *Advanced Drug Delivery Reviews*. https://doi.org/10.1016/j.addr.2019.01.005.

Dalvi, S. S., Gupta, K. C., Pohujani, S. M., Vaidya, A. B., & Satoskar, R. S. (1985). Bioavailability of aspirin after oral and rectal administration in volunteers and patients with fever. *Journal of Postgraduate Medicine*, *31*, 192–195.

Danhier, F., Ansorena, E., Silva, J. M., Coco, R., Le Breton, A., & Préat, V. (2012). PLGA-based nanoparticles: An overview of biomedical applications. *Journal of Controlled Release*. https://doi.org/10.1016/j.jconrel.2012.01.043.

Desborough, M. J. R., & Keeling, D. M. (2017). The aspirin story—From willow to wonder drug. *British Journal of Haematology*. https://doi.org/10.1111/bjh.14520.

Dhand, C., Dwivedi, N., Loh, X. J., Jie Ying, A. N., Verma, N. K., Beuerman, R. W., et al. (2015). Methods and strategies for the synthesis of diverse nanoparticles and their applications: A comprehensive overview. *RSC Advances*. https://doi.org/10.1039/c5ra19388e.

Du, G. H. (2018). *Natural small molecule drugs from plants*. https://doi.org/10.1007/978-981-10-8022-7.

Duan, H., Wang, D., & Li, Y. (2015). Green chemistry for nanoparticle synthesis. *Chemical Society Reviews, 44*, 5778–5792. https://doi.org/10.1039/c4cs00363b.

Duffin, J. (2000). Poisoning the spindle: Serendipity and discovery of the anti-tumor properties of the Vinca alkaloids. *Canadian Bulletin of Medical History, 17*, 155–192. https://doi.org/10.3138/cbmh.17.1.155.

Ekor, M. (2014). The growing use of herbal medicines: Issues relating to adverse reactions and challenges in monitoring safety. *Frontiers in Neurology*. https://doi.org/10.3389/fphar.2013.00177.

Ellingsrud, C., & Agewall, S. (2016). Morphine in the treatment of acute pulmonary oedema—Why? *International Journal of Cardiology*. https://doi.org/10.1016/j.ijcard.2015.10.014.

Farlow, M. R. (2003). Clinical pharmacokinetics of galantamine. *Clinical Pharmacokinetics*. https://doi.org/10.2165/00003088-200342150-00005.

Fattinger, K., Benowitz, N. L., Jones, R. T., & Verotta, D. (2000). Nasal mucosal versus gastrointestinal absorption of nasally administered cocaine. *European Journal of Clinical Pharmacology, 56*, 305–310. https://doi.org/10.1007/s002280000147.

Feliú, A., Moreno-Martet, M., Mecha, M., Carrillo-Salinas, F. J., De Lago, E., Fernández-Ruiz, J., et al. (2015). A Sativex®-like combination of phytocannabinoids as a disease-modifying therapy in a viral model of multiple sclerosis. *British Journal of Pharmacology, 172*, 3579–3595. https://doi.org/10.1111/bph.13159.

Gazaliev, A. N., Zhurinov, M. Z., Fazylov, S. D., & Balitskii, S. N. (1989). Isolation, analysis, and synthesis of ephedrine and its derivatives. *Chemistry of Natural Compounds, 25*, 261–271. https://doi.org/10.1007/BF00597698.

Gong, L. L., Da Du, L., & Du, G. H. (2018). Theophylline. In *Natural small molecule drugs from plants* (pp. 469–478). https://doi.org/10.1007/978-981-10-8022-7_78.

Griffin, J. P., Posner, J., & Barker, G. R. (2013). *The textbook of pharmaceutical medicine*. https://doi.org/10.1002/9781118532331.

Gupta, M. (2010). Pharmacological properties and traditional therapeutic uses of important Indian spices: A review. *International Journal of Food Properties, 13*, 1092–1116. https://doi.org/10.1080/10942910902963271.

Guzmán-Rodríguez, J. J., Ochoa-Zarzosa, A., López-Gómez, R., & López-Meza, J. E. (2015). Plant antimicrobial peptides as potential anticancer agents. *BioMed Research International*. https://doi.org/10.1155/2015/735087.

Hampson, P., Chahal, H., Khanim, F., Hayden, R., Mulder, A., Assi, L. K., et al. (2005). PEP005, a selective small-molecule activator of protein kinase C, has potent antileukemic activity mediated via the delta isoform of PKC. *Blood, 106*, 1362–1368. https://doi.org/10.1182/blood-2004-10-4117.

Hollman, A. (1996). Drugs for atrial fibrillation. Digoxin comes from Digitalis lanata. *BMJ*. https://doi.org/10.1136/bmj.312.7035.912.

Hussain, I., Singh, N. B., Singh, A., Singh, H., & Singh, S. C. (2016). Green synthesis of nanoparticles and its potential application. *Biotechnology Letters*. https://doi.org/10.1007/s10529-015-2026-7.

Ikram, N. K. B. K., & Simonsen, H. T. (2017). A review of biotechnological artemisinin production in plants. *Frontiers in Plant Science*. https://doi.org/10.3389/fpls.2017.01966.

Izawa, K., Amino, Y., Kohmura, M., Ueda, Y., & Kuroda, M. (2010). Human-environment interactions—Taste. In H.-W.(Ben) Liu, & M. Lew (Eds.), *Vol. 4. Comprehensive natural products II: Chemistry and biology* (pp. 631–671). Amsterdam, Netherlands: Elsevier.

Jain, P. K., Huang, X., El-Sayed, I. H., & El-Sayed, M. A. (2008). Noble metals on the nanoscale: Optical and photothermal properties and some applications in imaging, sensing, biology, and medicine. *Accounts of Chemical Research, 41,* 1578–1586. https://doi.org/10.1021/ar7002804.

Jeong, H. H., Choi, E., Ellis, E., & Lee, T. C. (2019). Recent advances in gold nanoparticles for biomedical applications: From hybrid structures to multi-functionality. *Journal of Materials Chemistry B.* https://doi.org/10.1039/c9tb00557a.

Jiang, J., Pi, J., & Cai, J. (2018). The advancing of zinc oxide nanoparticles for biomedical applications. *Bioinorganic Chemistry and Applications.* https://doi.org/10.1155/2018/1062562.

Karunamoorthi, K., Jegajeevanram, K., Vijayalakshmi, J., & Mengistie, E. (2013). Traditional medicinal plants: A source of phytotherapeutic modality in resource-constrained health care settings. *Evidence-based Complementary and Alternative Medicine, 18,* 67–74. https://doi.org/10.1177/2156587212460241.

Katiyar, C., Kanjilal, S., Gupta, A., & Katiyar, S. (2012). Drug discovery from plant sources: An integrated approach. *Ayu, 33,* 10. https://doi.org/10.4103/0974-8520.100295.

Kesarwani, K., & Gupta, R. (2013). Bioavailability enhancers of herbal origin: An overview. *Asian Pacific Journal of Tropical Biomedicine, 3,* 253–266. https://doi.org/10.1016/S2221-1691(13)60060-X.

Khanna, C., Rosenberg, M., & Vail, D. M. (2015). A review of paclitaxel and novel formulations including those suitable for use in dogs. *Journal of Veterinary Internal Medicine.* https://doi.org/10.1111/jvim.12596.

Kircik, L., Sung, J. C., Stein-Gold, L., & Goldenberg, G. (2017). United States food and drug administration product label changes. *The Journal of Clinical and Aesthetic Dermatology, 10,* 20–29.

Kong, L. L., Wang, J. H., & Du, G. H. (2018). Morphine. In *Natural small molecule drugs from plants* (pp. 295–302). https://doi.org/10.1007/978-981-10-8022-7_49.

Krishna, S., Bustamante, L., Haynes, R. K., & Staines, H. M. (2008). Artemisinins: Their growing importance in medicine. *Trends in Pharmacological Sciences.* https://doi.org/10.1016/j.tips.2008.07.004.

Krishnamurti, C., & Rao, S. S. C. C. (2016). The isolation of morphine by serturner. *Indian Journal of Anaesthesia.* https://doi.org/10.4103/0019-5049.193696.

Lee, S. H., & Jun, B. H. (2019). Silver nanoparticles: Synthesis and application for nanomedicine. *International Journal of Molecular Sciences.* https://doi.org/10.3390/ijms20040865.

Lei, T. T., Wang, J. H., & Du, G. H. (2018). Galantamine. In *Natural small molecule drugs from plants.* https://doi.org/10.1007/978-981-10-8022-7_42.

Lemaire-Hurtel, A. S., & Alvarez, J. C. (2014). Drugs involved in drug-facilitated crime-pharmacological aspects. In *Toxicological aspects of drug-facilitated crimes* (pp. 47–91). https://doi.org/10.1016/B978-0-12-416748-3.00003-7.

Leung, Y. Y., Yao Hui, L. L., & Kraus, V. B. (2015). Colchicine-update on mechanisms of action and therapeutic uses. *Seminars in Arthritis and Rheumatism.* https://doi.org/10.1016/j.semarthrit.2015.06.013.

Li, F., Jiang, T., Li, Q., & Ling, X. (2017). Camptothecin (CPT) and its derivatives are known to target topoisomerase I (Top1) as their mechanism of action: Did we miss something in CPT analogue molecular targets for treating human disease such as cancer? *American Journal of Cancer Research.*

Li, X., Wang, L., Fan, Y., Feng, Q., & Cui, F. Z. (2012). Biocompatibility and toxicity of nanoparticles and nanotubes. *Journal of Nanomaterials.* https://doi.org/10.1155/2012/548389.

Li, F., Wang, Y., Li, D., Chen, Y., Qiao, X., Fardous, R., et al. (2018). Perspectives on the recent developments with green tea polyphenols in drug discovery. *Expert Opinion on Drug Discovery.* https://doi.org/10.1080/17460441.2018.1465923.

Lilienfeld, S. (2002). Galantamine—A novel cholinergic drug with a unique dual mode of action for the treatment of patients with Alzheimer's disease. *CNS Drug Reviews.* https://doi.org/10.1111/j.1527-3458.2002.tb00221.x.

Limberger, R. P., Jacques, A. L. B., Schmitt, G. C., & Arbo, M. D. (2013). Pharmacological effects of ephedrine. In *Natural products: Phytochemistry, botany and metabolism of alkaloids, phenolics and terpenes* (pp. 1217–1237). https://doi.org/10.1007/978-3-642-22144-6_41.

Lingbeck, J. M., O'Bryan, C. A., Martin, E. M., Adams, J. P., & Crandall, P. G. (2015). Sweetgum: An ancient source of beneficial compounds with modern benefits. *Pharmacognosy Reviews.* https://doi.org/10.4103/0973-7847.156307.

López-Meza, J. E., Ochoa-Zarzosa, A., Barboza-Corona, J. E., & Bideshi, D. K. (2015). Antimicrobial peptides: Current and potential applications in biomedical therapies. *BioMed Research International.* https://doi.org/10.1155/2015/367243.

Lü, S., & Wang, J. (2014). Homoharringtonine and omacetaxine for myeloid hematological malignancies. *Journal of Hematology & Oncology.* https://doi.org/10.1186/1756-8722-7-2.

Makarov, V. V., Love, A. J., Sinitsyna, O. V., Makarova, S. S., Yaminsky, I. V., Taliansky, M. E., et al. (2014). "Green" nanotechnologies: Synthesis of metal nanoparticles using plants. *Acta Naturae.* https://doi.org/10.32607/20758251-2014-6-1-35-44.

Malingré, M. M., Beijnen, J. H., Rosing, H., Koopman, F. J., Jewell, R. C., Paul, E. M., et al. (2001). Co-administration of GF120918 significantly increases the systemic exposure to oral paclitaxel in cancer patients. *British Journal of Cancer, 84*, 42–47. https://doi.org/10.1054/bjoc.2000.1543.

Manatunga, D. C., de Silva, R. M., de Silva, K. M. N., Wijeratne, D. T., Malavige, G. N., & Williams, G. (2018). Fabrication of 6-gingerol, doxorubicin and alginate hydroxyapatite into a bio-compatible formulation: Enhanced anti-proliferative effect on breast and liver cancer cells. *Chemistry Central Journal, 12.* https://doi.org/10.1186/s13065-018-0482-6.

Marslin, G., Siram, K., Maqbool, Q., Selvakesavan, R. K., Kruszka, D., Kachlicki, P., et al. (2018). Secondary metabolites in the green synthesis of metallic nanoparticles. *Materials (Basel).* https://doi.org/10.3390/ma11060940.

Maude, R. J., Woodrow, C. J., & White, L. J. (2010). Artemisinin antimalarials: Preserving the "magic bullet". *Drug Development Research.* https://doi.org/10.1002/ddr.20344.

Medhi, B., Patyar, S., Rao, R. S., Byrav Ds, P., & Prakash, A. (2009). Pharmacokinetic and toxicological profile of artemisinin compounds: An update. *Pharmacology.* https://doi.org/10.1159/000252658.

Mohanpuria, P., Rana, N. K., & Yadav, S. K. (2008). Biosynthesis of nanoparticles: Technological concepts and future applications. *Journal of Nanoparticle Research.* https://doi.org/10.1007/s11051-007-9275-x.

Narayanan, R., & El-Sayed, M. A. (2004). Shape-dependent catalytic activity of platinum nanoparticles in colloidal solution. *Nano Letters, 4*, 1343–1348. https://doi.org/10.1021/nl0495256.

Nasri, H., Baradaran, A., Shirzad, H., & Kopaei, M. R. (2014). New concepts in nutraceuticals as alternative for pharmaceuticals. *International Journal of Preventive Medicine, 5*, 1487–1499.

Nemunaitis, J., Mita, A., Stephenson, J., Mita, M. M., Sarantopoulos, J., Padmanabhan-Iyer, S., et al. (2013). Pharmacokinetic study of omacetaxine mepesuccinate administered subcutaneously to patients with advanced solid and hematologic tumors. *Cancer Chemotherapy and Pharmacology, 71*, 35–41. https://doi.org/10.1007/s00280-012-1963-2.

Nomicos, E. Y. H. (2007). Myrrh: Medical marvel or myth of the magi? *Holistic Nursing Practice, 21*, 308–323. https://doi.org/10.1097/01.HNP.0000298616.32846.34.

Padil, V. V. T., Wacławek, S., Černík, M., & Varma, R. S. (2018). Tree gum-based renewable materials: Sustainable applications in nanotechnology, biomedical and environmental fields. *Biotechnology Advances*. https://doi.org/10.1016/j.biotechadv.2018.08.008.

Pandey, N., Meena, R. P., Rai, S. K., & Pandey-Rai, S. (2011). Medicinal plants derived nutraceuticals: A re-emerging health aid. *International Journal of Pharma and Bio Sciences*, 2(4), 420–441.

Pang, X. C., Zhang, L., & Du, G. H. (2018). Podophyllotoxin. In *Natural small molecule drugs from plants* (pp. 545–550). https://doi.org/10.1007/978-981-10-8022-7_90.

Patil, S. P., Goswami, A., Kalia, K., & Kate, A. S. (2019). Plant-derived bioactive peptides: A treatment to cure diabetes. *International Journal of Peptide Research and Therapeutics*. https://doi.org/10.1007/s10989-019-09899-z.

Patra, J. K., Das, G., Fraceto, L. F., Campos, E. V. R., Rodriguez-Torres, M. D. P., Acosta-Torres, L. S., et al. (2018). Nano based drug delivery systems: Recent developments and future prospects. *Journal of Nanbiotechnology*, 16. https://doi.org/10.1186/s12951-018-0392-8.

Pereira, L., Mehboob, F., Stams, A. J. M., Mota, M. M., Rijnaarts, H. H. M., & Alves, M. M. (2015). Metallic nanoparticles: Microbial synthesis and unique properties for biotechnological applications, bioavailability and biotransformation. *Critical Reviews in Biotechnology*. https://doi.org/10.3109/07388551.2013.819484.

Peterson, C. T., Denniston, K., & Chopra, D. (2017). Therapeutic uses of triphala in ayurvedic medicine. *Journal of Alternative and Complementary Medicine*, 23, 607–614. https://doi.org/10.1089/acm.2017.0083.

Phan, T. T. V., Huynh, T. C., Manivasagan, P., Mondal, S., & Oh, J. (2020). An up-to-date review on biomedical applications of palladium nanoparticles. *Nanomaterials*. https://doi.org/10.3390/nano10010066.

Pooja, B., Joginder, S. D., & Suresh, K. G. (2014). Biogenesis of nanoparticles: A review. *African Journal of Biotechnology*, 13, 2778–2785. https://doi.org/10.5897/ajb2013.13458.

Pool, H., Campos-Vega, R., Herrera-Hernández, M. G., García-Solis, P., García-Gasca, T., Sánchez, I. C., et al. (2018). Development of genistein-PEGylated silica hybrid nanomaterials with enhanced antioxidant and antiproliferative properties on HT29 human colon cancer cells. *American Journal of Translational Research*, 10, 2306–2323.

Powell, R. G., Weisleder, D., & Smith, C. R. (1972). Antitumor alkaloids from Cephalotaxus harringtonia: Structure and activity. *Journal of Pharmaceutical Sciences*, 61, 1227–1230. https://doi.org/10.1002/jps.2600610812.

Prakash, J., Srivastava, S., Ray, R. S., Singh, N., Rajpali, R., & Singh, G. N. (2017). Current status of herbal drug standards in the Indian pharmacopoeia. *Phytotherapy Research*. https://doi.org/10.1002/ptr.5933.

Prasad, S., & Aggarwal, B. B. (2011). Turmeric, the golden spice: From traditional medicine to modern medicine. In *Herbal medicine: Biomolecular and clinical aspects: Second edition* (pp. 263–288). CRC Press/Taylor & Francis.

Prathna, T. C., Mathew, L., Chandrasekaran, N., Raichur, A., & Mukherjee, A. (2010). Biomimetic synthesis of nanoparticles: Science, technology & amp; applicability. In *Biomimetics Learning from Nature*. https://doi.org/10.5772/8776.

Rai, M., Ingle, A. P., Birla, S., Yadav, A., & Santos, C. A. D. (2016). Strategic role of selected noble metal nanoparticles in medicine. *Critical Reviews in Microbiology*. https://doi.org/10.3109/1040841X.2015.1018131.

Rao, P., & Gan, S. (2014). Recent advances in nanotechnology-based diagnosis and treatments of diabetes. *Current Drug Metabolism*, 16, 371–375. https://doi.org/10.2174/1389200215666141125120215.

Rao, C. N. R., Kulkarni, G. U., John Thomas, P., & Edwards, P. P. (2002). Size-dependent chemistry: Properties of nanocrystals. *Chemistry—A European Journal.* https://doi.org/10.1002/1521-3765(20020104)8:1<28::AID-CHEM28>3.0.CO;2-B.

Rates, S. M. K. (2001). Plants as source of drugs. *Toxicon.* https://doi.org/10.1016/S0041-0101(00)00154-9.

Roy, P. (2018). Global pharma and local science: The untold tale of reserpine. *Indian Journal of Psychiatry*, 60, S277–S282. https://doi.org/10.4103/psychiatry.IndianJPsychiatry_444_17.

Rubilar, O., Rai, M., Tortella, G., Diez, M. C., Seabra, A. B., & Durán, N. (2013). Biogenic nanoparticles: Copper, copper oxides, copper sulphides, complex copper nanostructures and their applications. *Biotechnology Letters.* https://doi.org/10.1007/s10529-013-1239-x.

Samal, J. (2015). Situational analysis and future directions of Ayush: An assessment through five year plans of India. *Journal of Intercultural Ethnopharmacology*, 4, 348. https://doi.org/10.5455/jice.20151101093011.

Sapsford, K. E., Algar, W. R., Berti, L., Gemmill, K. B., Casey, B. J., Oh, E., et al. (2013). Functionalizing nanoparticles with biological molecules: Developing chemistries that facilitate nanotechnology. *Chemical Reviews.* https://doi.org/10.1021/cr300143v.

Sarup, P., Bala, S., & Kamboj, S. (2015). Pharmacology and phytochemistry of oleo-gum resin of Commiphora wightii (Guggulu). *Scientifica (Cairo)*, 2015, 1–14. https://doi.org/10.1155/2015/138039.

Sashidhar, R. B., Raju, D., & Karuna, R. (2014). Tree Gum: Gum Kondagogu. In K. Ramawat, & J. M. Mérillon (Eds.), *Polysaccharides.* Cham: Springer. https://doi.org/10.1007/978-3-319-03751-6_32-1.

Seca, A. M. L., & Pinto, D. C. G. A. (2018). Plant secondary metabolites as anticancer agents: Successes in clinical trials and therapeutic application. *International Journal of Molecular Sciences.* https://doi.org/10.3390/ijms19010263.

Souyoul, S. A., Saussy, K. P., & Lupo, M. P. (2018). Nutraceuticals: A review. *Dermatology and Therapy.* https://doi.org/10.1007/s13555-018-0221-x.

Sreeprasad, T. S., & Pradeep, T. (2013). Noble metal nanoparticles. In *Springer handbook of nanomaterials* (pp. 303–388). https://doi.org/10.1007/978-3-642-20595-8_9.

Stahl, W., & Sies, H. (2005). Bioactivity and protective effects of natural carotenoids. *Biochimica et Biophysica Acta: Molecular Basis of Disease*, 101–107. https://doi.org/10.1016/j.bbadis.2004.12.006.

Su, X. Z., & Miller, L. H. (2015). The discovery of artemisinin and the Nobel prize in physiology or medicine. *Science China. Life Sciences*, 58, 1175–1179. https://doi.org/10.1007/s11427-015-4948-7.

Sudo, K., Konno, K., Shigeta, S., & Yokota, T. (1998). Inhibitory effects of podophyllotoxin derivatives on herpes simplex virus replication. *Antiviral Chemistry & Chemotherapy*, 9, 263–267. https://doi.org/10.1177/095632029800900307.

Sun, J., Wei, Q., Zhou, Y., Wang, J., Liu, Q., & Xu, H. (2017). A systematic analysis of FDA-approved anticancer drugs. *BMC Systems Biology*, 11. https://doi.org/10.1186/s12918-017-0464-7.

Tewari, D., Stankiewicz, A. M., Mocan, A., Sah, A. N., Tzvetkov, N. T., Huminiecki, L., et al. (2018). Ethnopharmacological approaches for dementia therapy and significance of natural products and herbal drugs. *Frontiers in Aging Neuroscience.* https://doi.org/10.3389/fnagi.2018.00003.

Thakur, L., Ghodasra, U., Patel, N., & Dabhi, M. (2011). Novel approaches for stability improvement in natural medicines. *Pharmacognosy Reviews.* https://doi.org/10.4103/0973-7847.79099.

Thanh, N. T. K., & Green, L. A. W. (2010). Functionalisation of nanoparticles for biomedical applications. *Nano Today.* https://doi.org/10.1016/j.nantod.2010.05.003.

Verma, N., & Kumar, N. (2019). Synthesis and biomedical applications of copper oxide nanoparticles: An expanding horizon. *ACS Biomaterials Science & Engineering, 5*, 1170–1188. https://doi.org/10.1021/acsbiomaterials.8b01092.

Wang, L., & Du, G. H. (2018). Paclitaxel. In *Natural small molecule drugs from plants* (pp. 537–544). https://doi.org/10.1007/978-981-10-8022-7_89.

Wang, Z., Li, L., & Du, G. H. (2018). Homoharringtonine (HHT). In *Natural small molecule drugs from plants* (pp. 521–528). https://doi.org/10.1007/978-981-10-8022-7_86.

Wang, S., Su, R., Nie, S., Sun, M., Zhang, J., Wu, D., et al. (2014). Application of nanotechnology in improving bioavailability and bioactivity of diet-derived phytochemicals. *The Journal of Nutritional Biochemistry.* https://doi.org/10.1016/j.jnutbio.2013.10.002.

Wang, E. C., & Wang, A. Z. (2014). Nanoparticles and their applications in cell and molecular biology. *Integrative Biology.* https://doi.org/10.1039/c3ib40165k.

Wang, J., Xu, C., Wong, Y. K., Li, Y., Liao, F., Jiang, T., et al. (2019). Artemisinin, the magic drug discovered from traditional Chinese medicine. *Engineering, 5*, 32–39. https://doi.org/10.1016/j.eng.2018.11.011.

Wang, J. H., Yang, X. Y., & Du, G. H. (2018). Ephedrine. In *Natural small molecule drugs from plants* (pp. 231–236). https://doi.org/10.1007/978-981-10-8022-7_38.

Wani, S. A., & Kumar, P. (2018). Fenugreek: A review on its nutraceutical properties and utilization in various food products. *Journal of the Saudi Society of Agricultural Sciences.* https://doi.org/10.1016/j.jssas.2016.01.007.

Wani, M. C., Taylor, H. L., Wall, M. E., Coggon, P., & Mcphail, A. T. (1971). Plant antitumor agents. VI. The isolation and structure of taxol, a novel antileukemic and antitumor agent from Taxus brevifolia2. *Journal of the American Chemical Society, 93*, 2325–2327. https://doi.org/10.1021/ja00738a045.

Wei, Q., Yang, Q., Wang, Q., Sun, C., Zhu, Y., Niu, Y., et al. (2018). Formulation, characterization, and pharmacokinetic studies of 6-gingerol-loaded nanostructured lipid carriers. *AAPS PharmSciTech, 19*, 3661–3669. https://doi.org/10.1208/s12249-018-1165-2.

Wen, L., Yuan, Y. H., Kong, L. L., & Chen, N. H. (2018). Quinine. In *Natural small molecule drugs from plants* (pp. 613–618). https://doi.org/10.1007/978-981-10-8022-7_100.

WHO. (2019). World Health Organization model list of essential medicines, 21st list, 2019. Geneva: World Health Organization.

Wong, Y. K., Yang, J., & He, Y. (2020). Caution and clarity required in the use of chloroquine for COVID-19. *The Lancet Rheumatology.* https://doi.org/10.1016/s2665-9913(20)30093-x.

Xu, L. J., Liu, A. L., & Du, G. H. (2018). Scopolamine. In *Natural small molecule drugs from plants* (pp. 319–324). https://doi.org/10.1007/978-981-10-8022-7_53.

Yallapu, M. M., Nagesh, P. K. B., Jaggi, M., & Chauhan, S. C. (2015). Therapeutic applications of curcumin nanoformulations. *The AAPS Journal, 17*, 1341–1356. https://doi.org/10.1208/s12248-015-9811-z.

Yang, X. Y., & Du, G. H. (2018a). Capsaicin. In *Natural small molecule drugs from plants* (pp. 397–402). https://doi.org/10.1007/978-981-10-8022-7_66.

Yang, X. Y., & Du, G. H. (2018b). Metformin. In *Natural small molecule drugs from plants* (pp. 101–108). https://doi.org/10.1007/978-981-10-8022-7_16.

Yang, Z. H., Wang, S. B., & Du, G. H. (2018). Ajmaline. In *Natural small molecule drugs from plants* (pp. 5–12). https://doi.org/10.1007/978-981-10-8022-7_1.

Yao, J., Weng, Y., Dickey, A., & Wang, K. Y. (2015). Plants as factories for human pharmaceuticals: Applications and challenges. *International Journal of Molecular Sciences.* https://doi.org/10.3390/ijms161226122.

Zhang, W., Zhou, Q. M., & Du, G. H. (2018). Colchicine. In *Natural small molecule drugs from plants* (pp. 503–508). https://doi.org/10.1007/978-981-10-8022-7_83.

Zhao, Y., Wang, J. H., Yang, X. Y., & Du, G. H. (2018). Cocaine. In *Natural small molecule drugs from plants* (pp. 221–226). https://doi.org/10.1007/978-981-10-8022-7_36.

Ziental, D., Czarczynska-Goslinska, B., Mlynarczyk, D. T., Glowacka-Sobotta, A., Stanisz, B., Goslinski, T., et al. (2020). Titanium dioxide nanoparticles: Prospects and applications in medicine. *Nanomaterials.* https://doi.org/10.3390/nano10020387.

Chapter 5

Herbal medicine: Old practice and modern perspectives

Sami Ullah Qadir[a] and Vaseem Raja[b]
[a]*Department of Environmental Science and Engineering, Jamia Millia Islamia, New Delhi, India,*
[b]*Department of Botany, University of Kashmir, Srinagar, Jammu and Kashmir, India*

1. Introduction

For several thousands of years, plants and their associated products have been used in folk medicines for disease treatment in humans and animals. Traditional medicine refers to the application, knowledge, belief, and approaches of incorporating animal or plant-based properties in remedies, either alone or in amalgamation, to maintain the welfare of an individual and for preventing or treating the disease as well.

India enjoys almost 8% of the assessed inclusive biodiversity of about 126,000 species. India is a home to about 315 species of flowering plants out of 400 families found around the globe. At present, nearly 45,000 classes (almost 20% of total species diversity) originate in India, out of which approximately 3500 plant species possess medicinal values. Modern Ayurvedic industry makes use of around 500 medicinally important plant species for the development of modern pharmaceuticals; out of total plants with medicinal value, ~80% of plants are acquired from wilderness areas, and 10% are obtained from cultivated farmhouses that are actively involved in trade (Singh, 2006). The western Himalayan biodiversity hotspot provides about 80% of drugs obtained from natural herbs used in Ayurvedic, 46% in medicines of Unani, and 33% used in allopathic systems of medicine (Baragi, Patgri, & Prajapati, 2008). British Pharmacopeia comprises about 50% of drugs derived from medicinal plants growing in the Indian subcontinent. In the Indian subcontinent, folk and traditional system of medicines make use of approximately 25,000 plant products and their preparations for treatment of different ailments. Indian medicinal system uses around 2000 plant species in the Ayurvedic system, 1300 in medicines of Siddha, 1000 medicinal system of Unani, homeopathy makes use of around 800, Tibetan system uses 500, in contemporary 200 and folk medicine makes use of around 4500 medicinal plant species. Furthermore, the Indian system of medicine currently uses around 7500 plant species, which includes tonics; some

Phytomedicine: A Treasure of Pharmacologically Active Products from Plants
https://doi.org/10.1016/B978-0-12-824109-7.00001-7

149

are used in the treatment of malaria, to lower a body fever, sexual stimulants, in the treatment of cough and cold, to control heart diseases, and for treatment of rheumatism and kidney problems (Mukherjee & Wahile, 2006), as well as for the treatment of certain central nervous system (CNS) illnesses (Kumar & Khanum, 2012). Indian herbal medicines use either complete plant species or plant parts, which include the leaves, stem, bark, root, flower, seed, etc. Medications are also derived from a plant's secondary products namely gums, resins, and latex. In the Indian medicinal system, around 28 classes of chronic human diseases are treated by using spices, herbs, and herbal formulations (Sharma, Chandola, Singh, & Basisht, 2007). Exceptional herbal formulations, known as *Rasayana*, are used for aging retardation and rejuvenation process for stimulating longevity (Schippmann, Leaman, & Cunningham, 2002). Indian herbal industry uses about 700 plant species, 90% of which are procured from areas of wilderness. About half of the tropical forests commonly known as the treasure house of biodiversity have already been destroyed, which is the responsible factor for the various valuable medicinal plants for their extinction.

From time to time, information about plants of medicinal importance has been collected through familiarity, experimentation, and improvement. This unscientific and age-old method of obtaining acquaintance about these wild plant species does not encompass much understanding of how authentically "a cure" works. Even at present, their proper mechanism of work is still unknown on a large scale to the discipline of science, but there is no ambiguity in saying that these natural agents serve as an excellent raw material in the formulation of new medicines. These plant species of medicinal importance have an immeasurable capability to produce a plethora of secondary plant metabolites that have an extensive array of biological functions beneficial to humans (Vining, 1990). Thus, plant species are acknowledged as treasure houses for the production of subordinate secondary metabolites and are precious sources of pharmaceutical products (Siddiqui, Adhami, Saleem, & Mukhtar, 2006).

Since the last two to three decades, several organizations have introduced the broadcast of plants species aiming to find novel biologically active compounds gifted enough to be established in new remedies. This has led to the detection of thousands of phytochemicals with suppressive properties on diverse types of microbes. But it is projected that, of around 300,000 existing plant species, only about 30% have been examined chemically for pharmacological activities. Additionally, despite the superb achievements in chemo- and antibiotic therapy, only 30% of the existing 2000 ailments can be treated at present, and others can be cured with only 196 with known symptoms and some not at all. The necessity for effective, causally acting medicines to treat harmful diseases is, thus, a serious challenge, to which phytotherapy is likely to subsidize a lot.

The medicinal value of plants to humankind is recognized very well. Mankind has used many medicinally important species from old times for the treatment of various syndromes (Olalde Rangel, 2005). Plants produce complex secondary metabolites that have found numerous salutary uses in prescription. Modern medicine's initial history encompasses explanations of plant-based phytochemicals, countless of which still need to be tested for the treatment of many disarrays (Rishton, 2008). Traditional medicine denotes the application, familiarity, and belief, including an approach in integrating animal or plant-based properties in remedies, used alone or in a blend, to maintain the welfare of an individual and for averting or curing a disease as well (Pan et al., 2014).

The Himalayan region of India possesses an inordinate treasure of plants with medicinal importance, and the residents have good understanding about traditional medicinal knowledge. Garhwal Himalaya is one of the important regions of India with extraordinary floristic composition and is acknowledged to possess more than 300 documented plant species of medicinal importance (Pan et al., 2014). In Hindu culture, the ancient documentation of 67 medicinal plants is reported in the "Rigveda," an ancient book believed to be written between 4500 and 1600 BC. To India, the collection and trade of medicinal wealth is an ancient fact. There has been increasing interest during the last few decades in the traditional use and training of medicinal flora in different global regions (Rossato, Leitão-Filho, & Begossi, 1999). As per the documented reports of WHO, about 80% of the global community use traditional medicine in primary healthcare needs. Significant economic benefits are linked together in the expansion of aboriginal medicines and the use of herbal medicine for the handling of several maladies (Azaizeh, Fulder, Khalil, & Said, 2003).

Nature has bestowed medicinal plants to cure a limitless number of illnesses in humans. The analysis of extracts acquired from diverse traditional plants of medicinal importance as a novel foundations of drugs used in antimicrobial medicine has led to an accumulative awareness about the use of medicinal plants due to ample abundance and availability (Bonjar, 2004). Nowadays, excessive use of viable drugs used as antimicrobials for treatment of innumerable contagious diseases has led to multiple drug resistance (Davis, 1994). Supplemented to this problem, occasionally antibiotics are associated with hostile effects like immune system destruction, allergic reactions, and hypersensitivity on the host (Ahmad, Mehmood, & Mohammad, 1998). This condition enforced researchers to quest for some novel agents with promising agents for the control of microbial populations. Consequently, medicinal plants are the best alternative source for expansion of novel drugs for the treatment of contagious diseases caused by microorganisms (Clark, 1996; Cordell & Colvard, 2005). The screening of medicinal plants and herbal extracts in several parts of the globe have been carried out for drug development of antimicrobial agents (Chanda, Kaneria, & Nair, 2011; Nair, Narasimhan, Shiburaj, & Abraham, 2005). Chemical substances present in medicinal plants are responsible for certain bodily physiological actions in humans that include the substances alkaloids, tannins, flavonoids, and phenolic compounds (Edeoga, Okwu, & Mbaebie, 2005).

Cancer is a chief community health problem in both developing and developed economies. It was assessed that cancer caused at least 6.7 million deaths,

10.9 million fresh cases, and still 24.6 million individuals are living with carcinoma-like disease (Parkin, Bray, Ferlay, & Pisani, 2005). For prevention of this dreadful disease, the use of medicinal plant products is of current interest (Bauer & Mushler, 2007).

Regarding genetic resources, India is one of the richest countries with an emporium of medicinal plants in the world. It displays a varied choice in climate and topography, which has a great influence on floristic and vegetation configuration. Also, for domesticating and introducing new exotic plant varieties, the agroclimatic circumstances are encouraging (Martins et al., 2001). In various parts of the world, different medicinal products are used to cure numerous severe infectious diseases in humans (Caceres, Cabrera, Morales, Mollinedo, & Mendiab, 1991; Nweze, Okafor, & Njoku, 2004). The use of medicinal products and the plants they are derived from is well recognized as folk medicine in the countryside areas of many developing economies (Sandhu & Heinrich, 2005). Mitscher, Drake, Gollapudi, and Okwute (1987) reported that plants of higher order and microorganisms are a probable birthplace for developing new sources of natural antimicrobial agents (Zaika & Smith, 1975). Medicinal plant research overcomes the inadequacies of accessible drugs to boost the unearthing of novel pharmacotherapeutic means in the modern market (Cordell & Colvard, 2005).

2. Plant active ingredients for treatment

In traditional medicine, in the treatment of diseases globally, medicinal plants are of promising interest. As reported by WHO, more than 80% of inhabitants in the developing world depends on folk medicine for basic healthcare facilities. More than 50% of the global populace depends completely on medicinally important wild flora, as most significant and vigorous constituents of these old-style medical products are offered by plant species (Kumar & Navaratnam, 2013). The principal organ in the human body is the skin that forms a principal defense contour. Its three chief layers are the upper epidermis, middle dermis, and subcutaneous hypodermis. In skin homeostasis, individually each layer provides a unique role. The thickness of these layers differ from person-to-person all over the body (Tabassum & Hamdani, 2014). Microorganisms, harmful mechanical and chemical means, and resistant/autoimmune reactions exacerbate cutaneous swelling. It is a complex practice through which the body protects itself against harmful stimuli and repairs tissue damage. Indications like soreness, inflammation, burning, heat, and pain distinguish inflammation (Ikeda et al., 2008). Patients with atopic dermatitis have an elevated value of water loss through the epidermis, a condition called transepidermal water loss (TEWL), which is prominent in persons residing in zones of dry skin. Clinically, healthy skin is not affected by inflammation, which is linked to the loss of natural moisturizing factor (NMF), constituents, and depressed concentration of skin lipids in the body, predominantly those of ceramides (Murata, Gallese, Kaseda, & Sakata, 1996). For correction of atopic dermatitis, plant ingredients with anti-inflammatory properties and lipid normalizing capacity are used.

The readings of Pan et al. (2014) deals with a systematic assessment regarding the noteworthy influence to promotion of health in currently burgeoning populations and aging civilizations of history and status quo of herbal medicines like that of Chinese, Indian, and Arab cultures. Some scholars like Rahman, Mossa, Al-Said, and Al-Yahya (2004), Gamal, Khalifa, Gameel, and Emad (2010), and Daur (2015) have established that regular stress circumstances of drought and heat of Saudi Arabia are deliberated as constructive factors for medicinal plants growing in these regions of the world for treatment of eczema, like many skin diseases (Dawid-Pać, 2013; Khiljee et al., 2011; Radha & Laxmipriya, 2015; Tabassum & Hamdani, 2014; Zari & Zari, 2015).

3. Plants as sources of antimicrobial agents

To antifungal agents, the incidence of resistance has been considerably amplified in the past decade. The resistance of a particular community to antifungal medicines has important implications for indisposition, death, and healthcare facilities. Mushrooms (fungi) were not documented as significant disease-causing agents until 1950–70 because the mortality rate due candidiasis was stable from year to year (Wey, Mori, Pfaller, Woolson, & Wenzel, 2013). This mortality rate augmented considerably since 1970, owing to prevalent use of immune-suppressive agents for treatments of fungal diseases, the public use of indwelling circulatory maneuvers, immune-suppressive contagions such as AIDS, and extensive use of broad-spectrum antimicrobial agents. The connected intensification in fungal toxicities and these developments (Beck-Sagu, Jarvis, & System, 1993; Beck-Sague, Baneqee, & Janvis, 2000) demanded the hunt for harmless, new, and more potential means to fight these grave microbial and fungal toxicities. For treatment of these grave infections, amphotericin B, was the sole drug accessible for nearly 30 years, which was significantly responsible for nephrotoxicity. In the United States alone, fluconazole was the only used drug to treat more than 16 million people, which includes over 300,000 AIDS patients (Schulman, Kaiser, Polinsky, Srinivasan, & Baluarte, 1988). The widespread use and selective pressure of these antifungal drugs led to increased resistance of these few antifungal remedies (Rex, Rinaldi, & Pfaller, 1995).

Plants with medicinal importance are a good source with an extensive variety of bioactive complexes used broadly as pure amalgams or as raw material for the management of various diseases for many centuries. Of the estimated 250,000 to 500,000 plant species inhabiting the Earth, nearly 1%–10% are used by people (Borris, 1996). These plants possess an enormous array of metabolites of a secondary order, for choosing the molecule of anticipated bioactivity and relatively cheap source of biomaterials. One biologically active compound isolated of plant origin, namely hydroxydihydrocornin aglycones, 2-decanone (Lee, Park, Lee, Kim, & Park, 2007), various indole derivatives (Ruszkowska, Chrobak, Wróbel, & Czarnocki, 2003), and isoflavones are stated to possess antifungal activities.

Plants have limitless capacity to produce compounds of an aromatic nature with diverse functional groups, maximum of which include phenols or their oxygen-substituted by-products (Geissman, 1963). Out of 13,000 compounds, most of which include secondary metabolites, less than 10% of the total have been isolated (Schutzki & Cregg, 2007). These substances in plants act as defense molecules against insects, microbes, and herbivores. The plants endowed with various metabolites with medicinal properties like terpenoids (capsaicin from chili), quinones, and tannins, etc. are used for their flavor. Most of the spices derived from plants and herbs produce useful compounds with immense medicinal importance and are used as food stuffs by humans.

4. Herbal medicine

A plant part or entire plant used for its odor, aroma, or healing properties is known as an herb. Herbs with medicinal importance are used by humans to sustain or increase their status of health. Herbal medicines are dietary supplements used as pills, capsules, concentrates, extracts, brews, and desiccated or fresh plant species and their products. Herbal medicine, also known as phytomedicine or botanical medicine, involves the use of plant parts such as seeds, rhizomes, grasses, twigs, berries, or flowers or a whole plant for the purpose of medicine.

The term "traditional medicine" refers to acquaintance, abilities, and observations based on concepts, philosophies, and individual knowledge native to diverse cultures used to sustain well-being, avert spread of disease, reduce disease, or complete treatment in both physical and mental health. In both underdeveloped and developed countries, at least 75%–80% of people trust the remedies of herbal origin for their healthcare, as it is the most prevalent and oldest form of medicine with a long history. The evidence is communicated generation to generation. Therapeutic response of herbs vary from plant to plant (Khiljee et al., 2011). The use and study of plants for medicinal purposes is called herbology, herbalism, or herbal medicine. Throughout human history, the use of plant species form a chief source for medical treatments, and such conventional medicine is still extensively practiced at present. Outside conventional medicine, herbal medicines are also used for traditional use. It is becoming more normal as developments in quality control and analysis, along with progresses in medical investigation, represent the value of herbal medicine in disease prevention and treatment.

Herbalism as a form of substitute medicine is acknowledged by the present system of medicine, as the practice of using different plants for purpose of medicine is not sternly grounded on suggestion assembled through scientific technique. Use of plants and their derived compounds, however, as contemporary medicine as the basis for evidence-based therapeutic drugs, and plant-based therapy works to muddy current criterions of efficiency testing for medicines and herbs from natural environments. Sometimes the opportunity of herbal medicine is stretched to include fungal diversity and products obtained from bees, shells from aquatic animals, minerals, and some parts of animals.

5. Herbal medicine history

Plants have been used for medicinal devotions long before the documented past. Egyptian papyrus and ancient Chinese literature describe the use of medicinal plants before 3000 BC. The use of medicinal plants has been documented as archeological evidence dates to at least 60,000 years ago in the Paleolithic era. The written evidence of herbal remedies provided by Sumerians dates to over 5000 years ago, who first shaped lists of medicinal plants. The evidence on the use of medicinal plants for various purposes is also mentioned in several prehistoric cultures. Herbs are mentioned in Egyptian medical papyri, and ancient Egypt depicted illustrations on rare plant species in tombs found in homeopathic vessels comprising trace quantities of herbs. The Ebers Papyrus, a medical papyri of ancient Egypt dating from about 1550 BC, is the ancient, longest, and most significant papyri, encompassing more than 700 medicines of plant origin (Atanasov et al., 2015). The initially known Greek herbals were identified as early as in 4th century B.C. appearing in *Historia Plantarum* in Greek by Theophrastus of Eresos, a Krateuas writer in the 1st century B.C., and of Diocles of Carystus, who wrote in the 3rd century B.C. Only a few remains of these works have persisted intact, but from what remains, scholars have reported a large extent of connection with the herbals of Egypt (Robson, Dodd, & Thomas, 2009). The history of seeds used for herbalism has been found in the archeological sites of Bronze Age and in China socializing from the Shang Dynasty. In the *Huangdi Neijing* documents at least 224 drugs, in which over a hundred cited in an early Chinese medical literature are herbal products. In ancient Indian medicine, herbs were also mentioned very commonly, where the primary management for diseases was nutrition. In both Western and Eastern medical literature, the certification of herbs and their uses was an essential part through to the 1600s, and these works occupied a noteworthy part in the advancement of botany as a science.

Morphine, the first medicinally active and pure compound, was produced some 200 years ago, from the seed extract of opium from poppy *Papaver somniferum* pods. The drugs obtained from plants shown by this innovation can be refined and administered in fixed amounts irrespective of the age and material source (Hartmann, 2007; Rousseaux & Schachter, 2003). The discovery of penicillin enhanced this approach (Li & Vederas, 2009). Products derived from natural sources and plants (marine microorganisms and fungi) or analogs stimulated by them with this sustained trend have donated significantly to viable drug findings at present. Noteworthy to mention are antibiotics like erythromycin and penicillin; salicylic acid, an antecedent of aspirin, resulting from bark of a willow tree (*Salix spp.*); the cardiac intoxicant dioxin from *Digitalis purpurea*; an antimalarial drug "quinine" from the bark of *Cinchona* plant; lipid-controlling agents (lovastatin) from a yeast and reserpine from *Rauwolfia* spp. *Viola odorata*, a drug with antipsychotic and antihypertensive properties (Feher & Schmidt, 2003; Li & Vederas, 2009; Rishton, 2008).

Herbs in their remedial sacraments were used by aboriginal beliefs (African and Native American), while herbal therapies were used in other advanced and conventional systems of medicine (Traditional Chinese Medicine and Ayurveda). People residing in different parts of the globe, as established by researchers are inclined to use similar plants for identical determinations. Plants are a repository of compounds that include aromatic substances or secondary metabolites, which include phenols or their substituted oxygenated derivatives such as tannins (Hartmann, 2007). Not only when plant ingredients are used openly as healing agents, ethnobotanicals prove to be significant for medicinal investigation and drug advance, and also as preliminary resources as replicas for pharmacologically active complexes or the synthesis of new drugs (Li & Vederas, 2009). In testing or on the market, more than 60% of cancer therapeutics are natural plant or animal products. Of the 177 drugs approved globally for cancer treatment, 70% or more are produced from natural origin or mimetics, most of which are developed with chemical combinations. From plants, the medicines used for cancer therapies include camptothecin, derived from Chinese plant species popularly called as "happy tree" *Camptotheca acuminate* and used for the preparation of drugs like irinotecan and topotecan; Pacific yew tree sequestered drug paclitaxel and combretastatin, a South African bush willow derivative (Brower, 2008). About 25% of the drugs recommended globally are assessed to be clichéd of plants, and currently 121 such bioactive compounds are in use (Sahoo, Manchikanti, & Dey, 2010). United States approved the use of 13 drugs derived from natural products between 2005 and 2007. In medical studies, more than 100 drugs are of natural derivation (Li & Vederas, 2009), and out of a total 252 drugs in the WHO's indispensable prescription list 11% are exclusively plant-based (Sahoo et al., 2010).

6. Herbal medicine nowadays

Of the world's total populace, 60% rely on conventional medicine and 80% of the population of developing nations almost completely depend on old-age medical observations, in particular, plant-based medicinal system for their primary healthcare needs as reported by WHO. Drugs approved between 1981 and 2002, which includes 60% anticancerous and 75% antiinfectious are products of natural origin according to Newman and Cragg (2012). During the same era, 61% new chemical agents introduced globally as drugs could be traced back to or were motivated by products of a natural source (Gupta, Siddique, Saxena, & Chowdhuri, 2005).

As part of human culture, herbal medicine has played a significant role. The awareness that some plants had healing properties, certainly that they possess what would be considered antimicrobial philosophies, were well acknowledged from prehistoric times when humans revealed the existence of microorganisms as cited by Ríos and Recio (2005). The incidence of compounds called secondary metabolites, which fluctuate from plant to plant, is the basis for the curative stuff of these medicinal plant species, which is typically interconnected with these secondary products. A considerable fraction (38%) of preparation has been stated to contain one or more natural plant products as the beneficial means (Nanos & Ilias, 2007).

As reported earlier (Rates, 2001), more than 75 pure compounds, most of which are produced synthetically and are used in modern medicine, are derived from higher plants used in herbal remedies. Extract from various parts of medicinal plants in current readings possess broad-spectrum antimicrobial activities against harmful disease-causing microorganisms (Karamenderes, Khan, Tekwani, Jacob, & Khan, 2006; Oyetayo, Oyetayo, & Ajewole, 2007; Sudhakar, Kuotsu, & Bandyopadhyay, 2006). Herbal drugs are easily available to most people. Of the total population of developing world economies, about 60%–85% of the inhabitants trust herbal medicine or aboriginal forms of prescription. The problematic resistance toward different antibiotics in developing countries and high effective cost are factors responsible for the high investment in herbal drugs (Hack, Hoffmann, & Nelson, 2006; Okeke, Lamikanra, & Edelman, 1999). Worldwide biochemists, botanists, pharmacologists, microbiologists, and chemists are at present examining herbs for prime bioactive composites that could be recognized for control of detrimental diseases (Acharya & Shrivastava, 2007).

6.1   Antimicrobial potential of herbal medicines as a source of novel drugs

In the last few decades, there has been much awareness regarding the use of natural plant harvests as cradles of novel agents for treatment of microbial diseases. From conventional medicinal plants, different extracts have been used in the treatment of these microorganisms, and the effectiveness of these conventional herbs is documented by many reports. For the development of modern medicine, plants play pivotal roles (Evans, Banso, & Samuel, 2002). Plant-based by-products, such as phenolics and polyphenolics, used as antimicrobial agents are quite interesting (Cowan, 1999). In treatments, phytochemicals and the use of diverse plant excerpts with documented antimicrobial possessions can be of inordinate prominence. Several studies over the last years have been conducted to prove such efficiency. Because of their antimicrobial characters, numerous florae have been tested, which are due to plant-produced secondary metabolic compounds. These products are known active substances, such as phenolics and tannins, compounds that form a part of vital oils (Saxena & Kulshrestha, 2016; Saxena, Saxena, Nema, Singh, & Gupta, 2013).

Problems such as resistance, mutagenesis, carcinogenic effects, and increased resistance to antibiotics are offered by agents of microbial inoculants added to food and public reliance on food manufacturing industries are the main factors to avoid chemical preservatives, which justifies the research and expansion of new nature-oriented antimicrobials (Rauha et al., 2000). In the ethnic medicinal system, like that of Ayurveda, Siddha, and Unani, systematically about 1500 plants are used in the treatment of lowering these emerging problems. However, botanists, microbiologists, ethnopharmacologists, and chemists all over the world at present are working uninterruptedly in examining the medicinal efficiency of plant-derived medicines; based on existing data, plants with medicinal importance are reasonably insufficient in spite of the gigantic number of the vegetal populations. To treat infectious diseases, drug safety is still a global issue for the drugs already in use. Nearly 2.22 million admitted patients in world hospitals had adverse drug reactions (ADR) and about 106,000 died in a particular year 1968 in the U.S. Interestingly all over the globe, there are comparatively subordinate occurrences of adverse reactions due to herbal and natural products and plant preparations that find a place in old medicinal systems from olden days associated with contemporary conventional drugs. Herbal drugs are quite encouraging taken together with their economic cost for both public consumption and national healthcare organizations (Nair et al., 2005).

Antibiotic resistance is a global concern (Westh et al., 2004). Today, in human pathogenic microorganisms, there is growing frequency of multiple drug resistances, principally due to a haphazard use of profit-making drugs of antimicrobial activity for the treatment of contagious infections. This forced researchers to explore novel materials with antimicrobial properties from various plants species with medicinal importance (Iwu, 2002). Screening of plant products and plant extracts for antimicrobial activity has revealed that a possible source of novel antibiotic models is represented by plants of higher order (Afolayan, 2003).

7. Preference for herbal drugs in modern societies

Chemical elements obtained from plants of natural origin that yield a biological action on the human body are said to be medicinally important. Of these chemicals, the most significant and biologically active compounds are secondary plant products such as tannins, flavonoids, alkaloids, and phenolic compounds. In recent years, worldwide, particularly in developing nations, excessive use of profitable antibiotics for the prevention of communicable ailments, multiple drug resistance (MDR) in human microorganisms, and plant pathogens has been described frequently (Chattopadhyay, Bhattacharyya, Medda, Chanda, & Bag, 2009). From higher plants, with regard to the current methodology in the detection of innovative anti-infective means, phytochemical investigation based on ethnopharmacological evidence is normally deliberated (Duraipandiyan, Ayyanar, & Ignacimuthu, 2006). Also, among tribal groups of developing nations, traditional medicines still plays a central role for treatment of various ailments due to inefficient primary healthcare (Ali, Jülich, Kusnick, & Lindequist, 2001; Pandey, Rastogi, & Rawat, 2013). Since ancient times, the indispensable oils, also known as volatile or ethereal oils, are the standard bioactive compounds of numerous imperative herbal preparations (Guenther, 1948). The chemical composition of these volatile oils depends on factors such as period of collection, desiccation procedure, environmental conditions, condition of storage, and methods of isolation (Magiatis, Skaltsounis, Chinou, & Haroutounian, 2002). Although resistance development cannot be stopped, mortality and healthcare costs can be reduced by suitable action using plant-based, antibiotic-resistant inhibitors (Ahmad & Beg, 2001). Although some of the chief antibacterial substances have significant disadvantages, the quest for plant-based antimicrobials has been mostly inspired in terms of an imperfect antimicrobial continuum.

7.1 Phytochemistry

The study of phytochemicals, chemicals of plant origin, is known as phytochemistry. The secondary metabolites found in plants in huge number are specifically described by the discipline known as phytochemistry. To protect against plant diseases and pest attacks, many of these are known. For human consumers, they also exhibit several protective functions.

7.2 Medicinal plants and phytochemistry

Use of floras as therapeutic means undoubtedly can be traced back to the Akkadian and Sumerian cultures. Hippocrates (460–377 BC), one of the prehistoric writers, roughly registered about 400 diverse species of plants for the purpose of medicines and designated natural products of herbal and animal ancestries. In ancient customary medicinal systems of Chinese, Ayurvedic, and Egyptian cultures, products derived from natural plants and animals have played an integral part (Nahar, Hasanuzzaman, Alam, & Fujita, 2015; Sarkar & Li, 2006). Ethnopharmacology is the science dealing with the use of local or native medicinal remedies including plants in the management of diseases. In different forms, plants are applied in such as mixtures as teas or tinctures, poultices, concoctions of different plant mixtures, or as constituent concoctions in gruels and broths run in diverse means comprising oral, nasal (smoking, breathing, or steaming), bathing or rectal (enemas), and topicals (lotions, oils, or creams). In the treatment of urinary tract infections, respiratory system, intestinal, biliary systems, and in the skin, diverse plant parts (roots, rhizomes, leaves, barks, flowers or their mixtures, and volatile oils) and plant components have been engaged as shown in Table 1 (Adekunle & Adekunle, 2009; Ríos & Recio, 2005). The whole world is now being assimilated into conventional medicine, and ethnopharmacology has been the backbone of customary medicines. Undoubtedly for the rheostat of numerous communicable diseases such as fever, gonorrhea, and tuberculosis, owing to growing inefficacy of various contemporary drugs used and an amassed cost of available medications, for preservation of individual healthiness and hygiene, an increase in resistance by numerous microorganisms to several antibiotics medicinal flora are progressively gaining recognition even among the literates in urban areas (Levy, 1998; Smolinski & Pestka, 2003;

TABLE 1 Major plant compounds and their biological activity.

Common name	Scientific name	Compound	Class	Activity	Reference
Aloe	*Aloe barbadensis*	Latex	Complex	*Corynebacterium, Salmonella, Streptococcus, S. aureus*	Martinez, Betancourt, Alonso-Gonzalez, and Jauregui (1996)
Ashwagandha	*Withania somniferum*	Withafarin A	Lactone	Bacteria, fungi	Hunter and Hull (1993)
Barberry	*Berberis vulgaris*	Berberine	Alkaloid	Bacteria, protozoa	Ghosal, Prasad, and Lakshmi (1996)
Basil	*Ocimum basilicum*	Essential oils	Terpenoids	*Salmonella*, bacteria	Hunter and Hull (1993)
Bay	*Laurus nobilis*	Essential oils	Terpenoids	Bacteria, fungi	Hunter and Hull (1993)
Black pepper	*Piper nigrum*	Piperine	Alkaloid	Fungi, *Lactobacillus, Micrococcus, E. coli, E. faecalis*	Ghosal et al. (1996)
Cashew	*Anacardium pulsatilla*	Salicylic acids	Polyphenols	*P. acnes*, Bacteria, Fungi	Himejima and Kubo (1991)
Chili peppers	*Capsicum annuum*	Capsaicin	Terpenoid	Bacteria	Cichewicz and Thorpe (1996)
Dandelion	*Taraxacum officinale*		Essential oil	*C. albicans, S. cerevisiae*	Navaro, Villarreal, Rojas, and Lozoya (1996)

Dill	*Anethum graveolens*				Navaro et al. (1996)
Garlic	*Allium sativum*	Allicin, ajoene	Sulfoxide	General	Naganawa et al. (1996)
Ginseng	*Panax notoginseng*		Saponins	*E. coli, Sporothrixschenckii, Staphylococcus, Trichophyton*	Himejima and Kubo (1991)
Green tea	*Camellia sinensis*	Catechin	Flavonoid	General, *Shigella, Vibrio, S. mutans, Viruses*	Ooshima et al. (1993)
Onion	*Allium cepa*	Allicin	Sulfoxide	Bacteria, *Candida*	Vohora, Rizwan, and Khan (1973)

Van Den Bogaard & Stobberingh, 2000). Regrettably, to meet contemporary health services, a burgeoning human population has made it virtually difficult over the biosphere, thus tapping more stresses on the use of regular herbal crops as health medicines. The potential interest in plants with antimicrobial properties has been revitalized by existing complications connected to the indiscriminate use of antibiotics and augmented occurrence of MDR pathogen strains (Voravuthikunchai & Kitpipit, 2005).

Biologically important chemical compounds occurring naturally in plants provide numerous health benefits for humans and animals in addition to macronutrients and micronutrients, which are regarded as phytochemicals (Greek word phyto, meaning plant) (Hasler & Blumberg, 1999). Phytochemicals are classified by physical features, protective roles, and chemical features, and on the basis of these features, about 4000 phytochemicals have been classified (Mendoza & Silva, 2018). Phytochemicals in plants perform diverse functions; they give color, fragrance, and taste to the plants and defend plants from injuries and diseases caused by different microorganisms. In general, phytochemicals are important plant substances that shield plant species from hazardous environmental conditions such as drought, UV radiation exposure, disease attack, and stress caused by these agents (Gibson, Wardle, & Watts, 1998). Plant parts such as rhizomes, flowers, fruits, roots, stems, leaves, or seeds are important repositories for accumulation of phytochemicals (Costa, Xia, Davin, & Lewis, 1999). These phytochemicals in many plants are concentrated in the surface layers of several tissues, particularly those of pigment molecules. Phytochemicals fluctuate from species to species, depending on the plant diversity, method of food preparation, processing, and environments of growth (King & Young, 1999). These plant-derived chemicals are also available in additional forms, but due to lack of information, it is unknown whether will they offer same health assistance to humans as dietary materials. These mixtures of medicinal importance have immense biological properties as they help in the modulation of detoxification enzymes, have potential antioxidant and antimicrobial activities, decrease platelet clumping, strengthen the immune system, help in the inflection of hormone metabolism, and are used in the treatment of cancer. These properties are possessed by a particular group of plant medicines known as secondary plant metabolites. More than a thousand known chemicals are so far reported, but some are still unknown to humans. Plants using these secondary metabolites for their survival under adverse environmental conditions are well known, but many phytochemicals can also be used for treatment of human diseases as demonstrated by recent research.

Because of the wide diversity and depending on their role the exact classification of these phytochemicals is still lacking. Conversely, depending on their role performed in plant metabolism, phytochemicals may be either primary or secondary constituents. The amino acids, proteins, sugars, pyrimidines, purines of nucleic acids, and chlorophyll constitute the primary phytochemicals. The remaining ones, which include compounds like phenols, alkaloids, saponins, tannins, steroids, glucosides, lignans, etc., constitute the secondary phytochemicals (Gregory, 1998). Phenolics constitute the most important, numerous, and structurally diverse group of secondary plant phytoconstituents as suggested by current literature.

7.3 Conveyed medicinal properties of secondary herbal metabolites

In the case of herbs with medicinal properties, the presence of bioactivity of herbal secondary metabolites often determines the medicinal and pharmacological actions (Afolayan, 2003; Ahmad et al., 1998; Iqbal, Aref, & Khan, 2010; Kurkin, 2013; Salim, Chin, & Kinghorn, 2008). The medicinal property of these secondary plant metabolites is reported only in a few plant species (Salim et al., 2008). These chemical compounds are often used against disease-causing pathogens and herbivores as defense molecules. Some of these secondary herbal metabolites include alkaloids, terpenoids, waxes, fatty acids, phenolics, glycosides, and their derivative products (Eloff, 2001; Nazemiyeh et al., 2008). The brief description of some secondary herbal metabolites is outlined in the following text:

(I) *Phenolics* are a group of herbal secondary metabolites that are recognized due to the occurrence of one or more than one hydroxyl (−OH) group committed to a ring of benzene or with composite aromatic assemblies (Harborne & Williams, 2001; Salim et al., 2008). Phenolic secondary metabolites are accountable for development of color, fertilization, shielding against UV radiation and microorganisms, and are normally distributed in herbs (Salim et al., 2008). They also donate acidity and color to some food stuffs. Based on construction, phenolics can be grouped into two broad categories, viz., flavonoid-based and nonflavonoid-based phenolic complexes (Salim et al., 2008). Eugenol, hydroquinone, catechol, phloroglucinol hydroquinone, and p-anisaldehyde are a group of nonflavonoid phenolic compounds (Jadhav, Mallikarjuna, Rathore, & Pokle, 2012) as shown in Fig. 1. The C6-C3 phenylpropanoids and their by-products includes cinnamic acid, caffeic acid, ferulic acid, myristicin, and sinapoyl alcohol; the C6-C1 benzoic acids are vanillic acid, gallic acid, and protocatechuic acid; commonly used coumarin scopoletin; warfarin and dicoumarol; hydrolyzabletennis namely Gallo tannins and ellagitannins; and lignans and allied complexes (Kumar & Navaratnam, 2013). Flavones include six major groups of flavonoids, which

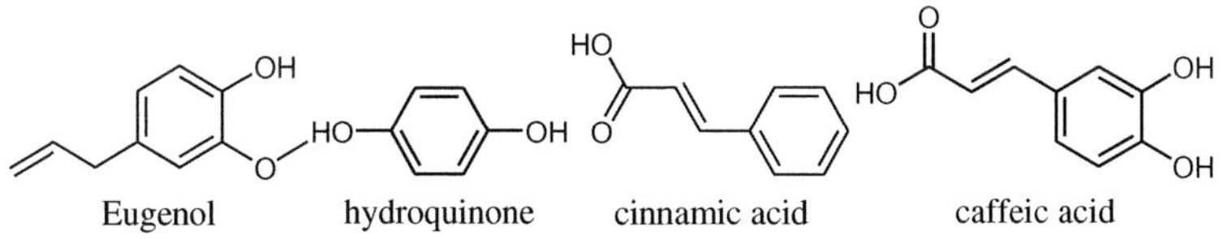

FIG. 1 Structures of some important phenolic compounds.

include flavonols, flavanones, catechins (flavanols), anthocyanidins, and isoflavones (Brahmachari & Gorai, 2006; Pietta, 2000; Scalbert, Manach, Morand, Rémésy, & Jiménez, 2005). The potential medicinal properties of these flavonoids include antioxidant, anti-inflammatory, anticancer, antibacterial, and antiviral (Cheynier, 2005; Harborne & Williams, 2001; Hollman & Katan, 1999; Manach, Scalbert, Morand, Rémésy, & Jiménez, 2004; Valsaraj, Pushpangadan, Smitt, Adsersen, & Nyman, 1997).

(II) *Alkaloids* are often characterized by the presence of one or multiple rings of carbon atoms, typically with an atom of nitrogen in the loop as shown in Fig. 2. Several alkaloids have established pharmacological activity (Harborne & Williams, 2001). The earlier reports suggest that medicinal herbs with several alkaloids were used by ancient man as recreational stimulants, pain killers, or in religious rituals to achieve a spiritual state to communicate with dynasties or divinity (Gurib-Fakim, 2006; Sandhu & Heinrich, 2005). Because of a strong bitter taste and toxic nature, most alkaloids are used by plants to defend themselves against pathogenic attacks, herbivore animals, and pest infusion (Harborne & Williams, 2001). On the basis of its ring system, alkaloids are categorized into numerous groups (e.g., isoquinoline, imidazole, atropine, indole, quinoline, piperidine alkaloids), plant-based (e.g., opium, belladonna, vinca, cinchona, and ergot alkaloids), or by their medicinal properties (e.g., analgesic, stimulant, or antimalarial alkaloids) (Harborne & Williams, 2001; Jones & Kinghorn, 2012; Sandhu & Heinrich, 2005) (see Table 2).

(III) *Glycosides* are comprised of two components: glycone, a component of carbohydrate, and a noncarbohydrate constituent called aglycone, which are herbal secondary metabolites. The aglycone may be among one of the secondary metabolites while the glycone constituent typically comprises of one or more units of glucose (Gurib-Fakim, 2006; Sandhu & Heinrich, 2005). The nature, number, and types of molecules of sugar attached to the aglycone are the three main factors that determine the solubility of glycosides (Starmans & Nijhuis, 1996). Aglycones are soluble in organic solvents and sugar components in aqueous solvents. In particular, glycosides can be removed with ethanol, acetone, or a mixture of aqueous/ethanol (Jones & Kinghorn, 2012). The glycosides of medicinal importance consist of coumarin glycosides, anthraquinone glycosides, and steroidal (cardiac) glycosides (Fig. 3).

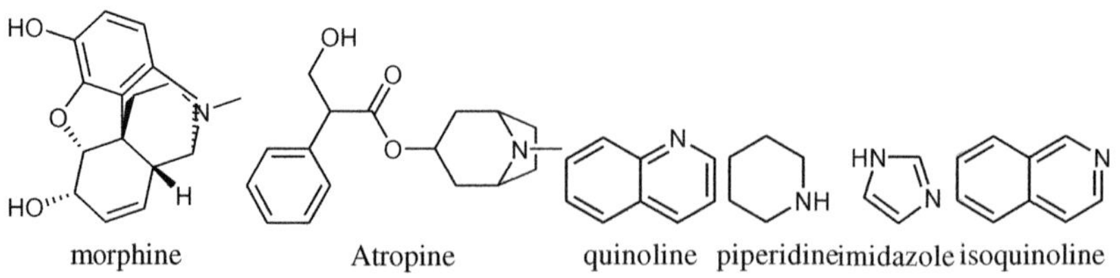

morphine   Atropine   quinoline piperidine imidazole isoquinoline

FIG. 2 Chemical structures of some important alkaloids.

TABLE 2 Major groups of antimicrobial compounds obtained from plants and their mechanism of action.

Class	Examples	Mechanism	Reference
Phenolics	Catechol Epicatechin Abyssinone	Substrate deprivation Membrane disruption Inactivate enzymes Inhibit HIV reverse transcriptase	Peres, Monache, Cruz, Pizzolatti, and Yunes (1997) Taniguchi and Kubo (2000) Ono, Nakane, Fukushima, Chermann, and Barre-Sinoussi (1989)
Terpenoids	Capsaicin	Membrane disruption	Cichewicz and Thorpe (1996)
Alkaloids	Berberine and Piperine	Intercalate into the cell wall and/or DNA	Atta-ur-Rahman and Choudhary (1995)
Lectins and polypeptides	Mannose-specific agglutinin Fabatin	Block viral fusion or adsorption Form disulfide bridges	Meyer, Afolayan, Taylor, and Erasmus (1997)
Polyacetylenes	8S-Heptadeca-2(Z),9(Z)-diene-4,6-diyne-1,8-diol	Enzyme inhibition Substrate deprivation	Estevez-Braun, Estevez-Reyes, Moujir, Ravelo, and Gonzalez (1994)

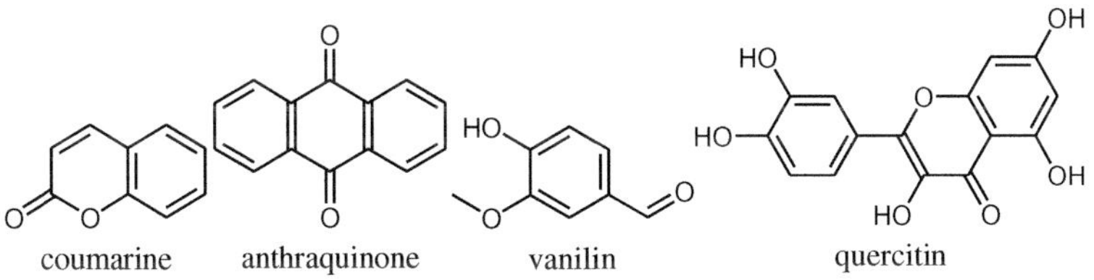

FIG. 3 Chemical structures of some important glycosides.

    coumarine   anthraquinone   vanilin   quercitin

(IV) *Saponin* is regarded as a high molecular weight compound in which, when a sugar molecule is united with triterpene or steroid aglycone, a high molecular weight compound is obtained, which is regarded as saponin. In saponin, the sugar is usually attached at C-3, and in maximum sapogenins, a hydroxyl group is attached at the C-3 position (Fig. 4). Saponin is a plant material of toxic nature, which proliferates in soapwort and makes foam when shaken with water, thus it is used as soap. Therapeutically, saponins are very essential because of

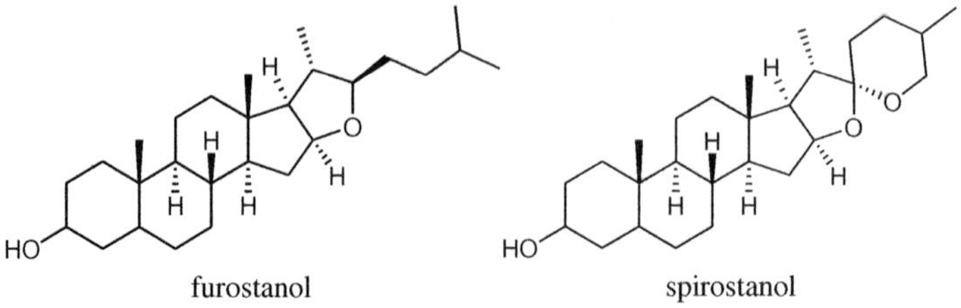

FIG. 4 Chemical structures of furostanol and spirostanol saponins.

their hypolipidemic and anticancer action. Saponins are also essential for the action of cardiac glycosides, and upon hydrolysis, an important product is obtained, an aglycone-shaped compound called sapogenin. The saponins so produced are of two categories recognized as triterpenoids and steroidal saponins. Saponins are enormously toxic, as they cause blood hemolysis and are acknowledged to cause poisoning in cattle (Kar et al., 2007; Karamenderes et al., 2006; Schulte-Elte et al., 1978).

(V) *Tannins* are phenolic compounds extensively distributed in plant flora with a high molecular weight that have a distinguishing feature to tan, i.e., conversion of things into the leather. They are acidic in nature, and the acidity to these compounds is accredited to the presence of carboxylic or phenolic groups (Kar et al., 2007). Tannins are soluble in water and alcohol and are mainly found in plant parts like rhizome, bark, stem, and outer tissues of plants. Tannins are classified into two types called hydrolyzable and condensed tannins. Tannins that produce ellagic and gallic acids on hydrolysis are called hydrolyzable (Fig. 5). Based on acid produced by tannins, they are recognized as ellagitannins or gallo tannins. These produce pyrogallic acid upon heating. Examples of some common hydrolyzable tannins include theaflavins (from tea), daidzein, genistein, and glycerin. Due to the presence of phenolic groups, they are used as antiseptics. Proteins, carbohydrates, gelatin, and alkaloids are the important compounds with which tannins form important multiplexes. Medicinal plants that serve as repositories of tannins act as healing agents in numerous ailments. In Ayurveda, diseases like leucorrhoea, rhinorrhea, and diarrhea have been treated with tannin-rich plant formulations.

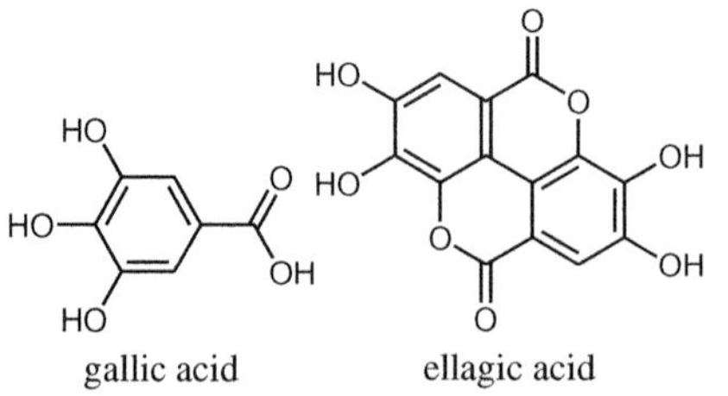

FIG. 5 Chemical structures of gallic acid and ellagic acid tannins.

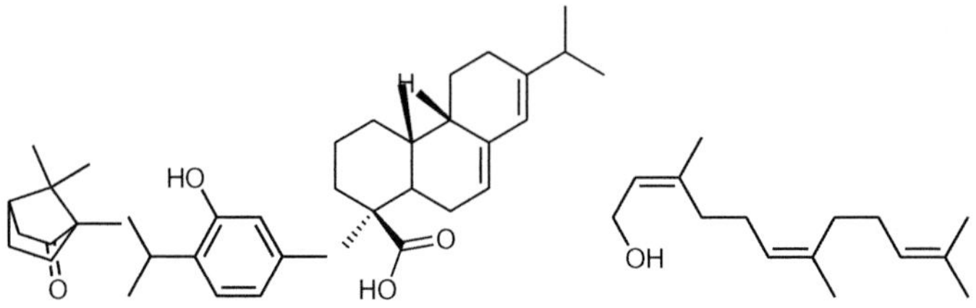

camphor   thymol   abietic acid   farnesol

FIG. 6   Chemical structures of some important mono- and diterpenoids.

(VI) *Terpenoids* are the largest assembly of herbal secondary metabolites constituted by isoprenoid family known as terpenoids. Terpenoids are convoluted in the fertilization of seed crops as well as in wound healing, resistance, and heat tolerance in plants (Sandhu & Heinrich, 2005). In quality of agricultural products, fruit flavors, and flower fragrances, the role of terpenoids can no longer be challenged. Terpenoids are classified as monoterpenes (C10), sesquiterpenes (C15), diterpene (C20), triterpenes (C30), and tetraterpenes (C40) based on some isoprene units (Fig. 6) (Alarcón, Pardo-De-Santayana, Priestley, Morales, & Heinrich, 2015; Gurib-Fakim, 2006; Sandhu & Heinrich, 2005). The chief components of important essential oils are the monoterpenes and sesquiterpenes normally found in families of plants like Labiatae, Myrtaceae, Pinaceae, and Rutaceae (Harborne & Williams, 2001; Sandhu & Heinrich, 2005). Gibberellins and resin acids are the diterpenes and are included in the group of plant hormones (Harborne & Williams, 2001). Many diterpenes are toxic, but the role played by some diterpenes like taxol, forskolin, and ginkgolides obtained from different plants used in cancer treatment, hypertension, and memory loss in contemporary medicine can no longer be shielded (Gurib-Fakim, 2006; Sandhu & Heinrich, 2005). Triterpenoids are a group of terpenes and are most abundantly derived from plants, forming essential components of saponins and steroidal glycosides, and are included as plant steroids (Harborne & Williams, 2001).

8.   Phytochemicals: The mechanism of action

Phytochemicals hamper with some essential metabolic processes, inhibit microorganisms, and may control the expression of specific genes and corridors of signal transduction (Kris-Etherton et al., 2002; Manson, 2003; Surh, 2003). Phytochemicals may either be used as chemopreventive or chemotherapeutic agents. Chemoprevention denotes the application of phytochemical agents to inverse, deter, or delay the incidence of tumors in humans or animals. Meanwhile, molecular contrivances may be similar in chemoprevention and cancer treatment; in cancer therapy, chemopreventive phytochemicals are widely used (D'Incalci, Steward, & Gescher, 2005; Sarkar & Li, 2006). Plant extracts and essential oils exhibit different approaches of action against bacterial strains,

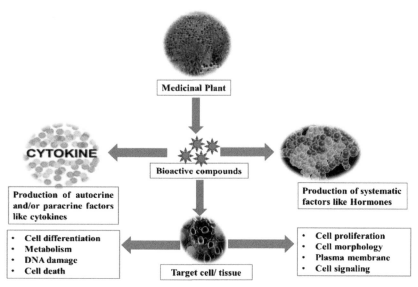

FIG. 7 Diagrammatic representation of toxic action of medicinally active compounds.

such as interference with cell membrane phospholipid bilayers, which proliferates the water permeability and helps in the impairment enzymes; damage of cellular components involved in the production of structural constituents; and cellular energy fabrication and suppression or annihilation of hereditary material (Fig. 7). Likewise, some other mechanistic pathways involved are a disturbance of the proton motive force, the disorder in cytoplasmic membrane integrity, active transport of materials, the flow of electrons, and cell content coagulation (Kotzekidou, Giannakidis, & Boulamatsis, 2008) (see Table 2).

9. Prospects of phytochemicals as a repository of antimicrobial and chemotherapeutic representatives

Generally, with natural product research, a few disadvantages are always associated. These include the inherent slowness of working with products of natural origin, problems in supply and access, and complex chemistry of these natural products. Despite these drawbacks, about 100 analogous schemes are in the infancy of development and about 100 natural product-based complexes are presently experiencing clinical judgments (Phillipson, 2007). The highest quantities of these products are contributed by natural plant diversity and others are produced with the help of microbial and animal sources. The schemes are principally being premeditated for their use in the treatment of various infections or cancer with nature-based products. With the global emergence of multidrug-resistant microorganisms and the discovery of successful drugs other than the multiplicity and assortment of synthetic compounds, the chemical diversity of natural products is an excellent match (Feher & Schmidt, 2003). Once again, interest in applying natural chemical diversity to drug discovery seems to be growing (Galm & Shen, 2007).

With progress in the development of techniques of fractionation to refine and separate natural products, e.g., countercurrent chromatography (Doughari, Human, Bennade, & Ndakidemi, 2009) and to elucidate structures in analytical methods (Sahoo et al., 2010), the anticipated timescale of high-throughput broadcast movements are now more attuned to a screening of natural product combinations. Sahoo et al. (2010) reported that it takes less than 2 weeks for pure and biologically active compounds to be sequestered from fermentation broths, and within 2 weeks about 90% or more new compound configurations can be elucidated. With the progress in NMR techniques, with less than 1 mg of parent material, these structures can be cracked. It is promising to organize a broadcast library of extremely varied complexes from plants as recently demonstrated with the compounds being selected previously from an examination of the Dictionary of Natural Products to be drug-like in their physicochemical assets (Doughari et al., 2009; Oleszek & Marston, 2000). Despite the popularity in chemical drugs, owing to an abundance of secondary metabolites such as alkaloids, tannins, flavonoids, and terpenoid-assured plants variations, herbs with medicinal properties in Africa and the rest of the biosphere continue to be experienced (Adekunle & Adekunle, 2009; Cowan, 1999; Lewis & Ausubel, 2006).

In current years, various scientific medical/pharmaceutical books have been printed aiming to provide the general public and healthcare professionals with confirmation of the welfare and dangers of medicines derived from different herbs of natural origin (Akhtar, Ihsan-ul-Haq, & Mirza, 2018; Phillipson, 2007). On specific medicinal herbs, pharmacopeia of European has circulated up to 125 articles with an additional 84 presently in groundwork, to provide up-to-date information of phytochemicals; the books are meant for describing the chemical outlines of medicinal herbs and thoughtful documentation of analytical tests of the herbs and any known potential components for numerical valuation (Phillipson, 2007). Numerous governing bodies including Medicines and Healthcare products Regulatory Agency (MHRA), Traditional Medicines Boards (TMBs), American Herbal Products Association (AHPA), and Herbal Medicines Advisory Committee (HMAC) (UK) and several other pharmacopeias (British, Chinese, German, Japanese) in some states provide information and strategies on the quality control, security levels, and consumption of herbal products (Yadav & Dixit, 2008). In the bodies of higher learning that deal explicitly with an examination in herbal drugs, scientific and research groups are presently involved in research of different phytochemicals and phytomedicine. In fruits, vegetables, and grains, it is projected that more than 5000 different phytochemicals are known, but a huge proportion remains unidentified and needs to be recognized before we can entirely fathom the health profits of phytomedicine (Liu, 2004).

10. Natural products: The future scaffolds for novel antimicrobials

In the past few years, global intensification in the rate of fungal infections have been detected and augmented in the resistance of some fungal species to diverse fungicides used in remedial practice. Fungi are one of the most ignored pathogens, as confirmed by the information that amphotericin B, a long-ago discovered polyene antibiotic in 1956, was quietly used as a "gold standard" for antifungal remedy. The last 20 years have observed a melodramatic increase in the frequency of life-threatening infections caused by systemic fungal strains. The challenge is to develop effective approaches for the treatment of fungal maladies, bearing in mind the increase in unscrupulous fungal toxicities in human immunodeficiency virus-positive patients and in those who are immunocompromised owing to chemotherapy for cancer treatment and the excessive use of antibiotics. The bulk of clinically tested antifungals have innumerable disadvantages in terms of harmfulness, effectiveness, and economic cost, and their recurrent use has led to the development of resistant fungal strains. The progression of resistance to antimicrobial drugs for control of different pathogens in medicine is a familiar problem. One longstanding apprehension is that the number of primarily different kinds of antifungal agents accessible for treatment remains exceptionally inadequate. Humans and their associated fungal pathogens belong to sister clades in the family tree, and possible drug targets that are exclusive and vital to the parasitic fungus, but not to the host, are very scarce. Resistance to dissimilar antimicrobial drugs has considerably amplified in the past years. Present-day antimicrobial means have essential consequences for indisposition, impermanence, and healthcare for the public. Originally, fungi were not documented as a significant disease-causing kingdom because the annual death rate due to candidiasis was very stable from 1950 to 1970 (Wey et al., 2013). From 1970 and onward, there was a dramatic increase in this rate due to the extensive use of immunosuppressive remedies, excessive use of broad-spectrum antibacterial drugs, the conjoint use of indwelling circulatory devices, and immunosuppressive virus-related contagions such as AIDS. These expansions and the accompanying intensification in both microbial and fungal infections (Beck-Sague et al., 2000; Beck-Sague, Banerjee, & Jarvis, 1993) demanded exploration of novel, safer, and potential drugs to combat grave contagions. AmB, which is extremely nephrotoxic, was the only drug accessible for about 30 years in the treatment of severe fungal infections. In the last half of the 1980s and early 1990s, the imidazoles and triazoles were chief developments in the safe and actual management of indigenous and universal fungal contaminations. Owing to the higher safety of triazoles, fluconazole specifically has led to their widespread use of antifungal remedial treatment. Fluconazole has been used to treat more than 16 million patients, comprising over 300,000 patients suffering from AIDS, in the United States only since the innovation of this antifungal prescription (Schulman et al., 1988).

Due to discriminating stress and extensive use of these antifungal medications, there have been growing accounts of antifungal resistance (Rex et al., 1995). Therefore, there is a countless demand for new antimicrobials fitting a wide range of class structures, selectively performing on new objectives with fewer side effects. One method might be the testing of traditionally used plants species for antifungal actions as probable foundations for the development of new drugs. Medicinal plants are not only significant to the increasing population for folk medicines as a sole prospect for healthcare, and to those who use medicinally important plant species for several purposes, but also as a source of new drugs. Plant products, either as unadulterated or as homogenous plant extracts, afford limitless chances for novel and leading drugs, because of the unparalleled accessibility of chemical miscellany. Medicinal plants have been a foundation of a varied range of biologically active compounds for many years and have been used comprehensively as unfinished material or as pure complexes for treating disease conditions. Comparatively, 1–10% of plant species are used by individuals out of a projected 250,000–500,000 species of plants on Earth (Schürer, 2002). Plants are a fairly inexpensive cradle of an organic material having an enormous diversity of metabolites, whether primary or secondary, for choosing the iota of anticipated biological activity. Conventional prescription is progressively amenable to the use of antimicrobial and other drugs resulting from flora, as antibiotics used in traditional medicine become unsuccessful. The alternative energetic factor for the transformed attention in plant antimicrobials in the previous 20 years is due to the reckless extinction rate of various plant species (Pan et al., 2014; Ruszkowska et al., 2003). Ethnobotany exploits the inspiring collection of data collected by native people about the plant and animal products they have used to uphold their well-being (Rojas, Hernandez, Pereda-Miranda, & Mata, 1992). Finally, the dominance of the anthropological immunodeficiency virus (HIV) has encouraged concentrated examination into plant products, which may be operative, particularly for use in weak countries.

New varieties of humanoid fungal contagions are growing due to amplified cancer and AIDS patients. The amplified use of antifungal drugs also stemmed from the expansion of conflict to these medications. It's essential to ascertain new classes of antimicrobial compounds to treat infections caused by different species of fungal and bacterial strains. These molecules or plant extracts may be applied openly or deliberated as an ideal for developing superior molecules.

11.   Conclusion

Since immemorial times, disease is a principal reason of indisposition/transience and is linked with a substantial monetary encumbrance between people with diseases. Despite present-day advances in medicine and science, disease is quite a grave hazard to community health in industrialized and emerging economies, and all social groups living in urban and rural areas. To fight illness, prehistoric and contemporary people take drugs to maintain good health in case of sickness. Present-day modern drugs are chemically manufactured, and some are acquired from wild growing plant species for use in customary treatment. Nonetheless, our descendants took only guaranteed categories of natural remedial formulations to combat or avert a particular disease, whether microbial or fungal. Because current drug development is very expensive and is associated with increased risk incidence (and hence high failure), marketable attempts and artificial drugs have an extraordinary rate of antagonistic occasions. To keep these problems of modern medicines in mind, there is a widespread requirement of consuming medicinal herbs or associated goods for the treatment of diseases without causing economic depletion and risk to human life.

References

Acharya, D., & Shrivastava, A. (2007). Etanobotany, alternative medicine and conservation of Indian medicinal plants. In *Souvenir of Indian international convention on health for all in 21st century- medicine for future, 4th–5th August. 2007, Apollo Hospital, Gandhi Nagar, Gujarat* (pp. 22–25).

Adekunle, A. S., & Adekunle, O. C. (2009). Preliminary assessment of antimicrobial properties of aqueous extract of plants against infectious diseases. *Biology and Medicine, 1,* 20–24. https://doi.org/10.4172/0974-8369.1000029.

Afolayan, A. J. (2003). Extracts from the shoots of *Arctotis arctotoides* inhibit the growth of bacteria and fungi. *Pharmaceutical Biology, 41,* 22–25. https://doi.org/10.1076/phbi.41.1.22.14692.

Ahmad, I., & Beg, A. Z. (2001). Antimicrobial and phytochemical studies on 45 Indian medicinal plants against multi-drug resistant human pathogens. *Journal of Ethnopharmacology, 74,* 113–123. https://doi.org/10.1016/S0378-8741(00)00335-4.

Ahmad, I., Mehmood, Z., & Mohammad, F. (1998). Screening of some Indian medicinal plants for their antimicrobial properties. *Journal of Ethnopharmacology, 62,* 183–193. https://doi.org/10.1016/S0378-8741(98)00055-5.

Akhtar, N., Ihsan-ul-Haq, & Mirza, B. (2018). Phytochemical analysis and comprehensive evaluation of antimicrobial and antioxidant properties of 61 medicinal plant species. *Arabian Journal of Chemistry, 11,* 1223–1235. https://doi.org/10.1016/j.arabjc.2015.01.013.

Alarcón, R., Pardo-De-Santayana, M., Priestley, C., Morales, R., & Heinrich, M. (2015). Medicinal and local food plants in the south of Alava (Basque Country, Spain). *Journal of Ethnopharmacology, 176,* 207–224. https://doi.org/10.1016/j.jep.2015.10.022.

Ali, N. A. A., Jülich, W. D., Kusnick, C., & Lindequist, U. (2001). Screening of Yemeni medicinal plants for antibacterial and cytotoxic activities. *Journal of Ethnopharmacology, 74,* 173–179. https://doi.org/10.1016/S0378-8741(00)00364-0.

Atanasov, A. G., Waltenberger, B., Pferschy-Wenzig, E. M., Linder, T., Wawrosch, C., Uhrin, P., et al. (2015). Discovery and resupply of pharmacologically active plant-derived natural products: A review. *Biotechnology Advances, 33,* 1582–1614. https://doi.org/10.1016/j.biotechadv.2015.08.001.

Atta-ur-Rahman, & Choudhary, M. I. (1995). Diterpenoid and steroidal alkaloids. *Natural Product Reports,* 361–379.

Azaizeh, H., Fulder, S., Khalil, K., & Said, O. (2003). Ethnobotanical knowledge of local Arab practitioners in the middle eastern region. *Fitoterapia, 74,* 98–108. https://doi.org/10.1016/S0367-326X(02)00285-X.

Baragi, P. C., Patgri, B., & Prajapati, P. (2008). Neutraceuticals in Ayurveda with special reference to Avaleha Kalpana. *Ancient Science of Life*, *28*, 29–32.

Bauer, W. T., & Mushler, F. G. (2007). Bone graft materials. *Dental Clinics of North America*, *51*, 729–746. https://doi.org/10.1016/j.cden.2007.03.004.

Beck-Sagu, C. M., Jarvis, W. R., & System, N. N. I. S. (1993). Secular trends in the epidemiology of nosocomial fungal infections in the United States, 1980–1990. *Journal of Infectious Diseases*, *167*, 1247–1251.

Beck-Sague, C., Baneqee, S., & Janvis, W. R. (2000). Public health briefs infectious diseases and mortality among US nursing home residents. *American Journal of Public Health*, *83*, 1739–1742.

Beck-Sague, C., Banerjee, S., & Jarvis, W. R. (1993). Infectious diseases and mortality among US nursing home residents. *American Journal of Public Health*, *83*, 1739–1742. https://doi.org/10.2105/AJPH.83.12.1739.

Bonjar, S. G. (2004). Evaluation of antibacterial properties of Iranian medicinal-plants against *Micrococcus luteus*, *Serratia marcescens*, *Klebsiella pneumoniae* and *Bordetella bronchoseptica*. *Asian Journal of Plant Sciences*, *3*(1), 82–86.

Borris, R. P. (1996). Natural products research: Perspectives from a major pharmaceutical company. *Journal of Ethnopharmacology*, *51*, 29–38. https://doi.org/10.1016/0378-8741(95)01347-4.

Brahmachari, G., & Gorai, D. (2006). Progress in the research on naturally occurring flavones and flavonols: An overview. *Current Organic Chemistry*, *10*, 873–898. https://doi.org/10.2174/138527206776894438.

Brower, V. (2008). Back to nature: Extinction of medicinal plants threatens drug discovery. *Journal of the National Cancer Institute*, *100*, 838–839. https://doi.org/10.1093/jnci/djn199.

Caceres, A., Cabrera, O., Morales, O., Mollinedo, P., & Mendiab, P. (1991). Preliminary screening for antimicrobial activity. *Journal of Ethnopharmacology*, *33*, 216.

Chanda, S., Kaneria, M., & Nair, R. (2011). Chanda 2004.pdf. *Research Journal of Microbiology*, *6*, 124–131. https://doi.org/10.3923/jm.2011.124.131.

Chattopadhyay, R. R., Bhattacharyya, S. K., Medda, C., Chanda, S., & Bag, A. (2009). A comparative evaluation of antibacterial potential of some plants used in Indian traditional medicine for the treatment of microbial infections. *Brazilian Archives of Biology and Technology*, *52*, 1123–1128. https://doi.org/10.1590/S1516-89132009000500009.

Cheynier, V. (2005). Polyphenols in foods are more complex than often thought. *The American Journal of Clinical Nutrition*, *81*, 223–229. https://doi.org/10.1093/ajcn/81.1.223s.

Cichewicz, R. H., & Thorpe, P. A. (1996). The antimicrobial properties of Chile peppers (*Capsicum* species) and their uses in Mayan medicine. *Journal of Ethnopharmacology*, *8741*.

Clark, A. M. (1996). Natural products as a resource for new drugs. *Pharmaceutical Research*. https://doi.org/10.1023/A:1016091631721.

Cordell, G. A., & Colvard, M. D. (2005). Some thoughts on the future of ethnopharmacology. *Journal of Ethnopharmacology*, *100*, 5–14. https://doi.org/10.1016/j.jep.2005.05.027.

Costa, M. A., Xia, Z.-Q., Davin, L. B., & Lewis, N. G. (1999). Toward engineering the metabolic pathways of cancer-preventing lignans in cereal grains and other crops. *Phytochemicals in Human Health Protection, Nutrition, and Plant Defense*, 67–87. https://doi.org/10.1007/978-1-4615-4689-4_4.

Cowan, M. M. (1999). Plant products as antimicrobial agents. *Clinical Microbiology Reviews*, *12*, 564–582. https://doi.org/10.1128/cmr.12.4.564.

D'Incalci, M., Steward, W. P., & Gescher, A. J. (2005). Use of cancer chemopreventive phytochemicals as antineoplastic agents. *Lancet Oncology*, *6*, 899–904. https://doi.org/10.1016/S1470-2045(05)70425-3.

Daur, I. (2015). Chemical composition of selected Saudi medicinal plants. *Arabian Journal of Chemistry, 8*, 329–332. https://doi.org/10.1016/j.arabjc.2013.10.015.

Davis, R. J. (1994). MAPKs: New JNK expands the group. *Trends in Biochemical Sciences, 19*, 470–473. https://doi.org/10.1016/0968-0004(94)90132-5.

Dawid-Paċ, R. (2013). Medicinal plants used in treatment of inflammatory skin diseases. *Postepy Dermatologii i Alergologii, 30*, 170–177. https://doi.org/10.5114/pdia.2013.35620.

Doughari, J. H., Human, I. S., Bennade, S., & Ndakidemi, P. A. (2009). Phytochemicals as chemotherapeutic agents and antioxidants: Possible solution to the control of antibiotic resistant verocytotoxin producing bacteria. *Journal of Medicinal Plants Research, 3*, 839–848.

Duraipandiyan, V., Ayyanar, M., & Ignacimuthu, S. (2006). Antimicrobial activity of some ethnomedicinal plants used by Paliyar tribe from Tamil Nadu, India. *BMC Complementary and Alternative Medicine, 6*. https://doi.org/10.1186/1472-6882-6-35.

Edeoga, H. O., Okwu, D. E., & Mbaebie, B. O. (2005). Phytochemical constituents of some Nigerian medicinal plants. *African Journal of Biotechnology, 4*, 685–688. https://doi.org/10.5897/AJB2005.000-3127.

Eloff, J. N. (2001). Antibacterial activity of Marula (*Sclerocarya birrea*) (Anacardiaceae) bark and leaves. *Journal of Ethnopharmacology, 76*, 305–308.

Estevez-Braun, A., Estevez-Reyes, R., Moujir, L. M., Ravelo, A. G., & Gonzalez, A. (1994). Activity, antibiotic configuration, absolute configuration of SS-heptadeca-2(2),9(z)-diene-4,6-diyne-1,%diol from *Bupleurum salicifolium*. *Journal of Natural Medicines, 5*, 7–11.

Evans, C. E., Banso, A., & Samuel, O. A. (2002). Efficacy of some nupe medicinal plants against *Salmonella typhi*: An in vitro study. *Journal of Ethnopharmacology, 80*, 21–24. https://doi.org/10.1016/S0378-8741(01)00378-6.

Feher, M., & Schmidt, J. M. (2003). Property distributions: Differences between drugs, natural products, and molecules from combinatorial chemistry. *ChemInform, 34*, 218–227. https://doi.org/10.1002/chin.200317217.

Galm, U., & Shen, B. (2007). Natural product drug discovery: The times have never been better. *Chemistry and Biology, 14*, 1098–1104. https://doi.org/10.1016/j.chembiol.2007.10.004.

Gamal, E. E. G., Khalifa, S. A. K., Gameel, A. S., & Emad, M. A. (2010). Traditional medicinal plants indigenous to Al-Rass province, Saudi Arabia. *Journal of Medicinal Plants Research, 4*, 2680–2683. https://doi.org/10.5897/jmpr09.556.

Geissman, T. A. (1963). *Flavonoid compounds, tannins, lignins and, related compounds, comprehensive biochemistry*. Elsevier B.V. https://doi.org/10.1016/B978-1-4831-9718-0.50018-7.

Ghosal, S., Prasad, B. N. K., & Lakshmi, V. (1996). Antiamoebic activity of *Piper longum* fruits against *Entamoeba histolytica* in vitro and in vivo. *Journal of Ethnopharmacology, 8741*.

Gibson, E. L., Wardle, J., & Watts, C. J. (1998). Fruit and vegetable consumption, nutritional knowledge and beliefs in mothers and children. *Appetite, 31*, 205–228. https://doi.org/10.1006/appe.1998.0180.

Gregory, J. F. (1998). Nutritional properties and significance of vitamin glycosides. *Annual Review of Nutrition, 18*, 277–296. https://doi.org/10.1146/annurev.nutr.18.1.277.

Guenther, E. (1948). The production of essential oils: Methods of distillation, Enfleurage, maceration and extraction with volatile solvents. In *History-origin in plants production-analysis* (pp. 87–213). London: D. Van Nostrand Company, Inc.

Gupta, S. C., Siddique, H. R., Saxena, D. K., & Chowdhuri, D. K. (2005). Hazardous effect of organophosphate compound, dichlorvos in transgenic *Drosophila melanogaster* (hsp70-lacZ): Induction of hsp70, anti-oxidant enzymes and inhibition of acetylcholinesterase. *Biochimica et Biophysica Acta-General Subjects, 1725*, 81–92. https://doi.org/10.1016/j.bbagen.2005.04.033.

Gurib-Fakim, A. (2006). Medicinal plants: Traditions of yesterday and drugs of tomorrow. *Molecular Aspects of Medicine*, *27*, 1–93. https://doi.org/10.1016/j.mam.2005.07.008.

Hack, J. B., Hoffmann, R. S., & Nelson, L. S. (2006). Resistant alcohol withdrawal: Does an unexpectedly large sedative requirement identify these patients early? *Journal of Medical Toxicology: Official Journal of the American College of Medical Toxicology*, *2*, 55–60. https://doi.org/10.1007/BF03161171.

Harborne, J. B., & Williams, C. A. (2001). Anthocyanins and other flavonoids. *Natural Product Reports*, *18*, 310–333. https://doi.org/10.1039/b006257j.

Hartmann, T. (2007). From waste products to ecochemicals: Fifty years research of plant secondary metabolism. *Phytochemistry*, *68*, 2831–2846. https://doi.org/10.1016/j.phytochem.2007.09.017.

Hasler, C. M., & Blumberg, J. B. (1999). Symposium on phytochemicals: Biochemistry and physiology: Introduction. *Journal of Nutrition*, *129*, 756–757.

Himejima, M., & Kubo, I. (1991). Antibacterial agents from the cashew. *Journal of Agricultural and Food Chemistry*, 418–421.

Hollman, P. C. H., & Katan, M. B. (1999). Dietary flavonoids: Intake, health effects and bioavailability. *Food and Chemical Toxicology*, *37*, 937–942. https://doi.org/10.1016/S0278-6915(99)00079-4.

Hunter, D. M., & Hull, A.l. (1993). Variation in concentrations of phloridzin apple foliage and phloretin. *Phytochemistry*, *34*, 1251–1254.

Ikeda, Y., Yahata, N., Ito, I., Nagano, M., Toyota, T., Yoshikawa, T., et al. (2008). Low serum levels of brain-derived neurotrophic factor and epidermal growth factor in patients with chronic schizophrenia. *Schizophrenia Research*, *101*, 58–66. https://doi.org/10.1016/j.schres.2008.01.017.

Iqbal, M., Aref, I. M., & Khan, P. R. (2010). Behavioral responses of leaves and vascular cambium of *Prosopis cineraria* (L.) Druce to different regimes of coal-smoke pollution. *Journal of Plant Interactions*, *5*, 117–133. https://doi.org/10.1080/17429140903438084.

Iwu, M. (2002). Ethanomedicine and drug discovery. In M. M. Iwu, & J. Wooton (Eds.), *Advances in phytochemistry* (pp. 280–302). Elsevier.

Jadhav, D. R., Mallikarjuna, N., Rathore, A., & Pokle, D. (2012). Effect of some flavonoids on survival and development of *Helicoverpa armigera* (Hübner) and *Spodoptera litura* (Fab) (Lepidoptera: Noctuidae). *Asian Journal of Agricultural Sciences*, *4*, 298–307.

Jones, W. P., & Kinghorn, A. D. (2012). Extraction of plant secondary metabolites. *Methods in Molecular Biology*, *864*, 341–366. https://doi.org/10.1007/978-1-61779-624-1_13.

Kar, A., Mirkazemi, R., Singh, P., Potnis-lele, M., Lohade, S., Lalwani, A., et al. (2007). Disability in Indian patients with haemophilia. *Haemophilia*, *13*, 398–404. https://doi.org/10.1111/j.1365-2516.2007.01483.x.

Karamenderes, C., Khan, S., Tekwani, B. L., Jacob, M. R., & Khan, I. A. (2006). Antiprotozoal and antimicrobial activities of *Centaurea* species growing in Turkey. *Pharmaceutical Biology*, *44*, 534–539. https://doi.org/10.1080/13880200600883080.

Khiljee, S., Rehman, N. U., Khiljee, T., Ahmad, R. S., Khan, M. Y., & Qureshi, U. A. (2011). Use of traditional herbal medicines in the treatment of eczema. *Journal of Pakistan Association of Dermatologists*, *21*, 112–117.

King, A., & Young, G. (1999). Characteristics and occurrence of phenolic phytochemicals. *Journal of the American Dietetic Association*. https://doi.org/10.1016/S0002-8223(99)00051-6.

Kotzekidou, P., Giannakidis, P., & Boulamatsis, A. (2008). Antimicrobial activity of some plant extracts and essential oils against foodborne pathogens in vitro and on the fate of inoculated pathogens in chocolate. *LWT—Food Science and Technology*, *41*, 119–127. https://doi.org/10.1016/j.lwt.2007.01.016.

Kris-Etherton, P. M., Hecker, K. D., Bonanome, A., Coval, S. M., Binkoski, A. E., Hilpert, K. F., et al. (2002). Bioactive compounds in foods: Their role in the prevention of cardiovascular disease and cancer. *American Journal of Medicine, 113*, 71–88. https://doi.org/10.1016/s0002-9343(01)00995-0.

Kumar, G. P., & Khanum, F. (2012). Neuroprotective potential of phytochemicals. *Pharmacognosy Reviews, 6*, 81–90. https://doi.org/10.4103/0973-7847.99898.

Kumar, V. S., & Navaratnam, V. (2013). Neem (*Azadirachta indica*): Prehistory to contemporary medicinal uses to humankind. *Asian Pacific Journal of Tropical Biomedicine, 3*, 505–514. https://doi.org/10.1016/S2221-1691(13)60105-7.

Kurkin, V. A. (2013). Phenylpropanoids as the biologically active compounds of the medicinal plants and phytopharmaceuticals. *Advances in Biological Chemistry, 03*, 26–28. https://doi.org/10.4236/abc.2013.31004.

Lee, H., Park, H., Lee, Y., Kim, K., & Park, S. (2007). A practical procedure for producing silver nanocoated fabric and its antibacterial evaluation for biomedical applications. *Chemical Communications*, 2959–2961. https://doi.org/10.1039/b703034g.

Levy, S. B. (1998). *The challenge of antibiotic bacteria*. Scientific American.

Lewis, K., & Ausubel, F. M. (2006). Prospects for plant-derived antibacterials. *Nature Biotechnology, 24*, 1504–1507. https://doi.org/10.1038/nbt1206-1504.

Li, J. W., & Vederas, J. C. (2009). Drug discovery and natural products. *Science, 325*, 161–165. https://doi.org/10.1126/science.1168243.

Liu, R. H. (2004). Potential synergy of phytochemicals in cancer prevention: Mechanism of action. *The Journal of Nutrition, 134*, 3479S–3485S. https://doi.org/10.1093/jn/134.12.3479S.

Magiatis, P., Skaltsounis, A. L., Chinou, I., & Haroutounian, S. A. (2002). Chemical composition and in-vitro antimicrobial activity of the essential oils of three Greek Achillea species. *Zeitschrift fur Naturforschung—Section C Journal of Biosciences, 57*, 287–290. https://doi.org/10.1515/znc-2002-3-415.

Manach, C., Scalbert, A., Morand, C., Rémésy, C., & Jiménez, L. (2004). Polyphenols: Food sources and bioavailability. *American Journal of Clinical Nutrition, 79*, 727–747. https://doi.org/10.1093/ajcn/79.5.727.

Manson, M. M. (2003). Cancer prevention—The potential for diet to modulate molecular signalling. *Trends in Molecular Medicine, 9*, 11–18. https://doi.org/10.1016/S1471-4914(02)00002-3.

Martinez, M. J., Betancourt, J., Alonso-Gonzalez, N., & Jauregui, A. (1996). Screening of some Cuban medicinal plants for antimicrobial activity. *Journal of Ethnopharmacology, 8741*.

Martins, A. P., Salgueiro, L., Gonc, M. J., Cunha, Ë., Vila, R., Caæigueral, S., et al. (2001). As part of our work on the characterisation of aromatic and medicinal plants of S. TomØ and Príncipe (1), (2), (3), we re- port here the chemical composition and antimicrobial activity of the essential oils from two samples of fruit, that is studi. *Planta Medica, 67*, 580–584.

Mendoza, N., & Silva, E. M. E. (2018). Introduction to phytochemicals: Secondary metabolites from plants with active principles for pharmacological importance. *Phytochemicals—Source of Antioxidants and Role in Disease Prevention*. https://doi.org/10.5772/intechopen.78226.

Meyer, J. J., Afolayan, A., Taylor, M., & Erasmus, D. (1997). Antiviral activity of galangin isolated from the aerial parts of Helichrysum aureonitens. *Journal of Ethnopharmacology, 56*, 165–169.

Mitscher, L. A., Drake, S., Gollapudi, S. R., & Okwute, S. K. (1987). A modern look at folkloric use of anti-infective agents. *Journal of Natural Products, 50*, 1025–1040. https://doi.org/10.1021/np50054a003.

Mukherjee, P. K., & Wahile, A. (2006). Integrated approaches towards drug development from Ayurveda and other Indian system of medicines. *Journal of Ethnopharmacology, 103*, 25–35. https://doi.org/10.1016/j.jep.2005.09.024.

Murata, A., Gallese, V., Kaseda, M., & Sakata, H. (1996). Parietal neurons related to memory-guided hand manipulation. *Journal of Neurophysiology, 75*, 2180–2186. https://doi.org/10.1152/jn.1996.75.5.2180.

Naganawa, R., Iwata, N., Ishikawa, K., Fukuda, H., Fujino, T., & Suzuki, A. (1996). Inhibition of microbial growth by Ajoene, a sulfur-containing compound derived from garlic. *Applied and Environmental Microbiology, 62*, 4238–4242.

Nahar, K., Hasanuzzaman, M., Alam, M., & Fujita, M. (2015). Exogenous spermidine alleviates low temperature injury in mung bean (*Vigna radiata* L.) seedlings by modulating ascorbate-glutathione and exogenous. *International Journal of Molecular Sciences.* https://doi.org/10.3390/ijms161226220.

Nair, G. M., Narasimhan, S., Shiburaj, S., & Abraham, T. K. (2005). Antibacterial effects of *Coscinium fenestratum*. *Fitoterapia, 76*, 585–587. https://doi.org/10.1016/j.fitote.2005.04.005.

Nanos, G. D., & Ilias, I. F. (2007). Effects of inert dust on olive (*Olea europaea* L.) leaf physiological parameters. *Environmental Science and Pollution Research, 14*, 212–214. https://doi.org/10.1065/espr2006.08.327.

Navaro, V., Villarreal, M.l., Rojas, G., & Lozoya, X. (1996). Antimicrobial evaluation of some plants used in Mexican traditional medicine for the treatment of infectious diseases. *Journal of Ethnopharmacology, 8741.*

Nazemiyeh, H., Rahman, M. M., Gibbons, S., Nahar, L., Delazar, A., Ghahramani, M. A., et al. (2008). Assessment of the antibacterial activity of phenylethanoid glycosides from *Phlomis lanceolata* against multiple-drug-resistant strains of *Staphylococcus aureus*. *Journal of Natural Medicines, 62*, 91–95. https://doi.org/10.1007/s11418-007-0194-z.

Newman, D. J., & Cragg, G. M. (2012). Natural products as sources of new drugs over the 30 years from 1981 to 2010. *Journal of Natural Products, 75*, 311–335. https://doi.org/10.1021/np200906s.

Nweze, E., Okafor, J., & Njoku, O. (2004). Antimicrobial activities of methanolic extracts of *Trema guineensis* (schumm and thorn) and *Morinda lucida* Benth used in Nigerian. *Bio-Research.* https://doi.org/10.4314/br.v2i1.28540.

Okeke, I. N., Lamikanra, A., & Edelman, R. (1999). Socioeconomic and behavioral factors leading to acquired bacterial resistance to antibiotics in developing countries. *Emerging Infectious Diseases, 5*, 18–27. https://doi.org/10.3201/eid0501.990103.

Olalde Rangel, J. A. (2005). The systemic theory of living systems and relevance to CAM: The theory (Part III). *Evidence-based Complementary and Alternative Medicine, 2*, 267–275. https://doi.org/10.1093/ecam/neh119.

Oleszek, W., & Marston, A. (2000). Saponins in food, feedstuffs and medicinal plants. In W. Oleszek, & A. Marston (Eds.), *Saponins in food, feedstuffs and medicinal plants* (pp. 1–287). Springer Science.

Ono, K., Nakane, H., Fukushima, M., Chermann, J., & Barre-Sinoussi, F. (1989). Inhibition of reverse transcriptase activity by a flavonoid. *Biochemical and Biophysical Research Communications, 160*, 982–987.

Ooshima, T., Minami, T., Aono, W., Izumitani, A., Sobue, S., Fujiwara, T., et al. (1993). Oolong tea polyphenols inhibit experimental dental caries in SPF rats infected with mutans streptococci. *Caries Research, 27*, 124–129.

Oyetayo, F., Oyetayo, V., & Ajewole, V. (2007). Phytochemical profile and antibacterial properties of the seed and leaf of Luffa plant (*Luffa cylindrica*). *Journal of Pharamcology and Toxicology, 2*, 586–589.

Pan, S. Y., Litscher, G., Gao, S. H., Zhou, S. F., Yu, Z. L., Chen, H. Q., et al. (2014). Historical perspective of traditional indigenous medical practices: The current renaissance and conservation of herbal resources. *Evidence-based Complementary and Alternative Medicine, 2014.* https://doi.org/10.1155/2014/525340.

Pandey, M. M., Rastogi, S., & Rawat, A. K. S. (2013). Indian traditional ayurvedic system of medicine and nutritional supplementation. *Evidence-based Complementary and Alternative Medicine, 2013.* https://doi.org/10.1155/2013/376327.

Parkin, D. M., Bray, F., Ferlay, J., & Pisani, P. (2005). Global cancer statistics, 2002. *A Cancer Journal for Clinicians, 55,* 74–108.

Peres, M. T. L. P., Monache, F. D., Cruz, A. B., Pizzolatti, M. G., & Yunes, R. A. (1997). Chemical composition and antimicrobial activity of *Croton urucurana* Baillon (*Euphorbiaceae*). *Journal of Ethnopharmacology, 56,* 223–226.

Phillipson, J. D. (2007). Phytochemistry and pharmacognosy. *Phytochemistry, 68,* 2960–2972. https://doi.org/10.1016/j.phytochem.2007.06.028.

Pietta, P. G. (2000). Flavonoids as antioxidants. *Journal of Natural Products, 63,* 1035–1042. https://doi.org/10.1021/np9904509.

Radha, M. H., & Laxmipriya, N. P. (2015). Evaluation of biological properties and clinical effectiveness of *Aloe vera*: A systematic review. *Journal of Traditional and Complementary Medicine, 5,* 21–26. https://doi.org/10.1016/j.jtcme.2014.10.006.

Rahman, M. A., Mossa, J. S., Al-Said, M. S., & Al-Yahya, M. A. (2004). Medicinal plant diversity in the flora of Saudi Arabia 1: A report on seven plant families. *Fitoterapia, 75,* 149–161. https://doi.org/10.1016/j.fitote.2003.12.012.

Rates, S. M. K. (2001). Plants as source of drugs. *Toxicon, 39,* 603–613. https://doi.org/10.1016/S0041-0101(00)00154-9.

Rauha, J.-P., Remes, S., Heinonen, M., Hopia, A., Marja, K., Kujala, T., et al. (2000). Antimicrobial effects of Finnish plant extracts containing flavonoids and other phenolic compounds Jussi-Pekka. *International Journal of Food Microbiology, 56,* 3–12.

Rex, J. H., Rinaldi, M. G., & Pfaller, M. A. (1995). Resistance of Candida species to fluconazole. *Antimicrobial Agents and Chemotherapy, 39,* 1–8. https://doi.org/10.1128/aac.39.1.1.

Ríos, J. L., & Recio, M. C. (2005). Medicinal plants and antimicrobial activity. *Journal of Ethnopharmacology, 100,* 80–84. https://doi.org/10.1016/j.jep.2005.04.025.

Rishton, G. M. (2008). Natural products as a robust source of new drugs and drug leads: Past successes and present day issues. *American Journal of Cardiology, 101.* https://doi.org/10.1016/j.amjcard.2008.02.007.

Robson, V., Dodd, S., & Thomas, S. (2009). Standardized antibacterial honey (Medihoney™) with standard therapy in wound care: Randomized clinical trial. *Journal of Advanced Nursing, 65,* 565–575. https://doi.org/10.1111/j.1365-2648.2008.04923.x.

Rojas, A., Hernandez, L., Pereda-Miranda, R., & Mata, R. (1992). Screening for antimicrobial activity of crude drug extracts and pure natural products from Mexican medicinal plants. *Journal of Ethnopharmacology, 35,* 275–283. https://doi.org/10.1016/0378-8741(92)90025-M.

Rossato, S. C., Leitão-Filho, H. D. F., & Begossi, A. (1999). Ethnobotany of Caicaras of the Atlantic Forest coast (Brazil). *Economic Botany, 53,* 387–395. https://doi.org/10.1007/BF02866716.

Rousseaux, C. G., & Schachter, H. (2003). Regulatory issues concerning the safety, efficacy and quality of herbal remedies. *Birth Defects Research Part B—Developmental and Reproductive Toxicology, 68,* 505–510. https://doi.org/10.1002/bdrb.10053.

Ruszkowska, J., Chrobak, R., Wróbel, J. T., & Czarnocki, Z. (2003). Novel bisindole derivatives of *Catharanthus* alkaloids with potential cytotoxic properties. *Advances in Experimental Medicine and Biology, 527,* 643–646. https://doi.org/10.1007/978-1-4615-0135-0_74.

Sahoo, N., Manchikanti, P., & Dey, S. (2010). Herbal drugs: Standards and regulation. *Fitoterapia, 81,* 462–471. https://doi.org/10.1016/j.fitote.2010.02.001.

Salim, A., Chin, Y., & Kinghorn, A. (2008). Drug discovery from plants. In R. KG, & M. JM (Eds.), *Bioactive molecules and medicinal plants* (pp. 1–21). Springer. https://doi.org/10.1007/978-3-540-74603-4_1.

Sandhu, D. S., & Heinrich, M. (2005). The use of health foods, spices and other botanicals in the sikh community in London. *Phytotherapy Research, 19*, 633–642. https://doi.org/10.1002/ptr.1714.

Sarkar, F. H., & Li, Y. (2006). Using chemopreventive agents to enhance the efficacy of cancer therapy. *Cancer Research, 66*, 3347–3350. https://doi.org/10.1158/0008-5472.CAN-05-4526.

Saxena, P., & Kulshrestha, U. (2016). Plant responses to air pollution. In U. Kulshrestha, & P. Saxena (Eds.), *Plant responses to air pollution* (pp. 1–165). New Delhi: Springer. https://doi.org/10.1007/978-981-10-1201-3.

Saxena, M., Saxena, J., Nema, R., Singh, D., & Gupta, A. (2013). Phytochemistry of medicinal plants. *Journal of Pharmacognosy and Phytochemistry, 1*, 13–14. https://doi.org/10.1007/978-1-4614-3912-7_4.

Scalbert, A., Manach, C., Morand, C., Rémésy, C., & Jiménez, L. (2005). Dietary polyphenols and the prevention of diseases. *Critical Reviews in Food Science and Nutrition, 45*, 287–306. https://doi.org/10.1080/1040869059096.

Schippmann, U., Leaman, D. J., & Cunningham, A. B. (2002). Impact of cultivation and gathering of medicinal plants on biodiversity : Global trends and issues. Biodiversity and ecosystem approach in agriculture, forestry. *Fisheries*, 1–21.

Schulman, S. L., Kaiser, B. A., Polinsky, M. S., Srinivasan, R., & Baluarte, H. J. (1988). Predicting the response to cytotoxic therapy for childhood nephrotic syndrome: Superiority of response to corticosteroid therapy over histopathologic patterns. *The Journal of Pediatrics, 113*, 996–1001. https://doi.org/10.1016/S0022-3476(88)80570-5.

Schulte-Elte, K. H., Gautschi, F., Renold, W., Hauser, A., Fankhauser, P., Limacher, J., et al. (1978). Vitispiranes, important constitutents of Vanilla aroma. *Helvetica Chimica Acta, 61*, 1125–1133. https://doi.org/10.1002/hlca.19780610326.

Schürer, N. Y. (2002). Implementation of fatty acid carriers to skin irritation and the epidermal barrier. *Contact Dermatitis, 47*, 199–205. https://doi.org/10.1034/j.1600-0536.2002.470402.x.

Schutzki, R. E., & Cregg, B. (2007). Abiotic plant disorders symptoms, signs and solutions. *Extension Bulletin.*

Sharma, H., Chandola, H., Singh, G., & Basisht, G. (2007). Utilization of Ayurveda in healthcare : An approach for prevention, health promotion, and treatment of diseases. Part 1—Ayurveda, the science of life. *Journal of Alternative and Complementary Medicine, 13*, 1011–1019. https://doi.org/10.1089/acm.2007.7017-A.

Siddiqui, I. A., Adhami, V. M., Saleem, M., & Mukhtar, H. (2006). Beneficial effects of tea and its polyphenols against prostate cancer. *Molecular Nutrition and Food Research, 50*, 130–143. https://doi.org/10.1002/mnfr.200500113.

Singh, H. (2006). Prospects and challenges for harnessing opportunities in medicinal plants sector in India. *Law Environment and Development, 2*, 198–210.

Smolinski, A. T., & Pestka, J. J. (2003). Modulation of lipopolysaccharide-induced proinflammatory cytokine production in vitro and in vivo by the herbal constituents apigenin (chamomile), ginsenoside Rb1 (ginseng) and parthenolide (feverfew). *Food and Chemical Toxicology, 41*, 1381–1390. https://doi.org/10.1016/S0278-6915(03)00146-7.

Starmans, D. A. J., & Nijhuis, H. H. (1996). Extraction of secondary metabolites from plant material: A review. *Trends in Food Science and Technology, 7*, 191–197. https://doi.org/10.1016/0924-2244(96)10020-0.

Sudhakar, Y., Kuotsu, K., & Bandyopadhyay, A. K. (2006). Buccal bioadhesive drug delivery—A promising option for orally less efficient drugs. *Journal of Controlled Release*, *114*, 15–40. https://doi.org/10.1016/j.jconrel.2006.04.012.

Surh, Y. J. (2003). Cancer chemoprevention with dietary phytochemicals. *Nature Reviews Cancer*, *3*, 768–780. https://doi.org/10.1038/nrc1189.

Tabassum, N., & Hamdani, M. (2014). Plants used to treat skin diseases. *Pharmacognosy Reviews*, *8*, 52–60. https://doi.org/10.4103/0973-7847.125531.

Taniguchi, M., & Kubo, I. (2000). Ethnobotanical drug discovery based on medicine men's trials in the African savanna : Screening of east. *Journal of Natural Prodvcts*, *56*, 1539–1546.

Valsaraj, R., Pushpangadan, P., Smitt, U. W., Adsersen, A., & Nyman, U. (1997). Antimicrobial screening of selected medicinal plants from India. *Journal of Ethnopharmacology*, *58*, 75–83. https://doi.org/10.1016/S0378-8741(97)00085-8.

Van Den Bogaard, A. E., & Stobberingh, E. E. (2000). Epidemiology of resistance to antibiotics: Links between animals and humans. *International Journal of Antimicrobial Agents*, *14*, 327–335. https://doi.org/10.1016/S0924-8579(00)00145-X.

Vining, L. C. (1990). Functions of secondary structural complexity of secondary metabolites waste products. *Annual Review of Microbiology*, *44*, 395–427.

Vohora, S., Rizwan, M., & Khan, M. J. (1973). Medicinal uses of common Indian vegetables. *Planta Medica*, *23*, 381–393.

Voravuthikunchai, S. P., & Kitpipit, L. (2005). Activity of medicinal plant extracts against hospital isolates of methicillin-resistant *Staphylococcus aureus*. *Clinical Microbiology and Infection*, *11*, 510–512. https://doi.org/10.1111/j.1469-0691.2005.01104.x.

Westh, H., Zinn, C. S., Rosdahl, V. T., Couto, E., Struelens, M., MacGowan, A., et al. (2004). An international multicenter study of antimicrobial resistance and typing of hospital *Staphylococcus aureus* isolates from 21 laboratories in 19 countries or states. *Microbial Drug Resistance*, *10*, 160–168. https://doi.org/10.1089/1076629041310055.

Wey, S. B., Mori, M., Pfaller, M. A., Woolson, R. F., & Wenzel, R. P. (2013). Hospital-acquired candidemia the attributable. *Archives of Internal Medicine*, *148*, 10–13.

Yadav, N. P., & Dixit, V. K. (2008). Recent approaches in herbal drug standardization. *International Journal of Integrative Biology*, *2*, 195–203.

Zaika, L. L., & Smith, J. L. (1975). Antioxidants and pigments of *Aspergillus niger*. *Journal of the Science of Food and Agriculture*, *26*, 1357–1369. https://doi.org/10.1002/jsfa.2740260915.

Zari, S., & Zari, T. A. (2015). A review of four common medicinal plants used to treat eczema. *Journal of Medicinal Plant Research*. https://doi.org/10.5897/JMPR2015.5831.

Chapter 6

Bioactive compounds obtained from plants, their pharmacological applications and encapsulation

Rocio Del Carmen Díaz-Torres[a], Angel Josabad Alonso-Castro[b], María Luisa Carrillo-Inungaray[c], and Candy Carranza-Alvarez[d]

[a]*Multidisciplinary Postgraduate Program in Environmental Sciences, Autonomous University of San Luis Potosi, San Luis Potosi, Mexico,* [b]*Department of Pharmacy, Division of Natural and Exact Sciences, University of Guanajuato, Guanajuato, Mexico,* [c]*Faculty of Professional Studies, Huasteca Zone and Multidisciplinary, Autonomous University of San Luis Potosí, San Luis Potosí, Mexico,* [d]*Faculty of Professional Studies, Huasteca Zone and Multidisciplinary Postgraduate Program in Environmental Sciences of the Autonomous University of San Luis Potosí, San Luis Potosí, Mexico*

1. Introduction

In the last years, many chronic and degenerative diseases have been increased worldwide. Human populations around the world demand effective therapeutic interventions. Therefore the obtention of novel, safe, and effective bioactive agents have become a need of great interest (Mehalaine, Belfadel, Menasria, & Messaili, 2017). To overcome the adverse reactions of synthetic drugs, the ancient use of medicinal herbs has gained popularity as the most sought-after alternative (Srivastava, Srivastav, Pandey, Khanna, & Pantalón, 2019). Medicinal and aromatic plants have been the basis of traditional medicine worldwide for centuries (Jin, Zhao, Li, & Zhang, 2019; Zouaoui, Chenchoun, Bouguerra, Massouras, & Barkat, 2020). In countries like Pakistan, China, Peru, and Mexico, the use of medicinal plants is a common practice among 40%–90% of the general population. In some of these countries, there are departments of traditional medicine in hospitals (García, 2019). In Mexico, population depends on herbal medicine for primary healthcare (Alonso-Castro et al., 2012). People trust in herbal medicine because of their efficacy and the fear to adverse effects produced by allopathic medicine. The use of complementary and alternative medicine, including medicinal plants, in Mexico, is recognized by the Secretariat of Health (World

Phytomedicine: A Treasure of Pharmacologically Active Products from Plants
https://doi.org/10.1016/B978-0-12-824109-7.00017-0
181

Health Organization, 2019). In Mexico the National Indigenous Institute supported the creation of an organization of traditional healers in Cuetzalan, Puebla, through a support program for traditional medicine. This program resulted in the formation of an organization of about 120 traditional therapists in this municipality and the creation of a Masehualpajti traditional medical center in 1991. The inclusion of traditional medicine in some health units from Mexico has represented an effort to achieve intercultural medicine in the Mexican health system (Duarte-Gómez, Brachet-Márquez, Campos-Navarro, & Nigenda, 2004). In Peru, through one of the departments of the Ministry of Health, INMETRA (National Institute of Traditional Medicine) has been making efforts to link traditional or complementary medicine with conventional or academic medicine, promoting responsible and professional use in the use of medicinal plants in primary health care (García, 2019). According to the Institute of Public Health of Mexico, complementary and conventional medicines can be complemented with an inclusive public health policy. The regulation of natural product trade should be implemented by national health systems (García, 2019).

The World Health Organization has stated that the trade of medicinal plants and herbal medicines is growing an annual growth rate of about 15% (Srivastava et al., 2019). In some cases medicinal plants contribute to the socioeconomic development in rural areas (Güney, 2019). Medicinal plants have offered a significant number of bioactive compounds with applications in the fields of medicine, pharmacy, and biology (Priya & Satheeshkumar, 2020). Bioactive compounds (secondary metabolites) produced by plants are synthesized by secondary metabolism, which protect plants from multiple biotic or abiotic stressful conditions and provide odor, color, and tastes (Bhattacharya, 2019; Zouaoui et al., 2020). The production of these compounds, influenced by multiple factors, is less than 1%–5% dry weight in plants (Krzyzanowska, Czubacka, & Oleszek, 2009; Rao & Ravishankar, 2002). Most of the secondary metabolites are stored in the vacuole (e.g., via glycosilation with the participation of H^+-ATPase), and their transport is via simple diffusion, vesicle-mediated transport, and carrier-mediated transport (Shitan, 2016). The metabolic pathways involved in the synthesis of secondary metabolites in plants include shikimic acid (for alkaloids, phenols, and tannins,), acetate malonate (phenols and alkaloids), and mevalonic acid (terpenes, steroids, and alkaloids) (Dewick, 2002). There are more than 100,000 different secondary metabolites (Hadacek, 2002), and approximately 5000–15,000 genes encode for enzymes involved in secondary metabolism, such as terpene synthases, phenylalanine ammonia lyase, NADPH-dependent dehydrogenases, polyketide synthases, and chalcone synthase (Somerville & Somerville, 1999). However, new enzymes, such as members of terpene synthase family, participating in secondary metabolism have been discovered (Pichersky & Gang, 2000). Secondary metabolites can help to prevent and treat various emergent diseases such as cancer, neurological diseases, diabetes, and atherosclerosis (Bonam et al., 2018; Zouaoui et al., 2020). The application of bioactive compounds from plants in medicine is very wide, and therefore alternatives should be sought to obtain better strategies for extraction, purification, and encapsulation of these compounds. The term phytomedicine can be defined as a medicine derived exclusively from a complete plant or parts of plants and manufactured in raw form or as a purified pharmaceutical formulation (Srivastava et al., 2019). The study of medicinal plants is a multidisciplinary field that involves the participation of experts in natural sciences such as chemistry, molecular biology, toxicology, pharmacology, botany, biochemistry, and biotechnology. This chapter summarizes general aspects of secondary metabolites obtained from medicinal plants and discusses the application of herbs in medicine and their encapsulation.

2.   Factors that influence the production of bioactive compounds

There are several factors that influence the composition and concentration of secondary metabolites in plants, including the following: (a) ecological habitat conditions, such as elevation, latitude, precipitation, humidity, sunlight, aridity, etc.; (b) characteristics of soil such as salinity, pH, and concentrations of phosphorus, potassium, calcium, nitrogen, sulfate, manganese, among other oligoelements; (c) plant-plant interactions; (d) plant-microbe interactions; (e) environmental stress; (f) genetic factors; and (g) plant-growing conditions (Mehalaine & Chenchouni, 2019; Zouaoui et al., 2020) (Fig. 1). Some of the main secondary metabolites of plants are alkaloids, flavonoids, phenols, glycosides, saponins, tannins, volatile oils, gums, and resins. These secondary metabolites are known to have various activities such as antimicrobials, chemopreventive, hypoglycemic, antihypertensive, antiinflammatory, and antioxidant (Altaf, Khan, & Ahmad, 2019). Different groups of secondary metabolites such as terpenes, alcohols, aldehydes, ketones, esters, and heterocyclic compounds are commonly found as volatile organic compounds due their low molecular weight (Zouaoui et al., 2020). These metabolites have demonstrated a wide spectrum of biological activities and can have a great potential for applications in various fields. Volatile compounds in plants, especially essential oils, have been reported to exert antibacterial, antifungal, and antioxidant effects (Mehalaine et al., 2017). Volatile compounds can be identified using gas chromatography-mass spectrometry. Additionally, analytical techniques such as HPLC and UPLC can be incorporated for the identification of secondary metabolites in medicinal plants.

Abiotic and biotic conditions influence the production and concentration of secondary metabolites in plants. For example, the content con flavonoids in vegetable crop and *Ilex paraguariensis* A. St.-Hil.is higher in plants exposed to higher sunlight intensity than plants growing in the shadow (Riachi et al., 2018; Schreiner, 2005). The content of flavonoids and polyphenols in *Ipomoea batatas* (L.) Lam. was increased due to longtime exposure to light irradiation (Carvalho, Cavaco, Carvalho, & Duque, 2010). Air temperature was the main

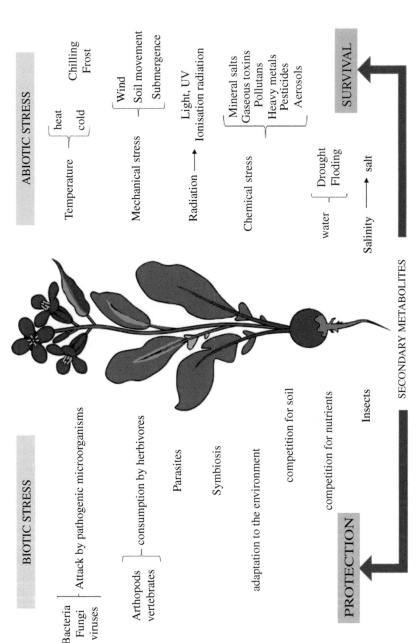

FIG. 1 Production of secondary metabolites in plants in response to biotic and abiotic stress. *Credit: Rocio Del Carmen Diaz-Torres.*

factor associated with the production of polyphenols and antioxidant activity in red amaranth growing in cold weather (Khandaker, Masum-Akond, Ali, & Oba, 2010). Water restriction conditions decreased the content of saponins in *Chenopodium quinoa* Willd. (Soliz-Guerrero, de Rodriguez, Rodriguez-Garcia, Angulo-Sanchez, & Mendez-Padilla, 2002) and the content of artemisinin in *Artemisia annua* L. (Charles, Simon, Shock, Feibert, & Smith, 1993).

On the contrary, drought conditions increased the levels of polyphenols, flavonoids, and alkaloids in *Trachyspermum ammi* (L.) Sprague (Azhar, Hussain, Ashraf, & Abbasi, 2011), *Crataegus laevigata* (Poir.) DC., and *Camptotheca acuminata* Decne. (Liu, 2000), respectively (Kirakosyan et al., 2004). The in vitro antimalarial activity of *Bidens pilosa* L. is affected by drying. Plants collected during the dry season showed better antimalarial activity than those plants collected in the wet season (Andrade-Neto et al., 2004). The polyphenol content in *Camellia sinensis* (L.) Kuntze, *Chelidonium majus* L., *Rosmarinus officinalis* L., *Salvia fruticosa* Mill, and *Cosmos caudatus* Kunth and the terpene content in *Ocimum basilicum* L. were higher during the dry season than the wet season (Anesini, Ferraro, & Filip, 2008; Hussain, Anwar, Hussain Sherazi, & Przybylski, 2008; Jakovljevic, Stankovic, & Topuzovic, 2013; Mediani, Abas, Ping, Khatib, & Lajis, 2012). The high sunlight intensity (including high UV pattern) during the wet season and the possible inactivation of phenolic acid degradation enzymes could explain this phenomenon. Therefore the season of the year and climatic conditions should be considered before collecting medicinal plants and evaluate their chemical composition. The effects of climate change in the composition of secondary metabolites in plants should be assessed. It is known that climate changes have affected the flowering time, photosynthesis rate, and crop yield (Jagadish, Craufurd, & Wheeler, 2008). For instance, elevated CO_2 levels affect the quantity of secondary metabolites in plants. The content of gallic acid in *Labisia pumila* Benth. & Hook. f., native of Malaysia, was increased due to high levels of CO_2 (Jaafar, Ibrahim, & Karimi, 2012), whereas the levels of flavonoids in ginkgo and *Gynostemma pentaphyllum* (Thunb.) Makino were decreased by the exposition to high levels of CO_2 (Chang et al., 2016; Huang, He, Liu, & Li, 2010). High UV radiation exposure increased the production of phenylpropanoid and monoterpenes in *Lonicera japonica* Thunb. (Ning et al., 2012). Elevated ozone concentrations increased the levels of phenolics and tannins in *Melissa officinalis* L. (Pellegrini, Carucci, Campanella, Lorenzini, & Nali, 2011). The adaptation mechanisms and the changes in metabolic pathways of medicinal plants against climate change should be considered in phytochemical studies.

3. Application of bioactive compounds from plants in medicine

Plant-derived compounds such as morphine, digoxin, artemisinin, codeine, and camptothecin are examples of the importance of medicinal plants in the isolation of bioactive compounds with beneficial effects for humans (Thomford et al., 2018).

This chapter highlights the diversity of bioactive compounds, obtained from medicinal plants, with pharmacological interest reported in scientific articles published in the last 3 years. Plant species cited in Table 1 belong to 18 botanical families. The plant parts used for the obtention of bioactive compounds are listed in the following order: aerial parts (24%), leaves (20%), roots (20%), fruits (16%), petal (5%), essential oil (5%), stem (5%), and in vitro (5%). Root harvesting is harmful to medicinal plants, especially in slow-growing plants. The harvesting of aerial parts of medicinal plants should be promoted as a sustainable use. Conservation studies of medicinal plants should be considered before carrying out chemical, pharmacological, and toxicological studies. Plants with slow growth might be at risk of extinction when roots are used for experimental studies. Plant tissue culture, including micropropagation, offers an attractive alternative to avoid the overexploitation of medicinal plants. As shown in Table 1, the use of in vitro plants has been increased. This technique avoids destructive harvesting methods. In addition, in vitro root culture is also a technique to avoid root harvesting. The use and study of essential oils from medicinal plants are current topics of great interest. However, the chemical composition of essential oils should be carried out to provide a certification of purity. As shown in Table 1, the main components found in essential oils are terpenes. The study of terpenes has been increased in the last decade. The study of natural antioxidants as a preventive strategy for cardiovascular and neurological diseases has gained attention.

Medicinal plants cited in Table 1 contain a wide range of bioactive compounds, including terpenes (e.g., oleanolic acid, camphor, Z-β-ocimene, borneol, isoborneol, α-bisabolol, viridiflorol, and trans-caryophyllene), flavonoids (e.g., kaempferol, naringenin, and quercetin), phenolic acids (protocatechuic acid), hydroxycinnamic acids (caffeic acid and chlorogenic acid), organic acids (gallic acid), polyphenols (catechin), and phytosterols (β-sitosterol) (Fig. 2). These bioactive compounds are responsible for therapeutic activities such as analgesic, hypoglycemic, antidiabetic, antioxidant, antimicrobial, antiinflammatory, anticancer, antimalarial, antidiarrheal, osteogenic, and antiinfluenza (Table 1). Current research on medicinal plants and their bioactive compounds is focusing on finding treatments for diseases like cancer, diabetes, infections, and inflammatory diseases (rheumatoid arthritis). Most of the active compounds cited in Table 1 have been studied for their pharmacological studies. However, the isolation of novel compounds should be encouraged. It is also necessary to perform clinical trials with active compounds obtained from medicinal plants to demonstrate their effectiveness, comparing the pharmacological with reference drugs.

TABLE 1 Main phytochemicals produced by medicinal plants and their applications.

Scientific name	Used part	Medicinal properties	Bioactive compounds	Reference
Argemone mexicana L. (Papaveraceae)	Aerial parts	Analgesic, antidiarrheal, and antibacterial	Alkaloids, flavonoids, phenolic compounds, tannins, and glycosides	Sarkar, Mitra, Acharyya, and Sadhu (2019)
Artemisia judaica L. (Asteraceae)	Aerial parts	Antioxidant activity and hypoglycemic effect	Phenolics, saponins, flavonoids, tannins, and alkaloids	Bhat, Ullah, and Abu-Duhier (2019)
Carica papaya L. (Caricaceae)	Ripe fruit	Antioxidant activity	carotenoids and bioactive compounds such as flavonoids, hydroxycinnamic acids, and others	Asuquo, Edagha, Ekandem, and Peter (2019)
Echinacea purpurea var. *laevigata* (Asteraceae)	Plant material derived from in vitro cultures	Immunomodulatory and antitumor effects	Phenols, cathartic acid, chlorogenic acid, chichoric acid, and caffeic acid	Lema-Rumińska, Kulus, Tymoszuk, Varejão, and Bahcevandziev (2019)
Euphorbia subg. Tirucalli (Euphorbiaceae)	Root	Analgesic, acute inflammation, asthma, and arthritis	Terpenoid	Palit et al. (2018)
Hibiscus rosa-sinensis L. (Malvaceae)	Petal	Antioxidants	Phenolics, flavonoids, and anthocyanins	Anand and Sarkar (2017)
Kigelia pinnata (Jacq.) DC. (Bignoniaceae)	Leaf	Osteogenic activity	Polybotrin and scutellarin 7-O-β-D-glucopyranoside	Ramakrishna et al. (2017)

Continued

TABLE 1 Main phytochemicals produced by medicinal plants and their applications—cont'd

Scientific name	Used part	Medicinal properties	Bioactive compounds	Reference
Leea macrophylla Roxb. ex Hornem.(Vitaceae)	Root	Antioxidant	Oleanolic acid and its derivatives, 7α, 28-olean diol and stigmasterol	Mahmud et al. (2017)
Monotheca buxifolia (Falc.) A. DC. (Sapotaceae)	Leaf	Antioxidant and antifungal	Polyphenols and flavonoids	Ali, Khan, and Zia (2019)
Oenocarpus bacaba var. *bacaba.* (Arecaceae)	Fruit	Anticancer effect, apoptotic effect, and antioxidant	Polyphenolics	Abadio-Finco, Kloss, and Graeve (2016)
Paeonia lactiflora Pall. (Paeoniaceae)	Root	Antiinfluenza activity	Gallic acid, methyl gallate, and pentagalloylglucose	Zhang et al. (2020)
Pituranthos scoparius Benth. & Hook. f. (Apiaceae)	Aerial parts	Antioxidant	Flavonoids, tannins, coumarins, saponins, and steroids	Belyagoubi-Benhammou, Belyagoubi, Bechlaghem, Chembaza, and Atik-Bekkara (2017)
Ricinus communis L. (Euphorbiaceae)	Leaf	Antibacterial and antifungal	Tannins, saponins, terpenoids, polyuronoids, reducing sugars, flavonoids, alkaloids, and anthraquinones	Suurbaar, Mosobil, and Donkor (2017)
Rosa multiflora Thunb (Rosaceae)	Dry fruits	Purgative	Quercitrin, multinoside A, multiflorin B, and multinoside A acetate	Kitahiro, Ikeda, Im, Kodaira, and Shibano (2019)

Species (Family)	Part	Activity	Compounds	Reference
Rosmarinus officinalis L. (Lamiaceae)	Aerial parts	Antimicrobial, antirheumatic, antiinflammatory, insecticidal, and antioxidant	Essential oils (camphor, Z-β-ocimene, isoborneol, α-bisabolol, borneol, trans-caryophyllene, and α-terpineol	Mehalaine et al. (2017)
Scandix pecten-veneris L. (Apiaceae)	Leaf	Antibacterial, antifungal and antioxidant	Alkaloids, flavonoids, polyphenols y tannins	Wahab et al. (2018)
Spiraea chamaedryfolia L. (Ranunculaceae)	Root	Antibacterial	Alkaloids	Kiss et al. (2017)
Syzygium aromaticum (L.) Merr. & L.M. Perry (Myrtaceae)	Essential oil	Antioxidant	Polyphenols and flavonoids Eugenol, β-Caryophyllene, eugenyl acetate, and α-humulene	El Ghallab et al. (2019)
Teucrium pseudoscorodonia Desf. (Lamiaceae)	Leaf	Antioxidants	Methylated flavonoids or hydroxyflavonoles	Belarbi et al. (2017)
Thymus algeriensis Boiss. & Reut. (Lamiaceae)	Aerial parts	Treat respiratory and digestive disorders	essential oils (camphor, 1.8-cineol, borneol, viridiflorol, linalool, α-terpineol and caroyophyllene oxide)	Mehalaine et al. (2017)
Traganum nudatum Delile (Amaranthaceae)	Aerial parts	Antioxidant and the antimicrobial activities	Sinapic, syringic, ferulic, p-coumaric acids, catechin, naringenin, and vanillin	Mouderas, El Haci, and Lahfa (2019)

Continued

TABLE 1 Main phytochemicals produced by medicinal plants and their applications—cont'd

Scientific name	Used part	Medicinal properties	Bioactive compounds	Reference
Tripterygium regelii Sprague & Takeda (Celastraceae)	Stem	Rheumatoid arthritis and antiinflammatory activity	Sesquiterpenoids, diterpenoids (Abietanes diterpenoids, abietanes with benzenoid rings, diterpene quinoids, and diterpene quinoids with lactone rings.) and triterpenoids	Chen, Liu, Xiao, Yang, and Li (2018)
Salvia officinalis L. (Lamiaceae)	Aerial parts	Antioxidant, antiseptic, bactericidal, and fungicidal	Essential oils (β-thujone, viridiflorol, camphor, 1,8-cineol, trans-caryophyllene, α-humulene, caryophyllene oxide, αthujone, and borneol)	Mehalaine et al. (2017)
Oryza sativa L. (Poaceae)	Caryopsis fruit	Anticancer, hypoglycemic effect and lipid lowering effect	Vitamin E (tocotrienols), anthocyanins (cyanidin-3-O-glucoside, and peonidin-3-O-glucoside) and phytosterols (γ-oryzanol, β-sitosterol, and triterpene alcohol)	Poosri, Thilavech, Pasukamonset, Suparprom, and Adisakwattana (2019)

Credit: Rocio Del Carmen Díaz-Torres.

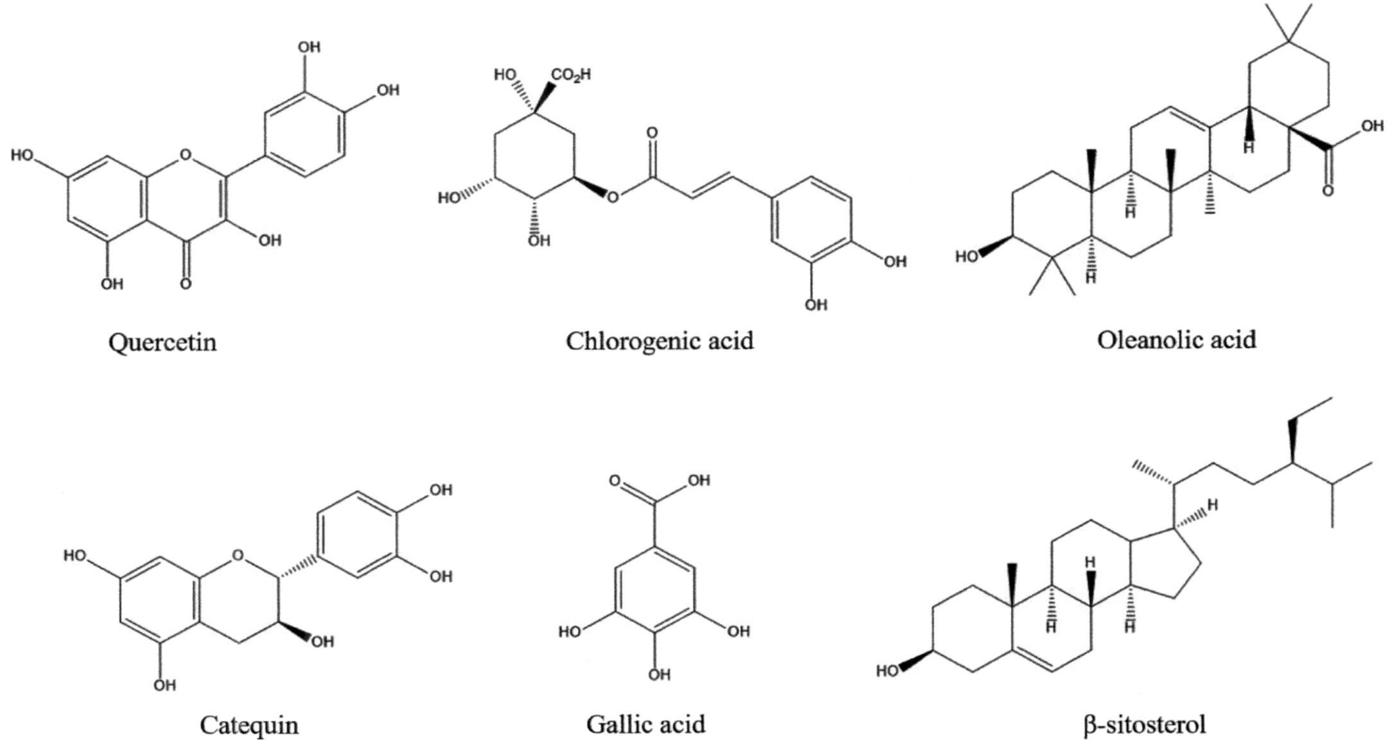

FIG. 2 Secondary metabolites isolated from medicinal plants. *Credit: Rocío Del Carmen Díaz-Torres.*

4. Encapsulation of phytochemicals

Plants and fruits, as well as other plant resources, contain bioactive components that have a beneficial effect on human health. However, they have less stability in environmental conditions, so their encapsulation emerged to decrease their sensitivity to degradation. The encapsulation process is defined as the coating of small particles of liquids, solids, or gases with a thin protective layer of barrier or wall material, which inhibits volatilization and protects the encapsulated material against adverse environmental conditions, such as the effects of light, moisture, and oxygen. In addition, encapsulation provides protection against ultraviolet radiation and exposure to dangerous or toxic products, helps to the prolongation of chemical stability and functionality of the bioactive compounds, and promotes the controlled release of bioactive compounds (Ishwarya & Anandharamakrishnan, 2017; Rezvankhah, Emam-Djomeh, & Gholamreza, 2020). Encapsulation is considered as an important alternative to the solution of problems that may arise either due to physical or chemical instability of the active ingredient or due to incompatibility between the active ingredient and the matrix of the drug to be manufactured. The selection of the encapsulation method is based on the size of required particle, the physicochemical properties of the wall material, and the substance to encapsulate material applications and release mechanisms and costs. There are several encapsulation methods, which can be divided into three groups: (a) physical processes such as spray drying, extrusion, and crystallization; (b) chemical processes such as interfacial polymerization and molecular inclusion; and (c) physicochemical processes such as simple or complex coacervation (separation of the aqueous phase), evaporation of solvent in the emulsion (separation of the organic phase), and emulsion solidification (Liu, Huang, Chen, Lin, & Wang, 2019).

4.1 Encapsulation methods

4.1.1 Complex coacervation

Coacervation is the process during which a homogeneous solution of charged macromolecules and separates into two liquid phases in equilibrium, in which the most concentrated colloid phase is known as the "phase coacervated" and the other is known as "equilibrium phase" (Ifeduba & Akoh, 2015). Currently, coacervate complexes are used as an ingredient in food products to protect one or more hydrophobic substances in beverages, dry foods, and semiwet foods (Zhang & Mutilangi, 2013). This method presents the possibility of controlled release.

4.1.2 Spray drying

Spray drying is one of the most common encapsulation methods used in food and pharmaceutical industries due its low processing costs. However, the use of spray drying is limited in heat-sensitive bioactive compounds. A possible alternative in these circumstances is the use of lyophilization. However, this alternative requires a longtime process (Rezvankhah et al., 2020). Spray drying produces dry powders from a low-viscosity liquid, suspension, emulsion, suspension, or paste. This method has unique characteristics, such as rapid evaporation of water, continuous operation, various drying settings, multiscale operating capacity, and high production rate (Ishwarya & Anandharamakrishnan, 2015).

4.1.3 Hydrogels

Hydrogels are defined as dual or multicomponent systems with networks of three-dimensional polymer chains, where the empty spaces between the macromolecules that make it up can be occupied by water, increasing its volume and/ or modifying its shape (Ahmed, 2015; Gaharwar, Peppas, & Khademhosseini, 2014). There are many materials that can be incorporated into this type of systems such as amino acids, nucleic acids of therapeutic base for the generation of tissues, drugs, and hydrophilic bioactive compounds such as antioxidants (Rehman et al., 2019).

4.2 Materials used for encapsulation

Polysaccharides are widely used as encapsulating agents because of their high chemical stability, high abundance in nature, and low cost (Fathi, Martín, & McClements, 2014). Furthermore the use of pectin in combination with other compounds such as proteins and lipids is found to be the most promising wall materials for encapsulating bioactive compounds from plants. Sodium alginate and chitosan have shown good biocompatibility (Rehman et al., 2019). Surfactant molecules can be macromolecules such as polysaccharides or proteins, added to pharmaceuticals such as phospholipids, mono-/diglycerides, polysorbates, and sorbitan esters. Polysaccharides and proteins can be used individually or together to protect the functional components. The ability of these particles to emulsify, protect, and release the active component depends on the molecular and physicochemical factors such as its composition, structure internal, polarity, electric charge, and physical dimensions (Adjonu, Doran, Torley, & Agboola, 2014).

4.3 Encapsulation of medicinal plants and bioactive compounds

Medicinal plants are commonly used as infusions or decoctions. However, many of the plant compounds can be degraded by oxidation, by incorrect storage conditions, or during these extraction procedures. Many plant extracts or their active compounds can be hydrophobic (essential oils, somopolyphenols, etc.), and bioavailability can be reduced. Table 2 shows medicinal plants and/ or their bioactive compounds encapsulated by different methods. Encapsulation shows an improvement in the pharmacological actions of plant extracts, and their active compounds increases, in most of the cases, the chemical stability of bioactive compounds and decreases the adverse effects. Most of the studies evaluated the release of essential oils, plant extracts, or bioactive compounds in intestinal fluids. Many encapsulation methods avoid the release of bioactive components in stomach and esophagus, minimizing their chemical degradation.

Table 2 clearly indicates that there is a need to carry out more studies about the encapsulation of medicinal plants. The extensive use of medicinal plants requires studies incorporating pharmaceutical technology. Stability tests should also be carried out during encapsulation of medicinal plants. However, many encapsulation methods need to be scaled up for industrial production.

5. Future perspectives of phytomedicines

The global market for herbal medicines is approximately US $83 billion annually, and this rate will reach more than US $115 billion in 2020 (Gunjan et al., 2015). Therefore quality analysis on herbal products should be carried out. Some medicinal plants lack scientific evidence of their safety and efficacy in clinical trials (Bonam et al., 2018). Some aspects that should be considered in the production of phytomedicines include (a) the standardization of plant extracts, using HPLC and/or gas chromatography-mass spectrometry; (b) quality assessment (stability proofs), (c) safety assessment, including the evaluation of contaminants (i.e., pesticides, pathogenic microorganisms, and heavy metals); (d) intrinsic toxicity of medicinal plants; (e) assessment of adulterants; and (f) assessment of the pharmacological effects in clinical trials. In Mexico, herbal products are categorized as herbal medicines, herbal remedies, and dietary supplements. In 2009 154 herbal medicines (79 from parts of plants or extracts and 75 presented in pharmaceutical form) were registered. The national regulation on herbal medicines was most recently updated in 2013 to reference the Mexican herbal pharmacopeia (Farmacopea herbolaria de los Estados Unidos Mexicanos). Nevertheless, most of herbal medicines are sold as nonprescription medicines. Many countries lack regulation on the use of herbal medicines.

The standardization of extraction procedures in medicinal plants should be carried out. It is known that the quantitative composition of phytomedicines can vary depending on the extraction methods and the conditions of plant conservation. For instance, the amount of volatile bioactive compounds in air-dried samples away from sunlight is much higher compared with artificial drying methods such as microwave oven drying (Zouaoui et al., 2020). Plant metabolomics can be a useful tool in the identification of pathways involved in the synthesis of secondary metabolites.

Pharmacokinetics studies should be carried out with at least one bioactive compound found in the phytomedicine. These studies can help to provide information to avoid toxicity and prevent possible interactions with other medicinal plants and prescription drugs.

TABLE 2 Encapsulation of medicinal plants and bioactive compounds.

Medicinal plant(s) or bioactive compound(s)	Method of encapsulation	Materials used for encapsulation	Findings	Reference
Anthocyanins	Encapsulation	Sodium alginate (polymer matrix) and CaCl (gelling agent)	Over 85% of the anthocyanins present in the blueberry extracts were encapsulated	Acciarri (2017)
Curcumin	Hydrogel	Liposomes/thiolated chitosan	Drug-loaded liposomal hydrogels deliver curcumin with a tumoricidal effect. These hydrogels are a promising new drug delivery vehicle	Li et al. (2020)
Melissa officinalis L. (Lamiaceae) oil	Hydrogel	Multicellulose	Hydrogels successfully inhibited the growth of *Candida albicans* demonstrating the potential of a new antimicrobial therapy	Serra et al. (2020)
Cinnamomum verum J. Presl (Lauraceae) essential oil	Spray drying	Sodium alginate	A controlled release of oil was obtained. Encapsulation decreased toxic effects of oil, but increased antifungal activity	Makimori et al. (2020)
Piper nigrum L. (Piperaceae)essential oil	complex coacervates	Lactoferrin and sodium alginate	Encapsulation preserved secondary metabolites in the essential oil and increased bioavailability	Heckert Bastos, Correados Santos, Geraldode Carvalho, and Garcia-Rojasac (2020)

Continued

TABLE 2 Encapsulation of medicinal plants and bioactive compounds—cont'd

Medicinal plant(s) or bioactive compound(s)	Method of encapsulation	Materials used for encapsulation	Findings	Reference
Plukenetia volubilis L. (Euphorbiaceae) oil	complex coacervation	Ovalbumin and sodium alginate	These results demonstrate that it is possible to use OVA/AL as encapsulating agents to protect bioactive compounds and to improve the thermal behavior of microcapsules	Silva Soares, Pinto Siqueira, Geraldo de Carvalho, Vicente, and Garcia-Rojasad (2019)
Rubus fruticosus L. (Rosaceae)	spray-drying, freeze-drying and supercritical antisolvent (SAS)	Polyvinylpyrrolidone	The encapsulation processes did not lead to great degradation of antioxidant	da Fonseca-Machado et al. (2018)
Allium ursinum L. (Amaryllidaceae)	spray congealing	Microparticles (MPs)	The encapsulation of *Allium ursinum* extract by spray congealing is a promising approach to improve the biopharmaceutical properties of the extract, without affecting its antibacterial activity	Tomšik et al. (2019)
Stevia rebaudiana Bert. (Asteraceae)	Ionic gelation	Alginate	Encapsulation efficiency was increased by using alginate blends	Aceval-Arriola et al. (2019)
Vaccinium myrtillus L. (Ericaceae) anthocyanins	Encapsulation	Whey protein capsules	Encapsulation increased short-term bioavailability and stability of bilberry extract	Mueller et al. (2018)

Smallanthus sonchifolius (Poepp.) H. Rob. (Asteraceae) leaf extract	Encapsulation	Supercritical fluid extraction of emulsions	Encapsulation increased the antioxidant activity of the plant extract. A stable suspension was obtained	Cruz, Lima Reis, Ferreira, Masson, and Corazza (2020)
Myrciaria cauliflora (Mart.) O. Berg (Myrtaceae) anthocyanin	Encapsulation	Calcium alginate	Encapsulation increased the stability of anthocyanins and the antioxidant was unaltered	Santos, Albarelli, Beppu, and Meirelesa (2013)
Momordica charantia L. (Cucurbitaceae) medicinal fruit	Spray-drying	Maltodextrin and gum Arabic	Encapsulation did not affect the stability of plant extract and its bioactive compounds	Tan, Kha, Parksa, Stathopoulos, and Roach (2015)
Piper sarmentosum Roxb. (Piperaceae) aqueous extract	Hydrogel	Calcium alginate	Encapsulation did not alter the stability of plant compounds	Eng-Seng, Zhi-Hui, Soon-Hock, Mansa, and Ravindra (2010)
Cinnamomum zeylanicum Blume (Lauraceae) proanthocyanidin	Complex coacervation	Gum arabic, pectin, cashew gum, carboxymethylcellulose, and κ-carrageenan	Dry extracts presented high antioxidant capacity and high potential to inhibit digestive enzymes, but showed a lost of the antimicrobial capacity in comparison with the liquid extract	de Souza et al. (2018)

Credit: Rocío Del Carmen Díaz-Torres.

The toxicity of many medicinal plants has not been studied despite their high use and consumption by the general population. Some medicinal plants can trigger undesirable effects on human health due to (a) pharmacodynamic interaction with prescription drugs, (b) intrinsic effects, (c) pharmacokinetic interaction with prescription drugs, (d) the presence of contaminants and/ or pathogenic microorganisms, and (e) other factors such as the age of the patient, the nutritional status, and the presence of chronic diseases (Alonso-Castro et al., 2017). Recently, Alonso-Castro, Domínguez, et al. (2017) registered a total of 216 medicinal plants belonging to 77 families reported as toxic. Of these plants, 76 had been studied, and 140 plants lacked studies on their toxicological effects. Of the studied plants the parts considered toxic are listed in the following order: aerial parts including branches, leaves, and flowers (22%); whole plant (22%); leaves exclusively (15%); seeds (14%); roots (8%); fruits (8%); bark (4%); latex (3%); and other parts of the plant. The signs and symptoms of toxicity induced by medicinal plants were as follows: nausea and vomiting (20%), dermatitis (14%), gastritis (9%), abdominal pain (9%), abortions (8%), skin burns (8%), hepatotoxicity (7%), severe diarrhea (6%), cardiotoxicity (5%), nephrotoxicity (2%), numbness (2%), dizziness (2%) and hallucinations (2%). Phytopharmacovigilance programs should also be implemented in Mexico to record possible adverse effects induced by herbal medicines. Health professionals should receive academic information about medicinal plants to provide information to the patient about possible herb-drug interactions and avoid self-medication (Alonso-Castro et al., 2017). Finally, approximately 20% of medicinal plants are exposed to a risk of extinction due to overexploitation, habitat fragmentation, spread of invasive species, etc. (Bentley, 2010). Conservation programs should be implemented in countries to preserve local biodiversity. Large-scale cultivation of medicinal plants and the use of plant tissue culture are options to achieve the demand of medicinal plants in the market.

6. Conclusions

Many bioactive compounds present in plants are labile and require protection by the medium of its encapsulation. Encapsulation by complex coacervation, spray drying, and hydrogels have shown to be useful methodologies for preserving the physicochemical characteristics of bioactive compounds. The use of biomaterials (polysaccharides and proteins) as encapsulating agents has been an excellent strategy for the release of active ingredients to along the gastrointestinal tract.

References

Abadio-Finco, F. D. B., Kloss, L., & Graeve, L. (2016). Bacaba (*Oenocarpus bacaba*) phenolic extract induces apoptosis in the MCF-7 breast cancer cell line via the mitochondria-dependent pathway. *NFS Journal, 5,* 5–15. https://doi.org/10.1016/j.nfs.2016.11.001.

Acciarri, G. (2017). *Estabilización de antioxidantes naturales por encapsulación y su incorporación a derivados lácteos con valor agregado. Área Fisicoquímica.* Facultad de Ciencias Bioquímicas y Farmacéuticas. URN.

Aceval-Arriola, N. D., Chater, P. I., Wilcox, M., Lucini, L., Rocchetti, G., Dalmina, M., et al. (2019). Encapsulation of *Stevia rebaudiana* Bertoni aqueous crude extracts by ionic gelation—Effects of alginate blends and gelling solutions on the polyphenolic profile. *Food Chemistry, 275*, 123–134. https://doi.org/10.1016/j.foodchem.2018.09.086.

Adjonu, R., Doran, G., Torley, P., & Agboola, S. (2014). Whey protein peptides as components of nanoemulsions: A review of emulsifying and biological functionalities. *Journal of Food Engineering, 122*, 15–27. https://doi.org/10.1016/j.jfoodeng.2013.08.034.

Ahmed, E. M. (2015). Hidrogel preparation, characterization, and applications: A review. *Journal of Advanced Research, 105*, 21. https://doi.org/10.1016/j.jare.2013.07.006.

Ali, J. S., Khan, I., & Zia, M. (2019). Antimicrobial, cytotoxic, phytochemical and biological properties of crude extract and solid phase fractions of *Monotheca buxifolia*. *Oriental Pharmacy and Experimental Medicine*. https://doi.org/10.1007/s13596-019-00409-6.

Alonso-Castro, A. J., Domínguez, F., Maldonado-Miranda, J. J., Castillo-Pérez, L. J., Carranza-Álvarez, C., Solano, E., et al. (2017). Use of medicinal plants by health professionals in Mexico. *Journal of Ethnopharmacology, 198*, 81–86. https://doi.org/10.1016/j.jep.2016.12.038.

Alonso-Castro, A. J., Dominguez, F., Ruíz-Padilla, A. J., Campos-Xolalpa, N., Zapata-Morales, J. R., Carranza-Alvarez, C., et al. (2017). Review article medicinal plants from north and Central America and the Caribbean considered toxic for humans: The other side of the coin. *Evidence-based Complementary and Alternative Medicine, 28*. https://doi.org/10.1155/2017/9439868.

Alonso-Castro, A. J., Maldonado-Miranda, J. J., Zarate-Martínez, A., Jacobo-Salcedo, M. R., Fernandez-Galicia, C., Figueroa-Zuñiga, L. A., et al. (2012). Medicinal plants used in the Huasteca Potosina, México. *Journal of Ethnopharmacology, 143*, 292–298. https://doi.org/10.1016/j.jep.2012.06.035.

Altaf, M. M., Khan, M. S. A., & Ahmad, I. (2019). Chapter 2. Diversity of bioactive compounds and their therapeutic potential. In *New look to phytomedicine advancements in herbal products as novel drug leads* (pp. 15–34)., ISBN:9780128146194. https://doi.org/10.1016/B978-0-12-814619-4.00002-1.

Anand, A., & Sarkar, B. (2017). Phytochemical screening and antioxidant property of Anthocyanins extracts from *Hibiscus rosa-sinensis*. In K. Mukhopadhyay, A. Sachan, & M. Kumar (Eds.), *Applications of biotechnology for sustainable development* (pp. 139–147). Singapore: Springer. https://doi.org/10.1007/978-981-10-5538-6_17.

Andrade-Neto, V. F., Brandão, M. G. L., Oliveira, F. Q., Casali, V. W. D., Njaine, B., Zalis, M. G., et al. (2004). Antimalarial activity of *Bidens pilosa* L. (Asteraceae) ethanol extracts from wild plants collected in various localities or plants cultivated in humus soil. *Phytotherapy Research, 18*, 634–639. https://doi.org/10.1002/ptr.1510.

Anesini, C., Ferraro, G. E., & Filip, R. (2008). Total polyphenol content and antioxidant capacity of commercially available tea (*Camellia sinensis*) in Argentina. *Journal of Agricultural and Food Chemistry, 56*(19), 9225–9229. https://doi.org/10.1021/jf8022782.

Asuquo, I. E., Edagha, I. A., Ekandem, G. J., & Peter, A. I. (2019). *Carica papaya* attenuates testicular histomorphological and hormonal alterations following alcohol-induced gonado toxicity in male rats. *Toxicological Research*. https://doi.org/10.1007/s43188-019-00017-1. ISSN 2234-2753.

Azhar, N., Hussain, B., Ashraf, M., & Abbasi, K. (2011). Water stress mediated changes in growth, physiology and secondary metabolites of desi ajwain (*Trachyspermum ammi* L.). *Pakistan Journal of Botany, 43*, 15–19.

Belarbi, K., Atik-Bekkara, F., El Haci, I. A., Bensaid, I., Beddou, F., & Bekhechi, C. (2017). In vitro antioxidant activity and phytochemical analysis of *Teucrium pseudo-Scorodonia* Desf. Collected from Algeria. *Oriental Pharmacy and Experimental Medicine, 17*, 151–160. https://doi.org/10.1007/s13596-017-0260-3.

Belyagoubi-Benhammou, N., Belyagoubi, L., Bechlaghem, N., Ghembaza, N., & Atik-Bekkara, F. (2017). Evaluación del potencial antioxidante y análisis fitoquímico del extracto crudo de *Pituranthos scoparius* y sus fracciones. *Oriental Pharmacy and Experimental Medicine, 17*, 51–57. https://doi.org/10.1007/s13596-016-0253-7.

Bentley, R. E. (2010). *Medicinal plants* (pp. 23–46). London: Domville-Fife Press.

Bhat, S. H., Ullah, M. F., & Abu-Duhier, F. M. (2019). Bioactive extract of *Artemisia judaica* causes in vitro inhibition of dipeptidyl peptidase IV and pancreatic/intestinal enzymes of the carbohydrate absorption cascade: Implication for anti-diabetic new molecular entities (NMEs). *Oriental Pharmacy and Experimental Medicine, 19*, 71–80. https://doi.org/10.1007/s13596-018-0347-5.

Bhattacharya, A. (2019). Chapter 5—High-temperature stress and metabolism of secondary metabolites in plants. In *Effect of high temperature on crop productivity and metabolism of macro molecules* (pp. 391–484). https://doi.org/10.1016/B978-0-12-817562-0.00005-7.

Bonam, S. R., Wu, Y. S., Tunki, L., Chellian, R., Halmuthur, M. S. K., Muller, S., et al. (2018). What has come out from phytomedicines and herbal edibles for the treatment of cancer? *Medicinal Chemistry in India, 13*(18), 1854–1872. https://doi.org/10.1002/cmdc.201800343.

Carvalho, I. S., Cavaco, T., Carvalho, L. M., & Duque, P. (2010). Effect of photoperiod on flavonoid pathway activity in sweet potato (*Ipomoea batatas* (L.) lam.) leaves. *Food Chemistry, 118*, 384–390.

Chang, J. D., Mantri, N., Sun, B., Jiang, L., Chen, P., Jiang, B., et al. (2016). Effects of elevated CO_2 and temperature on *Gynostemma pentaphyllum* physiology and bioactive compounds. *Journal of Plant Physiology, 196-197*, 41–52.

Charles, D. J., Simon, J. E., Shock, C. C., Feibert, E. B. G., & Smith, R. M. (1993). Effect of water stress and post-harvest handling on artemisinin content in the leaves of *Artemisia annua* L. In J. Janick, & J. E. Simon (Eds.), *Proceedings of the second national symposium: New crops, exploration, research and commercialization* (pp. 640–643). New York: John Wiley and Sons Inc.

Chen, X. L., Liu, F., Xiao, X. R., Yang, X. W., & Li, F. (2018). Anti-inflammatory abietanes diterpenoids isolated from *Tripterygium hypoglaucum*. *Phytochemistry, 156*, 167–175. https://doi.org/10.1016/j.phytochem.2018.10.001.

Cruz, P. N., Lima Reis, P. M. C., Ferreira, S. R. S., Masson, M. L., & Corazza, M. L. (2020). Encapsulation of yacon (*Smallanthus sonchifolius*) leaf extract by supercritical fluid extraction of emulsions. *The Journal of Supercritical Fluids, 160*, 1–11. https://doi.org/10.1016/j.supflu.2020.104815.

da Fonseca-Machado, A. P., Alves Rezende, C., Alexandre Rodrigues, R., Fernández Barbero, G., Vieira e Rosa, P. T., & Martínez, J. (2018). Encapsulación de extracto rico en antocianinas a partir de residuos de mora por secado por pulverización, liofilización y antidisolvente supercrítico. *Tecnología de polvo, 340*, 553–562. https://doi.org/10.1016/j.powtec.2018.09.063.

de Souza, V. B., Thomazini, M., Echalar Barrientos, M. A., Nalin, C. M., Ferro-Furtado, R., Genovese, M. I., et al. (2018). Functional properties and encapsulation of a proanthocyanidin-rich cinnamon extract (*Cinnamomum zeylanicum*) by complex coacervation using gelatin and different polysaccharides. *Food Hydrocolloids, 77*, 297–306. https://doi.org/10.1016/j.foodhyd.2017.09.040.

Dewick, P. M. (2002). *Medicinal natural products* (p. 495). New York: Jonh Wiley & Sons Ltd.

Duarte-Gómez, M. B., Brachet-Márquez, V., Campos-Navarro, R., & Nigenda, G. (2004). National health policies and local decisions in Mexico: The case of an intercultural hospital in

Cuetzalan, Puebla. *Salud publica de Mexico*, *46*(5), 388–389. https://doi.org/10.1590/s0036-36342004000500005.

El Ghallab, Y., Al Jahid, A., Jamal Eddine, J., Haj Said, A. A., Zarayby, L., & Derfoufi, S. (2019). *Syzygium aromaticum* L.: Phytochemical investigation and comparison of the scavenging activity of essential oil, extracts and eugenol. *Oriental Pharmacy and Experimental Medicine*. https://doi.org/10.1007/s13596-019-00416-7.

Eng-Seng, C., Zhi-Hui, Y., Soon-Hock, P., Mansa, R. F., & Ravindra, P. (2010). Encapsulation of herbal aqueous extract through absorption with ca-alginate hydrogel beads. *Food and Bioproducts Processing*, *88*(2–3), 195–201. https://doi.org/10.1016/j.fbp.2009.09.005.

Fathi, M., Martín, Á., & McClements, D. J. (2014). Nanoencapsulation of food ingredients using carbohydrate based delivery systems. *Trends in Food Science and Technology*, *39*, 18–39. https://doi.org/10.1016/j.tifs.2014.06.007.

Gaharwar, A. K., Peppas, N. A., & Khademhosseini, A. (2014). Nanocomposite hydrogels for biomedical applications. *Biotechnology and Bioengineering*, *111*(3), 441–453. https://doi.org/10.1002/bit.25160.

García, R. (2019). Medicina tradicional o complementaria: pacientes que lo utilizan al mismo tiempo que su tratamientofarmacológico. *Ciencia y Desarrollo*, *22*(1), 25–30. https://doi.org/10.21503/cyd.v22i1.1735.

Güney, O. I. (2019). Consumption attributes and preferences on medicinal and aromatic plants: A consumer segmentation analysis. *Ciência Rural*, *49*(5). https://doi.org/10.1590/0103-8478cr20180840.

Gunjan, M., Naing, T. W., Saini, R., Ahmad, D. A., Naidu, D. J., & Kumar, I. (2015). Marketing trends and future prospects of herbal medicine in treatment of various diseases. *World Journal of Pharmaceutical Research*, *4*(9), 132–155.

Hadacek, F. (2002). Secondary metabolites as plant traits: Current assessment and future perspectives. *Critical Reviews in Plant Sciences*, *21*, 273–322.

Heckert Bastos, L. P., Correados Santos, C. H., Geraldode Carvalho, M., & Garcia-Rojasac, E. E. (2020). Encapsulation of the black pepper (*Piper nigrum L.*) essential oil by lactoferrin-sodium alginate complex coacervates: Structural characterization and simulated gastrointestinal conditions. *Food Chemistry*, *316*, 1–33. https://doi.org/10.1016/j.foodchem.2020.126345.

Huang, W., He, X. Y., Liu, C. B., & Li, D. W. (2010). Effects of elevated carbon dioxide and ozone on foliar flavonoids of *Ginkgo biloba*. *Advanced Materials Research*, *113–116*, 165–169. https://doi.org/10.4028/www.scientific.net/AMR.113-116.165.

Hussain, A. I., Anwar, F., Hussain Sherazi, S. T., & Przybylski, R. (2008). Chemical composition, antioxidant and antimicrobial activities of basil (*Ocimum basilicum*) essential oils depends on seasonal variations. *Food Chemistry*, *108*(3), 986–995. https://doi.org/10.1016/j.foodchem.2007.12.010.

Ifeduba, E. A., & Akoh, C. C. (2015). Microencapsulation of stearidonic acid soybean oil in complex coacervates modified for enhanced stability. *Food Hydrocolloids*, *51*, 136–145. https://doi.org/10.1016/j.foodchem.2015.12.011.

Ishwarya, S. P., & Anandharamakrishnan, C. (2015). Spray-freeze-drying approach for soluble coffee processing and its effect on quality characteristics. *Journal of Food Engineering*, *149*(3), 171–180.

Ishwarya, S. P., & Anandharamakrishnan, C. (2017). Spray drying. In *Handbook of drying for dairy products* (pp. 57–94). Wiley Online Library.

Jaafar, H. Z., Ibrahim, M. H., & Karimi, E. (2012). Phenolics and flavonoids compounds, phenylalanine ammonia lyase and antioxidant activity responses to elevated CO_2 in *Labisia pumila* (Myrisinaceae). *Molecules*, *17*, 6331–6347. https://doi.org/10.3390/molecules17066331.

Jagadish, S. V. K., Craufurd, P. Q., & Wheeler, T. R. (2008). Phenotyping parents of mapping populations of rice (*Oryza sativa* L.) for heat tolerance during anthesis. *Crop Science, 48*, 1140–1146.

Jakovljevic, Z. D., Stankovic, S. M., & Topuzovic, D. M. (2013). Seasonal variability of *Chelidonium majus* L. secondary metabolites content and antioxidant activity. *EXCLI Journal, 12*, 260–268.

Jin, C., Zhao, L., Li, Y., & Zhang, M. (2019). Traditional herbal medicines. Clinical, pharmaceutical and biological analysis. *Encyclopedia of Analytical Science, 3*, 148–151. https://doi.org/10.1016/B978-0-12-409547-2.14026-0.

Khandaker, L., Masum-Akond, A. S. M. G., Ali, M. B., & Oba, S. (2010). Biomass yield and accumulations of bioactive compounds in red amaranth (*Amaranthus tricolor* L.) grown under different colored shade polyethylene in spring season. *Scientia Horticulturae, 123*, 289–294. https://doi.org/10.1016/j.scienta.2009.09.012.

Kirakosyan, A., Kaufman, P., Warber, S., Zick, S., Aaronson, K., Bolling, S., et al. (2004). Applied environmental stresses to enhance the levels of polyphenolics in leaves of hawthorn plants. *Physiologia Plantarum, 121*, 182–186.

Kiss, T., Cank, K. B., Orbán-Gyapai, O., Liktor-Busa, E., Zomborszki, Z. P., Rutkovska, S., et al. (2017). Phytochemical and pharmacological investigation of *Spiraea chamaedryfolia*: A contribution to the chemotaxonomy of *Spiraea* genus. *BMC Research Notes, 10*, 762. https://doi.org/10.1186/s13104-017-3013-y.

Kitahiro, Y., Ikeda, H., Im, H., Kodaira, E., & Shibano, M. (2019). Phytochemical characterization of *Rosa multiflora* Thunb. (Rosaceae) in Japan and South Korea, with a focus on the bioactive flavonol glycoside 'multiflorin A'. *Journal of Natural Medicines, 73*, 555–565. https://doi.org/10.1007/s11418-019-01302-x.

Krzyzanowska, J., Czubacka, A., & Oleszek, W. (2009). Dietary phytochemicals and human health. In M. T. Giardi, G. Rea, & B. Berra (Eds.), *Bio-farms for nutraceuticals: Functional food and safety control by biosensors*. Georgetown: Landes BioScience.

Lema-Rumińska, J., Kulus, D., Tymoszuk, A., Varejão, J. M. T. B., & Bahcevandziev, K. (2019). Profile of secondary metabolites and genetic stability analysis in new lines of *Echinacea purpurea* (L.) Moench micropropagated via somatic embryogenesis. *Industrial Crops and Products, 142*. https://doi.org/10.1016/j.indcrop.2019.111851.

Li, R., Lin, Z., Zhang, Q., Zhang, Y., Liu, Y., Lyu, Y., et al. (2020). Injectable and in-situ formable thiolated chitosan coated liposomal hydrogels as curcumin carriers for prevention of in vivo breast cancer recurrence. *ACS Applied Materials & Interfaces*, 1–38. https://doi.org/10.1021/acsami.9b21528.

Liu, Z. (2000). Drought-induced in vivo synthesis of camptothecin in *Camptotheca acuminata* seedlings. *Physiologia Plantarum, 110*, 483–488.

Liu, Q., Huang, H., Chen, H., Lin, J., & Wang, Q. (2019). Food-grade nanoemulsions: Preparation, stability and application in encapsulation of bioactive compounds. *Molecules, 24*, 4242. https://doi.org/10.3390/molecules24234242.

Mahmud, Z. A., Bachar, S. C., Hasan, C. M., Emran, T. B., Qais, N., & Nasir-Uddin, M. M. (2017). Phytochemical investigations and antioxidant potential of roots of *Leea macrophylla* (Roxb.). *BMC Research Notes, 10*, 245. https://doi.org/10.1186/s13104-017-2503-2.

Makimori, R. Y., Endo, E. H., Makimori, J. W., Zanqueta, E. B., Ueda. Nakamura, T., Leimann, F. V., et al. (2020). Preparation, characterization and antidermatophytic activity of free- and microencapsulated cinnamon essential oil. *Journal de Mycologie Médicale*. https://doi.org/10.1016/j.mycmed.2020.100933.

Mediani, A., Abas, F., Ping, T. C., Khatib, A., & Lajis, N. H. (2012). Influence of growth stage and season on the antioxidant constituents of Cosmos caudatus. *Plant Foods for Human Nutrition, 67*(4), 344–350. https://doi.org/10.1007/s11130-012-0317-x.

Mehalaine, S., Belfadel, O., Menasria, T., & Messaili, A. (2017). Chemical composition and anti-bacterial activity of essential oils of three medicinal plants from Algerian semi-arid climatic zone. *Phytothérapie*, 1–9. https://doi.org/10.1007/s10298-017-1143-y.

Mehalaine, S., & Chenchouni, H. (2019). Effect of climatic factors on essential oil accumulation in two lamiaceae species from algerian semiarid lands. In H. Chenchouni, E. Errami, F. Rocha, & L. Sabato (Eds.), *Advances in science, technology & innovation (IEREK interdisciplinary series for sustainable development)*. *Exploring the nexus of geoecology, geography, geoarcheology and geotourism: Advances and applications for sustainable development in environmental sciences and agroforestry research. CAJG 2018* (pp. 57–60). Cham: Springer, ISBN:9783030016838. https://doi.org/10.1007/978-3-030-01683-8_12.

Mouderas, F., El Haci, I. A., & Lahfa, F. B. (2019). Phytochemical profile, antioxidant and antimicrobial activities of *Traganum nudatum* Delile aerial parts organic extracts collected from Algerian Sahara's flora. *Oriental Pharmacy and Experimental Medicine*, *19*, 299–310. https://doi.org/10.1007/s13596-019-00365-1.

Mueller, D., Jung, K., Winter, M., Rogoll, D., Melcher, R., Kulozik, U., et al. (2018). Encapsulation of anthocyanins from bilberries—Effects on bioavailability and intestinal accessibility in humans. *Food Chemistry*, *248*, 217–224. https://doi.org/10.1016/j.foodchem.2017.12.058.

Ning, W., Peng, X., Ma, L., Cui, L., Lu, X., Wang, J., et al. (2012). Enhanced secondary metabolites production and antioxidant activity in postharvest *Lonicera japonica* Thunb. in response to UV radiation. *Innovative Food Science & Emerging Technologies*, *13*, 231–243. https://doi.org/10.1016/j.ifset.2011.10.005.

Palit, P., Mukherjee, D., Mahanta, P., Shadab, M., Ali, N., Roychoudhury, S., et al. (2018). Attenuation of nociceptive pain and inflammatory disorders by total steroid and terpenoid fraction of *Euphorbia tirucalli* Linn root in experimental in vitro and in vivo model. *Inflammopharmacology*, *26*, 235–250. https://doi.org/10.1007/s10787-017-0403-7.

Pellegrini, E., Carucci, G., Campanella, A., Lorenzini, G., & Nali, C. (2011). Ozone stress in *Melissa officinalis* plants assessed by photosynthetic function. *Environmental and Experimental Botany*, *73*, 94–101.

Pichersky, E., & Gang, D. R. (2000). Genetics and biochemistry of secondary metabolites in plants: An evolutionary perspective. *Trends in Plant Science*, *5*(10), 439–445. https://doi.org/10.1016/s1360-1385(00)01741-6.

Poosri, S., Thilavech, T., Pasukamonset, P., Suparpprom, C., & Adisakwattana, S. (2019). Studies on Riceberry rice (*Oryza sativa* L.) extract on the key steps related to carbohydrate and lipid digestion and absorption: A new source of natural bioactive substances. *NFS Journal*, *17*, 17–23. https://doi.org/10.1016/j.nfs.2019.10.002.

Priya, S., & Satheeshkumar, P. K. (2020). Chapter 5—Natural products from plants: Recent developments in phytochemicals, phytopharmaceuticals, and plant-based neutraceuticals as anticancer agents. In *Functional and preservative properties of phytochemicals* (pp. 145–163). https://doi.org/10.1016/B978-0-12-818593-3.00005-1.

Ramakrishna, E., Dev, K., Kothari, P., Kumar Tripati, A., Trivedi, R., & Maurya, R. (2017). Phytochemical investigation of *Kigelia pinnata* leaves and identification of osteogenic agents. *Medicinal Chemistry Research*, *26*, 940–946. https://doi.org/10.1007/s00044-017-1807-z.

Rao, S. R., & Ravishankar, G. A. (2002). Plant cell cultures: Chemical factories of secondary metabolites. *Biotechnology Advances*, *20*, 101–153. https://doi.org/10.1016/S0734-9750(02)00007-1.

Rehman, A., Ahmad, T., Aadil, R. M., Spotti, M. J., Bakry, A. M., Khan, I. M., et al. (2019). Pectin polymers as wall materials for the nano-encapsulation of bioactive compounds. *Trends in Food Science and Technology*, *90*, 35–46. https://doi.org/10.1016/j.tifs.2019.05.015.

Rezvankhah, A., Emam-Djomeh, Z., & Gholamreza, A. (2020). Encapsulation and delivery of bio-active compounds using spray and freeze-drying techniques: A review. *Drying Technology*, *38*(1), 235–258. https://doi.org/10.1080/07373937.2019.1653906.

Riachi, L. G., Simas, D., Coelho, G. C., Marcellini, P. S., Ribeiro da Silva, A. J., & Bastos de Maria, C. A. (2018). Effect of light intensity and processing conditions on bioactive compounds in maté extracted from yerba mate (*Ilex paraguariensis* A. St.-Hil.). *Food Chemistry*, *266*, 317–322. https://doi.org/10.1016/j.foodchem.2018.06.028.

Santos, D. T., Albarelli, J. Q., Beppu, M. M., & Meirelesa, A. A. (2013). Stabilization of anthocyanin extract from jabuticaba skins by encapsulation using supercritical CO2 as solvent. *Food Research International*, *50*, 617–624. https://doi.org/10.1016/j.foodres.2011.04.019.

Sarkar, K. K., Mitra, T., Acharyya, R. N., & Sadhu, S. K. (2019). Phytochemical screening and evaluation of the pharmacological activities of ethanolic extract of *Argemone mexicana* Linn. aerial parts. *Oriental Pharmacy and Experimental Medicine*, *19*, 91–106. https://doi.org/10.1007/s13596-018-0357-3.

Schreiner, M. (2005). Vegetable crop management strategies to increase the quantity of phytochemicals. *European Journal of Nutrition*, *44*, 85–94. https://doi.org/10.1007/s00394-004-0498-7.

Serra, E., Saubade, F., Ligorio, C., Whitehead, K., Sloan, A., Williams, D. W., et al. (2020). Methylcellulose hydrogel with *Melissa officinalis* essential oil as a potential treatment for Oral candidiasis. *Microorganisms*, *8*(2), 2015. https://doi.org/10.3390/microorganisms8020215.

Shitan, N. (2016). Secondary metabolites in plants: Transport and self-tolerance mechanisms. *Bioscience, Biotechnology, and Biochemistry*, *80*(7), 1283–1293. https://doi.org/10.1080/0916845 1.2016.1151344.

Silva Soares, B., Pinto Siqueira, R., Geraldo de Carvalho, M., Vicente, J., & Garcia-Rojasad, E. E. (2019). Microencapsulation of sacha inchi oil (*Plukenetia volubilis* L.) using complex coacervation: Formation and structural characterization. *Food Chemistry*, *298*, 1–8. https://doi.org/10.1016/j.foodchem.2019.125045.

Soliz-Guerrero, J. B., de Rodriguez, D. J., Rodriguez-Garcia, R., Angulo-Sanchez, J. L., & Mendez-Padilla, G. (2002). Quinoasaponins: concentration and compositionanalysis. In J. Janick, & A. Whipkey (Eds.), *Trends in new crops and new uses* (p. 110). Alexandria: ASHS Press.

Somerville, C., & Somerville, S. (1999). Plant functional genomics. *Science*, *285*, 380–383.

Srivastava, A., Srivastav, P., Pandey, A., Khanna, V. K., & Pantalón, A. B. (2019). Chapter 24—Phytomedicine: A potential alternative medicine in controlling neurological disorders. In *New look to phytomedicine advancements in herbal products as novel drug leads* (pp. 625–655). ISBN 9780128146194 https://doi.org/10.1016/B978-0-12-814619-4.00025-2.

Suurbaar, J., Mosobil, R., & Donkor, A. (2017). Antibacterial and antifungal activities and phytochemical profile of leaf extract from different extractants of *Ricinus communis* against selected pathogens. *BMC Research Notes*, *10*, 660. https://doi.org/10.1186/s13104-017-3001-2.

Tan, S. P., Kha, T. C., Parksa, S. E., Stathopoulos, C. E., & Roach, P. D. (2015). Effects of the spray-drying temperatures on the physiochemical properties of an encapsulated bitter melon aqueous extract powder. *Powder Technology*, *281*, 65–75. https://doi.org/10.1016/j.powtec.2015.04.074.

Thomford, N. E., Senthebane, D. A., Rowe, A., Munro, D., Seele, P., Maroyi, A., et al. (2018). Natural products for drug discovery in the 21st century: Innovations for novel drug discovery. *International Journal of Molecular Sciences*, *19*(6), 1578. https://doi.org/10.3390/ijms19061578.

Tomšik, A., Šarić, L., Bertoni, S., Protti, M., Albertini, B., Mercolini, L., et al. (2019). Encapsulations of wild garlic (*Allium ursinum* L.) extract using spray congealing technology. *Food Research International*, *119*, 941–950. https://doi.org/10.1016/j.foodres.2018.10.081.

Wahab, A., Jan, S. A., Rauf, A., Rehman, Z. U., Khan, Z., Ahmed, A., et al. (2018). Phytochemical composition, biological potential and enzyme inhibition activity of *Scandix pecten-veneris* L. *Journal of Zhejiang University Science B*, *19*, 120–129. https://doi.org/10.1631/jzus.B1600443.

World Health Organization. (2019). *WHO global report on traditional and complementary medicine*. https://www.who.int/traditionalcomplementaryintegrativemedicine/WhoGlobalReportOnTraditionalAndComplementaryMedicine2019.pdf?ua=1. Accessed 27.03.20.

Zhang, T., Lo, C., Xiao, M., Cheng, L., Mokc, C. K., & Shaw, P. (2020). Anti-influenza virus phytochemicals from *Radix Paeoniae* Alba and characterization of their neuraminidase inhibitory activities. *Journal of Ethnopharmacology*, *253*. https://doi.org/10.1016/j.jep.2020.112671.

Zhang, N., & Mutilangi, W. (2013). *Coacervate complexes, methods and food products: Google patents*. https://patentimages.storage.googleapis.com/aa/ef/fb/9d50ea4eb157ff/US20130004617A1.pdf (Accessed 27.03.20).

Zouaoui, N., Chenchoun, H., Bouguerra, A., Massouras, T., & Barkat, M. (2020). Characterization of volatile organic compounds from six aromatic and medicinal plant species growing wild in north African drylands. *NFS Journal*, *18*, 19–28. https://doi.org/10.1016/j.nfs.2019.12.001.

Chapter 7

Medicinal plants and their traditional uses in different locations

Juan José Maldonado Miranda
Faculty of Professional Studies, Huasteca Zone, Autonomous University of San Luis Potosí, San Luis Potosí, Mexico

1. Introduction

Since ancient times, humankind in various cultures have implemented the use of plants for therapeutic and medicinal purposes. Medicinal plants are an important source of current medicines, as approximately 25% of the drugs prescribed worldwide come from plants (Alonso Castro et al., 2012). According to the World Health Organization (2004), the use of medicinal plants continues to expand rapidly worldwide, and many people take medicines or plant-based products for their medical care. The use of herbal medicine has increased around the world due to its presumptive efficiency, availability, and general acceptance (Alonso Castro et al., 2017).

In Mexico, of the total vascular species available, 16% is used for medicinal purposes (CONABIO, 2006). In addition, 90% of Mexicans are considered to use medicinal plants, herbal remedies, or some plant-based preparation for the empirical treatment of various diseases due to the efficacy, tradition, and low cost of plants. However, they do not often inform their doctors about the use of medicinal plants; so in many cases it is not possible to report the exact number of plants used (Alonso Castro et al., 2012; Poss, Pierce, & Prieto, 2005).

In Huasteca Potosina, Mexico, a great wealth of plant species with medicinal properties is preserved, as reported by Alonso Castro et al. (2012). Due to the great diversity of flora present in the region, there is great potential to research new compounds to treat various conditions (Alonso-Castro et al., 2011). In the Huasteca Potosina, plants represent an important natural resource for traditional medicine.

Therefore the objective of this book chapter was to conduct a review on the basis of scientific data on the traditional use of medicinal plants globally and

Phytomedicine: A Treasure of Pharmacologically Active Products from Plants
https://doi.org/10.1016/B978-0-12-824109-7.00014-5

in particular in a locality with an indigenous population, so that the traditional knowledge about the consumption of medicinal plants can be documented.

2. Traditional medicine

Traditional medicine is the set of knowledge, skills (ability to employ empirical knowledge), and practices based on theories, beliefs, and experiences of different cultures, whether they are explicable or not and used for the maintenance of health and for the prevention, diagnosis, improvement, or treatment of physical or mental illness (WHO, 2017). Generally the use of herbal medicine is highly frequent in traditional medicine for the treatment of diseases. However, traditional medicine is a wider area, where the use of animals, fungi, or other components of nature (rocks, minerals, etc.) can also be included for the treatment of conditions or diseases.

Across the world, traditional medicine either is the mainstay of healthcare delivery or serves as a complement to it. In some countries, traditional medicine or nonconventional medicine may be termed complementary medicine (WHO, 2013) for its, more recently, popular use in parallel with allopathic medicine, especially for the treatment and control of chronically diseases.

Diffusion and the increasing use of traditional medicine have created challenges in public health from the point of view of politics, safety, efficacy, quality, access, and rational use (WHO, 2013). Mainly because for some ethnic groups, traditional medicine has represented the only option for disease prevention and cure; this is mainly due to exclusion and extreme poverty in which they live, as well as the lack of health services (http://www.udg.mx/, 2011).

Traditional medicine is an important healthcare component in low-income countries. The prevalence of traditional medicine use in low-income countries is estimated to be between 40% and 71% (Bodeker & Kronenberg, 2002), for example, in sub-Saharan Africa, it is estimated at 58.2% on average in the general population, but prevalence rates vary widely among studies (James, Wardle, Steel, & Adams, 2018).

The high prevalence of traditional medicine use in low-income countries has important clinical implications, especially when traditional medicine and conventional treatments are used concurrently. Traditional medicine use has important historical and cultural significance in diverse settings and populations and may provide benefit when used safely and appropriately (Hill et al., 2019).

3. Ethnopharmacology

The term ethnopharmacology is defined as the interdisciplinary field that studies natural medicine derived from plants, animals, fungi, and other natural resources that have traditionally been used by cultural groups of people to treat various diseases (Bannister, 2009). Two aspects are particularly important in ethnopharmaceutology: (1) the potential health benefits and (2) the potentially toxic risks of traditional remedies. Empirical knowledge of phytotherapy was transmitted orally and occasionally recorded in herbariums (Heinrich, Barnes, Prieto-Garcia, Gibbons, & Williamson, 2017). Today ethnopharmacology has focused on the development of medicinal plant–based drugs used in indigenous communities. The purpose of the International Society for Ethnopharmacology (ISE) is to contribute to the development of new pharmaceutical products; to examine, in situ, the importance of medicinal plant extracts, pure compounds, and other products derived from nature; and to better understand the drugs and help improve the use of those products (Bannister, 2009; Elisabetsky & Etkin, 2009).

4. Medicinal plants

The term medicinal plants was first used in 1967 in the context of the study of hallucinogenic plants. A medicinal plant is that species of the plant kingdom, whose parts (flowers, leaves, roots, stems, fruits, or seeds) are directly used or used in some preparation as a medicine to treat a condition or disease. Knowledge of the beneficial properties of medicinal plants to treat diseases represents a valuable resource to preserve the biological and cultural diversity of different ethnicities (Heinrich et al., 2017).

Medicinal plants have been used by indigenous physicians since pre-Hispanic times and are part of the traditional knowledge of humanity (Heinrich, Ankli, Frei, Wiemann, & Sticher, 1998). The use of medicinal plants derives from having secondary metabolites with pharmacological properties, and some are an important source of components for antitumors, antivirals, antiepileptics, antibiotics, antiinflammatories, antinociceptives, among others (Alonso-Castro et al., 2011; Le Rhun, Devos, & Bourg, 2019; Sharma, Flores-Vallejodel, Cardoso-Taketa, & Villarreal, 2017; Wang et al., 2019).

The most current reports indicate that the planet is home to a diversity of 391,000 vascular plants, of which at least 35,000 species have a potential medicinal use. In addition, 25% of the bioactive compounds used in various medicines come from plants; there are also reports indicating that at least 80% of the world's population depend on traditional herbal remedies (García de Alba García et al., 2012; Kew, Quinn, Quon, & Ducharme, 2016). Worldwide the use of natural products for primary health care has increased during the last decade (Ekor, Adeyemi, & Otuechere, 2013). The global market for medicinal plants and plant-derived drugs in 2015 was estimated at 25.6 billion dollars and is expected to rise to 35.4 billion dollars in 2020 (Board of Trustees of the Royal Botanic Gardens, Kew et al., 2016). This clearly indicates that the consumption of medicinal plants is a current topic of interest (Alonso Castro et al., 2017). Nevertheless, many medicinal plants lack scientific evidence for their toxicity, chemical composition, and pharmacological effects (Alonso-Castro et al., 2018). Despite improving access to health care in urban areas, many people rely on medicinal plants (Alonso Castro et al., 2012; Juárez Vázquez et al., 2013).

The preparation and consumption of medicinal plants depend on the place where they are used. However, there are some methods for the extraction of its bioactive compounds. The selection of the extraction method depends on the time and economic resources available to them.

In Fig. 1 the different treatments that are traditionally used and at the research level in a specialized laboratory in San Luis Potosí, Mexico, for the extraction of active compounds from medicinal plants are represented.

5. Herbal medicine

Herbal medicine is conceived from different points of view depending on the place where it is used, but its definitions have the same conception, that is, it is the use of plants or parts of the plant to treat a disease and achieve health wellness. The herbal medicines include herbs, herbal materials, herbal preparations, and finished herbal products that contain as active ingredient parts of plants, or other plant materials, or combinations (WHO, 2019).

Humankind in its early years used nature to implement suitable conditions for its survival, improving its quality of life over the years. Since those ancient years the knowledge of herbal medicine based on tradition has served to cure the ailments of the humankind (González Rodríguez & Cardentey García, 2016). Herbal medicine includes numerous substances extracted from plants, homemade infusions, plants harvested for medicinal purposes, and products that have to be approved by government regulatory bodies (Carrillo Esper, Lara Caldera, & Ruiz Morales, 2010). Much of the herbal remedies are obtained by alcoholic, acetonic, or aqueous extraction and sometimes without manipulation of seeds, leaves, stem, bark, or roots of these plants (Carrillo Esper et al., 2010).

The use of herbal medicine has increased around the world due to its presumptive efficiency, availability, and general acceptance. Approximately 80% of the general population, especially in developing countries, use medicinal herbs for primary health care (Taddei Bringas, Santillana Macedo, Romero Cancio, & Romero Tellez, 1999; WHO, 1993).

6. Global use of herbal medicine

In the early civilizations of Mesopotamia, Egypt, China, and India, written evidence was left on the use of plants to treat numerous physiological and spiritual diseases of different populations. In Mesopotamia, reports on tablets (cuneiform system) of plant use date from approximately 2600 BCE, where substances obtained from about 1000 plants such as *Cupressus sempervirens* (cyprus), *Glycyrrhiza glabra* (licorice), and *Papaver somniferum* (opium poppy) were used for cough treatments, colds, parasitic infections, and inflammations. In Egypt, there is the pharmaceutical report "Ebers Papyrus" from 1500 BCE, which contains about 700 medicines (most of which come from plants, although animal organs and some minerals are included).

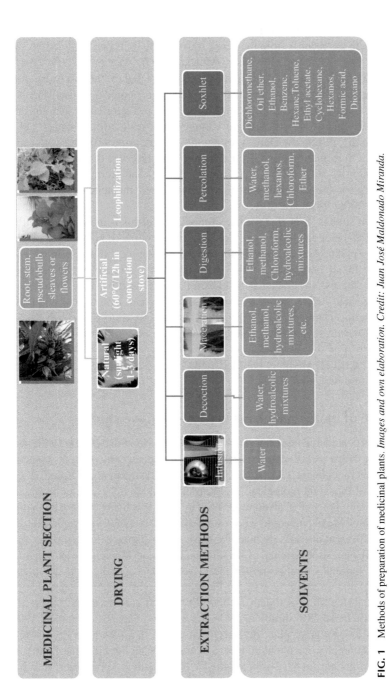

MEDICINAL PLANT SECTION

Root, stem, pseudobulb sleaves or flowers

DRYING

Natural (sunlight 1-3 days)

Artificial (60°C/12h in convection stove)

Leophilization

EXTRACTION METHODS

Infusion

Decoction

Maceration

Digestion

Percolation

Soxhlet

SOLVENTS

Water

Water, hydroalcolic mixtures

Ethanol, methanol, hydroalcolic mixtures, etc.

Ethanol, methanol, Chloroform, hydroalcolic mixtures

Water, methanol, hexanos, Chloroform, Ether

Dichloromethane, Oil ether, Ethanol, Benzene, Hexane, Toluene, Ethyl acetate, Cyclohexane, Hexanos, Formic acid, Dioxano

FIG. 1 Methods of preparation of medicinal plants. *Images and own elaboration. Credit: Juan José Maldonado Miranda.*

On the other hand, in China, medicine refers to the document "Materia Medica" written in 1100 BCE, which describes 52 herbal prescriptions; in addition, there are records from the years 100 and 659 BCE that included 365 and 850 medicines, respectively (Castro Méndez, 2006; Wiseman, 2004; Zhou & Nunes, 2015). In Central Africa and America, since before the discovery of the New World, it has been proven that all indigenous people also used plants as food, for their varied health conditions, and rites to their deities, among other uses. Native plants are still used for similar purposes (Castro Méndez, 2006). Subsequently, during the Dark Ages and the Middle Ages (5th and 12th centuries), basic information on medicinal plants was preserved thanks to monasteries in countries such as England, Ireland, France, and Germany, which acted as centers for the production of preparations from medicinal plants (Castro Méndez, 2006; Wiseman, 2004).

Recent studies show that the use of medicinal plants is increasingly settling as a global practice, since there are scientific reports showing that different medicinal plants are used to treat the most common local diseases in different countries in all continents of the world. In Latin American and European countries, the use of medicinal herbalism has increased by more than 12% in recent decades, while in Asian countries, more than 70% of the population use medicinal plants (Villar López, Ballinas Sueldo, Soto Franco, & Medina Tejada, 2016). A review was conducted on the use of different medicinal plants worldwide. For this review, academic databases such as SCOPUS, Web of Science, SCIELO, Medline, PubMed, and Google Scholar were consulted; this report is summarized in Table 1.

7.   Use of medicinal plants in Mexico: Case study

Mexico is considered one of the countries in the world with the greatest diversity of plants, with a record of around 30,000 vascular plants, of which 4000 species have medicinal effects (CONABIO, 2006; Ocegueda, Moreno, & Koleff, 2005). Because of this great biological wealth, traditional medicine is used in Mexico as a recurring practice (Caballero Uribe et al., 2002), which can be seen in more than 90% of Mexican people who use medicinal plants for the empirical treatment of different diseases (Schlaepfer & Mendoza-Espinoza, 2010). To register medicinal plants the Library of Traditional Mexican Medicine was integrated into the National Commission for the Development of Indigenous Peoples in 1990 in Mexico, setting the base for the Digital Library of Traditional Mexican Medicine, which brings together about 2000 plant species from 183 families (about 28% trees, 28% shrubs, and 44% herbs) (Bye, Linares, & Estrada, 1995). Within Mexico's great plant diversity, some plant families have been studied from an ethnopharmacological approach, such as the Orchidaceae family, for which various species have been reported to have medicinal properties. However, most of the reports come from empirical knowledge provided by different ethnic groups in Mexico, and very few species have been scientifically vali-

TABLE 1 Ethnobotanical cultural use and exploitation of medicinal plants from different countries of the world.

Continents	Countries	Main diseases	Plant used	Bibliographic citations
America	Mexico	Digestive system	*Matricaria chamomilla* L., *Mentha x piperita* L., *Wormwood Artemisia absinthium* L., and *Psidium guajava* L.	Ortega Cala et al. (2019)
	Mexico	Dermatological affections, inflammation of the skin	Solanaceae, *Heterotheca inuloides*, *Aloe vera*, *Oenothera rosea*	Esquivel García, Pérez Calix, Ochoa Zarzosa, and García Pérez (2018)
	Mexico	Cancer, diabetes mellitus	*Justicia spicigera* (Mohuite)	Alonso Castro et al. (2017)
	Mexico	Gastro intestinal, gynecological, respiratory problems, deworming	*Juglans regia* L. (nuez), *Teloxys ambrosioides* L. Weber (epazote), and *Argemone ochroleuca* (chicatl)	Barrera Catalán, Herrera Castro, Catalán Heverástico, and Ávila Sánchez (2015)
	Mexico	Renal problems	*Mentha piperita*, *Marrubium vulgare*, *Mimosa albida*, and *Psidium guajava*	Juárez Vázquez et al. (2013)
	United States	Cancer, diabetes mellitus, hypertension	*Cymbidium*, *Dendrobium*, *Phalaenopsis*	Sranjeet-Kaurt (2018)
	United States	Rejuvenation	*Echinacea purpurea*, *Valeriana officinalis*, *Ginkgo biloba*, *Hypericum perforatum*, and *Trigonella foenum-graecum*	Ivanova, Kuzmina, Braukmann, Borisenko, and Zakharov (2016)
	United States	Diabetes mellitus, hypertension	*Spartina alterniflora*	Murphy, Wimp, Lewis, and Denno (2012)
	United States	Diabetes mellitus, hypertension, arthritis	*Morinda citrifolia* L. Noni	Palu, Kauvaka, Kalisi, Palu, and Hifo (2010)

Continued

TABLE 1 Ethnobotanical cultural use and exploitation of medicinal plants from different countries of the world—cont'd

Continents	Countries	Main diseases	Plant used	Bibliographic citations
	Ecuador	Renal problems, respiratory and skin problems, diabetes mellitus, hypertension, arthritis, cancer	*Melissa officinalis* (Toronjil), *Glycyrrhiza glabra* (Zaragoza), *Mentha spicata* (Menta), *Origanum vulgare* L. (orégano), *Chenopodium ambrosioides* (paico), *Aloe vera* (sábila), *Scoparia dulcis* (teatina), *Chamaemelum nobile* (manzanilla), *Eucalyptus urograndis* (eucalipto), *Plantago major* (llantén), *Allium sativum* (ajo), *Kalanchoe pinnata* (hoja del aire), *Citrus limon* (limón), *Averrhoa carambola* (carambola)	Gallegos Zurita (2016)
	Brazil	Analgesics	*Cocos nulifera* L.	Heisler et al. (2015)
	Colombia	Gastrointestinal, hypertension, renal problems	*Nicotiana tabacum* (tobacco)	Cardona Arias and Palomino Rivera (2012)
Africa	Cameroon	Gout, gastrointestinal, hypertension, renal problems	*Amaranthus spinosus*, *Termitomyces clypeatus*, *Irvingia gabonensis*, *Ricinodendron heudelotii*, *Aframomum* sp.	Fedoung Fongnzossie et al. (2020)
	Nigeria	Cancer Mellitus diabetics, hypertension	*Gnetum africanum* (Igbo name: Ukazi), *Gongronema latifolium*, (Igbo name: Ugu), *Telfairia occidentalis* (Igbo name: Nchoanwu), *Ocimum gratissimum*	Iweala Emeka (2009)

Continent	Country	Use	Plant species	Reference
Asia	India	Diabetes mellitus, hypertension	*Brassaiopsis hainla* Seem., *Gnetum gnemon* L., *Pilea scripta* (Buch.-Ham. ex D. Don) Wedd., *Rhynchotechum ellipticum* (Wall. ex D. Dietr.) A. D.C., *Sarcochlamys pulcherrima* Gaudich. Voy. Bonite.	Panmei, Gajurel, and Singh (2019)
	India	Hypertension, gastrointestinal, anti-HIV, antiinflammatory, antioxidant	*Ocimum basilicum* L. (sweet basil)	Biswas and Mukhopadhyay (2018)
	India	Diabetes mellitus diabetics, hypertension	*Capparis zeylanica* (CZ)	Balekari and Veeresham (2015)
	India	Cancer	*Morinda citrifolia* L. (noni)	Westendorf and Mettlich (2009)
	China	Cancer	(*Hedyotis diffusa*) snake needle grass	Chen (2010)
Europe	Spain	Therapeutic uses	*Malva sylvestris*, *Juglans regia Mentha* spp., *Sambucus nigra*, and *Urtica dioica*	Herrero and Santos (2009)
	Spain	Abortifacient, digestive, circulatory, antirheumatic, antiseptic, and antiinflammatory, analgesic and repellent against parasites	(*Ruta* L., Rutaceae) Rue.	San Miguel (2003)
	Spain	Therapeutic uses	*Rosmarinus officinalis*, and *Sambucus nigra*, *Ruta graveolens*, *Urtica dioica*, *Malva* sp., *Juglans regia*, *Allium sativum*, *Taxus baccata*, *Saxifraga spathularis*, *Taraxacum officinale*, *Osmunda regalis*, and *Silybum marianum*	González Rodríguez and Cardentey García (2016)

Credit: Iuan José Maldonado Miranda based on various research queries.

dated (Castillo-Pérez, Martínez-Soto, Maldonado-Miranda, Alonso-Castro, & Candy Carranza-Álvarez, 2019).

7.1 Description of the traditional use of medicinal plants in Huasteca Potosina, Mexico

The state of San Luis Potosi ranks ninth nationwide in biodiversity of flora and fauna, due to its climate, soils, hydrology, geology, and location (Reyes-Pérez, Vázquez-Solís, Reyes-Hernández, & Nicolás-Carettay Rivera-González, 2012). The state is divided into four regions, one of which is the Huasteca Potosina. This region comprises 20 of the 58 municipalities of the state of San Luis Potosi and is mainly inhabited by indigenous groups Teenek, Nahuas, Pames, and Mestizos (Gallardo Arias, 2004). This region is located at the eastern end of the state, on the coastal plain of the Gulf of Mexico. According to Luna-Vargas (2014), Huasteca Potosina is considered a rich region in biological diversity, and some plants are used for their medicinal properties for the treatment of diseases. The prescription of medicinal plants in rural areas of Huasteca Potosina is mainly given by healers called "shamanes" (Alonso Castro et al., 2012). In a quantitative study carried out in the municipality of Aquismon, S.L.P., it was found that a total of 73 plants belonging to 37 families are used for medical purposes, which are used to treat 52 diseases and 2 cultural affiliations. Among the main diseases for which these plants are used are those of the respiratory and gastrointestinal type (Alonso Castro et al., 2012). In Fig. 2 the relationship between society and medicinal plants is presented. The figure was made with information collected in a case study.

7.1.1 Data collection in the case study

The present case study was conducted in popular markets in Huasteca Potosina, which represent the main point of sale of medicinal plants in that region (Fig. 3). Ten markets from different municipalities of Huasteca Potosina were visited to carry out this study. For data collection a cross-sectional descriptive and participatory methodology was followed. The study group were informants (sellers), connoisseurs of medicinal plants in each market visited, selected according to the following inclusion criteria: (i) sellers of medicinal plants in stalls in each market, (ii) positive response to participate and answer interview questions, and (iii) people over the age of 18. The collection of information on the traditional use of medicinal plants was done through a semistructured interview applied to sellers. Sociodemographic variables and knowledge of practices related to the traditional use of medicinal plants were consulted and measured.

7.1.2 Results of the case study

In the 10 markets visited, a total of 35 sites were found selling medicinal plants; however, not all vendors agreed to provide information for fear of being questioned by the way of obtaining the plants or by lack of time to participate. The

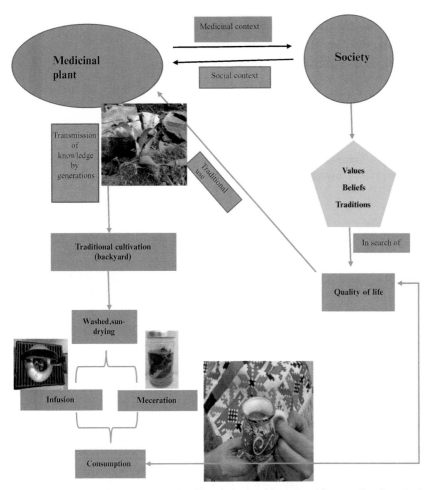

FIG. 2 Medicinal plant-society relationship. *Author's elaboration. Photo credit: Juan José Maldonado Miranda.*

FIG. 3 Image of sale point of medicinal plants at a local market in a Teenek community in Huasteca Potosina. *Photo credit: Juan José Maldonado Miranda.*

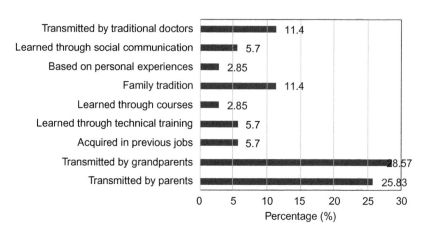

FIG. 4 Knowledge acquisition process on the use and properties of medicinal plants. *Credit: Juan José Maldonado Miranda.*

average age of the sellers was 58, of which 60% were females and 40% males. The educational level of the total number of participants was 20% with technical preparation, 70% with basic education, and 10% without any educational level. Fig. 4 shows the results on how vendors gained knowledge about traditional uses of medicinal plants.

Most of the interviewees indicated that knowledge was passed down from generation to generation (Fig. 4). In terms of obtaining medicinal plants, 40% commented that they collect plants directly in the field, and 60% commented that they obtain the plants from registered distributors. The traditional use attributed to medicinal plants is summarized in Fig. 5.

Total vendors (100%) expressed knowledge of the traditional uses of medicinal plants; 80% indicated how they were prepared and consumed. However, 100% indicated that the chemical compositions and the toxicity of the chemical compositions were unknown to them, and only 25% of them commented that they issued recommendations on extended use. In this regard the sellers were inquired whether according to their experience in the subject of medicinal plants, they would recommend their use as the sole option to treat a disease or if they would recommend the use of plants as a complement to prescribed drugs issued by top-level doctors. Eighty percent of the vendors responded that they would not recommend the exclusive use of a plant for a disease, and 20% said that they would recommend it based on personal experiences in which they have observed that plants can cure certain diseases on their own.

As for supplementing top-level medicine with the consumption of medicinal plants, most informants expressed that they do not have enough information about the additional drugs that people who buy plants consume, so they cannot make suggestions about the complementary use of both treatments. In addition, the study also asked about the perception that sellers have about the use that people give to the medicinal plants they buy. Comments were grouped into two factors: (1) social and (2) economic (Table 2).

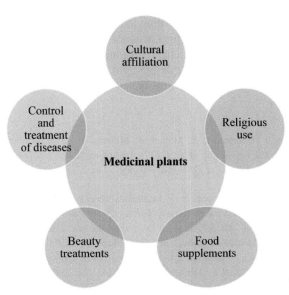

FIG. 5   Principal uses of medicinal plants in Huasteca Potosina. Self-creation with information collected in the field. *Credit: Juan José Maldonado Miranda.*

TABLE 2 Factors that influence the use of traditional medicinal plants.

Factor	Description
Social	Belief that medicinal plants have fewer side effects than medicinal drugs
	Family tradition
	Use and custom
	Recommendation by people who had positive results
Economic	The low cost of medicinal plants in comparison with the price of medicinal drugs
	Lack of public services
	Easy access to medicinal plants
	No need of a prescription for purchasing medicinal plants

Credit: Juan José Maldonado Miranda.

8.   Conclusion and perspectives

Medicinal plants are an essential natural resource for the treatment of more persistent diseases. Various medicinal plants can be used to treat similar diseases, depending on the country in which the disease occurs. In some localities, medicinal plants are perceived according to their traditional uses and represent a low-cost alternative to treat various diseases. However, more ethnobotanical studies are still needed to quantitatively document the use of medicinal plants and their beneficial effects when they are used as the only option to treat a disease, as well as their toxicological effects.

References

Alonso Castro, A. J., Domínguez, F., Maldonado Miranda, J. J., Castillo Pérez, L. J., Carranza Álvarez, C., Solano, E., et al. (2017). Use of medicinal plants by health professionals in Mexico. *Journal of Ethnopharmacology, 198,* 81–86. https://doi.org/10.1016/j. jep.2016.12.038.

Alonso-Castro, A. J., Carranza-Álvarez, C., Maldonado-Miranda, J. J., Jacobo-Salcedo, M. R., Quezada-Rivera, D. A., Lorenzo-Márquez, H., et al. (2011). Zootherapeutic practices in Aquismón, San Luis Potosí, México. *Journal of Ethnopharmacology, 138,* 233–237.

Alonso Castro, A., Maldonado Miranda, J. J., Zarate Martinez, A., del Jacobo Salcedo, M. R., Fernández Galicia, C., Figueroa Zuñiga, A. L., et al. (2012). Medicinal used in the Huasteca Potosina, México. *Journal of Ethnopharmacology, 143,* 292–298. https://doi.org/10.1016/j. jep.2012.06.035.

Alonso-Castro AJ, Zapata-Morales JR, Arana-Argáez V, Torres-Romero JC, Ramírez-Villanueva E, Pérez-Medina SE, et al. (2018). Pharmacological and toxicological study of a chemical-standardized ethanol extract of the branches and leaves from Eysenhardtia polystachya (Ortega) Sarg. (Fabaceae). Journal of Ethnopharmacology 224, 314–322. doi:10.1016/j.jep.2018.06.016. Epub 2018 Jun 18. PMID: 29913299.

Balekari, U., & Veeresham, C. (2015). In vivo and in vitro evaluation of anti diabetic and insulin secretagogue activities of Capparis zeylanica. *Pharmacology & Pharmacy, 6,* 311–320. https:// doi.org/10.4236/pp.2015.67033.

Bannister, K. (2009). Instrumentos no legales para la protección del patrimonio cultural inmaterial: Roles clave para los códigos éticos y protocolos comunitarios. In C. Bell, & RK Paterson (Eds.), *Protección del patrimonio cultural de las primeras naciones: leyes, políticas y reformas.* Toronto, ON: UBC Press.

Barrera Catalán, E. D., Herrera Castro, N., Catalán Heverástico, C., & Ávila Sánchez, P. (2015). Medicinal plants in Tixtla de Guerrero, Mexico. *Revista Fitotecnia Mexicana, 38,* 109–111.

Biswas, T., & Mukhopadhyay, S. (2018). Estimation of rosmarinic acid content of genetically stable clones of *Ocimumbasilicum* L. (sweet basil) regenerated through in vitro shoot bud multiplication and root culture. *Open Access Journal of Medicinal and aromatic Plants, 9,* 34–41.

Board of Trustees of the Royal Botanic Gardens, Kew. (2016). Introduction to the State of the World's Plants. *The State of the World's Plants Report. Foundation would like to thank the Sfumato Foundation for generously funding the State of the Word's Plants Project.* (84 pp). Kew: Royal Botanic Gardens. Available from: https://ceiba.org.mx/publicaciones/plantsreport2016. pdf; (2010) (Accessed 20 March 2010).

Bodeker, G., & Kronenberg, F. A. (2002). Public health agenda for traditional, complementary, and alternative medicine. *American Public Health Association, 92,* 1582–1591. https://doi. org/10.2105/ajph.92.10.1582. 12356597.

Bye, R., Linares, E., & Estrada, E. (1995). Biological diversity of medicinal plants in México. In J. T. Arnason, R. Mata, & J. T. Romeo (Eds.), *Recent advances in phytochemistry (Proceedings of the phytochemical society of North America): Vol. 29. Phytochemistry of medicinal plants.* Boston, MA: Springer. https://doi.org/10.1007/978-1-4899-1778-2_4.

Caballero Uribe, CV., Wilches, H., Wilches, K., Rojas, J., Salas Siado, J., Forero, E., et al. (2002). Utilización de medicinas alternativas en la consulta reumatológica en la ciudad de Barranquilla. *Revista colombiana de reumatología, 9*(3), 194–200.

Cardona Arias, J. A., & Palomino Rivera, Y. (2012). Social representations about traditional medicine and foreign diseases in Embera Chami natives in Colombia. *Revista Cubana de Salud Pública, 38,* 471–483.

Carrillo Esper, R., Lara Caldera, B., & Ruiz Morales, M. J. (2010). Hierbas, medicinaherbolaria y suimpactoen la prácticaclínica. *Revista de Investigación Médica Sur México, 17,* 124–130. Available from: https://www.medigraphic.com/pdfs/medsur/ms-2010/ms103a.pdf (Accessed 20 March 2015).

Castillo-Pérez, L. J., Martínez-Soto, D., Maldonado-Miranda, J. J., Alonso-Castro, A. J., & Candy Carranza-Álvarez, C. (2019). The endemic orquids of México: A review. *Biologia, 74,* 1–13. https://doi.org/10.2478/s11756-018-0147-x.

Castro Méndez, I. (2006). Actualidad de la medicina tradicional herbolaria. *Revista Cubana de Plantas Medicinales, 11*(2). Recuperadoen 06 de abril de 2020, de http://scielo.sld.cu/scielo.php?script=sci_arttext&pid=S1028-47962006000200001&lng=es&tlng=es.

CONABIO. (2006). *Capital Natural y Bienestar Social. Mexico, D.F.: Comisión Nacional para el Conocimiento y Uso de la Biodiversidad.*

Chen, S. (2010). Snake needle grass (Hedyotisdiffusa): A mini review. *Journal of Medicinal Food Plants, 2,* 10–11.

Ekor, M., Adeyemi, O. S., & Otuechere, C. A. (2013). Management of anxiety and sleep disorders: Role of complementary and alternative medicine and challenges of integration with conventional orthodox care. *Chinese Journal of Integrative Medicine, 19,* 5–14. https://doi.org/10.1007/s11655-013-1197-5.

Elisabetsky, E., & Etkin, N. L. (Eds.). (2009). Ethnopharmacology. In *Vol. 1. Encyclopedia of life support systems* (p. 290). United Nations Educational, Scientific and Cultural Organization. https://www.eolss.net/ebooklib/bookinfo/ethnopharmacology.aspx.

Esquivel García, R., Pérez Calix, E., Ochoa Zarzosa, A., & García Pérez, M. E. (2018). Ethnomedicinal plants used for the treatment of dermatological affections on the Purépecha plateau, Michoacán, Mexico. *Actabotanica Mexicana, 125,* 95–132. https://doi.org/10.21829/abm125.2018.1339.

Fedoung Fongnzossie, E., Biyegue Nyangono, C. F., Biwole, A. B., Besong Ebai, P. N., Ndifongwa, N. B., Motove, J., et al. (2020). Wild edible plants and mushrooms of the Bamenda highlands in Cameroon: Ethnobotanical assessment and potentials for enhancing food security. *Journal of Ethnobiology and Ethnomedicine, 16,* 1–12. https://doi.org/10.1186/s13002-020-00362-8.

Gallardo Arias, P. (2004). Huastecos de San Luis Potosí. Pueblos indígenasdel México contemporáneo. Huastecos de San Luis Potosí, México. In *Vol. 1. Comisión Nacional para el Desarrollo de los Pueblos Indígenas* (Primeraedición, p. 34). Programa de las Naciones Unidas para el desarrollo. http://www.cdi.gob.mx/dmdocuments/huastecos.pdf.

Gallegos Zurita, M. (2016). Medicinal plants: Main alternative for health care, in the rural town of Babahoyo, Ecuador. *Magazine Scielo. Anales de la Facultad de Medicina, 77,* 327–332.

García de Alba García, J. E., Ramírez Hernández, B. C., Robles Arellano, G., Zañudo Hernández, J., Salcedo Rocha, A.L., & García de Alba Verduzco, J.E. (2012). Conocimiento y uso de las plantas medicinales en la zona metropolitana de Guadalajara (pp. 29–44). Desacatos, núm. 39. México: Centro de Investigaciones y Estudios Superiores en Antropología Social Distrito Federal. ISSN: 1607-050x. Available from: https://www.redalyc.org/pdf/139/13923111003.pdf.

González Rodríguez, R., & Cardentey García, J. (2016). La medicina herbolaria como terapéutica en un consultorio. *Revista de Ciencias Médicas de Pinar del Río, 20,* 20–27. Available from: http://www.revcmpinar.sld.cu/index.php/publicaciones/article/view/2490/html (Accessed 20 March 2018).

Heinrich, M., Ankli, A., Frei, B., Wiemann, C., & Sticher, O. (1998). Medicinal plants in Mexico: Healer's consensus and cultural importance. *Social Science and Medicine*, *47*, 1859–1871.

Heinrich, M., Barnes, J., Prieto-Garcia, J., Gibbons, S., & Williamson, E. M. (Eds.). (2017). *Vol. 1. Fundamentals of pharmacognosy and phytotherapy* (p. 360). Elsevier Health Sciences. https://www.elsevier.com/books/fundamentals-of-pharmacognosy-and-phytotherapy/heinrich/978-0-7020-7008-2.

Heisler, E. V., de Budó Denardin, M. L., Schimith, M. D., Badke, M. R., Ceolin, S., & Heck, R. M. (2015). Use of medicinal plants in health care: Scientific production of theses and dissertation Brazilian nursing. *Magazine Scielo. Enfermería Global*, *14*, 390–403.

Herrero, B., & Santos, L. (2009). Medicinal plants of traditional use in Castilla y León (Spain). *Journal Acta Horticultura*, *826*, 229–236.

Hill, J., Seguin, R., Phanga, T., Manda, A., Chikasema, M., & Gopal, S. (2019). Facilitators and barriers to traditional medicine use among cancer patients in Malawi. *PLoS One*, *14*, 10. https://doi.org/10.1371/journal.pone.0223853.

Ivanova, N. V., Kuzmina, M. L., Braukmann, T. W. A., Borisenko, A. V., & Zakharov, E. V. (2016). Authentication of herbal supplements using next-generation sequencing. *PLoS One*, *11*, 1–5. https://doi.org/10.1371/journal.pone.0156426.

Iweala Emeka, E. J. (2009). Preliminary qualitative screening for cancer chemopreventive agents in Telfairia occidentalis Hook.f., Gnetum africanum Welw., Gongronema latifolium Benth. and Ocimum gratissimum L. from Nigeria. *Journal of Medicinal Food Plants*, *1*, 58–63.

James, P. B., Wardle, J., Steel, A., & Adams, J. (2018). Traditional, complementary and alternative medicine use in sub-Saharan Africa: A systematic review. *BMJ Global Health*, *3*. https://doi.org/10.1136/bmjgh-2018-000895, e000895.

Juárez Vázquez, M. C., Carranza Álvarez, C., Alonso Castro, A. J., González Alcarez, V. F., Bravo-Acevedo, E., Chamarro Tinajero, J., et al. (2013). Ethnobotany of medicinal plants used in Xalpatlahuac, Guerrero, México. *Journal of Ethnopharmacology*, *148*, 521–527. https://doi.org/10.1016/j.jep.2013.04.048.

Kew, K. M., Quinn, M., Quon, B. S., & Ducharme, F. M. (2016). Increased versus stable doses of inhaled corticosteroids for exacerbations of chronic asthma in adults and children. *Cochrane Database of Systematic Reviews*, *6*. https://doi.org/10.1002/14651858.CD007524.pub4.

Le Rhun, E., Devos, P., & Bourg, V. (2019). Uso de la medicina complementaria y alternativa en pacientes con glioma en Francia. *Journal of Neuro-Oncology*, *145*, 487–499. https://doi.org/10.1007/s11060-019-03315-8.

Luna-Vargas, S. (2014). Biodiversidad de la Huasteca Potosina. In: Naturaleza, cultura y desarrollo endógeno: un nuevo paradigma del turismo sustentable. Eumed. Net. Available from: http://www.eumed.net/libros-gratis/2014/1377/#indice.

Murphy, S. M., Wimp, G. M., Lewis, D., & Denno, R. F. (2012). Nutrient presses and pulses differentially impact plants, herbivores, detritivores and their natural enemies, authentication of herbal supplements using next-generation sequencing. *PLoS One*, *7*, 1–8. https://doi.org/10.1371/journal.pone.0043929.

Ocegueda, S., Moreno, E, & Koleff, P. (2005). Plantas utilizadas en la medicina tradicional y su identificación científica. *CONABIO. Biodiversitas*, *62*, 12–15.

Ortega Cala, L. L., Monroy Ortiz, C., Monroy Martínez, R., Colín Bahena, H., Flores Franco, G., Luna Cavazos, M., et al. (2019). Medicinal plants used for diseases of the digestive system in Teteladel Volcan, State of Morelos, Mexico. *Bulletin Latin American Caribbean Plants Medicinal Aromatic*, *6*, 106–129.

Palu, A. K., Kauvaka, P., Kalisi, T., Palu, A., & Hifo, T. (2010). Fakatafe: A forgotten art of traditional wound-healing using nonu leaves in the Friendly Islands of Tonga. *Journal of Medicinal Food Plants*, *2*, 27–33.

Panmei, R., Gajurel, P. R., & Singh, B. (2019). Ethnobotany and nutritional values of some selected wild edible plants used by rongmei tribe of Manipur Northeast India. *International Journal of Applied Biology and Pharmaceutical Technology*, 7, 1–11. https://doi.org/10.21276/ijabpt.2016.7.4.1.

Poss, J., Pierce, R., & Prieto, V. (2005). Herbal remedies used by selected migrantameworkers in El Paso, Texas. *Journal of Rural Health*, 21, 187–191.

Reyes-Pérez, O., Vázquez-Solís, V., Reyes-Hernández, H., & Nicolás-Carettay Rivera-González, M. J. G. (2012). Potencial turístico de la región Huasteca del estado de San Luis Potosí, México. *Economía, Sociedad y Territorio*, 12(38), 249–275.

San Miguel, E. (2003). Rue (Ruta L., Rutaceae) in traditional Spain: Frequency and distribution of its medicinal and symbolic applications. *Economic Botany*, 57, 231–244. https://doi.org/10.1663/0013-0001(2003)057[0231:RRLRIT]2.0.CO;2.

Schlaepfer, L., & Mendoza-Espinoza, J. (2010). Las plantas medicinales en la lucha contra el cáncer, relevancia para México. *Revista Mexicana de Ciencias Farmacéuticas*, 41(4), 18–27.

Sharma, A., Flores-Vallejodel, R. del C., Cardoso-Taketa, A., & Villarreal, M. L. (2017). Antibacterial activities of medicinal plants used in Mexican traditional medicine. *Journal of Ethnopharmacology*, 208, 264–329.

Sranjeet-Kaurt. (2018). Conservation of orchids—A review. *Open Access Journal of Medicinal and Aromatic Plants*, 9, 52–62.

Taddei Bringas, G. A., Santillana Macedo, M. A., Romero Cancio, J. A., & Romero Tellez, M. B. (1999). Acceptance and use of therapeutic medical plants in family medical care. *Salud Pública de México*, 41, 216–220. Available from: https://www.ncbi.nlm.nih.gov/pubmed/10420791 (Accessed 20 March 2018).

The International Union for Conservation of Nature and Natural Resources (IUCN), Gland, Switzerland, in partnership with The World Health Organization (WHO), Geneva, Switzerland, & WWF—World Wide Fund for Nature, Gland, Switzerland. (1993). *Guidelines on the conservation of medicinal plants*. Vol. 1 (p. 38). World Health Organization (OMS). https://apps.who.int/medicinedocs/documents/s7150e/s7150e.pdf.

Villar López, M., Ballinas Sueldo, Y., Soto Franco, J., & Medina Tejada, N. (2016). Conocimiento, aceptación y uso de la medicinatradicional, alternativa y/o complementariapormédicos del Seguro Social de Salud. *Revista Peruana de Medicina Integrativa*, 1, 13–18.

Wang, J., Xu, J., Gong, X., Yang, M., Zhang, C., & Li, M. (2019). Biosynthesis, chemistry, and pharmacology of polyphenols from Chinese *Salvia* species: A review. *Molecules*, 24, 155. https://doi.org/10.3390/molecules24010155.

Westendorf, J., & Mettlich, C. (2009). The benefits of noni juice: An epidemiological evaluation in Europe. *Journal of Medicinal Food Plants*, 1, 64–79.

Wiseman, N. (2004). Designations of medicines. *Evidence-Based Complementary and Alternative Medicine*, 1, 327–329. https://doi.org/10.1093/ecam/neh053.

World Health Organization. (2004). *Guidelines on safety monitoring of herbal medicines in pharmacovigilance systems*. WHO: Geneva. ISBN 92 41592214. Available from: https://apps.who.int/iris/bitstream/handle/10665/43034/9241592214_eng.pdf.

World Health Organization (WHO). (2013). *Traditional medicine strategy 2014–2023*. Available from: http://apps.who.int/iris/bitstream/10665/92455/1/9789241506090_eng.pdf (Accessed 20 March 2018).

World Health Organization. (2017). *Report on traditional and complementary medicine.*

World Health Organization. (2019). WHO global report on traditional and complementary medicine. Available from: https://www.who.int/traditionalcomplementaryintegrativemedicine/Who-GlobalReportOnTraditionalAndComplementaryMedicine2019.pdf?ua=1.

Zhou, L., & Nunes, J. M. B. (2015). *Knowledge sharing in Chinese hospitals, Berlin, Heidelberg* (p. 221). Berlin, Heidelberg: Springer.

Chapter 8

Perspectives of phytotherapeutics: Diagnosis and cure

Lubna Azmi[a,b] and Ila Shukla[c]

[a]Hygia Institute of Pharmaceutical Education & Research, Lucknow, India, [b]Department of Chemistry, University of Lucknow, Lucknow, India, [c]Harsha Institute of Pharmacy, Itaunja, Lucknow, India

1. Overview of phytotherapeutics

An ancient form of remedies practiced by people to treat different ailments is collectively called herbal medicines. Although modern medicine has made remarkable advances in the healthcare system, plants still have a huge contribution. About 120 plant substances form present drugs of high significance (Fabricant & Farnsworth, 2001). Often, all these substances work together to give a synergistic effect in the finished product, but sometimes these may have antagonistic effects that may reduce the overall effect of the finished preparation. Some herbal medicines may offer polypharmacology not possible with conventional drugs. This feature of herbal medicines is due to their chemical constituents that possess a large range of bioactivities. Usually, natural products are not evaluated under clinical trials, and hence they have less systematic evidence to support their efficiency, although these therapeutic substances are chemical only, synthesized by plants as a part of their defense mechanisms or manufactured synthetically. Therefore, they are considered to have comparable quality, medicinal efficiency, as well as safety. This measure will be an independent source of these substances. Here, quality refers to identity, purity, and stability of the given substance (Angell & Kassirer, 1998).

According to the World Health Organization (WHO), the term *herbal medicines* refer to important and easily available means, but the need is to form an organized record of therapeutic plants sources and to develop some regulatory procedures. WHO also stated that good manufacturing practices should be applied so that these herbal medicines may find a place in the standard pharmacopeia (Robinson & Zhang, 2011). In the recent past, the population has expressed dissatisfaction toward the cost of prescription drugs. This fact has

Phytomedicine: A Treasure of Pharmacologically Active Products from Plants
https://doi.org/10.1016/B978-0-12-824109-7.00010-8

favored increased attention of returning toward naturally found medications. For this reason, industrialized countries reassessed the use of preparations of natural origin. Usually any preparation or manufactured good that solely possesses one or more than one substance of natural origin as their active constituent, or if they are a combination of one or more such formulations, is termed as an *herbal medicinal product* (HMP) (World Health Organization, 2013).

HMP have composite combinations of substances, many times referred to as a phytocomplex. This name suggests that they often act mutually to exert their effect on the body. It is rarely completely recognized. Usually the action of any phytocomplex is more powerful compared with the summation of effect of solo active components. Many times, it's seen that substances with no particular action often show a remarkable synergistic effect (Williamson, 2001). One beautiful example of such phenomena is *Ginkgo biloba*. It possesses a number of activities ranging from showing antagonistic effect on release of oxygen radical, corticosteroid production (antianxiety activity); escalating uptake of glucose and its utilization and adenosine triphosphate (ATP) synthesis; reducing red blood cells (RBCs) aggregation; and initiating nitric oxide (NO) synthesis (Chan, Xia, & Fu, 2007).

Phytotherapy is a discipline that includes the study of substances extracted from natural origins and their use as medicines or in the promotion of health. Because the effects of phytotherapy are analyzed based on causes and symptoms of an ailment, it may be taken as an allopathic discipline. Phytotherapy is categorized as a standard discipline in Germany just like any other science-oriented medicine. Hence, any health and medicine publication has to abide by the scientific requirements that form the basis of any chemically defined substances, i.e., its safety, qualitative nature, and effectiveness (Peschel, 2014). Differences arise when it comes to mass manufacturing of natural products as nutraceuticals or herbal medicine, because the basis of defining the safety and efficacy of herbal medicines becomes difficult because there is a great variation in dosage form, methods of administration, ingredients of medicines, manufacturing methods, and indications. The possible advantages received by the consumption of herbal medicine are occasionally undermined by the clinical risks that could be associated with them. These risks come into picture when there is a lack of prescribed levels of medicinally active component from plants (Ernst, 1998). To resolve this situation, WHO has given monographs to maintain the quality, efficacy, and safety of the chosen medicinal plants. These monographs also present some guidelines about the cultivation of medicinally important plants and maintenance of their quality, efficacy, and safety (World Health Organization, 1993).

2. Preclinical and clinical research on phytotherapeutics

2.1 Preclinical research on phytotherapeutics

2.1.1 Quality standardization of phytotherapeutics

Preclinical study is defined as "a study to test a drug, a procedure, or another medical treatment in animals". This study is usually aimed at collecting data in relation to the safety of the new treatment. These studies are required before clinical trials in humans can be initiated (Yau, Goh, & Koh, 2015).

The branch of science that studies medicinally important plants and their important medicaments from past to present times is called pharmacognosy. It covers the structural, physical, chemical, biochemical, and sensory characteristics of the medicinal plant or the natural drug (Yau et al., 2015). It also comprises the study of its history, distribution, collection, cultivation, identification, evaluation, and preservation. Scope of pharmacognosy includes: (1) Chief source of medicinally important substances (from past to now), e.g., morphine, ergotamine, hyoscine, ouabain, etc.; and (2) Making available a guide for discovering newer medicaments, e.g., Pathidine (analgesic drug) designed from morphine. Mainly, pharmacognosy provides information about the sources and constituents of medicinally important plants or natural drugs (Alamgir, 2017).

The foundation of a good safety profile and clinical efficiency can be placed on the proper recognition of a good quality herb (Ramzan & Li, 2015). Testing labs, herbal drug dispensing centers, and relevant regulatory bodies depend upon this information. National and international pharmacopeias include this discipline as identification of species, macroscopic identification, microscopic identification, and quality control. An example of such comprehensive monographs are WHO monographs (Allard, Wenner, Greten, & Efferth, 2013), although statutory standards are found in *British Pharmacopeia* while chemical analysis is used for quality control in Australia (Cartwright, 2016). Standardization of quality of herbal medicines is based on bioequivalence. Bioequivalence is defined as a condition when the active substance is same as its bioavailability, due to its clinical effectiveness and safety (European Medicines Agency, 2012; García-Arieta & Gordon, 2012). For any two products to be declared bioequivalent, the following tests are to be conducted: (i) Standardization of quality, which refers to pharmaceutical equivalence; (ii) To prove equivalence in pharmacokinetic parameters, bioavailability and time-to-peak concentration should be the same; (iii) When products yield similar results in vivo and in vitro, they are termed as pharmacodynamically equivalent; (iv) Clinical studies of any product define its therapeutic equivalence. An example of testing bioequivalence is a study on *Ginkgo biloba* to determine the bioavailability of ginkgolide B, ginkgolide A, and bilobalide. These were taken from two different commercial brands (Kressmann et al., 2002). The in vitro dissolution of the product was very slow, which indicated a huge decline in its bioavailability. A broad spectrum platform to evaluate herbal products is referred to as the bioequivalence concept (Osimani et al., 2018). Several techniques are employed for maintaining the quality of these herbal products. These techniques involve HPLC using UV for detection. This concurrently quantifies eight major phenolic compounds in Chinese propolis. Another technique, TLC is used in combination with chemometric fingerprinting, which is employed to differentiate between Chinese propolis and poplar tree gum (Fan & He, 2006).

2.1.2 Pharmacological studies and identification of bioactive compounds

The discipline that deals with the mechanism of action of medicines in living systems (pharmacodynamics) and their pharmacokinetic nature is called pharmacology. This discipline defines the clinical applications of a medicinal product. Possessing a huge range of compounds and having multiple pharmacological targets is exclusive to herbal medicines (Rainsford, 2016). Majorly in vitro studies are employed for pharmacological research of herbal products at the cellular or tissue levels. These procedures help in unraveling the mode of action of components present in the product. For example, cytotoxicity is studied at the cellular level in cancer cell lines; after that, animal models are employed to analyze the preclinical properties of the product and acquire pharmacokinetic data about the same. Pentacyclic triterpenoids (oleanane, ursane, and lupane groups) are an example, which are analyzed by the previously given procedures. The preclinical (i.e., streptozotocin-induced diabetes in rat model) and some clinical data have indicated toward their biological activity, and they may be incorporated in traditional medicines for treating diabetes and its complications (Mgbeahuruike, Yrjönen, Vuorela, & Holm, 2017).

Probable health benefits of food and food supplements, e.g., as shown by common chemical components like gallic acid, is very well supported by growing evidence for them (Nassiri-Asl & Hosseinzadeh, 2016). Quercetin and berberine are used as nutraceuticals for cardiovascular ailments and for managing diabetes, respectively. St. John's wort is an herb that is researched upon for chemistry and pharmacological potential. Flavonoids, e.g., quercetin; phloroglucinols, e.g., hyperforin; and naphthodianthrones, e.g., hypericins are some of the active compounds present in the plant. These may be found independently or in mixture. Extract from this plant interacts with many neurotransmitter systems and hence found its application for treating depression and psychiatric illness. The mode of action lying behind it is an uptake of serotonin, noradrenaline, and dopamine, and interaction with γ-aminobutyric acid (GABA) receptors, monoamine oxidases, and dopamine-beta hydroxylase (Nahrstedt & Butterweck, 2010), athough the precise mode of action and the compounds behind it are not defined completely. Lavender flower (*Lavandula officinalis*) also possesses multiple actions like antianxiety, antimicrobial activity, curing dyspepsia, and healing wounds and sores. Oil of *Lavandula officinalis* possesses noteworthy anxiolytic activity, reported in rats and mice in a dose-dependent manner. These results were found to be on par with a standard anxiolytic agent, e.g., lorazepam. This lavender oil is also found to decrease the mean heart rate in dogs (Komiya, Sugiyama, Tanabe, Uchino, & Takeuchi, 2009) and increase the pentobarbital-induced sleeping time (Kumar, 2013). Thorough studies for analyzing the complete mechanism of action showed that it slowed down voltage-dependent Ca^{++} channels present in synaptosomes and primary neurons of the hippocampus (Schuwald et al., 2013), and in the olfactory bulb of mice it amplified the dopamine D3 receptor subtype (Kim, Kim, Kim, & Kim, 2009).

Largely, the mode of action and the properties of active components are yet not properly defined for most herbal medicines. Moreover, broadly the research upon herbal medicines focuses on finding the biological activities of single compounds or crude extracts in absence of a distinct fingerprint of the crude extract or formulation. To link the biological activities of a single component or multiple components to various standardized extracts, there is a need to establish a novel multidisciplinary research platform (Schuwald et al., 2013).

2.2 Clinical research on phytotherapies

2.2.1 Effectiveness of phytotherapeutics

Generally, clinical data or case reports form the clinical evidence about herbal formulations. Around 200 proper systematic reviews about herbal medicines have been published in last 5 years. These can be found in the Cochrane Library. This library has records about the most popular herbs, e.g., valerian (*Valeriana officinalis*), echinacea (*Echinacea* species), milk thistle (*Silybum marianum*), bitter melon (*Momordica charantia*), ginkgo (*Ginkgo biloba*), hawthorn (*Crataegus monogyna*), and black cohosh (*Cimicifuga* species). The results of a huge number of trials gave positive outcomes, but no decisive proof was obtained about the effectiveness of the popular herbs like ginkgo, ginseng, milk thistle, bitter melon, and black cohosh. The ginkgo, ginseng, milk thistle, bitter melon, and black cohosh were reported for cognitive impairment and dementia (Birks & Evans, 2009); cognition and preventing and treating the common cold (Karsch-Völk et al., 2014); alcohol or nonalcohol hepatitis, and other liver diseases (Woelkart, Linde, & Bauer, 2008); type 2 diabetes mellitus (Ooi, Yassin, & Hamid, 2012); and menopausal symptoms (Leach & Moore, 2012), respectively. Adding up, hawthorn extract acted as an adjunct in controlling symptoms of chronic heart failure treatment (Pittler, Guo, & Ernst, 2008).

Another popular phytotherapy having noteworthy effect on the management of anxiety includes lavender. To obtain reliable results about lavender's properties, a randomized, double-blind, double-dummy trial was conducted. In this trial, 539 adults having generalized anxiety ailment were given 160 or 80 mg preparation of lavender (SilexanR), 20 mg of paroxetine, or placebo once daily for 10 weeks. Silexan was found to be more effective than placebo (Kasper et al., 2014). A methodical assessment of seven trials concluded that Silexan was considerably better than placebo where patients were having subsyndromal anxiety disorder, and also its activity was similar to that of lorazepam (Kasper, 2013).

2.2.2 Translational research in relation to food nutrition

As many phytotherapies are given as food and almost all food items derived from plants contain phytochemicals, it's really difficult to maintain a boundary between food and phytotherapy. In the management of metabolic disorders and problems related to cardiovascular illnesses, new perspectives have been added by analyzing the effect of food on health and duration of life. To demonstrate this, mice were fed one of 25 diets ad lib. When carbohydrates replaced protein, longevity and health were found to be most favorable (Yi et al., 2010). Although intake of a diet high in protein activated hepatic mammalian target of rapamycin (mTOR), it also augmented the circulating branched-chain amino acids and glucose (Solon-Biet et al., 2014). Also, the function of micronutrients in food has been of great interest of researchers, anthocyanin in fruits and rice being one such example. One of the popular pharmacological activities of anthocyanin is arthrosclerosis prevention by suppression of oxidative stress-induced endothelial injury in endothelial cells, mouse peritoneal macrophages (Xia et al., 2007), apolipoprotein E-deficient mice, and dyslipidemic subjects (Chawla et al., 2001).

3.   Quality control and standardization of phytotherapeutics

Standardization and quality maintenance of HMP engages quite a few steps: (i) Availability of top quality unprocessed material and (ii) Framing of standards to accurately identify the components of every manufactured item, along with certification of the role of the components in combinations. These steps will optimize and control the circumstances until pharmaceutical-grade HMP is achieved. Lastly, it is indispensable to place its efficacy through biological assays and outline its adverse effects through text or from toxicological studies (both short-term and long-term) followed by clinical trials (Li et al., 2010). The deficient pharmacological and clinical data about most of the herbal medicinal products show a main obstruction to the receipt of natural products by usual medicine.

The authenticity of herbal raw material may be affected by unintentional botanical replacement (i.e., wrong recognition of plant species) or deliberate botanical replacement (i.e., knowingly swapping with other, maybe more toxic, plant species) (Qin et al., 2009). Many times, some unidentified chemical or synthetic substances are also responsible for the variation in the quality of the collected material. Different concentrations of active ingredients in herbal preparation, e.g., 15 to 200 times dissimilarity was reported in the amount of ginsenosides and eleutherosides (active components of ginseng); by analyzing the 25 ginseng products, there have also been a reason of variation in quality. Another common problem regarding this is the presence of toxic heavy metals in herbal preparations beyond the permissible limit set by national regulatory authorities. This could be intentional or accidental (Qin et al., 2009). The contamination of raw material can happen starting from collection to manufacturing (Li et al., 2010). A need for managing and regulating the use of pesticides has been highlighted by the presence of their residues in herbal material. Proper methods for detection of pesticides in cultivated or wild medicinal plants has been given by WHO, which also prescribed the maximum residue limit for pesticides (Patel, 2011). Chromatographic fingerprinting should be adopted for assessing the chemical consistency of herbal medicines at different stages of development. This would ensure the identity and purity of any product. This would also help in maintaining the accurate quantification of active components or marker compounds. Furthermore, regulatory authorities have touched upon the significance of qualitative and quantitative analysis, as these methods are of utmost importance for sample characterization, biomarker/chemical marker quantification, and recording fingerprint profiles (European Medicines Agency, 2012).

To overcome the problem of identification of all of a plant's constituents, due to their complex nature or nonavailability of analytical methods or reference compounds, the German Commission E has presented elaborated plant monographs (Schellekens, Perrinjaquet-Moccetti, Verbruggen, & Dimpfel, 2009). Also, to recognize and manufacture the chemical reference substances for quality control and assurance of herbal drugs and formulations, explaining the analytical methods and quality specifications for each product, *European Pharmacopeia* has initiated a program covering all of these aspects. A similar approach is being followed by the *United States Pharmacopeia* stressing upon the need to prepare guidelines for a rising number of herbal formulations (Schellekens et al., 2009).

4. Unevenness in phytotherapeutics: Variation in the concentration of biologically active constituents

Initially, macroscopic and microscopic evaluations are carried out to identify herbal drugs. This comparison is done with authentic herbal material or with precise descriptions of genuine herbs. Also, for further investigation, tests like thin-layer chromatography may be carried out. It is made indispensable that herbal components should be mentioned according to their binomial Latin names of genus and species. Additional noteworthy information is the procedure of extraction and standardization. Concentration of a particular bioactive component in any herbal formulation is often dependent on them. A few phytochemicals are more water soluble, whereas others are not. They could be more soluble in alcohol or oil. Hence, methods of extraction differ from plant to plant, defined by the types of active constituents (Becker, 2013).

Standardized extracts are defined as the "processed extracts that possess some specific and identified compounds that are documented to add further to remedial activity than others". These are regulated to give a quantity containing an acceptable tolerance. Consistency is attained by adjusting the herbal formulations using excipients or by blending a group of herbal components or formulations. Every other constituent is there in the extract so far, so the activity of the plant could be an outcome of synergistic action of some components. Extracts with distinct properties are yielded by different methods of standardization. This makes transferability of clinical data from one extract to another roughly unattainable. Although this can be made possible, only bioequivalence (or phytoequivalence) and bioavailability studies can prove necessary resemblance (Barnes, 2003). Herbal formulations are pharmaceutically analogous if they possess a comparable amount of the similar active components in similar dosage forms. These dosage forms should meet comparable or equivalent standards (Leslie, 2013).

Specific parameters for active components of extract:

1. Quality of herbal substances.
2. Solvent used for extraction.
3. Procedures used for extraction.
4. Ratio of drug to extract.

Specific parameters for a finished product:

1. Native extracts weight (per dosage form).
2. Weight of pure medicinally constituents (accountable for the therapeutic effect per dosage form).
3. Weight of pharmacologically active constituents (per dosage form).
4. Type of excipients used per dosage form.
5. Type of dosage form.

Germany developed the concept of phytoequivalence to ensure the consistency of herbal formulations (Williamson, 2001). A clinically proven reference product should be employed to develop chromatographic fingerprinting for an herbal formulation, and its profile should be compared with the reference product. This helps in determining the reliability and repeatability of pharmacological and clinical research.

Regrettably, phytoequivalence is extremely complex to be obtained because phytochemical outlines are very complicated and intimately connected to the method "seed to patient" process that has been followed. Medicinal plants possess varying chemical constituents due to the following factors (Cañigueral et al., 2008):

1. Variation in different species or within a single species, because it is controlled genetically.
2. Conditions at the time of cultivation causes variation in synthesis of herbal constituents.
3. Time of harvest also affects the concentration of active components, because it is related to the growth cycle.
4. Plant part being used, because different plant parts store different components.
5. After harvesting, storage conditions also affect the quality of herbal component. Improper storage often lead to microbial contamination (WHO, 1998).

5. Phytotherapeutics for treating major ailments involving vital organs

Several herbal medicines are used to treat a number of ailments. A few of them are listed in the following tables (Principles and Practice of Phytotherapy, 2013):

5.1 For treating anxiety or insomnia

Common name	Latin name	Part(s) of plant used	Key constituents	Daily dose
German chamomile	*Matricaria recutita*	Flower heads	Coumarins, flavonoids	5 g
Hops	*Humulus lupulus*	Glandular hairs	Bitter acids, flavonoids	1–2 g
Kava	*Piper methysticum*	Rhizome	Kava lactones, flavonoids,	1.5–3 g
Lavender	*Lavandula angustifolia*	Flower	Volatile oil, tannins, coumarins, caffeic acid derivatives	20–80 mg of the oil

5.2 For treating depression

Common name	Latin name	Part(s) of plant used	Key constituents	Daily dose
Corydalis	Corydalis cava	Tubers	Isoquinoline alkaloids	1 g
Lemon balms	*Melissa officinalis*	Leaves	Volatile oil, glycosides, caffeic acid derivatives	1.5–4.5 g
Mugwort	*Artemisia vulgaris*	Roots	Volatile oil, sesquiterpene lactones, flavonoids,	1.5–6 g
Passionflower	*Passiflora incarnata*	Aerial parts	Volatile oil, cyanogenic glycosides, volatile oil	4–8 g

5.3   For treating hypertension

Common name	Latin name	Part(s) of plant used	Key constituents	Daily dose
Evodia	Evodia rutaecarpa	Fruits	Alkaloids (rutaecarpine), tannins	3–9 g
Garlic	*Allium sativum*	Bulb	Alliins	4 g
Onion	*Allium cepa*	Bulb	Alliins, Flavonoids	20 g
Scotch broom	*Cytisus scoparius*	Aerial parts	Quinolizidine alkaloids, biogenic amines, flavonoids	1–2 g

5.4   For treating chronic venous insufficiency

Common name	Latin name	Part(s) of plant used	Key constituents	Daily dose
Bilberry	*Vaccinium myrtillus*	Fruits	Tannins, anthocyanin, flavonoids	20–50 g
Buckwheat	*Fagopyrum esculentum*	Aerial parts	Flavonoids, anthracene derivatives	1.5–2 g
Gotu kola	*Centella asiatica*	Leaves, stem	Triterpenes, flavonoids, volatile oil	1.8 g

5.5   For treating hyperlipidemia

Common name	Latin name	Part(s) of plant used	Key constituents	Daily dose
Artichoke	*Cynara scolymus*	Leaves	Caffeic acid derivatives, flavonoids sesquiterpene lactones	4–9 g
Fenugreek	*Trigonella foenum graecum*	Seeds	Steroids saponins, flavonoids, fibers	6 g

Common name	Latin name	Part(s) of plant used	Key constituents	Daily dose
Garlic	Allium sativum	Bulb	Alliins	4 g
Ginseng	*Panax ginseng*	Roots	Triterpene saponins	1–2 g

5.6   For treating diabetes

Common name	Latin name	Part(s) of plant used	Key constituents	Daily dose
Aloe	*Aloe vera*	Dried juice of leaves	Anthraquinone derivatives, flavonoids	0.1–0.2 g
Fenugreek	Trigonella foenum graecum	Seeds	Steroids saponins, flavonoids, fibers	6–12 g
Garlic	Allium sativum	Bulb	Alliins	4 g
Ginseng	Panax ginseng	Roots	Triterpene saponins	1–2 g

5.7   Use as diuretics

Common name	Latin name	Part(s) of plant used	Key constituents	Daily dose
Birch	Betula spp.	leaves	Flavonoids, proanthocyanins, triterpene alcohol ester	2–3 g
Corn silk	*Zea mays*	Styles	Fatty oil, amines, tannins, potassium, saponins, sterol	40 g
Java Tea	Orthosiphon spicatus	Leaves	Volatile oil, flavonoids, caffeic acid derivatives, triterpenes saponins	6–12 g
Witch grass	Agropyron repens	Rhizome	Carbohydrates, volatile oil, flavonoids, saponin, minerals	6–9 g

5.8   For treating inflammation

Common name	Latin name	Part(s) of plant used	Key constituents	Daily dose
Birch	Betula spp.	Leaves	Flavonoids, proanthocyanins, triterpene alcohol ester	6–9 g
Black currant	*Ribes nigrum*	Oil from the seeds	Fatty oil (gamma-linolenic acid, alpha-linolenic acid)	10 g
Eucalyptus	*Eucalyptus globulus*	Oil from the leaves	Cineole	0.3–0.6 g
Ginger	Gingiber officinalis	Root	Volatile oil, gingerols, shogaols, ginger diols, starch	–

5.9   For treating headache or migraine

Common name	Latin name	Part(s) of plant used	Key constituents	Daily dose
Butterbur	*Petasites hybridus*	Leaves	Sesquiterpene, pyrrolizidine alkaloids, volatile oil	4.5–7 g
Catnip	*Nepeta cataria*	Aerial parts	Volatile oil	1 g
Lemon balm	Melissa officinalis	Leaves	Volatile oil, glycosides, caffeic acid derivatives, flavonoids	1.5–4.5 g
Sweet violet	*Viola odorata*	Rhizome	Volatile oil, saponins, alkaloids	–

5.10   For treating bronchial spasms or asthma

Common name	Latin name	Part(s) of plant used	Key constituents	Daily dose
Belladonna	*Atropa belladonna*	Leaves	Tropane alkaloids (atropine, scopolamine), flavonoids	0.2–0.4 g

Common name	Latin name	Part(s) of plant used	Key constituents	Daily dose
Gingko	Gingko biloba	Leaves	Flavonoids, ginkgolides	3–6 g
Datura	Datura	Leaves	Tropane alkaloids, flavonoids	1.5–4.5 g
Henbane	*Hyoscyamus niger*	Leaves	Tropane alkaloids, flavonoids	0.5–1 g

5.11   Use as expectorants

Common name	Latin name	Part(s) of plant used	Key constituents
Anise	*Pimpinella anisum*	Oil from fruits	Anethole
Citronella	*Cymbopogon nardus*	Oil from leaves	Citronellal, geraniol, geranyl acetate
Thyme	*Thymus vulgaris*	Oil from herb	Thymol, carvacrol
Eucalyptus	Eucalyptus globulus	Oil from leaves	Cineole (eucalyptol)

5.12   For treating ulcers

Common name	Latin name	Part(s) of plant used	Key constituents	Daily dose
Chili	*Capsicum annuum*	Fruits	Capsaicin	5–10 g
German chamomile	Matricaria recutita	Flower heads	Volatile oil, flavonoids, coumarins	5 g
Liquorice	*Glycyrrhiza glabra*	Roots	Triterpene saponins, flavonoids, coumarins	2–4 g
Iceland moss	*Cetraria islandica*	Thallus	Mucilages, lichen acids	4–6 g

5.13   Use as hepatoprotectant

Common name	Latin name	Part(s) of plant used	Key constituents	Daily dose
Turmeric	Curcuma dosmetica	Rhizome	Volatile oil, curcuminoids	1.5–3 g
Milk Thistle	*Silybum marianum*	Seeds	Flavonolignans (silymarin), flavonoids, fatty oil	5 g
Liquorice	Glycyrrhiza glabra	Roots	Triterpene saponins, flavonoids, coumarins	5–15 g
Schizandra	Schizandra chinensis	Fruit	Volatile oil, ascorbic acid, lignans	1.5–6 g

6.   New drug delivery systems and formulations for phytotherapies

The effects of a given drug can be improved drastically by incorporating it in an herbal novel drug delivery system. Examples of such systems are mucoadhesive systems, dosage forms for transdermal release, tablets with high oral dissolution, etc. Many researchers are trying to fabricate such formulations. Some of the formulations are at the stage of testing, and many have found their place in the market (De Jong & Borm, 2008).

Traditional medicines of natural origin affect a living system at multiple levels, upon multiple targets, and show coordinated intervention. This practice has been lost as present delivery systems of herbal formulations mainly focus on a single active component; the reports about total extracts of herbal medicines are insufficient (Verma & Singh, 2008). This is due to the extraordinarily complicated chemical composition of medicinal herbs. It is challenging to efficiently formulate these constituents with varying properties and retain the integrity of an herb or prescription. Although challenging, it is very important to resolve the limitations of novel herbal nanomedicines via the oral route as it is the most popular delivery method for herbal medicines (Li, Zhang, Liu, & Lu, 2013). For example, nanocarriers assembled using degradable polymers should avoid drug leakage in the gastrointestinal tract. Meanwhile, the particle size determines the diffusion of nanoparticles through the mucus layer and the elimination by mucociliary clearance. Novel formulations of herbal extracts have received much attention from the pharmaceutical industry (Ajazuddin, 2010) (see Fig. 1).

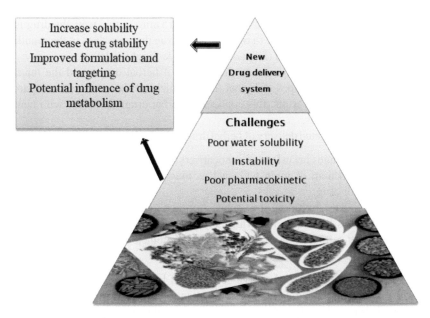

FIG. 1 Effect of novel drug delivery system on herbal formulation.

Currently, there are several companies such as Cosmetochem International AG that specialize in formulating liposomal herbal extracts. Indena has developed a series of herbal products via Phytosome® technology. Absorption of active constituents in these novel formulations is enhanced when administrated transdermally, and systemic bioavailability is improved for the oral dosage forms. As an example, absorption of standardized, decaffeinated green tea catechin (GTC) extract in healthy human subjects was compared with the GCT produced by Phytosome (Bhagyashree, 2015). Volunteers were given a single dose of GTC (400 mg) for 6 h, and the epigallocatechin-3-gallate (EGCG) levels in plasma were measured. The phytosome formulation achieved EGCG peak plasma concentration of approximately 4.0 mM, which was significantly higher than that of the standardized extract (around 2.0 mM). Meriva®, a patented complex of curcuminoids formulated with soy phosphatidylcholine, demonstrated superior bioavailability and stability (Rasaie, Ghanbarzadeh, Mohammadi, & Hamishehkar, 2014).

Res-Q is theworld's first polyherbal (tablet) that dissolves rapidly in the mouth. It was developed by Asoka Life Science Limited. It possesses a new system for delivering medicament, which increases its efficacy (Saedi, Md Noor, Ismail, & Othman, 2014). This is the first attempt to make medicines more efficient in managing chronic diseases in the ayurvedic medicine system. This formulation is really effective for lung problems and other respiratory ailments like asthma. The drug directly reaches blood bypassing the first pass metabolism.

This formulation relieves respiratory distress in 15 min. In this way, the drug is similar in efficacy as sorbitrate (a revolutionary drug that dissolves in the mouth and relieves cardiac distress) (Dhyani, Chander, & Singh, 2019).

Herbal formulation requires detailed study of herbal drugs, and the main purpose is to use such types of herbal drugs, which are direct targets to a particular site. For better results, this concept of herbal drugs coincides with new target delivery systems.

Strategies for targeted delivery of herbal drugs (Hu, Li, Wang, Gu, & Fan, 2019):

(i) Exterior changes in surface of nanocarriers, with help of active constituents like glycyrrhizin of main herb, act as magic-shot tissue-based targeting delivery of drug.

(ii) To minimize multidrug resistance, co-administration of active herbal constituents like curcumin with therapeutic agents. To achieve particular therapeutic efficacy, passive targets were combined with nanoparticles and actively targeted with herbal drugs.

The past has witnessed the large-scale use of herbal medicines worldwide, acknowledged by hospital doctors and patients due to effective significance of herbal drugs. Herbal medicines possess less unfavorable effects compared with the modern system of medicines. These medicaments originating from Ayurveda have been employed with an improved type of superior effectiveness by integrating them in a new drug dose concentration (Pawar, Kalamkar, Jain, & Shoyab, 2015). Phytotherapeutics require a scientific move toward delivering constituents in a newer way to augment compliance of the patient with repeated drug administration. A novel drug delivery system is a method that can accomplish such objectives. This delivery system is helpful to boost the therapeutic range by dropping toxicity and rising the bioavailability as well as better administration to overcome compliance problems (Dhyani et al., 2019) (see Table 1).

7. Misuse of phytotherapies: Consideration of side effects

Consumers have a prevalent mistaken belief that "natural" means "harmless", and one of the general faiths is that medicines from a natural source are not dangerous and have no hazard. Conversely, a few remedial plants are naturally poisonous. Just like syntactical drugs, herbal drugs are anticipated to possess adverse effects in the body. A few unfavorable conditions are accounted to be associated with plant-based drugs due to differences in quality aspect of herbal medicines (Adeyemi, Gbolade, Moody, Ogbole, & Fasanya, 2010). A chief reason of such happenings is adulteration of various adulterants in herbal formulations, strong pharmaceutical additives, and unknown supplementary drugs, e.g., NSAIDS and steroidal drugs (Bacchi, Palumbo, Sponta, & Coppolino, 2012). An unfavorable condition arises sometimes when mistaken species are used and a wrong dose is given to patients (Klein, 2017).

TABLE 1 List of herbal formulations, their source, active constituents, pharmacological property, delivery system, limit, and route of administration.

Herbal drugs	Source	Active constituents	Pharmacological activity	Drug formulation	Limitation	Route	Reference
Catechins	*Camellia sinensis*	(+)-Catechin (−)-Epicatechin	Antioxidant Anticancer Antiobesity Antiviral	Liposomes Nanoparticles Chitosan	Bioavailability is <5%	Oral	Jiang et al. (2019)
Camptothecin	*Camptotheca acuminate*	Irinotecan Topotecan Rubitecan	Cancer treatment	Conjugations nanoparticles liposome	Toxic effects Neutropenia thrombocytopenia Anemia	Intravenous	Gavvala, Sengupta, and Hazra (2013)
Curcumin	*Curcuma longa (turmeric)*	Curcumin Demethoxycurcumin, Bisdemethoxy-curcumin	Antitumor Antioxidant Antiamyloidh antiplatelet aggregation Anti-inflammatory	Microemulsion Liposomes, nanoparticles	Low aqueous solubility and bioavailability	Intravenous	Shen and Ji (2007)
Silymarin	*Silybus marianum*	Silybin Taxifolin Isosilybin Silydianin Silychristin	Hepatoprotective agent	Liposomes, proliposomes matrix tablets	Owing to degradation in gastrointestinal tract	Oral	Woo, Min, Chae, Chun, and Kwon (2015)

Continued

TABLE 1 List of herbal formulations, their source, active constituents, pharmacological property, delivery system, limit, and route of administration—cont'd

Herbal drugs	Source	Active constituents	Pharmacological activity	Drug formulation	Limitation	Route	Reference
Vincristin	*Catharanthus roseus*	Vincristine Vinblastine	Anticancer	Liposomes	Neurotoxicity	Intravenous	Sikorska, Słomkowski, Maślanka, Konopka, and Górski (2004)
Cuscuta chinensis	*Cuscuta chinensis lam*	Quercetin Kaempferol	Liver and kidney tonic Anticancer Antioxidant Antiaging Immuno-stimulatory	Nanoparticles	Poor water solubility	Oral	Yen, Wu, Lin, Cham, and Lin (2008)
Hypericin	*Hypericum perforatum*	Hypericin	Photoche-motherapy	Polymeric Nanoparticles	Problematic and restricts diagnostic applications	Intravenous, topical, and oral admin-istration	Agostinis, Vantieghem, Merlevede, and De Witte (2002)
Triptolide	*Tripterygium wilfondil hookf*	Sesquiterpenoid Diterpenoid Triterpenoid	Antineoplastic Antipsoriatic	Nanoparticles, Microemulsions, Polymeric nanoparticles	Adverse drug reaction	Transdermal	Liu (2011)

Name	Plant source	Compounds	Activities	Delivery system	Limitation	Route	Reference
Tetrandrine	*Stephania tetrandria*	Tetrandrine Fangchinoline	Anti-inflammatory antiplatelet aggregation Antiarthritis Ca^{2+} channel block Immuno-suppressive	Microparticles scaffold Nanoparticles	Accumulates in the liver and causes hepatic damage	Intravenous inhaled	Meng et al. (2004)
Podophyllotoxin	*Podophyllum peltatum*	Podophyllin	Antivirus Anticancer	Polymer conjugations nanoparticle	Insoluble and unpredictable in systemic behavior	Intravenous epidermal	Pang, Zhang, and Du (2018)
Guarana	*Paullinia cupana var. Sorbilis (mart)*	Caffeine Theophylline Theobromin	Bronchial muscle relaxation CNS stimulant Antidiuretics CVS stimulants	Transdermal patch		Transdermal	Schimpl, Da Silva, Gonçalves, and Mazzafera (2013)
Khellin	*Ammi visnaga*	Khellin	Anticancer Antipsoriatic	Transdermal gel	Nausea and Hepatic toxicity	Oral	Travaini et al. (2016)
Breviscapine	*Erigeron breviscapus*	Scutellarin	Treatment of cerebral infarction Heart disease	Liposomes	Shorter systemic circulation	Injectable multivesicular	Pennock et al. (1998)

Subsequent instances reveal a series of some troubles that arise due to herbal drugs and products:

- 0.1 to 0.3 mg dose of betamethasone in herbal capsule form shows a corticosteroid-type advance effect in patients.
- Some herbal products cause severe side effects like kidney damage due to improper characterization and toxicity study of plants.
- Information found from drug safety supervising organization of increased subcutaneous hematomas, augmented coagulation time, prothrombin time, and intracranial hemorrhage linked due to effective use of Ginkgo.
- Few well-developed herbal products show some cases of gastrointestinal interstitial pneumonia due to interaction of herbal drug with interferon.

Adverse events reported so far are related to products of herbal plants due to deprived quality of product or inappropriate utilization. Hence, it is hard to differentiate authentic adverse reactions to herbal medicines and herbal products awaiting the reason for such events to be identified.

8.   Recent developments in phytotherapy

The large-scale utilization of various herbal drugs are not that much limited to Asian and African countries. According to estimated data, nearly 70% of health workers in France and Germany prescribed herbal-based drugs and their formulations. In recent years, herbal-based drug therapy approaches have raised significantly (Romano et al., 2013). The market of herbal medicine has increased due to relaxation of U.S. FDA guidelines for herbal medicines. It is evident from the records that European countries have an herbal medicine market nearly about $7 billion to $22 billion in year of 1991, in which in which $3 billion of marketing of herbal medicine was done by Germany, and France and Italy covers $1.6 billion and $0.6 billion, respectively (Mucke, 2001).

In 1996, the American herbal medicine market covers nearly about $4.5 billion, while it is twofold at the present time; $1.5 billion of medicines were available in the Indian herbal market, and it exports about $80 million of herbal crude extract. In the recent past, an interesting thing has occurred to herbal medicine (Muthuswamy, 2011). It has made a comeback in a situation in which medical science and pharmaceutical chemistry had almost removed it from the healthcare system. Herbal medicine has been established to have a few striking certifications. These are developed by observation or experience by using trial and error methods, even though several herbal treatments are extraordinarily effective (Robinson & Zhang, 2011).

Latest survey about market of drug has revealed nearly about 30% of every 520 novel market drugs were obtained from natural source where they were synthetically derived from a plant source; 75% of anticancer and antibacterial drugs and their lead compounds were obtained from natural sources. A strong antibiotic named penicillin interacted with mercury for syphilis treatment was obtained from a plant source (Shah, Shah, Acharya, & Acharya, 2013). For GIT-related problems, belladonna was used as an antiseptic. Reserpine, an active ingredient of *Rauvolfia serpentina* (Indian snake root), was the fundamental ingredient in a range of tranquilizers initially used for the treatment of mental disorders in 1950. Today, reserpine is rarely used for treatment of mental illness, and that's why it works as one of the major components for an herbal formulation for the treatment of hypertension (Kim, 2015).

8.1 Limitations and risks of phytotherapeutics

Phytotherapeutics' reputation as "natural medicine" is not free from adverse effects, so it creates an idea of innocuousness. From a literature study, it was found that herbal drugs caused various side effects, in which a few of them are serious and leads to mortality (Baratta, Di Lascio, Tarditi, Petieau, & Brusa, 2016). Consumption of herbal medicine aggravates some risk factors, as follows:

1. Various marketed plant's clinical and biochemical properties are still not determined to solve this problem until a scientific database is developed and is available in ISI database; studies showed only 156 plants were undergoing clinical trial studies (Baratta et al., 2016).
2. Secondary risk factor is a behavior of patients for herbal drugs, as they were taking these drugs as supplementary treatment and never mentioning it to their consultant. In 2011, a study was performed that showed 44% cases were regularly using herbal drugs, but there was no data of such type of medicines in the record file of the doctors. Some patients were suffering from problems of side effects (Hugh et al., 2014).

Nearly about 500 medicinal plants are concerned with various allergic reactions in the human body starting from simple redness to shock (Pathak, 1986). Sometimes, ginkgo causes headache, and millepertuis induces dizziness. Some medicinal plants affect the property of synthetic drugs like a high dose of ginkgo reduces the effect of antiulcer drugs like omeprazole (Moreira, Teixeira, Monteiro, De-Oliveira, & Paumgartten, 2014). Due to such reasons, when patient surgery is scheduled, doctors are strictly advised to remove herbal drugs and supplement drugs.

9. Conclusion

The concept of phytotherapy is described as the therapeutic application of medicinal plants and their various parts. Scientific evidence for phytotherapy remains unknown, and their effects on ADME remains lacking. Data published in various articles regarding treatment with phytomedicine worldwide needs awareness in the general population. Phytotherapy works as an alternative for classical treatment therapies in clinical practice. It is very impotent that there is strong interaction relationship with conventional therapy of treatment to identify such types of medicinal plants that are clinically advisable for surgery and alternative medicine for chemotherapy.

References

Adeyemi, A. A., Gbolade, A. A., Moody, J. O., Ogbole, O. O., & Fasanya, M. T. (2010). Traditional anti-fever phytotherapies in Sagamu and Remo north districts in Ogun state, Nigeria. *Journal of Herbs Spices & Medicinal Plants*. https://doi.org/10.1080/10496475.2010.511075.

Agostinis, P., Vantieghem, A., Merlevede, W., & De Witte, P. A. M. (2002). Hypericin in cancer treatment: More light on the way. *The International Journal of Biochemistry & Cell Biology*. https://doi.org/10.1016/S1357-2725(01)00126-1.

Ajazuddin, S. S. (2010). Applications of novel drug delivery system for herbal formulations. *Fitoterapia*. https://doi.org/10.1016/j.fitote.2010.05.001.

Alamgir, A. N. M. (2017). Origin, definition, scope and area, subject matter, importance, and history of development of pharmacognosy. In *Progress in drug research*. https://doi.org/10.1007/978-3-319-63862-1_2.

Allard, T., Wenner, T., Greten, H., & Efferth, T. (2013). Mechanisms of herb-induced nephrotoxicity. *Current Medicinal Chemistry*. https://doi.org/10.2174/0929867311320220006.

Angell, M., & Kassirer, J. P. (1998). Alternative medicine—The risks of untested and unregulated remedies. *The New England Journal of Medicine*. https://doi.org/10.1056/NEJM199809173391210.

Bacchi, S., Palumbo, P., Sponta, A., & Coppolino, M. F. (2012). Clinical pharmacology of nonsteroidal anti-inflammatory drugs: A review. *Anti-Inflammatory & Anti-Allergy Agents in Medicinal Chemistry*. https://doi.org/10.2174/187152312803476255.

Baratta, F., Di Lascio, G., Tarditi, F., Petieau, R., & Brusa, P. (2016). Galenic formulations to fight the phenomenon of counterfeiting in developing countries. *Journal of Drug Delivery Science and Technology*. https://doi.org/10.1016/j.jddst.2015.10.006.

Barnes, J. (2003). Quality, efficacy and safety of complementary medicines: Fashions, facts and the future. Part II: Efficacy and safety. *British Journal of Clinical Pharmacology*. https://doi.org/10.1046/j.1365-2125.2003.01811.x.

Becker, A. (2013). Virus-induced gene silencing: Methods and protocols. *Methods in Molecular Biology*. https://doi.org/10.1007/978-1-62703-278-0.

Bhagyashree, H. A. P. (2015). Phytosome as a novel biomedicine: A microencapsulated drug delivery system. *Journal of Bioanalysis and Biomedicine*. https://doi.org/10.4172/1948-593x.1000116.

Birks, J., & Evans, J. G. (2009). *Ginkgo biloba* for cognitive impairment and dementia. *Cochrane Database of Systematic Reviews*. https://doi.org/10.1002/14651858.CD003120.pub3.

Cañigueral, S., Tschopp, R., Ambrosetti, L., Vignutelli, A., Scaglione, F., & Petrini, O. (2008). The development of herbal medicinal products: Quality, safety and efficacy as key factors. *Pharmaceutical Medicine*. https://doi.org/10.1007/BF03256690.

Cartwright, A. C. (2016). *The British pharmacopoeia, 1864 to 2014: Medicines, international standards and the state*. https://doi.org/10.4324/9781315614182.

Chan, P. C., Xia, Q., & Fu, P. P. (2007). *Ginkgo biloba* leave extract: Biological, medicinal, and toxicological effects. *Journal of Environmental Science and Health, Part C: Environmental Carcinogenesis & Ecotoxicology Reviews*. https://doi.org/10.1080/10590500701569414.

Chawla, A., Boisvert, W. A., Lee, C. H., Laffitte, B. A., Barak, Y., Joseph, S. B., et al. (2001). A PPARγ-LXR-ABCA1 pathway in macrophages is involved in cholesterol efflux and atherogenesis. *Molecular Cell*. https://doi.org/10.1016/S1097-2765(01)00164-2.

De Jong, W. H., & Borm, P. J. A. (2008). Drug delivery and nanoparticles: Applications and hazards. *International Journal of Nanomedicine*. https://doi.org/10.2147/ijn.s596.

Dhyani, A., Chander, V., & Singh, N. (2019). Formulation and evaluation of multipurpose herbal cream. *Journal of Drug Delivery and Therapeutics*. https://doi.org/10.22270/jddt.v9i2.2540.

Ernst, E. (1998). Harmless herbs? A review of the recent literature. *The American Journal of Medicine*. https://doi.org/10.1016/S0002-9343(97)00397-5.

European Medicines Agency. (2012). *Guideline on the investigation of drug interactions*. European Medicines Agency. https://doi.org/10.1093/deafed/ens058.

Fabricant, D. S., & Farnsworth, N. R. (2001). The value of plants used in traditional medicine for drug discovery. *Environmental Health Perspectives*. https://doi.org/10.1289/ehp.01109s169.

Fan, J. P., & He, C. H. (2006). Simultaneous quantification of three major bioactive triterpene acids in the leaves of *Diospyros kaki* by high-performance liquid chromatography method. *Journal of Pharmaceutical and Biomedical Analysis*. https://doi.org/10.1016/j.jpba.2006.01.044.

García-Arieta, A., & Gordon, J. (2012). Bioequivalence requirements in the European Union: Critical discussion. *The AAPS Journal, 14*(4), 738–748.

Gavvala, K., Sengupta, A., & Hazra, P. (2013). Modulation of photophysics and pKa shift of the anti-cancer drug camptothecin in the nanocavities of supramolecular hosts. *ChemPhysChem*. https://doi.org/10.1002/cphc.201200879.

Hu, Q., Li, H., Wang, L., Gu, H., & Fan, C. (2019). DNA nanotechnology-enabled drug delivery systems. *Chemical Reviews*. https://doi.org/10.1021/acs.chemrev.7b00663.

Hugh, J., Van Voorhees, A. S., Nijhawan, R. I., Bagel, J., Lebwohl, M., Blauvelt, A., et al. (2014). From the Medical Board of the National Psoriasis Foundation: The risk of cardiovascular disease in individuals with psoriasis and the potential impact of current therapies. *Journal of the American Academy of Dermatology, 70*(1), 168–177.

Jiang, Y., Ding, S., Li, F., Zhang, C., Sun-Waterhouse, D., Chen, Y., et al. (2019). Effects of (+)-catechin on the differentiation and lipid metabolism of 3T3-L1 adipocytes. *Journal of Functional Foods*. https://doi.org/10.1016/j.jff.2019.103558.

Karsch-Völk, M., Barrett, B., Kiefer, D., Bauer, R., Ardjomand-Woelkart, K., & Linde, K. (2014). Echinacea for preventing and treating the common cold. *Cochrane Database of Systematic Reviews*. https://doi.org/10.1002/14651858.CD000530.pub3.

Kasper, S. (2013). An orally administered lavandula oil preparation (Silexan) for anxiety disorder and related conditions: An evidence based review. *International Journal of Psychiatry in Clinical Practice*. https://doi.org/10.3109/13651501.2013.813555.

Kasper, S., Gastpar, M., Müller, W. E., Volz, H. P., Möller, H. J., Schläfke, S., et al. (2014). Lavender oil preparation Silexan is effective in generalized anxiety disorder—A randomized, double-blind comparison to placebo and paroxetine. *The International Journal of Neuropsychopharmacology*. https://doi.org/10.1017/S1461145714000017.

Kim, S. K. (2015). *Handbook of anticancer drugs from marine origin*. https://doi.org/10.1007/978-3-319-07145-9.

Kim, Y., Kim, M., Kim, H., & Kim, K. (2009). Effect of lavender oil on motor function and dopamine receptor expression in the olfactory bulb of mice. *Journal of Ethnopharmacology*. https://doi.org/10.1016/j.jep.2009.06.017.

Klein, E. (2017). Phytotherapie—Studienlage und Einsatzgebiete. *Der Gynäkologe*. https://doi.org/10.1007/s00129-016-4001-3.

Komiya, M., Sugiyama, A., Tanabe, K., Uchino, T., & Takeuchi, T. (2009). Evaluation of the effect of topical application of lavender oil on autonomic nerve activity in dogs. *American Journal of Veterinary Research*. https://doi.org/10.2460/ajvr.70.6.764.

Kressmann, S., Biber, A., Wonnemann, M., Schug, B., Blume, H. H., & Müller, W. E. (2002). Influence of pharmaceutical quality on the bioavailability of active components from Ginkgo biloba preparations. *The Journal of Pharmacy and Pharmacology*. https://doi.org/10.1211/002235702199.

Kumar, V. (2013). Characterization of anxiolytic and neuropharmacological activities of Silexan. *Wiener Medizinische Wochenschrift.* https://doi.org/10.1007/s10354-012-0164-2.

Leach, M. J., & Moore, V. (2012). Black cohosh (Cimicifuga spp.) for menopausal symptoms. *Cochrane Database of Systematic Reviews.* https://doi.org/10.1002/14651858.cd007244.pub2.

Leslie, S. J. (2013). Generics. In *Routledge companion to philosophy of language.* https://doi.org/10.4324/9780203206966-38.

Li, D., Wang, D., Wang, Y., Ling, W., Feng, X., & Xia, M. (2010). Adenosine monophosphate-activated protein kinase induces cholesterol efflux from macrophage-derived foam cells and alleviates atherosclerosis in apolipoprotein E-deficient mice. *The Journal of Biological Chemistry.* https://doi.org/10.1074/jbc.M110.159772.

Li, X. Z., Zhang, S. N., Liu, S. M., & Lu, F. (2013). Recent advances in herbal medicines treating Parkinson's disease. *Fitoterapia.* https://doi.org/10.1016/j.fitote.2012.12.009.

Liu, Q. (2011). Triptolide and its expanding multiple pharmacological functions. *International Immunopharmacology.* https://doi.org/10.1016/j.intimp.2011.01.012.

Meng, L. H., Zhang, H., Hayward, L., Takemura, H., Shao, R. G., & Pommier, Y. (2004). Tetrandrine induces early G1 arrest in human colon carcinoma cells by down-regulating the activity and inducing the degradation of G 1-S-specific cyclin-dependent kinases and by inducing p53 and p21Cip1. *Cancer Research.* https://doi.org/10.1158/0008-5472.CAN-04-0313.

Mgbeahuruike, E. E., Yrjönen, T., Vuorela, H., & Holm, Y. (2017). Bioactive compounds from medicinal plants: Focus on piper species. *South African Journal of Botany.* https://doi.org/10.1016/j.sajb.2017.05.007.

Moreira, D.d. L., Teixeira, S. S., Monteiro, M. H. D., De-Oliveira, A. C. A. X., & Paumgartten, F. J. R. (2014). Traditional use and safety of herbal medicines. *Brazilian Journal of Pharmacognosy.* https://doi.org/10.1016/j.bjp.2014.03.006.

Mucke, H. (2001). Phytotherapy: Europe's tradition is a hope for the US market. *D & Md Newsletter, 12*(7), 216–220.

Muthuswamy, V. (2011). Challenges in herbal materials (herbs, herbal materials, herbal products, natural sources). *Toxicology International, 18.*

Nahrstedt, A., & Butterweck, V. (2010). Lessons learned from herbal medicinal products: The example of St. John's wort. *Journal of Natural Products.* https://doi.org/10.1021/np1000329.

Nassiri-Asl, M., & Hosseinzadeh, H. (2016). Review of the pharmacological effects of Vitis vinifera (grape) and its bioactive constituents: An update. *Phytotherapy Research.* https://doi.org/10.1002/ptr.5644.

Ooi, C. P., Yassin, Z., & Hamid, T. A. (2012). Momordica charantia for type 2 diabetes mellitus. *Cochrane Database of Systematic Reviews.* https://doi.org/10.1002/14651858.CD007845.pub3.

Osimani, B., Aronson, J., Anjum, R., Crupi, V., Edwards, R., Holman, B., et al. (2018). 9 Roundtable on philosophy of evidence: 'dimensions of evidence and criteria for standards improvement.'. *BMJ Evidence-Based Medicine.* https://doi.org/10.1136/bmjebm-2018-111024.9.

Pang, X. C., Zhang, L., & Du, G. H. (2018). Podophyllotoxin. In *Natural small molecule drugs from plants.* https://doi.org/10.1007/978-981-10-8022-7_90.

Patel, P. (2011). WHO guidelines on quality control of herbal medicines. *International Journal of Research in Ayurveda and Pharmacy, 2.*

Pathak, M. A. (1986). Phytophotodermatitis. *Clinics in Dermatology.* https://doi.org/10.1016/0738-081X(86)90069-6.

Pawar, P., Kalamkar, R., Jain, A., & Shoyab, A. (2015). Ethosomes: A novel tool for herbal drug delivery. *IJPPR Human, 3.*

Pennock, J. R., Boyer, J. N., Herrerea-Silveira, J. A., Iverson, R. L., Whitledge, T. E., Mortazavi, B., et al. (1998). Nutrient behavior and phytoplankton production in Gulf of Mexico estuaries. In *Biogeochemistry of Gulf of Mexico estuaries*. https://doi.org/10.1016/j.phymed.2013.03.027.

Peschel, W. (2014). The use of community herbal monographs to facilitate registrations and authorisations of herbal medicinal products in the European Union 2004–2012. *Journal of Ethnopharmacology*. https://doi.org/10.1016/j.jep.2014.07.015.

Pittler, M. H., Guo, R., & Ernst, E. (2008). Hawthorn extract for treating chronic heart failure. *Cochrane Database of Systematic Reviews*. https://doi.org/10.1002/14651858.CD005312.pub2.

Principles and Practice of Phytotherapy. (2013). *Principles and practice of phytotherapy*. https://doi.org/10.1016/c2009-0-48725-7.

Qin, Y., Xia, M., Ma, J., Hao, Y. T., Liu, J., Mou, H. Y., et al. (2009). Anthocyanin supplementation improves serum LDL- and HDL-cholesterol concentrations associated with the inhibition of cholesteryl ester transfer protein in dyslipidemic subjects. *The American Journal of Clinical Nutrition*. https://doi.org/10.3945/ajcn.2009.27814.

Rainsford, K. D. (2016). *Pharmacological assays of plant-based natural products*. International Publishing Switzerland. https://doi.org/10.1007/978-3-319-26811-8.

Ramzan, I., & Li, G. Q. (2015). Phytotherapies-past, present, and future. In *Phytotherapies: Efficacy, safety, and regulation*. https://doi.org/10.1002/9781119006039.ch1.

Rasaie, S., Ghanbarzadeh, S., Mohammadi, M., & Hamishehkar, H. (2014). Nano phytosomes of quercetin: A promising formulation for fortification of food products with antioxidants. *Pharmaceutical Sciences*.

Robinson, M. M., & Zhang, X. (2011). *The world medicines situation 2011 traditional medicines : Global situation, issues and challenges*. World Health Organization.

Romano, B., Pagano, E., Montanaro, V., Fortunato, A. L., Milic, N., & Borrelli, F. (2013). Novel insights into the pharmacology of flavonoids. *Phytotherapy Research*. https://doi.org/10.1002/ptr.5023.

Saedi, T. A., Md Noor, S., Ismail, P., & Othman, F. (2014). The effects of herbs and fruits on leukaemia. *Evidence-based Complementary and Alternative Medicine*. https://doi.org/10.1155/2014/494136.

Schellekens, C., Perrinjaquet-Moccetti, T., Verbruggen, M., & Dimpfel, W. (2009). Neuravena® and the brain: First clinical data. *Zeitschrift für Phyther*. https://doi.org/10.1055/s-0029-1239918.

Schimpl, F. C., Da Silva, J. F., Gonçalves, J. F. D. C., & Mazzafera, P. (2013). Guarana: Revisiting a highly caffeinated plant from the Amazon. *Journal of Ethnopharmacology*. https://doi.org/10.1016/j.jep.2013.08.023.

Schuwald, A. M., Nöldner, M., Wilmes, T., Klugbauer, N., Leuner, K., & Müller, W. E. (2013). Lavender oil-potent anxiolytic properties via modulating voltage dependent calcium channels. *PLoS One*. https://doi.org/10.1371/journal.pone.0059998.

Shah, U., Shah, R., Acharya, S., & Acharya, N. (2013). Novel anticancer agents from plant sources. *Chinese Journal of Natural Medicines*. https://doi.org/10.1016/S1875-5364(13)60002-3.

Shen, L., & Ji, H. F. (2007). Theoretical study on physicochemical properties of curcumin. *Spectrochimica Acta Part A: Molecular and Biomolecular Spectroscopy*. https://doi.org/10.1016/j.saa.2006.08.018.

Sikorska, A., Słomkowski, M., Maślanka, K., Konopka, L., & Górski, T. (2004). The use of vinca alkaloids in adult patients with refractory chronic idiopathic thrombocytopenia. *Clinical and Laboratory Haematology*. https://doi.org/10.1111/j.1365-2257.2004.00643.x.

Solon-Biet, S. M., McMahon, A. C., Ballard, J. W. O., Ruohonen, K., Wu, L. E., Cogger, V. C., et al. (2014). The ratio of macronutrients, not caloric intake, dictates cardiometabolic health, aging, and longevity in ad libitum-fed mice. *Cell Metabolism*. https://doi.org/10.1016/j.cmet.2014.02.009.

Travaini, M. L., Sosa, G. M., Ceccarelli, E. A., Walter, H., Cantrell, C. L., Carrillo, N. J., et al. (2016). Khellin and Visnagin, Furanochromones from Ammi visnaga (L.) lam., as potential bioherbicides. *Journal of Agricultural and Food Chemistry.* https://doi.org/10.1021/acs.jafc.6b02462.

Verma, S., & Singh, S. P. (2008). Current and future status of herbal medicines. *Veterinary World.* https://doi.org/10.5455/vetworld.2008.347-350.

WHO. (1998). *Quality control methods for medicinal plant materials.* Geneva: World Health Organization. https://doi.org/10.1016/j.pbi.2015.01.003.

Williamson, E. M. (2001). Synergy and other interactions in phytomedicines. *Phytomedicine.* https://doi.org/10.1078/0944-7113-00060.

Woelkart, K., Linde, K., & Bauer, R. (2008). Echinacea for preventing and treating the common cold. *Planta Medica.* https://doi.org/10.1055/s-2007-993766.

Woo, S. M., Min, K. J., Chae, I. G., Chun, K. S., & Kwon, T. K. (2015). Silymarin suppresses the PGE2-induced cell migration through inhibition of EP2 activation; G protein-dependent PKA-CREB and G protein-independent Src-STAT3 signal pathways. *Molecular Carcinogenesis.* https://doi.org/10.1002/mc.22092.

World Health Organization. (1993). *Research guidelines for evaluating the safety and efficacy of herbal medicines.* World Health Organization. Res. Guidel. Eval. Saf. Effic. Herb. Med.

World Health Organization. (2013). WHO expert committee on specifications for pharmaceutical preparations. *World Health Organization Technical Report Series.* https://doi.org/10.1590/s0036-46652008000300013.

Xia, M., Ling, W., Zhu, H., Wang, Q., Ma, J., Hou, M., et al. (2007). Anthocyanin prevents CD40-activated proinflammatory signaling in endothelial cells by regulating cholesterol distribution. *Arteriosclerosis, Thrombosis, and Vascular Biology.* https://doi.org/10.1161/01.ATV.0000254672.04573.2d.

Yau, W. P., Goh, C. H., & Koh, H. L. (2015). Quality control and quality assurance of phytomedicines: Key considerations, methods, and analytical challenges. In *Phytotherapies: Efficacy, safety, and regulation.* https://doi.org/10.1002/9781119006039.ch2.

Yen, F. L., Wu, T. H., Lin, L. T., Cham, T. M., & Lin, C. C. (2008). Nanoparticles formulation of Cuscuta chinensis prevents acetaminophen-induced hepatotoxicity in rats. *Food and Chemical Toxicology.* https://doi.org/10.1016/j.fct.2008.01.021.

Yi, L., Chen, C.y., Jin, X., Mi, M.t., Yu, B., Chang, H., et al. (2010). Structural requirements of anthocyanins in relation to inhibition of endothelial injury induced by oxidized low-density lipoprotein and correlation with radical scavenging activity. *FEBS Letters.* https://doi.org/10.1016/j.febslet.2009.12.006.

Chapter 9

Ethnobotanical perspectives in the treatment of communicable and noncommunicable diseases

Sateesh Suthari[a], Srinivas Kota[b], Omkar Kanneboyena[c], Mir Zahoor Gul[d], and Sadanandam Abbagani[b]

[a]*Department of Botany, Vaagdevi Degree & PG College, Hanamkonda, Telangana, India,* [b]*Department of Biotechnology, Kakatiya University, Warangal, Telangana, India,* [c]*Department of Botany, Kakatiya Government Degree College, Hanamkonda, Telangana, India,* [d]*Department of Biochemistry, University College of Science, Osmania University, Hyderabad, Telangana, India*

1. Introduction

Health is wealth. Keeping healthy is highly important to any living being, and health is "a condition of perfect physical, mental, and social well-being and not merely the absence of disease or infirmity" as defined by the World Health Organization (WHO, 1946). Multisectoral and interdisciplinary activities will establish a complete healthcare system and is an important public factor to note. The basic health factors decide health status, which includes; (i) pure drinking water, effective sanitary waste management, and environmental conditions; (ii) appropriate food intake with high nutritious value and supply chain; (iii) regulation on transmitted (communicable) diseases; and (iv) a healthy lifestyle that changes the manifestation of nontransmittable (noncommunicable) diseases. A "disease" is an abnormal status that influences the organisms. It is a pathological aspect that indicates distinct characteristic features or expressions. Pathogens are either external or internal factors, and they interrupt the normal body functions. Finally, it leads to an abnormal change in temperature of the body that indicates the health condition of a person. Generally, diseases are categorized into three types: (a) *localized* or *target-oriented* diseases that affect particular and targeted organ of the body, (b) *disseminated* or *dispersal* diseases that spread easily to other organs in the body, and (c) *systemic* diseases that affect the whole body. Each disease has its origin and effects on the body, but some diseases perform differently and it is difficult to diagnose them (e.g., HIV, COVID-19). This leads to the severity of the disease, which spreads and sometimes causes death.

Phytomedicine: A Treasure of Pharmacologically Active Products from Plants
https://doi.org/10.1016/B978-0-12-824109-7.00016-9

2. State of disease

The disease responds to certain internal or external factors; it is a physiological response whereas a disorder is an abnormality and intrinsic. These have variations and similarities in some characteristics. The major difference between these two is that a disease may be transmittable from person to person, whereas a disorder is not transmittable. Notably, all diseases are not transmittable; usually, this type of disease is called a nontransmittable disease. Diseases occur due to pathogens, which include fungi, bacteria, viruses, and nematodes, and these spread and multiply on the host.

Based on transmission, diseases are categorized into two types. They are:

1. Communicable, transmissible, or infectious diseases
2. Noncommunicable, nontransmissible, or noninfectious diseases

2.1 Communicable diseases

Communicable disease can be imparted from one person to another or from a source to a host via transmitting agents. This transmission is through vector mediate or nonvector mediate. Sometimes, the illness symptoms may not be exhibited by the infectious organism that plays a key role in transmitting the disease to other organisms. Public health is of higher importance than any other factor in the world. The long-term infectious traces show a huge effect on human health apart from societal, economic, and environmental factors. Some of the prominent communicable diseases in India are flu, tuberculosis (TB), hepatitis-A, hepatitis-B, hepatitis-C, malaria, cholera, HIV/AIDS, sexually transmitted disease (STD), coronavirus disease-2019 (COVID-19), anthrax, botulism, conjunctivitis, cryptosporidiosis, diphtheria-pertussis-tetanus (DPT), ebola, gonorrhea, influenza, leprosy, herpes, ringworm, rubella, rotavirus, norovirus, rabies, scabies, severe acute respiratory syndrome (SARS), smallpox, chickenpox, zika, etc.

These transmissible diseases spread easily by physical contact between the infectious person and normal person or by contacting a contaminated surface, consumption of insects or animals that act as reservoirs for the pathogen, or via airborne or waterborne pathogens.

2.1.1 Trilogy among host, agent, and environment

Favorable environmental conditions act as a reservoir for the infectious organism by which it influences the different hosts. The reservoir acts as a normal environmental provider that provides favorable conditions for the infectious agents to reach the infectious stage by completing other biological cycles. This is called an incubation time/period. These reservoirs may be soil, plants, arthropods, animals, people, etc., and infectious agents use these reservoirs for survival and multiplication. Microbial-contaminated objects may also transmit diseases. An infectious agent may develop up to certain stages in the host but do no harm; these types of hosts are called intermediate hosts. Clinically, to cause disease, a required quantity of agents is needed; minimal causing agents may cause little illness but not a disease. Technically, effective virulence of the infectious agent determines the pathogenic activity of that particular organism by which it will enter the host and complete its lifecycle. Later, it damages the host tissue and causes disease. Virulence depends on the type of strains that cause severe infection or disease and is toxic to the host by modifying its defense system.

2.1.2 Classification of communicable diseases

There are various methods used to categorize infectious diseases. They are transmission mode, medical pattern, preventive methods, and infective agents like fungus, bacteria, and viruses. Protein-coated encapsulation of nucleic acids is called "virus particles"; these cannot be called "living particles" because they complete their biological activity within the cell. Lipid-contained capsid usually act as a "protective layer", as it enables the attachment of a viral particle to the cellular surface and also helps to penetrate and inject its nucleic acid molecule into the cell. Viral particles replicate within the host cell by hijacking its host's cellular mechanism. Misfolded proteins having the ability to cause disease are called "prions"; discovered by Stanley Prusiner, he received the Nobel prize in 1997.

Bacteria are single-cell microorganisms that replicate by asexual and sexual methods. Some bacteria that use oxygen for metabolic activities are called "aerobic bacteria", and others that use CO_2 are called "anaerobic bacteria." Bacteria are categorized based on various characters such as complex structure, developmental requirements, Gram strain, type of causing disease, etc. Two types of bacteria are usually found; one is indigenous (exists within the body) and the other is pathogenic (disease-causing) bacteria. Pathogenic bacteria cause illness by conquering the natural immune system and developing within the body. Fungi are multicellular organisms. Clinical appearances of fungal diseases are generally peripheral infections and hostile life situations, which threaten human life when their immune system is fragile. In HIV-positive patients, *Aspergillus*, *Cryptococcus*, *Mucor*, and *Candida* are most commonly appearing fungal infections. It also causes epidemic infections. Protozoans are one of the major unicellular organisms causing various infectious diseases.

2.2 Noncommunicable diseases

NCDs are also treated as chronic or lifestyle diseases that are not transmitted from source to host, and these can be sustained for a long time in the organism and have minimal progression (WHO, 2018). In the top 10 causes for deaths in the world, the five main NCDs are lung cancer, bronchus, trachea, diabetes,

Alzheimer's disease (WHO, 2015, 2018). This disease is primarily associated with modern (western) lifestyle, urbanization, globalization, and higher expectations to sustain life. The major causes for NCDs are tobacco, laziness, consumption of alcohol, lack of physical exercise, food habits (unhealthy diet), cultural, political, and environmental determinants apart from urbanization, globalization, and population aging. Irrespective of geographical region, age, gender, and community, these noninfectious diseases influence all people on Earth. Mostly it is associated with older people because of their slow physiological activities. The common NCDs in the country are heart disease, psoriasis, sickle cell anemia, stress, stroke, blood pressure, diabetes, obesity, kidney stones, depression, vision impairment, Alzheimer's disease, liver disease, jaundice, epilepsy, eczema, bleeding disorder, thalassemia, cancer, arthritis, etc.

2.3 Status of communicable and no-communicable diseases

According to the assessment report of the WHO (2018), in 2016, out of 56.9 million deaths on Earth, only the top 10 diseases caused about 54% of deaths (Fig. 1). Of these, about 40.5 million people died (71%) of noncommunicable disease whereas the rest died of communicable diseases (15.9 million; 29%). About 78% of global NCD deaths occurred in middle- and low-income countries. Approximately 15.2 million lives were lost due to the world's top-most diseases ischemic heart disease and stroke, whereas chronic obstructive pulmonary and lower respiratory infections killed 3 million people each. Alzheimer's disease and other dementias caused 1.9 million deaths in 2016 at a global level. Lung, bronchus, and trachea cancers accounted for 1.7 million deaths, while

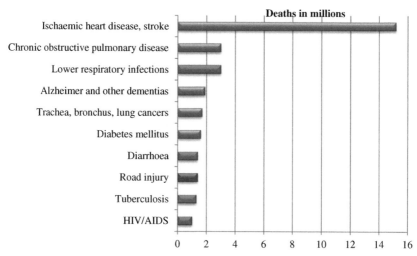

FIG. 1 Top 10 causes of mortality in the world, 2016. *Source: WHO (2018) World Health Statistics 2018: Monitoring health for the SDGs. World Health Organization.*

diabetes mellitus caused 1.65 million deaths, and diarrhea and road injury killed 1.6 million people globally. TB and HIV/AIDS also stood among the top 10 causes for deaths with 1.3 and 1 million people, respectively (WHO, 2018). As per the assumption of WHO (2018), the number of deaths from NCDs will increase from 40.5 to 55 million people by 2030, if timely measurements are not taken by the authorities to control NCDs.

2.3.1   Status of communicable diseases

Communicable diseases or infectious diseases are caused by fungi, bacteria, viruses, and other parasites that can transmit diseases from a source to host either directly or indirectly. A variety of disease-producing microorganisms can establish their shelters in various parts of the body and can be found in the mouth, nose, respiratory tract, and throat. Primarily four types of diseases are found under this category that are among the top 10 causes for death; they are lower respiratory infections, diarrhea, TB, and HIV/AIDS. About 15.9 million people have died from various communicable diseases, which accounts for 29% of total deaths at the global level in 2016. About 3 million people were killed due to lower respiratory infections, while 1.6 million, 1.3 million, and 1 million civilians lost their lives due to diarrhea, TB, and HIV/AIDS, respectively (WHO, 2018). Diseases like leprosy, TB, and influenza (flu) strains can easily spread by cough, saliva, runny nose, sneezing, congestion, and unwashed hands, while HIV/AIDS and hepatitis can be transmitted through semen, blood, and vaginal secretions.

The bacterium *Mycobacterium tuberculosis* causes TB, and it often affects the lungs. Globally, TB is still an effective disease; even though the fight against TB is splendid, it took 16 years to achieve a 19% drop in the rate of occurrence of disease. At the same time, the death rate due to TB in non-HIV patients dropped to 39%. More than 90% of adults affected by TB are found in the 10.4 million HIV-infected population. Meanwhile, TB-diagnosed people are undergoing medication, and millions of people are recovering from that, whereas most HIV-infected TB patients are in danger. Furthermore, the risk remains in TB in the form of drug resistance (Global Tuberculosis Report, 2017).

The human immunodeficiency syndrome (AIDS) is caused by human immunodeficiency virus (HIV), which targets the immune system and weakens the ability of the body to fight against infections. Worldwide, reporting of HIV cases have dropped down from 0.4 to 0.26 in 11 years from 2005 to 2016 (AIDS info-online database). WHO Aftrican Continent Report (2016) revealed that the region stays highly influenced by HIV occurrence, with a proportion of 1.24 of every 1000 uninfected people. Application of antiretroviral therapy brought down the HIV death rate to 48% in 2017, whereas in 2005 it climbed to 53% due to the availability of antiretroviral therapy (ART) drugs. HIV is a global threat because there is no proper medication or vaccine in the world; it has caused more than 35 million deaths, and 37.9 million infected people are living in different corners of the world. In 2018, it killed 7.7 lakh people and infected 1.7 million people on Earth.

In 2015, slow progressing hepatitis-B/hepatitis-C virus infections were noticed in 325 million people across the world; a time delay can cause severe pancreatic damage leading to death. Usually hepatitis occurs in children before the age of 5. Extensive usage of vaccines in infants has notably controlled this. In the meantime, the occurrence of hepatitis-B infection in people was reduced from 4.3% to 3.5% (Global Hepatitis Report, 2017).

Malaria can be considered a "poor man's disease", as almost half of modern civilization in more than 100 countries is affected. Globally, WHO estimated 228 million cases and 405,000 deaths occurred in year 2018. In India, 40 million people are suffering from malaria, and every year, the National Vector Borne Disease Control Programme (NVBDCP) reports more than 1.5 million confirmed cases in India of which 40%–50% are caused by *Plasmodium falciparum* (Pf).

Diarrhea is also one of the infectious diseases in the world; the bacterium, *Vibrio cholera,* causes acute diarrhea through contaminated food or water. Cholera is now endemic in many countries. The disease is most common in places with poor sanitation, unhealthy food habits, urbanization, conflict zones, and famine. At a global level, each year there are approximately 1.3 to 4.0 million cases recorded, and the death cases range from 21,000 to 143,000 people.

Socioeconomic conditions along with tropical environmental situations trigger a few diseases in the individual, and these groups are known as neglected tropical diseases (NTDs) (Neglected Tropical Diseases, 2013). In 2016, reports stated that about 1.5 billion civilians needed specific or mass treatment for NTDs by reducing 0.5 billion people compared to 2010. In 2016, improved prevention of different diseases in the various countries was observed. Worldwide, in the same circumstances, more than 1 billion individuals needed medical care in developing and developed countries for NTDs.

2.3.2 Status of noncommunicable diseases

NCDs are also termed as "chronic diseases" and are the result of a combination of physiological, genetic, and environmental behaviors and socioeconomic conditions. These have low progress and long duration. NCDs can also lead to deaths in the world. About 71%, approximately 41 million deaths out of 57 million, occurred due to these diseases in 2016 (WHO, 2018). Globally in 10 years' span, NCD's death risk rate increased by 18% in the 30 to 70 years of age group. One-third of the death rate caused by NCDs can be reduced by 2030 by restricting risk factors like consumption of tobacco, alcohol, lack of proper food, absence of physical exercise, along with proper diagnosing and disease treatment. As per WHO (2018), consumption levels of tobacco and alcohol quantities varies from one region to the other. Improved measurements are needed to avail for the coverage of alcohol consumption and its respective drug treatment because the global data available is not sufficient to make a proper decision on analysis.

Globally, more than 1.5 billion people aged 15 and above smokes tobacco. Smoking is injurious to health and is one of the major risk factors for various diseases like cancer, cardiovascular, and pulmonary diseases along with adverse effects on socioeconomic and ecological life of the respective patients. WHO had approved tobacco control to the 181 respective authorities in the concerned countries. In 2015–16, 98 countries of WHO took initiation to implement and strengthen their efforts to restrict tobacco products by placing danger symbols and warning quotes. Right now, almost 146 counties are presently observing their own people's smoking activities, and out of 146 countries just 109 countries are monitoring all types of tobacco goods (WHO, 2018).

In India, NCDs are treated as a major health problem for many people that accounts for 62% of the total burden and causes 53% of total deaths in 2018. In India, a study report stated that one person is affected and dies of cardiovascular diseases in 25% of families and 50% of families with cancer, and these diseases caused economic losses of about 10% and 25%, respectively, for the treatment. In India, nearly 5.8 million people die from NCDs every year, that is, 1 in 4 Indian nationals have a risk of dying from an NCD before they reach the age of 70.

2.4 Risk factors

A risk factor is a condition or characteristic feature that promotes acquiring a disease or disorder. Risk factors are often presented alone, but in practice, they do not occur individually. They often coexist and interact with many other features. In recent times, consciousness on health has been increasing enormously, placing this health wing under increasing budget pressure, and it is also very important to be aware of healthcare and risk factors behind any disease, disorder, or injury.

Various types of risk factors can involve many diseases. These are categorized into:

(a) Behavioral—consumption of tobacco, alcohol, improper diet, lack of physical exercise, lack of vaccination, unprotected sex, etc.
(b) Demographical—age, gender, population, etc.
(c) Environmental—pollution, social distance, improper sanitation, etc.
(d) Physiological—obesity, BP, diabetes, cholesterol, etc.
(e) Genetic—genetic history, mutations, etc.

2.4.1 Risk factors of noncommunicable diseases

Few major risk factors influence the increase in communicable diseases. They are:

Safe drinking water and maintaining hygiene conditions: People residing in flooding areas are at high and moderate risk of getting waterborne diseases.

Dislocation of overcrowding population: High population density displacement also leads to a high-risk rate of the occurrence of transmission diseases like measles and respiratory infections.

Disease carrier reproduction: Flooding influences the drastic growth of vector reproduction sites and causes a risk of dengue, malaria, and other endemic arthropod-borne viruses like malaria.

Lack of health infrastructures: Without them, people cannot get greater access to health services.

Lack of hygienic diet: This is the greatest risk factor for the transmission of diseases. A lack of nutritious food supplements governs the deficiency in natural immunity and places them on the path for the reoccurrence of infectious diseases.

2.4.2 Risk factors of noncommunicable diseases

NCDs risk factors are of two types, namely modifiable and nonmodifiable. (i) *Modifiable risk factors* are habituated risk factors that can be altered by behavioral changes that minimizes the possibility of disease occurrence. According to the WHO, they include physical inactivity, tobacco use, alcohol consumption, and abnormal food habitat. (ii) *Nonmodifiable risk factors* cannot be changed. Risk factors that fall under this category include age, gender, race, and genetic history of the family. In addition to these, there are metabolic risk factors that involve biochemical processes and also cause various diseases and damage human health. Behavioral risk factors also fall under this category. According to WHO, there are four metabolic risk factors. They are abnormal blood pressure, unusual cholesterol present in the body, excess or lower glucose levels in the body, and obesity.

3. Ethnobotany perspectives

All over the world, despite advances in the western system of medicine and medical technology, it is increasingly being realized that, if we have to support the healthcare requirements of our ever-increasing population, we will have to resort to cheaper, yet effective alternatives. And there is no better alternative than herbal drugs, which have an ancient history of safe utility in different parts across the planet, especially in India. The traditional wisdom generated following years of painstaking practical examination and usage by ancient rishis, saints, prophets, seers, and scholars led to the development of a rich traditional medicine, Ayurveda, based in our country. However, this traditional knowledge base is slowly eroding due to a lack of awareness and a modern lifestyle. History is replete with examples wherein herbs have saved numerous lives. To quote one, during the age of Ramayana (Treta Yuga), the plant "*Sanjeevani*" was brought by Lord Hanuman at the behest of Lord Rama from the Himalayas to bring back to life Laxmana who fell during the war.

Ancient people were fully aware of the rich potential of plants for curing different types of ailments. Plants are primary sources for many materials needed for all living beings, and these are a rich source of compounds, nutritional values, and used to cure various types of ailments due to their therapeutic properties. In recent times, herbal medicine plays a key role in curing a variety of disorders or diseases compared to synthetic drugs. There has been a striking improvement in the utility of herbal medicines in almost all low-, middle-, and high-income countries due to their natural origin, cost-effectiveness, and minimal or no side effects. Over the decades, several drugs have been sourced and isolated from medicinal plants, e.g., reserpine and turmeric. The plants may be used directly as teas or extracts, and they may be used in the production of drugs, for the prevention and treatment of many diseases, or to promote the health of people or animal husbandry.

Since ancient times, people had a close association with the surrounding plants, and they gained everything from various plant species then available. The relationship among the aboriginal people and their surrounding regions is known as "ethnobotany." Ethnobotany is an interdisciplinary science that includes the study of the plants used by indigenous people for food, shelter, medicine, and clothing (Jain & De, 1966). Harshberger (1896) was the first person to introduce the term "ethnobotany" and defined it as "the study of plants used by primitive and aboriginal people." Later, many eminent scientists defined ethnobotany in other ways. The study and evaluation of the knowledge of all phases of plant life among primitive societies and the effect of the vegetal environment upon their life, opined by Robbins, Harrington, and Freire-Marreco (1916), was where the "study of the interrelations of the primitive man and plants", as defined by Jones (1941). Rao and Henry (1995) explained it as "the study of the past and present interrelationships that exist between the tribal societies and their surrounding vegetal environment." Today, ethnobotany has become more quantitative, qualitative, analytical, multidimensional, multidisciplinary, and multi-institutional (Hamilton et al., 2003), and it is attracting more academicians with interest and growing rapidly (McDonald, 2009). Ethnobotany shows a relation between plants and man in a dynamic ecosystem (Rawat & Chowdhury, 1998).

Ethnobotany is an interdisciplinary aspect that incorporates many disciplines of science and sociology (Jain, 1989). Most importantly, plants and plant extracts have been administered to cure various ailments since ancient times and decrease human mortality and improve the life span. Ethnomedicine has thrived because of the cost-effectiveness, acceptability, and biomedical benefits.

Medicinal plants form the primary and raw material source for the preparation of traditional crude drugs for a long time (WHO, 2002). The basic needs such as food, clothes, shelter, and medicines for the good health of mankind are fulfilled by plants and plant products. Herbal remedies are an integral part of our lives and are considered to be the oldest forms of healthcare of people throughout the globe (Dangwal & Sharma, 2011).

About 422,000 flowering plants have been reported so far (Govaerts, 2001), and of these more than 50,000 plant taxa have been used for medicinal purposes (Schippmann, Leaman, & Cunningham, 2002). India is endowed with about 49,000 plant taxa (Hajra & Mudgal, 1997), and of these about 8000 species are considered to have medicinal values (Tiwari, 2000). More than 43% of the total flowering plants have medicinal importance (Pushpangadan, 1995) and approximately 90% of these plants grow in wild conditions. In India, indigenous and rural people depend on medicinal plants to meet their healthcare and livestock needs. The knowledge of tribal people on surrounding plants has attracted many scientists and botanists to document their oral knowledge. India is rich in biodiversity, and about 8% of the estimated flora and fauna are found here. It is one of the 12 megabiodiversity countries with two hotspots of biodiversity, i.e., the Western Ghats and Northeastern Himalayan regions (Raju, Reddy, & Suthari, 2010). The lifestyle, behavior, stature, and dialect of ethnic tribes are distinct from other people, and they tend to be confined to specific geographical areas that very remote areas of the forests, culturally homogenous, and associated with nature and a natural environment (Gupta, 1974). The tribal people of India mostly live in the forests, foothills, plateaus, and naturally isolated regions. They are differently named as "Adivasi" (original settlers), "aboriginal" (indigenous), "Adimjati" (primitive caste), "adim niwasi" (oldest ethnological sector of the population), "anusuchit janjati" (scheduled tribe), "Girijan" (hillsmen), "janjati" (folk communities), "vanjati" (forest caste), and "Vanvasi" (forest inhabitants). Among these terms, the most popular is *"adivasi"*, whereas the constitutional term is "Scheduled Tribes" (Anusuchit Janjati) who reside in the Union India. India consists of one of the largest groups of tribal communities in the world, a total of 10.43 crore tribal people constitute about 8.6% of the total population pertaining to 227 ethnic groups in 573 tribal communities with 116 different dialects; 10.03% are in urban areas and 89.97% live in rural areas, in about 5000 villages in forests occupying about 15% of land and sporadically distributed in different geographic locations of the country (Anonymous, 2011). Each tribal community has its own social and cultural identity.

India has a centuries-old heritage of benefiting from medicinal and aromatic plants. Herbal medicines are used in the curing of human illnesses and the promotion of health in remote areas. Medicinal plants are often the only accessible healthcare alternative for most of the population, and traditional medicines are an integral part of tribal healthcare. Indigenous people have shown evidence of historical continuity of resource use and possess a broad knowledge of the complex ecological system existing in the vicinity of their habitat.

3.1 Ethnomedicine

The word "ethnomedicine" was coined by Hughes in 1968. Generally, ethnomedicine is the study of traditional medical practices concerned with the cultural interpretation of health, diseases, and illness. It is a complex system and acts as a bridge between traditional knowledge and plant science (Sharma & Mujumdar, 2003). It is highly important as recognized by WHO, and it is one of the prominent healing sources for people for millennia, playing a key part in the treatment of ethnic tribes and rural people (Suthari, Kandagalta, Ragan, & Raju, 2016). The WHO (2008), realizing the significance of ethnomedicinal knowledge, implied that it supports the medicinal needs of some populations. The botanical knowledge of indigenous people is highly interesting and helpful to pharmaceutical explorations for drug discovery.

Tribal communities have accumulated knowledge on plants over the centuries, and they live as forest dwellers in the forest fringes, particularly on the use of plant parts and products as herbal remedies for different ailments (Sreeramulu, Suthari, Ragan, & Raju, 2013). In India, the utility of medicinal plants to cure common ailments has also been practiced for a long time. This medicinal knowledge is mostly undocumented, passed on verbally from one generation to the next as part of their cultural heritage (Mohan & Divya, 2017). About 6000 plants were estimated to be explored in the traditional system of medicine, of which 2500 plant taxa were expected to have medicinal value (Huxley, 1984). Traditional practice as well ethnobotany play vital roles in the conservation of biodiversity and sustainable development (Rajasekaran & Warren, 1994).

3.2   Ethnobotanical explorations in India

To meet survival, medicinal, and multifarious needs, a huge number of wild, indigenous plants have been used by the ethnic tribes in the country, but there is no proper documentation about plant distribution, availability of plants, preparation, and administration of crude drugs. For the first time in 1954, the Botanical Survey of India (Howrah) took up an official program on ethnobotanical exploration studies in India that led by Janaki Ammal. She is considered to be the founder of ethnobotanical studies in the Indian subcontinent and explored the food plants of certain tribes of south India (Ammal, 1956). Inspired by her, Jain and his associates started their extensive field studies by interviewing ethnic tribes in central India and documented the ethnobotanical knowledge (Jain, 1963a, 1963b, 1963c; Jain, 1964a, 1964b, 1964c; Jain, 1965). The initial steps triggered the young botanists', anthropologists', and ayurvedic medical practitioners' minds in the early 1960s to concentrate on ethnobotanical explorations. Mudgal (1987) recorded an account on ethnobotanical plants, and Binu, Nayar, and Pushpangadan (1992) reported on ethnobotanical studies in India. Later, the work on the ethnobotany of the world with special reference to the Indian subcontinent was reviewed by Lalramnghinglova and Jha (1999). Over a period of 6 decades, many field botanists and ethnobotanists searched for ethnomedicinal plants in each and every corner of the country and published their observations. Listed here are some of the noteworthy field observations and documentations provided for understanding: Jain and De (1966);

Jain and Tarafder (1970); Jain, Banerjee, and Pal (1973); Jain and Banerjee (1974); Kirtikar and Basu (1975); Jain and Dam (1979); Jain and Borthakur (1980); Jain (1981, 1986, 1988, 1994, 2000, 2004, 2019); Jain and Saklani (1991); Ramachandran and Manian (1991); Singh, Bhuyan, and Ahmed (1996); Ignacimuthu, Sankarasivaraman, and Kesavan (1998), Ignacimuthu, Ayyanar, and Sankarasivaraman (2008); Jain and Srivastava (1999); Kala (2005, 2006, 2007, 2009); Chakraborty and Bhattacharjee (2006); Ganesan, Venkateshan, and Banumathy (2006); Kingston et al. (2006); Sajem and Gosai (2006, 2010); Uniyal, Singh, Jamwal, and Lal (2006); Rani, Devi, Rajani, Padmavathi, and Maiti (2007); (Udayan, Tushar, George, & Balachandran, 2007); Vijayan, John, Parthipan, and Renuka (2007); Yesodharan and Sujana (2007); Das, Dutta, and Sharma (2008); Jagtap, Deokule, and Bhosle (2008); Shukla, Chakravarty, and Gautam (2008); Borah, Das, Saikia, and Borah (2009); Augustine, Sreejesh, and Bijeshmon (2010); Binu (2010); Jain and Singh (2010); Jaiswal (2010); Kadhirvel et al. (2010); Nanjunda (2010); Phondani et al. (2010); Rout, Panda, Mishra, and Panda (2010); Sutha, Mohan, Kumaresan, Murugan, and Athiperumalsami (2010); Ayyanar and Ignacimuthu (2009); Deepa, Pradeepa, Anjana, and Mohan (2011); Dey and De (2011); Jeyaprakash, Ayyanar, Geetha, and Sekar (2011); Narayanan et al. (2011); Panduranga, Prasanthi, and Reddi (2011); Rao, Seetharami, and Kumar (2011); Sahu, Pattnaik, Sahoo, Lenka, and Dhal (2011); Sen, Chakraborty, and De (2011); Simon, Jebaraj, Suresh, and Ramachandran (2011); Ajesh and Kumuthakalavalli (2012); Kapale (2012); Khatoon, Das, Dutta, and Singh (2014); Kumar, Partap, Sharma, and Jha (2012); Naseef, Ajesh, and Kumuthakalavalli (2012); Prabu and Kumuthakalavalli (2012); Samydurai, Thangapandian, and Aravinthan (2012); Smita, Sangeeta, Kumar, Soumya, and Deepak (2012); Sudeesh (2012); Wangyal (2012); Azad and Bhat (2013); Ballabha, Tiwari, and Tiwari (2013); Dhatchanamoorthy, Kumar, and Karthik (2013); Murthy and Vidyasagar (2013); Prasad, Shyma, and Raghavendra (2013); Rajkumari, Singh, Das, and Dutta (2013); Rani, Rana, Jeelani, Gupta, and Kumari (2013); Senthilkumar, Aravindhan, and Rajendran (2013); Shankar, Lavekar, Deb, and Sharma (2012); Shrivastava and Kanungo (2013a, 2013b); Shukla, Srivastava, and Rawat (2013); Thomas and Rajendran (2013); Vaidyanathan, Senthilkumar, and Basha (2013); Vidyasagar and Siddalinga (2013); Chatterjee (2014); Das, Talukdar, and Choudhury (2014); Kumar and Bharati (2014); Lone, Bhardwaj, Shah, and Tabasum (2014); Bose, Aron, and Mehalingam (2014); Islam et al. (2014); Mishra and Patil (2014); Pfoze, Kehie, Kayang, and Mao (2014); Sarkhel (2014); Sarvalingam, Rajendran, Sivalingam, and Jayanthi (2014); Sharma, Sharma, and Debbarma (2014); Sharma and Samant (2014); Shrestha, Prasai, Shrestha, Shrestha, and Zhang (2014); Tiwari and Tiwari (2014); Bandana and Debabrata (2015); Das and Khatoon (2015); Kumar and Pandey (2015); Modak, Gorai, Dhan, Mukherjee, and Dey (2015); Omkar, Suthari, and Raju (2015); Padhan and Panda (2015); Pattanayak and Dhal (2015); Rahaman and Karmakar (2015); Satyavathi, Deepika, Padal, and Rao (2015); Sen and Chakraborty (2015); Xavier, Kannan,

Lija, Auxillia, and Rose (2014), Xavier, Kannan, and Auxilia (2015); Bala and Singh (2016); Chakraborty, Mondal, and Mukherjee (2016); Chanthru, Gomthi, and Manian (2016); Joshi and Pande (2016); Kumar, Mallikarjuna, Prasad, and Naidu (2016); Prakash, Unnikrishnan, and Hariramamurthi (2016); Sasidhar, Brahmajirao, and Kumar (2016); Talukdar and De (2016); Basha and Priydarsini (2017); Dharm and Pramod (2017); Narayan and Singh (2017); Pandey, Mishra, and Hemalatha (2017); Pullaiah, Krishnamurthy, and Bahadur (2017); Ramesh, Pratap, Nagaraju, and Sudarsanam (2017); Reddy, Paul, and Basha (2017); Sripriya (2017a, 2017b); Kekuda, Vinayaka, and Sachin (2018); Kundu, Hazra, Pal, and Bhattacharya (2018); Raj et al. (2018); Rupani and Chavez (2018); Satyavathi and Padal (2018); Vishnuvardhan, Jyothirmayi, and Jyothi (2018); Reddy, Babu, and Rao (2019); and Naidu, Suthari, Kumar, and Venkaiah (2020).

In addition to the ethnobotanical studies in the country, in Telangana there was also extensive ethnobotanical work carried out by various scientists since 1953. In Telangana, for the first time, the ethnobotanical information on some plants was mentioned in Forest Flora of Hyderabad by Khan (1953). Later, Kapoor and Kapoor (1980), Ramarao (1988), and Ravisankar (1990) documented about the wealth of medicinal plants from the Karimnagar, Warangal, and Adilabad districts, respectively. Ravishankar and Henry (1992) documented an account on the Ethnobotany of Gonds of Adilabad district. Furthermore, notes on ethnomedicinal uses and modes of administration to treat various ailments using important medicinal plant taxa were documented by many botanists, ethnobotanists, and scientists from different areas of Telangana state (Kumar & Pullaiah, 1999; Mohan, Suthari, Kandagatla, & Ragan, 2017; Mohan, Suthari, & Ragan, 2017; Murthy, Reddy, Reddy, & Raju, 2007; Naqvi, 2001; Omkar, Suthari, Alluri, Ragan, & Raju, 2012; Padmarao & Reddy, 1999; Priyadarshini & Ragan, 2019; Pullaiah & Kumar, 1996; Raju & Reddy, 2005; Reddy, 2002; Reddy, 2008; Reddy, Bhanja, & Raju, 1998; Reddy & Raju, 2002; Reddy & Rao, 2002; Reddy, Trimurthulu, & Reddy, 2010; Saidulu, 2014; Saidulu, Suthari, Kandagatla, Ajmeera, & Vatsavaya, 2015; Sangameshwar, 2018; Sreeramulu, 2008; Sreeramulu et al., 2013; Sudharani, Umadevi, Rajani, Padmavathi, & Maiti, 2007; Suthari, 2020; Suthari et al., 2016; Suthari & Raju, 2016; Suthari, Raju, & Majeti, 2018; Suthari, Sreeramulu, Omkar, & Raju, 2014; Suthari, Sreeramulu, Omkar, Reddy, & Raju, 2014; Upadhyay & Chauhan, 2000).

3.3 Ethnobotanical explorations in abroad

Attention has been paid throughout the world to carry out ethnobotanical studies for the welfare of people. Gunther (1945) surveyed the use of 150 taxa as food and medicine by the Native Americans of western Washington. Schultes (1960) studied the ethnobotanical practices of the indigenous people of northwest Amazon. Turner and Bell (1971) studied the ethnobotanical aspects of Coast Salish Indians of Vancouver Island and recorded 122 taxa used as food and medicine. Liengme (1983) conducted exploration studies on ethnobotany among Khoi Bantu and San tribes of South Africa. Bhargava surveyed the Onge tribe of Andaman and Nicobar Island and recorded 40 ethnomedicinal plant taxa. Rodin (1985) studied the culture of the Kwanyama Ovambos tribe of South Africa. Pakia, Cooke, and Van Staden (2003) conducted ethnobotanical studies in Southern Africa and documented that 125 plant taxa are being used to treat 26 ailment categories by the Midzichenda tribes of Kenya. Gazzaneo, De Lucena, and de Albuquerque (2005) recorded 125 plants distributed among 61 flowering angiospermous families used by the Tres Ladeiras community residing in the Atlantic forest in Pernambuco State, Brazil. Reyes-García, Huanca, Vadez, Leonard, and Wilkie (2006) quantitatively analyzed a total of 114 plants used by the Tsimane of Bolivian Amazon for cultural, practical, and economic uses. Giday, Asfaw, and Woldu (2009) revealed 51 medicinal plants used by the Meinit ethnic groups of Ethiopia. Oladunmoye and Kehinde (2011) recorded a total of 208 medicinal plants used by the Yoruba tribe of southwestern Nigeria for treating viral infections. Srithi, Trisonthi, Wangpakapattanawong, and Balslev (2012) documented 79 medicinal plants used by the Hmong, an ethnic group of Thailand, to treat gynecological ailments. Lawal, Amao, Lawal, Alamu, and Sowunmi (2013) studied nine plant taxa used by the women from Oyo State, Nigeria, to treat gynecological disorders. Morilla, Sumaya, Rivero, and Madamba (2014) recorded 60 medicinal plants belong to 29 families used to treat minor ailments by the Subanens tribe, Philippines. Hundreds of ethnobotanists and plant scientists have explored the ethnomedicinal plants from different nations of the world to treat different ailments (Adnan et al., 2015; Andrade, Mosquera, & Armijos, 2017; Batsatsashvili, Kikvidze, & Bussmann, 2020; Bhat, Hegde, Hegde, & Mulgund, 2014; Bussmann, 2019; Bussmann et al., 2017, 2016, 2018; Bussmann, Paniagua-Zambrana, & Romero, 2020; Catarino, Havik, Indjai, & Romeiras, 2016; de Natale & Pollio, 2007; de Santana, Voeks, & Funch, 2016; Dwivedi & Dwivedi, 2015; Faruque et al., 2018; Giday et al., 2016; González & Marioli, 2010; Khatun, Parlak, Polat, & Cakilcioglu, 2012; Mesfin, Tekle, & Tesfay, 2013; Nadembega, Boussim, Nikiema, Poli, & Antognoni, 2011; Ong & Kim, 2014; Rahmatullah et al., 2015; Randrianarivony et al., 2016; Sadeghi & Mahmood, 2014; Sağıroğlu, Dalgıç, & Toksoy, 2013; Shiracko, Owuor, Gakuubi, & Wanzala, 2016; Shisanya, 2017; Tugume et al., 2016; Voeks, 2007; Wangyal, 2012; Zambrana & Bussmann, 2020).

3.4 Ethnobotany—A multidisciplinary science

Ethnobotany gives valid information about the utility of plant species by indigenous people, thus, it is a combination of two branches, i.e., botany and anthropology. It is interdisciplinary between plant science and ethnology. The botanical knowledge of an ethnic people can be useful in many aspects, so it is called a "multidisciplinary science." In recent years, several disciplines have come to be used in connection with ethnic people and their knowledge, namely ethnomedicine, ethnotaxonomy, ethnotoxicology, ethnoecology, ethnogynaecology, ethnopharmacology, ethnopadiatrics, ethnoforestry, etc. (Fig. 2).

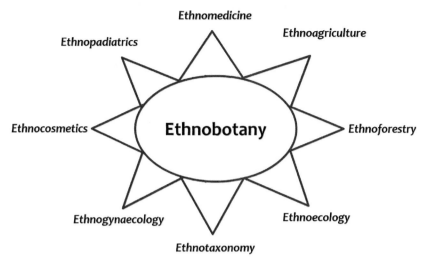

FIG. 2 Schematic presentation of prominent utility patterns of ethnobotany in various disciplines of science.

3.5   Ailments categories

Based on the International Classification of Diseases and Related Problems (ICD), WHO (2015) classified ailment or therapeutic indications into 19 categories. The ailments categories are: (i) circulatory system/cardiovascular, (ii) dental, (iii) dermatological infections, (iv) ear, nose, throat, (v) endocrine disorders, (vi) eye, (vii) fever, (viii) gastrointestinal, (ix) genitourinary, (x) gynecological, (xi) hemorrhoids, (xii) infectious and parasitic, (xiii) liver, (xiv) neurological, (xv) oncological, (xvi) poisonous bites, (xvii) respiratory system, (xviii) scalp and hair, and (xix) skeletomuscular. A brief compilation of ethnomedicinal plant data is provided in Table 1 for the treatment of various ailment categories.

3.6   Traditional medicine and Indian society

Traditional botanical knowledge (TBK) is the overall knowledge of local people regarding plants. In TBK, there are many steps involved such as identification, processing, and management of plants used for survival and medicine. In this, sociological, cultural, and spiritual aspects are also included. Cotton (1996) explained the utility pattern of plants in many aspects such as ecological, cognitive, and vegetation management. The new techniques in biotechnology, biochemistry, and pharmacy indicate the importance and significance of traditional knowledge for the commercial production of new compounds for the benefit of humankind and animal husbandry. The rapid development of equipment and techniques in isolation of phytocomponents and screening techniques for active compounds and testing tools have evolved in recent times for the formulation of drugs to control the diseases or disorders that have a profound effect on traditional medicine.

TABLE 1 Ethnomedicinal plants use report in the management of various diseases as categorized by WHO (2015).

	Ailment category (Genera/Species)	Scientific Names
1	Circulatory system/ Cardiovascular (25/25)	*Capparis spinosa, Cerascioides turgida, Cichorium intybus, Cinnamomum zeylanicum, Citrullus colocynthis, Cocculus hirsutus, Coriandrum sativum, Equisetum arvense, Erythrina suberosa, Caruga pinnata, Hemidesmus indicus, Hordeum vulgare, Litsea glutinosa, Memecylon umbellatum, Nerium oleander, Ocimum basilicum, Portulaca oleracea, Scleria corymbosa, Selaginella bryopteris, Sesamum indicum, Solanum nigrum, Vitis vinifera, Woodfiordia fruticosa, Ziziphus mauritiana, Zingiber officinale*
2	Dental (31/35)	*Acacia catechu, Acacia nilotica, Achyranthes aspera, Acmella paniculata, Anacardium occidentale, Annona squamosa, Azadirachta indica, Boehmeria pentaphylla, Buchanania cochinchinensis, Ficus benghalensis, Ficus racemosa, Jatropha curcas, Madhuca longifolia var. latifolia, Melia azedarach, Mimosa pudica, Mimusops elengi, Murraya koenigii, Murraya paniculata, Nicotiana tabacum, Phyllanthus emblica, Phyllanthus reticulatus, Pterocarpus marsupium, Ricinus communis, Senna occidentalis, Shorea robusta, Sida acuta, Solanum virginianum, Streblus asper, Tephrosia purpurea, Terminalia alata, Terminalia bellirica, Tragia involucrata, Vitex negundo, Wrightia tinctoria, Ziziphus rugosa*
3	Dermatological and Infections (28/28)	*Acalypha indica, Albizia amara, Argemone mexicana, Artemisia nilagirica, Asparagus gonocladus, Calotropis gigantea, Cannabis sativa, Cassia fistula, Centella asiatica, Cyclea peltata, Chaemesyce hirta, Cryptolepis buchananii, Datura metel, Drymaria cordata, Euphorbia tirucalli, Ficus virens, Holoptelea integrifolia, Ipomoea eriocarpa, Jatropha aureus, Litsea cubeba, Mimosa pudica, Lygodium flexuosum, Plantago major, Plumbago zeylanica, Plumeria acutifolia, Pongamia pinnata, Senna alata, Trichosanthes cucumerina*
4	Ear, Nose, Throat (42/43)	*Aegle marmelos, Aerva lanata, Albizia lebbeck, Allium sativum, Alstonia scholaris, Artemisia afra, Aristolochia indica, Artocarpus heterophyllus, Barleria prionitis, Bidens pilosa, Borreria latifolia, Bryophyllum pinnatum, Calamus rotang, Calotropis gigantean, Capsicum annuum, Carica papaya, Catharanthus roseus, Cheilocostus speciosus, Clitoria ternatea, Coccinia indica, Cocculus hirsutus, Curcuma longa, Euphorbia hirta, Gmelina asiatica, Justicia adhatoda, Leucas aspera, Lippia javanica, Merremia turpethum, Nerium oleander, Ocimum basilicum, Ocimum tenuiflorum, Pavetta indica, Phyllanthus emblica, Pseudarthria viscida, Sansevieria roxburghiana, Shorea robusta, Sida acuta, Syzygium cumini, Terminalia arjuna, Tragia involucrata, Vanda tesellata, Vitex negundo, Zingiber officinale*

5	Endocrine disorders (32/34)	*Acacia arabica, Adansonia digitata, Aegle marmelos, Allium cepa, Allium sativum, Aloe vera, Azadirachta indica, Benincasa hispida, Beta vulgaris, Capparis zeylanica, Citrullus colocynthis, Coccinia grandis, Eucalyptus globulus, Ficus benghalensis, Gymnema sylvestre, Hemidesmus indicus var. pubescens, Ipomoea batatas, Mangifera indica, Momordica charantia, Moringa oleifera, Morus alba, Mucuna pruriens, Ocimum tenuiflorum, Phyllanthus emblica, Piper nigrum, Pterocarpus marsupium, Punica granatum, Syzygium cumini, Tamarindus indica, Terminalia bellirica, Terminalia chebula, Tinospora cordifolia, Trigonella foenum-graecum, Ziziphus mauritiana*
6	Eye (25/27)	*Achyranthes aspera, Alternanthera philoxeroides, Alternanthera sessilis, Amaranthus spinosus, Amaranthus viridis, Argemone mexicana, Berberis asiatica, Biophytum sensitivum, Boerhavia diffusa, Butea monosperma, Cassia fistula, Clitorea ternatea, Cyanthillium cinereum, Cynodon dactylon, Dalbergia sissoo, Eclipta prostrata, Hemidesmus indicus, Hyptis suaveolens, Nelumbo nucifera, Pavetta indica, Phyllanthus emblica, Piper longum, Pseudarthria viscida, Pterocarpus santalinus, Tabernaemontana divaricata, Terminalia bellirica, Trianthema portulacastrum*
7	Fever (28/28)	*Acacia leucophloea, Agave americana, Andrographis paniculata, Anisomeles indica, Annona squamosa, Azadirachta indica, Acacia chundra, Calotropis gigantea, Diplocyclos palmatus, Vernonia cinerea, Balanites roxburghii, Bauhinia malabarica, Caesalpinia bonduc, Cardiospermum halicacabum, Casaeria elliptica, Crotalaria verrucosa, Curculigo orchioides, Cyperus rotundus, Drimia indica, Evolvulus alsinoides, Gymnema sylvestre, Phyllanthus amarus, Pueraria tuberosa, Terminalia arjuna. Vanda tasellata. Vitex negundo, Wrightia tinctoria. Ziziphus xylopyrus*
8	Gastrointestinal (30/30)	*Abutilon indicum, Acacia nilotica. Albizia lebbeck, Allium cepa, Azadirachta indica, Boerhavia diffusa, Carissa carandas, Cassia fistula, Cedrus deodara, Cinnamomum zeylanicum, Citrullus colocynthis. Coriandrum sativum, Curcuma longa, Cyperus rotundus, Euphorbia hirta, Ficus religiosa, Foeniculum vulgare, Melia azedarach, Mentha viridis, Opuntia dillenii, Oxalis corniculata, Phyllanthus emblica, Punica granatum, Rosa indica, Saccharum officinarum, Solanum surattense, Tamarindus indica, Withania somnifera, Woodfordia fruticosa, Ziziphus mauritiana*
9	Genitourinary (39/40)	*Acacia torta, Ailanthus excelsa, Aloe vera, Butea monosperma, Butea superba, Caesalpinia bonduc, Cajanus cajan, Capparis sepiaria. Cardiospermum halicacabum, Careya arborea. Ceriscioides turgida, Cheilocostus speciosus, Cocculus hirsutus, Cochlospermum religiosum, Combretum albidum, Crotalaria albida, Cryptolepis buchananii; Curculigo orchioides, Dioscorea alata. Diospyros montana, Firmiana simplex, Gardenia latifolia, Gloriosa superba. Heliotropium indicum, Hybanthus ennaespermis, Hymenodictyon orixense, Ledebouria hyacinthina. Leptadenia reticulata. Litsea glutinosa, Moringa pterygosperma, Naringi crenulata, Pedalium murex, Pterocarpus marsupium, Saraca indica, Smilax perfoliata, Streblus asper; Tacca leontopetalaoides, Tamarindus indica, Tinospora cordifolia, Wrightia arborea*

Continued

TABLE 1 Ethnomedicinal plants use report in the management of various diseases as categorized by WHO (2015) — cont'd

	Ailment category (Genera/Species)	Scientific Names
10	Gynecological (24/24)	*Achyranthes aspera, Allium cepa, Aloe vera, Borassus flabellifer, Brassica juncea, Curcuma longa, Erythrina variegata, Firmiana simplex, Foeniculum vulgare, Ipomoea obscura, Lawsonia inermis, Mangifera indica, Moringa oleifera, Nerium oleander, Papaver somniferum, Physalis angulata, Punica granatum, Ricinus communis, Sesbania grandiflora, Sida acuta, Smilax zeylanica, Tribulus lanuginosus, Trigonella foenum-graecum, Zingiber officinale*
11	Hemorrhoids (40/44)	*Abutilon indicum, Acalypha indica, Achyranthes aspera, Aegle marmelos, Allium cepa, Aloe vera, Alpinia galanga, Amaranthus spinosus, Azadirachta indica, Bauhinia variegata, Calotropis procera, Capparis spinosa, Carica papaya, Chaemesyce hirta, Cissus quadrangularis, Citrus limon, Cocos nucifera, Coriandrum sativum, Cynodon dactylon, Cyperus rotundus, Datura metel, Datura stramonium, Eclipta prostrata, Ficus carica, Ficus racemosa, Gloriosa superba, Mangifera indica, Mimosa pudica, Lantana camara, Madhuca longifolia var. latifolia, Phoenix dactylifera, Portulaca oleracea, Ricinus communis, Semicarpus anacardium, Senna auriculata, Senna tora, Sesamum indicum, Solanum melongena, Solanum nigrum, Tamarindus indica, Terminalia chebula, Trigonella foenum-graecum, Vitis vinifera, Zingiber officinale*
12	Infectious and Parasitic (62/65)	*Abrus precatorius, Abutilon indicum, Aegle marmelos, Ampelocissus tomentosa, Anogeissus latifolia, Aristolochia bracteolata, Asparagus racemosus, Azima tetracantha, Barringtonia acutangula, Bauhinia racemosa, Bombax ceiba, Boswellia serrata, Byttneria herbacea, Caesalpinia bonduc, Canavalia gladiata, Capparis sepiaria, Cerascioides turgida, Cissus quadrangularis, Clerodendrum phlomides, Clitoria ternatea, Coccinia grandis, Combretum latifolium, Commelina benghalensis, Curcuma pseudomontana, Cyphostemma setosum, Derris scandens, Dichrostachys cinerea, Dillenia pentagyna, Dioscorea bulbifera, Dioscorea pentaphylla, Diospyros chloroxylon, Dolichandrone falcata, Dregea volubilis, Eleusine coracana, Ficus benghalensis, Ficus talboti, Gloriosa superba, Gymnema sylvestre, Hemidesmus indicus, Holarrhena pubescens, Jasminum auriculatum, Justicia adhatoda, Lannea coromandelica, Lawsonia inermis, Mimosa pudica, Moringa concanensis, Olax scandens, Oroxylum indicum, Paederia foetida, Pavetta indica, Phyllanthus amarus, Phyllanthus emblica, Phyllanthus reticulatus, Plumbago zeylanicum, Pueraria tuberosa, Senna occidentalis, Strychnos nux-vomica, Syzygium cumini, Tectona grandis, Tephrosia purpurea, Vitex negundo, Withania somnifera, Wrightia tinctoria, Xylia xylocarpa, Ziziphus oenopolia*

Continued

13	Liver (35/37)	Amaranthus spinosus, Aloe vera, Apium graveolens, Asparagus racemosus, Azadirachta indica, Baliospermum montanum, Berberis aristata, Berberis lyceum, Boerhavia diffusa, Cassia fistula, Calotropis procera, Decalepis hamiltonii, Euphorbia fusiformis, Hedychium spicatum, Hygrophila auriculata, Melochia corchorifolia, Momordica dioica, Ocimum tenuiflorum, Phyllanthus amarus, Phyllanthus emblica, Picrorrhiza kurroo, Piper longum, Podophyllum hexandrum, Pterocarpus marsupium, Ricinus communis, Saussurea obvallata, Scoparia dulcis, Silybum marianum, Taraxacum officinale, Terminalia arjuna, Tylophora indica, Trichosanthes cucumerina, Tridax procumbens, Urtica parviflora, Vitex negundo, Withania somnifera, Woodfordia fruticosa
14	Neurological (32/34)	Ageratum conyzoides, Allium sativum, Arachys hypogea, Azadirachta indica, Bacopa monnieri, Bidens pilosa, Centella asiatica, Cinnamomum zeylanicum, Citrus aurantium, Cocos nucifera, Coffea arabica, Corchorus olitorius, Coriandrum sativum, Cymbopogon citrates, Cyperus rotundus, Eucalyptus globulus, Ficus carica, Glycyrrhiza glabra, Lantana camara, Mangifera indica, Momordica charantia, Musa paradisiacal, Ocimum basilicum, Ocimum gratissimum, Pistacia atlantica, Phyllanthus amarus, Pterocarpus marsupium, Senna occidentalis, Sesamum indicum, Scoparia dulcis, Sida acuta, Terminalia catappa, Terminalia chebula, Vitex negundo
15	Oncological (24/24)	Aegle marmelos, Allium sativum, Andrographis paniculata, Artemisia asiatica, Azadirachta indica, Cannabis sativa, Cathranthus roseus, Centella asiatica, Cistanche tubulosa, Coriandrum sativum, Dendrophthoe falcata, Eclipta prostrata, Elephantopus scaber, Glycyrrhiza glabra, Mangifera indica, Operculina turpethum, Oroxylum indicum, Phyllanthus amarus, Phyllanthus emblica, Plumbago zeylanica, Scutellaria barbata, Semecarpus anacardium, Withania somnifera, Zingiber officinale
16	Poisonous bites (44/47)	Abrus precatorius, Acacia chundra, Acacia farnesiana, Acalypha indica, Abrus precatorius, Abutilon indicum, Achyranthes aspera, Alangium salviifolium, Aloe vera, Ampelocissus tomentosa, Andrographis paniculata, Anogeissus latifolia, Aristolochia bracteolata, Aristolochia indica, Asparagus racemosus, Boerhavia diffusa, Boswellia serrata, Canthium parviflorum, Capparis divaricata, Crinum asiaticum, Crotolaria verrucosa, Curculigo orchioides, Cyclosorus unitus, Cynodon dactylon, Derris scandens, Diospyros chloroxylon, Erythrina variegata, Gloriosa superba, Helicteres isora, Jatropha curcas, Leea asiatica, Marsdenia tenacissima, Mentha spicata, Mimosa pudica, Pavetta indica, Phyllanthus reticulatus, Rauvolfia serpentina, Sarcostemma acidum, Semecarpus anacardium, Senna obtusifolia, Senna tora, Strychnos nux-vomica, Tamarindus indica, Tephrosia purpurea, Tinospora cordifolia, Vanda tesellata, Ziziphus xylopyrus

TABLE 1 Ethnomedicinal plants use report in the management of various diseases as categorized by WHO (2015)—cont'd

	Ailment category (Genera/Species)	Scientific Names
17	Respiratory system (50/52)	*Abelmoschus esculentus, Abrus precatorius, Acacia arabica, Acacia chundra, Acacia nilotica, Achyranthes aspera, Ageratum conyzoides, Albizia lebbeck, Allium cepa, Allium sativum, Aloe vera, Alstonia scholaris, Amaranthus viridis, Anogeissus latifolia, Azadirachta indica, Balanites roxburghii, Bambusa bambos, Bauhinia variegata, Blumea mollis, Breynia retusa, Calotropis gigantea, Capparis spinosa, Cissus quadrangularis, Datura metel, Dendrophthoe falcata, Dioscorea pentaphylla, Dodonaea viscosa, Drypetes roxburghii; Ephedra ciliata, Eucalyptus globulus, Ficus benghalensis, Holarrhena pubescens, Jasminum auriculatum, Justicia adhatoda, Lawsonia inermis, Mimosa pudica, Moringa concanensis, Nyctanthes arbor-tristis, Opuntia dillenii, Papaver somniferum, Phyllanthus emblica, Polygonum hydropiper, Portulaca oleracea, Ricinus communis, Solanum nigrum, Tephrosia purpurea, Terminalia chebula, Trianthema portulacastrum, Vitex negundo, Withania somnifera, Zingiber officinale, Ziziphus mauritiana*
18	Scalp and Hair (35/37)	*Abrus precatorius, Acacia caesia, Acacia sinuata, Achyranthes aspera, Alangium salviifolium, Aloe vera, Annona squamosa, Anisomeles indica, Azadirachta indica, Bacopa monnieri, Caryota urens, Centella asiatica, Citrullus colocynthis, Citrus aurantium, Cocos nucifera, Dalbergia lanceolaria subsp. paniculata, Delonix elata, Eclipta prostrata, Eucalyptus globulus, Ficus benghalensis, Hibiscus rosa-sinensis, Indigofera tinctoria, Lawsonia inermis, Madhuca longifolia var. latifolia, Melia azedarach, Merremia hederacea, Ocimum gratissimum, Ocimum tenuiflorum, Phyllanthus emblica, Ricinus communis, Salvia officinalis, Santalum album, Senna angustifolia, Sesamum indicum, Terminalia bellirica, Trigonella foenum-graecum, Vitex negundo*
19	Skeletomuscular (10/10)	*Asparagus racemosus, Hibiscus rosa-sinensis, Morinda pubescens, Moringa oleifera, Morus alba, Pergularia daemia, Syzygium cumini, Tamarindus indica, Tinospora cordifolia, Vitex negundo*

Ref. Suthari et al. (2018).

Everyone knows that no drug is an ideal one, and every drug can show side effects on the human body. To avoid these harmful effects on patients, the scientific community is in search of effective alternatives from plant products. Traditional knowledge was not properly documented in any form, as it is only passed orally from person to person for generations in the form of folklore, stories, songs, proverbs, rituals, legends, etc. Because of all these, WHO and the Rio Earth Summit have directed all nations toward the proper utility of traditional knowledge of indigenous or local people for the management of healthcare.

In the Indian subcontinent, various previous shreds of evidence proved that, for more than 7000 years, there has been medical practice on dental and other diseases. Archeological findings proved that, in the Middle Gangetic region, the use of medicinal plants to cure diseases occurred earlier than the 2nd millennium, and the use of those medicinal plant taxa is also seen in current Ayurvedic medicine (National Centre for Biological Sciences, 2015). A diversified medicinal system was developed in ancient India because the Indian subcontinent is composed of various ethnic groups with different languages and cultures. This made Indian society reliably dependent on traditional medicine practice, because once the British ruler's allopathic medicine entered, it raised questions on the existence of traditional medicine. Vast scientific research development in allopathic medicine restricted further development in the utilization of traditional medicine. After all these hurdles, up to today, a lot of rural and some urban people believe in traditional medicine as an initial healthcare system (Mafuva & Marima-Matarira, 2014; Pandey, Rastogi, & Rawat, 2013).

An Indian traditional system of medicine, Ayurveda suggests that drinking water containers must be silver or copper, yet today copper and silver pots are replaced by plastic and steel containers. As per Ayurveda, water stored in copper and silver pots prevents waterborne disease because *Escherichia coli-, Salmonella typhi-,* and *Vibrio cholerae*-causing organisms cannot survive in water stored in silver and copper pots. These two metals show lethal effects on microorganisms (Bodeker & Graz, 2013).

Native inhabitants traditionally use tools including smoke from burning plants, hanging plants inside houses, and crushed plant parts applied on the body. Through traditional knowledge, the local people make their own repellents for many insects such as mosquitoes and houseflies. These plant species contain terpenoids and phenols, and several investigations have also proved their insecticidal properties. These are excellent natural remedies, and traditional knowledge plays a vital role in the prevention and control of microorganisms without harming the environment.

3.6.1 Role of traditional health practitioners

Traditional health practitioners (THPs) are persons familiar to a group of people, who treats local people to cure various health problems using different ingredients from animals and plants in respect to cultural and social beliefs and holy information depending on geographical context, beliefs, and approaches to physical, psychological, and communal welfare by the cautious knowledge of diseases and cures for their society. Traditional therapists contain vast knowledge for the utilization of surrounding vegetation, achieving respective nutritional and medicinal aspects that they developed via various medicinal methods from different natural components. They depend on bush medicines with an effective lifestyle, behavior, diet habits, and all other preliminary information and understand the condition of the patient to treat diseases. In remote and interior areas, these local healers are the only source to save the lives of these ethnic people by treating various ailments due to lack of hospitals, proper settlements, awareness, and all other facilities.

3.6.2 Relevance of traditional botanical knowledge today

At the global level, there are wonderful techniques and exploration studies that have evolved to understand the human body, biochemistry, and physiology in the last two centuries. Due to rapid screening techniques, there was isolation, screening, and identification of many bioactive principles of the medicinal plants in pure form and preparation of synthetic compounds as drugs to treat various diseases and disorders. Modern drug use is often associated with harmful side effects. Currently, microbial resistance and side effects to modern allopathic compounds used by herbal products are increasing rapidly to cure various diseases and infections. In India, around 70% of new drugs are invented from naturally available resources, and a notable number of allopathic compound analogs are also isolated from plant extracts (Sen et al., 2011; Sen, Chakraborty, De, & Mazumder, 2009). The alternate medicines to substitute for natural (traditional) ones weakened the interrelationship between humans and other biological species that have evolved over generations. Furthermore, the prices of drugs have increased, and the marginalized sections of the society find it burdensome to buy. Meanwhile, WHO recognized the role of traditional systems of medicine, as part of a strategy to provide primary healthcare to the masses in developing and underdeveloped countries. Today, about 80% of the world's population is dependent on traditional therapies, many of which involve greater reliance on medicinal plants. The ethnodirected research has obvious benefits in national as well as global interests as directed by WHO and other health organizations.

3.7 Challenges

Humans have learned many lessons by observing nature and the surrounding environment. Humans have learned to use plant parts directly or as a mixture of extracts and plant-derived products to cure different diseases or disorders as remedies. Knowledge of all these techniques was not acquired in a single night. Ancient man observed all these things through trial and error methods and practiced efficiently for safety. Many changes have occurred in civilization due to advanced developments in science and technologies, which helped to discover many inventories for the provision of facilities. Urbanization, modernization, and globalization led to many adaptations in the lifestyle of humans. Many techniques evolved to discover new synthetic drugs that show an adverse effect on the healthcare system. These synthetic drugs have the capability of undesirable reactions that can cause threats to life, injuries, and sometimes lead to death. To reduce the side effects of synthetic drugs, interest has increased toward the usage of herbal products, and everyone is looking for plant-based products as remedies. WHO has also directed all the countries of the world toward the proper utilization of indigenous knowledge for new drug discovery. Biopiracy and intellectual property rights (IPRs) are important ones to protect indigenous knowledge and natural resources. Lack of proper scientific and efficient technological approaches to indicate pharmacodynamic properties and mechanisms of ethnobotanical-based formulations lead to a serious problem in the country. To overcome all these challenges, modern and ethnobotanical knowledge often subsist side-by-side. Challenges in the identification knowledge of the plant specimen is lacking in younger generations of ethnic groups due to negligence, lack of income generation, lack of awareness about properties of medicinal plants, lack of interest, provision of modern technologies to waste time even in some remote areas, and THPs.

Communicable diseases cause 14 million deaths annually and are interrelated to socioeconomic, environmental, and behavioral factors and population movements that promote the spreading of these communicable diseases. The big challenge is to meet the resources for public health and monitor the spreading of CDs due to globalization and rapid economic activity. Poverty, inequalities, and the inability of people are major challenges to control infectious diseases. The huge population ca. 138 crore and poor literacy are also reasons that affect and spread diseases in India. As such, India accounts for over 5.9 million of the global 38 million NCD deaths (WHO, 2014). To combat the NCDs occurrence in India, the government initiated the National Programme for Prevention and Control of Cancer, Diabetes, Cardiovascular, and Stroke (NPCDCS) and set specific targets to bring down the burden of NCD mortality to 25% by 2025. There is an urgent need for massive screening, treatment, and periodical followup in the country. The education and wealth were inversely associated with disease occurrence. In undeveloped and developed countries like India, there are many social, cultural, and functional barriers to access healthcare services.

3.8 Prevention and control methods

Ethnobotanical knowledge plays vital roles and reveals emerging perspectives to prevent diseases and is used to fit into all existing prevention strategies. Cognizant efforts need to be taken for proper recognition, identification, and mode of administration of ethnomedicinal plants and strict implementation of these strategies. Nongkynrih, Patro, and Pandav (2014) recommended a comprehensive five-point methodology to control and manage diseases in India. The strategies include; (i) strengthening of general healthcare services, (ii) improve proper environmental health, sanitation, and hygiene programs in healthcare settings, (iii) information, education, and communication (IEC) campaign concerning specific problems, (iv) establishment of epidemiological services (surveillance, study distribution, analytical studies), and (v) disease-specific measures (detection, quantification). The disease control programs should be exclusively implemented through nongovernmental organizations (NGOs), educational institutions, mass media, and other civil institutions that should be envisaged. The government has to undertake a concerted drive to empower decentralized local self-government institutions to take responsibilities in social sector development activities and healthcare services. The present chapter recommends following the "5 T" strategy for the prevention and control of diseases in India. The strategies include tracing, testing, treatment, tracking, teamwork, and regular monitoring of the health status of society.

There are so many steps to be taken for a healthy society that supply a nutritious diet to improve the immune system, which can inhibit the chance of contracting many diseases. Seasonal fruits, leaves, tubers, kernels, nuts, seeds, and other NTFPs (non-timber forest products) are very important energetic food resources for minerals, vitamins, and other essential nutrients for the security, growth, and development of the body. Generally, local people and THPs recommend plant-based natural food or food products to cure many ailments. The constitution of technical and authentic resource groups for the collection and processing of plant-based materials is highly important for that knowledge makes it available in all seasons and to preserve the knowledge of indigenous people for plant identification, locality, collection, mode of administration, and sociocultural and magicoreligious beliefs. Detection epidemics in an early stage should be met with strategy to prevent the spread in the future for a safe, healthy, and hygienic society. Making of formulations to control the emergence and spread of disease transmission in the population is highly encouraged.

Control of any communicable disease requires information on the pathogen, source of infection, incubation period, transmission type, period of infectiousness, its impact on the health, and method of target-specific treatment. Infectious agents include microorganisms such as bacteria, viruses, fungi, protozoans, nematodes, etc. and sources of infection refer to reservoirs where the microorganisms live, which include insects (mosquitoes), livestock (pigs, cows, monkeys, etc.). Soil mode of transmission refers to the pathogens that carry the microbes from one place to another by air, droplet, or direct contact. Hosts are vulnerable to diseases. For example, infants, young children, elderly persons, and patients with chronic diseases are more susceptible to infection because of weakened immunity. Develop the personal immune system via immunization, by physical exercise, providing nutritious food, and changing lifestyle. Wearing surgical or homemade masks is highly promoted when infectious diseases have respiratory symptoms like cough, cold, flu, runny nose, sneezing, etc.

Self-quarantine of suspected and sick personnel is highly essential to escape from the threat of infectious diseases like COVID-19 and should avoid physical contact with infected persons and pet animals.

Public and political awareness, understanding, and practice about prevention and control of diseases are highly important. The government should take necessary steps for proper implementation to make strict policies. It is highly essential to congregate international cooperation for resource mobilization, financial cooperation, capacity building, training, and exchange of information on experienced matters and best practices. Mobilize civil societies and various private sectors to raise awareness and support the implementation of the instructions at global, national, regional, and local levels. To support local communities, educated personalities and local managers should prevent and reduce the harmful use of alcohol, alcohol disorders, behavior of alcohol addicts, and associated conditions. Encourage warning people about the effect of tobacco and tobacco-related products through mass and social media, avoid smoking, and make available smoke-free environment zones in all indoor workplaces, public places, and other areas. Awareness campaigns about a balanced diet and its nutrients, physical activity, increasing consumption of fruit and vegetables, and reducing salt intake is also highly important in day-to-day life.

4. Conclusions

India is a holy land with its culture and traditions. India is famous for its tremendous biodiversity and rich heritage of traditional botanical knowledge through its cultural and ethnic background since ancient times. Four Vedas, scriptures, and Upanishads have evolved in the country to behave ourselves with nature. Nature trains us in every aspect of the country. Since ancient times, the locally available (native) medicinal plants have been used to cure various ailments by rishis, saints, scholars, and prophets. Local healers, traditional practitioners, and orthodox health practitioners are an integral part of the healthcare system in India, and approximately 80% of the population depends on these practitioners for instant treatment of any disease or disorder. Regular surveillance and monitoring are essential steps to identify and understand the status of different diseases at national, regional, and local levels for the development of need-based health interventions. Surveillance and monitoring also provide baseline data on financial, cultural, infrastructural, local governance, and availability of human resources for proper policy implementation to prevent and control infectious and noninfectious diseases. The government should also promote researchers and academicians to interact with ethnic people by providing financial support to document indigenous knowledge for new drug discoveries in the future and promote the gradual integration of ethnic knowledge about plants into the modern healthcare system. Care and proper measurements should also be taken by the government to scale up prevention and case management approaches for the effective control of transmittable and nontransmittable diseases.

Acknowledgments

S. Suthari acknowledges the support from the principal and management of Vaagdevi Degree and PG College, Hanamkonda. Authors S. Kota and S. Abbagani are grateful to the Head, Department of Biotechnology, Kakatiya University, Warangal, for facilities. K. Omkar is obliged to the Principal, Kakatiya Govt. Degree College, Hanamkonda, for support. Author Mir Zahoor Gul is grateful to the CSIR, New Delhi, for Research Associate Fellowship (No.09/132(0883)/2019-EMR-1). MZG also thanks Prof. B. Sashidhar Rao, Department of Biochemistry, for his guidance, support, and encouragement.

Conflict of interest

The authors declare no conflict of interest for the publication.

References

Adnan, M., Bibi, R., Azizullah, A., Andaleeb, R., Mussarat, S., Tariq, A., et al. (2015). Ethnomedicinal plants used against common digestive problems. *African Journal of Traditional, Complementary, and Alternative Medicines, 12*, 99–117.

Ajesh, T., & Kumuthakalavalli, R. (2012). Ethnic herbal practices for gynaecological disorders from urali tribes of Idukki district of Kerala, India. *International Journal of Pharmacy and Life Sciences, 3*, 2213–2219.

Ammal, E. J. (1956). An introduction to the subsistence economy of India. In T. William (Ed.), *International symposium on 'Man's role in changing the face of the earth* (pp. 324–335). Chicago: Wenner-Gren Foundation University of Chicago Press.

Andrade, J. M., Mosquera, H. L., & Armijos, C. (2017). Ethnobotany of indigenous Saraguros: Medicinal plants used by community healers "Hampiyachakkuna" in the San Lucas Parish, Southern Ecuador. *BioMed Research International*. https://doi.org/10.1155/2017/9343724, 9343724.

Anonymous. (2011). *Census of India 2011: provisional population totals-India data sheet Office of the Registrar General Census Commissioner*. India: Indian Census Bureau.

Augustine, J., Sreejesh, K., & Bijeshmon, P. (2010). Ethnogynecological uses of plants prevalent among the tribes of Periyar Tiger Reserve, Western Ghats. *Indian Journal of Traditional Knowledge, 9*.

Ayyanar, M., & Ignacimuthu, S. (2009). Herbal medicines for wound healing among tribal people in Southern India: Ethnobotanical and scientific evidences. *International Journal of Applied Research in Natural Products, 2*, 29–42.

Azad, S. A., & Bhat, A. R. (2013). Ethnomedicinal plants recorded from Rajouri-Poonch districts of J&K state. *Indian Journal of Life Sciences, 2*, 71–74.

Bala, L. D., & Singh, R. (2016). Ethnobotanical study of traditional medicinal plants used by Mawasi tribe of Chitrakoot region Distt. Satna (MP) India. *International Journal of Biological Research, 1*, 13–15.

Ballabha, R., Tiwari, J., & Tiwari, P. (2013). Medicinal plant diversity in Dhundsir Gad Watershed of Garhwal Himalaya, Uttarakhand, India. *The Journal of Ethnobiology and Traditional Medicine Photon, 119*, 424–433.

Bandana, P., & Debabrata, P. (2015). Wild edible plant diversity and its ethnomedicinal use by indigenous tribes of Koraput, Odisha, India. *Research Journal of Agriculture and Forestry Sciences, 3*(9), 1–10.

Basha, S. K. M., & Priydarsini, A. I. (2017). Ethnobotanical study on rapur–chitvelghat, Eastern Ghats, Andhra Pradesh. *International Journal of Engineering and Technical Research, 6*, 69–72.

Batsatsashvili, K., Kikvidze, Z., & Bussmann, R. W. (2020). *Ethnobotany of the mountain regions of Central Aisa and Altai.* Springer International Publishing.

Bhat, P., Hegde, G. R., Hegde, G., & Mulgund, G. S. (2014). Ethnomedicinal plants to cure skin diseases—An account of the traditional knowledge in the coastal parts of Central Western Ghats, Karnataka, India. *Journal of Ethnopharmacology, 151*, 493–502.

Binu, S. (2010). Medicinal plants used for treating snake bite by the tribals of Pathanamthitta district, Kerala, India. *Indian Journal of Botanical Research, 6*(1&2), 17–22.

Binu, S., Nayar, T., & Pushpangadan, P. (1992). An outline of ethnobotanical research in India. *Journal of Economic and Taxonomic Botany, 10*, 405–428.

Bodeker, G., & Graz, B. (2013). Traditional medicine. In *Hunter's tropical medicine and emerging infectious diseases* (pp. 194–199). Elsevier.

Borah, S., Das, A., Saikia, D., & Borah, J. (2009). A note on the use of ethnomedicine in treatment of diabetes by mishing communities in Assam. *India Ethnobotanical Leaflets, 2009,* 2.

Bose, M. F. J. N., Aron, S., & Mehalingam, P. (2014). An ethnobotanical study of medicinal plants used by the Paliyars aboriginal community in Virudhunagar district, Tamil Nadu, India. *Indian Journal of Traditional Knowledge, 13.*

Bussmann, R. W. (2019). Making friends in the field: How to become an ethnobotanist. A personal reflection. *Ethnobotany Research and Applications, 18*, 1–13.

Bussmann, R. W., Paniagua, Z., Narel, Y., Sikharulidze, S., Kikvidze, Z., Kikodze, D., et al. (2017). Plants in the spa—The medicinal plant market of Borjomi, Sakartvelo (Republic of Georgia), Caucasus. *Indian Journal of Traditional Knowledge, 16.*

Bussmann, R., Paniagua Zambrana, N., Sikharulidze, S., Kikvidze, Z., Kikodze, D., Tchelidze, D., et al. (2016). Medicinal and food plants of Svaneti and Lechkhumi, Sakartvelo (Republic of Georgia), Caucasus. *Medicinal & Aromatic Plants, 5.* 2167–0412.1000266.

Bussmann, R. W., Paniagua-Zambrana, N. Y., & Romero, C. (2020). *Hyptis capitata* Jacq. L amiaceae. In *Ethnobotany of the Andes* (pp. 1–4).

Bussmann, R. W., Paniagua-Zambrana, N. Y., Wood, N., Njapit, S. O., Njapit, J. N. O., Osoi, G. S. E., et al. (2018). Knowledge loss and change between 2002 and 2017—A revisit of plant use of the Maasai of Sekenani Valley, Maasai Mara, Kenya. *Economic Botany, 72*, 207–216.

Catarino, L., Havik, P. J., Indjai, B., & Romeiras, M. M. (2016). Ecological data in support of an analysis of Guinea-Bissau's medicinal flora. *Data in Brief, 7*, 1078–1097.

Chakraborty, M., & Bhattacharjee, A. (2006). Some common ethnomedicinal uses for various diseases in Purulia district, West Bengal. *Indian Journal of Traditional Knowledge, 5.*

Chakraborty, R., Mondal, M., & Mukherjee, S. K. (2016). Ethnobotanical information on some aquatic plants of South 24 Parganas, West Bengal. *Plant Science Today, 3*, 109–114.

Chanthru, K., Gomthi, R., & Manian, S. (2016). Ethnomedicinal uses of medicinal plants used by tribes of Valparai, Coimbatore, tamil Nadu, India. *Indian Forester, 14*(2), 607–612.

Chatterjee, A. K. (2014). Study of ethno-medicinal plants among the tribals of Surguja region (C.G.). *International Journal on Advanced Computer Theory and Engineering, 3*(2), 56–60.

Cotton, C. M. (1996). *Ethnobotany: Principles and applications* (p. 424). Chichester, New York: John Willy & Sons.

Dangwal, L., & Sharma, A. (2011). Indigenous traditional knowledge recorded on some medicinal plants in Narendra Nagar Block (Tehri Garhwal), Uttarakhand. *Indian Journal of Natural Products and Resources, 2.*

Das, A. K., Dutta, B., & Sharma, G. (2008). Medicinal plants used by different tribes of Cachar district, Assam. *Indian Journal of Traditional Knowledge, 7*.

Das, A. K., & Khatoon, R. (2015). Plants used in gynaecological problem by the Kim tribe of Manipur, India. *International Journal of Current Research, 7*(3), 13146–13148.

Das, B., Talukdar, A., & Choudhury, M. D. (2014). A few traditional medicinal plants used as antifertility agents by ethnic people of Tripura, India. *International Journal of Pharmacy and Pharmaceutical Sciences, 6*, 47–53.

de Natale, A., & Pollio, A. (2007). Plants species in the folk medicine of Montecorvino Rovella (inland Campania, Italy). *Journal of Ethnopharmacology, 109*, 295–303.

de Santana, B. F., Voeks, R. A., & Funch, L. S. (2016). Ethnomedicinal survey of a maroon community in Brazil's Atlantic tropical forest. *Journal of Ethnopharmacology, 181*, 37–49.

Deepa, M., Pradeepa, R., Anjana, R., & Mohan, V. (2011). Noncommunicable diseases risk factor surveillance: Experience and challenge from India. *Indian Journal of Community Medicine, 36*, S50.

Dey, A., & De, J. (2011). Ethnobotanical aspects of *Rauvolfia serpentina* (L). Benth. ex Kurz. in India, Nepal and Bangladesh. *Journal of Medicinal Plant Research, 5*, 144–150.

Dharm, N., & Pramod, K. S. (2017). Ethnobotanical importance and herbal medicine in Vindhya region of Eastern Uttar Pradesh, India. *Journal of Medicinal Plant Research, 11*(25), 403–413.

Dhatchanamoorthy, N., Kumar, N. A., & Karthik, K. (2013). Ethnomedicinal plants used by Irular tribes Javadhu hills of Southern Eastern Ghats, Tamil Nadu, India. *International Journal of Current Research and Development, 2*(1), 31–37.

Dwivedi, S. D. S., & Dwivedi, A. (2015). Herbal remedies for respiratory diseases among the Natives of Madhya Pradesh, India. *American Journal of Life Science Researches, 3*.

Faruque, M. D. O., Uddin, S. B., Barlow, J. W., Hu, S., Dong, S., Cai, Q., et al. (2018). Quantitative ethnobotany of medicinal plants used by indigenous communities in the Bandarban district of Bangladesh. *Frontiers in Pharmacology*. https://doi.org/10.3389/fphar.2018.00040. eCollection 2018.

Ganesan, S., Venkateshan, G., & Banumathy, N. (2006). Medicinal plants used by ethnic group Thottianaickans of Semmalai hills (reserved forest), Tiruchirappalli district, Tamil Nadu. *Indian Journal of Traditional Knowledge, 5*, 245–252.

Gazzaneo, L. R. S., De Lucena, R. F. P., & de Albuquerque, U. P. (2005). Knowledge and use of medicinal plants by local specialists in an region of Atlantic Forest in the state of Pernambuco (Northeastern Brazil). *Journal of Ethnobiology and Ethnomedicine, 1*, 9.

Giday, M., Asfaw, Z., & Woldu, Z. (2009). Medicinal plants of the Meinit ethnic group of Ethiopia: An ethnobotanical study. *Journal of Ethnopharmacology, 124*, 513–521.

Giday, K., Lenaerts, L., Gebrehiwot, K., Yirga, G., Verbist, B., & Muys, B. (2016). Ethnobotanical study of medicinal plants from degraded dry afromontane forest in northern Ethiopia: Species, uses and conservation challenges. *Journal of Herbal Medicine, 6*, 96–104.

Global Hepatitis Report. (2017). *Geneva: World Health Organization.* http://apps.who.int/iris/bitstream/handle/10665/255016/9789241565455-eng.pdf?sequence=1. Accessed 2 April 2020.

Global Tuberculosis Report. (2017). *Geneva: World Health Organization.* http://www.who.int/tb/publications/global_report/en/. Accessed 2 April 2020.

González, M., & Marioli, J. (2010). Antibacterial activity of water extracts and essential oils of various aromatic plants against Paenibacillus larvae, the causative agent of American Foulbrood. *Journal of Invertebrate Pathology, 104*, 209–213.

Govaerts, R. (2001). How many species of seed plants are there? *Journal of Taxonomy, 50*, 1085–1090.

Gunther, M. (1945). Sore nipples causes and prevention. *The Lancet, 246*, 590–593.

Gupta, S. P. (1974). *Tribes of Chotanagpur Plateau: An ethno-nutritional & pharmacological cross-section*. Govt. of Bihar, Welfare Dept.

Hajra, P. K., & Mudgal, V. (1997). *Plant Diversity: Hot spots in India-an overview* (pp. 1–179). Calcutta: Botanical Survey of India, Ministry of Environment and Forests.

Hamilton, A., Shengji, P., Kessy, J., Khan, A. A., Lagos-Witte, S., & Shinwari, Z. K. (2003). *The purposes and teaching of applied ethnobotany*. United Nations Educational, Scientific and Cultural Organization (UNESCO).

Harshberger, J. W. (1896). The purpose of ethno-botany. *Botanical Gazaette, 21*, 146–158.

Huxley, A. (1984). *Green inheritance*. Collins/Harvel, London: The World Wildlife Fund Book of India.

Ignacimuthu, S., Ayyanar, M., & Sankarasivaraman, K. (2008). Ethnobotanical study of medicinal plants used by Paliyar tribals in Theni district of Tamil Nadu, India. *Fitoterapia, 79*, 562–568.

Ignacimuthu, S., Sankarasivaraman, K., & Kesavan, L. (1998). Medico-ethnobotanical survey among Kanikar tribals of Mundanthurai Sanctuary, western ghats, India. *Fitoterapia (Milano), 69*, 409–414.

Islam, M. K., Saha, S., Mahmud, I., Mohamad, K., Awang, K., Uddin, S. J., et al. (2014). An ethnobotanical study of medicinal plants used by tribal and native people of Madhupur forest area, Bangladesh. *Journal of Ethnopharmacology, 151*, 921–930.

Jagtap, S., Deokule, S., & Bhosle, S. (2008). Ethnobotanical uses of endemic and RET plants by Pawra tribe of Nandurbar district, Maharashtra. *Indian Journal of Traditional Knowledge, 7*.

Jain, S. K. (1963a). Studies in Indian Ethnobotany-less known uses of fifty common plants from the tribal areas of Madhya Pradesh. *Bulletin of Botanical Survey of India, 5*, 223–226.

Jain, S. K. (1963b). Observations of Ethnobotany of the tribals of Madhya Pradesh. *Vanyajati, 11*, 177–183.

Jain, S. K. (1963c). *Studies in Indian Ethnobotany-Plants used in medicine by the tribals of Madhya Pradesh. Vol. 1* (pp. 126–128). Jammu: Bulletin of Regional Research Laboratory.

Jain, S. K. (1964a). The role of botanist in folklore research. *Folklore, 5*, 145–150.

Jain, S. K. (1964b). Wild plant foods of the tribals of Bastar. *Gramodyog, 10*, 557–561.

Jain, S. K. (1964c). Native plant remedies for snake-bite among Adivasis of central India. *Indian Medical Journal, 57*, 307–369.

Jain, S. K. (1965). Wooden musical instruments of the Gonds of Central India. *Ethnomusicology, 9*, 39–42.

Jain, S. K. (1981). *Observations on ethnobotany of the tribal of Central India* (pp. 193–198). New Delhi: Glimpses of Indian Ethnobotany Oxford & IBH.

Jain, S. K. (1986). Ethnobotany. *Interdisciplinary Science Reviews, 11*, 285–292.

Jain, S. K. (1988). Role of ethnobotany in conservation of plant genetic resources. In *Plant genetic resources: Indian perspective: Proceedings of the national symposium on plant genetic resources* (p. 59). National Bureau of Plant Genetic Resources.

Jain, S. K. (1989). Ethnobotany: An interdisciplinary science for holistic approach to man plant relationships. In *Jodhpur Methods Approaches in Ethnobotany* (pp. 9–12).

Jain, S. K. (1994). Ethnobotany and research in medicinal plants in India. In *Vol. 185. Ethnobotany and the search for new drugs* (pp. 153–168).

Jain, S. K. (2000). Human aspects of plant diversity. *Economic Botany, 54*, 459.

Jain, S. K. (2004). Credibility of traditional knowledge—The criterion of multilocational and multiethnic use. *Indian Journal of Traditional Knowledge, 3*.

Jain, S. K. (2019). The widening panorama of ethnobotany in India. *The Journal of Indian Botanical Society, 98*, 98–102.

Jain, S. K., & Banerjee, D. K. (1974). Preliminary observations on the ethnobotany of the genus Coix. *Economic Botany, 28*, 38–42.

Jain, S. K., Banerjee, D., & Pal, D. (1973). Medicinal plants among certain Adibasis in India. *Nelumbo, 15*, 85–91.

Jain, S. K., & Borthakur, S. (1980). Ethnobotany of the Mikirs of India. *Economic Botany, 34*, 264–272.

Jain, S. K., & Dam, N. (1979). Some ethnobotanical notes from Northeastern India. *Economic Botany, 33*, 52–56.

Jain, S. K., & De, J. (1966). Observations on ethnobotany of Purulia, West Bengal. *Journal of Nelumbo, 8*, 237–251.

Jain, S. K., & Saklani, A. (1991). Observations on the ethnobotany of the Tons valley region in the Uttarkashi district of the northwest Himalaya, India. *Mountain Research and Development, 11*, 157–161.

Jain, S. K., & Singh, J. (2010). Traditional medicinal practices among the tribal people of Raigarh (Chhatisgarh), India. *Indian Journal of Natural Products and Resources, 1*.

Jain, S. K., & Srivastava, S. (1999). *Dictionary of ethnoveterinary plants of India*. Deep Publications.

Jain, S. K., & Tarafder, C. (1970). Medicinal plant-lore of the santals (A revival of PO Bodding's work). *Economic Botany, 24*, 241–278.

Jaiswal, V. (2010). Culture and ethnobotany of Jaintia tribal community of Meghalaya, Northeast India—A mini review. *Indian Journal of Traditional Knowledge, 9*.

Jeyaprakash, K., Ayyanar, M., Geetha, K., & Sekar, T. (2011). Traditional uses of medicinal plants among the tribal people in Theni District (Western Ghats), Southern India. *Asian Pacific Journal of Tropical Biomedicine, 1*, S20–S25.

Jones, V. H. (1941). The nature and status of ethnobotany. *Chronicle of Botany, 6*(10), 219–221.

Joshi, A., & Pande, N. (2016). Indigenous practices used by bhotia tribe in kumaun for prenatal and postnatal care. *Economic Botany, 25*, 414–424.

Kadhirvel, K., Ramya, S., Sudha, T. P. S., Ravi, A. V., Rajasekaran, C., Selvi, R. V., et al. (2010). Ethnomedicinal survey on plants used by tribals in Chitteri Hills. *Environment & We: An International Journal of Science & Technology, 5*, 35–46.

Kala, C. P. (2005). Current status of medicinal plants used by traditional Vaidyas in Uttaranchal state of India. *Ethnobotany Research and Applications, 3*.

Kala, C. P. (2006). Ethnobotany and ethnoconservation of *Aegle marmelos* (L.) Correa. *Indian Journal of Traditional Knowledge, 5*.

Kala, C. P. (2007). Local preferences of ethnobotanical species in the Indian Himalaya: Implications for environmental conservation. *Current Science*, 1828–1834.

Kala, C. P. (2009). Aboriginal uses and management of ethnobotanical species in deciduous forests of Chhattisgarh state in India. *Journal of Ethnobiology and Ethnomedicine, 5*, 20.

Kapale, R. (2012). Ethnobotany of Baiga tribals with references to utilization of forest resources in Amarkantak Biosphere Reserve (India). *Bulletin of Environment, Pharmacology and Life Sciences, 1*, 73–76.

Kapoor, S., & Kapoor, L. (1980). Medicinal plant wealth of the Karimnagar district of Andhra Pradesh. *Bulletin of Medico-Ethnobotanical Research, 1*, 120–144.

Kekuda, T. P., Vinayaka, K., & Sachin, M. (2018). Chemistry, ethnobotanical uses and biological activities of the lichen genus Heterodermia Trevis.(Physciaceae; Lecanorales; Ascomycota): A comprehensive review. *Journal of Applied Pharmaceutical Science, 8*, 148–155.

Khan, M. S. (1953). *Forest flora of Hyderabad*. Hyderabad: State Govt Press.

Khatoon, R., Das, A., Dutta, B., & Singh, P. (2014). Study of Traditional handloom weaving by the Kom tribe of Manipur. *Indian Journal of Traditional Knowledge, 2*.

Khatun, S., Parlak, K. U., Polat, R., & Cakilcioglu, U. (2012). The endemic and rare plants of Maden (Elazig) and their uses in traditional medicine. *Journal of Herbal Medicine, 2*, 68–75.

Kingston, C., Jeeva, S., Shajini, R., Febreena, G. L., Jasmine, T. S., Laloo, R., et al. (2006). Antivenom drugs used by indigenous community in traditional healthcare system. *Journal for Nature Conservation, 18*, 137–143.

Kirtikar, K. R., & Basu, B. D. (1975). *Indian medicinal plants* (pp. 75–80). Dehradun, India: M/s Bishen Singh Mahendra Pal Singh.

Kumar, R., & Bharati, K. A. (2014). Ethnomedicines of Tharu tribes of Dudhwa national park, India. *Ethnobotany Research and Applications, 12*, 001–013.

Kumar, O. A., Mallikarjuna, K., Prasad, S. S. D., & Naidu, L. M. (2016). Ethnomedicine used for asthma by trubes of Papikondalu forest, Andhra Pradesh, India. *Current Botany, 7*, 1–4.

Kumar, S., & Pandey, S. (2015). An ethnobotanical study of local plants and their medicinal importance in Tons river area, Dehradun, Uttarakhand. *Indian Journal of Tropical Biodiversity, 23*, 227–231.

Kumar, A., Partap, S., Sharma, N. K., & Jha, K. (2012). Phytochemical, ethnobotanical and pharmacological profile of Lagenaria siceraria: A review. *Journal of Pharmacognosy and Phytochemistry, 1*, 24–31.

Kumar, T., & Pullaiah, T. (1999). Ethno-medicinal uses of some plants of Mahabubnagar District, Andhra Pradesh, India. *Journal of Economic and Taxonomic Botany, 23*, 341–345.

Kundu, M. K., Hazra, S., Pal, D., & Bhattacharya, M. (2018). A review on noncommunicable diseases (NCDs) burden, its socio-economic impact and the strategies for prevention and control of NCDs in India. *Indian Journal of Public Health, 62*(4), 302–304.

Lalramnghinglova, H., & Jha, L. (1999). Ethnobotany: A review. *Journal of Economic and Taxonomic Botany, 23*, 1–27.

Lawal, I., Amao, A., Lawal, K., Alamu, O., & Sowunmi, I. (2013). Phytotherapy approach for the treatment of gynaecological disorder among women in ido local government area of Ibadan, Oyo state, Nigeria. *Journal of Advanced Scientific Research, 4*(3), 41–44.

Liengme, C. (1983). A survey of ethnobotanical research in southern Africa. *Bothalia, 14*, 621–629.

Lone, P. A., Bhardwaj, A. K., Shah, K. W., & Tabasum, S. (2014). Ethnobotanical survey of some threatened medicinal plants of Kashmir Himalaya, India. *Journal of Medicinal Plant Research, 8*, 1362–1373.

Mafuva, C., & Marima-Matarira, H. T. (2014). Toward professionalization of traditional medicine in Zimbabwe: A comparative analysis to the South African policy on traditional medicine and the Indian Ayurvedic system. *International Journal of Herbal Medicine, 2*, 154–161.

McDonald, J. H. (2009). *Handbook of biological statistics. Vol. 2.* Baltimore, MD: Sparky House Publishing.

Mesfin, K., Tekle, G., & Tesfay, T. (2013). Ethnobotanical study of traditional medicinal plants used by indigenous people of Gemad District, Northern Ethiopia. *Journal of Medicinal Plants Studies, 1*.

Mishra, N. P., & Patil, P. (2014). Ethnomedicinal plants used for curing skin diseases by tribal of Betul district, Madhya Pradesh. *Journal of Biological Sciences, 3*, 731–734.

Modak, B. K., Gorai, P., Dhan, R., Mukherjee, A., & Dey, A. (2015). Tradition in treating taboo: Folkloric medicinal wisdom of the aboriginals of Purulia district, West Bengal, India against sexual, gynaecological and related disorders. *Journal of Ethnopharmacology, 169*, 370–386.

Mohan, A. C., & Divya, S. (2017). Phytochemical analysis and screening of total flavonoid, tannin and phenolic contents in Croton bonplandianum leaf and stem. *World Journal of Pharmaceutical Research, 6*.

Mohan, A. C., Suthari, S., Kandagatla, R., & Ragan, A. (2017). Antirheumatic plants used by indigenous people of Kawal Wildlife sanctuary region, Telangana. *International Journal of Advanced Research in Science and Technology, 6*, 666–669.

Mohan, A., Suthari, S., & Ragan, A. (2017). Ethnomedicinal plants of Kawal wildlife sanctuary, Telangana, India. *Annals of Plant Science, 6*, 1537–1542.

Morilla, L. J. G., Sumaya, N. H. N., Rivero, H. I., & Madamba, M. (2014). Medicinal plants of the Subanens in Dumingag, Zamboanga del Sur, Philippines. In *International Conference on Food, Biological and Medical Sciences* (pp. 28–29).

Mudgal, V. (1987). *Recent ethnobotanical works on different states/tribes of India–A synoptic treatment. A manual of ethno-botany* (pp. 43–58). Jodhpur: Scientific Publishers.

Murthy, E. N., Reddy, C. S., Reddy, K., & Raju, V. S. (2007). Plants used in ethnoveterinary practices by Koyas of Pakhal wildlife sanctuary, Andhra Pradesh. *India Ethnobotanical Leaflets, 2007*, 1.

Murthy, S. M. S., & Vidyasagar, G. M. (2013). Traditional knowledge on medicinal plant used in the treatment of respiratory disorders in Bellary district, Karnataka, India. *Indian Journal of Natural Products and Resources, 4*(2), 189–193.

Nadembega, P., Boussim, J. I., Nikiema, J. B., Poli, F., & Antognoni, F. (2011). Medicinal plants in baskoure, kourittenga province, Burkina Faso: An ethnobotanical study. *Journal of Ethnopharmacology, 133*, 378–395.

Naidu, M. T., Suthari, S., Kumar, O. A., & Venkaiah, M. (2020). Ethno-botanico-medicine in the treatment of diabetes by the tribal groups of Visakhapatnam district, Andhra Pradesh, India. In S. M. Khasim, et al. (Eds.), *Vol. II. Medicinal plants: Biodiversity, sustainable utilization and conservation* Springer Nature Singapore Pvt Ltd.

Nanjunda, D. (2010). Ethno-medico-botanical investigation of Jenu Kuruba ethnic group of Karnataka state, India. *Bangladesh Journal of Medical Science, 9*, 161–169.

Naqvi, A. H. (2001). *Flora of Karimnagar District, Andhra Pradesh, India*. Ph. D. Thesis Warangal, India: Kakatiya University.

Narayan, D., & Singh, P. K. (2017). Ethno botanical importance and herbal medicine in Vindhya region of Eastern Uttar Pradesh, India. *Journal of Medicinal Plant Research, 11*, 403–413.

Narayanan, M. R., Anilkumar, N., Balakrishnan, V., Sivadasan, M., Alfarhan, H. A., & Alatar, A. (2011). Wild edible plants used by the Kattunaikka, Paniya and Kuruma tribes of Wayanad District, Kerala, India. *Journal of Medicinal Plant Research, 5*, 3520–3529.

Naseef, S. A., Ajesh, T. P., & Kumuthakalavalli, R. (2012). Study of folklore medicinal practices of Paniya tribes for gynaecological ailments. *International Journal of Pharma and Bio Sciences, 3*(4), 493–501.

National Centre for Biological Sciences. (2015). *Overview of Indian healing traditions.* Accessed from: https://www.ncbs.res.in/HistoryScienceSociety/content/overview-indian-healing-traditions.

Neglected Tropical Diseases. (2013). Prevention, control, elimination and eradication. In *Report by the Secretariat to the Sixty-sixth World Health Assembly, Geneva, 20–28 May 2013*. Geneva: World Health Organization. Provisional agenda item 16.2 http://apps.who.int/gb/ebwha/pdf_files/WHA66/A66_20-en.pdf?ua=1. Accessed 2 April 2020.

Nongkynrih, B., Patro, B. K., & Pandav, C. S. (2014). Current status of communicable and noncommunicable diseases in India. *Journal of the Association of Physicians of India, 52*, 118–123.

Oladunmoye, M., & Kehinde, F. (2011). Ethnobotanical survey of medicinal plants used in treating viral infections among Yoruba tribe of South Western Nigeria. *African Journal of Microbiology Research, 5*, 2991–3004.

Omkar, K., Suthari, S., Alluri, S., Ragan, A., & Raju, V. S. (2012). Diversity of NTFPs and their utilization in Adilabad district of Andhra Pradesh, India. *Journal of Plant Studies, 1*, 33.

Omkar, K., Suthari, S., & Raju, V. S. (2015). Ethnobotanical knowedge of inhabitants from Gund-labrahmeswaram wildlife sanctuary (Eastern Ghats), Andhra Pradesh, India. *American Journal of Ethnomedicine*, 2(6), 333–346.

Ong, H. G., & Kim, Y.-D. (2014). Quantitative ethnobotanical study of the medicinal plants used by the Ati Negrito indigenous group in Guimaras island, Philippines. *Journal of Ethnopharmacology*, 157, 228–242.

Padhan, B., & Panda, D. (2015). *Wild edible plant diversity and its ethno-medicinal use by indigenous tribes of Koraput, Odisha, India*. International Science Congress Association.

Padmarao, P., & Reddy, P. (1999). A note on folk treatment of bone fractures in Ranga Reddy district, Andhra Pradesh. *Ethnobotany*, 11, 107–108.

Pakia, M., Cooke, J., & Van Staden, J. (2003). The ethnobotany of the Midzichenda tribes of the coastal forest areas in Kenya: 2. Medicinal plant uses. *South African Journal of Botany*, 69, 382–395.

Pandey, D., Mishra, S., & Hemalatha, S. (2017). Antiurolithic activity of different fractions of Aganosma dichotoma as folk medicine of Andhra Pradesh. *International Journal of Green Pharmacy*, 11(2), S324.

Pandey, M., Rastogi, S., & Rawat, A. (2013). Indian traditional ayurvedic system of medicine and nutritional supplementation. *Evidence-based Complementary and Alternative Medicine*, 2013, 1–12.

Panduranga, R. M., Prasanthi, S., & Reddi, S. T. (2011). Medicinal plants in Folk medicine for Women's diseases in use by Konda Reddis. *Indian Journal of Traditional Knowledge*, 10.

Pattanayak, B., & Dhal, N. (2015). Plants having mosquito repellant activity: An ethnobotanical survey. *International Journal of Research and Development in Pharmacy and Life Sciences*, 4, 1760–1765.

Pfoze, N. L., Kehie, M., Kayang, H., & Mao, A. A. (2014). Estimation of ethnobotanical plants of the Naga of North East India. *Journal of Medicinal Plants Studies*, 2, 92–104.

Phondani, P. C., Maikhuri, R. K., Rawat, L. S., Farooquee, N. A., Kala, C. P., Vishvakarma, S. R., et al. (2010). Ethnobotanical uses of plants among the Bhotiya tribal communities of Niti Valley in Central Himalaya, India. *Ethnobotany Research & Applications*, 8.

Prabu, M., & Kumuthakalavalli, R. (2012). Folk remedies of medicinal plants for snake bites, scorpion stings and dog bites in Eastern Ghats of Kolli Hills, Tamil Nadu, India. *International Journal of Research in Ayurveda & Pharmacy*, 3.

Prakash, B., Unnikrishnan, P., & Hariramamurthi, G. (2016). Medicinal flora and related traditional knowledge of Western Ghats: A potential source for community-based malaria management through endogenous approach. In *Vol. 2. Ethnobotany of India* (pp. 259–274). Apple Academic Press.

Prasad, A. D., Shyma, T., & Raghavendra, M. (2013). Plants used by the tribes for the treatment of digestive system disorders in Wayanad district, Kerala. *Journal of Applied Pharmaceutical Science*, 3, 171.

Priyadarshini, E. S., & Ragan, A. (2019). Survey on folklore medicinal plants knowledge of inhabitants of Khammam District, Telangana, India. *Annals of Plant Sciences*, 8(8), 3583–3590.

Pullaiah, T., Krishnamurthy, K., & Bahadur, B. (2017). *Ethnobotany of India. Vol. 3*. North-East India and the Andaman and Nicobar Islands: CRC Press.

Pullaiah, T., & Kumar, D. (1996). Herbal plants in Mannanur forest of Mahabubnagar district, Andhra Pradesh. *Journal of Economic and Taxonomic Botany*, 12, 218–220.

Pushpangadan, P. (1995). *Ethnobiology in India: A status report*. New Delhi: Government of India.

Rahaman, C. H., & Karmakar, S. (2015). Ethnomedicine of Santal tribe living around Susunia hill of Bankura district, West Bengal, India: The quantitative approach. *Journal of Applied Pharmaceutical Science*, 5, 127–136.

Rahmatullah, M. D., MDNK, A., MDM, R., Seraj, S., Mahal, M. J., Mou, S. M., et al. (2015). A survey of medicinal plants used by Garo and non-Garo traditional medicinal practitioners in two villages of Tangail district, Bangladesh. *American-Eurasian Journal of Sustainable Agriculture, 5*(3), 350–357.

Raj, A. J., Biswakarma, S., Pala, N. A., Shukla, G., Vineeta, K. M., Chakravarty, S., et al. (2018). Indigenous uses of ethnomedicinal plants among forest-dependent communities of Northern Bengal, India. *Journal of Ethnobiology and Ethnomedicine, 14*(1), 8. https://doi.org/10.1186/s13002-018-0208-9.

Rajasekaran, B., & Warren, D. (1994). Indigenous knowledge for socio-economic development and biodiversity conservation: The Kolli Hills. *Indigenous Knowledge and Development Monitor, 2*, 13–17.

Rajkumari, R., Singh, P., Das, A. K., & Dutta, B. (2013). Ethnobotanical investigation of wild edible and medicinal plants used by the Chiru Tribe of Manipur, India. *Pleione, 7*, 167–174.

Raju, V. S., & Reddy, K. (2005). Ethnomedicine for dysentery and diarrhoea from Khammam district of Andhra Pradesh. *Indian Journal of Traditional Knowledge, 4*.

Raju, V. S., Reddy, C. S., & Suthari, S. (2010). Flowering plant diversity and endemism in India: An overview. *ANU Journal of Natural Sciences, 2*, 27–39.

Ramachandran, V., & Manian, S. (1991). Ethnobotanical studies on the Irulas, the Koravas and the Puliyas of Coimbatore district, Tamil Nadu. *Indian Botanical Reporter, 8*, 85–91.

Ramarao, N. (1988). *The Ethnobotany of Eastern Ghats in Andhra Pradesh, India.* Ph. D. thesis Coimbatore: Bharathiar University.

Ramesh, C., Pratap, G. P., Nagaraju, V., & Sudarsanam, G. (2017). Ethno-botanical claims collected from tribal and rural people of Kadapa district, Andhra Pradesh. *International Journal of Ayurved and Pharma Research, 5*(8), 1–9.

Randrianarivony, T. N., Andriamihajarivo, T. H., Ramarosandratana, A. V., Rakotoarivony, F., Jeannoda, V. H., Kuhlman, A., et al. (2016). Value of useful goods and ecosystem services from Agnalavelo sacred forest and their relationships with forest conservation. *Madagascar Conservation & Development, 11*, 44–51.

Rani, T., Devi, M., Rajani, B., Padmavathi, V., & Maiti, R. (2007). Ethnobotanical survey of Nalgonda district, Andhra Pradesh, India. *Research on Crops, 8*, 700.

Rani, S., Rana, J., Jeelani, S., Gupta, R., & Kumari, S. (2013). Ethnobotanical notes on 30 medicinal polypetalous plants of district Kangra of Himachal Pradesh. *Journal of Medicinal Plant Research, 7*, 1362–1369.

Rao, N. R., & Henry, A. N. (1995). *The Ethnobotany of Eastern Ghats in Andhra Pradesh, India.* Botanical Survey of India.

Rao, J. K., Seetharami, T., & Kumar, O. A. (2011). Ethnobotany of stem bark of certain plants of Visakhapatnam district, Andhra Pradesh. *Current Botany, 2*.

Ravisankar, T. (1990). *Ethnobotanical studies in Adilabad and Karimnagar districts of Andhra Pradesh India.* Coimbatore: Bharathiar University (Ph.D. thesis).

Ravishankar, T., & Henry, A. (1992). Ethnobotany of Adilabad district, Andhra Pradesh, India. *Ethnobotany, 4*, 45–52.

Rawat, M. S., & Chowdhury, S. (1998). *Ethno medico botany of Arunachal Pradesh (Nishi & Apatani tribes).* Bishen Singh Mahendra Pal Singh.

Reddy, K. N. (2002). *Ethnobotany of Khammam district, Andhra Pradesh.* Ph.D. thesis Warangal: Kakatiya University.

Reddy, A. V. B. (2008). Use of various bio-fencing plants in the control of human disease by Lambada tribe inhabiting Nalgonda district, Andhra Pradesh, India. *Ethnobotanical Leaflets, 12*, 520–523.

Reddy, A. M., Babu, M. V. S., & Rao, R. R. (2019). Ethnobotanical study of traditional herbal plants used by local people of Seshachalam Biosphere Reserve in Eastern Ghats, India. *Herba Polonica*, *65*, 40–54.

Reddy, K., Bhanja, M., & Raju, V. S. (1998). Plants used in ethnoveterinary practices in Warangal district, Andhra Pradesh, India. *Ethnobotany*, *10*, 75–84.

Reddy, K. V., Paul, M. J., & Basha, S. M. (2017). Ethnobotanical study of durgamkonda of veligonda hill range, Eastern Ghats, Andhra Pradesh. *International Journal of Latest Trends in Engineering and Technology*, *8*(3), 258–262.

Reddy, K. N., & Raju, V. S. (2002). Ethnobotanical observations on Konda reddis of Mothugudem in Khammam district, Andhra Pradesh. In *Abstractpublished in national seminar on conservation of Eastern Ghats*.

Reddy, P. R., & Rao, P. P. (2002). A survey of plant crude drugs in folklore from Ranga Reddy district, Andhra Pradesh, India. *Indian Journal of Traditional Knowledge*, *1*(1), 20–25.

Reddy, K., Trimurthulu, G., & Reddy, C. S. (2010). Medicinal plants used by ethnic people of Medak district, Andhra Pradesh. *Indian Journal of Traditional Knowledge*, *9*.

Reyes-García, V., Huanca, T., Vadez, V., Leonard, W., & Wilkie, D. (2006). Cultural, practical, and economic value of wild plants: A quantitative study in the Bolivian Amazon. *Economic Botany*, *60*, 62–74.

Robbins, W. W., Harrington, J. P., & Freire-Marreco, B. W. (1916). *Ethnobotany of the Tewa Indians. Vol. 55*. US Government Printing Office.

Rodin, J. (1985). Insulin levels, hunger, and food intake: An example of feedback loops in body weight regulation. *Health Psychology*, *4*, 1.

Rout, S., Panda, S., Mishra, N., & Panda, T. (2010). Role of tribals in collection of commercial non-timber forest products in Mayurbhanj District, Orissa. *Studies of Tribes Tribals*, *8*, 21–25.

Rupani, R., & Chavez, A. (2018). Medicinal plants with traditional use: Ethnobotany in the Indian subcontinent. *Clinics in Dermatology*, *36*, 306–309.

Sadeghi, Z., & Mahmood, A. (2014). Ethno-gynecological knowledge of medicinal plants used by Baluch tribes, southeast of Baluchistan, Iran. *Revista Brasileira de Farmacognosia*, *24*, 706–715.

Sağıroğlu, M., Dalgıç, S., & Toksoy, S. (2013). Medicinal plants used in Dalaman (Muğla). *Turkey Academic Journals*, *7*, 2053–2066.

Sahu, S., Pattnaik, S., Sahoo, S., Lenka, S., & Dhal, N. (2011). Ethnobotanical study of medicinal plants in the coastal districts of Odisha. *Current Botany*, *2*.

Saidulu, P. (2014). *Ethnobotany of Pocharm wildlife sanctuary, Andhra Pradesh*. M.Phil. dissertation Warangal: Department of Botany, Kakatiya University.

Saidulu, P., Suthari, S., Kandagatla, R., Ajmeera, R., & Vatsavaya, R. S. (2015). Ethnobotanical knowledge studied in Pocharam wildlife sanctuary, Telangana, India. *Notulae Scientia Biologicae*, *7*, 164–170.

Sajem, A. L., & Gosai, K. (2006). Traditional use of medicinal plants by the Jaintia tribes in North Cachar Hills district of Assam, northeast India. *Journal of Ethnobiology and Ethnomedicine*, *2*, 33.

Sajem, A. L., & Gosai, K. (2010). Ethnobotanical investigations among the Lushai tribes in North Cachar hills district of Assam, northeast India. *Indian Journal of Traditional Knowledge*, *9*.

Samydurai, P., Thangapandian, V., & Aravinthan, V. (2012). *Wild habits of Kolli Hills being staple food of inhabitant tribes of eastern Ghats, Tamil Nadu, India*. NISCAIR-CSIR.

Sangameshwar, M. (2018). *Ethnobotany of north Telangana*. Ph.D. thesis Warangal: Department of Botany, Kakatiya University.

Sarkhel, S. (2014). Ethnobotanical survey of folklore plants used in treatment of snakebite in Paschim Medinipur district, West Bengal. *Asian Pacific Journal of Tropical Biomedicine, 4*, 416–420.

Sarvalingam, A., Rajendran, A., Sivalingam, R., & Jayanthi, P. (2014). Ipomoea muelleri Benth. (Convolvulaceae)-a new record for Asian continent. *Jordan Journal of Biological Sciences, 7*, 299–300.

Sasidhar, K., Brahmajirao, P., & Kumar, S. A. (2016). Ethnobotanical studies on medicinal plant utilization by the Yandhi tribe of Ananthasagaram mandal, Nellore district, Andhra Pradesh, India. *International Advanced Research Journal of Science, Engineering and Technology, 3*(3), 75–80.

Satyavathi, K., Deepika, D. S., Padal, S. B., & Rao, J. P. (2015). Ethnomedicinal plants used for leucorrhoea by tribes of Srikakulam district, Andhra Pradesh, India. *Malaya Journal of Sciences, 2*(4), 194–197.

Satyavathi, K., & Padal, S. (2018). Folklore medicine of primitive tribals in Dumbriguda Mandal, Visakhapatnam District, Andhra Pradesh, India. *International Journal of Life Sciences, 6*, 523–528.

Schippmann, U., Leaman, D. J., Cunningham, A. B., & Walter, S. (2005). Impact of cultivation and collection on the conservation of medicinal plants: Global trends and issues. *In III WOCMAP congress on medicinal and aromatic plants: Conservation, cultivation and sustainable use of medicinal and aromatic plants, Chiang Mai.*

Schultes, R. E. (1960). Tapping our heritage of ethnobotanical lore. *Economic Botany, 14*, 257–262.

Sen, S., & Chakraborty, R. (2015). Toward the integration and advancement of herbal medicine: A focus on traditional Indian medicine. *Biologics: Targets & Therapy, 5*, 33–44.

Sen, S., Chakraborty, R., & De, B. (2011). Challenges and opportunities in the advancement of herbal medicine: India's position and role in a global context. *Journal of Herbal Medicine, 1*, 67–75.

Sen, S., Chakraborty, R., De, B., & Mazumder, J. (2009). Plants and phytochemicals for peptic ulcer: An overview. *Pharmacognosy Reviews, 3*, 270–279.

Senthilkumar, K., Aravindhan, V., & Rajendran, A. (2013). Ethnobotanical survey of medicinal plants used by Malayali tribes in Yercaud hills of Eastern Ghats, India. *Journal of Natural Remedies, 13*, 118–132.

Shankar, R., Lavekar, G., Deb, S., & Sharma, B. (2012). Traditional healing practice and folk medicines used by Mishing community of North East India. *Journal of Ayurveda and Integrative Medicine, 3*, 124.

Sharma, P., & Mujumdar, A. (2003). Traditional knowledge on plants from Toranmal Plateau of Maharashtra. *Indian Journal of Traditional Knowledge, 2*.

Sharma, P., & Samant, S. (2014). Diversity, distribution and indigenous uses of medicinal plants in Parbati Valley of Kullu district in Himachal Pradesh, Northwestern Himalaya. *Asian Journal of Advanced Basic Science, 2*, 77–98.

Sharma, M., Sharma, C. L., & Debbarma, J. (2014). Ethnootanical studies of some plants used by Tripuri tribe of Tripura, India with special reference to Magico religious beliefs. *International Journal of Plant, Animal and Environmental Sciences, 4*(3), 518–528.

Shiracko, N., Owuor, B. O., Gakuubi, M. M., & Wanzala, W. (2016). A survey of ethnobotany of the AbaWanga people in Kakamega county, western province of Kenya. *Indian Journal of Traditional Knowledge, 15*.

Shisanya, C. A. (2017). *Role of traditional ethnobotanical knowledge and indigenous institutions in sustainable land management in western highlands of Kenya indigenous people* (p. 159).

Shrestha, N., Prasai, D., Shrestha, K. K., Shrestha, S., & Zhang, X.-C. (2014). Ethnomedicinal practices in the highlands of central Nepal: A case study of Syaphru and Langtang village in Rasuwa district. *Journal of Ethnopharmacology, 155*, 1204–1213.

Shrivastava, S., & Kanungo, V. (2013a). Ethnobotanical survey of Surguja district with special reference to plants used by Uraon tribe in treatment of diabetes. *International Journal of Herbal Medicine, 1,* 127–130.

Shrivastava, S., & Kanungo, V. (2013b). Ethnobotanical survey of Surguja District with special reference to plants used by uraon tribe in treatment of respiratory diseases. *International Journal of Herbal Medicine, 1,* 131–134.

Shukla, R., Chakravarty, M., & Gautam, M. (2008). Indigenous medicine used for treatment of gynecological disorders by tribal of Chhattisgarh, India. *Journal of Medicinal Plant Research, 2,* 356–360.

Shukla, A. N., Srivastava, S., & Rawat, A. (2013). A survey of traditional medicinal plants of Uttar Pradesh (India)-used in treatment of infectious diseases. *Natural Science, 11,* 24–36.

Simon, S., Jebaraj, N., Suresh, K., & Ramachandran, V. (2011). Ethnobotanical knowledge on single drug remedies from Idukki district, Kerala for the treatment of some chronic diseases. *International Journal of Research in Ayurveda and Pharmacy, 2,* 531–534.

Singh, J., Bhuyan, T. C., & Ahmed, A. (1996). Ethnobotanical studies on the Mishing tribes of Assam with reference to food and medicinal plants-I. *Journal of Economic and Taxonomic Botany, 12,* 350–356.

Smita, R., Sangeeta, R., Kumar, S. S., Soumya, S., & Deepak, P. (2012). An ethnobotanical survey of medicinal plants in Semiliguda of Koraput District, Odisha, India. *Research Journal of Recent Sciences, 2,* 20–30.

Sreeramulu, N. (2008). *Traditional botanical knowledge of local people in Nalgonda and Warangal Districts of Telangana, Andhra Pradesh, India.* Ph.D. Thesis Warangal: Kakatiya University.

Sreeramulu, N., Suthari, S., Ragan, A., & Raju, V. S. (2013). Ethno-botanico-medicine for common human ailments in Nalgonda and Warangal districts of Andhra Pradesh, India. *Annals of Plant Sciences, 2*(7), 220–229.

Sripriya, D. (2017a). Ethno-botanical uses of some plants by tribes in AP, India. *International Journal of Pharma and Bio Sciences, 7,* 67–74.

Sripriya, D. (2017b). Traditionally used Medicinal Plants for Wound Healing in the Chittoor District, Andhra Pradesh (India). *International Journal of Life Sciences, 5,* 706–708.

Srithi, K., Trisonthi, C., Wangpakapattanawong, P., & Balslev, H. (2012). Medicinal plants used in Hmong women's healthcare in northern Thailand. *Journal of Ethnopharmacology, 139,* 119–135.

Sudeesh, S. (2012). Ethnomedicinal plants used by Malayaraya tribes of Vannapuram village in Idukki, Kerala, India. *Age, 15,* 30.

Sudharani, T., Umadevi, M., Rajani, B., Padmavathi, V., & Maiti, R. K. (2007). Ethnobotanical survey of Nalgonda district, Andhra Pradesh, India. *Research on Crops, 8*(3), 700–715.

Sutha, S., Mohan, V., Kumaresan, S., Murugan, C., & Athiperumalsami, T. (2010). *Ethnomedicinal plants used by the tribals of Kalakad-Mundanthurai Tiger Reserve (KMTR), Western Ghats.* Tamil Nadu for the Treatment of Rheumatism.

Suthari, S. (2020). Utility of wild food plants by indigenous tribes from Telangana, India: An ethnobotanical perspective. *American Journal of Ethnomedicine* (in press).

Suthari, S., Kandagalta, R., Ragan, A., & Raju, V. S. (2016). Plant wealth of a sacred grove: Mallur Gutta, Telangana state, India. *International Journal of General Medicine, 9,* 369–381.

Suthari, S., & Raju, V. S. (2016). Antidote botanicals for snake bites from Koyas of Warangal district, Telangana, India. *Journal of Herbs, Spices & Medicinal Plants, 22,* 57–68.

Suthari, S., Raju, V. S., & Majeti, N. V. P. (2018). Ethnobotanical explorations in Telangana, The youngest state in union of India: A synoptic account. In M. Ozturk, & K. R. Hakeem (Eds.), *Vol. 1. Plant and human health* (pp. 65–123). Springer.

Suthari, S., Sreeramulu, N., Omkar, K., & Raju, V. (2014). The climbing plants of northern Telangana in India and their ethnomedicinal and economic uses Indian. *Journal of Plant Sciences*, *3*, 86–100.

Suthari, S., Sreeramulu, N., Omkar, K., Reddy, C., & Raju, V. S. (2014). Intracultural cognizance of medicinal plants of Warangal North Forest Division, Northern Telangana, India. *Ethnobotany Research and Applications*, *12*, 211–235.

Talukdar, N., & De, A. (2016). Ethnobotanical Knowledge used for primary healthcare in Loharbond region of innerline reserve forest. *International Journal of Recent Scientific Research*, *7*, 11200–11206.

Thomas, B., & Rajendran, A. (2013). Less known ethnomedicinal plants used by Kurichar tribe of Wayanad district, Southern Western Ghats Kerala, India. *Botany Research International*, *6*, 32–35.

Tiwari, D. N. (2000). *Report of the task force on conservation and sustainable use of medicinal plants*. New Delhi, India: Government of India.

Tiwari, N., & Tiwari, S. (2014). Assessment of traditional medicinal plants in Balaghat district (M.P.). *Golden Research Thoughts*, *4*(6), 91–112.

Tugume, P., Esezah, K. K., Mukadasi, B., Justine, N., Maud, K., Patrick, M., et al. (2016). Ethnobotanical survey of medicinal plant species used by communities around Mabira Central Forest Reserve, Uganda. *Journal of Ethnobiology and Ethnomedicine*, 1.

Turner, N. C., & Bell, M. A. (1971). The ethnobotany of the coast Salish Indians of Vancouver Island. *Economic Botany*, *25*, 63–99.

Udayan, P., Tushar, K., George, S., & Balachandran, I. (2007). Ethnomedicinal information from Kattunayakas tribes of Mudumalai wildlife sanctuary, Nilgiris district, Tamil Nadu. *Indian Journal of Traditional Knowledge*, *6*(4), 574–578.

Uniyal, S. K., Singh, K., Jamwal, P., & Lal, B. (2006). Traditional use of medicinal plants among the tribal communities of Chhota Bhangal, Western Himalaya. *Journal of Ethnobiology and Ethnomedicine*, *2*, 14.

Upadhyay, R., & Chauhan, S. V. S. (2000). Ethnobotanical observations on Koya tribe of Gundala Mandal of Khammam district, Andhra Pradesh. *Ethnobotany*, *12*, 93–99.

Vaidyanathan, D., Senthilkumar, M. S., & Basha, M. G. (2013). Studies on ethnomedicinal plants used by malayali tribals in Kolli hills of Eastern ghats, Tamilnadu, India. *Asian Journal of Plant Science & Research*, *3*, 29–45.

Vidyasagar, G., & Siddalinga, M. S. (2013). *Medicinal plants used in the treatment of Diabetes mellitus in Bellary district, Karnataka*. NISCAIR-CSIR.

Vijayan, A., John, J., Parthipan, B., & Renuka, C. (2007). *Traditional remedies of Kani tribes of Kottoor reserve forest, Agasthyavanam, Thiruvananthapuram, Kerala*. CSIR.

Vishnuvardhan, Z., Jyothirmayi, G., & Jyothi, D. (2018). Medicinal plants of tribal traditional system from Guntur District, Andhra Pradesh, India. *International Journal of Life Sciences (Kathmandu)*, *6*, 194–204.

Voeks, R. A. (2007). Are women reservoirs of traditional plant knowledge? Gender, ethnobotany and globalization in northeast Brazil. *Singapore Journal of Tropical Geography*, *28*, 7–20.

Wangyal, J. T. (2012). *Ethnobotanical knowledge of local communities of Bumdeling wildlife sanctuary, Trashiyangtse, Bhutan*. NISCAIR-CSIR.

WHO. (1946). *World Health Organization, Geneva*. Official Records of WHO, No. 2 (p. 100). https://www.who.int/about/.

WHO. (2002). *The world health report 2002: reducing risks, promoting healthy life*. World Health Organization.

WHO. (2008). *WHO report on the global tobacco epidemic, 2008: The MPOWER package*. World Health Organization.

WHO. (2014). *Global status report on non-communicable diseases*. World Health Organization.

WHO. (2015). *Trends in maternal Mortality: 1990–2015: Estimates from WHO*. UNICEF, UNFPA, World Bank Group and the United Nations Population Division.

WHO. (2018). *World Health Statistics 2018: Monitoring health for the SDGs*. World Health Organization.

WHO Aftrican Continent Report. (2016). *Health of the people: The African Continent Health Report 2016*. WHO Regional Office for Africa, AFRO Publication.

Xavier, T. F., Kannan, M., & Auxilia, A. (2015). Traditional Medicinal Plants Used in the treatment of different skin diseases. *International Journal of Current Microbiology and Applied Sciences, 4*, 1043–1053.

Xavier, T. F., Kannan, M., Lija, L., Auxillia, A., & Rose, A. K. F. (2014). Ethnobotanical study of Kani tribes in Thoduhills of Kerala, South India. *Journal of Ethnopharmacology, 152*, 78–90.

Yesodharan, K., & Sujana, K. (2007). *Ethnomedicinal knowledge among Malamalasar tribe of Parambikulam wildlife sanctuary, Kerala*. CSIR.

Zambrana, N. Y. P., & Bussmann, R. W. (2020). *Ethnobotnay of mountain regions-ethnobotany of the Andes*. Springer International Publishing.

Chapter 10

Health benefits of bioactive compounds from microalgae

Dig Vijay Singh, Atul Kumar Upadhyay, Ranjan Singh, and D.P. Singh
Department of Environmental Science, Babasahib Bhimrao Ambedkar University, Lucknow, India

1. Introduction

Rapid population growth (Singh, Bhat, Bhat, Dervash, & Ganei, 2018), industrialization (Leon, 2008), advancement in agricultural practices (Sarkar, Aronson, Patil, & Hugar, 2012), increasing pollution level (Ghorani-Azam, Riahi-Zanjani, & Balali-Mood, 2016), and faulty food habits are the major factors responsible for deterioration of human health. The health deterioration (Searle, 2006) is further enhanced by the use of synthetic medicine that has numerous side effects on the body (Abdel-Aziz, Aeron, & Kahil, 2016) and also eventually depletes the health of the environment (Boxall, 2004). Synthetic medicines have certain harmful impact on the body, but exploitation of the bioactive compounds for improvement of human health can minimize their impacts on the body (Bhattacharya, 2013). Developing nutrient-rich pharmaceuticals and food products supplemented with bioactive compounds not only minimize the use of medicines but also improve the immunity and protect humans from chronic diseases in elegant manner (Nasri, Baradaran, Shirzad, & Rafieian-Kopaei, 2014). Bioactive compounds can also be obtained from several plants (Altemimi, Lakhssassi, Baharlouei, Watson, & Lightfoot, 2017), but microalgae have shown tremendous potential to enhance their production (Khan, Shin, & Kim, 2018) in cost-effective and eco-friendly manner (de Jesus Raposo, de Morais, & de Morais, 2013).

Microalgae are ubiquitous and are efficient converters of solar energy into biomass (Hussian, 2018). Microalgae include prokaryotic and eukaryotic organisms (Urtubia, Betanzo, & Vásquez, 2016) inhibiting different habitats that have more photosynthetic potential compared with terrestrial plants (Barkia, Saari, & Manning, 2019). Globally, microalgae are responsible for two-fifth part of photosynthesis on the earth (Moreno-Garrido, 2008). Microalgae produce several compounds in good amount, which are mostly lacking in higher plants (Sathasivam, Radhakrishnan, Hashem, & Abd Allah, 2019). Microalgae are also reliable and cost-effective source for bioactive compound production

Phytomedicine: A Treasure of Pharmacologically Active Products from Plants
https://doi.org/10.1016/B978-0-12-824109-7.00015-7

as this system does not compete with agricultural land and synthetic chemicals (Sharma & Sharma, 2017). Thousands of algal species exist on earth, but merely one-fourth are collected and identified. Huge microalgae resource still remains unexplored (Shalaby, 2011) and can be exploited for production of valuable compounds having distinctive roles in different fields (Bilal, Rasheed, Ahmed, Iqbal, & Sada, 2017). The enormous diversity of microalgae can be the potent source of diverse compounds having magnificent role in medical and pharmaceutical industry (Sathasivam et al., 2019). Bioactive compounds isolated from microalgae biomass have unique properties that can be helpful in maintaining proper growth of different organisms (Barkia et al., 2019). Bioactive compounds have high nutritional value that can easily accomplish the growing nutrient demand (Ghosh & Smarta, 2016). Microalgal biomass obtained can be either used as feed to animals or feedstock for production of unique compounds having extensive medicinal importance (Hamed, 2016). Bioactive compounds of pharmaceutical importance are of utmost significance and are gaining impetus globally (Takshak, 2018). Bioactive compounds like carotenoids, proteins, vitamins, lipids, fatty acids, and sterol are the several compounds produced by microalgae in an eco-friendly manner (Galasso et al., 2019). These compounds play pivotal role by maintaining the essential nutrient supply to the human body (Liu, 2013). Microalgae generate ample compounds by producing biomass or releasing them into the culture media (de Morais, Vaz, de Morais, & Costa, 2015). The incomparable benefits of microalgae have made them valuable source of bioactive compounds on the earth (Martínez-Francés & Escudero-Oñate, 2018).

Incorporation of vital compounds from microalgae into food and pharmaceutical products can have several positive health effects on humans (Caporgno & Mathys, 2018). Countries like Japan are consuming microalgae (*Chlorella*) as potent nutrient source and also use it for medical treatments (Sathasivam et al., 2019). Consuming microalgae helps to fight against cancer, cardiovascular diseases, atherosclerosis, aging (Ku, Yang, Park, & Lee, 2013), and digestion problems (El-Hack et al., 2019). Bioactive compounds can also pave ways for the development of new products having immense importance in pharmaceutical industry (Atanasov et al., 2015). Pharmaceutically important bioactive compounds from microalgae and its commercialization on wide scale are still at initial stage (Bhattacharjee, 2016). Ability of microalgae to produce several valuable compounds has put microalgae in limelight for biotechnological purpose (Martínez-Francés & Escudero-Oñate, 2018). There is tremendous surge of interest in using bioactive compounds for improving and maintaining normal health by incorporating such compound in food and pharmaceutical products (Canizales et al., 2018). The wide array of compounds synthesized by microalgae through various metabolic pathways can become the promising source of natural compounds (Sathasivam et al., 2019). Bioactive compounds produced by microalgae are investigated to understand their possible applications at industrial level (de Morais et al., 2015). The main obstacle in the exploitation of "bioactive compound"–rich products is lack of financial support and less awareness among people about the health benefits of these products. But utilizing microalgae biomass as feedstock of essential compounds can help to meet the nutritional requirement of the expanding human population and also assist in tackling the issue of climate change.

2. Microalgae and bioactive compounds production and synthesis

Microalgae have great potential to acclimatize in varying culture conditions (Benavente-Valdés, Aguilar, Contreras-Esquivel, Méndez-Zavala, & Montañez, 2016; Hamidi et al., 2020) and simultaneously produce variety of bioactive compounds that are lacking in other organisms (de Morais et al., 2015). The accumulation of bioactive compounds enhances under different stress conditions like salinity, light, and temperature (Kumar & Sharma, 2018). These compounds have positive effects on human health by boosting immune system (El-Hack et al., 2019) that enhances the capability to fight against several diseases (Walia, Gupta, & Sharma, 2019). Microalgae can grow easily in regions and climates not suitable for agricultural cultivation (Baliga & Powers, 2010). In addition of their potential to grow profusely in hostile conditions, microalgae have short life span, quick growth, high rate of photosynthesis and produce high yield compared to other terrestrial plants (Mostafa, 2012). Studies on different microalgae species have led to the identification and extraction of compounds having antimicrobial, antioxidant, and antitumor activity (de Morais et al., 2015). The several incomparable benefits have made microalgae one of the most versatile organisms for production of valuable bioactive compounds in economical manner (Coêlho et al., 2019). Several microalgae species studied globally for bioactive compound production are as follows:

Dunaliella is unicellular and salt-tolerant microalgae that have essential functions in diverse ecosystem (Ramos et al., 2011). Because of unique physiology and tolerance to extreme conditions, it also acts as the affluent source of several bioactive compounds (Sukla, Subudhi, & Pradhan, 2019). β-Carotene, lipids, proteins, and fatty acids are the essential compounds easily obtained from *Dunaliella* (de Morais et al., 2015). Production of these compounds enhances with the change in culture conditions like nutrients, salinity, temperature, and light (Benavente-Valdés et al., 2016). The biomass produced by *Dunaliella* acts as feedstock for antioxidant production, and the extracts obtained have antimicrobial and hepatoprotective roles in the human body (Camacho, Macedo, & Malcata, 2019). Studies on *Dunaliella* cultivation for β-carotene have been conducted in both developing (India) and developed countries (the United States).

Nostoc forms spherical colonies and has heterocyst for atmospheric nitrogen fixation. Nostoc, an abundant source of different compounds, has been extensively used as dietary supplement and also in pharmaceutical industry (Panjiar, Mishra, Yadav, & Verma, 2017). *Nostoc* is rich in vitamins, proteins (Li & Guo, 2018), docosahexaenoic acid (DHA), and eicosapentaenoic acid

(EPA) (Wang, Xu, Jiang, Li, & Kuang, 2000) that have several beneficial roles in the human body. Biomass obtained from *Nostoc* assists in digestion, boosts immune system, and controls blood pressure in elegant manner. Biomass from *Nostoc* contains essential fatty acids that are precursors of prostaglandins, thereby attaining special attention from pharmaceutical industry (Bhattacharjee, 2016). *Nostoc* also produces protein cyanovirin, which have significant roles in the treatment of HIV and H1N1 (Mostafa, 2012). The affirmative role of *Nostoc* in improving health has encouraged for large-scale cultivation that can certainly improve economic condition and boost exploitation of sustainable source of bioactive compound (Deng, Hu, Lu, Liu, & Hu, 2008).

Spirulina is freshwater and multicellular cyanobacteria and is one of the most studied microalgae for effective treatment of tumors, obesity, and blood pressure (Habib, 2008). *Spirulina* has been legally approved as food supplement without posing any risk to human health (Karkos, Leong, Karkos, Sivaji, & Assimakopoulos, 2011). The plentiful presence of essential compounds along with high digestibility makes *Spirulina* a cheap source of valuable compounds (Falquet & Hurni, 1997). *Spirulina* is rich in proteins (50%–70%, w/w), carbohydrates (10%–20%), lipid (5%–10%), vitamins (B12), pigments, and essential macro- and micronutrients, and due to abundance of these compounds, *Spirulina* is one of the most investigated microalgae in the world (Sathasivam et al., 2019; Sharoba, 2014).

Chlorella sp. is unicellular microalgae containing essential compounds like proteins (53% w/w), carbohydrates (23%), lipids (9%), and minerals (Costa & Morais, 2013). Investigation of *Chlorella* evidenced their role in antitumor, antimicrobial activity, liver protection, and boosting of immune system (de Morais et al., 2015). *Chlorella* is also abundant in antioxidants, which are helpful in protection against oxidative stress (Sikiru et al., 2019). *Chlorella* grown under suitable and hygienic conditions can also be used as dietary supplement without having any impact on human health (de Morais et al., 2015). Biomass obtained from *Chlorella* is excellent source of essential compounds whose content varies with species and culture conditions (Khan et al., 2018). Medicinal properties of *Chlorella* have been proved experimentally, and evidences are available about their role in controlling pathogen growth and preventing tumor formation and fat accumulation in human body (Li, Li, Kim, & Lee, 2013; Medina-Jaritz, Carmona-Ugalde, Lopez-Cedillo, & Ruiloba-De Leon, 2013). Several compounds from "*Chlorella*" are used as natural colorants and are essential for muscular health (Zhao & Sweet, 2008). β-1,3-Glucan produced by *Chlorella* can help by promoting growth and boosting of immune system in the body (de Morais et al., 2015). Vitamin B12 is another important compound for blood cell regeneration, which can also be easily obtained from *Chlorella* (Costa & de Morais, 2014). More than 70 countries are involved in *Chlorella* cultivation, and leading among them is Japan, which produces 4-lakh tons of biomass annually (Rösch & Posten, 2012). Biomass from *Spirulina* and *Chlorella* is major contributors in the biomass market having production of 3000 and 4000 tons annually. Both *Chlorella* and *Spirulina* are important microalgae species rich in various metabolites having multiple beneficial roles in human body (Barkia et al., 2019).

3.   Bioactive compounds

Bioactive compounds (Fig. 1) from microalgae have several functional roles (de Morais et al., 2015), and their extraction is increasing due to various scientific processes that have made extraction of these compounds easy. Microalgae can become the source of novel drugs that have capacity to fight or control chronic diseases (Galasso et al., 2019). Cultivation of microalgae for production of bioactive compounds depends upon culture conditions as culture conditions affect metabolic pathways leading to build up of certain compounds in microalgae (Barkia et al., 2019). The various compounds produced are as follows.

3.1   Polysaccharides

Microalgae produce various compounds having protective roles in the body, which can be used directly as food or as medicines (de Morais et al., 2015). Polysaccharides are polymeric carbohydrates (de Souza et al., 2019) produced by microalgae that can be used in several industries (Arad & Levy-Ontman, 2010). Microalgal polysaccharides are structurally complex, which can be exploited by industries for formation of various products that can be helpful in improving human health (Pierre et al., 2019). Sulfated polysaccharides are very essential for pharmaceutical industry, and microalgae can increase their availability owing

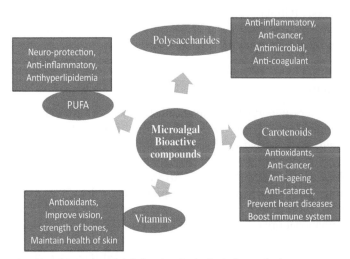

FIG. 1   Bioactive compounds and their functional roles in the human body.

to the simple process of extraction (Cunha & Grenha, 2016). Sulfated polysaccharides like carrageenan from microalgae possess distinctive properties and are also known to have antiviral activities (Ahmadi, Zorofchian Moghadamtousi, Abubakar, & Zandi, 2015). Polysaccharides from microalgae can also inhibit activity of virus and prevent the entry of virus into microalgae cell, which is assumed to be due to positive charge on the surface of the cell (Wang, Wang, & Guan, 2012). *Spirulina* also produce certain volatile compounds such as heptadecane and tetradecane that are known to have antimicrobial activity (Elshouny, El-Sheekh, Sabae, Khalil, & Badr, 2017). Long-chain PUFAs, sulfurized polysaccharides, and pigments are well known for their antiinflammatory activities (Bhattacharjee, 2016; de Morais et al., 2015). The antiinflammatory activities of microalgae are considered important for tissue reconstitution especially in patients with total skin lost due to burning (de Morais et al., 2010). Polysaccharides from microalgae enhance enzymatic activity and improve functioning of the immune system (Barkia et al., 2019). Polysaccharides also have strong antioxidant properties by enhancing quenching of superoxide, hydroxyl, and peroxide radical and protect from the damage posed by reactive oxygen species (Tan, Norhaizan, Liew, & Sulaiman Rahman, 2018). Polysaccharides (immulina and immurella) extracted from *Spirulina platensis* and *Chlorella pyrenoidosa* are also known to have anticancer activity (Andrade, Andrade, Dias, Nascimento, & Mendes, 2018). The main polysaccharide produced by *Spirulina* is rhamnose, which constitutes about 52.3% of total polysaccharides (Mišurcováa, Orsavováb, & Ambrožováa, 2014; Pugh, Ross, ElSohly, ElSohly, & Pasco, 2001). Polysaccharides have several beneficial effects on health by acting as dietary fiber (Lunn & Buttriss, 2007). Insoluble fiber of cellulose and lignin promotes movement of food in digestive system and also enhances the growth of microflora in the gut region, while soluble fiber adjusts blood sugar level and also reduces the fat accumulation in the body (Grundy et al., 2016). *Chlorella*, *Dunaliella*, and *Scenedesmus* are some of the microalgal species producing polysaccharides in good amount (Sathasivam et al., 2019; Wells et al., 2017).

3.2 Proteins

Proteins are very important constituent of the body and are vital for growth, repair, and maintenance of cellular activities in the body. Proteins are synthesized within the body and can be supplied externally in the form of dietary supplements if human body is not capable of synthesizing proteins (Stokes, Hector, Morton, McGlory, & Phillips, 2018). Microalgae contain a good quantity of protein, and some of them contain protein more than 50% of dry weight (Bleakley & Hayes, 2017). Microalgae are the safe and cheap source of proteins. International organizations (WHO and FAO) have recommended amino acid profile of proteins for humans, and proteins from microalgae were found to have batter amino acid profile than recommended (Gorissen et al., 2018). Microalgae species have huge protein production potential, which is helpful in protection against several diseases (Barkia et al., 2019). Some microalgal species like *Chlorella*, *Arthrospira*, and *Nostoc* produce proteins comparable with other protein-rich sources like soybean, milk, meat, and egg (Koyande et al., 2019). Microalgal biomass can be used as functional food that can reduce the incidence of several diseases like hypertension and cholesterol (de Morais et al., 2015) and protect cells from damage by increasing the availability of proteins in the body (Matos, Cardoso, Bandarra, & Afonso, 2017). Proteins extracted from *Chlorella* biomass can be used as dietary supplements that might be helpful in preventing the diseases induced by oxidative stress (Sathasivam et al., 2019). Another important product from microalgae is bioactive peptides, which are having antioxidant (protection of DNA from ROS), anticancer, and antimicrobial properties inside human body (Barkia et al., 2019). Proteins from microalgae biomass are the best alternative compared with traditional sources that can be easily supplemented into the food material and pharmaceutical products (Caporgno & Mathys, 2018) to improve human health.

3.3 Pigments

Microalgae are the superior pigment source owing to their higher rate of photosynthesis leading to elevated pigment production as compared with terrestrial plants (Benedetti, Vecchi, Barera, & Dall'Osto, 2018; Hu, Nagarajan, Zhang, Chang, & Lee, 2018). Utilization of microalgae as the feasible source of bioactive compounds like chlorophyll and carotenoids is viable alternative and eco-friendly method (Khan et al., 2018). Pigments like chlorophyll and carotenoids are most important compounds present in microalgae in plentiful amount (Alam, Najam, & Al-Harrasi, 2018). Chlorophyll helps in photosynthesis, while carotenoid provides protection from intense light (Dall'Osto et al., 2012). Pigments are essential for microalgal growth, and these pigments act as antimicrobial, antiaging, anticarcinogen, and antiinflammatory agents in human body (Gujar, Cui, Ji, Kubar, & Li, 2019). Pigments have numerous therapeutic properties such as it acts as efficient antioxidants, protects retina, and regulates blood cholesterol in the body (Martín, Kuskoski, Navas, & Asuero, 2017).

Carotenoids have several health benefits that involve treatment of degenerative diseases like cataract, cancer, and inflammation (Bungau et al., 2019). β-Carotene, lutein, tocopherols, zeaxanthin, lycopene phycocyanin, and phycoerythrin are different types of essential carotenoids that have diverse application in food and pharmaceutical industries (Dasgupta, 2015; Jaswir, Noviendri, Hasrini, & Octavianti, 2011). In *Spirulina*, β-carotene is predominant carotenoids (Park et al., 2018) composing approximately four-fifth of total carotenoids, while in *Dunaliella salina*, β-carotene constitutes 10%–14% of its total dry mass (Xu, Ibrahim, Wosu, Ben-Amotz, & Harvey, 2018; Hosseini Tafreshi & Shariati, 2009). β-Carotene improves vision, enhances immune system, and prevents lipid peroxidation, thus providing protection from several lethal diseases (De Carvalho & Caramujo, 2017; Gammone, Riccioni, & D'Orazio, 2015).

Lutein extracted from microalgal biomass is also helpful in prevention of several diseases (Sathasivam et al., 2019). The loss of vision with age can be due to deficiency of lutein and zeaxanthin in the body (Buscemi et al., 2018). Lutein and zeaxanthin prevent photooxidative damage and maintain vision of human eye by absorbing UV light (Mares, 2016; Roberts & Dennison, 2015). Lycopene prevents myocardial infarction, reduces the risk of heart stroke, and decreases cholesterol level by lowering lipoprotein density (Gammone, Pluchinotta, Bergante, Tettamanti, & D'Orazio, 2017; Riccioni, Mancini, Di Ilio, Bucciarelli, & D'Orazio, 2008). C-phycocyanin is accessory blue color pigment belonging to phycobilliprotein family found in cyanobacteria (Dasgupta, 2015; Jiang et al., 2017). Phycobilliproteins are also major pigment from microalgae and have vital roles in the food industry (Mysliwa-Kurdziel & Solymosi, 2017). Phycobilliprotein protects liver and is also involved in antiinflammatory and anticancer activities (Jiang et al., 2017; Mysliwa-Kurdziel & Solymosi, 2017).

Fucoxanthin is known to have higher antioxidant activity as compared with β-carotene and is responsible for preventing blood cell proliferation, obesity, and skin melanogenesis (Irvani, Hajiaghaee, & Zarekarizi, 2018; Peng, Yuan, Wu, & Wang, 2011). Fucoxanthin protects liver, skin, and heart and also improves strength of bones (Peng et al., 2011). Fucoxanthin inhibits angiogenesis by repressing the growth and development of microvessels (Sugawara, Matsubara, Akagi, Mori, & Hirata, 2006). Natural compounds from microalgal biomass act as antiangiogenic (Khan et al., 2018). Angiogenesis is the process of new blood vessel development from the existing one and occurs rapidly during the process of uterus development, embryo formation, and healing process (Felmeden, Blann, & Lip, 2003). Angiogenesis is normal process but becomes threatening in the case of the person suffering from cancer, arthritis, and atherosclerosis (Bisht, Dhasmana, & Bist, 2010). Angiogenesis further promotes the growth of the cancerous cell in the body, but microalgae can be helpful in treating cancer by synthesizing siphonaxanthin, which suppresses the process of angiogenesis (Ganesan, Matsubara, Sugawara, & Hirata, 2013; Khan et al., 2018). Bioactive compounds like siphonaxanthin derived from microalgal biomass can destroy cancerous cells by inducing apoptosis that leads to death of cells (El-Hack et al., 2019; Khan et al., 2018).

Like other pigments, astaxanthin also improves health by boosting immune system responses and prevents lipid peroxidation and the accumulation of fats in the body (Higuera-Ciapara, Felix-Valenzuela, & Goycoolea, 2006; Sztretye et al., 2019). Astaxanthin incorporated in pharmaceutical compounds and dietary supplements provides protection against oxidative damage, photooxidation, and antiaging (Dhankhar, Kadian, & Sharma, 2012; Guerin, Huntley, & Olaizola, 2003). Thus microalgae are promising source of different compounds that can provide natural strength to fight against several chronic diseases.

3.4 Phenolic compounds

Phenolic compounds from microalgae consist of aromatic rings and hydroxyl groups, which can become vital component of dietary supplements

(Barkia et al., 2019). Phenolic compounds can be halogenated that lead to various activities of these compounds (Greenwood, 2012). Phenolic compound accumulation in microalgae is the results of stress conditions, which protect microalgae from metal and ultraviolet radiation stress (Generalić Mekinić et al., 2019). The content of phenolic compounds in microalgae is either less or equal to that of terrestrial plants (Haoujar et al., 2019). Microalgae are capable of producing complex phenolic compounds that need identification and characterization, and microalgae may also contain some novel phenolic substances that can be beneficial for human health (Barkia et al., 2019; de Morais et al., 2015). Phenolic compounds from microalgae protect liver from damage caused by ROS, which is the easy target of oxidative stress (Montero-Lobato et al., 2018; Tan et al., 2018). Phenolic compounds neutralize free radicals and protect human body from damage (Pham-Huy, He, & Pham-Huy, 2008). Phenolic compounds are also well known for providing protection against cancer, inflammation, microbial attack, and degenerative diseases (Oliveira, Carvalho, & Melo, 2014; Tungmunnithum, Thongboonyou, Pholboon, & Yangsabai, 2018). Gallic acid, vanillin acid, salicylic acid, and cinnamic acid are the few phenolic compounds found in *Spirulina* (Safafar, Van Wagenen, Møller, & Jacobsen, 2015). Among phenolic compounds, one-third of total phenolic compound intake is in the form phenolic acids, while remaining two-third is in the form of flavonoids (Atanassova, Georgieva, & Ivancheva, 2011). *Spongiochloris spongiosa, Anabaena doliolum, Spirulina, Porphyra tenera, Spirogyra* sp., and *Ankistrodesmus* are the few species of microalgae containing good quantity of phenolic compounds (Jerez-Martel et al., 2017; Wang, 2016).

3.5 Vitamins

Vitamins have fundamental role and are essential for the proper functioning of the human body (World Health Organization, 2004). Microalgae have vitamin profile that is vital for growth and development of the human body (Koyande et al., 2019). Diverse compounds produced by microalgal species help to increase the production of vital metabolites in green and sustainable manner (Bhalamurugan, Valerie, & Mark, 2018). Vitamins from microalgae are important for human nourishment, metabolism, cell regeneration, skin health, membrane protection, and normal bone development; regulate blood sugar; and provide strength to immune system (Andrade et al., 2018; Schagen, Zampeli, Makrantonaki, & Zouboulis, 2012). Vitamins enhance circulation of blood in the body and also assist in controlling muscle functions (Rizvi, Raza, Faizal Ahmed, Abbas, & Mahdi, 2014). Vitamin A and vitamin B complex produced by microalgae in abundant quantity is responsible for vision, good enzyme activity, and healthy skin (Wells et al., 2017). Vitamins (C and E) also have antioxidant properties that protect from oxidative damage induced by free radicals (Lobo, Patil, Phatak, & Chandra, 2010). Incorporation of microalgal biomass in dietary supplements can fulfill vitamin requirement of the body in cheap manner. High vitamin content in microalgae can be due to alteration in the photosynthetic apparatus that is related to changes in cellular component (Barkia et al., 2019). Production of vitamins from microalgae changes with environmental condition and also with method of harvesting (Tandon, Jin, & Huang, 2017).

3.6 Polyunsaturated fatty acids (PUFA)

PUFAs are vital bioactive compounds that help to maintain tissue integrity, protect against high blood pressure and cardiovascular diseases, and also act as active antiinflammatory agents (Schuchardt, Huss, Stauss-Grabo, & Hahn, 2010). Microalgae are the promising source of PUFA compared with fish oil, which is not able to attain the rising demands (Adarme-Vega et al., 2012). Microalgae are the cheap and cost-effective source of these essential fatty acids, which are helpful in prevention of several diseases in human body (Khan et al., 2018). Fatty acids (omega 3 and omega 6) cannot be synthesized in the human body (Kaur, Chugh, & Gupta, 2014) but have numerous beneficial functions inside the body and should be taken externally in food supplement and pharmaceutical products (da Silva Vaz, Moreira, de Morais, & Costa, 2016; Kaur, Allahbadia, & Singh, 2018). Deficiency of essential fatty acids in humans can led to depression, schizophrenia, rough skin (Perica & Delaš, 2011), dermatitis (Kaczmarski, Cudowska, Sawicka-Żukowska, & Bobrus-Chociej, 2013), and sensitivity to infections (Mogensen, 2017); prevent wound healing (Silva et al., 2018); and restrict growth of the body (Hibbeln & Gow, 2014).

Essential fatty acids (linolenic acid) are vital for cosmetic industry as it helps to regenerate skin and prevent aging (Mourelle, Gómez, & Legido, 2017). Linoleic acid is essential for regeneration process and enhances response of immune system toward various diseases. Arachidonic acid (AA) is vital for development of babies, and low intake can lead to several problems in babies (Hadley, Ryan, Forsyth, Gautier, & Salem, 2016). AA plays the vital role in muscle development and increases stamina in person doing physical exercises (Hashimoto & Hossain, 2018). Deficiency of AA can lead to neurological problems in humans (Alzheimer's and autism) (Yui, Imataka, Nakamura, Ohara, & Naito, 2015).

EPA and DHA are vital fatty acids having multiple roles in human body (Swanson, Block, & Mousa, 2012). Intake of EPA ensures heart health and prevents from strokes and also blood clot formation. DHA is essential for retina formation and is also vital for brain health. DHA improves memory, imagination power, reasoning, and judgment ability (Andrade et al., 2018). DHA and EPA are also known to produce several substances that have the antiinflammatory and neuroprotective role in the body (Reimers & Ljung, 2019). These substances produced in the brain also prevent the formation of harmful substances and assist in reducing the risks of neurodegenerative diseases (Dyall, 2015). The production of essential fatty acids in microalgae can be enhanced commercially by using genetic manipulation (Sprague, Betancor, & Tocher, 2017). By genetic manipulation, *Phaeodactylum tricornutum* produces 36.5% DHA and 23.6% EPA of total fatty acid, thus making it best alternative for EPA and DHA production commercially (Hamilton, Powers, Napier, & Sayanova, 2016). Selection of potent strains, optimum environmental conditions, and efficient extraction techniques are important for increasing the production of these valuable compounds (Serive et al., 2012). Microalgal species like *Chlorella, Spirulina, Porphyridium cruentum, Phaeodactylum tricornutum, Pavlova lutheri,* and *Arthrospira platensis* have shown potential of high PUFA accumulation and thus can be used in production of novel dietary and pharmaceutical supplement (Sathasivam et al., 2019).

3.7 Antioxidants

Antioxidants are vital compounds responsible for sequestration of reactive oxygen species (ROS) produced under several stress conditions (Das & Roychoudhury, 2014). Oxidative damage to macromolecules can lead to cancer, arthritis, aging, and cardiovascular diseases in humans (Liguori et al., 2018). Antioxidant scavenges free radicals, protects macromolecules from injury, and repairs the oxidative damage, thereby helping in protection of biomolecules from oxidative damage (Engwa, 2018; Tan et al., 2018). Microalgal biomass rich in antioxidants has multifarious role in food and pharmaceutical industries. Antioxidants (carotenoids and vitamins) (Plaza, Herrero, Alejandro Cifuentes, & Ibanez, 2009) are present in microalgae in good amount and can be easily supplied in diet to protect the human body from several chronic diseases (Tan et al., 2018). Natural and synthetic antioxidants incorporated in food and pharmaceutical supplements can prevent the peroxidation of macromolecules (Taghvaei & Jafari, 2015). Antioxidants like dimethylsulfoniopropionate (de Morais et al., 2015) and mycosporine amino acids (Shick & Dunlap, 2002) produced from *Porphyra columbina* (Korbee Peinado, Abdala Díaz, Figueroa, & Helbling, 2004), *Isochrysis galbana* (Spielmeyer, Gebser, & Pohnert, 2011) efficiently serves as UV radiation blockers (Bhatia et al., 2011) and thus help in protecting body from UV radiation damage. Huge potential of microalgae to synthesize bioactive compounds (Khan et al., 2018) not only increases the nutritional value (Byrd, 2001) but also increases their economic feasibility for pharmaceutical, food, cosmetic, and nutraceutical industries (Fu et al., 2017).

3.8 Volatile compounds (VOCs)

The interest on the use of volatile compounds is increasing because of their diverse function in the biological system García, de Vicente, & Galán, 2017). More than 105 volatile compounds were produced by *Chlorella vulgaris,* and only few of them were identified (30) (Abdel-Baky, Shallan, El-Baroty, & El-Baz, 2002). Volatile compounds produced by microalgae are used as flavoring agents and also act as the natural alternative and are known to have antimicrobial and anticancer properties (Andrade et al., 2018). *Spirulina strains* were also studied for volatile compound production, and on chemical profile analysis, hydrocarbons were majorly found in *Spirulina* (Milovanović et al., 2015). Both *Chlorella* and *Spirulina* contain sufficient quantity of VOCs, which can be added in food products to perform biological functions in the body (Andrade et al., 2018).

3.9 Phytosterol

Phytosterol in microalgae has fundamental role in maintaining membrane integrity and is gaining interest because of simple production from microalgae (Ahmed, 2015; Levasseur, Perré, & Pozzobon, 2020). Humans cannot synthesize phytosterol, but several microalgae have shown potential to synthesize sterol, which can become an important component of human diet (Luo, Su, & Zhang, 2015). Phytosterols from microalgae are known to maintain membrane permeability; reduce blood cholesterol; prevent inflammation, cancer, and diabetes; and also offset heart diseases in humans (Hamed, Özogul, Özogul, & Regenstein, 2015). Ergosterol and 7-dehydroporiferasterol are the two phytosterol representing 45% of total sterol production in microalgae (Andrade et al., 2018; Chu et al., 2008). Total cholesterol level decreases in humans as phytosterol prevents the absorption of cholesterol from the intestine (Jesch & Carr, 2017). Phytosterols because of different health benefits can be used in both food and pharmaceutical industries (Fernandes & Cabral, 2007). *Isochrysis galbana*, *Nannochloropsis gaditana*, *Phaeodactylum tricornutum*, *Pavlova lutheri*, *Chaetoceros*, *Thalassiosira*, and *Tetraselmis* are the microalgae species commonly used for the production of phytosterol (Ahmed, Zhou, & Schenk, 2015; García et al., 2017). Microalgae serve as the cost-effective source of several compounds and assist in eliminating the use of multiple food and pharmaceutical supplements for proficient growth of the body (Luo et al., 2015).

4. Role of bioactive compound in health promotion

Microalgae as a source of functional metabolites are important for maintaining proper health of organisms (Barkia et al., 2019). Compounds produced by microalgae reduce heart-related diseases and have antioxidant, antiinflammation, and antiaging properties (Barkia et al., 2019). The diverse functional roles performed by bioactive compounds are as follows.

4.1 Anticancer activity

Cancer is chronic disease that is responsible for large number of deaths globally. The problem of cancer is rising globally (Bray et al., 2018), but use of bioactive compounds from microalgae can help to change the outlook against the disease. Various types (200) of cancer diseases have been identified, but efficient treatment is still lacking in both developing and developed countries

(Brown, Goldie, Draisma, Harford, & Lipscomb, 2006). The challenge is to discover the drugs that are reliable and effectively act against the disease in elegant manner. Bioactive compound, stigmasterol produced by *Navicula incerta*, inhibits the activity of cancerous cells by promoting cell apoptosis (Kim, Li, Kang, Ryu, & Kim, 2014). Bioactive compounds such as lutein, zeaxanthin, fucoxanthin, β- carotene (Sathasivam & Ki, 2018), polysaccharides (El-Hack et al., 2019), EPA, DHA (Andrade et al., 2018), vitamin E (Talero et al., 2015), phytosterol (Sanjeewa et al., 2016), and polyphenols (James & Thomas, 2019) from different microalgae (Table 1) control the growth of malignant cells in humans (Martínez Andrade, Lauritano, Romano, & Ianora, 2018; Talero et al., 2015). Anticancer activity of bioactive compounds is due to several mechanisms viz., antioxidants, resist cell cycle, inhibit angiogenesis, prevent spread of cancerous cells from the point of formation to other body parts, and also stimulate death of the cell (Diaconeasa, Frond, Ştirbu, Rugina, & Socaciu, 2018; Subramaniam, Selvaduray, & Radhakrishnan, 2019). Bioactive compounds incorporated into pharmaceutical and dietary supplements can enhance the action of killer cell, induce cell-mediated cytotoxicity, boost defense responses of immune system, and arrest growth of tumor cells (Abotaleb et al., 2019; Trejo-Solís et al., 2013). Fucoxanthin from microalgae arrests growth of malignant cells and also suppresses genes responsible for cancer cell growth without affecting the process of tumor cell apoptosis (Kumar, Hosokawa, & Miyashita, 2013; Martin, 2015). Microalgae being the storehouse of myriad of bioactive compound can certainly overcome or inhibit the growth of cancer disease, but the need is to exploit their use on wide scale to make it more reliable and feasible for chronic disease treatment (Alves et al., 2018; Galasso et al., 2019). Thus utilizing bioactive compounds for cancer treatment not only averts the side effects but also can decline the cost of treatment process and improve health in effectual manner.

4.2 Antimicrobial activities

Pathogenic organisms are responsible for severe disease in plants, animals, and human beings (Wilkinson et al., 2011). These organisms hinder several metabolic processes and disturb the balance of the body (Eisenreich, Heesemann, Rudel, & Goebel, 2013). To promote body growth, antibiotics are used, which has ultimately led to increased resistance of pathogens toward antibiotics (Manyi-Loh, Mamphweli, Meyer, & Okoh, 2018). Synthetic antibiotics because of several side effects, huge cost of production, and development of resistant pathogenic strains have led to extensive search for natural products. The search for natural metabolites that have antimicrobial activity is rising so that dependence on artificial antibiotics can be minimized. Natural metabolites mainly from microalgae can help to minimize the use of artificial antibiotics that have several side effects on human health. Bioactive compounds with no side effects and ability to hamper pathogens

TABLE 1 Production of several bioactive compounds from diverse microalgal species.

Bioactive compounds	Microalgal species	Functional roles	References
Vitamin A	*Porphyridium cruentum, Dunaliella salina*	Vitamin A: healthy eye sight and teeth and immune system	Mus et al. (2013), Hosseini Tafreshi and Shariati (2009), and Sathasivam and Juntawong (2013)
Vitamin B1, Vitamin B2	*Spirulina, Dunaliella tertiolecta, Dunaliella salina*	Vitamin B1: maintains health of brain and heart and prevents beriberi disease. Vitamin B2: energy production, prevents DNA damage and lung cancer	Richmond (1988), Fabregas and Herrero (1990), and Hosseini Tafreshi and Shariati (2009)
Vitamin B3, Vitamin B5	*Tetraselmis suecica*	Vitamin B3: prevents migraine headache and nonmelanoma skin cancer and promotes development of heart and nervous system. Vitamin B5: maintains health of skin hairs and digestive system and promotes blood cell formation	Fabregas and Herrero (1990)
Vitamin B6	*Tetraselmis suecica, Dunaliella salina*	Vitamin B6: promotes red blood cell formation and brain functions, maintains health of spleen and lymph nodes, and reduces depression	Fabregas and Herrero (1990) and Hosseini Tafreshi and Shariati (2009)
Vitamin B12	*Spirulina, Dunaliella tertiolecta, Chlorella*	Vitamin B12: genetic material formation, synthesis of red blood cells and fatty acid	Richmond (1988), Fabregas and Herrero (1990), and Watanabe, Takenaka, Kittaka-Katsura, Ebara, and Miyamoto (2002)
Vitamin C	*Chlorella, Dunaliella, Tetraselmis suecica*	Vitamin C: reduces oxidative damage, regulates health of bones and teeth, and accelerates	Barbosa et al. (2005), Running, Severson, and Schneider (2002), and Fabregas and Herrero (1990)
	Porphyridium cruentum	wound healing	Mus et al. (2013)
Vitamin E	*Dunaliella tertiolecta Haslea ostrearia, Porphyridium cruentum Chlorella sorokiniana*	Vitamin E: protects from heart disease and cancer and also acts as antioxidant	Fabregas and Herrero (1990), Khan et al. (2018), and Mus et al. (2013) Koyande et al. (2019)

Compound	Microalgae species	Health benefits	References
Fatty acids	*Cryptecodinium, Schizochytrium, Thraustochytrids, and Ulkenia*	Prevent cardiovascular disease, atherosclerosis and inflammation maintain immune system health and brain formation	Wynn, Behrens, Sundararajan, Hansen, and Apt (2010)
EPA	*Phaeodactylum tricornutum*	Reduce depression, prevent inflammation, improve eye and brain health	Hamilton, Haslam, Napier, and Sayanova (2014), Draaisma et al. (2013), Chauton, Kjell, Niels, Ragnar, and Hans (2015), and Koller, Muhr, and Braunegg (2014)
DHA	*Phaeodactylum tricornutum*	Prevent Alzheimer's disease, reduce heart disease, blood pressure, and risk of preterm birth	Chauton et al. (2015)
Fucoxanthin	*Cylindrotheca closterium and Phaeodactylum tricornutum*	Protect from depression, cancer, ageing, diabetes, obesity and inflammation	Kim, Jung, Kwon, Cha, and Um (2012)
Phycocynin	*Spirulina sp.*	Protect liver and kidney, promote immunity, and enhance functioning of immune system	Spolaore, Joannis-Cassan, Duran, and Isambert (2006)
Astaxanthin	*Haematococcus pluvialis*	Protect from cholesterol, hypertension, diabetes inflammation, and Alzheimer's disease	Lorenz and Cysewski (2000) and Sathasivam, Radhakrishnan, Hashem, and Abd Allah (2017)
Xanthophyll	*Spirulina sp.*	Maintain skin health, protect retina, and reduce effects of Alzheimer's disease	Richmond (1988)
Lutein	*Chlorella sorokiniana*	Maintain eye health, protect from cancer, atherosclerosis, diabetes and heart disease	Koyande et al. (2019)
Phycobiliproteins	*Porphyridium sp., Spirulia sp., and Aphanizomenon flos-aquae*	Protect liver and enhance immune function	Gouveia, Batista, Sousa, Raymundo, and Bandarra (2008)
Phytosterol	*Dunaliella tertiolecta and Dunaliella salina Pavlova and Thalassiosira Chaetoceros sp. and Nannochloropsis sp., Pavlova lutheri, Tetraselimis sp.*	Reduce cholesterol, atherosclerosis, inflammation and heart diseases	Borowitzka (2013) Luo et al. (2015) and Santhosh, Dhandapani, and Hemalatha (2016) Ahmed et al. (2015)

growth should be utilized for the proper growth of the body (Mostafa, 2012). Discovering of the novel compounds having antimicrobial activity can improve human health in efficient manner as resistance to antibiotic in humans is rising due to continuous use of antibiotics (Cheesman, Ilanko, Blonk, & Cock, 2017; Fair & Tor, 2014). The first antimicrobial compound "chlorellin" isolated from *Chlorella* inhibits the activity of both Gram-positive and Gram-negative bacteria (Ghasemi, Moradian, Mohagheghzadeh, Shokravi, & Morowvat, 2007). Bioactive compounds attack specific sites by damaging cell membrane that leads to outflow of material from the cell causing death of the bacterial cell (de Morais et al., 2015). Bioactive compounds having antimicrobial properties hinder cell wall formation and prevent bacteria from peptidoglycan formation, which is essential for the survival of bacteria (Bush, 2012). Bioactive compounds repress transpeptidation, destabilize the cytoplasmic membrane of bacterial cell (Liu & Breukink, 2016), and also hinder the DNA gyrase synthesis having essential role in uncoiling of DNA strands. Antimicrobial activity of various compounds like lutein, α-carotene, zeaxanthin (Kilic, Erdem, & Donmez, 2019), fucoxanthin (Karpiński & Adamczak, 2019) fatty acid, and polysaccharides obtained from microalgae can reduce the use of antibiotics against pathogens (de Morais et al., 2015; Falaise et al., 2016; Martínez-Francés & Escudero-Oñate, 2018). Bioactive compounds such as fatty acid are responsible for repressing the growth of pathogenic organisms in the body. Fatty acid's ability to arrest growth of bacteria depends upon the degree of unsaturation and length of fatty acid chain (Thangavel et al., 2018; Thevenieau & Nicaud, 2013). Fatty acid having more than 10 carbon atoms leads to protoplast lysis in bacteria and ultimately leads to death of bacteria (Mostafa, 2012) and thus can assist in protecting human body from pathogenic organisms.

4.3 Antiinflammatory activity

Inflammation is quick reaction of cell, which can be caused by pathogens or some agents. Body responds to the stimuli by recognizing the agent and endeavor to neutralize the agent quickly (Hilleman, 2004). Inflammation leads to redness, swelling, and pain at the site of attack by the pathogen (Chen et al., 2018). Antiinflammatory compounds reduce swelling, assist in healing, and improve immune system for protection against pain (de Morais et al., 2015). Pigments like carotenoids, fatty acids, and fucoxanthin show antiinflammatory role by reduce swelling, assist in healing, and improve immune system for protection against pain (Sathasivam & Ki, 2018). Phycocyanin also shows antiinflammatory action by preventing the release of histidine in the body. Bioactive compound like cyanovirin from microalgae is also known for antimicrobial and antiinflammatory activities with enough capacity to reduce and prevent numerous other diseases.

4.4 Protection against degenerative diseases

Microalgae consist of broad range of bioactive compounds having potent roles in augmenting proper health (Galasso et al., 2019) and are also attaining wide consideration at industrial level (Camacho et al., 2019). Oxidative damage posed by reactive oxygen species can cause severe diseases like obesity, high blood pressure, aging, cancer, and cardiovascular disease (Liguori et al., 2018). Chronic inflammation can led to severe neurodegenerative diseases in the human body (Amor, Puentes, Baker, & Van Der Valk, 2010). Vitamins (Pooja & Sunita, 2014), PUFA, and carotenoids prevent macromolecules from damage by scavenging of free radicals, thereby protecting human body from degenerative and cardiovascular diseases (Raposo, De Morais, & De Morais, 2015). Thus bioactive compounds from microalgae are neuroprotective and can efficiently cure or prevent neurodegenerative diseases in humans (Olasehinde, Olaniran, & Okoh, 2017).

5. Conclusion

Microalgae known as tiny factories are the frugal and sustainable feedstock for the production of several bioactive compounds. Utilization of microalgae biomass for bioactive compound production can surely reduce the dependence on synthetic compounds, which are costly and have side effects on the body. Escalating population, changing food habits, and increasing pollution and workload stress are boosting the chronic diseases incidence in the human beings. Bioactive compounds are vital for the proper functioning of body and are also helpful in averting numerous chronic diseases in human body. Bioactive compounds from microalgae enhance body growth, protect skin, improve blood circulation and vision, promote growth of brain, and also provide protection from oxidative stress. Further research is required for isolation and identification of metabolites from microalgae and mechanism involved in promoting growth and prevention of several diseases in humans.

References

Abdel-Aziz, S. M., Aeron, A., & Kahil, T. A. (2016). Health benefits and possible risks of herbal medicine. In *Microbes in food and health* (pp. 97–116). Cham: Springer.

Abdel-Baky, H. H., Shallan, M. A., El-Baroty, G., & El-Baz, F. K. (2002). Volatile compounds of the microalga *Chlorella vulgaris* and their phytotoxic effect. *Pakistan Journal of Biological Sciences*, 5(1), 61–65.

Abotaleb, M., Samuel, S. M., Varghese, E., Varghese, S., Kubatka, P., Liskova, A., et al. (2019). Flavonoids in cancer and apoptosis. *Cancers*, 11(1), 28.

Adarme-Vega, T. C., Lim, D. K., Timmins, M., Vernen, F., Li, Y., & Schenk, P. M. (2012). Microalgal biofactories: A promising approach towards sustainable omega-3 fatty acid production. *Microbial Cell Factories*, 11(1), 96.

Ahmadi, A., Zorofchian Moghadamtousi, S., Abubakar, S., & Zandi, K. (2015). Antiviral potential of algae polysaccharides isolated from marine sources: A review. *BioMed Research International, 2015*, 10. 825203.

Ahmed, F. (2015). *Induction of carotenoid and phytosterol accumulation in microalgae* (M.Sc. thesis). School of Agriculture and Food Sciences, The University of Queensland.

Ahmed, F., Zhou, W., & Schenk, P. M. (2015). *Pavlova lutheri* is a high-level producer of phytosterols. *Algal Research, 10*, 210–217.

Alam, T., Najam, L., & Al-Harrasi, A. (2018). Extraction of natural pigments from marine algae. *Journal of Agricultural and Marine Sciences, 23*(1), 81–91.

Altemimi, A., Lakhssassi, N., Baharlouei, A., Watson, D. G., & Lightfoot, D. A. (2017). Phytochemicals: Extraction, isolation, and identification of bioactive compounds from plant extracts. *Plants, 6*(4), 42.

Alves, C., Silva, J., Pinteus, S., Gaspar, H., Alpoim, M. C., Botana, L. M., et al. (2018). From marine origin to therapeutics: The antitumor potential of marine algae-derived compounds. *Frontiers in Pharmacology, 9*, 777.

Amor, S., Puentes, F., Baker, D., & Van Der Valk, P. (2010). Inflammation in neurodegenerative diseases. *Immunology, 129*(2), 154–169.

Andrade, L. M., Andrade, C. J., Dias, M., Nascimento, C. A. O., & Mendes, M. A. (2018). *Chlorella* and *Spirulina* microalgae as sources of functional foods. *Nutraceuticals, and Food Supplements*, 45–58.

Arad, S. M., & Levy-Ontman, O. (2010). Red microalgal cell-wall polysaccharides: Biotechnological aspects. *Current Opinion in Biotechnology, 21*(3), 358–364.

Atanasov, A. G., Waltenberger, B., Pferschy-Wenzig, E. M., Linder, T., Wawrosch, C., Uhrin, P., et al. (2015). Discovery and resupply of pharmacologically active plant-derived natural products: A review. *Biotechnology Advances, 33*(8), 1582–1614.

Atanassova, M., Georgieva, S., & Ivancheva, K. (2011). Total phenolic and total flavonoid contents, antioxidant capacity and biological contaminants in medicinal herbs. *Journal of the University of Chemical Technology & Metallurgy, 46*(1).

Baliga, R., & Powers, S. E. (2010). Sustainable algae biodiesel production in cold climates. *International Journal of Chemical Engineering, 2010*, 13. 102179.

Barbosa, M. J., Zijffers, J. W., Nisworo, A., Vaes, W., van Schoonhoven, J., & Wijffels, R. H. (2005). Optimization of biomass, vitamins, and carotenoid yield on light energy in a flat-panel reactor using the A-stat technique. *Biotechnology and Bioengineering, 89*, 233–242.

Barkia, I., Saari, N., & Manning, S. R. (2019). Microalgae for high-value products towards human health and nutrition. *Marine Drugs, 17*(5), 304.

Benavente-Valdés, J. R., Aguilar, C., Contreras-Esquivel, J. C., Méndez-Zavala, A., & Montañez, J. (2016). Strategies to enhance the production of photosynthetic pigments and lipids in chlorophycae species. *Biotechnology Reports, 10*, 117–125.

Benedetti, M., Vecchi, V., Barera, S., & Dall'Osto, L. (2018). Biomass from microalgae: The potential of domestication towards sustainable biofactories. *Microbial Cell Factories, 17*(1), 1–18.

Bhalamurugan, G. L., Valerie, O., & Mark, L. (2018). Valuable bioproducts obtained from microalgal biomass and their commercial applications: A review. *Environmental Engineering Research, 23*(3), 229–241.

Bhatia, S., Garg, A., Sharma, K., Kumar, S., Sharma, A., & Purohit, A. P. (2011). Mycosporine and mycosporine-like amino acids: A paramount tool against ultra violet irradiation. *Pharmacognosy Reviews, 5*(10), 138.

Bhattacharjee, M. (2016). Pharmaceutically valuable bioactive compounds of algae. *Asian Journal of Pharmaceutical and Clinical Research, 9*, 43–47.

Bhattacharya, S. (2013). *Health promoting effects of bioactive compounds in plants: Targeting type 2 diabetes* (Ph.D. thesis). Department of food Science Faculty of Science and Technology, Aarhus University. July.

Bilal, M., Rasheed, T., Ahmed, I., Iqbal, H. M. N., & Sada, E. G. (2017). High-value compounds from microalgae with industrial exploitability—A review. *Frontiers in Bioscience, 9*, 319–342.

Bisht, M., Dhasmana, D. C., & Bist, S. S. (2010). Angiogenesis: Future of pharmacological modulation. *Indian Journal of Pharmacology, 42*(1), 2.

Bleakley, S., & Hayes, M. (2017). Algal proteins: Extraction, application, and challenges concerning production. *Foods, 6*(5), 33.

Borowitzka, M. A. (2013). High-value products from microalgae—Their development and commercialization. *Journal of Applied Phycology, 25*, 743–756.

Boxall, A. B. (2004). The environmental side effects of medication. *EMBO Reports, 5*(12), 1110–1116.

Bray, F., Ferlay, J., Soerjomataram, I., Siegel, R. L., Torre, L. A., & Jemal, A. (2018). Global cancer statistics 2018: GLOBOCAN estimates of incidence and mortality worldwide for 36 cancers in 185 countries. *CA: A Cancer Journal for Clinicians, 68*(6), 394–424.

Brown, M. L., Goldie, S. J., Draisma, G., Harford, J., & Lipscomb, J. (2006). Health service interventions for cancer control in developing countries. *Disease Control Priorities in Developing Countries, 2*, 569–589.

Bungau, S., Abdel-Daim, M. M., Tit, D. M., Ghanem, E., Sato, S., Maruyama-Inoue, M., et al. (2019). Health benefits of polyphenols and carotenoids in age-related eye diseases. *Oxidative Medicine and Cellular Longevity, 2019*, 22. 9783429.

Buscemi, S., Corleo, D., Di Pace, F., Petroni, M. L., Satriano, A., & Marchesini, G. (2018). The effect of lutein on eye and extra-eye health. *Nutrients, 10*(9), 1321.

Bush, K. (2012). Antimicrobial agents targeting bacterial cell walls and cell membranes. *Revue Scientifique et Technique, 31*(1), 43–56.

Byrd, S. J. (2001). Using antioxidants to increase shelf life of food products. *Cereal Foods World, 46*(2), 48–53.

Camacho, F., Macedo, A., & Malcata, F. (2019). Potential industrial applications and commercialization of microalgae in the functional food and feed industries: A short review. *Marine Drugs, 17*(6), 312.

Canizales, J. R., Rodríguez, G. R., Avila, J. A., Saldaña, A. M., Parrilla, E. A., Ochoa, M. A., et al. (2018). Encapsulation to protect different bioactives to be used as nutraceuticals and food ingredients. In *Bioactive molecules in food* (pp. 1–20). Cham: Springer.

Caporgno, M. P., & Mathys, A. (2018). Trends in microalgae incorporation into innovative food products with potential health benefits. *Frontiers in Nutrition, 5*, 58.

Chauton, M. S., Kjell, I. R., Niels, H. N., Ragnar, T., & Hans, T. K. (2015). A techno-economic analysis of industrial production of marine microalgae as a source of EPA and DHA-rich raw material for aquafeed: Research challenges and possibilities. *Aquaculture, 436*, 95–103.

Cheesman, M. J., Ilanko, A., Blonk, B., & Cock, I. E. (2017). Developing new antimicrobial therapies: Are synergistic combinations of plant extracts/compounds with conventional antibiotics the solution? *Pharmacognosy Reviews, 11*(22), 57.

Chen, L., Deng, H., Cui, H., Fang, J., Zuo, Z., Deng, J., et al. (2018). Inflammatory responses and inflammation-associated diseases in organs. *OncoTarget, 9*(6), 7204.

Chu, F. L. E., Lund, E. D., Littreal, P. R., Ruck, K. E., Harvey, E., Le Coz, J. R., et al. (2008). Sterol production and phytosterol bioconversion in two species of heterotrophic protists, *Oxyrrhis marina* and *Gyrodinium dominans*. *Marine Biology, 156*(2), 155–169.

Coêlho, D. D. F., Tundisi, L. L., Cerqueira, K. S., Rodrigues, J. R. D. S., Mazzola, P. G., Tambourgi, E. B., et al. (2019). Microalgae: Cultivation aspects and bioactive compounds. *Brazilian Archives of Biology and Technology, 62*.

Costa, J. A. V., & de Morais, M. G. (2014). An open pond system for microalgal cultivation. In *Biofuels from algae* (pp. 1–22). Elsevier.

Costa, J. A., & Morais, M. G. (2013). Microalgae for food production. In C. R. Soccol, A. Pandey, & C. Larroche (Eds.), *Fermentation process engineering in the food industry* (p. 486). Boca Raton, USA: Taylor & Francis.

Cunha, L., & Grenha, A. (2016). Sulfated seaweed polysaccharides as multifunctional materials in drug delivery applications. *Marine Drugs, 14*(3), 42.

da Silva Vaz, B., Moreira, J. B., de Morais, M. G., & Costa, J. A. V. (2016). Microalgae as a new source of bioactive compounds in food supplements. *Current Opinion in Food Science, 7*, 73–77.

Dall'Osto, L., Holt, N. E., Kaligotla, S., Fuciman, M., Cazzaniga, S., Carbonera, D., et al. (2012). Zeaxanthin protects plant photosynthesis by modulating chlorophyll triplet yield in specific light-harvesting antenna subunits. *Journal of Biological Chemistry, 287*(50), 41820–41834.

Das, K., & Roychoudhury, A. (2014). Reactive oxygen species (ROS) and response of antioxidants as ROS-scavengers during environmental stress in plants. *Frontiers in Environmental Science, 2*, 53.

Dasgupta, C. N. (2015). Algae as a source of phycocyanin and other industrially important pigments. In *Algal biorefinery: An integrated approach* (pp. 253–276). Cham: Springer.

De Carvalho, C. C., & Caramujo, M. J. (2017). Carotenoids in aquatic ecosystems and aquaculture: A colorful business with implications for human health. *Frontiers in Marine Science, 4*, 93.

de Jesus Raposo, M. F., de Morais, R. M. S. C., & de Morais, A. M. M. B. (2013). Health applications of bioactive compounds from marine microalgae. *Life Sciences, 93*(15), 479–486.

de Morais, M. G., Stillings, C., Dersch, R., Rudisile, M., Pranke, P., Costa, J. A. V., et al. (2010). Preparation of nanofibers containing the microalga *Spirulina* (*Arthrospira*). *Bioresource Technology, 101*(8), 2872–2876.

de Morais, M. G., Vaz, B. D. S., de Morais, E. G., & Costa, J. A. V. (2015). Biologically active metabolites synthesized by microalgae. *BioMed Research International, 2015*.

de Souza, M. P., Sanchez-Barrios, A., Rizzetti, T. M., Benitez, L. B., Hoeltz, M., de Souza Schneider, R. D. C., et al. (2019). Concepts and trends for extraction and application of microalgae carbohydrates. In *Microalgae-from physiology to application* IntechOpen.

Deng, Z., Hu, Q., Lu, F., Liu, G., & Hu, Z. (2008). Colony development and physiological characterization of the edible blue-green alga, *Nostoc sphaeroides* (Nostocaceae, Cyanophyta). *Progress in Natural Science, 18*(12), 1475–1484.

Dhankhar, J., Kadian, S. S., & Sharma, A. (2012). Astaxanthin: A potential carotenoid. *International Journal of Pharmaceutical Sciences and Research, 3*(5), 1246.

Diaconeasa, Z. M., Frond, A. D., Ştirbu, I., Rugina, D., & Socaciu, C. (2018). Anthocyanins-smart molecules for cancer prevention. In T. Asao, & M. D. Asaduzzaman (Eds.), *Phytochemicals: Source of antioxidants and role in disease prevention* (p. 75). IntechOpen. https://doi.org/10.5772/intechopen.79613.

Draaisma, R. B., Wijffels, R. H., Slegers, P. M., Brentner, L. B., Roy, A., & Barbosa, M. J. (2013). Food commodities from microalgae. *Current Opinion in Biotechnology, 24*, 169–177.

Dyall, S. C. (2015). Long-chain omega-3 fatty acids and the brain: A review of the independent and shared effects of EPA, DPA and DHA. *Frontiers in Aging Neuroscience, 7*, 52.

Eisenreich, W., Heesemann, J., Rudel, T., & Goebel, W. (2013). Metabolic host responses to infection by intracellular bacterial pathogens. *Frontiers in Cellular and Infection Microbiology, 3*, 24.

El-Hack, M. E. A., Abdelnour, S., Alagawany, M., Abdo, M., Sakr, M. A., Khafaga, A. F., et al. (2019). Microalgae in modern cancer therapy: Current knowledge. *Biomedicine & Pharmacotherapy, 111*, 42–50.

Elshouny, W. A. E. F., El-Sheekh, M. M., Sabae, S. Z., Khalil, M. A., & Badr, H. M. (2017). Antimicrobial activity of *Spirulina platensis* against aquatic bacterial isolates. *Journal of Microbiology, Biotechnology and Food Sciences*, 6(5), 1203.

Engwa, G. A. (2018). Free radicals and the role of plant phytochemicals as antioxidants against oxidative stress-related diseases. *Phytochemicals: Source of Antioxidants and Role in Disease Prevention*, 49–74. BoD–Books on Demand.

Fabregas, J., & Herrero, C. (1990). Vitamin content of four marine microalgae. Potential use as source of vitamins in nutrition. *Journal of Industrial Microbiology*, 5, 259–263.

Fair, R. J., & Tor, Y. (2014). Antibiotics and bacterial resistance in the 21st century. *Perspectives in Medicinal Chemistry*, 6, PMC–S14459.

Falaise, C., François, C., Travers, M. A., Morga, B., Haure, J., Tremblay, R., et al. (2016). Antimicrobial compounds from eukaryotic microalgae against human pathogens and diseases in aquaculture. *Marine Drugs*, 14(9), 159.

Falquet, J., & Hurni, J. P. (1997). *The nutritional aspects of Spirulina*. Antenna Foundation. Available online at: https://www.antenna.ch/wp-content/uploads/2017/03/AspectNut_UK.pdf (Accessed 25 July 2017).

Felmeden, D. C., Blann, A. D., & Lip, G. Y. H. (2003). Angiogenesis: Basic pathophysiology and implications for disease. *European Heart Journal*, 24(7), 586–603.

Fernandes, P., & Cabral, J. M. S. (2007). Phytosterols: Applications and recovery methods. *Bioresource Technology*, 98(12), 2335–2350.

Fu, W., Nelson, D. R., Yi, Z., Xu, M., Khraiwesh, B., Jijakli, K., et al. (2017). Bioactive compounds from microalgae: Current development and prospects. In *Vol. 54. Studies in natural products chemistry* (pp. 199–225). Elsevier.

Galasso, C., Gentile, A., Orefice, I., Ianora, A., Bruno, A., Noonan, D. M., et al. (2019). Microalgal derivatives as potential nutraceutical and food supplements for human health: A focus on cancer prevention and interception. *Nutrients*, 11(6), 1226.

Gammone, M. A., Pluchinotta, F. R., Bergante, S., Tettamanti, G., & D'Orazio, N. (2017). Prevention of cardiovascular diseases with carotenoids. *Frontiers in Bioscience (Scholar Edition)*, 9, 165–171.

Gammone, M. A., Riccioni, G., & D'Orazio, N. (2015). Marine carotenoids against oxidative stress: Effects on human health. *Marine Drugs*, 13(10), 6226–6246.

Ganesan, P., Matsubara, K., Sugawara, T., & Hirata, T. (2013). Marine algal carotenoids inhibit angiogenesis by down-regulating FGF-2-mediated intracellular signals in vascular endothelial cells. *Molecular and Cellular Biochemistry*, 380(1–2), 1–9.

García, J. L., de Vicente, M., & Galán, B. (2017). Microalgae, old sustainable food and fashion nutraceuticals. *Microbial Biotechnology*, 10(5), 1017–1024.

Generalić Mekinić, I., Skroza, D., Šimat, V., Hamed, I., Čagalj, M., & Popović Perković, Z. (2019). Phenolic content of brown algae (pheophyceae) species: Extraction, identification, and quantification. *Biomolecules*, 9(6), 244.

Ghasemi, Y., Moradian, A., Mohagheghzadeh, A., Shokravi, S., & Morowvat, M. H. (2007). Antifungal and antibacterial activity of the microalgae collected from paddy fields of Iran: Characterization of antimicrobial activity of *Chroococcus dispersus*. *Journal of Biological Sciences*, 7(6), 904–910.

Ghorani-Azam, A., Riahi-Zanjani, B., & Balali-Mood, M. (2016). Effects of air pollution on human health and practical measures for prevention in Iran. *Journal of Research in Medical Sciences: The Official Journal of Isfahan University of Medical Sciences*, 21.

Ghosh, D., & Smarta, R. B. (2016). *Pharmaceuticals to nutraceuticals: A shift in disease prevention*. CRC Press.

Gorissen, S. H., Crombag, J. J., Senden, J. M., Waterval, W. H., Bierau, J., Verdijk, L. B., et al. (2018). Protein content and amino acid composition of commercially available plant-based protein isolates. *Amino Acids, 50*(12), 1685–1695.

Gouveia, L., Batista, A. P., Sousa, I., Raymundo, A., & Bandarra, N. M. (2008). Microalgae in novel food product. In *Food chemistry research developments* Nova Science Publishers, Inc.

Greenwood, D. (Ed.). (2012). Medical microbiology, with student consult online access. In *Vol. 18. Medical microbiology* Elsevier Health Sciences.

Grundy, M. M. L., Edwards, C. H., Mackie, A. R., Gidley, M. J., Butterworth, P. J., & Ellis, P. R. (2016). Re-evaluation of the mechanisms of dietary fibre and implications for macronutrient bioaccessibility, digestion and postprandial metabolism. *British Journal of Nutrition, 116*(5), 816–833.

Guerin, M., Huntley, M. E., & Olaizola, M. (2003). *Haematococcus astaxanthin*: Applications for human health and nutrition. *Trends in Biotechnology, 21*(5), 210–216.

Gujar, A., Cui, H., Ji, C., Kubar, S., & Li, R. (2019). Development, production and market value of microalgae products. *Applied Microbiology: Open Access, 5*, 162.

Habib, M. A. B. (2008). *Review on culture, production and use of Spirulina as food for humans and feeds for domestic animals and fish.* Food and Agriculture Organization of the United Nations.

Hadley, K. B., Ryan, A. S., Forsyth, S., Gautier, S., & Salem, N. (2016). The essentiality of arachidonic acid in infant development. *Nutrients, 8*(4), 216.

Hamed, I. (2016). The evolution and versatility of microalgal biotechnology: A review. *Comprehensive Reviews in Food Science and Food Safety, 15*(6), 1104–1123.

Hamed, I., Özogul, F., Özogul, Y., & Regenstein, J. M. (2015). Marine bioactive compounds and their health benefits: A review. *Comprehensive Reviews in Food Science and Food Safety, 14*(4), 446–465.

Hamidi, M., Kozani, P. S., Kozani, P. S., Pierre, G., Michaud, P., & Delattre, C. (2020). Marine bacteria versus microalgae: Who is the best for biotechnological production of bioactive compounds with antioxidant properties and other biological applications? *Marine Drugs, 18*(1), 28.

Hamilton, M., Haslam, R., Napier, J., & Sayanova, O. (2014). Metabolic engineering of microalgae for enhanced production of omega-3 long chain polyunsaturated fatty acids. *Metabolic Engineering, 22*, 3–9.

Hamilton, M. L., Powers, S., Napier, J. A., & Sayanova, O. (2016). Heterotrophic production of omega-3 long-chain polyunsaturated fatty acids by trophically converted marine diatom *Phaeodactylum tricornutum*. *Marine Drugs, 14*(3), 53.

Haoujar, I., Cacciola, F., Abrini, J., Mangraviti, D., Giuffrida, D., Oulad El Majdoub, Y., et al. (2019). The contribution of carotenoids, phenolic compounds, and flavonoids to the antioxidative properties of marine microalgae isolated from mediterranean morocco. *Molecules, 24*(22), 4037.

Hashimoto, M., & Hossain, S. (2018). Fatty acids: From membrane ingredients to signaling molecules. In *Biochemistry and health benefits of fatty acids* IntechOpen.

Hibbeln, J. R., & Gow, R. V. (2014). Omega-3 fatty acid and nutrient deficits in adverse neurodevelopment and childhood behaviors. *Child and Adolescent Psychiatric Clinics of North America, 23*(3), 555.

Higuera-Ciapara, I., Felix-Valenzuela, L., & Goycoolea, F. M. (2006). Astaxanthin: A review of its chemistry and applications. *Critical Reviews in Food Science and Nutrition, 46*(2), 185–196.

Hilleman, M. R. (2004). Strategies and mechanisms for host and pathogen survival in acute and persistent viral infections. *Proceedings of the National Academy of Sciences, 101*(suppl 2), 14560–14566.

Hosseini Tafreshi, A., & Shariati, M. (2009). *Dunaliella* biotechnology: Methods and applications. *Journal of Applied Microbiology, 107*(1), 14–35.

Hu, J., Nagarajan, D., Zhang, Q., Chang, J. S., & Lee, D. J. (2018). Heterotrophic cultivation of microalgae for pigment production: A review. *Biotechnology Advances, 36*(1), 54–67.

Hussian, A. E. M. (2018). The role of microalgae in renewable energy production: Challenges and opportunities. In M. Türkoğlu, U. Önal, & A. Ismen (Eds.), *Marine ecology: Biotic and abiotic interactions* (p. 257). IntechOpen. https://doi.org/10.5772/intechopen.73573.

Irvani, N., Hajiaghaee, R., & Zarekarizi, A. R. (2018). A review on biosynthesis, health benefits and extraction methods of fucoxanthin, particular marine carotenoids in algae. *Journal of Medicinal Plants, 3*(67), 6–30.

James, J., & Thomas, J. (2019). Anticancer activity of microalgae extract on human cancer cell line (MG-63). *Asian Journal of Pharmaceutical and Clinical Research, 12*(1), 139–142.

Jaswir, I., Noviendri, D., Hasrini, R. F., & Octavianti, F. (2011). Carotenoids: Sources, medicinal properties and their application in food and nutraceutical industry. *Journal of Medicinal Plant Research: Planta Medica, 5*(33), 7119–7131.

Jerez-Martel, I., García-Poza, S., Rodríguez-Martel, G., Rico, M., Afonso-Olivares, C., & Gómez-Pinchetti, J. L. (2017). Phenolic profile and antioxidant activity of crude extracts from microalgae and cyanobacteria strains. *Journal of Food Quality, 2017*.

Jesch, E. D., & Carr, T. P. (2017). Food ingredients that inhibit cholesterol absorption. *Preventive Nutrition and Food Science, 22*(2), 67.

Jiang, L., Wang, Y., Yin, Q., Liu, G., Liu, H., Huang, Y., et al. (2017). Phycocyanin: A potential drug for cancer treatment. *Journal of Cancer, 8*(17), 3416.

Kaczmarski, M., Cudowska, B., Sawicka-Żukowska, M., & Bobrus-Chociej, A. (2013). Supplementation with long chain polyunsaturated fatty acids in treatment of atopic dermatitis in children. *Advances in Dermatology and Allergology/Postępy Dermatologii I Alergologii, 30*(2), 103.

Karkos, P. D., Leong, S. C., Karkos, C. D., Sivaji, N., & Assimakopoulos, D. A. (2011). Spirulina in clinical practice: Evidence-based human applications. *Evidence-Based Complementary and Alternative Medicine, 2011*, 4. 531053.

Karpiński, T. M., & Adamczak, A. (2019). Fucoxanthin—An antibacterial carotenoid. *Antioxidants, 8*(8), 239.

Kaur, K. K., Allhbadia, G., & Singh, M. (2018). Synthesis and functional significance of poly unsaturated fatty acids (PUFA's) in body. *Acta Scientific Nutritional Health, 2*(4), 8.

Kaur, N., Chugh, V., & Gupta, A. K. (2014). Essential fatty acids as functional components of foods—A review. *Journal of Food Science and Technology, 51*(10), 2289–2303.

Khan, M. I., Shin, J. H., & Kim, J. D. (2018). The promising future of microalgae: Current status, challenges, and optimization of a sustainable and renewable industry for biofuels, feed, and other products. *Microbial Cell Factories, 17*(1), 36.

Kilic, N. K., Erdem, K., & Donmez, G. (2019). Bioactive compounds produced by *Dunaliella* species, antimicrobial effects and optimization of the efficiency. *Turkish Journal of Fisheries and Aquatic Sciences, 19*(11), 923–933.

Kim, S. M., Jung, Y. H., Kwon, O., Cha, K. H., & Um, B. H. (2012). A potential commercial source of fucoxanthin extracted from the microalga *Phaeodactylum tricornutum*. *Applied Biochemistry and Biotechnology, 166*, 1843–1855.

Kim, Y. S., Li, X. F., Kang, K. H., Ryu, B., & Kim, S. K. (2014). Stigmasterol isolated from marine microalgae *Navicula incerta* induces apoptosis in human hepatoma HepG2 cells. *BMB Reports, 47*(8), 433.

Koller, M., Muhr, A., & Braunegg, G. (2014). Microalgae as versatile cellular factories for valued products. *Algal Research, 6*, 52–63.199.

Korbee Peinado, N., Abdala Díaz, R. T., Figueroa, F. L., & Helbling, E. W. (2004). Ammonium and UV radiation stimulate the accumulation of mycosporine-like amino acids in *Porphyra columbina* (Rhodophyta) from Patagonia, Argentina. *Journal of Physiology, 40*, 248–259.

Koyande, A. K., Chew, K. W., Rambabu, K., Tao, Y., Chu, D. T., & Show, P. L. (2019). Microalgae: A potential alternative to health supplementation for humans. *Food Science and Human Wellness, 8,* 16–24.

Ku, C. S., Yang, Y., Park, Y., & Lee, J. (2013). Health benefits of blue-green algae: Prevention of cardiovascular disease and nonalcoholic fatty liver disease. *Journal of Medicinal Food, 16*(2), 103–111.

Kumar, S. R., Hosokawa, M., & Miyashita, K. (2013). Fucoxanthin: A marine carotenoid exerting anti-cancer effects by affecting multiple mechanisms. *Marine Drugs, 11*(12), 5130–5147.

Kumar, I., & Sharma, R. K. (2018). Production of secondary metabolites in plants under abiotic stress: An overview. *Significances of Bioengineering & Biosciences, 2,* 1–5.

Leon, D. A. (2008). Cities, urbanization and health. *International Journal of Epidemiology,* 4–8.

Levasseur, W., Perré, P., & Pozzobon, V. (2020). A review of high value-added molecules production by microalgae in light of the classification. *Biotechnology Advances,* 107545.

Li, Z., & Guo, M. (2018). Healthy efficacy of *Nostoc commune Vaucher. OncoTarget, 9*(18), 14669.

Li, L., Li, W., Kim, Y. H., & Lee, Y. W. (2013). 2013. *Chlorella vulgaris* extract ameliorates carbon tetrachloride-induced acute hepatic injury in mice. *Experimental and Toxicologic Pathology, 65*(1–2), 73–80.

Liguori, I., Russo, G., Curcio, F., Bulli, G., Aran, L., Della-Morte, D., et al. (2018). Oxidative stress, aging, and diseases. *Clinical Interventions in Aging, 13,* 757.

Liu, R. H. (2013). Dietary bioactive compounds and their health implications. *Journal of Food Science, 78*(s1), A18–A25.

Liu, Y., & Breukink, E. (2016). The membrane steps of bacterial cell wall synthesis as antibiotic targets. *Antibiotics, 5*(3), 28.

Lobo, V., Patil, A., Phatak, A., & Chandra, N. (2010). Free radicals, antioxidants and functional foods: Impact on human health. *Pharmacognosy Reviews, 4*(8), 118.

Lorenz, R. T., & Cysewski, G. R. (2000). Commercial potential for *Haematococcus* microalgae as a natural source of astaxanthin. *Trends in Biotechnology, 18,* 160–167.

Lunn, J., & Buttriss, J. L. (2007). Carbohydrates and dietary fibre. *Nutrition Bulletin, 32*(1), 21–64.

Luo, X., Su, P., & Zhang, W. (2015). Advances in microalgae-derived phytosterols for functional food and pharmaceutical applications. *Marine Drugs, 13*(7), 4231–4254.

Manyi-Loh, C., Mamphweli, S., Meyer, E., & Okoh, A. (2018). Antibiotic use in agriculture and its consequential resistance in environmental sources: Potential public health implications. *Molecules, 23*(4), 795.

Mares, J. (2016). Lutein and zeaxanthin isomers in eye health and disease. *Annual Review of Nutrition, 36,* 571–602.

Martin, L. J. (2015). Fucoxanthin and its metabolite fucoxanthinol in cancer prevention and treatment. *Marine Drugs, 13*(8), 4784–4798.

Martín, J., Kuskoski, E. M., Navas, M. J., & Asuero, A. G. (2017). Antioxidant capacity of anthocyanin pigments. In *Vol. 3. Flavonoids-from biosynthesis to human health* (pp. 205–255). IntechOpen.

Martínez Andrade, K. A., Lauritano, C., Romano, G., & Ianora, A. (2018). Marine microalgae with anti-cancer properties. *Marine Drugs, 16*(5), 165.

Martínez-Francés, E., & Escudero-Oñate, C. (2018). Cyanobacteria and microalgae in the production of valuable bioactive compounds. *Microalgal Biotechnology, 6,* 104–128.

Matos, J., Cardoso, C., Bandarra, N. M., & Afonso, C. (2017). Microalgae as healthy ingredients for functional food: A review. *Food & Function, 8*(8), 2672–2685.

Medina-Jaritz, N. B., Carmona-Ugalde, L. F., Lopez-Cedillo, J. C., & Ruiloba-De Leon, F. S. (2013). Antibacterial activity of methanolic extracts from *Dunaliella salina* and *Chlorella vulgaris. The FASEB Journal, 27*(S1), 1167.5. https://doi.org/10.1096/fasebj.27.1_supplement.1167.5.

Milovanović, I., Mišan, A., Simeunović, J., Kovač, D., Jambrec, D., & Mandić, A. (2015). Determination of volatile organic compounds in selected strains of cyanobacteria. *Journal of Chemistry*, *2015*.

Mišurcováa, L., Orsavováb, J., & Ambrožováa, J. V. (2014). Algal polysaccharides and health. In *Polysaccharides: Bioactivity and biotechnology*. Switzerland: Springer International Publishing.

Mogensen, K. M. (2017). Essential fatty acid deficiency. *Practical Gastroenterology*, *41*(6), 37–44.

Montero-Lobato, Z., Vázquez, M., Navarro, F., Fuentes, J. L., Bermejo, E., Garbayo, I., et al. (2018). Chemically-induced production of anti-inflammatory molecules in microalgae. *Marine Drugs*, *16*(12), 478.

Moreno-Garrido, I. (2008). Microalgae immobilization: Current techniques and uses. *Bioresource Technology*, *99*(10), 3949–3964.

Mostafa, S. S. (2012). Microalgal biotechnology: Prospects and applications. *Plant Science*, *12*, 276–314.

Mourelle, M. L., Gómez, C. P., & Legido, J. L. (2017). The potential use of marine microalgae and cyanobacteria in cosmetics and thalassotherapy. *Cosmetics*, *4*(4), 46.

Mus, F., Toussaint, J. P., Cooksey, K. E., Fields, M. W., Gerlach, R., Peyton, B. M., et al. (2013). Physiological and molecular analysis of carbon source supplementation and pH stress-induced lipid accumulation in the marine diatom *Phaeodactylum tricornutum*. *Applied Microbiology and Biotechnology*, *97*, 3625–3642.

Mysliwa-Kurdziel, B., & Solymosi, K. (2017). Phycobilins and phycobiliproteins used in food industry and medicine. *Mini Reviews in Medicinal Chemistry*, *17*(13), 1173–1193.

Nasri, H., Baradaran, A., Shirzad, H., & Rafieian-Kopaei, M. (2014). New concepts in nutraceuticals as alternative for pharmaceuticals. *International Journal of Preventive Medicine*, *5*(12), 1487.

Olasehinde, T. A., Olaniran, A. O., & Okoh, A. I. (2017). Therapeutic potentials of microalgae in the treatment of Alzheimer's disease. *Molecules*, *22*(3), 480.

Oliveira, L. D. L. D., Carvalho, M. V. D., & Melo, L. (2014). Health promoting and sensory properties of phenolic compounds in food. *Revista Ceres*, *61*, 764–779.

Panjiar, N., Mishra, S., Yadav, A. N., & Verma, P. (2017). Functional foods from cyanobacteria: An emerging source for functional food products of pharmaceutical importance. In *Microbial functional foods and nutraceuticals* (pp. 21–37). USA: Wiley-Blackwell.

Park, W. S., Kim, H. J., Li, M., Lim, D. H., Kim, J., Kwak, S. S., et al. (2018). Two classes of pigments, carotenoids and C-phycocyanin, in *Spirulina* powder and their antioxidant activities. *Molecules*, *23*(8), 2065.

Peng, J., Yuan, J. P., Wu, C. F., & Wang, J. H. (2011). Fucoxanthin, a marine carotenoid present in brown seaweeds and diatoms: Metabolism and bioactivities relevant to human health. *Marine Drugs*, *9*(10), 1806–1828.

Perica, M. M., & Delaš, I. (2011). Essential fatty acids and psychiatric disorders. *Nutrition in Clinical Practice*, *26*(4), 409–425.

Pham-Huy, L. A., He, H., & Pham-Huy, C. (2008). Free radicals, antioxidants in disease and health. *International Journal of Biomedical Science*, *4*(2), 89.

Pierre, G., Delattre, C., Dubessay, P., Jubeau, S., Vialleix, C., Cadoret, J. P., et al. (2019). What is in store for EPS microalgae in the next decade? *Molecules*, *24*(23), 4296.

Plaza, M., Herrero, M., Alejandro Cifuentes, A., & Ibanez, E. (2009). Innovative natural functional ingredients from microalgae. *Journal of Agricultural and Food Chemistry*, *57*(16), 7159–7170.

Pooja, V., & Sunita, M. (2014). Antioxidants and disease prevention. *International Journal of Advanced Scientific and Technical Research*, *4*(2), 903–911.

Pugh, N., Ross, S. A., ElSohly, H. N., ElSohly, M. A., & Pasco, D. S. (2001). Isolation of three high molecular weight polysaccharide preparations with potent immunostimulatory activity from *Spirulina platensis, Aphanizomenon flos-aquae* and *Chlorella pyrenoidosa*. *Planta Medica, 67*(08), 737–742.

Ramos, A. A., Polle, J., Tran, D., Cushman, J. C., Jin, E. S., & Varela, J. C. (2011). The unicellular green alga *Dunaliella salina* Teod. as a model for abiotic stress tolerance: Genetic advances and future perspectives. *Algae, 26*(1), 3–20.

Raposo, M. F. D. J., De Morais, A. M. M. B., & De Morais, R. M. S. C. (2015). Carotenoids from marine microalgae: A valuable natural source for the prevention of chronic diseases. *Marine Drugs, 13*(8), 5128–5155.

Reimers, A., & Ljung, H. (2019). The emerging role of omega-3 fatty acids as a therapeutic option in neuropsychiatric disorders. *Therapeutic Advances in Psychopharmacology, 9.* 2045125319858901.

Riccioni, G., Mancini, B., Di Ilio, E., Bucciarelli, T., & D'Orazio, N. (2008). Protective effect of lycopene in cardiovascular disease. *European Review for Medical and Pharmacological Sciences, 12*(3), 183.

Richmond, A. (1988). Spirulina. In A. Borowitzka, & L. Borowitzka (Eds.), *Microalgal biotechnology* (pp. 83–121). United Kingdom: Cambridge University Press.

Rizvi, S., Raza, S. T., Faizal Ahmed, A. A., Abbas, S., & Mahdi, F. (2014). The role of vitamin E in human health and some diseases. *Sultan Qaboos University Medical Journal, 14*(2), e157.

Roberts, J. E., & Dennison, J. (2015). The photobiology of lutein and zeaxanthin in the eye. *Journal of Ophthalmology, 2015.*

Rösch, C., & Posten, C. (2012). Challenges and perspectives of microalgae production. *Technikfolgenabschätzung—Theorie und Praxis, 21*(1).

Running, J. A., Severson, D. K., & Schneider, K. J. (2002). Extracellular production of L ascorbic acid by *Chlorella protothecoides, Prototheca* species, and mutants of *P. moriformis* during aerobic culturing at low pH. *Journal of Industrial Microbiology & Biotechnology, 29*, 93–98.

Safafar, H., Van Wagenen, J., Møller, P., & Jacobsen, C. (2015). Carotenoids, phenolic compounds and tocopherols contribute to the antioxidative properties of some microalgae species grown on industrial wastewater. *Marine Drugs, 13*(12), 7339–7356.

Sanjeewa, K. K. A., Fernando, I. P. S., Samarakoon, K. W., Lakmal, H. H. C., Kim, E. A., Kwon, O. N., et al. (2016). Anti-inflammatory and anti-cancer activities of sterol rich fraction of cultured marine microalga *Nannochloropsis oculata*. *Algae, 31*(3), 277–287.

Santhosh, S., Dhandapani, R., & Hemalatha, R. (2016). Bioactive compounds from microalgae and its different applications—A review. *Advances in Applied Science Research, 7*(4), 153–158.

Sarkar, A., Aronson, K. J., Patil, S., & Hugar, L. B. (2012). Emerging health risks associated with modern agriculture practices: A comprehensive study in India. *Environmental Research, 115,* 37–50.

Sathasivam, R., & Juntawong, N. (2013). Modified medium for enhanced growth of *Dunaliella* strains. *International Journal of Current Science, 5*, 67–73.

Sathasivam, R., & Ki, J. S. (2018). A review of the biological activities of microalgal carotenoids and their potential use in healthcare and cosmetic industries. *Marine Drugs, 16*(1), 26.

Sathasivam, R., Radhakrishnan, R., Hashem, A., & Abd Allah, E. F. (2017). Microalgae metabolites: A rich source for food and medicine. *Saudi Journal of Biological Sciences, 26*(4), 709–722.

Sathasivam, R., Radhakrishnan, R., Hashem, A., & Abd Allah, E. F. (2019). Microalgae metabolites: A rich source for food and medicine. *Saudi Journal of Biological Sciences, 26*(4), 709–722.

Schagen, S. K., Zampeli, V. A., Makrantonaki, E., & Zouboulis, C. C. (2012). Discovering the link between nutrition and skin aging. *Dermatoendocrinology, 4*(3), 298–307.

Schuchardt, J. P., Huss, M., Stauss-Grabo, M., & Hahn, A. (2010). Significance of long-chain polyunsaturated fatty acids (PUFAs) for the development and behaviour of children. *European Journal of Pediatrics, 169*(2), 149–164.

Searle, R. (2006). *Population growth, resource consumption, and the environment: Seeking a common vision for a troubled world*. Wilfrid Laurier Univ. Press.

Serive, B., Kaas, R., Bérard, J. B., Pasquet, V., Picot, L., & Cadoret, J. P. (2012). Selection and optimisation of a method for efficient metabolites extraction from microalgae. *Bioresource Technology, 124*, 311–320.

Shalaby, E. (2011). Algae as promising organisms for environment and health. *Plant Signaling & Behavior, 6*(9), 1338–1350.

Sharma, P., & Sharma, N. (2017). Industrial and biotechnological applications of algae: A review. *Journal of Advances in Plant Biology, 1*(1), 01.

Sharoba, A. M. (2014). Nutritional value of spirulina and its use in the preparation of some complementary baby food formulas. *Journal of Food and Dairy Sciences, 5*(8), 517–538.

Shick, J. M., & Dunlap, W. C. (2002). Mycosporine-like amino acids and related gadusols: Biosynthesis, accumulation, and UV-protective functions in aquatic organisms. *Annual Review of Physiology, 64*(1), 223–262.

Sikiru, A. B., Arangasamy, A., Alemede, I. C., Guvvala, P. R., Egena, S. S. A., Ippala, J. R., et al. (2019). *Chlorella vulgaris* supplementation effects on performances, oxidative stress and antioxidant genes expression in liver and ovaries of New Zealand White rabbits. *Heliyon, 5*(9), e02470.

Silva, J. R., Burger, B., Kühl, C., Candreva, T., dos Anjos, M. B., & Rodrigues, H. G. (2018). Wound healing and omega-6 fatty acids: From inflammation to repair. *Mediators of Inflammation, 2018*.

Singh, D. V., Bhat, J. I. A., Bhat, R. A., Dervash, M. A., & Ganei, S. A. (2018). Vehicular stress a cause for heavy metal accumulation and change in physico-chemical characteristics of road side soils in Pahalgam. *Environmental Monitoring and Assessment, 190*, 353.

Spielmeyer, A., Gebser, B., & Pohnert, G. (2011). Dimethylsulfide sources from microalgae: Improvement and application of a derivatization-based method for the determination of dimethylsulfoniopropionate and other zwitterionic osmolytes in phytoplankton. *Marine Chemistry, 124*(1–4), 48–56.

Spolaore, P., Joannis-Cassan, C., Duran, E., & Isambert, A. (2006). Commercial applications of microalgae. *Journal of Bioscience and Bioengineering, 101*, 87–96.

Sprague, M., Betancor, M. B., & Tocher, D. R. (2017). Microbial and genetically engineered oils as replacements for fish oil in aquaculture feeds. *Biotechnology Letters, 39*(11), 1599–1609.

Stokes, T., Hector, A. J., Morton, R. W., McGlory, C., & Phillips, S. M. (2018). Recent perspectives regarding the role of dietary protein for the promotion of muscle hypertrophy with resistance exercise training. *Nutrients, 10*(2), 180.

Subramaniam, S., Selvaduray, K. R., & Radhakrishnan, A. K. (2019). Bioactive compounds: Natural defense against cancer? *Biomolecules, 9*(12), 758.

Sugawara, T., Matsubara, K., Akagi, R., Mori, M., & Hirata, T. (2006). Antiangiogenic activity of brown algae fucoxanthin and its deacetylated product, fucoxanthinol. *Journal of Agricultural and Food Chemistry, 54*(26), 9805–9810.

Sukla, L. B., Subudhi, E., & Pradhan, D. (Eds.). (2019). *The role of microalgae in wastewater treatment* Springer Nature Springapore Pte Limited.

Swanson, D., Block, R., & Mousa, S. A. (2012). Omega-3 fatty acids EPA and DHA: Health benefits throughout life. *Advances in Nutrition, 3*(1), 1–7.

Sztretye, M., Dienes, B., Gönczi, M., Czirják, T., Csernoch, L., Dux, L., et al. (2019). Astaxanthin: A potential mitochondrial-targeted antioxidant treatment in diseases and with aging. *Oxidative Medicine and Cellular Longevity, 2019*.

Taghvaei, M., & Jafari, S. M. (2015). Application and stability of natural antioxidants in edible oils in order to substitute synthetic additives. *Journal of Food Science and Technology, 52*(3), 1272–1282.

Takshak, S. (2018). Bioactive compounds in medicinal plants: A condensed review. *SEJ Pharmacognosy and Natural Medicine, 1*, 1–35.

Talero, E., García-Mauriño, S., Ávila-Román, J., Rodríguez-Luna, A., Alcaide, A., & Motilva, V. (2015). Bioactive compounds isolated from microalgae in chronic inflammation and cancer. *Marine Drugs, 13*(10), 6152–6209.

Tan, B. L., Norhaizan, M. E., Liew, W. P. P., & Sulaiman Rahman, H. (2018). Antioxidant and oxidative stress: A mutual interplay in age-related diseases. *Frontiers in Pharmacology, 9*, 1162.

Tandon, P., Jin, Q., & Huang, L. (2017). A promising approach to enhance microalgae productivity by exogenous supply of vitamins. *Microbial Cell Factories, 16*(1), 219.

Thangavel, K., Krishnan, P. R., Nagaiah, S., Kuppusamy, S., Chinnasamy, S., Rajadorai, J. S., et al. (2018). Growth and metabolic characteristics of oleaginous microalgal isolates from Nilgiri biosphere Reserve of India. *BMC Microbiology, 18*(1), 1.

Thevenieau, F., & Nicaud, J. M. (2013). Microorganisms as sources of oils. *OCL, 20*(6), D603.

Trejo-Solís, C., Pedraza-Chaverrí, J., Torres-Ramos, M., Jiménez-Farfán, D., Cruz Salgado, A., Serrano-García, N., et al. (2013). Multiple molecular and cellular mechanisms of action of lycopene in cancer inhibition. *Evidence-based Complementary and Alternative Medicine, 2013*.

Tungmunnithum, D., Thongboonyou, A., Pholboon, A., & Yangsabai, A. (2018). Flavonoids and other phenolic compounds from medicinal plants for pharmaceutical and medical aspects: An overview. *Medicines, 5*(3), 93.

Urtubia, H. O., Betanzo, L. B., & Vásquez, M. (2016). Microalgae and cyanobacteria as green molecular factories: Tools and perspectives. In *Algae—Organisms for imminent biotechnology* (pp. 1–27). IntechOpen. https://doi.org/10.5772/63006.

Walia, A., Gupta, A. K., & Sharma, V. (2019). Role of bioactive compounds in human health. *Acta Scientific Medical Sciences, 3*(9), 25–33.

Wang, N. (2016). *Biological activities of tropical green algae from Australia (Ph.D. thesis)*. School of Chemical Engineering Faculty of Engineering, The University of New South Wales.

Wang, W., Wang, S. X., & Guan, H. S. (2012). The antiviral activities and mechanisms of marine polysaccharides: An overview. *Marine Drugs, 10*(12), 2795–2816.

Wang, M., Xu, Y. N., Jiang, G. Z., Li, L. B., & Kuang, T. Y. (2000). Membrane lipids and their fatty acid composition in *Nostoc flagelliforme* cells. *Acta Botanica Sinica, 42*(12), 1263–1266.

Watanabe, F., Takenaka, S., Kittaka-Katsura, H., Ebara, S., & Miyamoto, E. (2002). Characterization and bioavailability of vitamin B12-compounds from edible algae. *Journal of Nutritional Science and Vitaminology, 48*, 325–331.

Wells, M. L., Potin, P., Craigie, J. S., Raven, J. A., Merchant, S. S., Helliwell, K. E., et al. (2017). Algae as nutritional and functional food sources: Revisiting our understanding. *Journal of Applied Phycology, 29*(2), 949–982.

Wilkinson, K., Grant, W. P., Green, L. E., Hunter, S., Jeger, M. J., Lowe, P., et al. (2011). Infectious diseases of animals and plants: An interdisciplinary approach. *Philos. Trans. R. Soc. Lond. B Biol. Sci., 366*(1573), 1933–1942.

World Health Organization. (2004). *Vitamin and mineral requirements in human nutrition*. World Health Organization.

Wynn, J., Behrens, P., Sundararajan, A., Hansen, J., & Apt, K. (2010). Production of single cell oils from dinoflagellates. In Z. Cohen, & C. Ratledge (Eds.), *Microbial and algal oils. Single cell oils* (pp. 115–129). Champaign, Illinois: AOCS Press.

Xu, Y., Ibrahim, I. M., Wosu, C. I., Ben-Amotz, A., & Harvey, P. J. (2018). Potential of new isolates of *Dunaliella salina* for natural β-carotene production. *Biology, 7*(1), 14.

Yui, K., Imataka, G., Nakamura, H., Ohara, N., & Naito, Y. (2015). Eicosanoids derived from arachidonic acid and their family prostaglandins and cyclooxygenase in psychiatric disorders. *Current Neuropharmacology, 13*(6), 776–785.

Zhao, L., & Sweet, B. V. (2008). Lutein and zeaxanthin for macular degeneration. *American Journal of Health-System Pharmacy, 65*(13), 1232–1238.

Chapter 11

Some special diets used as neutraceuticals in Unani system of medicine with modern aspects

Huda Nafees[a], S. Nizamudeen[b], and Sana Nafees[c]

[a]Department of Saidla, Faculty of Unani Medicine, Aligarh Muslim University, Aligarh, India,
[b]GUMC, Chennai, India, [c]AIIMS, New Delhi, India

1. Introduction

Major advances in science and technology in recent decades have led to an increase in the development of highly expensive technologies related to medical and surgical procedures and drug therapies. At the same time, however, there has been an increase in the number of people turning to alternative medical therapies; Unani medicine is one such alternative medicine that emphasizes the importance of a "good diet" in maintaining and restoring health. The importance of diet is captured by this quote: "Let food be thy medicine and medicine be thy food", which was said by the great Persian polymath and father of medicine, Hippocrates (Glynn, Bhikha-Vallee, & Bhikha, 2013). This proclamation on food reflects the critical significance of food and lifestyle on one's health. According to Unani medicine, lifestyle diseases are the diseases that arise from an imbalance in the *Asbab-e-Sitta Zarooriyah* over a long period of time. The word *Asbab* (cause) in Unani terms refers to that which initiates a given state (health or disease) of humans. The basis of preventive medicine in the Unani system is based upon *Asbab-e-Sitta Zarooriah*. Food is one of the six essential prerequisites. It is mentioned that whenever a disease occurs, *Ilaj-bil-Ghiza* (dietotherapy) is one of the key principles for treating the patient (Baghdadi, 2005; Kabiruddin, 1930; Sina, 2010).

The importance of food and drink for healthy life of an individual and the selection of good and nutritional diets for the prevention and treatment of diseases are discussed by various ancient Unani physicians. Hippocrates stated, "Let food be your medicine and medicine be your food" and "Leave your drugs in the chemist's pot if you can heal the patient with food". Another statement of

Phytomedicine: A Treasure of Pharmacologically Active Products from Plants
https://doi.org/10.1016/B978-0-12-824109-7.00024-8

Hippocrates is "The cause of sickness is overeating, and the cause of health is eating like a bird". Avicenna said, "Stomach is the house of disease, and diet is the head of healing". Aristotle stated, "The person who takes grape juice, bread, and mutton; doing physical exercise; and taking adequate sleep won't be ill frequently" (Usaiba, 1990).

Razi was quoted saying, "Good nutrition, adequate rest, happiness, and best line of treatment are the pillars for curing of diseases". He also stated that "The amount of desired eatables should be less for a patient" and "Whenever possible, treatment of the diseases should be done by diets, not only by drugs" (Razi, 1991).

In the development of health and disease prevention, food and drinks have a paramount importance. Therefore, Unani scholars mention it under the heading of "*Taghzia*" (nutrition). Unani scholars believe that, before initiation of treatment through drugs, an individual must be treated with restrictions and alterations in the diet. Unani scholars not only introduced the concept of nutrition but also recommended specific diets for specific diseases (Avicenna, Gruner, Bakhtiar, & Shah, 1999; Sina, 2010).

In Unani medicine, *Ilaj-bil-Ghiza* (dietotherapy) is considered as the first important principle toward prevention and management of disease. Unani medicine lays great emphasis on treating certain ailments by administering a specific diet or by regulating the quality and quantity of food. The specific diet that can be included for both its nutritional value, having high nutritive content, and its pharmacological actions, for therapeutic purpose. We call these substances as *Dawa e Ghizai* (Nafees, 1954). Accordingly neutraceuticals has its roots since ancient times, defined by Unani scholars as *Dawa e Ghizai*. In Unani manuscripts, these modified diet regimens or special diets, separately or as an adjuvant with pharmacotherapy, is recommended for preventive measures and in many acute and chronic diseases.

In 1989, Stephen Defelice, MD, founder and chairman of the Foundation for Innovation in Medicine (FIM), Cranford, New Jersey, defined the term "neutraceuticals". According to him, neutraceuticals are substances that can be a food or a part of food that provides health benefits, including the prevention and treatment of disease by detoxifying the body, avoiding vitamin and mineral deficiencies, and restoring healthy digestion and dietary habit" (DeFelice, 1989; DeFelice, 1995).

2. Special diets used in Unani Medicine with modern approach

Ma-us-shaeer (Barley water): This is prepared with a good quality of *Jau* (Barley), which is thick; after boiling, it becomes soft and swollen (Kabeeruddin, 2010).

Ma-us-Shaeer is of different types:

1. *Maush Shaeer Sada*: Two methods are described in Unani Pharmacopeias.
   A. Fresh barley seeds are soaked in water for up to 4 hours, then their outer seed coat is removed by pestle and mortar. These seeds without seed coat are called *Jau Muqashshar* (Dehusked *Jau*). These 50-gm *Jau muqashshar*

are taken and boiled in 1 liter of water, i.e., barley in water in the ratio of 1:20. It has to cook until the water attains a thick consistency and reddish color, the jau gets torn up, then the boiling should be stopped. Then the water is filtered and that filtrate is known as *Maush shaeer*. Different ratios of barley and water are also mentioned in Unani literature such as 1:10, 1:20, or 1:24.

    B. Dehusked *Jau* is kept in a galvanized pot with 10-fold water and boiled at low flame; while cooking, the water in the pot can be changed seven times. Then it is filtered and used as *Maush shaeer* (Kabeeruddin, 2006).

2. *Maush shaeer Mulham*: To prepare *Maush shaeer mulham,* two methods have been mentioned in Unani literature.

    A. Mutton has to be cooked with spices then ghee (roast) is added in a small amount only for fragrance, and then barley water is added into it.

    B. After preparation of *Maush shaeer,* just add mutton soup to it.

3. *Maush shaeer Muhmaz*: The method of preparation of *Maush shaeer Mohmaz* is the same as that of *Maush shaeer Sada*. The only difference is that *Jau Muqashshar* has to be roasted before boiling it with water; then after the process, the filtrate is known as *Maus shaeer Mohmaz.*

4. *Kashkush-sheer*: Dehusked *Jau* has to boil with water until the *Jau* is ruptured, then throw out the boiled water and new water is added into it and boiled again. At last, *Jau* and water are mixed together. Then this final product is known as *Kashkush Shaeer* (Khan, 2006).

The major phytochemicals in barley that have shown health benefits include phenolic acids, flavonoids, lignans, vitamin E (tocols), sterols, and folates. The phenolic compounds provide important functions in reproduction and growth; act as an immune enhancer against pathogens, parasites, and predators; as well as contribute to the color of plants. Sterols and tocols are mainly components of plant oils that provide benefits such as protection against toxins, diabetes, and neurological diseases like Alzheimer's disease (Bartłomiej, Justyna, & Ewa, 2012; Shahidi & Zhong, 2010). Lignans have strong antioxidant properties similar to FA and better than vitamin E (Prasad, 2000). The phytosterols present in barley regulates cholesterol levels by micelle formation in the lumen of the intestine, consequently inhibiting cholesterol absorption and increasing secretion. Many studies speculated that the levels of total phytosterols in barley oils are sufficient enough (0.18e1.44 g/15 g oil) to significantly lower low-density lipoprotein (LDL) cholesterol at reasonable dosages of 15 mL/day (1 tablespoon/day) (Moreau, Flores, & Hicks, 2007). Barley and its products have bioactive compounds with antioxidative and immunomodulatory activities that are associated with cancer moderation. The dietary factors, especially those that reduce the impact of reactive oxygen species, can protect against DNA damage and stimulate the immune system, thus lowering cancer risks (Seifried, Anderson, Fisher, & Milner, 2007). Clinical studies have been done on the flavonoids and found that bioactive substances present in cereal grains are responsible for the moderation of many diseases including cancer and coronary heart diseases (Gani, SM, & FA, 2012; Tang et al., 2016). They have also been reported to lower the risk of CVD (Lucas et al., 2004).

Satto: Satto is another diet that is prepared by barley or corn flour. These are soaked in water then pounded with dry fruits to make satto. It should be given along with plenty of cold water to make it easily ingestible (Zaman, Basar, & Farah, 2013).

Maul-jubn **(Whey):** Maul-Jubn is prepared by milk curdling. To prepare it, the pregnant goat or any pregnant animal is fed with cold foods like palak (spinach) or khurfa (purslane). The animal shouldn't keep an empty stomach. When the lamb or new born is born, its milk should not be used for the purpose of curdling (Maul-Jubn) before 40 days. After then, the specific quantity of milk is taken and boiled in a tin-coated vessel. When it boiled well, then a little quantity of lemon juice or vinegar is added for curdling of milk. It is then removed from the fire and allowed to cool. After that, it is sieved through a thick cloth and thus the clean water is obtained and is known as *"Maul-Jubn"* (Arzani, 2009).

Milk is a nutrient physiological liquid containing bioactive compounds that have potential benefits to the newborn infant's growth and digestive system. It also improves the symbiotic microflora of the gut and the development of lymphoid tissues. A number of bioactive compounds are present in milk, remarkably in fermented milk products that are of great importance like curd, yogurt, cheese, butter, etc. and include certain specific proteins, vitamins, bioactive peptides, organic acids, and oligosaccharides (Akm, 2006).

Milk and milk products have a pronounced effect on body fat and body mass, and this might be due to presence of whey proteins, medium-chain fatty acids, high level of calcium, and other minerals. Other fermented milk components may have proteins, peptides, probiotic lactic acid bacteria, calcium, and other minerals that help in maintaining gut microbiota and may also be helpful in reduction of high blood pressure. Sphingolipids and their active metabolites are available in milk fat and have functional properties that may exert antimicrobial influences either directly or upon digestion (Mattila-Sandholm & Saarela, 2003).

Oligosaccharides   Lactic acid

Falooda: This is a special highly nutritious food item that is made with starch, cow's milk, and sugar. It is very much useful in general body weakness and cardiac weakness (Kabeeruddin, 2006).

Kavameekh: A diet is prepared with vegetables, milk, spices, and mint to increase appetite (Kabeeruddin, 1990).

Mazeera: It is prepared with curdled milk, which is useful in a hot season (Zaman et al., 2013).

Hareera: Harera is made with wheat or pieces of roti (bread), dry fruits, saffron, sugar, and cow's ghee. This diet is highly nutritious and increases a body's innate heat. Some recipes of hareera have been mentioned in Unani pharmacopoeias for the treatment of certain ailments such as cough, tuberculosis, hemoptysis, pleurisy, sexual dysfunctions, headache, etc.

Ma-ul-Asl (**Honey water**): It is prepared by taking 1 part honey and 2 parts water and boiled for a few minutes. Few herbs such as darchini (Cinnamomum zeylanicum), zinger (Zingiber officinale), mastagi (Pistacia lentiscus), safron (Crocus sativus), cardamom (Elettaria cardamomum), jaifal (Myristica fragrans), and javitri (Myristica fragrans) are mixed in the ma-ul-asl to increase its potency (Sina, 2010).

Honey is the mother of all natural sugars in which fructose constitutes 38% and glucose is about 31% (Nayik et al., 2014), and the other constituents are minerals, proteins, vitamins, organic acids, flavonoids, phenolic acids, enzymes, and other phytochemicals, which are responsible for dozens of pharmacological actions, and its consumption has a potential effect on human digestion. This effect is due to the presence of oligosaccharides in it (Curry, Curry, & Gomez, 1972; Gouyon et al., 2003; Jones, Butler, & Brooks, 2011).

Several in vitro and in vivo studies have been documented on the importance of dietary supplementation with natural honey and found that it increases the growth of essential and beneficial bacteria especially *Bifidobacteria* and *Lactobacilli* for the maintenance of good health and their prebiotic effect on the GIT (Gouyon et al., 2003; Jones et al., 2011). This can be justifiable with one such comparative study on natural (honey) and artificial (sucrose) sugars where it was found that honey increased *Lactobacilli* in both in vitro and in vivo (within the small and large intestine of experimental rats) while artificial sugar had no effect (Ushijima, Riby, Fujisawa, & Kretchmer, 1995).

Few studies enlightened that fructose reduces hyperglycemia or glucose levels in diabetic rodent models, healthy subjects, and patients with diabetes. It is also documented that fructose intake causes prolonged gastric emptying (Khalil, Alam, Moniruzzaman, Sulaiman, & Gan, 2011), which in turn reduces the intestinal absorption of gastric content (Topliss et al., 2002). Besides delaying absorption, fructose consumption lowers food intake, which is also attributed to the delayed gastric emptying (Vaisman, Niv, & Izkhakov, 2006). The slow absorption of fructose in the intestine might prolong the duration of contact and interaction between fructose and intestinal receptors that play a key role in satiety (Kwon, Kim, & Kim, 2008; Stanhope et al., 2011). In another study, it was found that the hypoglycemic effect of honey is due to fructose, as fructose neither increases plasma glucose nor requires insulin secretion for metabolism (Moran & McHugh, 1981). Dietary fructose is known to activate glucokinase (GKA) which is a key enzyme involved in the intracellular metabolism. Invertase (saccharase), diastase (amylase), and glucose oxidase are the main enzymes present in honey. Invertase is a carbohydrate-digestive enzyme that splits sucrose into glucose and fructose. It has an ability to hydrolyze the glycosidic bond between fructose and glucose, which makes it a vital part of the digestion of complex sugars into glucose and thereby it suffices as a prepared fuel source by the body (Simon et al., 2009). It also helps to reduce stomach toxicity, as it creates predigestive simple sugars (Busserolles, Gueux, Rock, Mazur, & Rayssiguier, 2002; Simon et al., 2009). Apart from that, invertase has been proven to have antimicrobial properties as it has the ability to pull moisture out of the body thereby subsiding bacterial manifestations (Busserolles et al., 2002). Hence, invertase sets up to play a key role not only in digestive processes but also in overall human disease prevention, physical rejuvenation, and antiaging processes.

A phenolic compound, syringic acid is found in honey and has antioxidant properties, free radicals, and ROS (Paramás, Bárez, Marcos, García-Villanova, & Sánchez, 2006; Sakaguchi, Inoue, & Ogihara, 1998). Another chemical constituent, aldose reductase inhibitors (ARIs), are potent therapeutic agents for the prevention and treatment of DC in a number of animal model-based studies (Jung et al., 2010; Locatelli et al., 2008; Meyskens Jr, Buckmeier, McNulty, & Tohidian, 1999).

Flavonoids, namely chrysin and kaempferol, present in honey (Levites, Amit, Youdim, & Mandel, 2002) are responsible for inhibiting the replication of several herpes viruses, adenoviruses, and rotaviruses, while in some studies it is found that quercetin and rutin exerted antiviral activity against HSV, syncytial virus, poliovirus, and Sindbis virus (Joskova, Franova, & Sadlonova, 2011; Tripoli, Guardia, Giammanco, Majo, & Giammanco, 2007). These constituents exert their actions by inhibiting the viral polymerase and binding to the viral nucleic acids or viral capsid proteins (Tripoli et al., 2007). Few studies have also been conducted for their cardioprotective and antidiabetic potentials, and results suggest that the flavonoids in honey are responsible for improving coronary vasodilatation, decreasing the ability of platelets to clot and preventing LDLs from oxidizing and hence decrease the risk of CVDs (Levites et al., 2002).

Therefore, the pharmacological studies done on honey show that it could be a potential agent against oxidative stress disorders including diabetes, cardiovascular disease, cancer, hepatic and renal failure, and aging processes.

Sikanjabeen (Oxymel): It is generally prepared by boiling 1 part vinegar and 3 parts honey or sugar (Kabeeruddin, 2010). There are some specific sikanjabeen that have specific actions on the basis of presence of main drug in it such as *sikanjabeen-e-buzuri* used as diuretic, and *sikanjabeen-e-lemuni* and *sikanjabeen-e-nanaee* used in vomiting, acute hepatitis, and jaundice (Kabeeruddin, 2006).

The chemical constituents of honey that are responsible for health benefits have been described in Ma ul asal, and many chemical constituents have been reported from natural vinegar, which has medicinal properties and are good for health. They are carbohydrates; organic acid (like citric acid, lactic acid, acetic acid, formic acid, malic acid, tartaric and succinic acid); alcohols and amino acid and peptides; vitamins and mineral salts; and polyphenolic compounds like gallic acid, catechin, caffeic, and ferulic acid. Some pharmacological studies have been done due to these chemical constituents that have benefits on health such as antimicrobial properties, preventing inflammation and hypertension, lowering serum cholesterol levels, effects on reducing systolic blood pressure, enhanced calcium absorption and its retention, and also decreasing the glycemic index of carbohydrates in food for people with or without diabetes (Johnston & Gaas, 2006). Hippocrates prescribed the use of oxymel for persistent cough. Acetic acid is responsible for antibacterial action, and vitamin C is for antioxidant activity. The antioxidant properties of vinegars are mainly due to the presence of polyphenols and melanoidins, which are affected by the raw materials and fermentation conditions, respectively. The effects of some vinegars on control of blood glucose, regulation of lipid metabolism, and weight loss are due to the presence of acetic acid, which is produced by acetic acid bacteria (AAB) during fermentation.

The acetic acid in vinegars regulates the concentration of blood glucose in the following ways (Petsiou, Mitrou, Raptis, & Dimitriadis, 2014): (1) by delaying gastric emptying; (2) by inhibiting disaccharidase activity; (3) by improving insulin sensitivity; and (4) by promoting the production of glycogen. Acetic acid in vinegar affects weight loss through the following mechanisms: (1) by decreasing the synthesis of lipids; (2) by increasing the oxygenolysis and secretion of lipids; (3) by increasing postprandial satiety; and (4) by increasing energy consumption.

The antibacterial activities of the polyphenols present in vinegars are primarily achieved by destroying the integrity of the cell membrane and interfering with the activities of enzymes present in bacteria (Gradišar, Pristovšek,

Plaper, & Jerala, 2007; Sirk, Brown, Sum, & Friedman, 2008; Taguri, Tanaka, & Kouno, 2006; Yoda, Hu, Shimamura, & Zhao, 2004). Tryptophol is a novel anticancer compound that was isolated from Japanese black soybean vinegar by Inagaki et al. (2007). Cell experiments have shown that tryptophol denatures DNA repair enzymes by activating caspase-8 and -3, which inhibits the proliferation of human leukemia cells (U937) in vitro. Tryptophol was less toxic to normal lymphocytes and did not activate the caspase of lymphocytes (Inagaki et al., 2007).

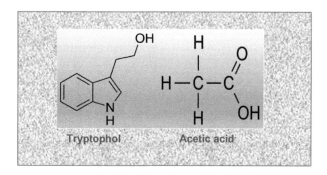

Ma-ul-Laham (**Meat distillate**): This is a form of meat/chicken soup prepared by method of distillation. Few drugs, viz., Cinnamomum tamala, Foeniculum vulgare, Coriandrum sativum, Mentha arvensis, Cinnamoum zylenicum, Lavandula stoechas, Pistacia lentiscus, Alpinia galanga, Curcuma zedoria, Valeriana jatamansi, etc. are mixed in ma-ul-laham to increase its potency and for other purposes (Khan, 2005). It is used for protein energy malnutrition, tuberculosis, cachexia, anemia, and cerebral and cardiac diseases (Kabeeruddin, 2006; Khan, 2005). Meat is rich in essential amino acids, whereas minerals are present in a lesser extent. Apart from it, essential fatty acids and vitamins are also present in it. Meat basically contains protein (26.1%); other components are water (61%) and fat (11.8 g). Animal protein contains all nine essential amino acids. Beef meat is mainly composed of saturated and monounsaturated fat present in almost equal amounts. The major fatty acids are oleic acid, stearic acid, and palmitic acid. The most common is conjugated linoleic acid (CLA), which is commonly found in beef and lamb. This CLA is responsible for several health benefits such as it helps in reducing weight (Schmid, Collomb, Sieber, & Bee, 2006). It contains vitamin B12, which is an important nutrient for formation of blood and for the brain and nervous system; it also contains vitamin B6, which is important for blood formation and energy metabolism. In minerals, it contains zinc, which is important for body growth and maintenance; and iron and selenium is found in high amounts, basically in heme form, which absorbs very efficiently. A few studies indicate that meat can increase the absorption of nonheme iron, which also comes from plants (Valenzuela, López de Romaña, Olivares, Morales, & Pizarro, 2009). Meat also contains some other bioactive compounds like creatine, which is found in an abundant amount. It serves as an energy source for muscles, and it may be beneficial for muscle growth and maintenance (Candow, Chilibeck, & Forbes, 2013). Meat also contain glutathione in higher amounts, which is an antioxidant that acts in antiaging and also is responsible for Increasing longevity, preventing illness, reducing the risk of chronic disease, and strengthening the immune system (Valencia, Marin, & Hardy, 2001).

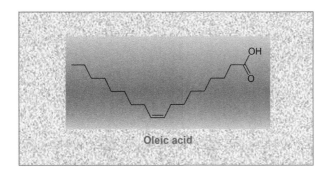

Asfeedaj: It is another special food in which meat is used. It is prepared by plain mutton soup without mixing spices and is useful in the winter season (Zaman et al., 2013).

Mazurat: This is prepared with mutton soup, vitis vinfera, sugar, and honey. It may be useful in chronic disease, as honey is the best immunomodulator for strengthening the immune system to fight against diseases.

Khageena: It is prepared by yolk of egg, aqueous extract of onion, ginger powder, Orchis latifolia, Cinnamomum zeylanicum, Syzygium aromaticum, and Elettaria cardamomum. It is very useful in sexual dysfunction (Kabeeruddin, 2006).

Ma-ul-Buqool (**Fresh vegetable juices**): *Ma-ul-Buqool* is prepared from fresh vegetables by crushing and pounding it. The effect on health by vegetables depends on the presence of chemical constituents, which vary in vegetables. Some of the vegetables that have effects on health promotion with their chemical constituents are given in Table 1.

TABLE 1 Chemical composition of *Ma ul buqool* with its pharmacological activities.

Part used	Chemical composition	Activities
Bulbs: Garlic, Leek, Onion, Shallot, Spring onion	Garlic contains allicin and carbohydrate, anthocyanins, proteins, amino acids, glycosides of kaempferol and quercetin, saponin, β-sitosterol, cholesterol, and campesterol, vitamins, polycarbohydrates, and prostaglandins A_2 and F_1 Onion contains vitamins, mineral salts, Na, K, phosphate and nitrate limestone, Fe, S, I, Si, phosphoric and acetic acids, allyl propyl disulfide, volatile oil, glucokinin, oxidase diastase (Bagiu, Vlaicu, & Butnariu, 2012; Farshori et al., 2013; Putnoky, Caunii, & Butnariu, 2013)	Hypolipidemic: Lowers low density lipoprotein Immunomodulatory: Improves and maintains healthy immune system of the body Antioxidant activity Antifungal Antibacterial Antiparasitic Antiviral Antithrombotic Anti-inflammatory (Mikaili, Maadirad, Moloudizargari, Aghajanshakeri, & Sarahroodi, 2013)
Flowers: Broccoli, Cauliflower, Choi sum, Artichoke, Broccoli Rabe, Gai john. courgette or other squash flowers	Q_{10} coenzyme, iron, phosphorus, vitamins A and C and riboflavin, potassium, calcium, phosphorus, and rich in fiber content, α-carotene, β-carotene, Flavonoids, α-tocopherol (Guo, Yuan, & Wang, 2013)	Antidiabetic. Analgesic. Anti-inflammatory. Antioxidant (Guo et al., 2013). Induces detoxification enzymes and reduces risk of cancer.
Fruits: Bitter melon, capsicum, chilli, choko, courgette, cucumber, eggplant, fuzzy melon, Indian marrow, marrow, plantain, pumpkin and squash, scallopini, tindora, tomatillo, tomato, turia	Iron, phosphorus, vitamins A and C and riboflavin, potassium, calcium and phosphorus, lycopene, Saponins, Carotene (Butnariu & Giuchici, 2011; Butnariu & Samfira, 2012)	Annihilate free radicals and preventing the formation of carcinogenic metabolites and hence reducing the risk of cancer Antidiarrheal. Antohelminthic Immunomodulatory Insecticidal (Butnariu & Giuchici, 2011)
Fungi (Mushrooms): Button white, Swiss brown, cup (opened not flat), enoki, oyster, Portabello (brown flat or cup), shiitake, truffle-black and white	Lignins, Polysaccharides, Scutigeral, Albaconol, Hispidin, Betulinan A, Australic acid.	Immunomodulator. Antibacterial. Antifungal. Antiviral. Antitumor. (Lindequist, Niedermeyer, & Jülich, 2005)

Plant part	Composition	Health effects
Leaves: Cabbage, lettuce, Spinach, sea kale, celtuce, romaine lettuce. Parsley	Rich in carotene, starch, minerals, vitamins, oxalates and folic acid minerals, calcium, vitamins (A, C, K, and riboflavin) and cellulose, anthocyanins, b-Carotene. And essentiall oil, ascorbic acid, chlorophyll, lactucin, leutin, amino acids (Metha & Belemkar, 2014; Park et al., 2013)	Antioxidant: preventing cancer of the lungs, oral cavity, stomach, and prostate Antimicrobial CNS depressant Hepatoprotective Protection against gamma radiation Antiproliferative Anti-inflammatory (Dong et al., 2013; Niu, Anstead, & Verchot, 2012)
Roots: Beetroot, carrot, celeriac, daikon, parsnip, radish, swede, turnip, Rutabaga	Good source of thiamin and minerals. Vitamin B, minerals, and rafanol, a sulfurglycoside with bactericidal potential. It also contains Ca and vitamin D, which is needed by the body for Ca absorption (Busserolles et al., 2002)	Antioxidant, increases resistance toward ultraviolet rays, smoothening of the skin and healthy coloring, provides nutrition to the body (Mech-Nowak, Swiderski, Kruczek, Luczak, & Kostecka-Gugala, 2012) Antiscorbutic, antisterility, increases Ca absorption (Roh et al., 2013) Healthy bones, promotes Fe absorption, accelerates wound healing (Butnariu & Caunii, 2013)
Seeds: Bean (green, French, butter, snake), broad bean, pea, pea, sweet corn	Vitamins B and macro- and micronutrients, carotene, vitamins, and minerals, Ca, Fe, Mg, proteins, (Mendes, Ramalho, & Abreu, 2012)	Decreases cholesterol and reduces risk of CVD (Aughey et al., 2013)
Stems: stalk is the main part of the vegetable Asparagus, celery, kohlrabi, fennel, artichoke	Carbohydrates, flavanoids, alkaloids, steroids, glycosides, phenols, furocoumarins, minerals and vitamins, fatty acids (AlSnafi, 2014)	Hypolipidemic: Lowers LDL, VLDL, Triglycerides and increases dHDL Anti-inflammatory Antiulcerogenic Antidepressant
Tubers: Potato, Jerusalem Artichoke, Yam, Crosne, Sweet potato, Taro, Jicama, Cassava, arrowroot	Saponins, riboflavin, β–carotene, thiamine, choline, niacin, serotonin, betulinic acid, tricontane, lupeol, stigmasterol, sitosterol, and its palmitate complex carbohydrates, fiber, K, Mg, Fe, vitamin C. vitamin K (Albiski, Najla, Sanoubar, Alkabani, & Murshed, 2012)	Immunomodulatory Improves gastrointestinal health Antiiinflammatory Antimicrobial Antidiarrhoeal (Pradeepika, Selvakumar, Krishnakumar, Nabi, & Sajeev, 2018)

Schematic representation of health benefits of *Ma ul buqool.*

Ma ul Usool: It is prepared by crushing and boiling of some roots like *Beekh e Kasni, Badyan, Karafs,* etc.

Ma ul Fawakeh (**Fruit Juices**): *Ma ul Fawakeh* are the juices extracted from fruits. It is one of the special diets given in Unani pharmacopeias. Fruit and fruit product consumption are known for promoting good general health and may also lower the risk of several noncommunicable diseases like cardiovascular diseases, stroke, gastrointestinal diseases, cancer, hypertension, and also degenerative diseases. Fruits contain dietary fibers in high amounts, certain vitamins (viz., ascorbic acid, vitamin A precursors, and folic acid), many minerals (viz., magnesium, potassium, iron, and calcium), and many other important phytochemicals possessing strong antioxidative properties. The phytochemicals present in different fruits are different and mainly these phytochemicals are responsible for the promotion of health. A brief description of phytochemicals in some fruits is given in Table 2.

TABLE 2 Chemical composition of *Ma ul Fawakeh* with its pharmacological activities.

Name of the fruit	Chemical composition	Activities
Pomegranate (*Punica granatum*)	Vitamin A, C, E, and Folic Acid, Granatin A, Granatin B, Pedunculagin, Castalagin, Corilagin, Mannitol, Pelleteirine, Pseudopelleteirine.	Lipid lowering agent Prevents heart attack Anthelmintic Antiaging (Rahimi, Arastoo, & Ostad, 2012)
Avocado (*Persea americana* Mill.)	Folic acid, Potassium, Oleic acid, terpenoid glycosides, flavonoids, coumarin (Yasir, Das, & Kharya, 2010)	Vasorelaxant Effect on growth and development of brain of the fetus, Anticancer, Analgesic, and anti-inflammatory activity
Strawberry (*Fragaria* x *ananassa*)	Ellagic acid, Phenols	Antioxidant. Prevents heart diseases. Removes pimples and wrinkles. Cleanses the digestive system. Prevents neurodegenerative diseases
Citrus Fruits	Vitamin C, potassium, folate, calcium, thiamin, niacin, vitamin B_6, phosphorus, magnesium, copper, riboflavin, pantothenic acid, both glycemic and nonglycemic carbohydrates (Whitney & Rolfes, 1999)	Lowering plasma cholesterol levels (United States National Academy of Sciences, Food and Nutrition Board, 1989a, 1989b; USDA, 1996) anticarcinogenic activity, antioxidant, wound healing activity, strong ligaments, tendons, dentin, skin, blood vessels, and bones, maintains the body's water and acid balance (Steinmetz & Potter, 1991)
Banana (*Musa acuminate*)	Dopamine, norepinephrine, carbohydrates, dietary fibres, αcarotene, β carotene, β cryptoxanthin, Lycopene, gallic acid, catechin, epicatechin, tannins, anthocyanins and ferulic acid, Flavanoids, sterols	Antioxidant Antimutagenic Antitumoral Hypolipidemic Decreases the risk of degenerative diseases (Sidhu & Zafar, 2018)

Continued

TABLE 2 Chemical composition of Ma ul Fawakeh with its pharmacological activities—cont'd

Name of the fruit	Chemical composition	Activities
Apple (*Malus domestica*)	Protein, vitamin, dietary fiber, Phenolic compounds like quercetin-3-galactoside, quercetin-3-glucoside, quercetin-3-rhamnoside, catechin, epicatechin, procyanidin, cyanidin-3-galactoside, coumaric acid, chlorogenic acid, gallic acid, and phloridzin (Lee, Kim, Kim, Lee, & Lee, 2003)	Antioxidant. Cardioprotective Anticancer. Antiobesity Antiasthmatic (Knekt et al., 2002)
Pineapple (*Ananas comosus.*L.)	Protein, lipid, carbohydrate, dietary fiber, potassium, magnesium, phosphorus, iron, manganese, vitamin C, asparagine, proline, aspartic acid, serine, glutamic acid, tyrosine, valine, and isoleucine (Gebhardt et al., 1983; U.S. Department of Agriculture (USDA), 2013; Krueger, Krueger, & Maciel, 1992; Elkins, Lyon, Huang, & Matthys, 1997; Gawler, 1962)	Antixodant Antimicrobial Anti-inflammatory Antitumor Debridement of third degree burn (Smith, 2003)
Grapes (*Vitis vinifera*)	Carbohydrate, proteins, fat, potassium, vitamin C, A, E, phenolics, flavonoids, anthocyanins, stilbenes, proanthocyanidins (Ali, Maltese, Choi, & Verpoorte, 2009; Ananga, Georgiev, & Tsolova, 2013; Nassiri-Asl & Hosseinzadeh, 2009)	Cardioprotective, Neuroprotective, Antimicrobial, Antiaging (Alzand & Mohamed, 2012; Raj, Sripal, Chaluvadi, & Krishna, 2001; Xia, Deng, Guo, & Li, 2010) Reducing the metabolic syndrome, Antiobesity (Chuang & McIntosh, 2011; Tsuda, 2011)

Brief description of some Important Diets given in the following table:

Diet forms	Chemical constituents	Indications
Ma us Shaeer	Phenolicacids, flavonoids, lignans, vitamin E	Tuberculosis, acute infections, diabetes, etc.
Ma ul Asl	Fructose, glucose, organic acids, flavonoids, phenolic acids, enzymes	Paralysis, facial palsy, cough, cerebrovascular disorders, etc.
Ma ul Jubn	Calcium, sphingolipids, proteins, vitamins, bioactive peptides, organic acids and oligosaccharides	Melancholia, Depression, Anxiety, Ascitis, Jaundice, etc.
Ma ul Fawakeh	Dietary fibers, ascorbic acid, folic acid, and vitamin A precursors, potassium, magnesium, iron, and calcium	Vitamin deficiency, constipation, general body weakness, etc.
Ma ul Buqool	Fiber, vitamin C, Mg, chlorophyll, vitamin K, carotenoids, B-complex vitamins	Cardiovascular diseases, general body weakness,
Sikanjbeen	Gallic acid, catechin, caffeic, ferulic acid, melanoidins	Vomiting, acute hepatitis and jaundice, cough, etc.
Ma-ul-Laham	Essential amino acids, vitamins, essential fatty acids, stearic acid, oleic acid, and palmitic acid, creatine	Protein energy malnutrition, tuberculosis, cachexia, anemia, cerebral and cardiac diseases.

3. Conclusion and future aspect

Dietotherapy has long been propounded and practiced as a specialized treatment in Unani medicine since the era of Hippocrates, Galen, Razi, and Avicenna, etc. The classical texts have extensive literature on this, which has been pointed out that the ancient Unani physicians were completely aware of the concept and importance of food and drink. And it is given that food is one of the prerequisites that come under the six essential prerequisites of life. Dietotherapy is a fundamental and vital prospect toward maintaining health, prevention, and treatment of many diseases ranging from communicable and noncommunicable disease. In the present scenario, lifestyle diseases (viz., hypertension, diabetes, dyslipidemia, etc.) has been become a major setback in the global disease burden with related risk and mortality rates, whereas a diet plan can enhance the lifestyle modules by selecting an appropriate diet either alone or as adjuvant with pharmacotherapy, which can reduce the disease burden and its mortality rate. This is why dietotherapy is widely acceptable, accessible, and above all affordable to patients with near nil adverse reactions. Several other diseases such as anemia, malnutrition, vitamin deficiency, etc. can also be treated by using an appropriate diet.

In developed countries, predictable factors have been largely responsible for encouraging the growth of the nutraceutical industry. High disposable incomes, changing lifestyles, unhealthy eating behaviors, increasing incidence and prevalence of health problems, increasingly larger aging population with exclusive dietary needs to maintain health, etc. have all prompted the development of new dietary solutions especially the use of neutraceuticals. Hence, there is a significant correlation between the growth of nutritional ingredients and demographic issues and confidence in the production and growth of nutraceutical products. Nowadays, this industry is also expanding in developing countries too.

Therefore, there is a gradual increase in understanding about the potential mechanism of active biological components in food, which could improve health and probably reduce the risk of disease and enhance overall wellbeing. Emerging fields in neutraceuticals are Nanoneutraceuticals, Nutrigenomics, Nutrigenetics, Molecular nutrition, and safety and efficacy of neutraceuticals. Huge demands of herbal products project the pharmaceutical industries to produce herbal nutritional supplements.

References

Akm, N. (2006). *Modfern Yogurt Bilimi ve Teknolojini*. Konya: Seluck Universitesi Ziraat Fakultesi Gida Muhendisligi Bolumu.

Albiski, F., Najla, S., Sanoubar, R., Alkabani, N., & Murshed, R. (2012). In vitro screening of potato lines for drought tolerance. *Physiology and Molecular Biology of Plants, 18*(4), 315–321. https://doi.org/10.1007/s12298-012-0127-5.

Ali, K., Maltese, F., Choi, Y., & Verpoorte, R. (2009). Metabolic constituents of grapevine and grape-derived products. *Phytochemistry Reviews, 9*(3), 357–378. https://doi.org/10.1007/s11101-009-9158-0.

AlSnafi, E. A. (2014). The pharmacology of Apium graveolens. A review. *International Journal for Pharmaceutical Research Scholars (IJPRS), 3*(1), 671–677.

Alzand, K. I., & Mohamed, M. A. (2012). Flavonoids: Chemistry, biochemistry and antioxidant activity. *Journal of Pharmacy Research, 5*, 4013–4020.

Ananga, A., Georgiev, V., & Tsolova, V. (2013). Manipulation and engineering of metabolic and biosynthetic pathway of plant polyphenols. *Current Pharmaceutical Design, 19*(34), 6186–6206. https://doi.org/10.2174/1381612811319340011.

Arzani, M. A. (2009). *Qarabadeen Qadri (Urdu Translation: CCRUM)*. New Delhi: CCRUM.

Aughey, R., Buchheit, M., Garvican-Lewis, L., Roach, G., Sargent, C., Billaut, F., et al. (2013). Yin and yang, or peas in a pod? Individual-sport versus team-sport athletes and altitude training. *British Journal of Sports Medicine, 47*(18), 1150–1154. https://doi.org/10.1136/bjsports-2013-092764.

Avicenna, O., Gruner, C., Bakhtiar, L., & Shah, M. H. (1999). *The canon of medicine of Avicenna: [volume 1]*. Chicago, IL: Kazi Publiations.

Baghdadi, I. H. (2005). *Kitabul Mukhtarat fit Tibb*. New Delhi: CCRUM.

Bagiu, R., Vlaicu, B., & Butnariu, M. (2012). Chemical composition and in vitro antifungal activity screening of the Allium ursinum L. (Liliaceae). *International Journal of Molecular Sciences, 13*(2), 1426–1436. https://doi.org/10.3390/ijms13021426.

Bartłomiej, S., Justyna, R., & Ewa, N. (2012). Bioactive compounds in cereal grains—Occurrence, structure, technological significance and nutritional benefits—A review. *Food Science and Technology International, 18*(6), 559–568. https://doi.org/10.1177/1082013211433079.

Busserolles, J., Gueux, E., Rock, E., Mazur, A., & Rayssiguier, Y. (2002). Substituting honey for refined carbohydrates protects rats from hypertriglyceridemic and prooxidative effects of fructose. *The Journal of Nutrition, 132*(11), 3379–3382. https://doi.org/10.1093/jn/132.11.3379.

Butnariu, M., & Caunii, A. (2013). Design management of functional foods for quality of life improvement. *Annals of Agricultural and Environmental Medicine, 20*(4), 736–741.

Butnariu, M., & Giuchici, C. (2011). The use of some nanoemulsions based on aqueous propolis and lycopene extract in the skin's protective mechanisms against UVA radiation. *Journal of Nanobiotechnology, 9*(1), 3. https://doi.org/10.1186/1477-3155-9-3.

Butnariu, M., & Samfira, I. (2012). Free radicals and oxidative stress. *Journal of Bioequivalence and Bioavailability, 04*(03). https://doi.org/10.4172/jbb.10000e13.

Candow, D., Chilibeck, P., & Forbes, S. (2013). Creatine supplementation and aging musculoskeletal health. *Endocrine, 45*(3), 354–361. https://doi.org/10.1007/s12020-013-0070-4.

Chuang, C., & McIntosh, M. (2011). Potential mechanisms by which polyphenol-rich grapes prevent obesity-mediated inflammation and metabolic diseases. *Annual Review of Nutrition, 31*(1), 155–176. https://doi.org/10.1146/annurev-nutr-072610-145149.

Curry, D., Curry, K., & Gomez, M. (1972). Fructose potentiation of insulin secretion. *Endocrinology, 91*(6), 1493–1498. https://doi.org/10.1210/endo-91-6-1493.

DeFelice, S. L. (1989). *The nutraceutical revolution: Fueling a powerful, new international market.* Como, Italy: Harvard University Advanced Management Program in Biomedical Research and Development.

DeFelice, S. (1995). The nutraceutical revolution: Its impact on food industry R&D. *Trends in Food Science and Technology, 6*(2), 59–61. https://doi.org/10.1016/s0924-2244(00)88944-x.

Dong, J., Yan, W., Bock, C., Nokhrina, K., Keller, W., & Georges, F. (2013). Perturbing the metabolic dynamics of myo-inositol in developing Brassica napus seeds through in vivo methylation impacts its utilization as phytate precursor and affects downstream metabolic pathways. *BMC Plant Biology, 13*(1), 84. https://doi.org/10.1186/1471-2229-13-84.

Elkins, E. R., Lyon, R., Huang, C. J., & Matthys, A. (1997). Characterization of commercially produced pineapple juice concentrate. *Journal of Food Composition and Analysis, 10*(4), 285–298. https://doi.org/10.1006/jfca.1997.0547.

Farshori, N. N., Al-Sheddi, E. S., Al-Oqail, M. M., Musarrat, J., Al-Khedhairy, A. A., & Siddiqui, M. A. (2013). Anticancer activity of Petroselinum sativum seed extracts on MCF-7 human breast cancer cells. *Asian Pacific Journal of Cancer Prevention, 14*(10), 5719–5723. https://doi.org/10.7314/apjcp.2013.14.10.5719.

Gani, A., SM, W., & FA, M. (2012). Whole-grain cereal bioactive compounds and their health benefits: A review. *Journal of Food Processing and Technology, 03*(03). https://doi.org/10.4172/2157-7110.1000146.

Gawler, J. H. (1962). Constituents of canned Malayan pineapple juice I—Amino-acids, non-volatile acids, sugars, volatile carbonyl compounds and volatile acids. *Journal of the Science of Food and Agriculture, 13*(1), 57–61. https://doi.org/10.1002/jsfa.2740130111.

Gebhardt, S. E., Cutrufelli, R., Matthews, R. H., & Consumer Nutrition Center (U.S.A). (1983). *Composition of foods—Fruits and fruit juices: Raw, processed, prepared.* Washington, D.C.: U.S. Dept. Of Agriculture, Human Nutrition Information Service.

Glynn, J., Bhikha-Vallee, N., & Bhikha, R. (2013). *Dietotherapy: Background and theory.* Ibn Sina Institute of Tibb. Reprint http://tibb.co.za/articles/Tibb-andDietotherapy.

Gouyon, F., Caillaud, L., Carrière, V., Klein, C., Dalet, V., Citadelle, D., et al. (2003). Simple-sugar meals target GLUT2 at enterocyte apical membranes to improve sugar absorption: A study in GLUT2-null mice. *The Journal of Physiology, 552*(3), 823–832. https://doi.org/10.1113/jphysiol.2003.049247.

Gradišar, H., Pristovšek, P., Plaper, A., & Jerala, R. (2007). Green tea catechins inhibit bacterial DNA Gyrase by interaction with its ATP binding site. *Journal of Medicinal Chemistry, 50*(2), 264–271. https://doi.org/10.1021/jm060817o.

Guo, R., Yuan, G., & Wang, Q. (2013). Effect of NaCl treatments on glucosinolate metabolism in broccoli sprouts. *Journal of Zhejiang University Science B, 14*(2), 124–131. https://doi.org/10.1631/jzus.b1200096.

Inagaki, S., Morimura, S., Gondo, K., Tang, Y., Akutagawa, H., & Kida, K. (2007). Isolation of tryptophol as an apoptosis-inducing component of vinegar produced from boiled extract of black soybean in human monoblastic leukemia U937 cells. *Bioscience, Biotechnology, and Biochemistry, 71*(2), 371–379. https://doi.org/10.1271/bbb.60336.

Johnston, C. S., & Gaas, C. A. (2006). Vinegar: Medicinal uses and antiglycemic effect. *Medscape General Medicine, 8*(2), 61.

Jones, H. F., Butler, R. N., & Brooks, D. A. (2011). Intestinal fructose transport and malabsorption in humans. *American Journal of Physiology. Gastrointestinal and Liver Physiology, 300*(2), G202–G206. https://doi.org/10.1152/ajpgi.00457.2010.

Joskova, M., Franova, S., & Sadlonova, V. (2011). Acutebronchodilatoreffectofquercetininexperimentalallergicasthma. *Bratislavské lekárske listy, 112*, 9–12.

Jung, H.-J., Kim, S.-J., Jeon, W.-K., Kim, B.-C., Ahn, K., Kim, K., et al. (2010). Anti-inflammatory activity of n-propyl gallate through down-regulation of NF-κB and JNK pathways. *Inflammation, 34*(5), 352–361. https://doi.org/10.1007/s10753-010-9241-0.

Kabeeruddin, M. (1990). *Sharah Asbab: Vol. II.* Hyderabad: Hikmat Book Depo.

Kabeeruddin, M. (2006). *Al-Qarabadeen. New Delhi: Central Council for Research in Unani Medicine.* Dept. of AYUSH, Ministry of H and FW, Govt. of India.

Kabeeruddin, M. (2010). *Bayaz-e-Kabeer Vol. III.* New Delhi: Idarah Kitab-us-Shifa.

Kabiruddin, M. (1930). *Translation and elaboration Kulliyate-Qanoon* (1st ed.). Karol Bagh, Delhi: Matba Darul Masih.

Khalil, M. I., Alam, N., Moniruzzaman, M., Sulaiman, S. A., & Gan, S. H. (2011). Phenolic acid composition and antioxidant properties of Malaysian honeys. *Journal of Food Science, 76*(6), C921–C928. https://doi.org/10.1111/j.1750-3841.2011.02282.x.

Khan, A. (2005). *Qarabadeen-e-Azam va Akmal (Urdu translation by CCRUM).* New Delhi: Dept. of AYUSH, Ministry of H and FW, Govt. of India.

Khan, S. (2006). *Elaj al-Amraz (Urdu translation by Kabeeruddin M).* New Delhi: Ejaz Publishing House.

Knekt, P., Kumpulainen, J., Järvinen, R., Rissanen, H., Heliövaara, M., Reunanen, A., et al. (2002). Flavonoid intake and risk of chronic diseases. *The American Journal of Clinical Nutrition, 76*(3), 560–568. https://doi.org/10.1093/ajcn/76.3.560.

Krueger, D. A., Krueger, R.-G., & Maciel, J. (1992). Composition of pineapple juice. *Journal of AOAC INTERNATIONAL, 75*(2), 280–282. https://doi.org/10.1093/jaoac/75.2.280.

Kwon, S., Kim, Y. J., & Kim, M. K. (2008). Effect of fructose or sucrose feeding with different levels on oral glucose tolerance test in normal and type 2 diabetic rats. *Nutrition Research and Practice, 2*(4), 252. https://doi.org/10.4162/nrp.2008.2.4.252.

Lee, K. W., Kim, Y. J., Kim, D.-O., Lee, H. J., & Lee, C. Y. (2003). Major phenolics in apple and their contribution to the total antioxidant capacity. *Journal of Agricultural and Food Chemistry, 51*(22), 6516–6520. https://doi.org/10.1021/jf034475w.

Levites, Y., Amit, T., Youdim, M. B. H., & Mandel, S. (2002). Involvement of protein kinase c activation and cell survival/cell cycle genes in green tea polyphenol (−)-epigallocatechin 3-gallate neuroprotective action. *Journal of Biological Chemistry, 277*(34), 30574–30580. https://doi.org/10.1074/jbc.m202832200.

Lindequist, U., Niedermeyer, T. H. J., & Jülich, W.-D. (2005). The pharmacological potential of mushrooms. *Evidence-based Complementary and Alternative Medicine, 2*(3), 285–299. https://doi.org/10.1093/ecam/neh107.

Locatelli, C., Rosso, R., Santos-Silva, M. C., de Souza, C. A., Licínio, M. A., Leal, P., et al. (2008). Ester derivatives of gallic acid with potential toxicity toward L1210 leukemia cells. *Bioorganic and Medicinal Chemistry, 16*(7), 3791–3799. https://doi.org/10.1016/j.bmc.2008.01.049.

Lucas, E. A., Lightfoot, S. A., Hammond, L. J., Devareddy, L., Khalil, D. A., Daggy, B. P., et al. (2004). Flaxseed reduces plasma cholesterol and atherosclerotic lesion formation in ovariectomized Golden Syrian hamsters. *Atherosclerosis, 173*(2), 223–229. https://doi.org/10.1016/j.atherosclerosis.2003.12.032.

Mattila-Sandholm, T., & Saarela, M. (2003). *Functional dairy products.* Boca Raton, FL: CRC Press. Cambridge.

Mech-Nowak, A., Swiderski, A., Kruczek, M., Luczak, I., & Kostecka-Gugała, A. (2012). Content of carotenoids in roots of seventeen cultivars of Daucus carota L. *Acta Biochimica Polonica, 59*(1). https://doi.org/10.18388/abp.2012_2190.

Mendes, M. P., Ramalho, M. A. P., & Abreu, A. F. B. (2012). Strategies in identifying individuals in a segregant population of common bean and implications of genotype x environment interaction in the success of selection. *Genetics and Molecular Research, 11*(2), 872–880. https://doi.org/10.4238/2012.april.10.3.

Metha, D., & Belemkar, S. (2014). Pharmacological activity of spinacia oleracea linn—A complete overview. *Asian Journal of Pharmaceutical Research and Development, 2*(1), 32–42.

Meyskens, F. L., Jr., Buckmeier, J. A., McNulty, S. E., & Tohidian, N. B. (1999). Activation of nuclear factor-kappa B in human metastatic melanoma cells and the effect of oxidative stress. *Clinical Cancer Research, 5*, 1197–1202.

Mikaili, P., Maadirad, S., Moloudizargari, M., Aghajanshakeri, S., & Sarahroodi, S. (2013). Therapeutic uses and pharmacological properties of garlic, shallot, and their biologically active compounds. *Iranian Journal of Basic Medical Sciences, 16*(10), 1031–1048.

Moran, T. H., & McHugh, P. R. (1981). Distinctions among three sugars in their effects on gastric emptying and satiety. *The American Journal of Physiology, 241*, R25–R30.

Moreau, R. A., Flores, R. A., & Hicks, K. B. (2007). Composition of functional lipids in Hulled and Hulless Barley in fractions obtained by scarification and in barley oil. *Cereal Chemistry Journal, 84*(1), 1–5. https://doi.org/10.1094/cchem-84-1-0001.

Nafees, B. (1954). *Tarjuma wa Sharae Kulliyate Nafeesi (urdu translation by Kabeeruddin M).* New Delhi: Idrare Kitabus Shifa. Qarshi AA. Ifadae Kabeer. New Delhi; Idrare Kitabus Shifa.

Nassiri-Asl, M., & Hosseinzadeh, H. (2009). Review of the pharmacological effects of Vitis vinifera(Grape) and its bioactive compounds. *Phytotherapy Research, 23*(9), 1197–1204. https://doi.org/10.1002/ptr.2761.

Nayik, G. A., Shah, T. R., Muzaffar, K., Wani, S. A., Gull, A., et al. (2014). Honey: Its history and religious significance: A review. *UJP, 3*, 5–8.

Niu, C., Anstead, J., & Verchot, J. (2012). Analysis of protein transport in the Brassica oleracea vasculature reveals protein-specific destinations. *Plant Signaling & Behavior, 7*(3), 361–374. https://doi.org/10.4161/psb.19020.

Paramás, A. M. G., Bárez, J. A. G., Marcos, C. C., García-Villanova, R. J., & Sánchez, J. S. (2006). HPLC-fluorimetric method for analysis of amino acids in products of the hive (honey and bee-pollen). *Food Chemistry, 95*(1), 148–156. https://doi.org/10.1016/j.foodchem.2005.02.008.

Park, S., Navratil, S., Gregory, A., Bauer, A., Srinath, I., Jun, M., et al. (2013). Generic Escherichia coli contamination of spinach at the preharvest stage: Effects of farm management and environmental factors. *Applied and Environmental Microbiology*, 79(14), 4347–4358. https://doi.org/10.1128/aem.00474-13.

Petsiou, E. I., Mitrou, P. I., Raptis, S. A., & Dimitriadis, G. D. (2014). Effect and mechanisms of action of vinegar on glucose metabolism, lipid profile, and body weight. *Nutrition Reviews*, 72(10), 651–661. https://doi.org/10.1111/nure.12125.

Pradeepika, C., Selvakumar, R., Krishnakumar, T., Nabi, S. U. N., & Sajeev, M. S. (2018). Pharmacology and Phytochemistry of underexploited tuber crops: A review. *Journal of Pharmacognosy and Phytochemistry*, 7(5), 1007–1019.

Prasad, K. (2000). Oxidative stress as a mechanism of diabetes in diabetic BB prone rats: Effect of secoisolariciresinol diglucoside (SDG). *Molecular and Cellular Biochemistry*, 209, 89–96.

Putnoky, S., Caunii, A., & Butnariu, M. (2013). Study on the stability and antioxidant effect of the *Allium ursinum* watery extract. *Chemistry Central Journal*, 7(1), 21.

Rahimi, H. R., Arastoo, M., & Ostad, S. N. (2012). A comprehensive review of Punica granatum (pomegranate) properties in toxicological, pharmacological, cellular and molecular biology researches. *Iranian Journal of Pharmaceutical Research: IJPR*, 11(2), 385–400.

Raj, N. K., Sripal, R. M., Chaluvadi, M. R., & Krishna, D. R. (2001). Flavonoids classification, pharmacological, biochemical effects and therapeutic potential. *Indian Journal of Pharmacology*, 33, 2–16.

Razi, Z. (1991). *Kitab al-Mansuri (Urdu translation by CCRUM)*. New Delhi: Dept. of AYUSH, Ministry of H and FW, Govt. of India.

Roh, S. S., Park, S. B., Park, S. M., Choi, B. W., Lee, M.-H., Hwang, Y. L., et al. (2013). A novel compound Rasatiol Isolated from Raphanus sativus has a potential to enhance extracellular matrix synthesis in dermal fibroblasts. *Annals of Dermatology*, 25(3), 315. https://doi.org/10.5021/ad.2013.25.3.315.

Sakaguchi, N., Inoue, M., & Ogihara, Y. (1998). Reactive oxygen species and intracellular Ca2, common signals for apoptosis induced by gallic acid. *Biochemical Pharmacology*, 55(12), 1973–1981. https://doi.org/10.1016/s0006-2952(98)00041-0.

Schmid, A., Collomb, M., Sieber, R., & Bee, G. (2006). Conjugated linoleic acid in meat and meat products: A review. *Meat Science*, 73(1), 29–41. https://doi.org/10.1016/j.meatsci.2005.10.010.

Seifried, H. E., Anderson, D. E., Fisher, E. I., & Milner, J. A. (2007). A review of the interaction among dietary antioxidants and reactive oxygen species. *The Journal of Nutritional Biochemistry*, 18(9), 567–579. https://doi.org/10.1016/j.jnutbio.2006.10.007.

Shahidi, F., & Zhong, Y. (2010). Lipid oxidation and improving the oxidative stability. *Chemical Society Reviews*, 39(11), 4067. https://doi.org/10.1039/b922183m.

Sidhu, J. S., & Zafar, T. A. (2018). Bioactive compounds in banana fruits and their health benefits. *Food Quality and Safety*, 2(4), 183–188. https://doi.org/10.1093/fqsafe/fyy019.

Simon, A., Traynor, K., Santos, K., Blaser, G., Bode, U., & Molan, P. (2009). Medical honey for wound care—Still the 'Latest Resort'? *Evidence-based Complementary and Alternative Medicine*, 6(2), 165–173. https://doi.org/10.1093/ecam/nem175.

Sina, I. (2010). *Al Qanoon fil Tibb, Urdu translation by Hkm Ghulam Hasnain Qantoori*. New Delhi: Idara-e-Kitab-us Shifa.

Sirk, T. W., Brown, E. F., Sum, A. K., & Friedman, M. (2008). Molecular dynamics study on the biophysical interactions of seven green tea catechins with lipid bilayers of cell membranes. *Journal of Agricultural and Food Chemistry*, 56(17), 7750–7758. https://doi.org/10.1021/jf8013298.

Smith, L. G. (2003). *Pineapples. Encyclopedia of food sciences and nutrition* (2nd ed., p. 4567).

Stanhope, K. L., Griffen, S. C., Bremer, A. A., Vink, R. G., Schaefer, E. J., Nakajima, K., et al. (2011). Metabolic responses to prolonged consumption of glucose- and fructose-sweetened beverages are not associated with postprandial or 24-h glucose and insulin excursions. *The American Journal of Clinical Nutrition*, *94*(1), 112–119. https://doi.org/10.3945/ajcn.110.002246.

Steinmetz, K. A., & Potter, J. D. (1991). Vegetables, fruit, and cancer. II. Mechanisms. *Cancer Causes and Control*, *2*(6), 427–442. https://doi.org/10.1007/bf00054304.

Taguri, T., Tanaka, T., & Kouno, I. (2006). Antibacterial spectrum of plant polyphenols and extracts depending upon hydroxyphenyl structure. *Biological and Pharmaceutical Bulletin*, *29*(11), 2226–2235. https://doi.org/10.1248/bpb.29.2226.

Tang, Y., Zhang, B., Li, X., Chen, P. X., Zhang, H., Liu, R., et al. (2016). Bound phenolics of quinoa seeds released by acid, alkaline, and enzymatic treatments and their antioxidant and α-glucosidase and pancreatic lipase inhibitory effects. *Journal of Agricultural and Food Chemistry*, *64*(8), 1712–1719. https://doi.org/10.1021/acs.jafc.5b05761.

Topliss, J. G., Clark, A. M., Ernst, E., Hufford, C. D., Johnston, G. A. R., Rimoldi, J. M., et al. (2002). Natural and synthetic substances related to human health (IUPAC Technical Report). *Pure and Applied Chemistry*, *74*(10), 1957–1985. https://doi.org/10.1351/pac200274101957.

Tripoli, E., Guardia, M. L., Giammanco, S., Majo, D. D., & Giammanco, M. (2007). Citrus flavonoids: Molecular structure, biological activity and nutritional properties: A review. *Food Chemistry*, *104*(2), 466–479. https://doi.org/10.1016/j.foodchem.2006.11.054.

Tsuda, T. (2011). Dietary anthocyanin-rich plants: Biochemical basis and recent progress in health benefits studies. *Molecular Nutrition & Food Research*, *56*(1), 159–170. https://doi.org/10.1002/mnfr.201100526.

U.S. Department of Agriculture (USDA). (2013). *National nutrient database for standard reference*. Release 26. Published online at: http://ndb.nal.usda.gov/. (Accessed 7 April 2014).

United States National Academy of Sciences, Food and Nutrition Board. (1989a). *Diet and health: Implications for reducing chronic disease risk*. Washington, DC: National Academy Press.

United States National Academy of Sciences, Food and Nutrition Board. (1989b). *Cleveland, Goldman and Borrud* (p. 1996).

Usaiba, I. A. (1990). *Uoyun al-Amba-Atibba Vol. I (Urdu translation by CCRUM)*. New Delhi: Dept. of AYUSH, Ministry of Health and FW. 74, 75, 97, 138, 220.

USDA. (1996). *Dietary guidelines for Americans*. Government Printing Office: Washington, DC.

Ushijima, K., Riby, J. E., Fujisawa, T., & Kretchmer, N. (1995). Absorption of fructose by isolated small intestine of rats is via a specific saturable carrier in the absence of glucose and by the disaccharidase-related transport system in the presence of glucose. *The Journal of Nutrition*, *125*(8), 2156–2164. https://doi.org/10.1093/jn/125.8.2156.

Vaisman, N., Niv, E., & Izkhakov, Y. (2006). Catalytic amounts of fructose may improve glucose tolerance in subjects with uncontrolled non-insulin-dependent diabetes. *Clinical Nutrition*, *25*(4), 617–621. https://doi.org/10.1016/j.clnu.2005.11.013.

Valencia, E., Marin, A., & Hardy, G. (2001). Glutathione—Nutritional and pharmacological viewpoints: Part II. *Nutrition*, *17*(6), 485–486. https://doi.org/10.1016/s0899-9007(01)00572-x.

Valenzuela, C., López de Romaña, D., Olivares, M., Morales, M. S., & Pizarro, F. (2009). Total iron and heme iron content and their distribution in beef meat and viscera. *Biological Trace Element Research*, *132*(1–3), 103–111. https://doi.org/10.1007/s12011-009-8400-3.

Whitney, E., & Rolfes, S. (1999). In W. Rolfes (Ed.), *Understanding nutrition* (8th ed.). Belmont, CA, USA: West/Wadsworth.

Xia, E.-Q., Deng, G.-F., Guo, Y.-J., & Li, H.-B. (2010). Biological activities of polyphenols from grapes. *International Journal of Molecular Sciences*, *11*(2), 622–646. https://doi.org/10.3390/ijms11020622.

Yasir, M., Das, S., & Kharya, M. (2010). The phytochemical and pharmacological profile of Persea americana Mill. *Pharmacognosy Reviews*, *4*(7), 77. https://doi.org/10.4103/0973-7847.65332.

Yoda, Y., Hu, Z.-Q., Shimamura, T., & Zhao, W.-H. (2004). Different susceptibilities of Staphylococcus and Gram-negative rods to epigallocatechin gallate. *Journal of Infection and Chemotherapy, 10*(1), 55–58. https://doi.org/10.1007/s10156-003-0284-0.

Zaman, R., Basar, S. N., & Farah, S. A. (2013). Dietotherapy in Unani system of medicine. *International Journal of Pharmaceutical, Chemical and Biological Sciences, 3*, 4.

Chapter 12

Phytomedicines: Synergistic and antagonistic phytometabolites-drug interactions

Monica Butnariu[a] and Marian Butu[b]

[a]*Banat's University of Agricultural Sciences and Veterinary Medicine "King Michael I of Romania" from Timisoara, Timis, Romania*, [b]*National Institute of Research and Development for Biological Sciences, Bucharest, Romania*

1. Background

Medicines are chemicals used for prophylaxis, diagnosis, or treatment of a disease. Even if these are natural extracts or products of chemical synthesis, besides the therapeutic effect, they have some undesirable effects on us. The effect of a medicament on a person may be different than expected, and this being due to an interaction with another drug (polymedication) or with phytometabolites, beverages, or phytometabolic supplements that are co-administered (Chan, 2013).

A medicament interaction is a situation where a substance affects the activity of a drug such that the effects increase or decrease. Interactions are responsible even for causing new effects. These interactions may occur through improper use of or due to lack of knowledge about the active ingredients existing in the respective substances. In the case of oral medication administration, the interactions can be multiple. This is because both the medication and the phytometabolites or natural juices go through the same stages in the digestive system (Ferreira Silva & Rita Carvalho Garbi Novaes, 2014).

In causing phytometabolite-drug interactions, several factors are involved, such as the order in which both drugs and phytometabolites are administered, the interval at which they are administered, the physicochemical properties of the medicament molecules, and the quality of the diet. Also, even the amount of water with which the medication is administered is important, as oral medicines are most often solid pharmaceutical forms—tablets, capsules, dragees—therefore, the administration is done with the help of water. The amount of water ingested could influence the effect of the oral drug (Thitz et al., 2020).

Phytomedicine: A Treasure of Pharmacologically Active Products from Plants
https://doi.org/10.1016/B978-0-12-824109-7.00021-2

To be absorbed, the drug in the solid pharmaceutical form should first disintegrate into the digestive tract, and the rate of dissolution of a substance categorically depends on the amount of water in which the substance is dissolved. The same applies to phytometabolites or juices. Digestive absorption is one of the always rigorously studied parameters before introducing drugs into therapy.

Medications have better absorption when given on an empty stomach, because ingestion of phytometabolites concomitantly will lead to decreased absorption of medication. But this is not possible in the case of anti-inflammatories. These are gastric irritants, and their administration is indicated by the meal (Vora et al., 2020).

There are situations when the absorption of medicaments is influenced only by certain components of phytometabolites such as tetracyclines that are inactivated by calcium ions present in dairy products. Dairy administration in this case is not indicated.

However, due to the fact that during the antibiotic treatment it is indicated that the lactate should be given for restocking the colon with *Lactobacilus acidophylus* or for gastric protection, the administration will be done at a time interval of 2–3 h with respect to the antibiotic and not in the same doses. The main side effects of some phytometabolites on medicaments include changes in the absorption of fats, proteins, and fibers (Chen et al., 2020).

Bioavailability is an important pharmacokinetic parameter that is correlated with the clinical effect of most medicaments. To evaluate the clinical relevance of a phytometabolite-medicament interaction, the effect that phytometabolite ingestion has on the clinical effect of medications should also be considered.

The most important interactions are those associated with a high risk of treatment failure resulting from a significantly reduced bioavailability during the meal. Such interactions are frequently caused by chelation with phytometabolite components.

The physiological response to phytometabolite intake, particularly gastric acid secretion, may decrease or increase the bioavailability of certain medicaments. Medicament interactions may alter the pharmacokinetics and pharmacodynamics of a medicament.

Pharmacodynamic interactions may be additive, synergistic, or antagonistic to the effects of a medicament. Medicament interactions are of great importance and are widely recognized as a source of medical errors (Kaufman-Szymczyk, Majda, Szuławska-Mroczek, Fabianowska-Majewska, & Lubecka, 2019).

Gastrointestinal uptake of medicaments may be affected by the concomitant use of other medicaments that have a large surface area on which the medicament can be absorbed.

Alteration of gastric pH, slowing of gastrointestinal motility, or impairment of transport proteins, such as glycoprotein P, and factors such as atypical kinetics, lower solubility, and different protein ratios may alter the kinetic behavior of an enzyme and subsequently extrapolate in vitro data to humans.

For example, coenzyme Q-10 is widely consumed by humans as a phytometabolic supplement due to its recognition by the public as an important nutrient in supporting human health. It interferes with the intestinal efflux transporter glycoprotein P, thus phytometabolite-medicament interactions occur.

Studies have been conducted on medicament interactions, for example, medicament-medicament, phytometabolite-medicament, plant-medicament products, and genetic factors that affect pharmacokinetics and pharmacodynamics that are expected to enhance the safety of medicament administration and allow individualization of medicament therapy (Ratriyanto & Prastowo, 2019).

2. Phytometabolite-drug interaction

Interactions between medicaments and nutrients may cause a change in the pharmacokinetics and pharmacodynamics of a medicament or compromise the nutritional status as a result of their interaction. This can be either harmful or beneficial. Common side effects include nutritional deficiencies, increased medicament toxicity, loss of therapeutic efficacy, and unwanted physiological changes.

The definition of medicament-phytometabolite interactions is an interaction resulting from a physical, chemical, physiological, or pathophysiological relationship between a medicament and a nutrient, nutritional products in general, or with the nutritional status of the body. The clinical consequences of an interaction are related to pharmacotherapeutic changes and the effect of the medicament or nutrient (Ogata, Imaoka, Akiyoshi, & Ohtani, 2019).

Both pharmaceuticals and nutrients are metabolized by the action of a diverse group of enzymes. The activity of these enzymes is affected by both diet and genotype (for example, lactose intolerance), and this may represent a risk to the patient. Given the number of nutrients and dietary components that affect the immune response, an association involving medicaments and nutrients could become an increasingly important issue.

Starting from the general aspects regarding the transformations that the medicinal substances undergo as well as phytometabolites, we can deduce the importance of knowing the phytometabolite-drug interactions (Oikeh, Sakkas, Blake, & Kyriazakis, 2019).

Several factors can influence the risk of developing phytometabolite-drug interactions such as age, and in this case, there is usually a polymedication as well as malnutrition, in addition to affecting the functions of organs and the genetic characteristics of the body where the risk of interactions is increased. Another category of patients targeted in the development of interactions are patients with chronic pathologies who take several medicaments simultaneously.

Children, unlike adults, are growing and developing and need special attention when it comes to a treatment scheme. In fact, in the first decade of life, the growth is irregular and does not follow a linear path, thus making the dosage of medication as well as feeding more complicated (Boullata, 2019).

Moreover, during childhood, the body weight changes due to changing constituents of the organism (proteins, intra- and extracellular water, fat). In newborns, muscle mass is reduced, and they have lower concentrations of albumin and intracellular fluid than adults.

Changing the water content in the body can significantly affect the volume of distribution of medicaments, particularly those that are extremely hydrophilic. Changes in serum proteins (e.g., albumin) may alter the concentration of certain medicaments in the circulation and thus change the pharmacodynamic response of the patient. In the second year of life, the fat mass is reduced with a corresponding increase in the protein mass. As the child grows, there is a corresponding change in the size of the liver and kidneys. In the period from 1 to 2 years, each organ reaches the maximum relative size, when the capacity of metabolization and elimination of medicaments is increased. The absorption surface of the small intestine is proportionally larger, while the gastrointestinal transit is slower (Bozkurt et al., 2019).

During adolescence, there is an increase of approximately 25% in height, while the weight is almost doubled. Prior to puberty in men and women, muscle mass, skeletal mass, and body fat are similar, but at maturity, women have twice as much fat per total body weight compared with a man. The differences that exist in each period of life are a very important aspect in determining the pharmacokinetics of a medicament. Phytometabolites and medicaments cross the gastrointestinal tract, where they must pass through the absorption phase before reaching the systemic circulation and occupying the site of action (Bitencourt, Oliveira, Sanches, Rossi, & Martinez-Rossi, 2019).

Absorption from the gastrointestinal lumen into the hepatic portal vein and subsequently into the systemic circulation is a series of complex processes, including dissolving the solid dosage form, which passes chemically along the intestinal tract (e.g., intestinal transit, gastric emptying), passive diffusion, active transport, and metabolism of the compounds.

Each of these processes may affect the pharmacokinetics of the medicament. The oral route continues to be the most widely used route in the administration of medicaments, in which case absorption is made in the small intestine. Phytometabolites present concomitantly in the digestive tract can lead to interactions by decreasing absorption and consequently leading to low therapeutic efficacy. Given the widespread use of medicaments combined with nutritional variability, eating habits, and phytometabolite composition, the number of phytometabolite-medicament interactions is overwhelming (Olukosi & Bedford, 2019).

The clinical consequences of an interaction are related to pharmacodynamic alteration and the effect of the medicament or phytometabolite. Several factors may influence the risk of developing phytometabolite-medicament interactions. These include children, the elderly, pregnant women, people with chronic diseases, and patients with polymedication.

It is very important that the history of the phytometabolite is investigated when developing a therapeutic scheme because a high phytometabolite rich in lipids or carbohydrates as well as an insufficient phytometabolite are factors that can also decrease the therapeutic efficacy.

In the case of the elderly, we can talk about an increased frequency of side effects and the development of phytometabolite-medicament interactions because the nutritional status is disturbed in this category of persons, as well as functions of the different organs (e.g., liver, kidneys) and the genetic characteristics of the body such as the polymorphism of the carriers and enzymes (Gröber, Schmidt, & Kisters, 2020).

The disturbance of the nutritional status refers to the imbalances that arise as a result of the metabolism of an inadequate dietary intake or by altering the metabolism of some nutritional principles. Phytometabolite-drug interactions can disrupt health status, and their effect refers to the action of nutrients as well as drugs at the cellular level.

Phytometabolite-drug interactions can be evaluated more as disturbances in drug pharmacokinetics and less frequently in pharmacodynamics.

Pharmacokinetic interactions concern drug uptake along the digestive tract and may involve enzymes involved in drug absorption, distribution, or elimination (Nagai et al., 2019).

Pharmacokinetic interactions are defined as changes in nutrients or drug parameters (bioavailability, clearance). Among the most important pharmacokinetic interactions are those that influence the amount of active principle (drugs with low therapeutic index).

2.1 The influence of drugs on nutritional status

Medicament-induced changes in nutritional status may be a direct or indirect consequence of a chemical class of drug or a specific drug. Acknowledgment and acceptance of medicament-induced changes in nutritional status are required to achieve desired therapeutic effects. Medications can interfere with a person's nutritional status. Some medicaments may disrupt the absorption of a nutrient, whereas others affect the use and/or excretion of nutrients, particularly vitamins and minerals in the body. These effects can lead to nutrient deficiency (de Leon, He, & Ullevig, 2018).

Changes to the overall nutritional status or specific nutrients may also be dependent on several factors. Medications can affect the nutrition status of the patient by altering body weight, changing taste perception (thus decreasing consumption), decreasing nutrient absorption, altering macronutrient metabolism, or leading to the use of essential vitamins and minerals. Whether this is a result of the mechanism of action of the medicament or its adverse effects, the patient's nutritional status may be impaired (Gezmen-Karadağ et al., 2018).

Changes in medicament-induced nutritional status may be considered a subclass of adverse effects. One of the known side effects is weight gain.

Medications that induce this alteration of nutritional status include psychotropic, oral antidiabetics, corticosteroids, beta-blocking estrogen, and oral contraceptives. Insulin and sulfonylurea derivatives administered in diabetes are associated with weight gain. Studies have reported weight gain from 0.8 to 6.6 kg since the initiation of treatment. But higher doses have been associated with greater weight gain.

Most commonly reported for this adverse effect are psychotropic drugs. Psychotropic drugs associated with weight gain include amitriptyline, chlorpromazine, clozapine, imipramine, lithium, mirtazapine, olanzapine, risperidone, and ziprasidone (Liu et al., 2018).

Weight gain has been reported for testosterone, testosterone derivatives, and selective estrogen receptor modulators. Subsequently, they were used intentionally to facilitate weight gain in undernourished patients. Oxandrolone, a testosterone derivative, has been approved by the Food and Drug Administration (FDA) for the effect of weight gain in patients who have suffered heavy weight loss due to chronic infections or severe trauma. Testosterone ($C_{19}H_{28}O_2$) treatment allows for a decrease in fat mass while causing muscle mass gains in HIV-positive patients with low testosterone and abdominal obesity (Quiñones, Mardare, Hassel, & Brüggemann, 2019).

Medications associated with weight loss are the central nervous system stimulant drugs. While anorexic properties have been desired in obese patients, it seems that there are undesirable effects in children who have been treated with these drugs for attention deficit. This treatment resulted in weight loss in children but did not affect the increase in height (Gurley, Tonsing-Carter, Thomas, & Fifer, 2018).

Serotoninergic drugs, selective serotonin ($C_{10}H_{12}N_2O$) reuptake inhibitors, and serotonin receptor agonists have been correlated with weight loss in obese and nonobese individuals. The effect is mediated by the functions of the hypothalamus. Lamotrigine and topiramate antiepileptics also lead to weight loss. Other drugs that may have anorexic side effects include epirubicin ($C_{27}H_{30}ClNO_{11}$), fluvoxamine ($C_{15}H_{21}F_3N_2O_2$), sibutramine ($C_{17}H_{26}ClN$), and antihistamines. As for caffeine, it is thought to produce greater thermogenesis, lipolysis, fat oxidation, and insulin ($C_{257}H_{383}N_{65}O_{77}S_6$) secretion. Another effect produced by drugs on nutritional status is changes in the gastrointestinal level (Beardmore et al., 2018).

There are drugs that increase gastrointestinal tract motility or cause gastrointestinal intolerance such as abdominal pain, cramps, or diarrhea. There are effects similar to vomiting, but these effects are prolonged or severe. This can lead to poor nutrient absorption. Patients experiencing abdominal pain and cramps decrease nutrient intake simply due to decreased appetite. Aspirin, anti-inflammatories, and iron are known to cause gastrointestinal irritation. Metoclopramide and erythromycin are included in the category of drugs that increase gastrointestinal motility (Schlessinger et al., 2018).

As mentioned previously, decreased gastrointestinal motility leads to inadequate delivery of nutrients. Anticholinergics decrease gastrointestinal motility by blocking the action of acetylcholine. Another adverse effect reported by many drugs is emesis. Prolonged or severe vomiting is one that becomes a concern and will alter the absorption of nutrients; this category includes chemotherapies (cisplatin, cyclophosphamide, epirubicin). Alterations of a patient's metabolic function or macronutrient status may also be attributed to drugs (Da Costa Silva et al., 2018).

Drug-induced changes in carbohydrate or lipid metabolism have been reported with several drugs. The changes can range from a transient effect to one that could endanger the patient's life. Osteoporosis and pancreatitis are the most serious consequences induced by some drugs.

Drug-induced hyperglycemia can lead to fluctuations in a patient's metabolism. Episodes of medication-induced hyperglycemia may worsen glucose control in the diabetic patient as well as the possibility of subsequent diabetes development (Evans, Piccio, & Cross, 2018).

The first medicaments that have reported changes in glucose metabolism are antipsychotic medicaments. It has been observed that medicaments used in schizophrenia have increased blood glucose levels and are responsible for the production of type II diabetes. Atypical antipsychotics such as clozapine, olanzapine, and risperidone have a stronger effect on diabetes than conventional antipsychotics (haloperidol). In a study conducted on 200 schizophrenic patients, 7% had adverse effect hyperglycemia and 5% had diabetes. Beta-blockers, a class of antihypertensive medicaments, cause hyperglycemia as an adverse effect, particularly in diabetic patients, which results in obstruction of insulin secretion. Corticosteroids as well as oral contraceptives are responsible for the hyperglycemic effect. Other medicaments commonly reported in the induction of hyperglycemia include alcohol, atypical antipsychotics, caffeine, calcium channel blockers, morphine ($C_{35}H_{44}O_{16}$), nicotine ($C_{10}H_{14}N_2$), phenytoin ($C_{15}H_{12}N_2O_2$), protease inhibitors, theophylline, thiazide diuretics, etc. Serotonin reuptake inhibitors have reported cases of hypoglycemia, as well as some oral antidiabetics (Reichenbach et al., 2018).

2.2 The influence of phytometabolites on drug metabolism

Phytometabolite ingestion has a complex influence on drug bioavailability. This can interfere not only with the process of drug disintegration, i.e., dissolution but also with the route of drugs through the gastrointestinal tract. It may affect the metabolic transformation of drugs in the intestinal wall and liver. Different components of phytometabolic products can have unexpected effects, and phytometabolic products can interact in opposite ways, even with chemically linked drugs. Therefore, the net effect of phytometabolites on the bioavailability of the medicament can only be predicted by direct clinical studies of the medicament in question (Nagai et al., 2018).

In treatment with anticonvulsants (lamotrigine, gabapentin), serum and sodium iron should be monitored. In addition, patients are advised to avoid the consumption of grapefruit juice within 1–2 h of their administration. Active ingredients in grapefruit include furanocoumarin and bioflavonoids, which are inhibitors of organic anion transporter polypeptides (OATP) when co-administered and may decrease the oral availability of medicaments that are OATP substrates, for example, fexofenadine ($C_{32}H_{40}ClNO_4$) (Li, An, Xu, & Tuo, 2019).

In general, a number of flavonoids present in grapefruit are identified as esterase inhibitors, of which narginin ($C_{15}H_{12}O_5$) and kaempferol ($C_{15}H_{10}O_6$) are responsible for pharmacokinetic interactions with several drugs: calcium channel blockers and statins, such as felodipine ($C_{18}H_{19}Cl_2NO_4$) and lovastatin ($C_{24}H_{36}O_5$). Lovastatin, a drug used to lower cholesterol ($C_{27}H_{46}O$), should be taken in combination with phytometabolites to enhance gastrointestinal uptake and bioavailability. The absorption of rosuvastatin, another hypolipidemic drug, was significantly reduced during meals compared with fasting conditions, suggesting that rosuvastatin should be administered on an empty stomach. Simvastatin and fluvastatin can be administered with phytometabolites without affecting them (Andrus, 2004).

Diets high in fiber can decrease the effectiveness of these medicaments. Co-administration of statins with phytometabolites may alter their pharmacokinetics or pharmacodynamics, increasing the risk of adverse effects such as rhabdomyolysis or reducing their pharmacological action. Consumption of pectin or oat bran along with lovastatin decreases its absorption, while alcohol consumption does not appear to affect the efficacy and safety of fluvastatin ($C_{24}H_{26}FNO_4$) treatment. Warfarin ($C_{19}H_{16}O_4$) is commonly used to treat or prevent thrombosis and thromboembolism. The possible interactions of this medicament are correlated with a diet rich in protein (Kim et al., 2006).

Some vegetables (parsley, Brussels sprouts, broccoli, spinach, cabbage, and others) are rich in vitamin K. Consumption in large quantities may interfere with the efficacy and safety of warfarin therapy. Administration of warfarin with cranberry juice appears to lead to an increase in International Normalized Ratio (INR) but no bleeding occurs. There have been a number of studies on the interaction between cranberry juice and warfarin. It appears that cranberry juice contains a flavonoid responsible for inhibiting the cytochrome P450 (CYP) iso-enzyme CYP2C9, the major enzyme involved in the metabolism of warfarin (Doostikhah, Panahpour, Nadian, & Gholami, 2020).

However, it has not been established whether the increase in INR was due solely to the administration of cranberry juice. Monoamine oxidase inhibitors (MAOIs) used in the treatment of depressive disorders are subject to interactions with phytometabolites containing tyramine, an amino acid that acts as a neurotransmitter.

The association of the MAOI with foods such as cheese, bananas, chocolate, yogurt, etc. can lead to hypertensive crises due to the tyramine content.

For patients treated with antihypertensives, a poor diet of sodium is required first (Oh et al., 2020).

There are particular restrictions on beta-blockers such as propranolol whose concentration can be increased if it is associated with a high protein diet. Celiprolol ($C_{20}H_{33}N_3O_4$), another beta-blocker, has low absorption when is taken with orange juice. It seems that the hesperidin contained in the orange juice is responsible for this low absorption.

Antibiotics are very often prescribed in medical practice, although many of them induce medicament interactions that may diminish their anti-infectious effect or may precipitate certain adverse effects (Holguín, Amariles, & Ospina, 2017).

Phytometabolic intake may affect the effectiveness of an antibiotic, interfering with its pharmacokinetics. An example of this would be the complexation of tetracyclines with calcium ions in dairy products, therefore the co-administration of milk products with tetracyclines is avoided to prevent their absorption. Numerous studies have shown that quinolones form soluble complexes with metal ions from ingested phytometabolites, leading to a dramatic decrease in bioavailability. A concrete example of this is the decrease in the absorption of ciprofloxacin by casein and calcium present in milk but also the interaction of dissolution and absorption of ciprofloxacin tablets with certain types of natural juices (García-Lino, Álvarez-Fernández, Blanco-Paniagua, Merino, & Álvarez, 2019).

The absorption of azithromycin on the other hand decreases with the co-administration of phytometabolites, resulting in a lower bioavailability by 43%. Phytometabolite-medicament interactions can greatly decrease the bioavailability of medicaments taken after meals, but the use of enterosoluble tablets that begin to break down in the lower region of the small intestine can decrease unwanted effects. Analgesics and antipyretics are used to treat mild to moderate pain and treat fever. For rapid enhancement, acetaminophen should be taken on an empty stomach because phytometabolites can slow its absorption. NSAIDs such as ibuprofen, naproxen, ketoprofen, and others may cause irritation of the stomach lining and therefore should be taken with phytometabolites or milk (Sharma, Kapoor, & Kaur, 2020).

Alcohol consumption is also avoided, as chronic alcohol intake may increase the risk of liver damage or the risk of gastric ulcer. Another phytometabolite-medicament interaction is with ibuprofen co-administration and cola when the maximum concentration of the medicament increases after single or repeated ingestion of juice, leading to a high degree of absorption of ibuprofen. Therefore, both the dosage of ibuprofen and the low dose of the medicament should be adjusted when consuming cola (Johnson & Seneviratne, 2014).

Bronchodilators such as theophylline, albuterol, and epinephrine have different effects when administered with phytometabolites.

The effect of phytometabolites on theophylline may vary widely, as high-fat meals may increase its absorption in the body, while high-carbohydrate meals may decrease it. It is recommended that alcohol be avoided during theophylline treatment, as it may increase the risk of side effects such as nausea, vomiting, headache, and irritability.

At the same time, avoid ingestion of large amounts of phytometabolites and beverages containing caffeine (e.g., chocolate, cola, coffee, and tea), as theophylline is a xanthine derivative and these substances also contain xanthine.

Therefore, the consumption of these substances in large quantities during theophylline treatment greatly increases the risk of theophylline toxicity, in addition, both bronchodilators and caffeine are central nervous system stimulants (Kotwal et al., 2020).

Mercaptopurine, a purine analog, is inactivated by xanthinoxidase, and concomitant use of phytometabolites containing this enzyme may decrease the bioavailability of the medicament. Cow's milk is known for its content in this enzyme, which is why it is recommended to space the administration of mercaptopurine from cow's milk. Tamoxifen is indicated not to be taken with sesame seeds because it interferes negatively with its pharmacodynamics.

The results indicated that phytometabolite-medicament interactions were avoided by separating the main site of medicament absorption from that of phytometabolites (Ogungbenro et al., 2014).

To enhance the effectiveness of antihistamines, it is recommended that their administration be done on an empty stomach. According to studies on rupatadine, it has been shown that the concomitant intake of phytometabolic at a dose of 20 mg of rupatadine significantly increases the bioavailability of the active substance.

Cimetidine on the other hand is indicated to be given with a phytometabolic because some of the ingested amount of medicament is absorbed in the presence of the phytometabolic, and the remaining fraction of medicament is eliminated, thus ensuring an effective therapeutic dose during administration (Nishiyama et al., 2019).

Antituberculosis medicaments such as isoniazid interact with molecules such as histamine and tyramine, because they inhibit monoamine oxidase and histaminase leading to numerous phytometabolite-medicament interactions, which is why phytometabolics greatly decrease the bioavailability of isoniazid. On the other hand, the widespread oleanolic acid ($C_{30}H_{48}O_3$) in phytometabolites and medicinal plants also has antituberculosis activity against *Mycobacterium tuberculosis*, and the synergistic therapeutic effect is manifested with isoniazid. As for the absorption of cycloserine, an antituberculostatic agent, this is considerably diminished in the presence of fat-rich phytometabolites, which can lead to therapeutic failure.

Glimepiride, an antidiabetic in the class of sulphonyl urea derivatives, should be given during a morning meal, because it has absolute bioavailability and does not interact at all with phytometabolites, so the therapeutic effect is provided every time.

Metformin should also be given during a meal, as it is decreased by the gastrointestinal adverse effects (nausea, vomiting, diarrhea) (Zsila, Bősze, & Beke-Somfai, 2020).

Immediate release antidiabetic glipizide should be taken 30 minutes before meals. Maximum efficiency of acarbose, an alpha-glucosidase inhibitor, is achieved when the medicament is given immediately at the beginning of each meal (not half an hour before or after), because it delays carbohydrate absorption by inhibiting the alpha-glucosidase enzyme.

Medications can only be shown to be effective if they are administered in an adequate amount with a correct combination of other medicaments and phytometabolites at the right time. A large number of medicaments are placed on the pharmaceutical market each year without any known pharmacovigilance elements.

Phytometabolite-medicament interactions can have adverse effects both on the safety and efficacy of medicament therapy, as well as on the nutrition status of the patient.

Generally speaking, medicament interactions are relatively easy to avoid due to the knowledge of their pharmacokinetic aspects. Like phytometabolics, oral medications should be absorbed through the lining of the stomach or small intestine (Naseeruddin et al., 2018).

The presence of phytometabolites in the digestive tract can decrease the absorption of a medicament. Often, these interactions can be avoided by administering the medication 1 h before or 2 h after a meal. On the other hand, phytometabolic medicaments are not sufficiently tested to fully understand their metabolism, so they can interact with prescribed or over-the-counter medicaments.

3.   Not recommended combinations of drugs with phytometabolites

Energizing drinks and blood pressure control medicaments.

Energizing drinks and those specially designed for athletes can be harmful when combined with blood pressure control medicaments. At the same time, juices containing bananas and kiwifruit can have adverse effects if used for the administration of high blood pressure medicaments due to their high potassium and magnesium content.

Coffee and medicaments containing ephedrine.

Coffee interacts and can have adverse effects if combined with medicaments containing ephedrine or other central nervous system stimulants. Coffee can block the effect of some medicaments in the treatment of asthma (Nomani et al., 2019).

Green tea and medicaments that control blood-clotting processes.

Green tea can lessen the effect of these medicaments (Albassam & Markowitz, 2017).

Chocolate and ritalin.

Chocolate contains a stimulant, called theobromine. The combination of these molecules in the human body can lead to incoherent behavior and various seizures when administered with ritalin. Ritalin is a psychostimulant recommended for patients suffering from hyperactivity or attention deficit disorder (Wells & Losin, 2008).

Citrus fruits and cough syrup.

Citrus fruits contain a significant amount of vitamin C. Some of them, such as green lemon, grapefruit, or certain varieties of oranges, are forbidden in combination with cough syrup. These fruits block the enzyme that metabolizes dextromethorphan, an antitussive medicament.

The molecule accumulates in the blood without being degraded, and the side effects increase, such as hallucinations and drowsiness. Citrus juices interact with more than 50 medicines, so the best extended solution is to never take citrus juice with medicines (Quirino et al., 2018).

Lactates and antibiotics.

Certain antibiotics, particularly ciprofloxacin, become ineffective if they come in contact with calcium, iron, and other dairy minerals. At the same time, milk, due to the calcium it contains, can interfere with thyroid medication.

It is recommended that between the time you take the medicines and the time when you have been drinking milk, you should spend at least 4 h (Stewart et al., 2019).

Smoked salmon and antidepressants.

Old-generation antidepressants interact with smoked foods, chocolate, beer, red wine, soy sauce, or aged cheese. Demonoamine, oxidase inhibitors from antidepressants, interact with tyramine in foods and can cause hypertension attacks with life-threatening potential (Cooper et al., 2018).

Fruit juice and allergy medicines.

Avoid apple, orange, or grapefruit juice if you are receiving fexofenadine treatment for pollen allergy at least 4 h after treatment. These juices inhibit the effect of the action of a peptide that transports the medicament's molecules from the stomach into the blood (Paśko et al., 2017).

Pomegranate juice and blood pressure-regulating medicaments.

Pomegranate juice is contraindicated in combination with blood pressure medicaments. Certain enzymes from pomegranate juice cancel the effect of these pills (Lu et al., 2015).

Red wine and antidepressants.

It is contraindicated to consume red wine during antidepressant treatment. The combination of red wine and antidepressants can lead to stroke, severe changes in blood pressure, and headaches.

A similar effect is found in a combination of antidepressants and energy drinks. In many cases, phytometabolites and therapeutic plants interact with the medication, and the consequences can be severe. This is because phytometabolites can alter the way the body metabolizes medicaments. Alcohol is a substance that interacts with virtually all medicaments, particularly antidepressant medication and antibiotics.

Also, the action of anticoagulant medicaments can be influenced by excessive consumption of spinach, broccoli, and green leafy vegetables containing vitamin K, which promotes blood coagulation. Excess dietary fiber can affect the absorption of painkillers (pain relievers - acetaminophen) and heart medication (digoxin) (Ndou et al., 2019).

Garlic has antihypertensive action and, when associated (in excess) with medication to treat high blood pressure, can lead to severe hypotension. Fermented cheese (with mold, parmesan) and alcoholic beverages interact with antidepressant medication, the main reaction felt at the level of blood pressure. Grapefruit juice or fruit consumed as such interacts with anticholesterol, contraceptive, and antiallergic medication (Ali, 2018).

Milk and orange juice may interact with the therapeutic effect of antibiotics. Caffeinated beverages can alter the action of antiasthma medication due to the excessive excitability that can be reached. Long-cooked meat on the grill can inhibit antiasthmatic medication, particularly ones containing theophylline. During the administration of a medicinal treatment, strict control of the diet to be followed is required, to maximize the therapeutic efficiency of the treatment and also to ensure the energy and nutritional needs of the food (McKetin, Coen, & Kaye, 2015).

4. *Citrus paradisi* (grapefruit) juice and drugs

Medicaments and phytometabolites are often taken together. The administration of medicines related to the time of the meal can enhance the patient's adherence to the management of regimen, particularly in the elderly. Certain phytometabolites may create an interaction that may increase or decrease the bioavailability of systemic medicaments resulting in modified clinical effects (Abdlekawy, Donia, & Elbarbry, 2017).

The natural compounds contained by *Citrus paradisi* may interact with drugs depending on the type of drug, as can be seen in Table 1.

Grapefruit (*Citrus paradisi*) juice contains furanocoumarins (bergamottin $[C_{21}H_{22}O_4]$, epoxybergamottin $[C_{21}H_{22}O_5]$, and 6′,7′-dihydroxybergamottin $[C_{21}H_{24}O_6]$), flavonoids (naringenin $[C_{15}H_{12}O_5]$, naringin $[C_{27}H_{32}O_{14}]$), and sesquiterpen (nootkatone $[C_{15}H_{22}O]$), bergapten (5-methoxypsoralen $[C_{12}H_8O_4]$), polyamines (e.g., putrescine $[C_4H_{12}N_2]$), and limonoids-furanolactone core structure (limonin $[C_{26}H_{30}O_8]$, nomilin $[C_{28}H_{34}O_9]$, azadirachtin $[C_{35}H_{44}O_{16}]$, nomilinic acid $[C_{34}H_{48}O_{16}]$). Major classes of phytometabolites from grapefruit (*Citrus paradisi*) are schematically represented in Fig. 1.

Research into the interactions between fruit juice and medicines began with a single crucial unanticipated secondary investigation in a clinical study that led to a very important inquiry over 25 years ago. It was discovered that *Citrus paradisi* juice could significantly increase the bioavailability of oral medicaments.

TABLE 1 Medicament influenced by *Citrus paradisi* juice

Medicament class	Major interactions	Minor interactions
Antiarrhythmic medication	Amiodarone ($C_{25}H_{29}I_2NO_3$)	–
	Dofetilide ($C_{19}H_{27}N_3O_5S_2$)	–
	Dronedarone ($C_{31}H_{44}N_2O_5S$)	–
Calcium channel blockers (CCB) drugs	–	Felodipine ($C_{18}H_{19}Cl_2NO_4$)
	–	Isradipine ($C_{19}H_{21}N_3O_5$)
	–	Nicardipine ($C_{26}H_{29}N_3O_6$)
	–	Nifedipine ($C_{17}H_{18}N_2O_6$)
	–	Nimodipine ($C_{21}H_{26}N_2O_7$)
	–	Nisoldipine ($C_{20}H_{24}N_2O_6$)
Cough suppressant/NMDA antagonist drugs	Dextromethorphan ($C_{18}H_{25}NO$)	–
Erectile dysfunction (ED) drugs	–	Sildenafil ($C_{22}H_{30}N_6O_4S$)
	–	Tadalafil ($C_{22}H_{19}N_3O_4$)
	–	Vardenafil ($C_{23}H_{32}N_6O_4S$)
HIV protease inhibitors (PIs) drugs	–	Amprenavir ($C_{25}H_{35}N_3O_6S$)
	–	Nelfinavir ($C_{32}H_{45}N_3O_4S$)
	–	Ritonavir ($C_{37}H_{48}N_6O_5S_2$)
	–	Saquinavir ($C_{38}H_{50}N_6O_5$)
Hormones drugs	–	Ethinylestradiol ($C_{20}H_{24}O_2$)
	–	Methylprednisolone ($C_{22}H_{30}O_5$)
Immunosuppressant drugs	–	Ciclosporin ($C_{62}H_{111}N_{11}O_{12}$)
	–	Mercaptopurine ($C_5H_4N_4S$)
	–	Sirolimus ($C_{51}H_{79}NO_{13}$)
	–	Tacrolimus ($C_{44}H_{69}NO_{12}$)

TABLE 1 Medicament influenced by *Citrus paradisi* juice—cont'd

Medicament class	Major interactions	Minor interactions
Statins (HMG-CoA reductase inhibitors) drugs	Lovastatin ($C_{24}H_{36}O_5$)	Atorvastatin ($C_{33}H_{35}FN_2O_5$)
	Simvastatin ($C_{25}H_{38}O_5$)	Cerivastatin ($C_{26}H_{34}FNO_5$)
Sedatives, hypnotics, and anxiolytics drugs	Buspirone ($C_{21}H_{31}N_5O_2$)	Alprazolam ($C_{17}H_{13}ClN_4$)
		Clonazepam ($C_{15}H_{10}ClN_3O_3$)
		Diazepam ($C_{16}H_{13}ClN_2O$)
		Ketamine ($C_{13}H_{16}ClNO$)
		Midazolam ($C_{18}H_{13}ClFN_3$)
		Triazolam ($C_{17}H_{12}Cl_2N_4$)
		Zaleplon ($C_{17}H_{15}N_5O$)
Other psychotropics drugs	–	Carbamazepine ($C_{15}H_{12}N_2O$)
	–	Fluvoxamine ($C_{15}H_{21}F_3N_2O_2$)
	–	Nefazodone ($C_{25}H_{32}ClN_5O_2$)
	–	Quetiapine ($C_{21}H_{25}N_3O_2S$)
	–	Trazodone ($C_{19}H_{22}ClN_5O$)
Other miscellaneous drugs	Cisapride ($C_{23}H_{29}ClFN_3O_4$)	–
	Ivabradine ($C_{27}H_{36}N_2O_5$)	–

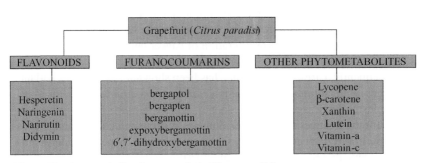

FIG. 1 The phytometabolites from grapefruit (*Citrus paradisi*).

The mechanism was obstruction of medicament metabolism, which was probably the first detecting of a phytometabolite capable of causing such an effect in humans. Subsequently, many scientists started from this finding, with hundreds of research articles investigating a number of issues related to this problem (Misaka et al., 2011).

4.1 Interactions of *Citrus paradisi* juice with drugs

Citrus paradisi primarily attenuates the intestinal activity of CYP and isoenzyme (CYP3A4). Moreover, the interaction of *Citrus paradisi* juice with over 50 medicaments has been demonstrated, some of which are widely used in medical practice by decreasing the concentration from the first liver passage. Orally administered medicaments have the same pathway in the digestive system as phytometabolites or natural juices (Baltes, Dubois, & Hanocq, 2001).

In causing the phytometabolite-medicament interactions, the physicochemical properties of the medicament molecules, the quality and composition of the food, the order in which the medicaments and the phytometabolites are ingested, and the time intervals between them becomes more and more imperative during the administration of a medicament, and control of the diet for the therapeutic efficiency of the treatment also ensures the energetic and nutritional needs of the phytometabolites (Fleisher, Unum, Shao, & An, 2015).

Regarding the natural juices, properly prepared and administered, they cannot supply the 20–25 g of fiber daily that adults need. Yet at the same time, for maximum benefit, it is recommended that the juices are prepared in the home and consumed immediately after preparation, not those in supermarkets (even if they are declared 100% natural), because pasteurization (practiced to increase the shelf life) destroys enzymes and vitamins in juices. For maximum benefit, it is recommended to use fresh fruits and vegetables grown in the country (they are fresher than imported) without pesticides (which we do not yet know all the long-term effects) and additives for growth and ripening (Johnson, Won, Köck, & Paine, 2017).

The widespread consumption of *Citrus paradisi* is not attributed solely to its taste or nutritional value. In fact, its consumption is due to medical research that has suggested that *Citrus paradisi* juice decreases the formation of atherosclerotic plaques and inhibits the proliferation of cancer cells in the breast, as well as breast tumorigenesis.

Subsequently, *Citrus paradisi* juice has been shown to contain antioxidants, substances with antinitrosamine action, antiseptic, cardiotonic, detoxifying, hypocholesterolemic agents, and sedatives. *Citrus paradisi* has long been indicated for the treatment of anorexia, diabetes, various infections, benign prostatic hypertrophy, cancers (prostate, skin, colon, breast), hypercholesterolemia, insomnia, and mycosis (Yamasaki et al., 2020).

Data from the literature indicate a significant decrease in cholesterol levels in people who consume one *Citrus paradisi* daily, which is why *Citrus paradisi* has entered the world diagram of foods that decrease the incidence of heart disease. In 1991, the first information on the interaction between *Citrus paradisi* juice and some medicaments, calcium channel blockers (felodipine and nifedipine), was reported.

This phenomenon was discovered by accident following a study of ethanol-medicament interaction in which *Citrus paradisi* juice was used as a vehicle for alcohol used in the study (to mask the taste of alcohol). Medicament interactions are a major source of therapeutic failures, particularly in polymedicine situations (Khuda et al., 2019).

Pharmacokinetic mechanisms that may induce medicament interactions are more difficult to predict for clinicians than pharmacodynamic interactions that can be anticipated based on the pharmacological action of the medicaments administered.

During the various phases of medicament kinetics, metabolism and in particular biotransformation catalyzed by CYP can lead to significant interactions that are manifested in intestinal uptake or renal excretion (Ushijima et al., 2018).

Most oral medications are absorbed mainly in the small intestine. Phytometabolites, present simultaneously in the digestive tract, can influence the effectiveness of the active medicament principles by diminishing or, on the contrary, intensifying the therapeutic effects.

Interactions often involve the absorption of medicaments along the digestive tract, but certain nutrients alter the metabolism and elimination of medicaments. Phytometabolites can influence both the rate of absorption and the bioavailability of a medicament (Chen, Zhou, Fabriaga, Zhang, & Zhou, 2018).

The *Citrus paradisi* fruit contains impressive amounts of ascorbic acid (vitamin C), a true value for the immune system; it decreases the symptoms of cold, relieves the symptoms of rheumatoid arthritis or asthma, and decreases the risk of heart attack, heart disease, and cancer.

Grapefruit contains native sugars and volatile chemical compounds (essential oils), such as limonene, which has anticancer effects, retinol, retinal (vitamin A), and pantothenic acid (vitamin B5). Pectin ($C_6H_{10}O_7$) from *Citrus paradisi* fruit is a soluble fiber that lowers low-density lipoprotein (LDL) cholesterol levels and decelerates the accumulation of plaque on the inner walls of blood vessels. Red *Citrus paradisi* contains lycopene, which impedes the destructive operation of low-density lipoprotein (LDL) cholesterol on the walls of blood vessels.

This is why *Citrus paradisi* fruit are included in foods that decrease the incidence of heart disease. In the mineral class, we find higher quantity of calcium, folic acid, phosphorus, and potassium. *Citrus paradisi* contains enzymes that burn fat and help in the absorption and reduction of starch and sugar in the body (Zargar, Al-Majed, & Wani, 2018).

Investigation has shown that there is a physiological link between *Citrus paradisi* and insulin ($C_{257}H_{383}N_{65}O_{77}S_6$), as the chemical properties of this fruit decrease insulin levels and support weight loss. This relationship is explained by the weight management hormone.

Although not its main function, insulin helps regulate fat metabolism, so that the lower the amount of insulin after a meal, the more efficient the body processes the food to use it as energy, and so less of is deposited in the form of fat. *Citrus paradisi* has a low glycemic index of 30, so metabolism can burn fat. *Citrus paradisi* has chemical properties that decrease insulin levels, leading to weight loss. Studies have shown that substances from citrus fruits can help decrease the risk of cancer. Researchers have shown that lyophilized *Citrus paradisi* can decrease the incidence of cancer-related injury in animals. Red *Citrus paradisi* is rich in lycopene, a substance with antitumor effects, which fights against free radicals that destroy cells (Grimm et al., 2018).

Specialists are still investigating this phenomenon because there are conflicting claims that *Citrus paradisi* contributes to increased levels of estrogen in the blood, a hormone associated with breast cancer risk. *Citrus paradisi* and *Citrus paradisi* juice have the potential to interact with many medicaments. Organic compounds that are derived from furocoumarin interfere with the hepatic and intestinal enzyme of CYP, the CYP 3A4 isoenzyme, and are considered to be the main substances responsible for the action of the fruit on the enzymes.

Bioactive compounds in *Citrus paradisi* juice can interfere with P-glycoprotein and organic polypeptide transporter anions (OPTAs) by increasing or decreasing the bioavailability of a number of medicaments (Shang et al., 2018).

Pomelo, an Asian fruit, was crossed with an orange to obtain the *Citrus paradisi*, which in turn contained a large amount of furanocoumarin derivatives (furan ring combined with coumarin). *Citrus paradisi* is obtained from hybrids between pomelo and different varieties of orange. The following medicaments are affected by CYP3A4 obstruction due to *Citrus paradisi* compounds: benzodiazepines, triazolam, orally administered midazolam, orally administered alprazolam, diazepam, and nitrazepam. CYP3A4 obstruction prevents the metabolism of protease inhibitors such as ritonavir and sertraline and some statins, such as lovastatin, simvastatin, and atorvastatin, but nonetheless rosuvastatin, pravastatin, and fluvastatin are not affected by *Citrus paradisi* (Dresser et al., 2017).

Citrus paradisi juice interacts with antiarrhythmic medicaments, including propafenone, amiodarone, disopyramide, carvedilol, and quinidine and antimigraine medicaments such as amitriptyline, nimodipine, and ergotamine. Blood levels are increased if cyclosporine is taken with *Citrus paradisi* juice. A plausible mechanism would be the combined obstruction of enteric CYP3A4 and glycoprotein P, which could lead to serious adverse events (e.g., nephrotoxicity). In the case of tacrolimus, blood levels can be equally affected for the same reason as cyclosporine, as both medicaments are calcineurin inhibitors (Ohkubo, Chida, Kikuchi, Tsuda, & Sunaga, 2017).

Although no official studies have been conducted with imatinib (a tyrosine kinase inhibitor used in chronic myeloid leukemia and the treatment of gastrointestinal stromal tumors), the fact that *Citrus paradisi* juice is a known inhibitor of CYP3A4 suggests that concomitant administration may increase plasma concentrations of imatinib. Although formal studies have been conducted, co-administration of imatinib with another type of citrus juice (citron, lime, lemon, grapefruit, and orange) or appointed *Seville orange* juice may lead to lifted plasma quantities of imatinib by inhibiting CYP3A isoenzymes. Seville orange juice is not currently consumed as a juice because of its bitter taste, but it is found in marmalade and other products. Consumption of this orange juice has been associated to be a feasible CYP3A enzyme inhibitor without affecting glycoprotein P when consumed at the same time with cyclosporine (Masuda, Watanabe, Tanaka, Tanaka, & Araki, 2018).

Anthelmintics (used to treat parasitic infections) include albendazole, mebendazole, carbamazepine, and praziquantel; *Citrus paradisi* or *Citrus paradisi* juice slows the breakdown of these medicaments resulting in increased blood levels. Blood concentrations of acetaminophen and paracetamol have been shown to increase with white and pink *Citrus paradisi* juice, with white *Citrus paradisi* juice acting faster. A single glass of *Citrus paradisi* juice has the potential to increase oral bioavailability to enhance the beneficial or negative effects of many medicaments even if the juice was consumed a few hours in advance. It has been found that *Citrus paradisi* juice has an effect for 72 h, and specialists recommend 24 h between the drinking of juice and administration of medicaments to achieve the desired therapeutic effect. *Citrus paradisi* juice acts by inhibiting the metabolism of the CYP3A isoenzyme-mediated presystemic medicaments in the small intestine (Kashihara et al., 2017).

The interaction appears to be particularly relevant, with at least a doubling of the plasma concentration of the medicament, with a steep concentration-response relationship or a narrow therapeutic index. Isolation of the active ingredient may lead to the identification of phytometabolites that produce this interaction or to incorporation into pharmaceutical formulations (Mouly, Lloret-Linares, Sellier, Sene, & Bergmann, 2017).

Further investigation is needed to better comprehend the interaction during *Citrus paradisi* juice consumption at quantities considered safe for administration with medicaments and with different patient types. Increased plasma concentrations of felodipine with *Citrus paradisi* juice have provided new fundamental insights to enhance pharmacotherapy and encourage investigation (Santes-Palacios, Romo-Mancillas, Camacho-Carranza, & Espinosa-Aguirre, 2016).

4.2 Study of the pharmacokinetic interaction between DCF and *Citrus paradisi* juice

Medicament interactions are a major source of therapeutic failures. The probability increases in the case of polymedication. The nutrition of patients as well as their ingredients can have a significant effect on the pharmacokinetics and pharmacodynamics of the medication.

These can affect the absorption, distribution, metabolism, transport, and excretion of medicaments. Body weight and other associated pathologies are also important factors that influence the correct dosage of the medication (Theile et al., 2017).

The various components of phytometabolites can have an effect on blood flow, enzymatic activity, gastric secretions, and gastrointestinal motility, which ultimately affect medicament metabolism. Interactions are highly variable, complex, and often difficult to predict. The incidence of phytometabolite-medicament interactions appears to be higher during periods with increased demands (e.g., during growth, pregnancy, and lactation) (Tsuji et al., 2016).

The effect that phytometabolites have on the action of the medication by modifying the pharmacokinetic and pharmacodynamic parameters should be considered when performing the therapeutic schemes. By definition, pharmacokinetics depicts the action of medicaments in the body, processes of absorption, distribution in tissues, metabolism, and elimination. This contrasts with pharmacodynamics, which is the study of biochemical and physiological repercussions, the mechanism of action, including the correlation of actions and repercussions with their chemical structure, as well as the repercussions of the action on another medicament or phytometabolite. The medicament interactions induced by the pharmacokinetic mechanisms are more difficult to predict than the pharmacodynamic interactions that can be anticipated based on the pharmacological action of the medicaments. Throughout the different phases of medicament kinetics, metabolism and biotransformation catalyzed by P450 cotrimes may produce interactions that are manifested in the intestinal absorption or renal excretion.

The presence of phytometabolites simultaneously with medicaments in the small intestine (in the case of oral administration) leads to intensification or diminution of therapeutic repercussions (Hu et al., 2016).

Nutrients can alter bioavailability and rate of absorption as well as elimination of medicaments. In contrast to medicament interactions, where access to information is easier, there are more studies done, and it is a more debated topic; in the case of phytometabolite-medicament interactions, there is much less information as well as studies, although the importance is probably just as great (Akamine et al., 2015).

This is a difficult and complex problem for accurately determining the repercussions that phytometabolites/nutrients can have in combination with medicaments. Most people under medicament therapy associate this therapy with a higher consumption of fruits, natural juices, and vegetables to speed up the healing process. It is a concept supported by many doctors who encourage the increased consumption of fruits and juices at the time of medicament therapy for their vitaminic effect and for strengthening the immune system, which certainly helps in achieving the desired therapeutic repercussions as quickly as possible (Holmberg et al., 2015).

For example, in the case of diabetes, the administration of natural juices and fruits is limited. Moreover, everything we ingest can interact with something else in the body, so phytometabolite-medicament interactions occur. It is also important that the consumption of natural juices is more beneficial when it is made in the house and not ones bought that contain many other ingredients that can lead to other possible interactions and that, as a result of the technological processes they go through, have less vitamins and enzymes. Of all the fruit juices, *Citrus paradisi* juice interacts with almost all classes of medicaments (Qinna, Ismail, Alhussainy, Idkaidek, & Arafat, 2016).

The juice changes the body's way of metabolizing medicaments and affects the liver's ability to do this step. A case of porphyria has been reported associated with concomitant ingestion of cilostazol, aspirin, and *Citrus paradisi* juice in a 79-year-old man. His porphyria disappeared when he stopped taking *Citrus paradisi* juice, although the medicaments were not changed (Takahashi, Onozawa, Ogawa, Uesawa, & Echizen, 2015).

The most likely cause of this is an increase in the level of cilostazol in the blood because of an obstruction of antiplatelet metabolism due to the components of *Citrus paradisi* juice.

Phytometabolite-medicament interaction is a common problem encountered in clinical practice and is based on the same pharmacokinetic and pharmacodynamic principles as with medicament interactions (Markert et al., 2014).

Several fruits and berries have recently been shown to contain substances that affect medicament metabolism by inhibiting the enzymes required for this process. *Citrus paradisi* is the best-known example but also other fruits such as Sevilla oranges, pomelo, and more "rare" fruits containing substances that inhibit CYP, 3A4 (CYP3A4), which is the most important enzyme in the metabolism of medicaments (Ainslie et al., 2014).

Numerous reports have demonstrated medicament interactions with *Citrus paradisi* juice that occur by inhibiting the CYP 3A4 enzyme. Furanocoumarin present in *Citrus paradisi* inhibits CYP3A4 and has been shown to increase the oral bioavailability of some medicaments that are substrates of the CYP 3A4 enzyme, such as felodipine, midazolam, and cyclosporine, and to increase their toxic concentrations. *Citrus paradisi* is generally contraindicated in patients who have taken psychotropic medicaments, and it is indicated that these patients should be informed about this aspect. In vitro data suggest that compounds from *Citrus paradisi* juice are capable of inhibiting P-glycoprotein activity by altering the arrangement of medicaments that are substrates of P-glycoprotein, such as talinolol. The administration of many medicaments with *Citrus paradisi* juice has been shown to lead to a fivefold increase in adverse repercussions. More and more such interactions are signaled by a large number of publications. DCF is one of the most widely used classes of medicaments (Kogure, Akiyoshi, Imaoka, & Ohtani, 2014).

4.3   General aspects regarding DCF pharmacokinetics

DCF 2-[2-(2,6-dichloro-4-hydroxyanilino) phenyl] acetic acid is a nonsteroidal anti-inflammatory medicament that, with a chemical structure of ($C_{14}H_{11}Cl_2NO_3$), carries a functional carboxylic acid group. DCF metabolism in humans occurs via acyl-glucuronidation and phenyl hydroxylation, the first reactions being mainly catalyzed by 5′-diphospho-glucuronosyl uridine transferase (UGTs) and phenyl hydroxylation being catalyzed by CYP (CYP2C9 and CYP3A4). Furthermore, hydroxylation of glucuronidated DCF was demonstrated in vitro using recombinant CYP2C8 and *Citrus paradisi* juice (Altan et al., 2020).

This hydroxylation may be of clinical significance by defining the main route of metabolism and elimination of DCF in humans. 4′-Hydroxylation of medicaments appears to be a characteristic reaction for CYP2C9 catalysis, and this regioselective oxidation is probably dictated by the interactions between the carboxyl group (C(=O)OH) of the substrate with supposed cationic groups of the enzyme. Several other groups of CYP2C9 have been identified in studies with site-directed mutants that influence delegation and substrate specificity, including Phe114, Arg97, Ser286, and Asn289. The influence of DCF-quinidine combination on CYP3A4 activity revealed enhanced DCF metabolism by quinidine, in vitro, attributed to an increase in maximal volume (Inoue, Mizuo, Ishida, Komori, & Kusano, 2020).

The therapeutic use of DCF is not commonly associated with hepatotoxicity, as it is characterized by delayed onset of symptoms and lack of a clear dose-response relationship.

DCF toxicity was classified as a metabolic idiosyncrasy; the acyl glucuronide conjugate of the medicament was confirmed as reactive and capable of modifying cellular proteins by covalently binding the liver proteins to the rat, dependent on the activity of "2-multimedicament resistant" hepatic channel transporters. One of the modified proteins was identified as dipeptidyl peptidase IV. The formation of protein adducts was evidenced by the oxidative metabolism of DCF, catalyzed by CYP enzymes. Thus, the reactive intermediates were probably DCF 1′,4′- and 2,5-quinone imine, both involved in conjugation with glutathione (GSH) and identified as glutathione adducts (Sharma et al., 2019).

These glutathione ($C_{10}H_{17}N_3O_6S$) adducts were detected in rats as well as in human hepatocytes treated with DCF; a derivative of mercapturic acid ($C_{22}H_{21}NO_3S$) has been identified in the urine of patients receiving the medicament. It is not excluded that acyl glucuronide-conjugate or benzoquinone imine, derived from covalent modification of the protein, are involved and thus the toxicity in patients susceptible to explanation, either by direct disruption of critical cellular functions or by eliciting immunological responses (Yang et al., 2019).

After oral administration, DCF uptake is rapid and complete; DCF binds extensively to plasma albumin. The area under the plasma concentration-time curve of DCF is proportional to the dose administered for oral doses from 25 to 150 mg. Substantial medicament concentrations are reached in the synovial fluid, which is the site of target action for NSAIDs. The concentration-effect relationships were established for the totally bound and unbound DCF and for the synovial fluid concentrations. DCF is eliminated after biotransformation into sulfate metabolites, respectively glucuronide-conjugate, which are excreted in the urine; a very small amount of the unmodified medicament is eliminated. Medicament interactions have been reported in combination with aspirin, lithium, methotrexate, digoxin, cyclosporine, cholestyramine, and colestipol but also *Citrus paradisi* juice (Al-Lawati, Vakili, Lavasanifar, Ahmed, & Jamali, 2019).

A review of human hepatic metabolism of DCF, with emphasis on the production of minor hydroxylated metabolites involved in idiosyncratic hepatotoxicity of the medicament, revealed that human hepatocytes form 3'-hydroxy-, 4'-hydroxy-, 5-hydroxy-, 4'-, 5-dihydroxy-, and N, 5-dihydroxy-DCF, together with certain lactams. Secondary metabolites were highlighted after incubation of 5-hydroxy-DCF with *Citrus paradisi* juice, resulting in 4', 5-dihydroxy-DCF and small amounts of N, 5-dihydroxy-DCF (Sultan et al., 2018).

Based on studies on *Citrus paradisi* juice, it was estimated that in vivo CYP2C9 enzymes would be solely responsible for 4'-hydroxylation of DCF (> 99.5%) as well as for 5-hydroxy-DCF formation (> 97%). In addition, CYP2C9 is solely responsible for the formation of 3'-hydroxy-DCF. Experimental research in which DCF was incubated with cell lines expressing different human CYP isoforms suggests that seven isoforms might be involved (Fig. 2).

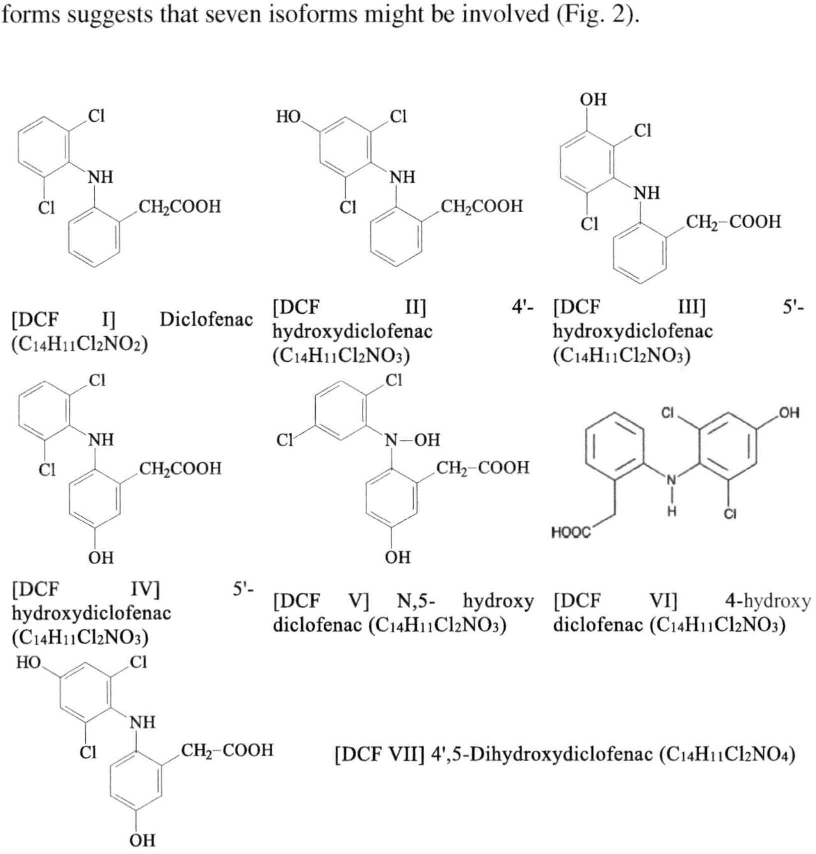

FIG. 2   Structure of diclofenac and seven known diclofenac metabolites (I to VII).

Comparison of the data obtained with cells expressing CYP, respectively with human hepatocytes, suggests that CYP2C8, CYP2C18, CYP2C19, and CYP2B6 are the main isoenzymes involved in 5-hydroxylation of DCF in vivo. DCF causes severe hepatotoxicity in rare cases, usually with a late onset (more than 1-3 months) (Kim et al., 2018).

Because there is no direct dose-liver injury relationship, these lesions cannot be replicated in known animal models; the individual sensitivity of the patients was evoked to explain the increased risk of these manifestations. While patient-related risk factors remained undefined, a number of molecular perturbations were characterized. These include metabolic factors (bioactivation by CYP2C9 or CYP3A4 in thiol-reactive quinone-imine, by activation of UGT2B7 in protein-reactive glucuronides, acyl glucuronide conjugates, and 4'-hydroxylation secondary to DCF), as well as kinetic factors (Shimura, Murayama, Tanaka, Onozeki, & Yamazaki, 2019).

Toxicodynamically, both oxidative stress (*Citrus paradisi* juice) and disturbances at the mitochondrial level (protonophoretic activity and permeability modification), separately or in combination, were implicated in DCF toxicity. CYP, the CYP2C11 isoform, is involved in DCF metabolism in rats in reactive products that covalently bind to this enzyme before it can diffuse further and react with other proteins. To determine whether P450 cytochromes present in human liver could catalyze a similar reaction, cvalent binding of DCF to hepatic microsomes has been studied in vitro (Wilson et al., 2018).

Only 3 of the 16 samples taken were identified as capable of activating appreciable DCF to form protein adducts in an NADPH-dependent pathway. CYP 2C9, which catalyzes the major oxidative metabolism pathway of DCF to produce 4'-hydroxy-DCF, does not appear to be responsible for the formation of protein adducts, because the enzyme inhibitor sulfaphenazole did not affect the formation of protein adducts.

Citrus paradisi juice like troleandomycin, an inhibitor of CYP, the CYP3A4 isoform, inhibits both protein adduct formation and DCF 5-hydroxylation. These findings were confirmed by the use of CYP2C9-dependent baculoviruses expressing human P450, the P450 3A4 isoform (Mohos et al., 2018).

P-benzoquinone imine, a biotransformation product of 5-hydroxy-DCF, is a possible reactive intermediate that would covalently bind to liver proteins.

This compound is formed by an oxidation apparently catalyzed by metal ions, which has been inhibited by EDTA, glutathione, and NADPH. P-benzoquinone imine, the product of 5-hydroxy-DCF metabolism, covalently binds to human liver microsomes in vitro in a reaction that is inhibited by GSH (Fig. 3).

On the contrary, glutathione did not inhibit covalent binding of DCF to human liver microsomes. These results suggest that appreciable bioactivation of DCF by P450 in vivo occurs both under the highly reactive CYP3A4 action and probably by the participation of other lower activity enzymes that catalyze alternative pathways of DCF metabolism. Moreover, the p-benzoquinone derivative of 5-hydroxy-DCF is involved in covalent attachment to the liver only when the

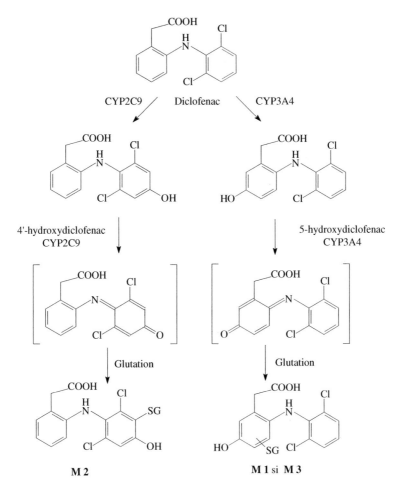

FIG. 3 Oxidative metabolism of DCF. M1 = 5-hydroxy-4-glutathione-S-yl DCF; M2 = 4-hydroxy-3-glutathione-S-yl DCF; M3 = 5-hydroxy-6-glutathione5-yl DCF.

levels of NADPH, GSH, and other reducing agents are decreased (Poór et al., 2018).

By visual analysis of plasma profiles, as expected, higher DCF concentrations were observed in the case of co-administration of *Citrus paradisi* juice.

However, great variability is observed in the samples taken at the same time but from different rats; the values of the pharmacokinetic parameters (area under the curve - AUC, T max, C max) highlight the tendency to increase the plasma concentrations of DCF administered simultaneously with *Citrus paradisi* juice. This can be explained both by the DCF pharmacokinetics, characterized by high interindividual variability, and also by the specificity of the clinical experiment - administration/animal slaughter at a certain time interval, which induces an additional variability (Nakanishi et al., 2018).

When analyzing the results of the pharmacokinetic analysis is performed, it can be observed that DCF has significantly higher plasma concentrations in the rat in the case of co-administration of *Citrus paradisi* juice than given alone. Because the dose administered (relative to kg body weight) was the same, this observation proves the existence of the pharmacokinetic interaction between DCF and *Citrus paradisi* juice.

Among the calculated pharmacokinetic parameters of DCF, Cmax increases from 1384.5 ng/mL (DCF administered alone) to 2262.6 ng/mL (DCF + *Citrus paradisi* juice).

These values correspond to a relative bioavailability of 163.4%, or in other words, the DCF concentrations co-administered with *Citrus paradisi* juice increase by 63.4% from normal due to the pharmacokinetic interaction. Because the increase in Cmax can be correlated with both the increase of the amount of the absorbed substance and the rate of absorption, the mechanism/place of pharmacokinetic interaction cannot be specified using only the Cmax values. Another changing pharmacokinetic parameter is the area under the curve, which increases from 7652.1 ng/mL·h to 9527.7 ng/mL·h. The AUC is a parameter that is correlated with the amount of substance that reaches the systemic circulation (Damkier, Hansen, & Brosen, 1999).

The relative bioavailability of DCF administered with *Citrus paradisi* juice versus DCF alone is 124.5%. In other words, in the case of administration with *Citrus paradisi* juice, the DCF will reach 24.5% more DCF than if it is administered alone (Misaka et al., 2013).

The increase of the AUC in the case of DCF co-administration with *Citrus paradisi* juice is also a proof of the pharmacokinetic interaction in this case. Although the difference between Tmax is from 3 h (DCF alone) to 2 h (DCF + *Citrus paradisi* juice), the difference is not relevant due to the discrete values of the Tmax parameter (between 2 and 3 h; it cannot take any other value because there were no samples). In such situations, the observed differences can be considered insignificant (Campbell, Teply, Mooss, & Hilleman, 2014).

The fact that the half-lives of DCF are similar regardless of the mode of administration (alone or with *Citrus paradisi* juice) proves that there is no interaction in the elimination of the substance; so, *Citrus paradisi* juice does not influence the body's ability to remove this substance. Since the elimination function of DCF is not affected by *Citrus paradisi* juice, but Cmax and AUC parameters prove the existence of the interaction, this can only be explained by the increased bioavailability due to the decrease of the effect of the first intestinal or hepatic passage or by the direct obstruction of CYP2A4 isozyme, (which exists in the proportion of 40% in the intestine), either by obstruction of glycoprotein P, as a "pump" that removes the medicament substance from the intestinal cell, preventing or diminishing its absorption (Shirasaka, Mori, Murata, & Nakanishi, 2014).

Citrus paradisi juice is involved in the biotransformation of medicaments, through its bioactive components and can modify their bioavailability as well as their therapeutic efficacy.

5. Conclusions and recommendations

There are situations when phytometabolites and medicaments conflict, mutually influencing their specific absorption and action or trying to impose their own effects. Most phytometabolites and medicaments are administered orally, which is why frequent interactions between them occur in the digestive tract and are influenced by the type and amount of food ingested, the period of time in contact, and the functioning optimum of the digestive system.

Medicaments that may influence nutrient absorption, metabolism, and excretion:

– *laxatives* accelerate the transit of phytometabolites through the intestine and can lead to loss of calcium and potassium
– *medicaments for lowering cholesterol* and neomycin (antibiotic) have the effect of reducing the sequestration of the bile in the gallbladder, which leads to poor digestion and absorption of food fats
– *antacids* used in the treatment of gastric ulcer increase the pH in the stomach and impede the absorption of vitamin B12; alkalinized gastric juice also leads to low absorption of minerals (calcium, iron, magnesium, zinc)
– *aspirin* and other acidic medicaments alter the digestive tract and impair its ability to absorb minerals, particularly calcium and iron
– *treatment against gout or tuberculosis* has a negative effect on the transport mechanisms of nutrients from the intestine, affecting the apoptotic intake of vitamin B12 and folic acid
– *the anticoagulants* used in cardiovascular diseases have as their basic mechanism the neutralization of vitamin K (coagulant vitamin)
– furosemide-type *diuretics* increase calcium excretion in the urine by decreasing its reabsorption in the kidney; chronic use of these medicaments causes increased renal excretion for other minerals (potassium, magnesium, zinc)
– *anticonvulsants* (phenobarbital) may cause deficiencies of vitamin D, folic acid, or biotin, as they alter the intestinal pH
– *oral contraceptives* with high doses of estrogen can lead to decreased serum B6 levels and increased iron and vitamin A levels in the blood
– *anti-inflammatory medicaments* in the class of glucocorticoids used for a long time have strong side effects: obesity, hyperglycemia, hypertension, bone decals with significant bone mass reduction, and osteopenia
– *hypertensive treatment* with beta-blockers increases the level of lipids in the blood and decreases the amount of HDL-cholesterol, the fraction with cardiovascular protective effect; beta-blockers also alter the body's tolerance to glucose, which may result in decreased efficacy of oral antidiabetic medicaments

The medical term "interaction" describes situations where phytometabolites accelerate, delay, or impede the absorption of a medicament; phytometabolites alter the distribution of a medicament in tissues or interfere with metabolism, transport, and elimination of substances in the body; phytometabolites block the effects of medicaments by altering pharmacometabolic or pharmacodynamic reactions that inhibits the side or natural effects of certain medicaments; change the taste or appetite for certain medicaments; or medicaments affect the metabolism, transport, and elimination of certain nutrients. There are phytometabolites and medicaments that may have similar chemical structures that, consumed simultaneously, cause confusion in the body and can cause absorption of one to the detriment of the other. The effect of phytometabolites on medicaments is influenced by gastric and intestinal pH, gastric motility and rate of progression of the food bowl, nutrient quantity and type, absorption capacity of intestinal cells, blood flow to the spleen and liver, transport rate of substances active through the blood from the intestine to the liver, the existence of other associated diseases and liver health, and the concomitant administration of other pharmaceutical preparations.

References

Abdlekawy, K. S., Donia, A. M., & Elbarbry, F. (2017). Effects of grapefruit and pomegranate juices on the pharmacokinetic properties of dapoxetine and midazolam in healthy subjects. *European Journal of Drug Metabolism and Pharmacokinetics, 42*(3), 397–405.

Ainslie, G. R., Wolf, K. K., Li, Y., Connolly, E. A., Scarlett, Y. V., Hull, J. H., et al. (2014). Assessment of a candidate marker constituent predictive of a dietary substance-drug interaction: Case study with grapefruit juice and CYP3A4 drug substrates. *The Journal of Pharmacology and Experimental Therapeutics, 351*(3), 576–584.

Akamine, Y., Miura, M., Komori, H., Tamai, I., Ieiri, I., Yasui-Furukori, N., et al. (2015). The change of pharmacokinetics of fexofenadine enantiomers through the single and simultaneous grapefruit juice ingestion. *Drug Metabolism and Pharmacokinetics, 30*(5), 352–357.

Albassam, A. A., & Markowitz, J. S. (2017). An appraisal of drug-drug interactions with green tea (*Camellia sinensis*). *Planta Medica, 83*(6), 496–508.

Ali, H. M. (2018). Mitigative role of garlic and vitamin E against cytotoxic, genotoxic, and apoptotic effects of lead acetate and mercury chloride on WI-38 cells. *Pharmacological Reports, 70*(4), 804–811.

Al-Lawati, H., Vakili, M. R., Lavasanifar, A., Ahmed, S., & Jamali, F. (2019). Delivery and biodistribution of traceable polymeric micellar diclofenac in the rat. *Journal of Pharmaceutical Sciences, 108*(8), 2698–2707.

Altan, F., Corum, O., Yildiz, R., Eser Faki, H., Ider, M., Ok, M., et al. (2020). Intravenous pharmacokinetics of moxifloxacin following simultaneous administration with flunixin meglumine or diclofenac in sheep. *Journal of Veterinary Pharmacology and Therapeutics, 43*(2), 108–114.

Andrus, M. R. (2004). Oral anticoagulant drug interactions with statins: Case report of fluvastatin and review of the literature. *Pharmacotherapy, 24*(2), 285–290.

Baltes, M. R., Dubois, J. G., & Hanocq, M. (2001). Application to drug-food interactions of living cells as in vitro model expressing cytochrome P450 activity: Enzyme inhibition by lemon juice. *Talanta, 54*(5), 983–987.

Beardmore, R. E., Cook, E., Nilsson, S., Smith, A. R., Tillmann, A., Esquivel, B. D., et al. (2018). Drug-mediated metabolic tipping between antibiotic resistant states in a mixed-species community. *Nature Ecology & Evolution*, 2(8), 1312–1320.

Bitencourt, T. A., Oliveira, F. B., Sanches, P. R., Rossi, A., & Martinez-Rossi, N. M. (2019). The prp4 kinase gene and related spliceosome factor genes in Trichophyton rubrum respond to nutrients and antifungals. *Journal of Medical Microbiology*, 68(4), 591–599.

Boullata, J. I. (2019). Drug-nutrition interactions and the brain: It's not all in your head. *Current Nutrition Reports*, 8(2), 92–98.

Bozkurt, M., Koçer, B., Ege, G., Tüzün, A. E., Bıyık, H. H., & Poyrazoğlu, E. (2019). Influence of the particle size and form of feed on growth performance, digestive tract traits and nutrient digestibility of white egg-laying pullets from 1 to 112 D of age. *Poultry Science*, 98(9), 4016–4029.

Campbell, J. A., Teply, R., Mooss, A. N., & Hilleman, D. E. (2014). Impact of grapefruit juice on the antiplatelet activity of loading and maintenance doses of clopidogrel in healthy volunteers. *Cardiology Research*, 5(1), 1–7.

Chan, L. (2013). Drug-nutrient interactions. *JPEN Journal of Parenteral and Enteral Nutrition*, 37(4), 450–459.

Chen, X., Luo, J., Liang, Z., Zhu, J., Li, L., & Wang, Q. (2020). Structural and physicochemical/ digestion characteristics of potato starch-amino acid complexes prepared under hydrothermal conditions. *International Journal of Biological Macromolecules*, 145, 1091–1098.

Chen, M., Zhou, S. Y., Fabriaga, E., Zhang, P. H., & Zhou, Q. (2018). Food-drug interactions precipitated by fruit juices other than grapefruit juice: An update review. *Journal of Food and Drug Analysis*, 26(2S), S61–S71.

Cooper, Z. D., Bedi, G., Ramesh, D., Balter, R., Comer, S. D., & Haney, M. (2018). Impact of co-administration of oxycodone and smoked cannabis on analgesia and abuse liability. *Neuropsychopharmacology*, 43(10), 2046–2055.

Da Costa Silva, B. Y., De Carvalho Sampaio, H. A., Shivappa, N., Hébert, J., Silva Albuquerque, L. D., Ferreira Carioca, A. A., et al. (2018). Interactions between dietary inflammatory index, nutritional state and Multiple Sclerosis clinical condition. *Clinical Nutrition ESPEN*, 26, 35–41.

Damkier, P., Hansen, L. L., & Brosen, K. (1999). Effect of diclofenac, disulfiram, itraconazole, grapefruit juice and erythromycin on the pharmacokinetics of quinidine. *British Journal of Clinical Pharmacology*, 48(6), 829–838.

de Leon, T. V., He, M., & Ullevig, S. L. (2018). Potential dietary supplement and medication interactions in a subset of the older adult population attending congregate sites. *Journal of Nutrition in Gerontology and Geriatrics*, 37(3-4), 218–230.

Doostikhah, N., Panahpour, E., Nadian, H., & Gholami, A. (2020). Tomato (*Lycopersicon esculentum* L.) nutrient and lead uptake affected by zeolite and DTPA in a lead-polluted soil. *Plant Biology (Stuttgart, Germany)*, 22(2), 317–322.

Dresser, G. K., Urquhart, B. L., Proniuk, J., Tieu, A., Freeman, D. J., Arnold, J. M., et al. (2017). Coffee inhibition of CYP3A4 in vitro was not translated to a grapefruit-like pharmacokinetic interaction clinically. *Pharmacology Research & Perspectives*, 5(5). https://doi.org/10.1002/prp2.346.

Evans, E., Piccio, L., & Cross, A. H. (2018). Use of vitamins and dietary supplements by patients with multiple sclerosis: A review. *JAMA Neurology*, 75(8), 1013–1021.

Ferreira Silva, R., & Rita Carvalho Garbi Novaes, M. (2014). Interactions between drugs and drug-nutrient in enteral nutrition: A review based on evidences. *Nutrición Hospitalaria*, 30(3), 514–518.

Fleisher, B., Unum, J., Shao, J., & An, G. (2015). Ingredients in fruit juices interact with dasatinib through inhibition of BCRP: A new mechanism of beverage-drug interaction. *Journal of Pharmaceutical Sciences*, 104(1), 266–275.

García-Lino, A. M., Álvarez-Fernández, I., Blanco-Paniagua, E., Merino, G., & Álvarez, A. I. (2019). Transporters in the mammary gland-contribution to presence of nutrients and drugs into milk. *Nutrients, 11*(10), E2372. https://doi.org/10.3390/nu11102372. Review.

Gezmen-Karadağ, M., Çelik, E., Kadayifçi, F. Z., Yeşildemir, Ö., Öztürk, Y. E., & Ağagündüz, D. (2018). Role of food-drug interactions in neurological and psychological diseases. *Acta Neurobiologiae Experimentalis (Wars), 78*(3), 187–197.

Grimm, M., Koziolek, M., Saleh, M., Schneider, F., Garbacz, G., Kühn, J. P., et al. (2018). Gastric emptying and small bowel water content after administration of grapefruit juice compared to water and isocaloric solutions of glucose and fructose: A four-way crossover MRI pilot study in healthy subjects. *Molecular Pharmaceutics, 15*(2), 548–559.

Gröber, U., Schmidt, J., & Kisters, K. (2020). Important drug-micronutrient interactions: A selection for clinical practice. *Critical Reviews in Food Science and Nutrition, 60*(2), 257–275.

Gurley, B. J., Tonsing-Carter, A., Thomas, S. L., & Fifer, E. K. (2018). Clinically relevant herb-micronutrient interactions: When botanicals, minerals, and vitamins collide. *Advances in Nutrition, 9*(4), 524S–532S.

Holguín, H., Amariles, P., & Ospina, W. (2017). Evolutionary interactions as a possible mechanism of drug interactions: An approach to the control of antibiotic-resistant bacteria. *Revista Chilena de Infectología, 34*(4), 307–313.

Holmberg, M. T., Tornio, A., Hyvärinen, H., Neuvonen, M., Neuvonen, P. J., Backman, J. T., et al. (2015). Effect of grapefruit juice on the bioactivation of prasugrel. *British Journal of Clinical Pharmacology, 80*(1), 139–145.

Hu, J., Shang, D., Xu, X., He, X., Ni, X., Zhang, M., et al. (2016). Effect of grapefruit juice and food on the pharmacokinetics of pirfenidone in healthy Chinese volunteers: A diet-drug interaction study. *Xenobiotica, 46*(6), 516–521.

Inoue, K., Mizuo, H., Ishida, T., Komori, T., & Kusano, K. (2020). Bioactivation of diclofenac in human hepatocytes and the proposed human hepatic proteins modified by reactive metabolites. *Xenobiotica, 18*, 1–10.

Johnson, B. A., & Seneviratne, C. (2014). Alcohol-medical drug interactions. *Handbook of Clinical Neurology, 125*, 543–559.

Johnson, E. J., Won, C. S., Köck, K., & Paine, M. F. (2017). Prioritizing pharmacokinetic drug interaction precipitants in natural products: Application to OATP inhibitors in grapefruit juice. *Biopharmaceutics & Drug Disposition, 38*(3), 251–259.

Kashihara, Y., Ieiri, I., Yoshikado, T., Maeda, K., Fukae, M., Kimura, M., et al. (2017). Small-dosing clinical study: Pharmacokinetic, pharmacogenomic (SLCO2B1 and ABCG2), and interaction (atorvastatin and grapefruit juice) profiles of 5 probes for OATP2B1 and BCRP. *Journal of Pharmaceutical Sciences, 106*(9), 2688–2694.

Kaufman-Szymczyk, A., Majda, K., Szuławska-Mroczek, A., Fabianowska-Majewska, K., & Lubecka, K. (2019). Clofarabine-phytochemical combination exposures in CML cells inhibit DNA methylation machinery, upregulate tumor suppressor genes and promote caspase-dependent apoptosis. *Molecular Medicine Reports, 20*(4), 3597–3608.

Khuda, F., Ovais, M., Zakiullah Khan, A., Ali, G., Ullah, S., Shah, W. A., et al. (2019). Mini-review: Drug-food interactions of commonly available juices of Pakistan. *Pakistan Journal of Pharmaceutical Sciences, 32*(5), 2189–2196.

Kim, S. B., Kim, K. S., Ryu, H. M., Hong, S. H., Kim, B. K., Kim, D. D., et al. (2018). Modulation of rat hepatic CYP1A and 2C Activity by Honokiol and Magnolol: Differential effects on phenacetin and diclofenac pharmacokinetics in vivo. *Molecules, 23*(6), E1470. https://doi.org/10.3390/molecules23061470.

Kim, M. J., Nafziger, A. N., Kashuba, A. D., Kirchheiner, J., Bauer, S., Gaedigk, A., et al. (2006). Effects of fluvastatin and cigarette smoking on CYP2C9 activity measured using the probe S-warfarin. *European Journal of Clinical Pharmacology*, *62*(6), 431–436.

Kogure, N., Akiyoshi, T., Imaoka, A., & Ohtani, H. (2014). Prediction of the extent and variation of grapefruit juice-drug interactions from the pharmacokinetic profile in the absence of grapefruit juice. *Biopharmaceutics & Drug Disposition*, *35*(7), 373–381.

Kotwal, P., Dogra, A., Sharma, A., Bhatt, S., Gour, A., Sharma, S., et al. (2020). Effect of natural phenolics on pharmacokinetic modulation of bedaquiline in rat to assess the likelihood of potential food-drug interaction. *Journal of Agricultural and Food Chemistry*, *68*(5), 1257–1265.

Li, T. T., An, J. X., xu, J. Y., & Tuo, B. G. (2019). Overview of organic anion transporters and organic anion transporter polypeptides and their roles in the liver. *World Journal of Clinical Cases*, *7*(23), 3915–3933.

Liu, D., Cheng, B., Li, D., Li, J., Wu, Q., & Pan, H. (2018). Investigations on the interactions between curcumin loaded vitamin E TPGS coated nanodiamond and Caco-2 cell monolayer. *International Journal of Pharmaceutics*, *551*(1-2), 177–183.

Lu, L. Y., Liu, Y., Zhang, Z. F., Gou, X. J., Jiang, J. H., Zhang, J. Z., et al. (2015). Pomegranate seed oil exerts synergistic effects with trans-resveratrol in a self-nanoemulsifying drug delivery system. *Biological & Pharmaceutical Bulletin*, *38*(10), 1658–1662.

Markert, C., Wirsching, T., Hellwig, R., Burhenne, J., Weiss, J., Riedel, K. D., et al. (2014). Lack of a clinically significant interaction of grapefruit juice with ambrisentan and bosentan in healthy adults. *International Journal of Clinical Pharmacology and Therapeutics*, *52*(11), 957–964.

Masuda, M., Watanabe, S., Tanaka, M., Tanaka, A., & Araki, H. (2018). Screening of furanocoumarin derivatives as cytochrome P450 3A4 inhibitors in citrus. *Journal of Clinical Pharmacy and Therapeutics*, *43*(1), 15–20.

McKetin, R., Coen, A., & Kaye, S. (2015). A comprehensive review of the effects of mixing caffeinated energy drinks with alcohol. *Drug and Alcohol Dependence*, *151*, 15–30.

Misaka, S., Miyazaki, N., Yatabe, M. S., Ono, T., Shikama, Y., Fukushima, T., et al. (2013). Pharmacokinetic and pharmacodynamic interaction of nadolol with itraconazole, rifampicin and grapefruit juice in healthy volunteers. *Journal of Clinical Pharmacology*, *53*(7), 738–745.

Misaka, S., Nakamura, R., Uchida, S., Takeuchi, K., Takahashi, N., Inui, N., et al. (2011). Effect of 2 weeks' consumption of pomegranate juice on the pharmacokinetics of a single dose of midazolam: An open-label, randomized, single-center, 2-period crossover study in healthy Japanese volunteers. *Clinical Therapeutics*, *33*(2), 246–252.

Mohos, V., Bencsik, T., Boda, G., Fliszár-Nyúl, E., Lemli, B., Kunsági-Máté, S., et al. (2018). Interactions of casticin, ipriflavone, and resveratrol with serum albumin and their inhibitory effects on CYP2C9 and CYP3A4 enzymes. *Biomedicine & Pharmacotherapy*, *107*, 777–784.

Mouly, S., Lloret-Linares, C., Sellier, P. O., Sene, D., & Bergmann, J. F. (2017). Is the clinical relevance of drug-food and drug-herb interactions limited to grapefruit juice and Saint-John's Wort? *Pharmacological Research*, *118*, 82–92.

Nagai, K., Omotani, S., Ito, A., Nishimura, I., Hatsuda, Y., Mukai, J., et al. (2019). Alterations in pharmacokinetics of orally administered carbamazepine in rats treated with sodium alginate: Possible interaction between therapeutic drugs and semi-solid enteral nutrients. *Drug Research*, *69*(3), 168–172.

Nagai, K., Omotani, S., Otani, M., Sasatani, M., Takashima, T., Hatsuda, Y., et al. (2018). In vitro and in vivo effects of selected fibers on the pharmacokinetics of orally administered carbamazepine: Possible interaction between therapeutic drugs and semisolid enteral nutrients. *Nutrition*, *46*, 44–47.

Nakanishi, K., Uehara, S., Kusama, T., Inoue, T., Shimura, K., Kamiya, Y., et al. (2018). In vivo and in vitro diclofenac 5-hydroxylation mediated primarily by cytochrome P450 3A enzymes in common marmoset livers genotyped for P450 2C19 variants. *Biochemical Pharmacology*, *152*, 272–278.

Naseeruddin, R., Sumathi, V., Prasad, T. N. V. K. V., Sudhakar, P., Chandrika, V., & Ravindra Reddy, B. (2018). Unprecedented synergistic effects of nanoscale nutrients on growth, productivity of sweet sorghum [*Sorghum bicolor* (L.) Moench], and nutrient biofortification. *Journal of Agricultural and Food Chemistry*, *66*(5), 1075–1084.

Ndou, S. P., Kiarie, E., Walsh, M. C., Ames, N., de Lange, C. F. M., & Nyachoti, C. M. (2019). Interactive effects of dietary fibre and lipid types modulate gastrointestinal flows and apparent digestibility of fatty acids in growing pigs. *The British Journal of Nutrition*, *121*(4), 469–480.

Nishiyama, K., Toshimoto, K., Lee, W., Ishiguro, N., Bister, B., & Sugiyama, Y. (2019). Physiologically-based pharmacokinetic modeling analysis for quantitative prediction of renal transporter-mediated interactions between metformin and cimetidine. *CPT: Pharmacometrics & Systems Pharmacology*, *8*(6), 396–406.

Nomani, H., Moghadam, A. T., Emami, S. A., Mohammadpour, A. H., Johnston, T. P., & Sahebkar, A. (2019). Drug interactions of cola-containing drinks. *Clinical Nutrition*, *38*(6), 2545–2551.

Ogata, Y., Imaoka, A., Akiyoshi, T., & Ohtani, H. (2019). Effects of food type on the extent of drug-drug interactions between activated charcoal and phenobarbital in rats. *Drug Metabolism and Pharmacokinetics*, *34*(4), 287–291.

Ogungbenro, K., Aarons, L., & CRESim & Epi-CRESim Project Groups. (2014). Physiologically based pharmacokinetic modelling of methotrexate and 6-mercaptopurine in adults and children. Part 2: 6-mercaptopurine and its interaction with methotrexate. *Journal of Pharmacokinetics and Pharmacodynamics*, *41*(2), 173–185.

Oh, J. M., Jang, H. J., Kim, W. J., Kang, M. G., Baek, S. C., Lee, J. P., et al. (2020). Calycosin and 8-O-methylretusin isolated from *Maackia amurensis* as potent and selective reversible inhibitors of human monoamine oxidase-B. *International Journal of Biological Macromolecules*, *151*, 441–448.

Ohkubo, A., Chida, T., Kikuchi, H., Tsuda, T., & Sunaga, K. (2017). Effects of tomato juice on the pharmacokinetics of CYP3A4-substrate drugs. *Asian Journal of Pharmaceutical Sciences*, *12*(5), 464–469.

Oikeh, I., Sakkas, P., Blake, D. P., & Kyriazakis, I. (2019). Interactions between dietary calcium and phosphorus level, and vitamin D source on bone mineralization, performance, and intestinal morphology of coccidia-infected broilers1. *Poultry Science*, *98*(11), 5679–5690.

Olukosi, O. A., & Bedford, M. R. (2019). Comparative effects of wheat varieties and xylanase supplementation on growth performance, nutrient utilization, net energy, and whole-body energy and nutrient partitioning in broilers at different ages. *Poultry Science*, *98*(5), 2179–2188.

Paśko, P., Rodacki, T., Domagała-Rodacka, R., Palimonka, K., Marcinkowska, M., & Owczarek, D. (2017). Second generation H1 - antihistamines interaction with food and alcohol—A systematic review. *Biomedicine & Pharmacotherapy*, *93*, 27–39.

Poór, M., Boda, G., Mohos, V., Kuzma, M., Bálint, M., Hetényi, C., et al. (2018). Pharmacokinetic interaction of diosmetin and silibinin with other drugs: Inhibition of CYP2C9-mediated biotransformation and displacement from serum albumin. *Biomedicine & Pharmacotherapy*, *102*, 912–921.

Qinna, N. A., Ismail, O. A., Alhussainy, T. M., Idkaidek, N. M., & Arafat, T. A. (2016). Evidence of reduced oral bioavailability of paracetamol in rats following multiple ingestion of grapefruit juice. *European Journal of Drug Metabolism and Pharmacokinetics*, *41*(2), 187–195.

Quiñones, J. P., Mardare, C. C., Hassel, A. W., & Brüggemann, O. (2019). Testosterone- and vitamin-grafted cellulose ethers for sustained release of camptothecin. *Carbohydrate Polymers, 206*, 641–652.

Quirino, A., Morelli, P., Capua, G., Arena, G., Matera, G., Liberto, M. C., et al. (2018). Synergistic and antagonistic effects of *Citrus bergamia* distilled extract and its major components on drug resistant clinical isolates. *Natural Product Research, 22*, 1–4.

Ratriyanto, A., & Prastowo, S. (2019). Floor space and betaine supplementation alter the nutrient digestibility and performance of Japanese quail in a tropical environment. *Journal of Thermal Biology, 83*, 80–86.

Reichenbach, A., Stark, R., Mequinion, M., Denis, R. R. G., Goularte, J. F., Clarke, R. E., et al. (2018). AgRP Neurons require carnitine acetyltransferase to regulate metabolic flexibility and peripheral nutrient partitioning. *Cell Reports, 22*(7), 1745–1759.

Santes-Palacios, R., Romo-Mancillas, A., Camacho-Carranza, R., & Espinosa-Aguirre, J. J. (2016). Inhibition of human and rat CYP1A1 enzyme by grapefruit juice compounds. *Toxicology Letters, 258*, 267–275.

Schlessinger, A., Welch, M. A., van Vlijmen, H., Korzekwa, K., Swaan, P. W., & Matsson, P. (2018). Molecular modeling of drug-transporter interactions—An international transporter consortium perspective. *Clinical Pharmacology and Therapeutics, 104*(5), 818–835.

Shang, D. W., Wang, Z. Z., Hu, H. T., Zhang, Y. F., Ni, X. J., Lu, H. Y., et al. (2018). Effects of food and grapefruit juice on single-dose pharmacokinetics of blonanserin in healthy Chinese subjects. *European Journal of Clinical Pharmacology, 74*(1), 61–67.

Sharma, A., Gour, A., Bhatt, S., Rath, S. K., Malik, T. A., Dogra, A., et al. (2019). Effect of IS01957, a para-coumaric acid derivative on pharmacokinetic modulation of diclofenac through oral route for augmented efficacy. *Drug Development Research, 80*(7), 948–957.

Sharma, A. K., Kapoor, V. K., & Kaur, G. (2020). Herb-drug interactions: A mechanistic approach. *Drug and Chemical Toxicology, 12*, 1–10.

Shimura, K., Murayama, N., Tanaka, S., Onozeki, S., & Yamazaki, H. (2019). Suitable albumin concentrations for enhanced drug oxidation activities mediated by human liver microsomal cytochrome P450 2C9 and other forms predicted with unbound fractions and partition/distribution coefficients of model substrates. *Xenobiotica, 49*(5), 557–562.

Shirasaka, Y., Mori, T., Murata, Y., Nakanishi, T., & Tamai, I. (2014). Substrate- and dose-dependent drug interactions with grapefruit juice caused by multiple binding sites on OATP2B1. *Pharmaceutical Research, 31*(8), 2035–2043.

Stewart, L. K., Smoak, P., Hydock, D. S., Hayward, R., O'Brien, K., Lisano, J. K., et al. (2019). Milk and kefir maintain aspects of health during doxorubicin treatment in rats. *Journal of Dairy Science, 102*(3), 1910–1917.

Sultan, T., Hamid, S., Hassan, S., Hussain, K., Ahmed, A., Bashir, L., et al. (2018). Development and evaluation of immediate release diclofenac sodium suppositories. *Pakistan Journal of Pharmaceutical Sciences, 31*(5), 1791–1795.

Takahashi, M., Onozawa, S., Ogawa, R., Uesawa, Y., & Echizen, H. (2015). Predictive performance of three practical approaches for grapefruit juice-induced 2-fold or greater increases in AUC of concomitantly administered drugs. *Journal of Clinical Pharmacy and Therapeutics, 40*(1), 91–97.

Theile, D., Hohmann, N., Kiemel, D., Gattuso, G., Barreca, D., Mikus, G., et al. (2017). Clementine juice has the potential for drug interactions—In vitro comparison with grapefruit and mandarin juice. *European Journal of Pharmaceutical Sciences, 97*, 247–256.

Thitz, P., Mehtätalo, L., Välimäki, P., Randriamanana, T., Lännenpää, M., Hagerman, A. E., et al. (2020). Phytochemical shift from condensed tannins to flavonoids in transgenic *Betula pendula* decreases consumption and growth but enhances growth efficiency of *Epirrita autumnata* larvae. *Journal of Chemical Ecology*, *46*(2), 217–231.

Tsuji, H., Ohmura, K., Nakashima, R., Hashimoto, M., Imura, Y., Yukawa, N., et al. (2016). Efficacy and safety of grapefruit juice intake accompanying Tacrolimus treatment in connective tissue disease patients. *Internal Medicine*, *55*(12), 1547–1552.

Ushijima, K., Mizuta, K., Otomo, S., Ogaki, K., Sanada, Y., Hirata, Y., et al. (2018). Increased tacrolimus blood concentration by *Beni-Madonna*—A new hybrid citrus cultivar categorized as 'Tangor', in a liver transplant patient: Likely furanocoumarin-mediated inhibition of CYP3A4 or P-glycoprotein. *British Journal of Clinical Pharmacology*, *84*(12), 2933–2935.

Vora, B., Green, E. A. E., Khuri, N., Ballgren, F., Sirota, M., & Giacomini, K. M. (2020). Drug-nutrient interactions: Discovering prescription drug inhibitors of the thiamine transporter ThTR-2 (SLC19A3). *The American Journal of Clinical Nutrition*, *111*(1), 110–121.

Wells, K. A., & Losin, W. G. (2008). In vitro stability, potency, and dissolution of duloxetine enteric-coated pellets after exposure to applesauce, apple juice, and chocolate pudding. *Clinical Therapeutics*, *30*(7), 1300–1308.

Wilson, C. E., Dickie, A. P., Schreiter, K., Wehr, R., Wilson, E. M., Bial, J., et al. (2018). The pharmacokinetics and metabolism of diclofenac in chimeric humanized and murinized FRG mice. *Archives of Toxicology*, *92*(6), 1953–1967.

Yamasaki, K., Iohara, D., Oyama, Y., Nishizaki, N., Kawazu, S., Nishi, K., et al. (2020). Processing grapefruit juice with γ-cyclodextrin attenuates its inhibitory effect on cytochrome P450 3A activity. *The Journal of Pharmacy and Pharmacology*, *72*(3), 356–363.

Yang, H. F., Li, Y. J., Li, Y. Y., Huang, C., Huang, L. X., & Bu, S. J. (2019). Pharmacokinetics of diclofenac sodium injection in swine. *Polish Journal of Veterinary Sciences*, *22*(2), 423–426.

Zargar, S., Al-Majed, A. A., & Wani, T. A. (2018). Potentiating and synergistic effect of grapefruit juice on the antioxidant and anti-inflammatory activity of aripiprazole against hydrogen peroxide induced oxidative stress in mice. *BMC Complementary and Alternative Medicine*, *18*(1), 106. https://doi.org/10.1186/s12906-018-2169-x.

Zsila, F., Bősze, S., & Beke-Somfai, T. (2020). Interaction of antitubercular drug candidates with α1-acid glycoprotein produced in pulmonary granulomas. *International Journal of Biological Macromolecules*, *147*, 1318–1327.

Chapter 13

Role of genetically modified plant repository in biopharmaceutical industries

Umair Riaz[a], Madiha Fatima[b], Laila Shehzad[c], and Hina Ahmed Malik[d]

[a]*Soil and Water Testing Laboratory for Research, Agriculture Department, Government of Punjab, Bahawalpur, Pakistan,* [b]*Department of Zoology, The University of Multan, Multan, Pakistan,* [c]*Sustainable Development Study Center, G.C. University, Lahore, Pakistan,* [d]*Institute of Soil and Environmental Sciences, University of Agriculture, Faisalabad, Pakistan*

1. Introduction

The plants have been widely used for pharmaceutical point since ancient times back thousands of years, but the exploiting genotypically engineered plants as a factory of desired biopharmaceuticals are a relatively new idea. Biopharmaceuticals are products of medicinal used manufactured in, extracted, or semisynthesized from living systems (Walsh, 2013). As the stipulation for biologically farmed therapeutics is anticipated to amplify with time, it would be wise to make their availability sure in radically more significant amounts to meet the future need on a cost-effective basis. At present the expense of biopharmaceutics puts a limit on their accessibility. Plant-based biopharmaceuticals are economical to make and stock up, are easy to bring up to industrial level for the production of larger quantities, and have low health risk as compared with those obtained from animals (Daniell, Streatfield, & Wycoff, 2001). "The use of transgenic plants for large scale production of recombinant compounds of pharmaceutical or industrial use is called molecular farming or biopharming" (Breyer, De Schrijver, Goossens, Pauwels, & Herman, 2012).

2. Genetically modified plants

Any organism having genetic sequences of any other organism incorporated into its genome can be referred to as *transgenic* or *genetically modified organisms*. A transgenic organism can be genetically engineered to get any protein product of interest. Although all transgenic organisms, including eukaryotes (animals and fungi) and prokaryotes, that is, bacteria, are used for the manufacture of

Phytomedicine: A Treasure of Pharmacologically Active Products from Plants
https://doi.org/10.1016/B978-0-12-824109-7.00026-1

377

any desired proteins, plants can prove to be a better choice (Wani & Sah, 2015). Plants are genetically modified to acquire valuable quality characteristics such as pest and pathogen resistance, herbicide and abiotic stress tolerance, high nutritional and yield potential, slow ripening, and increased ornamental value. GM plants can serve as bioreactors for the production of edible vaccines, therapeutic agents, antigens, monoclonal antibody fragments, and biopolymers (Rastogi Verma, 2013).

The plant-based production of recombinant biopharmaceutical proteins is expected to provide us with numerous apparent advantages, such as plant expression systems that are cheaper to run compared with running industrial operations based on the fermentation system and their production charges that are low than GM animals. Moreover, plants are easier to handle as compared with any other living being, and a piece of well-established machinery and expertise are already on hand for reaping and processing plants and plant harvest on a large scale. Plants are not known to be a vector for the human pathogen, so there is no or lesser possibility of contamination in the desired product health risks that are minimized (Lienard, Sourrouille, Gomord, & Faye, 2007). Proteins need to be extracted from bacteria or fungi to come into use; these refinement requirements can be brought down if the plant tissue containing the desired product is edible (Fischer, Stoger, Schillberg, Christou, & Twyman, 2004; Wani & Sah, 2015). Plants can be genetically programmed to translocate or even synthesize the proteins directly inside certain subcellular organelles where they are more stable. Finally, recombinant products could be formed in a desirable amount on industrial scale (Daniell et al., 2001).

Higher plants are preferred to use for the biopharming as they can synthesize eukaryotic proteins with correct final confirmation (bending and folding), posttranslational modification (glycosylation), and biological activity. Proteins pose more stability that is higher in plants compared with animals because plant cells have naturally evolved a capacity to store proteins in the lumens of subcellular organelles that reduce the risk of enzymatic degradation. For storage purposes, GM plants have the advantage of producing organs rich in a recombinant protein. Stably transformed plant lines can be used for subculturing in the field to get a considerable amount of biomass and, in turn, protein (Wani & Sah, 2015). Due to the reasons mentioned earlier, molecular plant farming has turn out to be more attractive for biotechnologists. The term "pharmaceutical crops" is used by scientist working in different areas of plant biotechnology in reference to different groups of plants and their consumption in biomedical science. Biologists characterize pharmaceutical crops as "genetically modified (GM) or engineered crops to produce vaccines, antibodies, and other therapeutic proteins" (Bauer, 2014), but occasionally, such group of the crop may be called as biopharma referring to plants that have been genetically altered for the assemblage of pharmaceuticals (antimicrobials, compounds used in diagnostic laboratories, plant bodies, and vaccine) or industrially valuable bimolecular (eco-friendly plastics, engine oils, and enzymes used in food industry) relative to those used for obtaining food, feed, or textile fibers. Conventionally biopharming refers to the technique of utilizing transgenic/engineered crops (e.g., tobacco, tomato, rice, potato, and safflower) as bioprocessing systems for the production of involved therapeutic agents.

However, the term pharmaceutical crops may occasionally be used by natural product chemists to use for a diverse category of plants possessing the capacity of producing specific micromolecules that are pharmaceutically active (Ma et al., 2005). These pharmaceutical ingredients occur as a single entity as secondary metabolites produced in natural metabolic pathways of the plant. For example, members of family *Taxaceae* (*Taxus* spp.) and Berberidaceae (*Podophyllum* spp.) yield anticancer drugs, while *Artemisia annua* L. belonging to family Asteraceae is used for extracting drugs against malaria. The different explanations for pharmaceutical crops not only can be a cause of confusion in researchers and manufacturers but also may be a source of misunderstanding in the layman (Marvier, 2007). Thus, pharmaceutical crops should be used for those domesticated plant species harvested for the purpose of obtaining active pharmaceutical ingredients (APIs), which can be constituent of pharmaceutical complexes, vaccines, antibodies, and other medically functional proteins. Pharmaceutical crops can be categorized into three discrete, however, interrelated groups on the basis of the type of pharmaceutical product, that is, crops yielding small therapeutic molecules (STMs), large therapeutic molecules (LTMs), and standardized therapeutic extracts (STEs). These crops can be present on land or water bodies. According to estimation, almost up to 500 plant types are being used for the production of STMs, fewer species are known to produce LTMs, however, and thousands of species are reported as reservoir crops for STEs (Li et al., 2010) (Tables 1 and 2).

TABLE 1 Plant expression hosts used for biopharmaceutical production.

Species	Example
Model plants	*Arabidopsis thaliana*
Simple plants	*Physcomitrella patens, Chlamydomonas reinhardtii, Lemna*
Leafy crops	Tobacco, alfalfa, clover, lettuce
Cereals	Maize, rice, wheat, barley
Legumes	Soybean, pea, pigeon pea
Fruits and vegetables	Potato, carrot, and tomato
Oil crops	Oilseed rape, *Camelina sativa*

From Fischer, R., Stoger, E., Schillberg, S., Christou, P., & Twyman, R. M. (2004). Plant-based production of biopharmaceuticals. Current Opinion in Plant Biology, 7(2), 152–158.

TABLE 2 Examples of various biopharmaceutical produced in transgenic plants.

Class	Protein	Host plant	Potential application
Vaccine	*Escherichia coli* heat labile toxin	Transgenic maize	For diarrhea
	Norwalk virus capsid protein	Transgenic potato	Norwalk virus infection
	Rabies glycoprotein	Viral vectors in spinach	Rabies
	Hepatitis B virus surface antigen	Transgenic lettuce, and potato	Hepatitis B
	SpaA	Tobacco	Dental caries
	HIV epitope (gp41)	Cow-pea	HIV
	Malarial B-cell epitope	Tobacco	Malaria
	Hemagglutinin	Tobacco	Influenza
	c-Myc	Tobacco	Cancer
Antibody	CaroRx	Transgenic tobacco	In dental caries
	Fv antibody fragments	Viral vector in tobacco	Non-Hodgkin's lymphoma
	NPI defensin	Tobacco	Produce IgM
	SIgA/G	Tobacco	Antiallergic response
Industrial enzyme	Lactase	Maize	Degrade phenolic compounds
Pharmaceutical enzyme	Glucocerebrosidase	Tobacco	For Gaucher's disease
	Gastric lipase	Transgenic maize	Pancreatitis
	Iduronidase	*Arabidopsis thaliana*	In mucopolysaccharidosis

Anticoagulants	Human protein C	Tobacco	Used in protein C pathway
	Human hirudin variant 2	Tobacco, oilseed	For indirect thrombin inhibitors
Recombinant hormones/proteins	Human epidermal growth factor	Tobacco	Wound repair and cell proliferation
	Human interferon-α	Rice, turnip	Used in Hepatitis C and B
	Human erythropoietin	Tobacco	In anemia
	Human serum albumin	Tobacco, potato	In liver cirrhosis
	Human hemoglobin	Tobacco	For blood substitute
	Insulin	Transgenic safflower	Diabetes
	CTB, cholera toxin B subunit-proinsulin	Tobacco, lettuce	Diabetes
Proteins/peptide inhibitors	Human α-1 antitrypsin	Rice	In cystic fibrosis
	Human aprotinin	Maize	Trypsin inhibitor for transplantation surgery
	Angiotensin I–converting enzyme	Tomato, tobacco	In hypertension
Nutraceuticals	Daffodil phytoene synthase	Rice	Used for provitamin A deficiency
	Amaranthus hypochondriacus Ama1 seed albumin	Potato	In amino acid deficiency
Dietary	Human intrinsic factor	Transgenic Arabidopsis	In vitamin B12 deficiency
	Lactoferrin	Transgenic maize	In gastrointestinal infection
	Lysozyme and human serum albumin	Transgenic rice	For diarrhea
Microbicide	Cyanoverin-N	Transgenic tobacco	In HIV

Data from Giddings, G., Allison, G., Brooks, D., & Carter, A. (2000). Transgenic plants as factories for biopharmaceuticals. Nature Biotechnology, 18(11) 1151–1155; Wani, S. H., & Sah. S. K. (2015). Transgenic plants as expression factories for bio pharmaceuticals. Journal of Botanical Sciences, 1, 1–4; Kermode, A. R. (2006). Plants as factories for production of biopharmaceutical and bioindustrial proteins: Lessons from cell biology. Botany, 84(4), 679–694; Cramer, C. L., Boothe, J. G., & Oishi, K. K. (2000). Transgenic plants for therapeutic proteins: Linking upstream and downstream strategies. In Plant biotechnology (pp. 95–118). Berlin, Heidelberg; Springer; Ma, J. K., Drake, P. M., & Christou, P. (2003). The production of recombinant pharmaceutical proteins in plants. Nature Reviews Genetics, 4(10), 794–805; Goldstein, D. A., & Thomas, J. A. (2004). Biopharmaceuticals derived from genetically modified plants. QJM, 97(11), 705–716. doi:10.1093/qjmed/hch121.

3. Production system for plant-derived biopharmaceuticals

Plants have provided humanity with a variety of valuable products for long times, but it has been possible only in the last two decades to manipulate plants for expression of heterogonous proteins of interest (Ma, Drake, & Christou, 2003). The technique of molecular farming or "biopharming" was introduced by Fischer, Liao, Hoffmann, Schillberg, and Emans (1999) to describe "the production of recombinant proteins in plants" established procedures for massive production of biopharmaceuticals, based on the bacterium *Escherichia coli*, yeast, and mammalian cell culture systems, is although well established. However, they possess a high cost to run, and there is always a health hazard associated. Taking such issues into considerations, a plant-based production system is a good substitute (Moon et al., 2020). The first medicinal protein derived from plants was the Hgh formed in the recombinant tobacco expression system in 1986. From that time, numerous additional human proteins have been expressed in a variety of crops. In 1989 the first immunoglobulin was produced in tobacco, which revealed that plants are capable of assembling complex functional polymeric glycoproteins. The anatomical accuracy of recombinant proteins extracted from plant was reported in 1992 in an experimental vaccine when the hepatitis B virus (HBV) surface antigen was successfully expressed in plants by Mason, Lam, and Arntzen (1992). Since then, many recombinant proteins, including pharmaceutical proteins such as recombinant antibodies and subunit vaccines with the therapeutic appliance, have been synthesized (Ma et al., 2003).The generation of genetically modified plants and their traits is shown in Fig. 1.

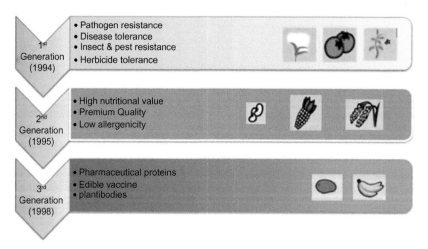

FIG. 1 Generations of GM plants with different traits.

Molecular farming mainly aims at the production of recombinant proteins at a larger rate and quantities. To meet this aim the expression-construct design should work at its maximum capacity at all phases of expression of the gene, from the production of nascent mRNA to posttranslational modification of a protein. Expression constructs are lab-designed chimeric nucleotide sequences, where transgene is flanked by heterogenic regulatory components known to be active in plants. For sophisticated transcription the two most fundamental and essential elements reported so far are the promoter and the polyadenylation site. Promoters that concede the temporal or spatial expression of a transgenic might also be useful. Such promoters result in improved constancy of the product and the shunning of protein amassing in vegetative organs, leading to prevent the toxicity to the expression host and nontarget organisms (Fischer et al., 2004; Ma et al., 2003). Subcellular targeting of the nascent recombinant polypeptide is also a critical factor affecting the yield of Rec. proteins, which influences the intertwined stages of polypeptide folding, protein assembly, and its posttranslational modification. Recombinant antibody-based experiments have revealed that targeting the proteins toward the secretory pathway provides a more appropriate atmosphere for the aforementioned stages of protein synthesis as compared with the cytosol, leading to high yield. Protein targeting toward the secretory pathway is accomplished through the insertion of a signal peptide toward the N-terminus of expression construct (Ma et al., 2003; Mason & Arntzen, 1995). Generally adopted methods for constructing transgenic plant lines for biofarming are agrobacterium-mediated transformation and particle bombardment. The choice of method depends on the selected host species, technical expertise of man force, and academic property issues. The pathway for protein synthesis is almost identical among higher eukaryotes, so anthropogenic genes that are incorporated into plants are expressed with an amino acid sequence identical to their native counterparts (Key, Ma, & Drake, 2008). The schematic representation of the stable transformation of plants by agrobacterium-mediated and particle bombardment method is shown in Fig. 2.

4. Plant expression host

At the research laboratory level, plant working as an expression host is mostly tobacco that is well known as a good model organism providing an excellent system for molecular farming for the fabrication of pharmaceutical proteins. But it cannot be used at an agricultural level as the recombinant protein product is relatively unstable in tobacco, which requires lyophilization of the leaf tissue for transport or processed at the farm (Ma et al., 2003). As compared with this the amassing of transgenic products in seeds helps in storage for longer time duration at atmospheric temperatures due to the higher stability of

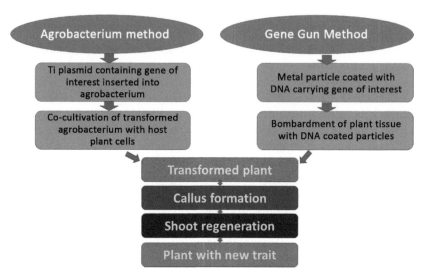

FIG. 2 Schematic representation of the stable transformation of plants by agrobacterium-mediated and particle bombardment method.

proteins in the seed. Seeds have specialized storage compartments in the form of integrated protein vesicles and vacuoles, derived from the secretory pathway, providing a suitable biochemical environment for protein accumulation. Seeds have less water content, which lessens the chances of enzymatic and nonenzymatic degradation of stored proteins, increasing the stability of the protein. Different studies have reported that plant bodies formed in seeds remain durable for a minimum of 3 years at ambient temperature with no apparent loss of biological activity (Rybicki, 2010; Stöger et al., 2000). Likewise, vegetables and fruits can also be used. The critical advantage of fruit, vegetable, and leafy salad crops is that they can be ingested in unprocessed or semiprocessed forms. They are predominantly appropriate for the synthesizing subunit vaccines, food additives, and antibodies for surface-based inert immunotherapy. For example, potatoes have widely been used for the edible manufacturing vaccines and in the majority of the human-based clinical assessment conducted so far (Artsaenko, Kettig, Fiedler, Conrad, & Düring, 1998) and have been investigated as a possible bulk production system for antibodies, antibody-fusion proteins, and human milk proteins (Chong & Langridge, 2000; De Wilde, Peeters, Jacobs, Peck, & Depicker, 2002; Schünmann, Coia, & Waterhouse, 2002). Plants offer the most practical approach to deliver bioencapsulated tolerogens because they are edible in raw form without being cooked. So, proteins that have been produced in them are not at risk of being heat denatured. On the other hand, new plant organs, that is, leaves or fruits, can be readily desiccated and frozen after being harvested, ground into a powder, and administered orally after being put into drinks or made into cookies for consumption or packaged in the form of capsules or tablets so can that it can be stored for a more extended period with increased therapeutic protein content, reliability, and homogeneous dose delivery (Ma, Liao, & M Jevnikar, 2015).

5. Mode of action

Vaccines have done near wonders in the struggle of humankind against contagious disease. Vaccines provoke the immune system to rapidly wipe out specific disease-causing agents or pathogens before they can proliferate up to an adequate number to cause a condition-indicating disease. Nowadays, biotechnologists are focusing on enhancing the use of edible flora to attain complete immunization against infectious disease-causing organisms (Carter III & Langridge, 2002). Plant vaccines, which protect against antigens adopting the intestinal route to enter the host, are a harmless and efficient substitute for high-cost vaccines (Yu & Langridge, 2000).

Most infectious agents initially get in touch with the host at the surface of epithelial membranes and enter at mucosal surfaces facing the lumen of alimentary canal/GI tract, air passage toward lungs, and genitourinary tracts. Mucosal immune system (MIS), a unified structure containing various tissues, cells, and macromolecules, is the most substantial immunologically active part of the body of an organism. MIS possesses the very first barrier against foreign invaders and also acts as the most efficient site for vaccination against them (Yu & Langridge, 2000). The most proficient way of mucosal immunization is administered through the oral or nasal route as compared with parental inoculation because here it will be able to produce mucosal resistance, Ig-mediated immune reaction, and cell-mediated immune response. Plant-derived edible vaccine, when taken orally, does not get hydrolyzed by gastric enzymes due to bioencapsulation, that is, sturdy cell wall. Instead, cells undergo degradation by the action of enteric or microbial enzymes, finally releasing the antigens in the intestines near Payer's patch (Jan et al., 2016). Payer's patches (PP) serve as mucosal immune effecter sites. They consist of 30–40 lymph nodules and are a prime source of IgA-secreting plasma cells. PP contains pocket-shaped follicles, filled with a cluster of B lymphocytes, T lymphocytes, and macrophages. The germinal center develops from follicles as a result of stimulation by a foreign invader. These follicles work as the site for antigen penetration into the epithelium of intestine, thereby amassing antigen inside the lymphoid structure. The antigens then penetrate M cells on Payer's patches and are diffused across the mucous membrane by M-cell toward activated B cells, within these lymphoid follicles due to the expression of class II MHC molecules by M cells. The B cells, activated by antigens, leave the lymphoid follicle and disperse into mucosal-associated lymphoid tissue (MALT), where they undergo maturation into plasma cells secreting the IgA class of secretory immunoglobulin. Secretory IgA neutralizes pathogenic toxins, inhibits microbial immigration, and eradicates antigens from the circulatory system (Langridge, 2000; Mishra, Gupta, Khatri, Goyal, & Vyas, 2008).

6. Future developments of GM plants in biopharmaceuticals

The genetically modified plants have become popular to produce new therapeutics with the advancement in biological sciences. As a result, undoubtedly, these plants will act as factories for the production of novel vaccines, antibodies, and therapeutic proteins. At one end, there is increased research, development, and commercialization on these genetically modified plants, and the other end is associated with risks to human health due to using these transgenic plants. As a result, there exist enormous challenges linked with the biosafety of these GM crops to the environment in general and public in specific. Biosafety trails should be conducted at all stages of GM crop development and release in the field for open trails. Potential risks are associated with the exposure of these crops to human health, biological diversity, and sustainable development of nations.

These "plantibodies" or biopharmaceuticals will be providing economic advantages by the increase in yields, improving quality, distribution, and handling of materials through preplanned biosafety procedures. With the improvement of technology and development in standard procedures, these biopharmaceuticals will play a significant role in significant therapeutics. There is, at the same time, much debate on the plant-based sources over nonplant sources of biopharmaceuticals. However, the recombinant plant-based DNA technology has become increasingly popular due to plasma proteins, the production of antibodies, and vaccines.

Another prospect of GM plants in biopharma comes from the public perception of using natural products; with the ever-increasing population, it is nearly impossible to produce organic food and medicines to overcome this demand. Here comes the role of GM food as technology has always played a vital role in the food industry. Genetically modified plants have advanced characteristics as compared with other plants and crops. Therefore these plants can be better in producing desire results in biopharmaceuticals.

References

Artsaenko, O., Kettig, B., Fiedler, U., Conrad, U., & Düring, K. (1998). Potato tubers as a biofactory for recombinant antibodies. *Molecular Breeding, 4*(4), 313–319.

Bauer, A. (2014). *Pharma crops, state of field trials worldwide.* http://www.umweltinstitut.org/download/field_trials_engl_september06_01-2.pdf. (Accessed 20 October 2014).

Breyer, D., De Schrijver, A., Goossens, M., Pauwels, K., & Herman, P. (2012). Biosafety of molecular farming in genetically modified plants. In *Molecular farming in plants: Recent advances and future prospects* (pp. 259–274). Dordrecht: Springer.

Carter, J. E., III, & Langridge, W. H. (2002). Plant-based vaccines for protection against infectious and autoimmune diseases. *Critical Reviews in Plant Sciences, 21*(2), 93–109.

Chong, D. K., & Langridge, W. H. (2000). Expression of full-length bioactive antimicrobial human lactoferrin in potato plants. *Transgenic Research, 9*(1), 71–78.

Daniell, H., Streatfield, S. J., & Wycoff, K. (2001). Medical molecular farming: Production of antibodies, biopharmaceuticals and edible vaccines in plants. *Trends in Plant Science, 6*(5), 219–226.

De Wilde, C., Peeters, K., Jacobs, A., Peck, I., & Depicker, A. (2002). Expression of antibodies and Fab fragments in transgenic potato plants: A case study for bulk production in crop plants. *Molecular Breeding, 9*(4), 271–282.

Fischer, R., Liao, Y. C., Hoffmann, K., Schillberg, S., & Emans, N. (1999). Molecular farming of recombinant antibodies in plants. *Biological Chemistry, 380*(7–8), 825–839.

Fischer, R., Stoger, E., Schillberg, S., Christou, P., & Twyman, R. M. (2004). Plant-based production of biopharmaceuticals. *Current Opinion in Plant Biology, 7*(2), 152–158.

Jan, N., Shafi, F., Hameed, O., Muzaffar, K., Dar, S., Majid, I., et al. (2016). An overview on edible vaccines and immunization. *Austin Journal of Nutrition and Food Sciences, 4*, 1078.

Key, S., Ma, J. K., & Drake, P. M. (2008). Genetically modified plants and human health. *Journal of the Royal Society of Medicine, 101*(6), 290–298.

Langridge, W. H. (2000). Edible vaccines. *Scientific American, 283*(3), 66–71.

Li, S., Yuan, W., Yang, P., Antoun, M. D., Balick, M. J., & Cragg, G. M. (2010). Pharmaceutical crops: An overview. *Pharmaceutical Crops, 1*, 1–17.

Lienard, D., Sourrouille, C., Gomord, V., & Faye, L. (2007). Pharming and transgenic plants. *Biotechnology Annual Review, 13*, 115–147.

Ma, J. K., Barros, E., Bock, R., Christou, P., Dale, P. J., Dix, P. J., et al. (2005). Molecular farming for new drugs and vaccines. Current perspectives on the production of pharmaceuticals in transgenic plants. *EMBO Reports, 6*(7), 593–599.

Ma, J. K., Drake, P. M., & Christou, P. (2003). The production of recombinant pharmaceutical proteins in plants. *Nature Reviews Genetics, 4*(10), 794–805.

Ma, S., Liao, Y. C., & M Jevnikar, A. (2015). Induction of oral tolerance with transgenic plants expressing antigens for prevention/treatment of autoimmune, allergic and inflammatory diseases. *Current Pharmaceutical Biotechnology, 16*(11), 1002–1011.

Marvier, M. (2007). Pharmaceutical crops have a mixed outlook in California. *California Agriculture, 61*(2), 59–66.

Mason, H. S., & Arntzen, C. J. (1995). Transgenic plants as vaccine production systems. *Trends in Biotechnology, 13*(9), 388–392.

Mason, H. S., Lam, D. M. K., & Arntzen, C. J. (1992). Expression of hepatitis B surface antigen in transgenic plants. *Proceedings of the National Academy of Sciences of the United States of America, 89*, 11745–11749.

Mishra, N., Gupta, P. N., Khatri, K., Goyal, A. K., & Vyas, S. P. (2008). Edible vaccines: A new approach to oral immunization. *Indian Journal of Biotechnology, 7*, 283–294.

Moon, K. B., Park, J. S., Park, Y. I., Song, I. J., Lee, H. J., Cho, H. S., et al. (2020). Development of systems for the production of plant-derived biopharmaceuticals. *Plants, 9*(1), 30.

Rastogi Verma, S. (2013). Genetically modified plants: Public and scientific perceptions. *ISRN Biotechnology, 2013*, 1–11.

Rybicki, E. P. (2010). Plant-made vaccines for humans and animals. *Plant Biotechnology Journal, 8*(5), 620–637.

Schünmann, P. H., Coia, G., & Waterhouse, P. M. (2002). Biopharming the SimpliRED™ HIV diagnostic reagent in barley, potato and tobacco. *Molecular Breeding, 9*(2), 113–121.

Stöger, E., Vaquero, C., Torres, E., Sack, M., Nicholson, L., Drossard, J., et al. (2000). Cereal crops as viable production and storage systems for pharmaceutical scFv antibodies. *Plant Molecular Biology, 42*(4), 583–590.

Walsh, G. (2013). *Biopharmaceuticals: Biochemistry and biotechnology.* John Wiley & Sons.

Wani, S. H., & Sah, S. K. (2015). Transgenic plants as expression factories for bio pharmaceuticals. *Journal of Botanical Sciences, 1*, 1–4.

Yu, J., & Langridge, W. H. (2000). Novel approaches to oral vaccines: Delivery of antigens by edible plants. *Current Infectious Disease Reports, 2*(1), 73–77.

Chapter 14

Role of phytomedicines in metabolic disorders

Takoua Ben Hlel[a], Ascensión Rueda Robles[b], Issam Smaali[a], and M. Nejib Marzouki[a]

[a]*LIP-MB Laboratory LR11ES24, National Institute of Applied Sciences and Technology, University of Carthage, Tunis, Tunisia*, [b]*Institute of Nutrition and Food Technology "José Mataix", Center of Biomedical Research, University of Granada, Granada, Spain*

1. Introduction

The medical research in the field of metabolic disorder (MD) treatment and prevention has undergone major advances, especially with the emergence of the new biochemical and molecular techniques. However, in spite of all the success of conventional medicine and synthetic drugs, many metabolic disorders are still incurable or hard to manage. Recent studies have shown that modern poor lifestyle choices are undeniably important risk factors for a lot of acquired metabolic diseases like obesity and type 2 diabetes. Besides, unbalanced and unhealthy diets can also trigger underlying inherited adult-onset metabolic disorders (Ki, Lee, Cho, Kim, & Kim, 2016). As a consequence the incidence and the mortality rate of metabolic disorders are unceasingly increasing (Saklayen, 2018).

They say prevention is better than cure. Luckily, some of the acquired metabolic diseases could be entirely avoided with the right anticipatory actions including a healthy diet, while other more stubborn health conditions could be at least managed. Traditional medical systems started the fight against these diseases long before modern medicine, which makes the human experience and collective heritage a precious tool in the journey of finding the right cure. Combining our traditional medical legacy with modern techniques could be the answer to finding new effective treatments to an unmet need like metabolic disorders. Traditional medicine includes a variety of practices such as using herbal remedies, cupping, acupuncture, and physical therapy (Park et al., 2012). Unlike the other methods, medicinal plants are considered as an unlimited source of raw material to prepare endless herbal formulas.

Phytomedicine: A Treasure of Pharmacologically Active Products from Plants
https://doi.org/10.1016/B978-0-12-824109-7.00022-4
389

In fact, plants are highly valued for their nutritional composition since they can contain proteins, vitamins, minerals, polyunsaturated fatty acids, and fiber while having a relatively low calorie intake (Bruneton, 2009). On the other hand, herbs represent a gold mine for researchers in metabolic diseases field, given their richness in bioactive compounds. Scientists have always been intrigued to precisely determine the actual effect of traditional herbal remedies on metabolic diseases. Starting by in vitro studies, a lot of herbal extracts showed therapeutic potential, mainly antidiabetic, antidyslipidemic, antiobesity, antioxidant, and modulatory activities (El Abed, 2014; Ko, Jang, & Jung, 2019; Yin, Zhang, & Ye, 2008). The next steps to prove its effectiveness were respectively animal models and clinical trials. With the evolution of analytical chemistry, biochemistry, genetics, and bioinformatics, the researchers gained the ability to identify major functional compounds and characterize the mechanisms of action. This chapter provides a general understanding of the contemporary role of traditional herbal medicine in treating metabolic disorders, with a realistic assessment of the currently available studies.

2. Metabolic disorders: Definition, causes, and prevalence

A metabolic disorder occurs when abnormal chemical reactions in the body interrupt the normal metabolic process. They often result in a deficit of an enzyme, its cofactor, or in other proteins involved in the metabolism of carbohydrates, amino acids, and fatty acids. As a consequence, these diseases are characterized by an inability to use and/or store energy properly. The metabolic abnormalities can be congenital or acquired (Gregersen & Brennser, 2001; Huynh, Schneider, & Gareau, 2016) (Fig. 1).

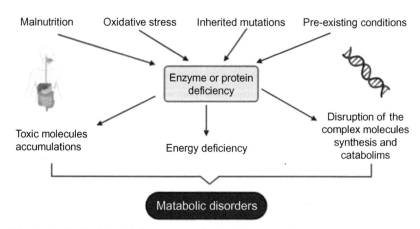

FIG. 1   A pathophysiological diagram presenting the risk factors of metabolic disorders.

Inherited metabolic diseases, also known as inborn errors of metabolism, can be divided into three major groups: (1) anomalies responsible for the accumulation of toxic substances proximal to the enzymatic bloc, which mainly consists on the enzyme deficiencies on the pathway of amino acid degradation (phenylketonuria, homocystinuria, and leucinosis), the urea cycle disorders mainly ornithine transcarbamylase OTC deficiency, sugar intolerances, and other conditions like Wilson's disease (lack of hepatic transport of copper) and porphyrias; (2) organelle dysfunctions (lysosomes, peroxisomes, endoplasmic reticulum, and Golgi) usually caused by deficiencies in the catabolism or synthesis of some complex molecules; and (3) intermediary metabolism anomalies with symptoms related to problems in the storage, production, or use of energy such as fatty acid oxidation, mitochondrial respiratory chain, and glycogen metabolism disorders (Fernandes, Saudubray, Van Den Berghe, & Walter, 2006; Jouvet & Touati, 2002).

The majority of MDs are single-gene disorders coming from an autosomal recessive inheritance. Although each one of these disorders has a low individual incidence of less than 1 per 100,000 births, their all-cause MD incidence is significantly higher. According to a systematic study published in 2018, which took into account 49 scientific reports, the global prevalence of collective MDs is estimated to be 50.9 in 100,000 live births. Only 500 disorders are currently identified, but the potential number is estimated to be around 4000–6000 when considering the number of genes coding for enzymes and transport proteins (Chabrol & de Lonlay, 2011; Sutton, 2019; Waters et al., 2018). The diagnosis of MDs is simple, and it can be done mostly through biochemical analysis. However, recognizing the signs and the symptoms can be challenging at first, due to the rarity of these disorders. Most of the MDs can be spotted and treated at the pediatric age, but lately, more severe conditions with neurologic syndromes are being discovered in adolescents and young adults. Depending on the type of the metabolic error, some conditions are treatable with medicines and special diets, while others are rarely treatable (Agana et al., 2018).

On the other hand, acquired metabolic disorders could occur under specific conditions, especially if an organ stops to function completely. They may also arise as complications of some surgeries or other diseases like hepatic, renal, and respiratory failure and cancer (Coller, 2014; Lameda & Koch, 2020). Other exogenous risk factors encompass malnutrition, unhealthy lifestyle, and alcohol and drug consumption (Ki et al., 2016). Diabetes, whether type 1 or 2, is a metabolic disease with a complex etiology, since there are many risk factors including a genetic predisposition. Diabetes prevalence is considered high with 8.5% of adults aged 18 years and older having diabetes in 2014, which makes it among the most common chronic metabolic diseases in the world (Emerging Risk Factors Collaboration, 2010; van Tilburg et al., 2001).

3. Traditional remedies used for the treatment of MDs

We cannot determine precisely the time when the practical experimentation and the use of natural remedies on an empirical basis to treat metabolic diseases began. The genetic nature of the majority of the MDs, the rarity of the cases, and the ambiguity of the syndromes made it very hard for ancient doctors and traditional healers to recognize, diagnose, or even understand the diseases. However, a quick glance into history reveal that several medical systems in ancient civilizations offered traditional remedies to a variety of symptoms of some medical conditions that could be metabolic-related diseases.

The traditional Chinese medicine, the Ayurveda, and the Greco-Arab and Islamic medicine all of these ancient systems were aware of the presence of metabolic disorders, even if the precise definition and understanding of the mechanisms came thousands of years later.

The metabolic syndrome is a combination of serious metabolic risk factors that include central obesity, high blood pressure, high blood sugar, low high-density lipoprotein (HDL) cholesterol levels, and a high triglyceride level (TG). The MetS increases significantly the risk of developing cardiovascular diseases, diabetes, and cancer (Ko, Jang, & Jung, 2019). Although MetS is a newly defined medical term, many of its symptoms including central obesity and its hallmark had been a focus of attention of many ancient medical systems.

Traditional Chinese medicine referred to diabetes as the disease that caused weight loss and excessive thirst, while obesity was the disease of overeating. To treat these conditions, patients received personalized herbal formulas composed of a variety of medicinal plants depending on the symptoms. *Aloe vera*, cinnamon, ginseng, ginger, and tea were one of the most used medicinal plants to treat MDs (Yin et al., 2008). Ancient Chinese practitioners of medicine also believed, along with the ancient Koreans, that metabolic diseases and obesity can be caused by blood stasis. Herbal formulas that contained red rooted sage, peach seed, safflower, and red peony root were used to promote blood circulation and remove the heat from the body (Ko et al., 2019; Liu, Yin, Shi, & Chen, 2012).

The ancient Ayurvedic texts described several clinical disorders characterized by polyuria, collectively called Prameha. The latter shares undeniable symptoms with medical conditions like obesity, diabetes mellitus, and metabolic syndrome. As a matter of a fact, Ayurveda reported two types of Prameha: the hereditary and the acquired that can be triggered by an unhealthy diet, psychological factors, and a sedentary lifestyle (Sharma & Chandola, 2011). The prognosis was different depending on the metabolic disease and its stage. The mild cases of obesity were considered as curable with the right herbal formula, whereas the advanced one with complications was deemed as hard to manage. The list of Indian herbs used to treat metabolic syndromes encompasses a large number of plants, namely, cedar wood, turmeric, bitter gourd pulp, holy basil leaves, tellicherry bark, black plum seeds, and Indian gooseberry fruit (Purkait & Bhattacharya, 2012; Rashmi, 2017).

In Islamic medicine, ancient doctors relied on the observation and recognized that some diseases have symptoms that appeared after the consumption of certain ingredients. They first treated patients with the prescription of a special diet, including fasting that was a holistic way of detoxifying the body, and when the treatment failed in some cases, they resorted to herbal remedies (Liu, 2011). Fenugreek, black seeds, garlic, walnut, saltbush, nettle, onions, and olive are among the popular plants used to treat obesity by Arab and Muslim ancient doctors (Said et al., 2008).

4. In vitro and preclinical studies of the use of herbal medicine in MDs

Herbal remedies or medicines consist of nonpurified plant portions or plant extracts containing various constituents, which often work together synergistically. The use of herbs as medicine is the oldest form of health care known to mankind and has been used in all cultures throughout history (Kunle, Egharevba, & Ahmadu, 2012). Many of the active ingredients they contain are still unknown, but numerous in vitro and in vivo studies have pointed to the role of herbal medicine in the treatment of MDs. Table 1 describes studies related to the effect of herbal medicines on some of the most important metabolic disorders that will be discussed here.

Insulin resistance is related to type 2 diabetes. *Scutellaria baicalensis Georgi* is a traditional Chinese herb. In its composition, it contains baicalin, one of the most potent and abundant polyphenols that exerts antiobesity and antidiabetic effects. An in vitro and in vivo study conducted by Fang et al. (2018) using obese mice and 3T3-L1 cells showed a decrease in food intake, body weight, homeostasis model of insulin resistance, and p-p38 MAPK and pERK levels. They also observed a reversal of high-fat diet-induced glucose intolerance, hyperglycemia, and insulin resistance (Fang et al., 2018). Kadan et al. (2018) assessed the chemical composition, cytotoxicity, and antidiabetic activity of *Gundelia tournefortii* extracts (Kadan et al., 2018). They proposed that *Gundelia tournefortii* exerts antidiabetic activity by enhancing GLUT4 translocation to the cell surface in skeletal muscle. Moreover, Tian et al. (2017) showed that *Wushenziye* formula (composed of *Radix Polygoni, Multiflori preparata, Mori fructus, Mori folium, and Cassiae semen*) regulates glucose metabolism management of type 2 diabetes Tian et al., 2017.

Luffa cylindrica, or bitter gourd, is a plant known for its antioxidant activity (Hlel et al., 2017). According to Akther et al. (2014), the methanolic extract of fruits reduced blood glucose levels when tested on Swiss albino mice (Akther et al., 2014). Another study reported similar results on alloxanized hyperglycemic rats in addition to the improvement of liver enzyme levels (Hazra et al., 2011).

To sum up, there are plenty of studies demonstrating the antidiabetic and antiobesity activities of plants, herbal formulas, and plant-derived constituents,

TABLE 1 Effect of selected herbal medicine in the most common metabolic disorders.

Metabolic disorder	Herbal medicine	Effects	References
Diabetes mellitus and obesity	Scutellaria baicalensis Georgi	Decrease in food intake in obese mice Decrease in p-p38 MAPK and pERK levels Reversed glucose intolerance, hyperglycemia, and insulin resistance	Fang et al. (2018)
	Gundelia tournefortii	Antidiabetic activity by enhancing GLUT4 translocation	Kadan, Sasson, Saad, and Zaid (2018)
	Wushenziye formula	Regulation of the glucose metabolism	Tian, Chang, La, and Li, (2017)
	Cynara scolymus	Delay of the carbohydrates and lipid digestion and absorption	Salem et al. (2017)
	Momordica charantia	Decrease in fasting blood glucose (FBG) Improvement in glucose tolerance Insulinomimetic and/or insulin releasing properties	Raman and Lau (1996)
	Olea europaea	Improvement in body weight and in metabolic markers level/increase in insulin level	Guex et al. (2019)
	Nigella sativa	Lowering body weight, food intake, glucose, serum cholesterol, triglycerides, and low-density lipoprotein (LDL) levels HDL level improvement	Bano et al. (2009)
	Cucurbita pepo	Antiobesity activity: decrease in body mass index	Nderitu, Mwenda, Macharia, Barasa, and Ngugi (2017)
	Brassica oleracea	Decrease in body weight, water and food intake, and blood glucose	Shah, Sarker, and Gousuddin (2016)
	Sonchus oleraceus	Blood sugar modulation through the downregulation of AMPK/Akt/GSK-3β pathway	Chen et al. (2020)
	Hamelia patens	Regulation of blood glucose	Tian et al. (2017)
	Coccinia indica	Hypoglycemic effect Improvement of the metabolic enzymes activity	Venkateswaran and Pari (2002)

Condition	Plant	Mechanism / Action	References
	Ocimum sanctum	Lowering blood sugar and cortisol level; Improvement in insulin release	Marcocci, Packer, Droy-Lefaix, Sekaki, and Gardes-Albert (1994) and Chattopadhyay (1993)
	Silybum marianum	Anti-α-amylase activity; Glucose level reduction	Shakeel and Yar (2014) and Abu-zaiton (2013)
	Trigonella foenum-graecum	Enhancement of the glucose homeostasis/delay of carbohydrate digestion and absorption action; Improvement of the activity of insulin	Hannan et al. (2007)
	Luffa cylindrica	Reduction of blood glucose level; Enhancement of liver enzymes levels	Akther et al. (2014) and Hazra et al. (2011)
Gout hyperuricemia	*Dendrobium candidum leaf*	Inhibition in uric acid production	Lou et al. (2020)
	Rhizoma Dioscoreae nipponicae	Reduction in uric acid level	Zhou, Yu, Zhang, Liu, and Lu (2014)
Dyslipidemia	*Garlic*	Suppress LDL oxidation; Reduce serum cholesterol, triglyceride, and LDL levels	Bayan, Koulivand, and Gorji (2014)
Mitochondrial dysfunctions	*Qiliqiangxin capsule*	Increase of the mitochondrial respiratory function and respiratory control ratio; Decrease in metabolic intermediate products	Zhang et al., 2013
	Panax ginseng	Improvement of the oxygen consumption rate during mitochondrial respiration; Increase in mitochondrial DNA content through activation of SIRT1	Huang et al. (2019a, 2019b) and Wang et al. (2013)
	Curcuma longa	Reduction in the mitochondrial oxidative damage by SIRT1 activation	Yang et al. (2013)

whether in vitro or using an animal model. Some of the other most known herbal medicines are depicted in Table 1 as mentioned earlier.

Usually related to cardiovascular diseases, dyslipidemia is defined as elevated LDL, TG, total cholesterol (TC), and low levels of HDL cholesterol in serum. It is an important risk factor for coronary heart disease (CHD) and stroke (Guo, Liu, Gao, & Shi, 2014). Garlic (*Allium sativum* L.) is a common spice consumed in the world, and it contains variety of bioactive compounds. In vitro effects have revealed that garlic administration suppress LDL oxidation and increase HDL levels. On the other hand, in vivo studies in rats suffering from induced hypercholesterolemia also indicate that the administration of garlic significantly reduces serum cholesterol, triglycerides, and LDL levels (Bayan et al., 2014).

Another important metabolic disorder is gout, an inherited disorder of purine metabolism that causes hyperuricemia resulting in an elevated uric acid level in the blood (Wolff, 2007). Lou et al. (2020) analyzed thee therapeutic effect of the extract of *Dendrobium candidum* leaf on rats with hyperuricemia. Its results indicate an inhibition of uric acid production and a decrease in inflammation by inhibiting the expression of the NF-κB and TLR4 proteins (Lou et al., 2020). Zhou, Liu, Yua, and Zhang (2016) studied the in vitro effect of saponins from *Rhizoma Dioscoreae Nipponicae* (RDN) on gouty arthritis and concluded that these secondary metabolites could regulate the NF-κB signal pathway and thus may help treat the disease. Another study using this time a mouse model showed that there is a significant reduction in the uric acid level using different doses of the total saponins of RDN. Researchers suggested that the decrease is due to the uricosuric effect of the plant saponins achieved through reversing the potassium oxonate-induced alterations in renal organic ion transporters (Zhou, et al., 2014).

As mitochondria are widely recognized as the mains source of endogenous reactive oxygen species, mitochondrial dysfunctions are usually observed in oxidative stress-related diseases like cancer and neurodegenerative diseases (De Moura, dos Santos, & Van Houten, 2010). Asian ginseng, a popular herbal remedy that has been already proven as a potent antioxidant (Huang et al., 2019a, 2019b), was tested for its ability to attenuate mitochondrial dysfunction. According to Huang et al. (2019a, 2019b), ginseng extracts improved spare respiratory capacity and ATP production of cardiomyoblasts. Moreover the extracts also enhanced the mitochondrial bioenergetics (basal respiration, ATP production, proton leak, maximal respiration, and spare respiration capacity) in the cells subjected to tBHP-induced oxidative stress (Huang et al., 2019a, 2019b). The enhancement of mitochondrial function in cardiomyoblasts is most likely due to ginsenosides, the major constituents of ginseng, capable of activating SIRT1 gene and recovering consumed oxygen Wang et al., 2013. Other herbal medicines, like the yellow spice, turmeric, and its active component curcumin, and Qiliqiangxin capsule, a Chinese medicine composed of 11 herbs, were also found to improve the mitochondrial function (Yang et al., 2013; Zhang et al., 2013).

The initial in vitro and in vivo screening of medicinal herbs that can be used for treatment of MDs requires further investigation. As a matter of fact, there are still plenty of unexplored herbs that have been used and recommended by the traditional medical systems and therefore might demonstrate positive effects in the treatment of several diseases.

5.   Clinical studies of the use of herbal medicines and natural compounds in MD: Scientific evidence

Using cell cultures and/or animal models is crucial to investigate the bioactivities of a medicinal plant and to have a primary look at its therapeutic properties. However, these techniques are insufficient to fully understand or anticipate the real effect on humans. The genetic, anatomic, and physiological differences between these models and the human body complicate the extrapolation of the results. Modern clinical trials and ancient medicine both shared the use of human subjects to test the effects of remedies, but the strict safety regulations and the much more reliable results rule in favor of clinical trials. Another obvious difference is depicted, which is the use of whole plant or plant organ extracts and herbal combinations to treat or manage metabolic diseases in folk medicine. Unlike ancient medicine, nowadays, many plant-derived compounds are being successfully characterized, especially with the progress of analytical chemistry and the development of new purification and identification techniques. Some may argue that the typical use of the whole plant or plant organ has better outcomes in the matter of efficacy, whereas others opted for the purified single compounds to avoid any unwanted effects that may exist from the other phytochemicals of the plant.

To treat metabolic disorders, only a limited number of medicinal plants were used in clinical trials in comparison with plants investigated in vitro. Most of these trials targeted acquired metabolic diseases, namely, diabetes and obesity.

5.1   Examples of clinical trials conducted on plants

Since *Momordica charantia* showed promising results in vitro and in preclinical assays, it was the subject of some clinical trials (Raman & Lau, 1996). Malik, Bashir, Khan, and Iqbal (2009) conducted a comparative study between *Momordica charantia* juice and an antibiotic drug, rosiglitazone. The study consisted in treating two groups of 25 type 2 diabetic subjects either with the plant juice or the standard drug. Another group of healthy subjects served as control. The group that have received *Momordica charantia* juice (55 mL/day, 5 months) showed significantly better results (decrease in FBG and TC levels) in comparison with the rosiglitazone-treated group. The authors also reported a decrease in sialic acid concentrations as compared with the control group. The latter is considered as a strong indicator of a high risk of fatal cardiovascular complications Malik et al., 2009. Another recent study published in 2018 showed that the administration of *Momordica charantia* in 24 patients of type 2 diabetic patients

(2000 mg/day, 3 months) resulted in significant decrease in the weight, body mass index (BMI), fat percentage, waist circumference, and in the levels of glycated hemoglobin (HbAlc), 2-h glucose tolerance, and an increase in insulin secretion (Cortez-Navarrete, Martinez-Abundis, Perez-Rubio, Gonzalez-Ortiz, & Mendez-Del Villar, 2018). However, Dans et al. (2007) indicated a limited effect in diabetic patients treated with a *Momordica charantia* (2 capsules 3 times/day, 3 months). The treated subjects showed a minor decrease in HbAlc levels (0.24%) and no significant effect regarding FBG and total cholesterol level (Dans et al., 2007).

Trigonella foenum-graecum, commonly known as fenugreek, is highly appreciated in traditional medicine as a slimming agent. The herb has been tested on human subjects in multiple occasions and has shown a hypoglycemic effect that affected FBG, 2-h glucose, and HbAlc levels. However, the majority of the trials administered medium or high doses of fenugreek (Hadi et al., 2020; Neelakantan, Narayanan, de Souza, & van Dam, 2014). A recent study published in 2020 showed that the consumption of fenugreek seed powder (5 g three times/day, 8 weeks) by the recruited type 2 diabetes patients along with their regular medication had beneficial results. Patients showed a decrease in fasting plasma glucose (FPG) and improvement in the levels of serum alanine aminotransferase (ALT), alkaline phosphatase (ALP), aspartate aminotransferase (AST), and systolic blood pressure (Hadi et al., 2020).

Cinnamomum cassia and *C. verum* are two species from the *Cinnamomum* genus, known for their hypoglycemic activity, and both are used as popular spices (cinnamon). Unfortunately, only *C. cassia* was subjected to human trials, and the obtained results are inconsistent and inconclusive. Some assays declared an improvement in the diabetic subjects' glucose and insulin levels, while others did not report any significant difference (Hasanzade, Toliat, Emami, & Emamimoghaadam, 2013; Medagama, 2015). Anderson et al. (2016) recruited 137 diabetic patients to be supplemented either with a placebo or a dietary product made from a water extract of cinnamon (CinSulin 500 mg/day) for 2 months. The assay revealed a significant decline of both FPG and 2-h glucose values after intervention. The authors also reported an enhancement in insulin sensitivity and in fructosamine concentrations (Anderson et al., 2016). This result is in line with another double-blind clinical trial conducted by Mang et al. (2006). The study took into account with data collected from 79 type 2 diabetic patients and announced a 10.3% improvement in plasma glucose levels after consumption of the aqueous cinnamon extract (3 g/day, 4 months). However, even the placebo group showed a nonnegligible decrease in the same parameter (3.4%) (Mang et al., 2006). According to Hasanzade et al. (2013) the administration of cinnamon in 72 type 2 diabetic subjects did not significantly reduce FBG or HbAlc levels Hasanzade et al., 2013. Several studies that were undertaken to determine the effect of cinnamon on cells response to insulin indicated that the herb did not improve insulin resistance and sensitivity (Gutierrez, Bowden, & Willoughby, 2016; Medagama, 2015; Roussel, Hininger, Benaraba, Ziegenfuss, & Anderson, 2009).

Ginkgo biloba or the legendary tree of Asia was the subject of a number of clinical trials as well. A study by Kudolo that came out in 2000 was about the insulin secretagogue capacity of the *G. biloba* extract. Twenty normal glucose-tolerant individuals were asked to consume it on daily basis. All participants underwent standard oral glucose tolerance tests before the start of the intervention and after it, as well as the measurement of the basic metabolic panel (electrolytes, plasma osmolality and anion gap), liver function, total lipid profile, insulin and C-peptide levels to indicate how much insulin is being produced by pancreatic β-Cells. The obtained data showed no significant changes in FPG and in the other metabolic parameters. Interestingly, though, an increase in fasting insulin and c-peptide level was recorded, along with a decrease in blood pressure. The author speculated that the increasing insulin resistance caused a *G. biloba*-induced accelerated pancreatic β-cells activity and as a consequence a rise in insulin/C-peptide levels (Kudolo, 2000). Another speculative statement was made by the author who proposed that the plant induced an increase in the insulin clearance but not in the c-peptide and justified it by a dissimilarity that existed between the insulin and C-peptide response curves to glucose ingestion. The same author conducted a follow-up trial with the same design; this time he recruited type 2 diabetic individuals. The *G. biloba* extract did not cause any increase in insulin levels since the pancreatic cells are already fully stimulated, but in subjects with pancreatic exhaustion, data showed an increase in insulin levels, which indicate its ability to improve the pancreatic function. However, even in this group, there was no decline in FPG levels. On the contrary, it was confirmed that the plant increased the rate of insulin clearance as well as the hypoglycemic drugs, which caused an FPG level rise (Kudolo, 2001).

Aloe vera has been regarded as a very valuable plant for centuries. In addition to diabetes and obesity, several clinical trials proved that the plant gel is useful in the treatment and management of dyslipidemia. For instance, Alinejad-Mofrad, Foadoddini, Saadatjoo, and Shayesteh (2015) administered capsules of *A. vera* (300 or 500 mg) to 45 prediabetic individuals for 2 months Alinejad-Mofrad et al., 2015. The subjects showed a significant decrease in the levels of FBG and HbA1C in comparison with the control group. Furthermore, TC, LDL-C, and TG levels decreased for the group who received the *A. vera* capsules of 500 mg. Another clinical study is in agreement with the latter, which consisted of 30 patients with type 2 diabetes supplemented with *A. vera* gel powder (100 or 200 mg) for 3 months. For both concentrations received, the results indicated a significant decline in the levels of FBG, postprandial plasma glucose, TC, TG, LDL, and very low-density lipoprotein (VLDL) in addition to an improvement in HDL level (Choudhary, Kochhar, & Sangha, 2014).

Traditional doctors recommended ginger root (*Zingiber officinale*) for a variety of health problems including diabetes and hyperlipidemia. This claim was supported by scientific proof. Jafarnejad et al. (2017) reviewed a total of nine randomized controlled trials and found that ginger supplementation significantly decreased FBG, TG, and TC and increased HDL level, but the changes depended on the clinical condition (Jafarnejad et al., 2017). The overall findings were in agreement with conclusions drawn by another review paper published in 2018 where the authors suggested considering ginger in the treatment of diabetes (Araujo, Jesus-Lima, Otoch, & Pessoa, 2018).

5.2 Prominent plant-derived compounds tested in clinical trials

Resveratrol is a polyphenol of the stilbene class that exists naturally in various plant species but in major contents in *Vitis vinifera* L (Fig. 2). The effect of resveratrol in diabetes mellitus has been investigated through clinical trials. The results of the various studies are conflicting. For instance, data collected from the treatment of 33 diabetic patients with resveratrol at a dose of 1 g/day for 45 days showed a decrease in FBG, HbAlc, insulin, and insulin resistance and an increase in HDL, in comparison with their baseline levels Movahed et al., 2013 While in another double-blind randomized clinical trial, the administration of 500 mg twice a day of resveratrol to 14 type 2 diabetic patients did not show any improvement on GLP-1 secretion, glycemic control, gastric emptying, body weight, or energy intake (Thazhath et al., 2016). Nevertheless, resveratrol might have other beneficial activities in diabetes management according to a study conducted by Goh et al., (2014) Authors reported that resveratrol regulates energy expenditure by increasing skeletal muscle SIRT1 and AMPK expression, which might indicate an exercise-mimetic effects in patients (Goh et al., 2014).

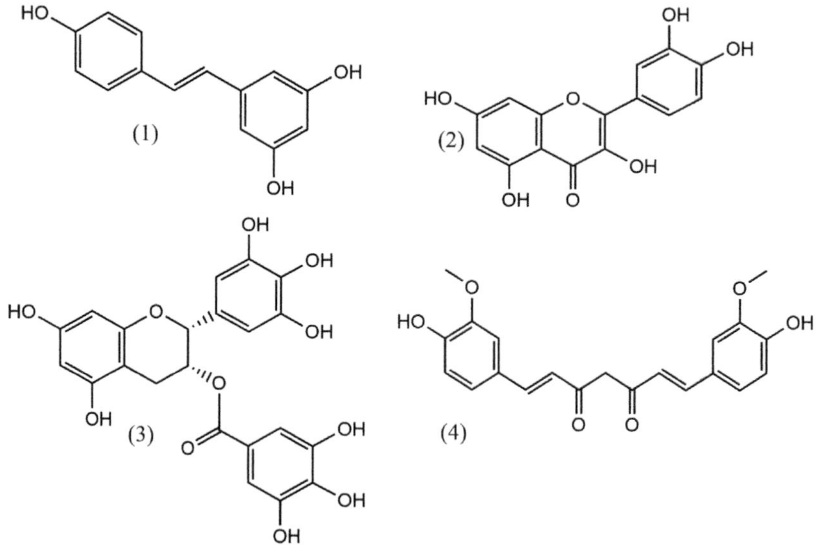

FIG. 2 Chemical structures of plant-derived polyphenols used in clinical trials for metabolic disorders. (1) Resveratrol, (2) quercetin, (3) epigallocatechin gallate, and (4) curcumin.

Quercetin and catechins are abundant flavonoids that can be found in vegetables, fruits, and teas; both have been linked to metabolic diseases treatments (Fig. 2). However, in spite of the positive results obtained in vitro and in animal model, quercetin was found to be ineffective in treating obesity and diabetes when it comes up to clinical trials (Huang et al., 2019a, 2019b; Sahebkar, 2017). So far the only worth mentioning beneficial effect that was detected in diabetic patients is the improvement of bone mineralization and formation biomarkers. Quercetin supplementation apparently caused an increase in the serum level of calcium, 25(OH) vitamin D, and osteocalcin (Hassan, Sharrad, & Sheri, 2018).

On the other hand, catechins, namely, epigallocatechin gallate (EGCG), which is the most abundant catechin in the green tea, have proven some promising results in human studies. According to a review written by Alipour et al. (2018), healthy subjects or patients who received EGCG, whether administered as pure compound or in green tea extracts, showed an improvement in their metabolic biomarkers and parameters. A lot of studies reported a decrease in body weight; total abdominal fat, in obese and normal subjects; decline in TC and TG in people with lipid abnormalities; and also a decrease in stress biomarkers levels in patients with the MetS (Alipour et al., 2018).

Many researchers were intrigued by curcumin, an active component found in the root of *Curcuma longa*, thus they explored its effect on glycemic and lipid parameters in human subjects. Several studies concluded that curcumin supplementation has positive results mainly on HbA1c level (Poolsup, Suksomboon, Kurnianta, & Deawjaroen, 2019). The most distinguished study is the one conducted by Chuengsamarn, Rattanamongkolgul, Luechapudiporn, Phisalaphong, and Jirawatnotai (2012), considering the high number of type 2 diabetic participants (237). The volunteers received either a curcumin capsule (1.5 g/day) or a placebo for 9 months. The authors reported that the levels of HbA1c, FPG, and 2-h postprandial glucose (2HPP) decreased in comparison with the control group Chuengsamarn et al. (2012). Another double-blind randomized clinical trial was conducted on 70 type 2 diabetic patients who received either curcumin (nanomicelle 80 mg/day) or placebo for 3 months. The data gathered after the end of the trial showed a significant decrease in the levels of HbA1C, FBG, TG, and BMI of the subjects who received the treatment (Rahimi et al., 2016).

Trigonelline is a natural pyridine alkaloid that was first extracted from the fenugreek seeds but exists in other plant species like coffee and green peas (Fig. 3) (Garg, 2016). The effect of the ingestion of 500-mg trigonelline and a placebo (1-g mannitol) on glucose and insulin concentrations during a 2-h oral glucose (75 g) tolerance test (OGTT) was explored in a randomized crossover trial that included 15 overweight men. Participants were asked to make a visit four times separated by at least days. Blood was collected before the supplementation and immediately after it. The second group of blood samples were collected at the start of the OGTT (30 min after the supplementation) and until 120 min after it. The authors reported that trigonelline reduced early glucose and insulin responses during an OGTT (15 min after). However, the treatments

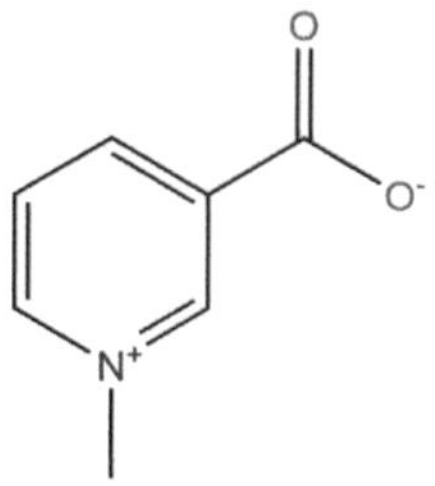

FIG. 3 Trigonelline chemical structure.

did not significantly affect insulin or glucose area under the curve values, in comparison with the placebo (Van Dijk et al., 2009). Moreover, another clinical trial revealed that the positive effects of trigonelline on glucose tolerance are not mediated by incretin hormones (GLP-1 and GIP) (Olthof, van Djik, Deacon, Heine, & van Dam, 2011). These results suggest that further clinical studies investigating the pharmacokinetics and mechanisms of action of trigonelline should be carried out.

6. Conclusion

The results of in vitro, preclinical, and clinical trials using herbal formulas and ingredients to treat metabolic disorders have been overall encouraging. Traditional medicine, mainly medicinal plant-based remedies, clearly still has a lot to offer to humanity in the fight against these serious diseases. Nevertheless, there is still a long way to go to confirm the efficacy and the safety of the treatments. More reliable clinical trials, defining acute and chronic side effects, performing pharmacokinetic and pharmacodynamic studies, and assessing the bioavailability and mechanisms of action of bioactive compounds, are needed for a better understanding and a wiser scientific judgment. Another big challenge is facing the use of traditional medicine, which is the genetic origin of many metabolic disorders. Using phytotherapy to regulate the gene expression is a new approach with little current applications yet looks very promising for the future.

References

Abu-zaiton, A. S. (2013). Evaluating the effect of Silybum marianum extract on blood glucose, liver and kidney functions in diabetic rats. *Advanced Studies in Biology*, 5(10), 447–454.

Agana, M., et al. (2018). Common metabolic disorder (inborn errors of metabolism) concerns in primary care practice. *Annals of Translational Medicine*, 6(24), 469.

Akther, F., et al. (2014). Methanolic extract of Luffa cylindrica fruits show anti hyperglycemic potential in Swiss Albino mice. *ANAS*, 8, 62–65.

Alinejad-Mofrad, S., Foadoddini, M., Saadatjoo, S. A., & Shayesteh, M. (2015). Improvement of glucose and lipid profile status with Aloe vera in pre-diabetic subjects: A randomized controlled-trial. *Journal of Diabetes & Metabolic Disorders*, 14(1), 22.

Alipour, M., et al. (2018). The effects of catechins on related risk factors with Type 2 diabetes: A review. *Progress in Nutrition, 20*(1), 12–20.

Anderson, R. A., et al. (2016). Cinnamon extract lowers glucose, insulin and cholesterol in people with elevated serum glucose. *Journal of Traditional and Complementary Medicine, 6*(4), 332–336.

Araujo, A., Jesus-Lima, J., Otoch, J., & Pessoa, A. (2018). Effect of ginger (Zingiber officinale) supplementation on diabetes: An update. *American Journal of Phytomedicine and Clinical Therapeutics, 6*(3), 13.

Bano, F., et al. (2009). Antiobesity, antihyperlipidemic and hypoglycemic effects of the aqueous extract of Nigella Sativa seeds (Kalongi). *Pakistan Journal of Biochemistry and Molecular Biology, 42*(4), 136–140.

Bayan, L., Koulivand, P. H., & Gorji, A. (2014). Garlic: A review of potential therapeutic effects. *Avicenna Journal of Phytomedicine, 4*(1), 1–14.

Bruneton, J. (2009). *Pharmacognosie, phytochimie, plantes médicinales* (4e éd.). Lavoisier.

Chabrol, B., & de Lonlay, P. (2011). *Maladies métaboliques héréditaires*. Doin.

Chattopadhyay, R. (1993). Hypoglycemic effect of Ocimum sanctum leaf extract in normal and streptozotocin diabetic rats. *Indian Journal of Experimental Biology, 31*(11), 891–893.

Chen, L., et al. (2020). Sonchus oleraceus Linn extract enhanced glucose homeostasis through the AMPK/Akt/GSK-3β signaling pathway in diabetic liver and HepG2 cell culture. *Food and Chemical Toxicology, 136*, 111072.

Choudhary, M., Kochhar, A., & Sangha, J. (2014). Hypoglycemic and hypolipidemic effect of Aloe vera L. in non-insulin dependent diabetics. *Journal of Food Science and Technology, 51*(1), 90–96.

Chuengsamarn, S., Rattanamongkolgul, S., Luechapudiporn, R., Phisalaphong, C., & Jirawatnotai, S. (2012). Curcumin extract for prevention of type 2 diabetes. *Diabetes Care, 35*(11), 2121–2127.

Coller, H. A. (2014). Is cancer a metabolic disease? *The American Journal of Pathology, 184*(1), 4–17.

Cortez-Navarrete, M., Martinez-Abundis, E., Perez-Rubio, K. G., Gonzalez-Ortiz, M., & Mendez-Del Villar, M. (2018). Momordica charantia administration improves insulin secretion in type 2 diabetes mellitus. *Journal of Medicinal Food, 21*(7), 672–677.

Dans, A. M. L., et al. (2007). The effect of Momordica charantia capsule preparation on glycemic control in type 2 diabetes mellitus needs further studies. *Journal of Clinical Epidemiology, 60*(6), 554–559.

De Moura, M. B., dos Santos, L. S., & Van Houten, B. (2010). Mitochondrial dysfunction in neurodegenerative diseases and cancer. *Environmental and Molecular Mutagenesis, 51*(5), 391–405.

Emerging Risk Factors Collaboration. (2010). Diabetes mellitus, fasting blood glucose concentration, and risk of vascular disease: A collaborative meta-analysis of 102 prospective studies. *The Lancet, 375*(9733), 2215–2222.

El Abed, N., et al. (2014). Chemical composition, antioxidant and antimicrobial activities of Thymus capitata essential oil with its preservative effect against Listeria monocytogenes inoculated in minced beef meat. *Evidence-based Complementary and Alternative Medicine, 2014*. https://doi.org/10.1155/2014/152487.

Fang, P., et al. (2018). Beneficial effect of baicalin on insulin sensitivity in adipocytes of diet-induced obese mice. *Diabetes Research and Clinical Practice, 139*, 262–271.

Fernandes, J., Saudubray, J. M., Van Den Berghe, G., & Walter, J. H. (2006). *Inborn metabolic diseases: Diagnosis and treatment*. Springer Science & Business Media.

Garg, R. C. (2016). *Fenugreek: Multiple health benefits in nutraceuticals* (pp. 599–617). Elsevier.

Goh, K. P., et al. (2014). Effects of resveratrol in patients with type 2 diabetes mellitus on skeletal muscle SIRT1 expression and energy expenditure. *International Journal of Sport Nutrition and Exercise Metabolism, 24*(1), 2–13.

Gregersen, N., & Brennser, S. (2001). Metabolic disorders, mutants. In *I Encyclopedia of genetics* (pp. 1187–1188). London: Academic Press.

Guex, C. G., et al. (2019). Antidiabetic effects of Olea europaea L. leaves in diabetic rats induced by high-fat diet and low-dose streptozotocin. *Journal of Ethnopharmacology, 235*, 1–7.

Guo, M., Liu, Y., Gao, Z. Y., & Shi, D. Z. (2014). Chinese herbal medicine on dyslipidemia: Progress and perspective. *Evidence-based Complementary and Alternative Medicine, 2014*, 163036.

Gutierrez, J. L., Bowden, R. G., & Willoughby, D. S. (2016). Cassia cinnamon supplementation reduces peak blood glucose responses but does not improve insulin resistance and sensitivity in young, sedentary, obese women. *Journal of Dietary Supplements, 13*(4), 461–471.

Hadi, A., et al. (2020). The effect of fenugreek seed supplementation on serum irisin levels, blood pressure, and liver and kidney function in patients with type 2 diabetes mellitus: A parallel randomized clinical trial. *Complementary Therapies in Medicine, 49*, 102315.

Hannan, J., et al. (2007). Soluble dietary fibre fraction of Trigonella foenum-graecum (fenugreek) seed improves glucose homeostasis in animal models of type 1 and type 2 diabetes by delaying carbohydrate digestion and absorption, and enhancing insulin action. *British Journal of Nutrition, 97*(3), 514–521.

Hasanzade, F., Toliat, M., Emami, S. A., & Emamimoghaadam, Z. (2013). The effect of cinnamon on glucose of type II diabetes patients. *Journal of Traditional and Complementary Medicine, 3*(3), 171–174.

Hassan, J. K., Sharrad, A. K., & Sheri, F. H. (2018). Effect of quercetin supplement on some bone mineralization biomarkers in diabetic type 2 patients. *Advances in Pharmacology and Pharmacy, 6*(2), 43–49.

Hazra, M., et al. (2011). Evaluation of hypoglycemic and antihyperglycemic effects of Luffa cylindrica fruit extract in rats. *Journal of Advanced Pharmacy Education and Research, 2*, 138–146.

Hlel, T. B., et al. (2017). Variations in the bioactive compounds composition and biological activities of Loofah (Luffa cylindrica) fruits in relation to maturation stages. *Chemistry & Biodiversity, 14*(10), e1700178.

Huang, Y., et al. (2019a). Ginseng extracts modulate mitochondrial bioenergetics of live cardiomyoblasts: A functional comparison of different extraction solvents. *Journal of Ginseng Research, 43*(4), 517–526.

Huang, H., et al. (2019b). Clinical effectiveness of quercetin supplementation in the management of weight loss: A pooled analysis of randomized controlled trials. *Diabetes, Metabolic Syndrome and Obesity: Targets and Therapy, 12*, 553.

Huynh, K., Schneider, M., & Gareau, M. G. (2016). Altering the gut microbiome for cognitive benefit? In N. Hyland, & C. Stanton (Eds.), *The gut-brain axis* (pp. 319–337). Academic Press.

Jafarnejad, S., et al. (2017). Effect of ginger (Zingiber officinale) on blood glucose and lipid concentrations in diabetic and hyperlipidemic subjects: A meta-analysis of randomized controlled trials. *Journal of Functional Foods, 29*, 127–134.

Jouvet, P., & Touati, G. (2002). Maladies héréditaires du métabolisme: ce que le réanimateur d'enfants peut transmettre au réanimateur d'adultes. *Réanimation, 11*(6), 433–439.

Kadan, S., Sasson, Y., Saad, B., & Zaid, H. (2018). Gundelia tournefortii antidiabetic efficacy: Chemical composition and GLUT4 translocation. *Evidence-based Complementary and Alternative Medicine, 2018*. https://doi.org/10.1155/2018/8294320.

Ki, N. K., Lee, H. K., Cho, J. H, Kim, S. C., & Kim, N. S. (2016). Factors affecting metabolic syndrome by lifestyle. *Journal of Physical Therapy Science, 28*(1), 38–45.

Ko, M. M., Jang, S., & Jung, J. (2019). Herbal medicines for metabolic diseases with blood stasis: A protocol for a systematic review and meta-analysis. *Medicine, 98*(8).

Kudolo, G. B. (2000). The effect of 3-month ingestion of Ginkgo biloba extract on pancreatic β-cell function in response to glucose loading in normal glucose tolerant individuals. *The Journal of Clinical Pharmacology, 40*(6), 647–654.

Kudolo, G. B. (2001). The effect of 3-month ingestion of ginkgo biloba extract (EGb 761) on pancreatic β-cell function in response to glucose loading in individuals with non-insulin-dependent diabetes mellitus. *The Journal of Clinical Pharmacology, 41*(6), 600–611.

Kunle, O. F., Egharevba, H. O., & Ahmadu, P. O. (2012). Standardization of herbal medicines—A review. *International Journal of Biodiversity and Conservation, 4*(3), 101–112.

Lameda, I. L. P., & Koch, T. R. (2020). *Acquired metabolic disorders in liver diseases* (pp. 107–116). Springer.

Liu, W. J. (2011). Introduction to traditional herbal medicines and their study. In *Traditional herbal medicine research methods: Identification, analysis, bioassay, and pharmaceutical and clinical studies* (pp. 1–26). Hoboken, USA: John Wiley & Sons, Inc.

Liu, Y., Yin, H. J., Shi, D. Z., & Chen, K. J. (2012). Chinese herb and formulas for promoting blood circulation and removing blood stasis and antiplatelet therapies. *Evidence-based Complementary and Alternative Medicine, 2012.* https://doi.org/10.1155/2012/184503.

Lou, X. J., et al. (2020). Beneficial effects of macroporous resin extract of dendrobium candidum leaves in rats with hyperuricemia induced by a high-purine diet. *Evidence-based Complementary and Alternative Medicine, 2020,* 3086106.

Malik, S. A., Bashir, M., Khan, R., & Iqbal, M. (2009). Serum sialic acid changes in non-insulin-dependant diabetes mellitus (NIDDM) patients following bitter melon (Momordica charantia) and rosiglitazone (Avandia) treatment. *Phytomedicine, 16*(5), 401–405.

Mang, B., et al. (2006). Effects of a cinnamon extract on plasma glucose, HbA1c, and serum lipids in diabetes mellitus type 2. *European Journal of Clinical Investigation, 36*(5), 340–344.

Marcocci, L., Packer, L., Droy-Lefaix, M. T., Sekaki, A., & Gardès-Albert, M. (1994). Antioxidant action of Ginkgo biloba extract EGb 761. In L. Packer (Ed.), *Methods in Enzymology* (pp. 462–475). San Diego: Academic Press. USA.

Medagama, A. B. (2015). The glycaemic outcomes of Cinnamon, a review of the experimental evidence and clinical trials. *Nutrition Journal, 14*(1), 108.

Movahed, A., et al. (2013). Antihyperglycemic effects of short term resveratrol supplementation in type 2 diabetic patients. *Evidence-based Complementary and Alternative Medicine, 2013.* https://doi.org/10.1155/2013/851267.

Nderitu, K. W., Mwenda, N. S., Macharia, N. J., Barasa, S. S., & Ngugi, M. P. (2017). Antiobesity activities of methanolic extracts of Amaranthus dubius, Cucurbita pepo, and Vigna unguiculata in progesterone-induced obese mice. *Evidence-based Complementary and Alternative Medicine, 2017.* https://doi.org/10.1155/2017/4317321.

Neelakantan, N., Narayanan, M., de Souza, R. J., & van Dam, R. M. (2014). Effect of fenugreek (Trigonella foenum-graecum L.) intake on glycemia: A meta-analysis of clinical trials. *Nutrition Journal, 13*(1), 7.

Olthof, M. R., van Djik, A. E., Deacon, C. F., Heine, R. J., & van Dam, R. M. (2011). Acute effects of decaffeinated coffee and the major coffee components chlorogenic acid and trigonelline on incretin hormones. *Nutrition and Metabolism, 8*(1), 10.

Park, H. L., et al. (2012). Traditional medicine in China, Korea, and Japan: A brief introduction and comparison. *Evidence-based Complementary and Alternative Medicine, 2012.* https://doi.org/10.1155/2012/429103.

Poolsup, N., Suksomboon, N., Kurnianta, P. D. M., & Deawjaroen, K. (2019). Effects of curcumin on glycemic control and lipid profile in prediabetes and type 2 diabetes mellitus: A systematic review and meta-analysis. *PLoS One, 14*(4).

Purkait, P., & Bhattacharya, M. (2012). Prameha and its ancient ayurvedic medicine in India. *Journal of the Anthropological Survey of India, 61*(2), 659–669.

Rahimi, H. R., et al. (2016). The effect of nano-curcumin on HbA1c, fasting blood glucose, and lipid profile in diabetic subjects: A randomized clinical trial. *Avicenna Journal of Phytomedicine, 6*(5), 567.

Raman, A., & Lau, C. (1996). Anti-diabetic properties and phytochemistry of Momordica charantia L.(Cucurbitaceae). *Phytomedicine, 2*(4), 349–362.

Rashmi, P. (2017). *Anti diabetic potential of some selected traditionally used medicinal plants in Western Ghats of India wsr to Prameha.* https://doi.org/10.18535/ijahm/v7i4.05.

Roussel, A. M., Hininger, I., Benaraba, N., Ziegenfuss, T. N., & Anderson, R. A. (2009). Antioxidant effects of a cinnamon extract in people with impaired fasting glucose that are overweight or obese. *Journal of the American College of Nutrition, 28*(1), 16–21.

Sahebkar, A. (2017). Effects of quercetin supplementation on lipid profile: A systematic review and meta-analysis of randomized controlled trials. *Critical Reviews in Food Science and Nutrition, 57*(4), 666–676.

Said, O., et al. (2008). Maintaining a physiological blood glucose level with 'glucolevel', a combination of four anti-diabetes plants used in the traditional Arab herbal medicine. *Evidence-based Complementary and Alternative Medicine, 5*(4), 421–428.

Saklayen, M. G. (2018). The global epidemic of the metabolic syndrome. *Current Hypertension Reports, 20*(2), 12.

Salem, M. B., et al. (2017). Protective effects of Cynara scolymus leaves extract on metabolic disorders and oxidative stress in alloxan-diabetic rats. *BMC Complementary and Alternative Medicine, 17*(1), 328.

Shah, M. A., Sarker, M., & Gousuddin, M. (2016). Antidiabetic potential of Brassica Oleracea Var. Italica in type 2 diabetic sprague dawley (sd) rats. *International Journal of Pharmacognosy and Phytochemical Research, 8*(3), 462–469.

Shakeel, A. H., & Yar, A. K. (2014). Effect of milk thistle (Silybum marianum) plant parts (seeds and leaves) to control the alloxan induced diabetes in rabbits. *Global Journal of Research on Medicinal Plants & Indigenous Medicine, 3*(1), 1.

Sharma, H., & Chandola, H. (2011). Prameha in ayurveda: Correlation with obesity, metabolic syndrome, and diabetes mellitus. Part 1–etiology, classification, and pathogenesis. *The Journal of Alternative and Complementary Medicine, 17*(6), 491–496.

Sutton, V. R. (2019). *Inborn errors of metabolism: Epidemiology, pathogenesis, and clinical features.* UpToDate [online serial] Waltham, MA: UpToDate. https://www.uptodate.com/legal/terms-of-use.

Thazhath, S. S., et al. (2016). Administration of resveratrol for 5 wk has no effect on glucagon-like peptide 1 secretion, gastric emptying, or glycemic control in type 2 diabetes: A randomized controlled trial. *The American Journal of Clinical Nutrition, 103*(1), 66–70.

Tian, C., Chang, H., La, X., & Li, J.-A. (2017). Wushenziye formula improves skeletal muscle insulin resistance in type 2 diabetes mellitus via PTP1B-IRS1-Akt-GLUT4 signaling pathway. *Evidence-based Complementary and Alternative Medicine, 2017.* https://doi.org/10.1155/2017/4393529.

Van Dijk, A. E., et al. (2009). Acute effects of decaffeinated coffee and the major coffee components chlorogenic acid and trigonelline on glucose tolerance. *Diabetes Care, 32*(6), 1023–1025.

van Tilburg, J., et al. (2001). Defining the genetic contribution of type 2 diabetes mellitus. *Journal of Medical Genetics, 38*(9), 569–578.

Venkateswaran, S., & Pari, L. (2002). Effect of Coccinia indica on blood glucose, insulin and key hepatic enzymes in experimental diabetes. *Pharmaceutical Biology, 40*(3), 165–170.

Wang, Y., et al. (2013). Ginsenoside Rd attenuates myocardial ischemia/reperfusion injury via Akt/GSK-3β signaling and inhibition of the mitochondria-dependent apoptotic pathway. *PLoS One, 8*(8).

Waters, D., Adeloye, D., Woolham, D., Wastnedge, E., Patel, S., & Rudan, I. (2018). Global birth prevalence and mortality from inborn errors of metabolism: A systematic analysis of the evidence. *Journal of Global Health, 8*(2).

Wolff, D. (2007). Gout. In S. J. Enna, & D. B. Bylund (Eds.), *xPharm: The comprehensive pharmacology reference* (pp. 1–8). New York: Elsevier.

Yang, Y., et al. (2013). SIRT1 activation by curcumin pretreatment attenuates mitochondrial oxidative damage induced by myocardial ischemia reperfusion injury. *Free Radical Biology and Medicine, 65,* 667–679.

Yin, J., Zhang, H., & Ye, J. (2008). Traditional chinese medicine in treatment of metabolic syndrome. *Endocrine, Metabolic & Immune Disorders-Drug Targets (Formerly Current Drug Targets-Immune, Endocrine & Metabolic Disorders), 8*(2), 99–111.

Zhang, J., et al. (2013). Protective effect of qiliqiangxin capsule on energy metabolism and myocardial mitochondria in pressure overload heart failure rats. *Evidence-based Complementary and Alternative Medicine, 2013.* https://doi.org/10.1155/2013/378298.

Zhou, Q., Liu, S., Yua, D., & Zhang, N. (2016). Therapeutic effect of total saponins from *Dioscorea nipponica Makino* on gouty arthritis based on the NF-κB signal pathway: An in vitro study. *Pharmacognosy Magazine, 12*(47), 235.

Zhou, Q., Yu, D. H., Zhang, C., Liu, S. M., & Lu, F. (2014). Total saponins from *Discorea nipponica* ameliorate urate excretion in hyperuricemic mice. *Planta Medica, 80*(15), 1259–1268.

Further reading

Zaid, H., & Saad, B. (2012). State of the art of diabetes treatment in Greco-Arab and islamic medicine. *Bioactive food as dietary interventions for diabetes* (pp. 327–335). London, UK: Academic Press.

Chapter 15

Artemisia amygdalina Decne.: A rich repository of pharmacologically vital phytoconstituents

Showkat Hamid Mir, Rezwana Assad, Aadil Farooq War, Irfan Rashid, and Zafar Ahmad Reshi

Department of Botany, University of Kashmir, Srinagar, Jammu and Kashmir, India

1. Introduction

Artemisia amygdalina Decne. is an endemic perennial rhizomatous herb, locally known as "veer tethven" in Kashmir valley, which occupies moist and resides along subalpine habitats of the dense pine forests of North Western Himalayas (Fig. 1). It grows alongside streams, preferably on sandy soil, rock crevices, and shady areas. Despite being pharmaceutically essential the species has got little attention and is now constrained to a small patch in northern parts of Kashmir valley.

A. *amygdalina* is generally utilized for the treatment of cold, cough, fever, helminthiasis, epilepsy, and nervous ailments (Rasool et al., 2012). Womankind exploits it for the treatment of dysmenorrhea and amenorrhea.

Phytochemical assay of extracts from different parts of A. *amygdalina* disclosed the incidence of alkaloids, cardiac glycosides, phenolics, steroids, tannins, terpenes, and essential oils, both in wild and in vitro raised plants (Taj, Khan, Ali, & Khan, 2019). Furthermore, HPLC studies of this species confirmed the occurrence of artemisinin in wild and tissue culture–raised plants (Rasool, Ganai, Akbar, & Kamili et al., 2013). Artemisinin is a chief drug used as cure of fever and malaria (White, 1997). The other major secondary metabolites reported from this plant are 1,8-cineole, p-cymene, and terpenes (Rather et al., 2012).

As many as 42 compounds have been identified in the leaf and stem extracts of A. *amygdalina* until now. Twenty-five chemical compounds were isolated from the leaf extracts belonging predominantly to the monoterpene

Phytomedicine: A Treasure of Pharmacologically Active Products from Plants
https://doi.org/10.1016/B978-0-12-824109-7.00006-6

409

FIG. 1 *Artemisia amygdalina.*

hydrocarbons and oxygenated monoterpenes, with 1,8 cineole, borneol, p-cymene, and sabinene being the major compounds. The stem extract showed the presence of 32 compounds dominated by monoterpene hydrocarbons, oxygenated monoterpenes, and sesquiterpene hydrocarbons, with borneol, camphene, α-pinene, and β-pinene as major constituents (Qadir, Shah, & Hussain, 2018).

As a result of overharvesting and deforestation, this particular species of genus *Artemisia* is now regarded as "critically endangered species" in Kashmir Himalaya.

2. Major therapeutic properties of *A. amygdalina*

A range of biological activities have been reported from the essential bioactive compounds extracted from *A. amygdalina*. A few of them are antibacterial, anticancerous, antidiabetic, antihelminthic, antiinflammatory, antimalarial, antioxidant, antiulcer, antiviral, and immunomodulatory activities (Fig. 2). In this chapter, major pharmaceutical properties of *A. amygdalina* are highlighted.

(a) Antibacterial effects

Antibacterial activity of the essential oils extracted from the leaves of *A. amygdalina* was ascertained by Qadir et al. (2018). Essential oils obtained from this plant demonstrated a wide range of antimicrobial effects against both gram-positive and gram-negative bacteria. However, it was found to be most effectual against *Bacillus subtilis*.

(b) Anticancerous activity

A range of essential oils have been reported from the *A. amygdalina* leaves with significant anticancerous activity. Qadir et al. (2018) conducted experiment on Lung and Hek cancer cell lines using essential oils obtained from

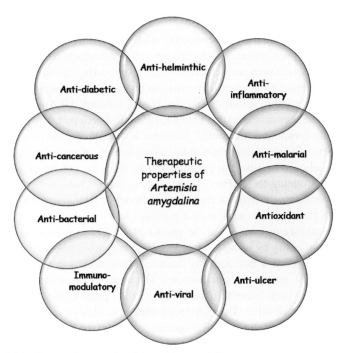

FIG. 2 Major therapeutic properties of *Artemisia amygdalina*.

this species. They found that the oil extracted from *A. amygdalina* showed high anticancerous activity against both the cell lines. However, the activity was more against Hek cancer cell lines, with an inhibition of 52.3% at a concentration of $10 \mu g\, mL^{-1}$, and IC50 value was reported to be at a concentration of $9.7 \mu g\, mL^{-1}$. As the anticancerous activity was very high, thus this species has a lot of potential in the food and pharmaceutical industries.

(c) Antidiabetic activity

A. amygdalina has been reported to have antidiabetic activity. Gazafar et al. (2014) studied the antidiabetic and antihyperlipidemic effects of this plant species. The extracts that were tested for the potential antidiabetic activity in diabetic rats were ethyl acetate, hydroethanolic, methanol, and petroleum ether extracts. Among the tested extracts the hydroethanolic and methanolic extracts showed significant results and reduced the glucose levels in diabetic rats. Both these extracts showed promising results in the reduction of some other biochemical factors such as alkaline phosphatase (ALP), cholesterol, low-density lipoproteins (LDL), serum creatinine, serum glutamate pyruvate transaminase (SGPT), serum glutamate oxaloacetate transaminase (SGOT), and triglycerides. In addition, the extract obtained from the mentioned plant significantly diminished consumption of feed and water in diabetic rats as compared with the control nondiabetic rats.

(d) Antihelminthic property

Helminthic ailments are exceptionally widespread, mainly in the beef- and pork-eating populations, and are recognized as the reason for chronic intestinal ailments. There are numerous studies supporting the fact that almost all the species of the genus *Artemisia* in one way or the other possess potent antihelminthic activity (Huassin, Hayat, Sahreen, Ul Ain, & Bokhari, 2017). The reason behind this valuable biological activity is the presence of principle compound "santonin." Santonin is a sesquiterpene present in the members of genus *Artemisia* that has a long history of antihelminthic usage and is believed to drive out parasitic worms from the gut, by either killing them or stunning them (Hiss, 1898). The actual mechanism of the drug is that it paralyzes the anterior end of the worm while stimulating the posterior end. As the result, these intestinal parasites lost their muscle coordination and the ability to hook up at the proper place. These worms can then be passed out easily by using a purgative. Furthermore, it has been confirmed by in vitro studies that aqueous extract of *Artemisia* species has the potential to act against intestinal worms. The aqueous extract of *A. amygdalina* has also been found to reduce the esterified propoxylated glycerol and worm burden in mice. Therefore such studies evidently specify the broad spectrum ability of the *Artemisia* species to prevent the helminthic diseases, thus offering great potential for the pharmaceutical industry.

(e) Antiinflammatory property

Among different extracts obtained from *A. amygdalina* that were tested for antiinflammation properties, methanolic extract obtained from leaves showed obstruction of paw edema development with an inhibition potential of 42.26% (Mubashir et al., 2013) in comparison with control, followed by aqueous (29.47%), while ethyl acetate and petroleum ether extracts illustrated an inhibition of 18.95% and 3.16%, respectively. The percent inhibition by methanolic extract is much comparable with a standard drug (Mubashir et al., 2013).

(f) Antimalarial activity

Malaria is a ruthless and highly transmissible infection and a widespread health problem faced by humans, leading to about 1 million deaths per annum. Malaria is the world's most hazardous parasitic disease in terms of sickness and death, the prominent causative agent of which is *Plasmodium falciparum*. The species of genus *Artemisia* including *A. amygdalina* are employed worldwide as antimalarial since prehistoric times. The pioneer in this work was a scientist from China who found antimalarial drug called artemisinin from *A. annua*. However, after the HPLC analysis, artemisinin was also found to be present in the petroleum ether extracts of wild aerial part, greenhouse acclimatized plants, and in vitro raised plants of *A. amygdalina* (Lone, Bhat, & Khuroo, 2015). Artemisinin is basically a sesquiterpenoid, obtained from the glandular trichomes of almost all the species of the genus *Artemisia* and is actually a current drug of choice against *Plasmodium falciparum*.

It contains endoperoxide bridge within the 1,2,3 trioxane system, which is cleaved to generate reactive oxygen species (such as superoxide anions, hydroxyl radicals, and carbon-centered radicals). The chief function of these radical species is to damage membranes and cause alkylation of proteins. At present, the World Health Organization (WHO) has suggested the use of artemisinin combination therapy (ACT), as the principal treatment to get rid of the spreading of this deadly disease. However, there were some disadvantages associated with artemisinin like short plasma half-life, low solubility, and poor bioavailability, which were later on surpassed by the semisynthetic/completely synthetic derivatives like artemisone, dihydroartemisinin, artemether, and artesunate (Li, Weina, & Milhous, 2007).

(g) Antioxidant activity

Antioxidant activity of tissue culture raised plants of *A. amygdalina* has been determined using several assays like, postmitochondrial supernatant (PMS) assay, riboflavin photooxidation assay, ferric thiocyanate (FTC) assay, DPPH assay, thiobarbituric acid (TBA) assay, and DNA damage assay (Rasool et al., 2013). From the DPPH assay the methanolic extracts from tissue culture-raised plants showed maximum inhibitory effect of about 92.11% (IC50 $26.06 \mu g\,mL^{-1}$), while as riboflavin photooxidation assay and ethyl acetate extract of both tissue culture grown and the aqueous extract of greenhouse acclimatized plants showed a significant free radical scavenging activity of 89.53% (IC50 = $16.2 \mu g\,mL^{-1}$) and 96.32% (IC50 = $22.04 \mu g\,mL^{-1}$), respectively.

Furthermore, in case of FTC assay, the ethyl acetate extract of the tissue culture-raised plant and the petroleum ether extract of the acclimatized plants revealed highest inhibitory activity of 76.57% (IC50 = $29.45 \mu g\,mL^{-1}$) and 88.28% (IC50 = $25.46 \mu g\,mL^{-1}$), respectively, against lipid peroxyl radicals, which were produced by linoleic acid oxidation. In the case of TBA assay, the highest inhibitory activity was observed by petroleum ether extract from both in vitro raised and the aqueous extract of the greenhouse acclimatized plants which was about 64.1% (IC50 = $25.34 \mu g\,mL^{-1}$) and 64.15% (IC50 = $26.93 \mu g\,mL^{-1}$), respectively, against TBA reactive species. Similarly, vitamin C (92.82% (IC50 = $4.2 \mu g\,mL^{-1}$) and vitamin E (50.13% (IC50 = $64.86 \mu g\,mL^{-1}$) of both tissue culture grown and greenhouse acclimatized plants showed a significant antioxidant activity against different free radicals.

In postmitochondrial supernatant (PMS) the methanolic extract of both tissue culture-raised and greenhouse acclimatized plants showed a significant antioxidant activity of 52.86% (IC50 $108.14 \mu g\,mL^{-1}$) and 67.04% (IC50 $82.11 \mu g\cdot mL^{-1}$), respectively, while in the case of deoxyribose assay, the methanolic extract of both tissue culture-raised and greenhouse acclimatized plants revealed highest percentage inhibitions of 46.17% and 69.33% (IC50 $13.96 \mu g\,mL^{-1}$), respectively against hydroxyl radicals produced by Fenton reaction.

(h) Antiulcer activity

Gastric ulcer is a chronic, inflammatory disease affecting stomach or duodenum of as many as 10% of the humans at some stage of their lives. Plant parts and products have been traditionally used as ethnomedicines by people to cure gastric ulcers. Plants have been found to contain numerous compounds with antiulcer activity that includes flavonoids (naringin, quercetin, silymarin, anthocyanosides, and sophoradin derivatives), saponins, and sesquiterpene lactones. However, sesquiterpene lactones of the eudesmanolide and guaianolide types are of particular interest in treatment of peptic and gastric ulcers as they are believed to play a role in regulation and prevention of oxidative damage and inflammation mediated damage (Repetto and Boveris, 2010). The reason behind the effectiveness of sesquiterpene lactones and their robustness in biological systems is the presence of α-methylene-*c*-lactone containing alkylating reaction center, molecular geometry, and lipophilicity. Ludartin, a sesquiterpene lactone of guaianolide type, isolated from the shoots of *A. amygdalina*, exercise in vivo cytoprotective actions against alcohol-induced gastric mucosal injury (Lone et al., 2013). Furthermore, aqueous extracts of this plant species may exhibit beneficial antiulcer effects through their antimicrobial activity. For instance, artemisinin, α-cadinol, caryophyllene oxide, and other sesquiterpenes exhibit strong antibacterial effects against *Helicobacter pylori*, the causative organism for the gastric ulcer disease.

(i) Antiviral potential

Artemisinin is the most hopeful compound to relieve the world from malaria burden besides having antiinflammatory, antiangiotensin, and antiproliferative activity against many tumor cell lines. Interestingly the biological activity and pharmaceutical property of artemisinin and its semisynthetic derivatives are even broader and show antagonistic action against many viral diseases. It is evident from studies that artemisinin possesses antiviral effect against an array of human herpes viruses like herpes simplex virus, Epstein-Barr virus and human cytomegalovirus (Efferth et al., 2008), hepatitis B virus, hepatitis C virus (Paeshuyse et al., 2006), bovine viral diarrhea virus (Romero et al., 2006), and influenza virus A (Krishna, Bustamante, Haynes, & Staines, 2008) in micromolar range. For instance, 50% effective concentration (EC_{50}) of artemisinin for inhibition of hepatitis C virus subgenomic replicon replication in Huh 5-2 cells has been found to be $78 \pm 21\,\mu M$. Furthermore, artesunate (semisynthetic derivative of artemisinin) has been proved to be successfully reducing cytomegalovirus number in immunosuppressed child without visible toxicity (Shapira et al., 2008).

(j) Immunomodulatory potential

Studies on immunomodulatory potential of different extracts from *A. amygdalina* revealed a significant immunosuppressive activity and a delayed humoral response by sheep RBCs. Among different extracts obtained from the mentioned species, the methanolic extract showed highest decline in humoral response followed by ethyl acetate and aqueous fractions. The methanolic and ethyl acetate fractions showed a maximum reduction in both primary and secondary antibody production. Petroleum ether fraction showed decline in primary response but a trivial increase in secondary response (Mubashir et al., 2013). Taking into consideration the potential of *A. amygdalina* to suppress both cell-mediated immunity and humoral immunity, it can serve as a potential therapeutic contender in several immunostimulant clinical conditions.

3.   Pharmacologically active phytoconstituents isolated from *A. amygdalina*

Scores of pharmacologically active phytoconstituents isolated from *A. amygdalina* along with their respective biological activities are summed up in Table 1.

TABLE 1 Phytochemical constituents of *Artemisia amygdalina*.

S. no	Chemical compound	Chemical group	Biological activity	References
1.	*Terpenes*			
	Limonene	Terpene	Anticancerous Antiinflammatory Antipyretic	Hirota et al. (2010) and Lone et al. (2015)
	Camphor	Terpene	Antiviral Myorelaxant	Lone et al. (2015) and Sokolova et al. (2015)
	Pinocarvyl acetate	Terpene	Antibacterial Antifungal	Lone et al. (2015) and Sharifi-Rad et al. (2015)
	β-Dehydroelsholtzia ketone	Terpene	Analgesic Antibacterial Antiinflammatory Antipyretic Antiviral	Liu et al. (2007) and Lone et al. (2015)
	En-in-dicycloether	Terpene	Antiinflammatory	Van and Arman (2013) and Lone et al. (2015)

Continued

TABLE 1 Phytochemical constituents of *Artemisia amygdalina*—cont'd

S. no	Chemical compound	Chemical group	Biological activity	References
2.	*Monoterpenes*			
	Camphene	Monoterpene	Antitumor	Lone et al. (2015) and Girola et al. (2015)
	3-Carene	Monoterpene	Antibiotic Antiinflammatory	Ocete et al. (1989) and Lone et al. (2015)
	p-Cymene	Monoterpene	Anticancerous Antipyretic	Beckford et al. (2009) and Lone et al. (2015)
	1,8-Cineole	Monoterpene	Analgesic Antiinflammatory	Asanova et al. (2003) and Lone et al. (2015)
	(Z)-β-ocimene	Monoterpenes	Antimicrobial	Tatsadjieu et al. (2003) and Lone et al. (2015)
	γ-Terpinene	Monoterpene	Antimicrobial Antimycotic Antipyretic	Foti and Ingold (2003) and Lone et al. (2015)
	α-Thujene	Monoterpene	Anticancerous Antipyretic	Hou et al. (2007) and Lone et al. (2015)
	α-Terpineol	Monoterpene	Antispasmodic Myorelaxant	Lone et al. (2015) and Francis and Thomas (2016)
	Terpenen-4-ol	Monoterpene	Antimicrobial	Halcón and Milkus (2004) and Lone et al. (2015)

TABLE 1 Phytochemical constituents of *Artemisia amygdalina*—cont'd

S. no	Chemical compound	Chemical group	Biological activity	References
	Borneol	Monoterpene	Analgesic Antimicrobial Antinflammtory Antiulcer Antiviral	Lone et al. (2015) and Sokolova et al. (2017)
	Piperitone	Monoterpene	Antibacterial Antifeedant Antifungal antipyretic	Grudniewska et al. (2011) and Lone et al. (2015)
	Linalool	Monoterpene	Anticonvulsant Antidepressant Anxiolytic Sedative	Sampaio et al. (2012) and Lone et al. (2015)
	Myrcene	Monoterpene	Prevention of peptic ulcers	Bonamin et al. (2014) and Lone et al. (2015)
	Myrtenyl acetate	Monoterpene	Antihemorrhagic Antimicrobial Hypoglycemic	Messaoud and Boussaid (2011) and Lone et al. (2015)
3.	*Sesquiterpenes*			
	α-Copane	Sesquiterpene	Anticonvulsant Antiinflammatory Antipyretic Antitubercular CNS depressant	Gupta et al. (2013) and Lone et al. (2015)
	α-Longipinene	Sesquiterpene	Antiinflammatory Antioxidant	Sakata and Miyazawa (2010) and Lone et al. (2015)
	β-Borbonene	Sesquiterpene	Anticancer Antiinflammatory Antimalarial Antinociceptive Antipyretic Atherosclerotic Hypoglycemia	Seyyedan et al. (2013) and Lone et al. (2015)

Continued

TABLE 1 Phytochemical constituents of *Artemisia amygdalina*—cont'd

S. no	Chemical compound	Chemical group	Biological activity	References
	β-Farnescene	Sesquiterpene	Antiinflammatory Antipyretic Aphidicidal	Seyyedan et al. (2013), Lone et al. (2015), and Qin et al. (2016)
	β-Caryophyllene	Sesquiterpene	Anticancer Anticonvulsant Antimicrobial Neuroprotective	Dahham et al. (2015), Lone et al. (2015), and de Oliveira et al. (2016)
	α-Bisabolol	Sesquiterpene	Anticancer Antimicrobial	da Silva et al. (2010) and Lone et al. (2015)
	Ledene oxide	Sesquiterpene	Antibacterial Antioxidant	Raja Rajeswari et al., 2011and Lone et al. (2015)
	α-Cadinol	Sesquiterpene	Antibacterial Antifungal Antiinflammatory Antimalarial Antioxidant Effective against cough and dysentery	Ogunwande et al. (2007) and Lone et al. (2015)
	Caryophyllene oxide	Sesquiterpenoid	Anticancer Anticonvulsant Antimicrobial Neuroprotective	Dahham et al. (2015), Lone et al. (2015), and de Oliveira et al. (2016)
	Spathulenol	Sesquiterpenoid	Anesthetic Antiinflammatory Antioxidant Treatment of cardiac impairments Vasodilator	Lone et al. (2015), Moreira et al. (2018), and Rajca Ferreira et al. (2018)

TABLE 1 Phytochemical constituents of *Artemisia amygdalina*—cont'd

S. no	Chemical compound	Chemical group	Biological activity	References
	Germacrene-D-4-ol	Sesquiterpenoid	Antiinflammatory Antimicrobial	Mevy et al. (2007) and Lone et al. (2015)
	E-Nerolidol	Sesquiterpene	Antiinflammatory Antimicrobial Antinociceptive Antiparasitic Antitumor Antiulcer Insecticidal Skin penetration enhancer	Lone et al. (2015)
	δ-Cadinene	Sesquiterpene	Analgesic Anticonvulsant Antidiabetic Antiinflammatory Proteolytic	Makhija et al. (2010) and Lone et al. (2015)
	BicycloGermacrene	Sesquiterpene	Analgesic Anticonvulsant Antidepressant Antidiabetic activities Antiinflammatory Antioxidant Cytotoxic	Asgarpanah (2012) and Lone et al. (2015)
	α-Humulene	Sesquiterpene	Anticancerous Antipyretic	Hou et al. (2007) and Lone et al. (2015)
	Germacrene D	Sesquiterpenoid	Antiinflammatory Antimicrobial	Mevy et al. (2007) and Lone et al. (2015)
	α-Curcumene	Sesquiterpene	Anticancer Demulcent Diuretic Expectorant Menstrual disorders Dyspepsia	Lobo et al. (2009) and Lone et al. (2015)

Continued

TABLE 1 Phytochemical constituents of *Artemisia amygdalina*—cont'd

S. no	Chemical compound	Chemical group	Biological activity	References
	α-Gurjunene	Sesquiterpene	Antidiabetic Antimicrobial	Lone et al. (2015) and Mosbah et al. (2018)
	α-muurolene	Sesquiterpene	Antimicrobial Antioxidant	Deng et al. (2009) and Lone et al. (2015)
	Artemisinin	Sesquiterpenoid	Antihelminthic Antimalarial	Haynes et al. (2004) and Lone et al. (2015)
4.	*Triterpenes*			
	Longiborneol	Triterpene	Anticancer Antiinflammatory Antimicrobial Antioxidant Antitumor	Mishra and Sree (2007) and Lone et al. (2015)
5.	*Fatty alcohol*			
	2-Nonene-1-ol	Fatty alcohol	Antifungal Antioxidant	Nihei et al. (2004) and Lone et al. (2015)

1,8-Cineole, 3-carene, artemisinin, artesunate, camphene, ludartin, p-cymene, santonin, α-cadinol, α-pinene, and β-pinene are the characteristic compounds isolated from *A. amygdalina* hitherto (Fig. 3).

4. Conclusions

A. amygdalina is as a rich repository of pharmacologically active constituents. Therapeutically active phytoconstituents of this plant species encompass alkaloids, cardiac glycosides, phenolics, steroids, tannins, terpenes, and other essential oils, with 1,8-cineole, 3-carene, artemisinin, artesunate, camphene, ludartin, p-cymene, santonin, α-cadinol, α-pinene, and β-pinene being the characteristic compounds isolated so far. Several therapeutic activities ascribed to these essential bioactive compounds extracted from this species include analgesic, antibacterial, anticancerous, anticonvulsant, antidepressant, antidiabetic, antifungal, antihelminthic, antiinflammatory, antimalarial, antioxidant, antispasmodic,

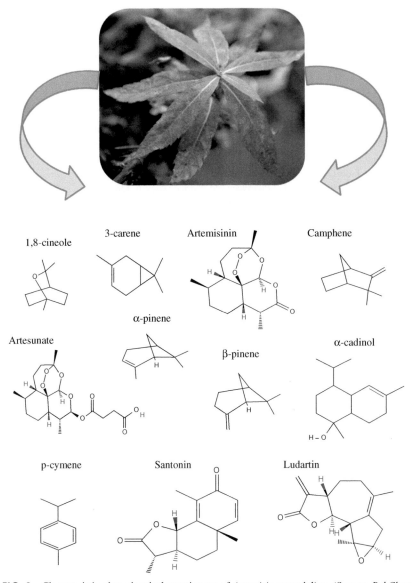

FIG. 3 Characteristic phytochemical constituents of *Artemisia amygdalina*. *(Source: PubChem Database. (2020). https://pubchem.ncbi.nlm.nih.gov. (Accessed 19 May 2020).)*

antiulcer, antiviral, anxiolytic, sedative, and immunomodulatory properties. Detailed phytochemical and pharmacological assays of this imperative *Artemisia* species demands further in-depth research. Harnessing of utmost benefit from this critically endangered plant species depends on the success of endeavors pertaining to its successful regeneration in the subalpine Himalayan habitats.

Acknowledgments

Authors acknowledge the funding support provided by University Grants Commission (UGC), India. We also state gratitude to Head, Department of Botany, University of Kashmir, India for facilitating this study.

References

Asanova, Z. K., Suleimenov, E. M., Atazhanova, G. A., Dembitskii, A. D., Pak, R. N., Dar, A., et al. (2003). Biological activity of 1, 8-cineole from levant wormwood. *Pharmaceutical Chemistry Journal, 37*, 28–30.

Asgarpanah, J. (2012). Phytochemistry, pharmacology and medicinal properties of *Hypericum perforatum* L. *African Journal of Pharmacy and Pharmacology, 6*, 1387–1394.

Beckford, F. A., Leblanc, G., Thessing, J., Shaloski, M., Jr., Frost, B. J., Li, L., et al. (2009). Organometallic ruthenium complexes with thiosemicarbazone ligands: Synthesis, structure and cytotoxicity of [(η6-p-cymene) Ru (NS) Cl]+(NS = 9-anthraldehyde thiosemicarbazones). *Inorganic Chemistry Communications, 12*, 1094–1098.

Bonamin, F., Moraes, T. M., Dos Santos, R. C., Kushima, H., Faria, F. M., Silva, M. A., et al. (2014). The effect of a minor constituent of essential oil from Citrus aurantium: The role of β-myrcene in preventing peptic ulcer disease. *Chemico-Biological Interactions, 212*, 11–19.

da Silva, A. P., Martini, M. V., de Oliveira, C. M., Cunha, S., de Carvalho, J. E., Ruiz, A. L., et al. (2010). Antitumor activity of (−)-α-bisabolol-based thiosemicarbazones against human tumor cell lines. *European Journal of Medicinal Chemistry, 45*, 2987–2993.

Dahham, S. S., Tabana, Y. M., Iqbal, M. A., Ahamed, M. B., Ezzat, M. O., Majid, A. S., et al. (2015). The anticancer, antioxidant and antimicrobial properties of the sesquiterpene β-caryophyllene from the essential oil of Aquilaria crassna. *Molecules, 20*, 11808–11829.

de Oliveira, C. C., de Oliveira, C. V., Grigoletto, J., Ribeiro, L. R., Funck, V. R., Grauncke, A. C. B., et al. (2016). Anticonvulsant activity of β-caryophyllene against pentylenetetrazol-induced seizures. *Epilepsy & Behavior, 56*, 26–31.

Deng, G. B., Zhang, H. B., Xue, H. F., Chen, S. N., & Chen, X. L. (2009). Chemical composition and biological activities of essential oil from the rhizomes of *Iris bulleyana*. *Agricultural Sciences in China, 8*, 691–696.

Efferth, T., Romero, M. R., Wolf, D. G., Stamminger, T., Marin, J. J., & Marschall, M. (2008). The antiviral activities of artemisinin and artesunate. *Clinical Infectious Diseases, 47*, 804–811.

Foti, M. C., & Ingold, K. U. (2003). Mechanism of inhibition of lipid peroxidation by γ-terpinene, an unusual and potentially useful hydrocarbon antioxidant. *Journal of Agricultural and Food Chemistry, 51*, 2758–2765.

Francis, S. C., & Thomas, M. T. (2016). Essential oil profiling of *Centella asiatica* (L.) Urb.-a medicinally important herb. *South Indian Journal of Biological Sciences, 2*, 169–173.

Gazafar, K., Ganaie, B. A., Seema, A., Mubashir, K., Showkat, A. D., Younis Dar, M., et al. (2014). Antidiabetic activity of *Artemisia amygdalina* decne in streptozotocin induced diabetic rats. *Biotechnology Research International, 2014*, 1–10.

Girola, N., Figueiredo, C. R., Farias, C. F., Azevedo, R. A., Ferreira, A. K., Teixeira, S. F., et al. (2015). Camphene isolated from essential oil of *Piper cernuum* (Piperaceae) induces intrinsic apoptosis in melanoma cells and displays antitumor activity in vivo. *Biochemical and Biophysical Research Communications, 467*, 928–934.

Grudniewska, A., Dancewicz, K., Białońska, A., Ciunik, Z., Gabryś, B., & Wawrzeńczyk, C. (2011). Synthesis of piperitone-derived halogenated lactones and their effect on aphid probing, feeding, and settling behavior. *RSC Advances, 1*, 498–510.

Gupta, A., Gupta, M., & Gupta, S. (2013). Isolation of piperine and few sesquiterpenes from the cold petroleum ether extract of *Piper nigrum* (black pepper) and its antibacterial activity. *International Journal of Pharmacognosy and Phytochemical Research*, *5*, 3–7.

Halcón, L., & Milkus, K. (2004). Staphylococcus aureus and wounds: A review of tea tree oil as a promising antimicrobial. *American Journal of Infection Control*, *32*, 402–408.

Haynes, R. K., Ho, W. Y., Chan, H. W., Fugmann, B., Stetter, J., Croft, S. L., et al. (2004). Highly antimalaria active artemisinin derivatives: Biological activity does not correlate with chemical reactivity. *Angewandte Chemie International Edition*, *43*, 1381–1385.

Hirota, R., Roger, N. N., Nakamura, H., Song, H. S., Sawamura, M., & Suganuma, N. (2010). Anti-inflammatory effects of limonene from yuzu (*Citrus junos* Tanaka) essential oil on eosinophils. *Journal of Food Science*, *75*, H87–H92.

Hiss, A. E. (1898). *Thesaurus of proprietary preparations and pharmaceutical specialties*. Chicago, IL: GP Engelhard & Co.

Hou, J., Sun, T., Hu, J., Chen, S., Cai, X., & Zou, G. (2007). Chemical composition, cytotoxic and antioxidant activity of the leaf essential oil of *Photinia serrulata*. *Food Chemistry*, *103*, 355–358.

Huassin, A., Hayat, M. Q., Sahreen, S., Ul Ain, Q., & Bokhari, S. A. (2017). Pharmacological promises of genus *Artemisia* (Asteraceae): A review. *Proceedings of the Pakistan Academy of Sciences: B. Life and Environmental Sciences*, *54*, 265–287.

Krishna, S., Bustamante, L., Haynes, R. K., & Staines, H. M. (2008). Artemisinins: Their growing importance in medicine. *Trends in Pharmacological Sciences*, *29*, 520–527.

Li, Q., Weina, P. J., & Milhous, W. K. (2007). Pharmacokinetic and pharmacodynamic profiles of rapid-acting artemisinins in the antimalarial therapy. *Current Drug Therapy*, *2*, 210–223.

Liu, A., Lee, S. M., Wang, Y., & Du, G. (2007). Elsholtzia: Review of traditional uses, chemistry and pharmacology. *Journal of Chinese Pharmaceutical Sciences*, *16*, 73.

Lobo, R., Prabhu, K. S., Shirwaikar, A., & Shirwaikar, A. (2009). Curcuma zedoaria Rosc.(white turmeric): A review of its chemical, pharmacological and ethnomedicinal properties. *Journal of Pharmacy and Pharmacology*, *61*, 13–21.

Lone, S. H., Bhat, K. A., & Khuroo, M. A. (2015). Phytochemical screening and HPLC analysis of *Artemisia amygdalina*. In *Chemical and pharmacological perspective of Artemisia amygdalina* (pp. 7–13). Cham: Springer.

Lone, S. H., Bhat, K. A., Naseer, S., Rather, R. A., Khuroo, M. A., & Tasduq, S. A. (2013). Isolation, cytotoxicity evaluation and HPLC-quantification of the chemical constituents from *Artemisia amygdalina* Decne. *Journal of Chromatography B*, *940*, 135–141.

Makhija, I. K., Sharma, I. P., & Khamar, D. (2010). Phytochemistry and pharmacological properties of Ficus religiosa: An overview. *Annals of Biological Research*, *1*, 171–180.

Messaoud, C., & Boussaid, M. (2011). Myrtus communis berry color morphs: A comparative analysis of essential oils, fatty acids, phenolic compounds, and antioxidant activities. *Chemistry & Biodiversity*, *8*, 300–310.

Mevy, J. P., Bessiere, J. M., Dherbomez, M., Millogo, J., & Viano, J. (2007). Chemical composition and some biological activities of the volatile oils of a chemotype of Lippia chevalieri Moldenke. *Food Chemistry*, *101*, 682–685.

Mishra, P. M., & Sree, A. (2007). Antibacterial activity and GCMS analysis of the extract of leaves of *Finlaysonia obovata* (a mangrove plant). *Asian Journal of Plant Sciences*, *6*, 168–172.

Moreira, C. M., Fernandes, M. B., Santos, K. T., Schneider, L. A., Da Silva, S. E. B., Sant'Anna, L. S., et al. (2018). Effects of essential oil of Blepharocalyx salicifolius on cardiovascular function of rats. *The FASEB Journal*, *32*, 715–717.

Mosbah, H., Louati, H., Boujbiha, M. A., Chahdoura, H., Snoussi, M., Flamini, G., et al. (2018). Phytochemical characterization, antioxidant, antimicrobial and pharmacological activities of *Feijoa sellowiana* leaves growing in Tunisia. *Industrial Crops and Products*, *112*, 521–531.

Mubashir, K., Ganaie, B. A., Gazafar, K., Seema, A., Akhter, H. M., & Akbar, M. (2013). Evaluation of *Artemisia amygdalina* D. for anti-inflammatory and immunomodulatory potential. *ISRN Inflammation, 2013*, 1–6.

Nihei, K. I., Nihei, A., & Kubo, I. (2004). Molecular design of multifunctional food additives: Antioxidative antifungal agents. *Journal of Agricultural and Food Chemistry, 52*, 5011–5020.

Ocete, M. A., Risco, S., Zarzuelo, A., & Jimenez, J. (1989). Pharmacological activity of the essential oil of *Bupleurum gibraltaricum*: Anti-inflammatory activity and effects on isolated rat uteri. *Journal of Ethnopharmacology, 25*, 305–313.

Ogunwande, I. A., Walker, T. M., & Setzer, W. N. (2007). A review of aromatic herbal plants of medicinal importance from Nigeria. *Natural Product Communications, 2*, 1311–1316.

Paeshuyse, J., Coelmont, L., Vliegen, I., Vandenkerckhove, J., Peys, E., Sas, B., et al. (2006). Hemin potentiates the anti-hepatitis C virus activity of the antimalarial drug artemisinin. *Biochemical and Biophysical Research Communications, 348*, 139–144.

Qadir, M., Shah, W. A., & Hussain, Z. (2018). Antibacterial and anticancer activity of essential oil of *Artemisia amygdalina* from Kashmir India. *International Journal of Advanced Research and Engineering, 7*, 47–55.

Qin, Y., Zhang, J., Song, D., Duan, H., Li, W., & Yang, X. (2016). Novel (E)-β-farnesene analogues containing 2-nitroiminohexahydro-1, 3, 5-triazine: Synthesis and biological activity evaluation. *Molecules, 21*, 825.

Raja Rajeswari, N., RamaLakshmi, S., & Muthuchelian, K. (2011). GC-MS analysis of bioactive components from the ethanolic leaf extract of Canthium dicoccum (Gaertn.) Teijsm & Binn. *Journal of Chemical and Pharmaceutical Research, 3*, 792–798.

Rajca Ferreira, A. K., Lourenço, F. R., Young, M. C. M., Lima, M. E. L., Cordeiro, I., Suffredini, I. B., et al. (2018). Chemical composition and biological activities of Guatteria elliptica RE Fries (Annonaceae) essential oils. *Journal of Essential Oil Research, 30*, 69–76.

Rasool, R., Ganai, B. A., Akbar, S., & Kamili, A. N. (2013). Free radical scavenging potential of *in vitro* raised and greenhouse acclimatized plants of *Artemisia amygdalina*. *Chinese Journal of Natural Medicines, 11*, 0377–0384.

Rasool, R., Ganai, B. A., Akbar, S., et al. (2012). Anti-oxidant activities of extracts of the inflorescences and aerial parts of the herb, *Artemisia amygdalina* Decne [J]. *The Journal of Horticultural Science and Biotechnology, 87*, 455–460.

Rather, M. A., Ganai, B. A., Kamili, A. N., Qayoom, M., Akbar, S., Masood, A., et al. (2012). Comparative GC–FID and GC–MS analysis of the mono and sesquiterpene secondary metabolites produced by the field grown and micropropagated plants of *Artemisia amygdalina* Decne. *Acta Physiologiae Plantarum, 34*(3), 885–890.

Repetto, M. G., & Boveris, A. (2010). Bioactivity of sesquiterpenes: Compounds that protect from alcohol-induced gastric mucosal lesions and oxidative damage. *Mini Reviews in Medicinal Chemistry, 10*, 615–623.

Romero, M. R., Serrano, M. A., Vallejo, M., Efferth, T., Alvarez, M., & Marin, J. J. (2006). Antiviral effect of artemisinin from Artemisia annua against a model member of the Flaviviridae family, the bovine viral diarrhoea virus (BVDV). *Planta Medica, 72*, 1169–1174.

Sakata, K., & Miyazawa, M. (2010). Regioselective oxidation of (+)-α-longipinene by Aspergillus Niger. *Journal of Oleo Science, 59*, 261–265.

Sampaio, L. F. S., Maia, J. G. S., de Parijós, A. M., de Souza, R. Z., & Barata, L. E. S. (2012). Linalool from rosewood (*Aniba rosaeodora* Ducke) oil inhibits adenylate cyclase in the retina, contributing to understanding its biological activity. *Phytotherapy Research, 26*, 73–77.

Seyyedan, A., Yahya, F., Kamarolzaman, M. F. F., Suhaili, Z., Desa, M. N. M., Khairi, H. M., et al. (2013). Review on the ethnomedicinal, phytochemical and pharmacological properties of Piper sarmentosum: Scientific justification of its traditional use. *TANG*, *3*, 1–32.

Shapira, M. Y., Resnick, I. B., Chou, S., Neumann, A. U., Lurain, N. S., Stamminger, T., et al. (2008). Artesunate as a potent antiviral agent in a patient with late drug-resistant cytomegalovirus infection after hematopoietic stem cell transplantation. *Clinical Infectious Diseases*, *46*, 1455–1457.

Sharifi-Rad, J., Hoseini-Alfatemi, S. M., Sharifi-Rad, M., & Setzer, W. N. (2015). Chemical, composition, antifungal and antibacterial activities of essential oil from Lallemantia royleana (Benth. in Wall.) Benth. *Journal of Food Safety*, *35*, 19–25.

Sokolova, A. S., Morozova, E. A., Vasilev, V. G., Yarovaya, O. I., Tolstikova, T. G., & Salakhutdinov, N. F. (2015). Curare-like camphor derivatives and their biological activity. *Russian Journal of Bioorganic Chemistry*, *41*, 178–185.

Sokolova, A. S., Yarovaya, O. I., Shtro, A. A., Borisova, M. S., Morozova, E. A., Tolstikova, T. G., et al. (2017). Synthesis and biological activity of heterocyclic borneol derivatives. *Chemistry of Heterocyclic Compounds*, *53*, 371–377.

Taj, F., Khan, M. A., Ali, H., & Khan, R. S. (2019). Improved production of industrially important essential oils through elicitation in the adventitious roots of *Artemisia amygdalina*. *Plants*, *8*, 430.

Tatsadjieu, L. N., Ngang, J. E., Ngassoum, M. B., & Etoa, F. X. (2003). Antibacterial and antifungal activity of Xylopia aethiopica, Monodora myristica, Zanthoxylum xanthoxyloıdes and Zanthoxylum leprieurii from Cameroon. *Fitoterapia*, *74*, 469–472.

Van, J. R., & Arman, C. G. (2013). *Anti-inflammatory drugs. Vol. 50*. Berlin, Heidelberg, New York: Springer-Verlag.

White, N. J. (1997). Assessment of the pharmacodynamic properties of antimalarial drugs in vivo. *Antimicrobial Agents and Chemotherapy*, *41*, 1413.

Chapter 16

Bioprospecting of endophytic fungi for antibacterial and antifungal activities

Bhat Mohd Skinder[a], Bashir Ahmad Ganai[a], and Abdul Hamid Wani[b]
[a]Centre of Research for Development/Department of Environmental Science, University of Kashmir, Srinagar, Jammu and Kashmir, India, [b]Department of Botony, University of Kashmir, Srinagar, Jammu and Kashmir, India

1. Introduction

The application of natural products in medicine obtained from plants can be traced back several eras in human history. The first written report of medicinal use of plants was discovered at Mesopotamia and dated from about 2600 BC (Cragg & Newman, 2005). Natural products are an integral part of several ethnic therapeutic systems (Liu, Qiu, Ding, & Yao, 2008). A lot of efforts have been deployed to isolate the novel compounds from plants and microorganisms that exhibit many clinical and pharmacological activities. Like plants, fungi have a long history for providing therapies to mankind in everyday life. It is reported that medicinal plants, besides having natural healing properties, also host some fungi involved in coproduction of bioactive metabolites, which have potential use in modern medicine (Alvin, Miller, & Neilan, 2014; Kaul, Gupta, Ahmed, & Dhar, 2012; Kusari, Pandey, & Spiteller, 2013). Fungi were used to treat intestinal ailments by Mayans about 3000 years ago (Strobel & Daisy, 2003; Strobel, Daisy, Castillo, & Harper, 2004). Since Pasteur's discovery, exploration of microbes as natural product resources sprung up. Fungi became a vital source of drugs for treatment of many diseases only after the discovery of penicillin isolated from the fungi *Penicillium chrysogenum*.

1.1 Endophytic fungi

Fungi are heterotrophs belonging to the eukaryotic group, which are plant-like organisms without chlorophyll that absorbs nutrients through its cell wall. They reproduce by spores and have a filamentous body called "thallus" (mycelium) composed of branching, microscopic tubular cells called "hyphae." Fungi are

Phytomedicine: A Treasure of Pharmacologically Active Products from Plants
https://doi.org/10.1016/B978-0-12-824109-7.00025-X

427

"biotrophs" when they get food from a living host, "saprotrophs" (saprobes, saprophytes) when they feed on a dead host, and "necrotrophs" when they infect and kill a living host to obtain their nutrients (Carris, Little, & Stiles, 2012). As per molecular data, fungi are almost more than 1 billion years old (Parfrey, Lahr, Knoll, & Katz, 2011), but fossil evidence record shows them as about 3.5 billion years old (Redecker, Kodner, & Graham, 2000). At least 99,000 fungal species have been labeled, and new species being designated at the rate of 1200 per day (Blackwell, 2011; Kirk, Cannon, Minter, & Stalpers, 2008). As per Hawksworth (2001), there are around 250,000 plant species worldwide considering there are six species of fungal per plant, which accounts for a total of 1.5 million fungal species $(1.5 \times 250,000)$. However, it is estimated per molecular studies that there are around 6 million soil fungi at the global level (Taylor et al., 2014). Fungi are ubiquitous-occurring heterotrophic organisms, often revealing symbiotic traits including mutualistic, antagonistic, or neutral symbiosis with different autotrophic organisms (Saar, Polans, Sørensen, & Duvall, 2001). A fungus is associated with both plants and animals. However, there is an ancient relationship between fungi and plants. Fungi on the plant surface are called "epiphytic fungi," and fungi residing within the plant tissues are called "endophytic fungi." Thus, these are fungal microorganisms that spend their entire or part of their life cycle residing inter- and/or intracellularly, inside the healthy plant tissues without causing apparent signs of any diseases (Petrini, 1991).

The term "endophyte" is from the Greek words "endo" or "endon" meaning within and "phyte" or "phyton" meaning plant, which was introduced by de Bary (1866), for fungi inhabiting plant tissue. An endophytic fungus lives in "mycelial" form in association with plant tissue. Thus, for a fungus to be termed endophyte, it should at least establish its "hyphae" in living tissue (Kaul et al., 2012). These are omnipresent in every plant, whether a plant found in the dessert or a plant found in a hotspot of global biodiversity. Medicinal plants of Western Ghats of India (a hotspot of global biodiversity) are a repository of diverse population of endophytic fungi (Raviraja, 2005). There is more than one endophyte inhibiting 300,000 plant species existing on Earth (Strobel & Daisy, 2003). They include all asymptomatic symbiotic associates of the eukaryotic group Plantae (Azevedo, Maccheroni, Pereira, & de Araújo, 2000; Bacon & White, 2000; Stone, Bacon, & White, 2000; Wilson, 1995), which is a vascular plant, or grasses all host endophytes (Zang, Becker, & Cheng, 2006). Fungi, bacteria, actinomycetes, and mycoplasma were found to be an endophytic organism in plants (Bandara, Seneviratne, & Kulasooriya, 2006). These are one of the most important elements in plant microecosystems, which have significant influences on growth and development of the host plants. However, a few of these plants have been studied for endophytic biology, but research on endophytes today are much more progressed and advanced. Different aspects of endophytic organisms could be investigated, to have a primary and elementary idea about the endophytic fungal population of particular plant species. There is an immense need of biodiversity, taxonomic, and molecular-based studies that include genomics, proteomics, and transcriptomics. The focus of studies also includes endophytes producing "secondary metabolites" and their various activities, antimicrobial, antifungal, antimalarial, antioxidant, anticancer, insecticidal, and pesticidal. The secondary metabolites were called natural products, first recognized by Sachs (1874). "Mycophenolic acid" isolated from *Penicillium glaucoma* is the first crystalline fungal secondary metabolite discovered by Gosio (1896). They have an ability to promote the accumulation of secondary metabolites of the host plants, which influenced the quantity and quality of drugs (Chen et al., 2016). Although endophytes play a very important role in affecting the quality and quantity of the crude drugs through a particular fungus-host interaction, our knowledge about the exact relationships between endophytic fungi and their host plants is still very limited. There is need to understand such relationships for the promotion of crude drug production (Faeth & Fagan, 2002). In the present context, bioprospecting of endophytic fungi from different medicinal plants for different bacterial and fungal strains in respect of antibacterial and antifungal activities are being highlighted that could possibly be used in the pharmaceutical industry to revolutionize the medicinal world in a sustainable way.

2. Bioprospecting of endophytic fungi

"Biodiversity-prospecting" (bioprospecting) can be explained in brief as "What are the ways to commercialize biodiversity"? The bioprospecting concept also includes research on knowledge of indigenous people related to proper management and utilization of natural resources. Bioprospecting is melded with sustainable use of natural resources, conservation, and rights of indigenous communities. There is a systematic approach for development of innovative sources of compounds, micro/macroorganisms, genes, and additional valuable products from the environment to grow and supply commercially valued products for different industrial applications like agriculture, pharmaceutical industry, cosmetics, etc. (Timmermans, 2001; UNDP, 2016). Bioprospecting of endophytic actinomycetes for biotechnological and pharmaceutical resolutions are vital for the exploration of novel drugs or substances for human therapeutics along with antibiotic, anticarcinogenics, and antimycotics (Berdy, 1995; Bi, Li, Song, Tan, & Ge, 2011; Yadav & Mishra, 2013). Research is being widely carried out to find new potentially vital secondary metabolites from endophytic fungi. However, there are major challenges in discovering new drugs based on endophytic fungi as it lies in well-organized approaches to recover bioactive strains. These are important components of biodiversity, having high potential for endophytes located in endemic areas with a high number of endemic plant species and high biodiversity (Strobel & Daisy, 2003). Among them, a repository for endophytes with novel compounds or metabolites of pharmaceuticals importance have been documented in medicinal plants (Strobel et al., 2004; Tan & Zou, 2001; Wiyakrutta et al., 2004). Some of the different endophytes isolated from different host plants are summarized in Table 1. It has been reported that more than 35% endophytes with antimicrobial activity are from medicinal plants (Fig. 1).

TABLE 1 Association of medicinal plants and endophytic fungi.

Medicinal plant	Endophytes fungi	References
Achillea fragrantissima	Dark sterile mycelia, white sterile mycelia, *Penicillium corylophilum*	Selim, El-Beih, AbdEl-Rahman, and El-Diwany (2011)
Adhathoda vasica	*Haplosporium* sp., *Alternaria alternate*, *Colletotrichum* sp., and *Phomopsis* sp.	Gautam (2014)
Adhatoda zeylanica	*Curvularia clavata* and *Phomopsis pittospori*	Raviraja, Maria, and Sridhar (2006)
Ageratum myriadenia	*Penicillium citrinum*, *P. griseofulvum*, *Alternaria arborescens*, and *Bipolaris* sp.	Rosa et al. (2010)
Artimisia herba alba	White sterile mycelia, dark sterile mycelia, *Aspergillus flavus*	Selim et al. (2011)
Atracty lancea	*Sclerotium* sp.	Chen, Dai, Li, Tian, and Xie (2008)
Atractylodes lancea	*Gilmaniella* sp.	Ren and Dai (2012)
Barringtonia acutangula	*Colletotrichum gloeosporioides* and *Phomopsis* species	Tilton, Aneesh Nair, Maany Ramanan, and Devi (2014)
Bauhinia phoenicea	*Curvularia lunata* and *Nigrospora sacchari*	Raviraja et al. (2006)
Buxus sp.	*Colletotrichum yunnanense*	Liu, Xie, and Duan (2007)
Cajanus cajan	*Fusarium oxysporum*, *F. solani*, *F.prolliferatum*, and *Neonectriamacrodidym*	Zhao et al. (2012)
Callicarpa tomentosa	*Alternaria longipes*, *Curvularia pallescens*, *Pithomyces* sp. *Nigrospora oryzae*	Raviraja et al. (2006)
Cannabis sativa	*Phoma* sp., *A. alternate*, *C. lunata*, *Colletotrichum* sp., *Cladosporium* sp., *A.niger*, *A. flavus*, *P. chrysogenum*, and *Mycelia sterilia*	Gautam (2014)
Capsicum annuum	*Penicillium resedanum*	Khan et al., 2013

Cassia spectabilis	Phomopsis cassiae	Silva et al. (2005, 2006)
Chiliadenus montanus	Nigrospora sphaerica, Dark sterile mycelia	Selim et al. (2011)
Chrysanthemummorifolium	Chaetomium globosum and Botrytis sp	Liu, Song, Zhang, and Ye (2011)
Cinchona ledgeriana	Diaporthe sp., Fomitopsis sp., Phomopsis sp., Schizophyllum sp., Penicillium sp. and Arthrinium sp.	Maehara et al. (2013)
Cinnamomum camphora	Cochliobolus nisikadoi	Chen et al. (2011)
Cirsium arvense	Chaetomium cochliodes, Cladosporium cladosporioides, and Trichoderma viride	Gange, Eschen, Wearn, Thawer, and Sutton (2012)
Coffea arabica L.	Penicillium coffeae and Alternaria alternata	Peterson, Vega, Posada, and Nagai (2005) and Fernandes et al. (2009)
Cucumis sativus	Phoma glomerata, Penicillium sp., and Chaetomium	Waqas et al. (2012) and Yan, Sikora, and Zheng (2011)
Curcuma wenyujin	Chaetomium globosum	Wang et al., 2012
Cymbopogon caesius	C. lunata	Avinash, Ashwini, Babu, & Krishnamurthy, 2015
Ephedra aphyla	Dark sterile mycelia, white sterile mycelia	Selim et al. (2011)
Eucommia ulmoides	Sordariomycete sp.	Chen, Sang, Li, Zhang, and Bai (2010)
Euphorbia sanctae catharinae	Phoma leveilleia, dark sterile mycelia, white sterile mycelia, Acremonium strictum	Selim et al. (2011)
Fritillaria unibracteata	Fusarium redolens	Pan et al. (2015)
Galium sinaicum	Pleospora tarda, Aspergillus sp., and Fusarium sp.	Selim et al. (2011)
Garcinia plant	Aspergillus, Botryosphaeria, Eutypella, Fusarium, Guignardia, Penicillium, Phomopsis, and Xylaria	Phongpaichit, Rungjindamai, Rukachaisirikul, and Sakayaroj (2006)

Continued

TABLE 1 Association of medicinal plants and endophytic fungi—cont'd

Medicinal plant	Endophytes fungi	References
Ginkgo biloba	F. oxysporum	Cui et al. (2012)
Glycinemax, Helianthus annuus, and Vigna radiate	Diaporthe miriciae	Thompson et al. (2015)
Hypericum sinaicum	Ulocladium chartarum, Pleospora tarda, Chaetomium globosum, white sterile mycelia	Selim et al. (2011)
Lannea coromandelica	A. flavus, A. niger, A. alternata, and C. gloeosporioides	Premjanu, Jaynthy, and Diviya (2016)
Launea spinosa	Acremonium strictum, Penicillium chrysogenoum, and Aspergillus niger	Selim et al. (2011)
Lavandula coronopifolia	Dark sterile mycelia	Selim et al. (2011)
Lippia sidoides	C. gloeosporioides, A. alternata, Guignardia bidwelli, and Phomopsis archeri	de Siqueira, Conti, de Araújo, and Souza-Motta (2011)
Lobelia nicotinifolia	F. oxysporum	Raviraja et al. (2006)
Luehea divaricate	Alternaria, Cochliobolus, Diaporthe, Epicoccum, Guignardia, Phoma, and Phomopsis	Bernardi-Wenzel, García, Celso Filho, Prioli, and Pamphile (2010)
Lycopodium serratum	P. chrysogenum	Zhou, Yang, Lan, Xu, and Hong (2009)
M. ilicifolia	Alternaria, Phyllosticta, Xylaria, Phomopsis, Pestalotiopsis sp., and Colletotrichum	Glienke et al. (2012)
Madhuka nerifolia (Moon) H. J. Lam.	Trichoderma koningi	Raviraja et al. (2006)
Manihot esculenta	Bacillus pumilus	Melo et al. (2009)
Maytenus ilicifolia	Pestalotiopsis sp., Pestalotiopis vismae, and Pestalotiopsis microspora	Gomes-Figueiredo et al. (2007)

Melia azedarach L.	Penicillium sp. and P. janthinellum	Marinho, Rodrigues-Filho, Moitinho, and Santos (2005)
Moringa oleifera Lam.	Aspergillus spp., A. flavus, A. niger, Aspergillus versicolor, Aspergillus terreus, Aspergillus ochraceus, and dematiaceous fungi namely Bipolaris spp.	Rajeswari, Umamaheswari, Arvind Prasanth, and Rajamanikandan (2016)
Murraya paniculata	Penicillium sp.	Pastre, Marinho, Rodrigues-Filho, Souza, and Pereira (2007)
Naregamia alata	Pestalotia disseminata	Raviraja et al. (2006)
Ocimum sanctum	A. niger, A. flavus, Nigrospora sp., A. alternate, Colletotrichum sp., C. lunata, Cladosporium sp., and Fusarium sp.	Gautam (2014)
Ocotea corymbosa	Curvalaria sp.	Teles et al. (2005)
Ophiopogon japonicus	F. oxysporum and Fusarium poae	Liu, Liu, Yuan, and Gu (2010)
Palicourea tetraphylla	Arthrium sp., F. oxysporum, Penicillium griseofulvum, and P. citrinum	Rosa et al. (2010)
Phlomis aurea	White sterile mycelia, dark sterile mycelia, Chaetomium spirale, Penicillium sp., A. flavus	Selim et al. (2011)
Picea rubens	150 foliar fungal endophytes	Sumarah, Puniani, Sorensen, Blackwell, and Miller (2010)
Piper aduncum	Xylaria sp.	Silva et al. (2010)
Piper nigrum	C. gloeosporioides	Chithra, Jasim, Sachidanandan, Jyothis, and Radhakrishnan (2014)
Piptadenia adiantoides	Arthrium sp. and Gibberella sp.	Rosa et al. (2010)
Pulicaria undulate	Dark sterile mycelia, Ulocladium chartarum, and Penicillium sp.	Selim et al. (2011)
Rehmannia glutinosa	Ceratobasidium sp.	Chen et al. (2011)

Continued

TABLE 1 Association of medicinal plants and endophytic fungi—cont'd

Medicinal plant	Endophytes fungi	References
Salvia miltiorrhiza	*Arbuscular mycorrhiza, Trichoderma atroviride*	Meng and He (2011)
Sapindus saponaria L.	*Cochliobolus intermedius* and *Phomopsis* sp.	Garcia et al. (2012))
Schinus terebinthifolius Raddi (Pepper tree)	*Alternaria* sp., *Phomopsis* sp., *Penicillium roseopurpureum*, basidiomycete, *Streptomyces* sp.	Glienke et al. (2012)
Sesbania sesban	*Funneliformis mosseae, Rhizophagus intraradices*, and *Claroideoglomus etunicatum*	Abd_Allah, Hashem, Alqarawi, Bahkali, and Alwhibi (2015)
Smallanthus sonchifolius	*Arthrinium arundinis, C. cladosporioides, Papulaspora immersa, Colletotrichum destructivum, C. gloeosporioides, Coniothyrium aleuritis* Teng, *F. oxysporum, Gibberella avenacea, N. sphaerica* (Sacc.), *Plectosphaerella cucumerina, Trichoderma asperellum,* and *Lecythophora* sp.	Rosa et al. (2012)
Smallanthus species	*Fusarium, Bionectria, Colletotrichum, Gibberella, Hypocrea, Lecythophora, Nigrospora, Cladosporium, Plectosphaerella,* and *Trichoderma*	Rosa et al. (2012)
Smallanthus uvedalius	*Bionectria ochroleuca, Colletotrichum* sp., *F. oxysporum, Hypocrea* sp.	Rosa et al. (2012)
Spondias mombin	*Spondias mombin* and *Guignardia* sp.	Rodrigues-Heerklotz, Drandarov, Heerklotz, Hesse, and Werner (2001)
Stachys aegyptiaca	*Mucor fuscus* and *A. flavus*	Selim et al. (2011)
Suaedamaritime and *Suaedamonoica*	*A. alternate, A. flavus, A. terreus, A. niger, Cladosporium* sp., *Fusarium* sp., *Penicillium* sp., sterile mycelium, and *Meyerozyma* sp.	Kalyanasundaram, Nagamuthu, and Muthukumaraswamy (2015)
Symphytum officinale L.	*Trichophyton* sp., *Chrysosporium* sp., *Candida pseudotropicalis*, and *Candida tropicalis*	Rocha et al. (2009)
Tanacetum sinaicum	White sterile mycelia, *Penicillium* sp.	Selim et al. (2011)

Plant species	Endophytes	References
Taxus chinensis	*Fusarium mairei*, *Ozonium* sp., *A. alternata*, *Botrytis* sp., *Ectostroma* sp., *Papulaspora* sp., and *Tubercularia* sp.	Zhou et al. (2007), Guo et al. (2006) and Wu et al. (2013)
Teucrium leucocladum	White sterile mycelia, dark sterile mycelia	Selim et al. (2011)
Teucrium polium	*A. alternate*, *N. sphaerica*, white sterile mycelia, *Penicillium corylophilum*, *P. chrysogenoum*, and *A. niger*	Selim et al. (2011)
Thymus sp.	*Alternaria* spp., *Phoma* spp., *Fusarium* spp., *Cladosporium* spp., *Stemphylium* spp., *Ulocladium* spp., *Colletotrichum* spp., *Cylindrocarpon* spp., *Drechslera* spp., *Curvularia* spp., *Aspergillus* spp., and yeast	Masumi, Mirzaei, Zafari, and Kalvandi (2015)
Tithonia diversifolia	*Phoma sorghina*	Borges and Pupo (2006)
Trichilia elegans	*Cordycepsmemorabilis*, *Phomopsis longicolla*, and *Dothideomycete* sp.	Rhoden, Garcia, Rubin Filho, Azevedo, and Pamphile (2012)
Trixis vauthieri	*Arthrium* sp. and *Xylaria* sp.	Rosa et al. (2010)
Vellozia gigantean	*Diaporthe miriciae*, *Trichoderma effusum*, *Penicillium herquei*, *P. adametzii*, and *P. quebecense*,	Ferreira et al. (2017)
Viguiera robusta,	*Chaetomium globosum*	Momesso et al. (2008)
Viola odorata	*Geotrichum* sp., *A. flavus*, *Phoma* sp., *A. alternate*, *Cladosporium* sp., and *Rhizopus* sp.	Gautam (2014)
Virola michelii	*Pestalotiopsis guepinii*	Oliveira et al. (2011)
Vitex negundo L.	*Chaetomium globosum* and *Papulospora* sp.	Raviraja et al. (2006)
Vochysia divergens (Cambará)	*Microbispora* sp., *Micromonospora* sp., and *Streptomyces sampsonii*	Glienke et al. (2012)
Withania somnifera (L.)	*Phoma* sp., *Alternaria* sp., *A. niger*, *A. flavus*, *A. clavatus*, *A. variecolor*, *Chaetomium bostrycodes*, *Eurotium rubrum*, *Melanospora fusispora*, *C. cladosporioides*, *Curvularia oryzae*, *Drechslera australiensis*, *Fusarium* sp., *Myrothecium roridum*, *Penicillium* sp., *Trichoderma* sp., *Stemphylium* sp., and *Colletotrichum* sp.	Khan et al. (2013) and Gautam (2014)

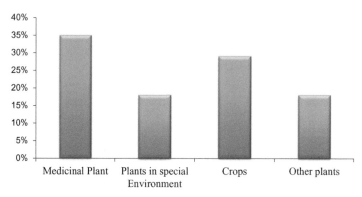

FIG. 1 Proportion of sources of endophytic fungi with antimicrobial activities (Yu et al., 2010).

Endophytes residing in medicinal plants are a vital source of bioactive compounds (secondary metabolites), contributing more than 80% to the natural drugs available in the markets (Singh & Dubey, 2015). The total number of microbial metabolites until 2005 was recognized to be around 50,000 (Berdy, 2005). As per the literature, the numbers still continue to increase over the years. Thus, countless drugs are being provided by medicinal plants and endophytes in the form of secondary metabolites that are selected for important therapeutic alternatives for numerous diseases. Endophytes are dynamic in nature and have the potential to produce pharmacologically active substances, for example, novel antibiotics, antifungal, anticancer, and immunosuppressant compounds (Firakova, Sturdikova, & Muckova, 2007; Mitchell, 2008; Strobel & Daisy, 2003). The antibacterial and antifungal activities of some of the isolated secondary metabolites from various endophytic fungi associated with different medicinal plants are depicted in Table 2. Endophytes are a rich source of functional metabolites and a potential source of finding new drugs for treating new diseases (Gouda, Das, Sen, Shin, & Patra, 2016; Kumar, Lau, Wan, Yang, & Hyde, 2005; Tejesvi, Kini, Prakash, Subbiah, & Shetty, 2007; Weber et al., 2004). The scientific study revealed more than 80% endophytic fungi produce biologically active compounds when screened for antifungal, antibacterial, and herbicidal activities (Schulz, Boyle, Draeger, Römmert, & Krohn, 2002). The success of various medicinal drugs from a microbial origin has become a source of innovations and shifted the focus of drug discovery from plants to microorganisms, for example, the β-lactam antibiotic "penicillin" from *Penicillium* sp., the antifungal agent "griseofulvin ($C_{17}H_{17}ClO_6$)" extracted from *P. griseofulvum* (Park et al., 2005), the "cyclosporine" immunosuppressant (*Tolypocladium inflatum* and *Cylindrocarponlucidum*), and the "lovastatin" (cholesterol biosynthesis inhibitor) obtained from *A. terreus*. It is evident that bioprospecting of endophytic microorganisms now becomes a fundamental emerging tool for pharmaceutical and biotechnological industries to discover new vital substances including antibiotics, antifungal, antioxidants, antimalarials, and anticarcinogenics for human

TABLE 2 Secondary metabolites from endophytic fungi associated with different medicinal plants.

Host plant	Endophyte	Compound	Biological activity	Reference
Artemisia mongolica	C. gloeosporioides	Colletotric acid	Antibacterial and antifungal activity	Zou et al. (2000)
Artemisia vulgaris	Chalara sp. (strain 6661)	Isofusidienol A, B, C, and D	Antifungal	Lösgen, Magull, Schulz, Draeger, and Zeeck (2008)
Azadirachta indica	Chloridium sp.	Javanicin	Antibacterial	Kharwar et al. (2009)
Cassia spectabilis	Phomopsis cassiae	3,12-Dihydroxycadalene	Antifungal	Silva et al. (2006)
Catunaregam tomentosa	Curvularia geniculata	Curvularides A–E	Antifungal	Chomcheon et al. (2010)
Cynodon dactylon	Aspergillus fumigatus,	Asperfumoid, Physcion, Fumigaclavine, Fumitremorgin, and Helvolic acid	Antifungal	Liu et al. (2004)
Daphnopsis americana	CR115 "unidentified fungus"	Guanacastepene	Antibacterial	Singh et al. (2000)
Dioscorea zingiberensis	Dzf12	Diepoxin κ	Antibacterial	Cai et al. (2009)
Ilex cornuta	Trichoderma harzianum	Trichodermin	Antifungal	Chen, Chen, Zheng, Zhang, and Yu (2007)

Continued

TABLE 2 Secondary metabolites from endophytic fungi associated with different medicinal plants—cont'd

Host plant	Endophyte	Compound	Biological activity	Reference
Juniperus cedrus and Erica arborea	Nodulisporium sp.	Nodulisporins A–C and Nodulisporins D–F	Antifungal, antibacterial, and antialgal	Dai, Krohn, Draeger, and Schulz (2009) and Dai et al. (2006)
Lannea coromandelica	A. flavus	Kojic acid, octadecanoic acid, n-hexadecanoic acid, diethyl phthlate, 3-phenyl propionic acid	Antifungal	Premjanu et al. (2016)
Laurencia sp.	P. chrysogenum	Conidiogenone B, conidiogenol	Antibacterial	Gao, Li, Li, Proksch, & Wang (2011) and Gao, Li, Zhang, Li, & Wang (2011)
Laurus azorica	Phomopsis sp.	Cycloepoxylactone, cycloepoxytriol A and B	Antifungal and antibacterial	Hussain et al. (2009)
Lycium intricatum	Microdiplodia sp.	Diversonol, ergosterol, microdiplodiasol, microdiplodiasone, and microdiplodiasolol	Antibacterial	Siddiqui et al. (2011)
Mentha pulegium	Stemphylium globuliferuman	Alterporriol N and E	Antibacterial	Debbab et al. (2009)
Paris polyphylla	Pichia guilliermondii	Helvolic acid	Antibacterial	Zhao et al. (2010)
Pinus sp.	Microdiplodia sp.	8-α-Acetoxyphomadecalin C Phomadecalin E	Antibacterial	Hatakeyama, Koseki, Murayama, and Shiono (2010)
Piper aduncum	Xylaria sp.	Phomenone	Antifungal	Silva et al. (2010)
Rosenvingea sp.	Pestalotia sp.	Pestalone	Antibacterial	Cueto et al. (2001)

Source	Endophyte	Compound	Activity	Reference
Sonneratia alba	*Alternaria* sp.	Xananteric aids I and II	Antibacterial	Kjer et al. (2009)
Stored sample	*Cladosporium cladosporiodes* (Fresen) de varies	Cladosporin-8-methyl ether, 5'-hydroxyasperentin, cladosporin, isocladosporin	Antifungal	Wang et al. (2012)
Taxus mairei	*Aspergillus clavatonanicus,*	Clavatol and patulin	Antifungal	Zhang et al. (2008)
Teucrium scorodonia	*Phomopsis* sp.	Phomosines A–C	Antibacterial	Krohn et al. (1995)
Theobroma cacao	*Epicoccum* sp.	Epicolactone, epicoccolides A and B	Antifungal	Talontsi, Dittrich, Schüffler, Sun, and Laatsch (2013)
Trixis vauthieri	*Alternaria* sp.	Altenusin	Antifungal and antiparasitic	Cota et al. (2008) and Johann et al. (2012)
Vernonia amygdalina	*Curvularia papendorfii*	Khartoumic acid (3,7,11,15-tetrahydroxy-18-hydroxymethyl-14,16,20,22,24-pentamethyl-hexacosa 4E,8E,12E,16,18-pentaenoic acid)	Antibacterial	Khiralla (2015) and Khiralla, Spina, Yagi, Mohamed, and Laurain-Mattar (2016)
Xylopia aromatica	*Periconia atropurpurea*	6,8-Dimethoxy-3-(2'-oxo-propyl)-coumarin and 2,4-dihydroxy-6-[(1'E, 3'E)-penta-1',3'-dienyl]-benzaldehyde, Periconicin	Antifungal	Teles et al. (2006)

therapeutics (Huang, Cai, Xing, Corke, & Sun, 2007; Kharwar, Mishra, Gond, Stierle, & Stierle, 2011; Strobel, 1998, 2002; Strobel & Daisy, 2003; Yu et al., 2010).

3. Antibacterial activity

Antibacterial resistance has now become a major health issue and globally affects our healthcare system (Aksoy & Unal, 2008; Ferri, Ranucci, Romagnoli, & Giaccone, 2017; Bungihan et al., 2011). Microbes have adopted various molecular mechanisms to resist antibiotics, which in turn downs the efficacy of the drug, for example, it prevents locating a target site and modifies the chemical nature of the drug (Blair, Webber, Baylay, Ogbolu, & Piddock, 2015). Thus to overcome this global problem, researchers are trying to introduce "endophytic fungi" isolated from different plant tissues as a potential source for new antibiotics (Radu & Kqueen, 2002; de Siqueira et al., 2011; Liang et al., 2012). The isolation and identification of endophytes is essential because the properties of a medicinal plant may be associated to its capacity to host endophytes to produce bioactive metabolites (Kaul et al., 2012; Kusari et al., 2013). The bioactive natural compounds produced by endophytic fungi such as alkaloids, flavonoids, steroids, terpenoids, phenolic acids, benzopyranones, chinones, quinone, tetralones, xanthones, etc. provide protection and survival to the plant (Carroll & Carroll, 1978; Tan & Zou, 2001).

There is one great advantage to using endophytic fungi in drug discovery. It is because their "diversity" offers a plentiful source of active metabolites, which can be the alternative source of compounds extracted from plants, thus reducing the large-scale harvesting of plant material in producing the compounds (Alvin et al., 2014). The fungal endophytes of *Ocimum* sp. (Tulsi) have potential antibacterial activity against *Pseudomonas aeroginosa, Mycobacterium smegmatis, and Salmonella typhimurium* (Pavithra, Sathish, & Ananda, 2012). About 33.60% of fungal endophytes isolated from the leaves of *Indigofera suffruticosa* Miller exhibited antibacterial activity; however, two species of fungi, *N. sphaerica* and *Pestalotiopsis maculans*, showed best results (Santos et al., 2015). More than 53% (8 of 15) endophytes revealed the production of antimicrobial compounds isolated from 15 species of endophytes isolated from eight medicinal plants of Western Ghats of India (*Adhatoda zeylanica, Bauhinia phoenicea, Callicarpa tomentosa, Clerodendrum serrate, Lobelia nicotinifolia, Madhuka nerifolia, Naregamia alata,* and *Vitex negundo*); among them, the top three endophytes (*Alternaria* sp., *N. oryzae,* and *Papulospora* sp.) revealed inhibitory action against both Gram +ve and Gram -ve bacteria (Raviraja et al., 2006). Although some plants have a large number of isolates, few of them show some activity. The study carried out by de Siqueira et al. (2011) revealed that only 16 endophytic fungi out of 203 endophytic isolates obtained from *Lippia sidoides* Cham exhibited antibacterial activity. There are times in such cases where all the isolated endophytic fungi from a host plant show activity against at least one tested pathogen. In the recent year, Marcellano, Collanto, and Fuentes (2017) isolated 12 endophytic fungi (*Colletotrichum* sp., *Fusarium* sp., *Cunninghamella* sp., *Mucor* sp., *Aspergillus* sp., *Penicillium* sp., *Pestalotiopsis* sp., *Phomopsis* sp., *Rhizoctonia* sp., and *Mycelia* sterilia) from the bark of medicinal tree *Cinnamomum mercadoi*, and interestingly all the endophytes exhibited antibacterial activity against at least one tested pathogen. However, after secondary screening, two endophytes, *Cunninghamella* sp. and *Fusarium* sp., exhibited broad spectrum antibacterial activity on all the test bacteria *Pseudomonas aeroginosa*, *Escherichia coli*, *Bacilluscereus*, and *Staphylococcusaureus*. In the forests of the Amazon, around 79 endophyte isolates were isolated from *Palicourea longiflora* and *Strychnos cogens*, but only 24% (19 of 79) exhibit the inhibitory action against at least one of the pathogenic microorganisms tested such as *S. aureus*, *E. coli*, *Bacillus* sp., and *B. subtilis* (de Souza et al., 2004). In Egypt, for the first time, a study was carried out on biodiversity of endophytic fungi and their antimicrobial activities. About 99 isolates of these fungi were revealed, and about 55.55% (55 of 99) endophytes exhibited a broad spectrum of inhibitory action against different pathogenic bacteria and yeasts (Selim et al., 2011). Similarly, for the first time, seven species of endophytic fungi (*Cladosporium* sp., *A. flavus*, *Aspergillus* sp., *C. lunata*, as well as three unknown species) were successfully isolated from *Kigelia africana* (Lam.). Most of the extracts of these endophytes showed in vitro inhibition of bacterial growth (*B. subtilis*, *S. aureus*, and *E. coli*) (Idris, Al-Tahir, & Idris, 2013). Sutjaritvorakul, Whalley, Sihanonth, and Roengsumran (2011) revealed metabolites from fungal extracts have positive antimicrobial activity against human pathogenic microbes (*E. coli*, *S. aureus*, *B. subtilis*, and *Pseudomonas aeruginosa*) and inhibition of Gram + ve bacteria was more than Gram −ve.

Researchers are now quite busy in the extraction of various metabolites from different endophytic fungi to boost the pharmaceutical industry in a conservative and sustainable way. Phongpaichit et al. (2006) worked on *five* medicinal Garcinia plants for extraction of endophytes and confirmed that bioactive metabolites produced by these fungi exhibited antibacterial action against *S. aureus*. The bioactive metabolite called Phomodione has inhibitory action against pathogenic bacterium *Staphylococcus aureus* produced by fungal endophyte "Phoma strain" isolated from *Saurauias caberrinae* (Hoffman et al., 2008). One of the endophytes, *Curvularia protubera*, isolated from Chilean native gymnosperms showed best inhibitory action against *S. aureus*, *Bacillus subtilis*, and *M. luteus* (Hormazabal & Piontelli, 2009). Exploration of antibacterial activities of different endophytes have been done from time to time like Alternaria (Kjer et al., 2009), Penicillium canescens (Bertinetti, Peña, & Cabrera, 2009), Phomopsis (Du et al., 2008), Pestalotiopsis (Liu et al., 2011), basidiomycete (Suay et al., 2000), and Streptomyces (Maruna et al., 2010) isolated from different species of plants. A decade ago in China, two new antimicrobial compounds, "10-oxo-10*H*-phenaleno[1,2,3-de] chromene-2-carboxylic acids" and xanalteric acids I and II, along with 11 already known secondary metabolites were obtained from the endophytic fungus *Alternaria* sp. isolated from the mangrove plant Sonneratia alba. These secondary metabolites were effective against *E. faecalis, P. aeroginosa*, and *S. epidermidis* (Kjer et al., 2009).

The endophyte *Penicillium janthinellum* produces a new modified anthraquinone, named "janthinone," from the fruits of a Chinaberry tree (Melia azedarach), together with known polyketides citrinin, emodin, and "1,6,8-trihydroxy-3-hydroxymethylanthraquinone" (Marinho et al., 2005). The metabolites azaphylones, citrinin, and citrinin H-1, which depict significant antibacterial activities, are also produced by *Penicillium* sp. isolated from *Melia azedarach* and *Murraya paniculata* (Pastre et al., 2007). Two novel anthraquinone derivatives, "dendryol E" and "dendryol F," produced by fungal endophyte *Phoma sorghina*, were isolated from the extract of medicinal plant *Tithonia diversifolia* (Borges & Pupo, 2006). Subsequent studies revealed another new anthraquinone derivative, "guepinone," produced by the fungus *Pestalotiopsis guepinii* along with the known compounds isosulochrin and chloroisosulochrin, isolated from *Virola michelii* (Oliveira et al., 2011). These quinone compounds have potential antibacterial efficiency against various pathogens *Enterococcus faecalis, S. aureus*, and *S. epidemidis* (Aly et al., 2008). The identified taxon, isolated from peppertree leaves, showed antibacterial activities against different bacterial strains *S. aureus, C. albicans, K. pneumonia, M. luteus, E. coli*, and methicillin-resistant *Staphylococcus aureus* (MRSA) and *P. aeruginosa* (Glienke et al., 2012). A biotechnological potential of crude extracts from fungal endophytes of a tree *Sapindus saponaria* L. was investigated by Garcia et al. (2012) who observed an extract of endophyte *Cochliobolus intermedius* was effective against all bacterial strains (*E. coli, S. aureus, S. typhi, M. luteus*, and *E. hirae*) but showed highest activity against *S. aureus*. In a similar approach, fungal endophytes *Cordycepsmemorabilis, Phomopsis longicolla*, and *Dothideomycete* sp. from *Trichilia elegans* exhibited antibacterial activity against five pathogenic bacteria (*Enterococus hirae, Micrococcus luteus, Escherichia coli, Salmonella typhi*, and *Staphylococcus aureus*). However, *Staphylococcus aureus* was unaffected by any endophyte extracted with ethyl acetate and no tested bacteria was inhibited by endophyte obtained by methanol extract (Rhoden et al., 2012), but *C. lunata* showed 92% inhibition against *Staphylococcus aureus* (Avinash et al., 2015). Furthermore, the two species *C. gloeosporioides* and *Phomopsis* of endophytic fungi from the leaves of *Barringtonia acutangula* (Indian oak or Samudraphal) showed significant activities against bacteria like *Bacillussubtillis* and MRSA (Tilton et al., 2014).

3.1 Mechanism of antibacterial activity

Antibacterial activity of active extracts of endophytes is due to the presence of novel compounds like alkaloids (Fig. 2) and phenolic compounds (Fig. 3) present in the plants. Other classes of compounds were also revealed in the extracts

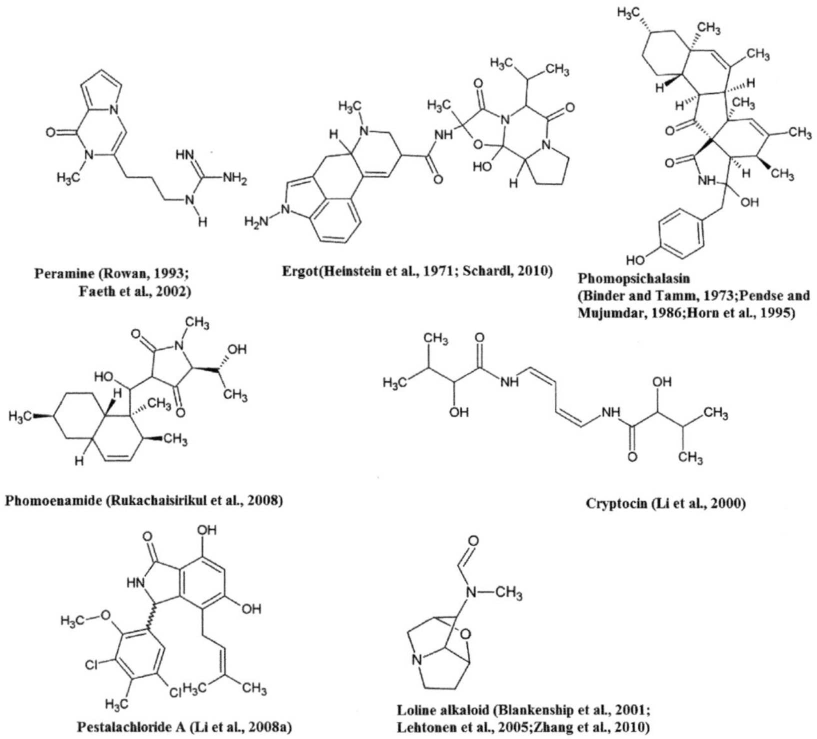

FIG. 2   Structure of alkaloid derivatives from fungal endophytes (Mousa & Raizada, 2013).

benzopyranones, chinones, phenolicacids, quinone, steroids, terpenoids, tetralones, and xanthones but less frequent. "Triterpenoid helvolic" acid is being produced by fungal endophyte *Cytonaema* sp. and has strong antibacterial activity (Kumar, Aharwal, Shukla, Rajak, & Sandhu, 2014). Anthraquinones from *Streptomyces* sp. and *Alternaria* sp. and terpenoids produced by *Phomopsis* sp. play important roles in antibacterial activities (Ceruks, Romoff, Fávero, & Lago, 2007; Degáspari, Waszczynskyj, & Prado, 2005; Glienke et al., 2012; Queires & Rodrigues, 1998). The production of a napthodianthrone derivative "hypericin ($C_{30}H_{16}O_8$)" and "emodin ($C_{15}H_{10}O_5$)" by endophytic fungi are believed to show both antibacterial (*Salmonella enterica, Staphylococcus aureus, Pseudomonas aeruginosa, Klebsiella pneumonia*, and *E. coli*) and antifungal activities (*C. albicans* and *A. niger*) (Kusari, Lamshöft, Zühlke, & Spiteller, 2008; Kusari, Zühlke, Kosuth, Cellarova, & Spiteller, 2009).

The mode of microbial action can be categorized in four ways.

1. Inhibition of cell wall synthesis: Bacterial cell wall is composed of peptidoglycan important for their survival. The drugs or secondary metabolites interfere in the biosynthesis of peptidoglycan, which affects sustainability of bacteria.

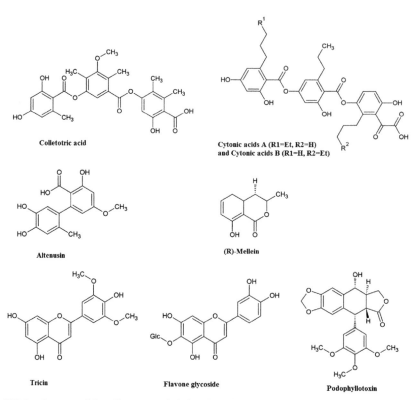

FIG. 3 Structure of phenolic compounds derived from fungal endophytes (Mousa & Raizada, 2013).

2. DNA synthesis inhibitor: The antibacterial activity of some compounds like Usnic acid produced by Lichens and closely related secondary metabolites produced by fungal endophytes "mycousnin" from *Mycosphaerella nawae* (Sassa & Igarashi, 1990), Usnic acid amide and Cercosporamide from *Cercosporidium henningsii* (Conover et al., 1992), and Phomodione and Cercosporamide from *Phoma* sp. (Hoffman et al., 2008) depicted strong inhibition of RNA and DNA synthesis (interference in elongation of DNA replication) in Gram + ve bacteria demonstrated by *S. aureus* and *B. subtilis* and slight inhibition also demonstrated by Gram −ve bacterium *Vibrio harveyi* (Maciąg-Dorszynska, Węgrzyn, & Guzow-Krzemińska, 2014). Twisting of long-stranded DNA into bundles is promoted by DNA gyrase and is important for the bacterial cell so that it completely fits into the cell. The inhibition of this enzyme by any DNA gyrase inhibitor like Quinolones does not allow folding of several millimeter-long DNA molecule. Thus, the power of cell DNA to synthesize proteins is halted and the bacteria dies.

3. Inhibitors of protein synthesis: Antibiotics or secondary metabolites interfere with the bacterial ribosome, which leads to the disruption of protein synthesis.

4. Metabolite inhibitor: Biosynthetic pathway of bacteria is disturbed by various inhibitors like trimethoprim, and sulfonamide inhibits synthesis of folic acids that are necessary for nucleic acid formation (Khan, 2007).

4. Antifungal activity

Fungal infection management now becomes more complex due to a limited number of effective antifungal drug sources, effectiveness and resistance of drug forms, side effects, and drug safety. Thus, during the last 15 years, fungal infections of various types have been growing at an alarming rate, which has been especially observed in immunocompromised/immunosuppressed persons with cancer, AIDS, and transplant patients. *Candida* sp. infection is one such example that has a significant pathogenic effect on immunocompromised patients and has resulted as a major cause of deaths in those being treated for malignant and immune deficiency diseases (Garcia-Ruiz, Amutio, & Pontón, 2004; Hayes & Denning, 2013). Now there is a need of better management through various techniques and *mechanisms* to find out new alternative sources of novel drugs, to improve present drugs, and to design new drugs for better treatment. The present new technique to find new sources of novel compounds is bioprospecting of endophytic fungi for various biological activities.

Scientists and scholars have revealed diverse antifungal properties of different secondary metabolites isolated from different fungal endophytes associated with different medicinal plants. *A. terreus* and other endophytes like *Penicillium* sp. and *Meyerozyma* sp. isolated from the leaves of *Suaedamaritime* and *Suaedamonoica* exhibited significant antifungal activity of different pathogenic fungi like *Candida albicans, Trichophyton rubrum, Microsporum canis, Trichophyton mentagrophyte*, and *Epidermophyton floccossum* (Kalyanasundaram et al., 2015). Metabolite extracts of *Alternaria* sp., inhibited the growth of *Candida albicans* as well (Raviraja et al., 2006); however, *C. lunata* associated with *Cymbopogeon caesius* showed 81% antifungal activity against *Candida albicans* (Avinash, Ashwini, Babu, & Krishnamurthy, 2015). Endophytic fungi obtained from *Palicourea longiflora* and *Strychnos cogens*, inhibited the growth of *Trichoderma* sp., *Candida albicans*, and *A. flavus* (de Souza et al., 2004). Phongpaichit et al. (2006) confirmed the most active endophytic fungi for production of secondary metabolites and antifungal activity (against *C. albicans, Cryptococcus neoformans*, and Microsporum gypseu) among Aspergillus, *Botryosphaeria, Eutypella, Fusarium, Guignardia, Penicillium*, and *Xylaria* obtained from the medicinal plant *Garcinia* sp. against *C. albicans*, were *Phomopsis* sp., *Botryo* sp., and *Haeria* sp. In 2004, Weber and coworkers had also confirmed the potential antifungal effect of active polyketide lactone "Phomol" from *Phomopsis* sp. Isolated from *Erythrina crista-galli*, which also has antiinflammatory and antibacterial effect (Weber et al., 2004). Scientific study further revealed two strains of *Penicillium* sp. isolated as endophytes from *Alibertia macrophylla* (Rubiaceae), which produces "orcinol" and 4-hydroxymellein, which showed inhibitory action against *C. cladospori-oides* and *C. sphaerospermum* (Oliveira et al., 2009). The antifungal metabo-lites produced by 39 different endophytes resides in *Artemisia annua* (Sweet wormwood), and 21 (53.84%) fungal metabolites are inhibitory to a few or all of the phytopathogenic fungi like *Fusarium graminearum*, *Gaeumannomyces graminis* var. *tritici*, *Helminthosporium satium*, *Rhizoctonia cerealis*, *Gerlachia nialis*, and *Phytophthora capsici* (Liu, Zou, Lu, & Tan, 2001). "Altenusin" iso-lated from the fungus endophyte Alternaria sp. is not only antimicrobial but also acts against pathogenic fungi (Kjer et al., 2009). The benzopyrans isolated from an endophytic fungus *Curvularia* sp., associated with *Ocotea corymbose*, tested against *C. sphaerospermum* and *C. cladosporioides*, and the results were convincing (Teles et al., 2005). Rosa et al. (2012) was working on isolation and identification of endophytic fungi from *Smallanthus* sp. and isolated 25 isolates of the genera *Trichoderma*, *Bionectria*, *Nigrospora*, *Cladosporium*, *Fusarium*, *Colletotrichum*, *Hypocrea*, *Gibberella*, *Lecythophora*, and *Plectosphaerella*. It was revealed that about 41.6% of the isolates displayed antifungal activi-ties (*F. oxysporum* and phylotypes *Lecythophora* sp.). The *Trichoderma lon-gibrachiatum* and *Syncephalastrum racemosum* isolated from *Markhamia tomentosa* exhibited significant antifungal inhibition (Ibrahim et al., 2017). The endophyte *Phomopsis cassiae* of *Cassia spectabilis* is a producer of pho-mopsilactone ($C_{13}H_{12}O_5$) and ethyl 2,4-dihydroxy-5,6-dimethylbenzoate. Both metabolites have strong potential to resist phytopathogenic fungi *C. cladospori-oides* and *C. sphaerospermum* (Silva et al., 2010, 2005). The fungal endophyte *Bacillus pumilus* produces the metabolite "pumilacidin" having potential of antifungal activity and was isolated from *Manihot esculenta* (Cassava) (Melo et al., 2009). Likewise, Cryptocin is a "tetramic acid" from endophytic fungus, *Cryptosporiopsis quercina*, an endophyte of *Tripterigeum wilfordii*, which is a strong antimycotic, possesses strong activity against the world's nastiest plant pests *Pyricularia oryzae*, and is considered as having possible agrochemical usage (Li, Strobel, Harper, Lobkovsky, & Clardy, 2000). Rajeswari et al. (2016) revealed fungal inhibition property of *A. flavus* isolated from *Moringa oleifera* (Lam) against *Staphylococcusaureus* and *Candida tropicalis*.

Antifungal activity is the result of bioactive metabolites produced by en-dophytes, and there are many such metabolites that are yet to be discovered and identified. Two new compounds, Asperentin-8-methyl ether and 5′-hy-droxyasperentin, have been isolated along with other known compounds Cladosporin, Isocladosporin, and 6,5′-diacetyl cladosporin (Fig. 4) from the endophytic fungus *C. cladosporiodes* (Fresen) varieties. It has been revealed that Cladosporin compound exhibited significant antifungal activity against *Colleotrichum fragariae*, *Colleotrichum acutatum*, *Colleotrichum gloeospori-oides*, and *Phomopsis viticola*. Isocladosporin compound was effective against *Colleotrichum fragariae*, *Colleotrichum acutatum*, and *Phomopsis viticola* (Wang et al., 2012). The differences in antifungal properties of compound Cladosporin produced by *C. cladosporiodes* depends upon and is influenced

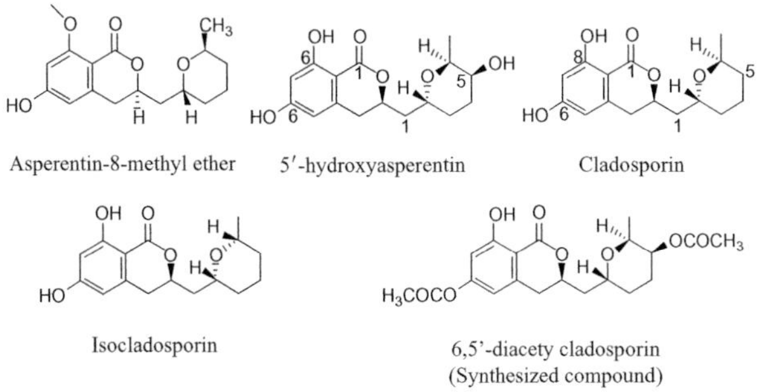

FIG. 4 Isolated compounds of *C. cladosporioides* from the endophytic fungi *C. cladosporiodes* (Wang et al., 2012).

by the introduction of functional group at C-6′, the stereochemical configuration of C-6, the openness of C-5′, and the hydroxyl group at C-8 (Fig. 4) (Wang et al., 2012). Song, Pongnak, and Soytong (2016) highlighted that the metabolites from *Nigrospors sphaerica* and *Fusarium falciforme* isolated from the host plant *Chrysalidocarpus lotescens* and *Mascarena lagencuulis* act as new antagonists against pathogenic fungi *Colletotrichum caffeanum*, which causes coffee anthracnose in the coffee plant. A few years back, new *N*-methoxypyridone analog (11*S*-hydroxy-1-methoxyfusaricide (1)) ($C_{18}H_{27}NO_4$) along with four known compounds epicoccarine B (2), (+)-epipyridobe (3), D8646-2-6 (4), and isoD8646-2-6 (5) isolated from the co-culture of Hawaiian endophytic fungi *Camporesia sambuci* and *Epicoccum sorghinum* and compounds 4 and 5 exhibited inhibitory action against pathogenic fungi (Li et al., 2017; Li, Sarotti, Yang, Turkson, & Cao, 2017).

4.1 Mechanism of antifungal activity

There are different layers in a fungal cell wall. The external part of the fungus cell, called mannoproteins, are supported by a β-glucan matrix consisting of β-(1,3)-glucan and β-(1,6)-glucan and chitin mixed within the β-glucans. Fungal cell membrane consists of bilayer of phospholipids in which "ergosterol" is the chief content, whereas a mammalian cell contains cholesterol and a plant cell contain phytosterols. The difference in sterol content of fungal cell leads to ergosterol, the foremost target of antifungal drugs and a vital target to search and study for antifungal agents (Walker & White, 2017; Tian et al., 2012). As per the structure of fungal cell (Fig. 5) there are six possible mechanisms of metabolites to show antifungal activity (Mc Clanahan, 2009; Walker & White, 2017).

1. First, formation of a fungal cell wall is inhibited as it is made up ofa a matrix of b-glucans and chitin, so if synthesis of these compounds is inhibited, the integrity of fungal cell wall will get disrupted (Hector, 1993; Mc Clanahan, 2009; Walker & White, 2017).

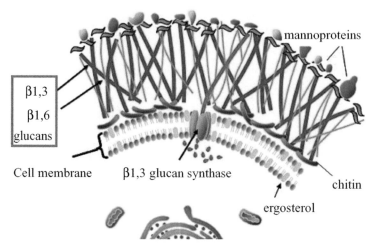

FIG. 5 Fungal cell wall composition. *(Adapted from Lagrouh, F., Dakka, N., & Bakri, Y. (2017). The antifungal activity of Moroccan plants and the mechanism of action of secondary metabolites from plants. Journal de Mycologie Medicale, 27 (3), 303–311.)*

2. Disruption of cell membrane due to inhibitors of ergosterol biosynthesis, as ergosterols are essential for the cell membrane, and these sterols are bound by antifungal drugs thus disrupting the cell membrane and making it leak (Ghannoum & Rice, 1999; Walker & White, 2017).

3. Mitochondria of the fungal cell becomes dysfunctional due to the inhibition of the electrotransport chain (ETC) thus reducing the potential of a mitochondrial membrane. There is a reduction in ATP generation due to inhibition of proton pumps in the respiratory chain and ultimately this leads to cell death (Costa, Vieira, Bizzo, Silveira, & Gimenes, 2012; Mc Clanahan, 2009).

4. Fungal cell division is inhibited via microtubule polymerization inhibition; it leads to inhibition of formation of the mitotic spindle (Mc Clanahan, 2009; Walker & White, 2017).

5. DNA/RNA or protein synthesis is inhibited due to interference of antifungal agents via active transport of ATPases and with RNA can cause defective synthesis of RNA and inhibition of DNA transcription. It also targets protein synthesis inhibition (Mc Clanahan, 2009).

6. Efflux pump inhibition: these pumps are present in all living organisms and helps them to clean out toxic substances from the cell. Drug resistance increases due to overexpression of efflux pumps, but inhibiting them can reduce drug resistance (Kang, Fong, & Tsang, 2010).

Endophytic fungi are diverse in nature, thereby different secondary metabolites, for example, Kojic acid, octadecanoic acid *n*-hexadecanoic acid, diethyl pthlate, and 3-phenyl propionic acid, are isolated with different techniques like Gas-Chromatography-Mass-spectrophotometry (GC-MS) from *A. flavus* associated with *Lannea coromandelica* depicting antimycotic activity against *Candida albicans* and *Malassezia pachydermis* by targeting the active site of the target protein and interacting amino acid, thus becoming an inhibitor similar to Flucanozole (Premjanu et al., 2016). Some secondary metabolites linked with both plants and endophytes are "nitrogen-containing substances" (alkaloids and glucosinolates), "phenolics" (phenolic acids, coumarins, lignans, stilbenes, flavonoids, tannins, and lignins), and "terpenes" (mono, di, tri, sesqui, and tetra-terpenes, saponins, steroids, cardiac glycosides, and sterols) (see Table 3). Some of the active phytochemical compounds belong to the different biosynthetic groups and are shown in the Table 4.

TABLE 3 Site of action for secondary metabolites.

Secondary metabolite	Site of action	Reference
Nitrogen-containing substances	Cell wall, cell membrane	Bagiu, Vlaicu, and Butnariu (2012) and Freiesleben and Jäger (2014)
Phenolics	Cell wall, cell membrane, mitochondria	Tian et al. (2012), Wu, Cheng, Sun, and Lou (2008), and Freiesleben and Jäger (2014)
Terpenes	Cell membrane, mitochondria	Walker and White (2017) and Tian et al. (2012)

TABLE 4 Examples of phytochemical compounds.

Biosynthetic groups	Example of active compound	Molecular formula
Terpenes	• Carvone • Limonene • Polygodial • Carvacrol	• $C_{10}H_{14}O$ • $C_{10}H_{16}$ • $C_{15}H_{22}O_2$ • $C_{10}H_{14}O$
Phenolics	• Apiol • Baicalein • Papyriflavonol • Eugenol • Reservatrol • Plagiochin E	• $C_{12}H_{14}O_4$ • $C_{15}H_{10}O_5$ • $C_{25}H_{26}O_7$ • $C_{10}H_{12}O_2$ • $C_{14}H_{12}O_3$ • $C_{28}H_{24}O_4$

Source: https://pubchem.ncbi.nlm.nih.gov/#.

5. Conclusion

Bioprospecting is a new way of getting natural compounds in a sustainable and innovative manner. There are various biological sources for natural compounds, but endophytic fungi from medicinal plants are considered most favorable because of their diversity in species and power in producing secondary metabolites. Countless drugs are being provided by medicinal plants and endophytes in the form of secondary metabolites, which are selected for important therapeutic alternatives for numerous diseases like antimicrobial, antifungal, antimalarial, antioxidant, anticancer, insecticidal, and pesticidal.

References

Abd_Allah, E. F., Hashem, A., Alqarawi, A. A., Bahkali, A. H., & Alwhibi, M. S. (2015). Enhancing growth performance and systemic acquired resistance of medicinal plant *Sesbania sesban* (L.) Merr using arbuscular mycorrhizal fungi under salt stress. *Saudi Journal of Biological Sciences*, *22*(3), 274–283.

Aksoy, D. Y., & Unal, S. (2008). New antimicrobial agents for the treatment of Gram-positive bacterial infections. *Clinical Microbiology and Infection*, *14*(5), 411–420.

Alvin, A., Miller, K. I., & Neilan, B. A. (2014). Exploring the potential of endophytes from medicinal plants as sources of antimycobacterial compounds. *Microbiological Research*, *169*(7–8), 483–495.

Aly, A. H., Edrada-Ebel, R., Wray, V., Müller, W. E., Kozytska, S., Hentschel, U., et al. (2008). Bioactive metabolites from the endophytic fungus *Ampelomyces* sp. isolated from the medicinal plant *Urospermum picroides*. *Phytochemistry*, *69*(8), 1716–1725.

Avinash, K. S., Ashwini, H. S., Babu, H. N., & Krishnamurthy, Y. L. (2015). Antimicrobial potential of crude extract of *Curvularia lunata*, an endophytic fungi isolated from *Cymbopogon caesius*. *Journal of Mycology*, *2015*, 1–4.

Azevedo, J. L., Maccheroni, W., Jr., Pereira, J. O., & de Araújo, W. L. (2000). Endophytic microorganisms: A review on insect control and recent advances on tropical plants. *Electronic Journal of Biotechnology*, *3*(1), 15–16.

Bacon, C. W., & White, J. F., Jr. (2000). Physiological adaptations in the evolution of endophytism in the Clavicipitaceae. In *Microbial endophytes* (pp. 251–276). CRC Press.

Bagiu, R. V., Vlaicu, B., & Butnariu, M. (2012). Chemical composition and in vitro antifungal activity screening of the *Allium ursinum* L. (Liliaceae). *International Journal of Molecular Sciences*, *13*(2), 1426–1436.

Bandara, W. M. M. S., Seneviratne, G., & Kulasooriya, S. A. (2006). Interactions among endophytic bacteria and fungi: Effects and potentials. *Journal of Biosciences*, *31*(5), 645–650.

Berdy, J. (1995). Are actinomycetes exhausted as a source of secondary metabolites?. Biotechnologija.

Berdy, J. (2005). Bioactive microbial metabolites. *Journal of Antibiotics*, *58*(1), 1–26.

Bernardi-Wenzel, J., García, A., Celso Filho, J. R., Prioli, A. J., & Pamphile, J. A. (2010). Evaluation of foliar fungal endophyte diversity and colonization of medicinal plant *Luehea divaricata* (Martius et Zuccarini). *Biological Research*, *43*(4), 375–384.

Bertinetti, B. V., Peña, N. I., & Cabrera, G. M. (2009). An antifungal tetrapeptide from the culture of *Penicillium canescens*. *Chemistry and Biodiversity*, *6*(8), 1178–1184.

Bi, S. F., Li, F., Song, Y. C., Tan, R. X., & Ge, H. M. (2011). New acrylamide and oxazolidin derivatives from a termite-associated *Streptomyces* sp. *Natural Product Communications*, *6*(3), 353–355. 1934578X1100600310.

Blackwell, M. (2011). The fungi: 1, 2, 3… 5.1 million species? *American Journal of Botany, 98*(3), 426–438.

Blair, J. M., Webber, M. A., Baylay, A. J., Ogbolu, D. O., & Piddock, L. J. (2015). Molecular mechanisms of antibiotic resistance. *Nature Reviews Microbiology, 13*(1), 42–51.

Borges, W. D. S., & Pupo, M. T. (2006). Novel anthraquinone derivatives produced by *Phoma sorghina*, an endophyte found in association with the medicinal plant *Tithonia diversifolia* (Asteraceae). *Journal of the Brazilian Chemical Society, 17*(5), 929–934.

Bungihan, M. E., Tan, M. A., Kitajima, M., Kogure, N., Franzblau, S. G., dela Cruz, T. E. E., et al. (2011). Bioactive metabolites of *Diaporthe* sp. P133, an endophytic fungus isolated from *Pandanus amaryllifolius*. *Journal of Natural Medicines, 65*(3–4), 606–609.

Cai, X., Shan, T., Li, P., Huang, Y., Xu, L., Zhou, L., et al. (2009). Spirobisnaphthalenes from the endophytic fungus Dzf12 of *Dioscorea zingiberensis* and their antimicrobial activities. *Natural Product Communications, 4*(11), 1934578X0900401105.

Carris, L. M., Little, C. R., & Stiles, C. M. (2012). Introduction to fungi. *The Plant Health Instructor*.

Carroll, G. C., & Carroll, F. E. (1978). Studies on the incidence of coniferous needle endophytes in the Pacific Northwest. *Canadian Journal of Botany, 56*(24), 3034–3043.

Ceruks, M., Romoff, P., Fávero, O. A., & Lago, J. H. G. (2007). Constituintes fenólicos polares de *Schinus terebinthifolius* Raddi (Anacardiaceae). *Quimica Nova, 30*(3), 597–599.

Chen, L. Z., Chen, J. M., Zheng, X. S., Zhang, J. F., & Yu, X. P. (2007). Identification and antifungal activity of the metabolite of endophytic fungi isolated from *Llex cornuta*. *Chinese Journal of Pesticide Science, 9*(2), 143–148.

Chen, J. X., Dai, C. C., Li, X., Tian, L. S., & Xie, H. (2008). Endophytic fungi screening from Atracty lancea and inoculating into the host plantlet. *Guihaia, 28*(2), 256–260.

Chen, X., Sang, X., Li, S., Zhang, S., & Bai, L. (2010). Studies on a chlorogenic acid-producing endophytic fungi isolated from *Eucommia ulmoides* Oliver. *Journal of Industrial Microbiology and Biotechnology, 37*(5), 447–454.

Chen, M., Yang, L., Li, Q., Shen, Y., Shao, A., Lin, S., et al. (2011). Volatile metabolites analysis and molecular identification of endophytic fungi bn12 from *Cinnamomum camphora* chvar. borneol. *Zhongguo Zhong yao za zhi = Zhongguo zhongyao zazhi = China Journal of Chinese Materia Medica, 36*(23), 3217–3221.

Chen, L., Zhang, Q. Y., Jia, M., Ming, Q. L., Yue, W., Rahman, K., et al. (2016). Endophytic fungi with antitumor activities: Their occurrence and anticancer compounds. *Critical Reviews in Microbiology, 42*(3), 454–473.

Chithra, S., Jasim, B., Sachidanandan, P., Jyothis, M., & Radhakrishnan, E. K. (2014). Piperine production by endophytic fungus Colletotrichum gloeosporioides isolated from *Piper nigrum*. *Phytomedicine, 21*(4), 534–540.

Chomcheon, P., Wiyakrutta, S., Aree, T., Sriubolmas, N., Ngamrojanavanich, N., Mahidol, C., et al. (2010). Curvularides A–E: Antifungal hybrid peptide–polyketides from the endophytic fungus *Curvularia geniculata*. *Chemistry—A European Journal, 16*(36), 11178–11185.

Conover, M. A., Mierzwa, R., King, A., Loebenberg, D., Bishop, W. R., & Puar, M. (1992). Usnic acid amide, a phytotoxin and antifungal agent from *Cercosporidium henningsii*. *Phytochemistry, 31*(9), 2999–3001.

Costa, T. D. S. A., Vieira, R. F., Bizzo, H. R., Silveira, D., & Gimenes, M. A. (2012). Secondary metabolites. In S. Dhanarasu (Ed.), *Chromatography and its applications*. Doctoral dissertation.

Cota, B. B., Rosa, L. H., Caligiorne, R. B., Rabello, A. L. T., Almeida Alves, T. M., Rosa, C. A., et al. (2008). Altenusin, a biphenyl isolated from the endophytic fungus *Alternaria* sp., inhibits trypanothione reductase from *Trypanosoma cruzi*. *FEMS Microbiology Letters, 285*(2), 177–182.

Cragg, G. M., & Newman, D. J. (2005). Plants as a source of anti-cancer agents. *Journal of Ethnopharmacology, 100*(1–2), 72–79.

Cueto, M., Jensen, P. R., Kauffman, C., Fenical, W., Lobkovsky, E., & Clardy, J. (2001). Pestalone, a new antibiotic produced by a marine fungus in response to bacterial challenge. *Journal of Natural Products, 64*(11), 1444–1446.

Cui, Y., Yi, D., Bai, X., Sun, B., Zhao, Y., & Zhang, Y. (2012). Ginkgolide B produced endophytic fungus (*Fusarium oxysporum*) isolated from Ginkgo biloba. *Fitoterapia, 83*(5), 913–920.

Dai, J., Krohn, K., Draeger, S., & Schulz, B. (2009). New naphthalene-chroman coupling products from the endophytic fungus, *Nodulisporium* sp. from Erica arborea. *European Journal of Organic Chemistry, 2009*(10), 1564–1569.

Dai, J., Krohn, K., Flörke, U., Draeger, S., Schulz, B., Kiss-Szikszai, A., et al. (2006). Metabolites from the endophytic fungus *Nodulisporium* sp. from *Juniperus cedre*. *European Journal of Organic Chemistry, 2006*(15), 3498–3506.

de Bary, A. (1866). *Morphologie und physiologie der pilze, flechten und myxomyceten.* Engelmann.

de Siqueira, V. M., Conti, R., de Araújo, J. M., & Souza-Motta, C. M. (2011). Endophytic fungi from the medicinal plant *Lippia sidoides* Cham. and their antimicrobial activity. *Symbiosis, 53*(2), 89–95.

de Souza, A. Q. L., de Souza, A. D. L., Astolffi Filho, S., Belém Pinheiro, M. L., de Moura Sarquis, M. I., & Pereira, J. O. (2004). Antimicrobial activity of endophytic fungi isolated from amazonian toxic plants: *Palicourea longiflora* (aubl.) rich and *Strychnos cogens* bentham. *Acta Amazonica, 34*(2), 185–196.

Debbab, A., Aly, A. H., Edrada-Ebel, R., Wray, V., Müller, W. E., Totzke, F., et al. (2009). Bioactive metabolites from the endophytic fungus *Stemphylium globuliferum* isolated from *Mentha pulegium*. *Journal of Natural Products, 72*(4), 626–631.

Degáspari, C. H., Waszczynskyj, N., & Prado, M. R. M. (2005). Atividade antimicrobiana de Schinus terebenthifolius Raddi. *Ciência e Agrotecnologia, 29*(3), 617–622.

Du, X., Lu, C., Li, Y., Zheng, Z., Su, W., & Shen, Y. (2008). Three new antimicrobial metabolites of *Phomopsis* sp. *Journal of Antibiotics, 61*(4), 250–253.

Faeth, S. H., & Fagan, W. F. (2002). Fungal endophytes: Common host plant symbionts but uncommon mutualists. *Integrative and Comparative Biology, 42*(2), 360–368.

Fernandes, M. D. R. V., Pfenning, L. H., Costa-Neto, C. M. D., Heinrich, T. A., Alencar, S. M. D., Lima, M. A. D., et al. (2009). Biological activities of the fermentation extract of the endophytic fungus *Alternaria alternata* isolated from *Coffea arabica* L. *Brazilian Journal of Pharmaceutical Sciences, 45*(4), 677–685.

Ferreira, M. C., Cantrell, C. L., Wedge, D. E., Gonçalves, V. N., Jacob, M. R., Khan, S., et al. (2017). Antimycobacterial and antimalarial activities of endophytic fungi associated with the ancient and narrowly endemic neotropical plant *Vellozia gigantea* from Brazil. *Memórias do Instituto Oswaldo Cruz, 112*(10), 692–697.

Ferri, M., Ranucci, E., Romagnoli, P., & Giaccone, V. (2017). Antimicrobial resistance: A global emerging threat to public health systems. *Critical Reviews in Food Science and Nutrition, 57*(13), 2857–2876.

Firakova, S., Sturdikova, M., & Muckova, M. (2007). Bioactive secondary metabolites produced by microorganisms associated with plants. *Biologia, 62*(3), 251–257.

Freiesleben, S. H., & Jäger, A. K. (2014). Correlation between plant secondary metabolites and their antifungal mechanisms—A review. *Medicinal and Aromatic Plants, 3*(2), 1–6.

Gange, A. C., Eschen, R., Wearn, J. A., Thawer, A., & Sutton, B. C. (2012). Differential effects of foliar endophytic fungi on insect herbivores attacking a herbaceous plant. *Oecologia, 168*(4), 1023–1031.

Gao, S. S., Li, X. M., Li, C. S., Proksch, P., & Wang, B. G. (2011). Penicisteroids A and B, antifungal and cytotoxic polyoxygenated steroids from the marine alga-derived endophytic fungus *Penicillium chrysogenum* QEN-24S. *Bioorganic and Medicinal Chemistry Letters, 21*(10), 2894–2897.

Gao, S. S., Li, X. M., Zhang, Y., Li, C. S., & Wang, B. G. (2011). Conidiogenones H and I, two new diterpenes of cyclopiane class from a marine-derived endophytic fungus *Penicillium chrysogenum* QEN-24S. *Chemistry and Biodiversity, 8*(9), 1748–1753.

Garcia, A., Rhoden, S. A., Bernardi-Wenzel, J., Orlandelli, R. C., Azevedo, J. L., & Pamphile, J. A. (2012). Antimicrobial activity of crude extracts of endophytic fungi isolated from medicinal plant *Sapindus saponaria* L. *Journal of Applied Pharmaceutical Science, 2*(10), 35.

Garcia-Ruiz, J. C., Amutio, E., & Pontón, J. (2004). Invasive fungal infection in immunocompromised patients. *Revista Iberoamericana de Micología, 21*, 55–62.

Gautam, A. K. (2014). Diversity of fungal endophytes in some medicinal plants of Himachal Pradesh, India. *Archives of Phytopathology and Plant Protection, 47*(5), 537–544.

Ghannoum, M. A., & Rice, L. B. (1999). Antifungal agents: Mode of action, mechanisms of resistance, and correlation of these mechanisms with bacterial resistance. *Clinical Microbiology Reviews, 12*(4), 501–517.

Glienke, C., Tonial, F., Gomes-Figueiredo, J., Savi, D., Vicente, V. A., Maia, B. H. S., et al. (2012). Antimicrobial activity of endophytes from Brazilian medicinal plants. In B. Varaprasad (Ed.), *Antibacterial agents* (1st ed., pp. 239–254). Rijeka: InTech.

Gomes-Figueiredo, J., Pimentel, I. C., Vicente, V. A., Pie, M. R., Kava-Cordeiro, V., Galli-Terasawa, L., et al. (2007). Bioprospecting highly diverse endophytic *Pestalotiopsis* spp. with antibacterial properties from *Maytenus ilicifolia*, a medicinal plant from Brazil. *Canadian Journal of Microbiology, 53*(10), 1123–1132.

Gosio, B. (1896). Richerche batteriologiche e chemiche sulle alterazoni del mais. *Rivista d'Igiene e Sanità pubblica, 7*, 825–849.

Gouda, S., Das, G., Sen, S. K., Shin, H. S., & Patra, J. K. (2016). Endophytes: A treasure house of bioactive compounds of medicinal importance. *Frontiers in Microbiology, 7*, 1538.

Guo, B. H., Wang, Y. C., Zhou, X. W., Hu, K., Tan, F., Miao, Z. Q., et al. (2006). An endophytic taxol-producing fungus BT2 isolated from *Taxus chinensis* var. mairei. *African Journal of Biotechnology, 5*(10), 875–877.

Hatakeyama, T., Koseki, T., Murayama, T., & Shiono, Y. (2010). Eremophilane sesquiterpenes from the endophyte *Microdiplodia* sp. KS 75-1 and revision of the stereochemistries of phomadecalins C and D. *Phytochemistry Letters, 3*(3), 148–151.

Hawksworth, D. L. (2001). The magnitude of fungal diversity: The 1.5 million species estimate revisited. *Mycological Research, 105*, 1422–1432.

Hayes, G. E., & Denning, D. W. (2013). Frequency, diagnosis and management of fungal respiratory infections. *Current Opinion in Pulmonary Medicine, 19*(3), 259–265.

Hector, R. F. (1993). Compounds active against cell walls of medically important fungi. *Clinical Microbiology Reviews, 6*(1), 1–21.

Hoffman, A. M., Mayer, S. G., Strobel, G. A., Hess, W. M., Sovocool, G. W., Grange, A. H., et al. (2008). Purification, identification and activity of phomodione, a furandione from an endophytic Phoma species. *Phytochemistry, 69*(4), 1049–1056.

Hormazabal, E., & Piontelli, E. (2009). Endophytic fungi from Chilean native gymnosperms: Antimicrobial activity against human and phytopathogenic fungi. *World Journal of Microbiology and Biotechnology, 25*(5), 813–819.

Huang, W. Y., Cai, Y. Z., Xing, J., Corke, H., & Sun, M. (2007). A potential antioxidant resource: Endophytic fungi from medicinal plants. *Economic Botany, 61*(1), 14.

Hussain, H., Akhtar, N., Draeger, S., Schulz, B., Pescitelli, G., Salvadori, P., et al. (2009). New bioactive 2, 3-epoxycyclohexenes and isocoumarins from the endophytic fungus *Phomopsis* sp. from *Laurus azorica*. *European Journal of Organic Chemistry, 2009*(5), 749–756.

Ibrahim, M., Kaushik, N., Sowemimo, A., Chhipa, H., Koekemoer, T., Van de Venter, M., et al. (2017). Antifungal and antiproliferative activities of endophytic fungi isolated from the leaves of *Markhamia tomentosa*. *Pharmaceutical Biology, 55*(1), 590–595.

Idris, A. M., Al-Tahir, I., & Idris, E. (2013). Antibacterial activity of endophytic fungi extracts from the medicinal plant *Kigelia africana*. *Egyptian Academic Journal of Biological Sciences, G. Microbiology, 5*(1), 1–9.

Johann, S., Rosa, L. H., Rosa, C. A., Perez, P., Cisalpino, P. S., Zani, C. L., et al. (2012). Antifungal activity of altenusin isolated from the endophytic fungus *Alternaria* sp. against the pathogenic fungus *Paracoccidioides brasiliensis*. *Revista Iberoamericana de Micología, 29*(4), 205–209.

Kalyanasundaram, I., Nagamuthu, J., & Muthukumaraswamy, S. (2015). Antimicrobial activity of endophytic fungi isolated and identified from salt marsh plant in Vellar Estuary. *Journal of Microbiology and Antimicrobials, 7*(2), 13–20.

Kang, K., Fong, W. P., & Tsang, P. W. K. (2010). Novel antifungal activity of purpurin against *Candida* species in vitro. *Medical Mycology, 48*(7), 904–911.

Kaul, S., Gupta, S., Ahmed, M., & Dhar, M. K. (2012). Endophytic fungi from medicinal plants: A treasure hunt for bioactive metabolites. *Phytochemistry Reviews, 11*(4), 487–505.

Khan, A. L., Waqas, M., Hamayun, M., Al-Harrasi, A., Al-Rawahi, A., & Lee, I. J. (2013). Cosynergism of endophyte *Penicillium resedanum* LK6 with salicylic acid helped *Capsicum annuum* in biomass recovery and osmotic stress mitigation. *BMC Microbiology, 13*(1), 1–13.

Khan, M. R. (2007). *Isolation, identification and cultivation of endophytic fungi from medicinal plants for the production and characterization of bioactive Fungal Metabolites*. Doctoral dissertation University of Karachi.

Kharwar, R. N., Mishra, A., Gond, S. K., Stierle, A., & Stierle, D. (2011). Anticancer compounds derived from fungal endophytes: Their importance and future challenges. *Natural Product Reports, 28*(7), 1208–1228.

Kharwar, R. N., Verma, V. C., Kumar, A., Gond, S. K., Harper, J. K., Hess, W. M., et al. (2009). Javanicin, an antibacterial naphthaquinone from an endophytic fungus of neem, *Chloridium* sp. *Current Microbiology, 58*(3), 233–238.

Khiralla, A. (2015). *Phytochemical study, cytotoxic and antibacterial potentialities of endophytic fungi from medicinal plants from Sudan*. Doctoral dissertation.

Khiralla, A., Spina, R., Yagi, S., Mohamed, L., & Laurain-Mattar, D. (2016). Endophytic fungi: Occurrence, classification, function and natural products. In *Endophytic fungi: Diversity, characterization and biocontrol* (pp. 1–19). Nova Science Publishers.

Kirk, P. M., Cannon, P. F., Minter, D. W., & Stalpers, J. A. (2008). *Ainsworth and Bisby's dictionary of the fungi*. Wallingford, United kingdom: CAB International.

Kjer, J., Wray, V., Edrada-Ebel, R., Ebel, R., Pretsch, A., Lin, W., et al. (2009). Xanaleric acids I and II and related phenolic compounds from an endophytic *Alternaria* sp. isolated from the mangrove plant *Sonneratia alba*. *Journal of Natural Products, 72*(11), 2053–2057.

Krohn, K., Michel, A., Römer, E., Flörke, U., Aust, H. J., Draeger, S., et al. (1995). Biologically active metabolites from fungi 61; phomosines AC three new biaryl ethers from *Phomopsis* sp. *Natural Product Letters, 6*(4), 309–314.

Kumar, S., Aharwal, R. P., Shukla, H., Rajak, R. C., & Sandhu, S. S. (2014). Endophytic fungi: As a source of antimicrobials bioactive compounds. *World Journal of Pharmaceutical Sciences, 3*(2), 1179–1197.

Kumar, D. S. S., Lau, C. S., Wan, J. M., Yang, D., & Hyde, K. D. (2005). Immunomodulatory compounds from *Pestalotiopsis leucothes*, an endophytic fungus from *Tripterygium wilfordii*. *Life Sciences*, *78*(2), 147–156.

Kusari, S., Lamshöft, M., Zühlke, S., & Spiteller, M. (2008). An endophytic fungus from *Hypericum perforatum* that produces hypericin. *Journal of Natural Products*, *71*(2), 159–162.

Kusari, S., Pandey, S. P., & Spiteller, M. (2013). Untapped mutualistic paradigms linking host plant and endophytic fungal production of similar bioactive secondary metabolites. *Phytochemistry*, *91*, 81–87.

Kusari, S., Zühlke, S., Kosuth, J., Cellarova, E., & Spiteller, M. (2009). Light-independent metabolomics of endophytic *Thielavia subthermophila* provides insight into microbial hypericin biosynthesis. *Journal of Natural Products*, *72*(10), 1825–1835.

Li, N., Chen, F., Cui, F., Sun, W., Zhang, J., Qian, L., et al. (2017). Improved postharvest quality and respiratory activity of straw mushroom (*Volvariella volvacea*) with ultrasound treatment and controlled relative humidity. *Scientia Horticulturae*, *225*, 56–64.

Li, C., Sarotti, A. M., Yang, B., Turkson, J., & Cao, S. (2017). A new N-methoxypyridone from the co-cultivation of Hawaiian endophytic fungi *Camporesia sambuci* FT1061 and *Epicoccum sorghinum* FT1062. *Molecules*, *22*(7), 1166.

Li, J. Y., Strobel, G., Harper, J., Lobkovsky, E., & Clardy, J. (2000). Cryptocin, a potent tetramic acid antimycotic from the endophytic fungus *Cryptosporiopsis* cf. *quercina*. *Organic Letters*, *2*(6), 767–770.

Liang, H., Xing, Y., Chen, J., Zhang, D., Guo, S., & Wang, C. (2012). Antimicrobial activities of endophytic fungi isolated from *Ophiopogon japonicus* (Liliaceae). *BMC Complementary and Alternative Medicine*, *12*(1), 238.

Liu, C., Liu, T., Yuan, F., & Gu, Y. (2010). Isolating endophytic fungi from evergreen plants and determining their antifungal activities. *African Journal of Microbiology Research*, *4*(21), 2243–2248.

Liu, H. Y., Qiu, N. X., Ding, H. H., & Yao, R. Q. (2008). Polyphenols content and antioxidant capacity of 68 Chinese herbals suitable for medical or food uses. *Food Research International*, *41*, 363–370.

Liu, J. Y., Song, Y. C., Zhang, Z., Wang, L., Guo, Z. J., Zou, W. X., et al. (2004). *Aspergillus fumigatus* CY018, an endophytic fungus in Cynodon dactylon as a versatile producer of new and bioactive metabolites. *Journal of Biotechnology*, *114*(3), 279–287.

Liu, X., Song, W., Zhang, K., & Ye, Y. (2011). Effects of two kinds of endophytic fungi infection on water stress of seedlings of *Chrysanthemum morifolium*. *Acta Horticulturae Sinica*, *38*(2), 335–342.

Liu, X., Xie, X., & Duan, J. (2007). *Colletotrichum yunnanense* sp. nov., a new endophytic species from *Buxus* sp. *Mycotaxon*, *100*, 137–144.

Liu, C. H., Zou, W. X., Lu, H., & Tan, R. X. (2001). Antifungal activity of *Artemisia annua* endophyte cultures against phytopathogenic fungi. *Journal of Biotechnology*, *88*(3), 277–282.

Lösgen, S., Magull, J., Schulz, B., Draeger, S., & Zeeck, A. (2008). Isofusidienols: Novel chromone-3-oxepines produced by the endophytic fungus *Chalara* sp. *European Journal of Organic Chemistry*, *2008*(4), 698–703.

Maciąg-Dorszynska, M., Węgrzyn, G., & Guzow-Krzemińska, B. (2014). Antibacterial activity of lichen secondary metabolite usnic acid is primarily caused by inhibition of RNA and DNA synthesis. *FEMS Microbiology Letters*, *353*(1), 57–62.

Maehara, S., Simanjuntak, P., Maetani, Y., Kitamura, C., Ohashi, K., & Shibuya, H. (2013). Ability of endophytic filamentous fungi associated with *Cinchona ledgeriana* to produce Cinchona alkaloids. *Journal of Natural Medicines*, *67*(2), 421–423.

Marcellano, J. P., Collanto, A. S., & Fuentes, R. G. (2017). Antibacterial activity of endophytic fungi isolated from the Bark of *Cinnamomum mercadoi*. *Pharmacognosy Journal*, *9*(3).

Marinho, A. M., Rodrigues-Filho, E., Moitinho, M. D. L. R., & Santos, L. S. (2005). Biologically active polyketides produced by *Penicillium janthinellum* isolated as an endophytic fungus from fruits of *Melia azedarach*. *Journal of the Brazilian Chemical Society*, *16*(2), 280–283.

Maruna, M., Sturdikova, M., Liptaj, T., Godany, A., Muckova, M., Certik, M., et al. (2010). Isolation, structure elucidation and biological activity of angucycline antibiotics from an epiphytic yew streptomycete. *Journal of Basic Microbiology*, *50*(2), 135–142.

Masumi, S., Mirzaei, S., Zafari, D., & Kalvandi, R. (2015). Isolation, identification and biodiversity of endophytic fungi o Thymus. *Progress in Biological Sciences*, *5*(1), 43–50.

Mc Clanahan, C. (2009). Antifungals. *BioFiles (Sigma-Aldrich)*, *4*(10).

Melo, F. M. P. D., Fiore, M. F., Moraes, L. A. B. D., Silva-Stenico, M. E., Scramin, S., Teixeira, M. D. A., et al. (2009). Antifungal compound produced by the cassava endophyte Bacillus pumilus MAIIIM4A. *Scientia Agricola*, *66*(5), 583–592.

Meng, J. J., & He, X. L. (2011). Effects of AM fungi on growth and nutritional contents of *Salvia miltiorrhiza* Bge. under drought stress. *Journal of Agricultural University of Hebei, 1*, 5721–5728.

Mitchell, A. (2008). Muscodor crispans, a novel endophyte from *Ananas ananassoides* in the Bolivian Amazon. *Fungal Diversity, 31*, 37–43.

Momesso, L. D. S., Kawano, C. Y., Ribeiro, P. H., Nomizo, A., Goldman, G. H., & Pupo, M. T. (2008). Chaetoglobosins produced by Chaetomium globosum, endophytic fungus found in association with *Viguiera robusta* Gardn (Asteraceae). *Quimica Nova, 31*(7), 1680–1685.

Mousa, W. K., & Raizada, M. N. (2013). The diversity of anti-microbial secondary metabolites produced by fungal endophytes: An interdisciplinary perspective. *Frontiers in Microbiology, 4*, 65.

Oliveira, M. N., Santos, L. S., Guilhon, G., Santos, A. S., Ferreira, I. C., Lopes-Junior, M. L., et al. (2011). Novel anthraquinone derivatives produced by *Pestalotiopsis guepinii*, an endophytic of the medicinal plant *Virola michelii* (Myristicaceae). *Journal of the Brazilian Chemical Society, 22*(5), 993–996.

Oliveira, C. M., Silva, G. H., Regasini, L. O., Zanardi, L. M., Evangelista, A. H., Young, M. C., et al. (2009). Bioactive metabolites produced by *Penicillium* sp. 1 and sp. 2, two endophytes associated with *Alibertia macrophylla* (Rubiaceae). *Zeitschrift Für Naturforschung C, 64*(11–12), 824–830.

Pan, F., Su, X., Hu, B., Yang, N., Chen, Q., & Wu, W. (2015). Fusarium redolens 6WBY3, an endophytic fungus isolated from *Fritillaria unibracteata* var. *wabuensis*, produces peimisine and imperialine-3β-D-glucoside. *Fitoterapia, 103*, 213–221.

Parfrey, L. W., Lahr, D. J., Knoll, A. H., & Katz, L. A. (2011). Estimating the timing of early eukaryotic diversification with multigene molecular clocks. *Proceedings of the National Academy of Sciences, 108*(33), 13624–13629.

Park, J. H., Choi, G. J., Lee, H. B., Kim, K. M., Jung, H. S., Lee, S. W., et al. (2005). Griseofulvin from *Xylaria* sp. strain F0010, an endophytic fungus of *Abies holophylla* and its antifungal activity against plant pathogenic fungi. *Journal of Microbiology and Biotechnology, 15*(1), 112–117.

Pastre, R., Marinho, A. M., Rodrigues-Filho, E., Souza, A. Q., & Pereira, J. O. (2007). Diversidade de policetídeos produzidos por espécies de *Penicillium* isoladas de *Melia azedarach* e *Murraya paniculata*. *Quimica Nova, 30*(8), 1867–1871.

Pavithra, N., Sathish, L., & Ananda, K. (2012). Antimicrobial and enzyme activity of endophytic fungi isolated from Tulsi. *Journal of Pharmaceutical and Biomedical Sciences (JPBMS), 16*(16), 2014.

Peterson, S. W., Vega, F. E., Posada, F., & Nagai, C. (2005). Penicillium coffeae, a new endophytic species isolated from a coffee plant and its phylogenetic relationship to *P. fellutanum*, *P. thiersii* and *P. brocae* based on parsimony analysis of multilocus DNA sequences. *Mycologia*, *97*(3), 659–666.

Petrini, O. (1991). Fungal endophytes of tree leaves. In *Microbial ecology of leaves* (pp. 179–197). New York, NY: Springer.

Phongpaichit, S., Rungjindamai, N., Rukachaisirikul, V., & Sakayaroj, J. (2006). Antimicrobial activity in cultures of endophytic fungi isolated from *Garcinia* species. *FEMS Immunology and Medical Microbiology*, *48*(3), 367–372.

Premjanu, N., Jaynthy, C., & Diviya, S. (2016). Antifungal activity of endophytic fungi isolated from lannea coromandelica—An insilico approach. *International Journal of Pharmacy and Pharmaceutical Sciences*, *8*(5), 207–210.

Queires, L. C. S., & Rodrigues, L. E. A. (1998). Quantificação das substâncias fenólicas totais em órgãos da aroeira Schinus Terebenthifolius (RADDI). *Brazilian Archives of Biology and Technology*, *41*(2), 247–253.

Radu, S., & Kqueen, C. Y. (2002). Preliminary screening of endophytic fungi from medicinal plants in Malaysia for antimicrobial and antitumor activity. *The Malaysian Journal of Medical Sciences*, *9*(2), 23.

Rajeswari, S., Umamaheswari, S., Arvind Prasanth, D., & Rajamanikandan, K. C. P. (2016). Bioactive potential of endophytic fungi *Aspergillus flavus* (SS03) against clinical isolates. *International Journal of Pharmacy and Pharmaceutical Sciences*, *8*(9), 37–40.

Raviraja, N. S. (2005). Fungal endophytes in five medicinal plant species from Kudremukh Range, Western Ghats of India. *Journal of Basic Microbiology: An International Journal on Biochemistry, Physiology, Genetics, Morphology, and Ecology of Microorganisms*, *45*(3), 230–235.

Raviraja, N. S., Maria, G. L., & Sridhar, K. R. (2006). Antimicrobial evaluation of endophytic fungi inhabiting medicinal plants of the Western Ghats of India. *Engineering in Life Sciences*, *6*(5), 515–520.

Redecker, D., Kodner, R., & Graham, L. E. (2000). Glomalean fungi from the Ordovician. *Science*, *289*(5486), 1920–1921.

Ren, C. G., & Dai, C. C. (2012). Jasmonic acid is involved in the signaling pathway for fungal endophyte-induced volatile oil accumulation of *Atractylodes lancea* plantlets. *BMC Plant Biology*, *12*(1), 128.

Rhoden, S. A., Garcia, A., Rubin Filho, C. J., Azevedo, J. L., & Pamphile, J. A. (2012). Phylogenetic diversity of endophytic leaf fungus isolates from the medicinal tree *Trichilia elegans* (Meliaceae). *Genetics and Molecular Research*, *11*(3), 2513–2522.

Rocha, R., Luz, D. E. D., Engels, C., Pileggi, S. A. V., Jaccoud Filho, D. D. S., Matiello, R. R., et al. (2009). Selection of endophytic fungi from comfrey (*Symphytum officinale* L.) for in vitro biological control of the phytopathogen *Sclerotinia sclerotiorum* (Lib.). *Brazilian Journal of Microbiology*, *40*(1), 73–78.

Rodrigues-Heerklotz, K. F., Drandarov, K., Heerklotz, J., Hesse, M., & Werner, C. (2001). Guignardic acid, a novel type of secondary metabolite produced by the endophytic fungus *Guignardia* sp.: Isolation, structure elucidation, and asymmetric synthesis. *Helvetica Chimica Acta*, *84*(12), 3766–3772.

Rosa, L. H., Gonçalves, V. N., Caligiorne, R. B., Alves, T., Rabello, A., Sales, P. A., et al. (2010). Leishmanicidal, trypanocidal, and cytotoxic activities of endophytic fungi associated with bioactive plants in Brazil. *Brazilian Journal of Microbiology*, *41*(2), 420–430.

Rosa, L. H., Tabanca, N., Techen, N., Pan, Z., Wedge, D. E., & Moraes, R. M. (2012). Antifungal activity of extracts from endophytic fungi associated with *Smallanthus* maintained in vitro as autotrophic cultures and as pot plants in the greenhouse. *Canadian Journal of Microbiology*, *58*(10), 1202–1211.

Saar, D. E., Polans, N. O., Sørensen, P. D., & Duvall, M. R. (2001). Angiosperm DNA contamination by endophytic fungi: Detection and methods of avoidance. *Plant Molecular Biology Reporter*, *19*(3), 249–260.

Sachs, J. (1874). *Lehrbuch der Botanik nach dem gegenwärtigen Stand der Wissenschaft. Vol. 1.* W. Engelmann.

Santos, I. P. D., Silva, L. C. N. D., Silva, M. V. D., Araújo, J. M. D., Cavalcanti, M. D. S., & Lima, V. L. D. M. (2015). Antibacterial activity of endophytic fungi from leaves of *Indigofera suffruticosa* Miller (Fabaceae). *Frontiers in Microbiology*, *6*, 350.

Sassa, T., & Igarashi, M. (1990). Structures of (−)-mycousnine,(+)-isomycousnine and (+)-oxymycousnine, new usnic acid derivatives from phytopathogenic *Mycosphaerella nawae*. *Agricultural and Biological Chemistry*, *54*(9), 2231–2237.

Schulz, B., Boyle, C., Draeger, S., Römmert, A. K., & Krohn, K. (2002). Endophytic fungi: A source of novel biologically active secondary metabolites. *Mycological Research*, *106*(9), 996–1004.

Selim, K. A., El-Beih, A. A., AbdEl-Rahman, T. M., & El-Diwany, A. I. (2011). Biodiversity and antimicrobial activity of endophytes associated with Egyptian medicinal plants. *Mycosphere*, *2*(6), 669–678.

Siddiqui, I. N., Zahoor, A., Hussain, H., Ahmed, I., Ahmad, V. U., Padula, D., et al. (2011). Diversonol and blennolide derivatives from the endophytic fungus *Microdiplodia* sp.: Absolute configuration of diversonol. *Journal of Natural Products*, *74*(3), 365–373.

Silva, G. H., Oliveira, C. M. D., Teles, H. L., Bolzani, V. D. S., Araujo, A. R., Pfenning, L. H., et al. (2010). Citocalasinas produzidas por *Xylaria* sp., um fungo endofítico de *Piper aduncum* (piperaceae). *Quimica Nova*, *33*(10), 2038–2041.

Silva, G. H., Teles, H. L., Trevisan, H. C., Bolzani, V. D. S., Young, M., Pfenning, L. H., et al. (2005). New bioactive metabolites produced by *Phomopsis cassiae*, an endophytic fungus in *Cassia spectabilis*. *Journal of the Brazilian Chemical Society*, *16*(6B), 1463–1466.

Silva, G. H., Teles, H. L., Zanardi, L. M., Young, M. C. M., Eberlin, M. N., Hadad, R., et al. (2006). Cadinane sesquiterpenoids of *Phomopsis cassiae*, an endophytic fungus associated with *Cassia spectabilis* (Leguminosae). *Phytochemistry*, *67*(17), 1964–1969.

Singh, R., & Dubey, A. K. (2015). Endophytic actinomycetes as emerging source for therapeutic compounds. *Indo Global Journal of Pharmaceutical Sciences*, *5*(2), 106–116.

Singh, M. P., Janso, J. E., Luckman, S. W., Brady, S. F., Clardy, J., Greenstein, M., et al. (2000). Biological activity of guanacastepene, a novel diterpenoid antibiotic produced by an unidentified fungus CR115. *Journal of Antibiotics*, *53*(3), 256–261.

Song, J., Pongnak, W., & Soytong, K. (2016). Antifungal activity of endophytic fungi from palm trees against coffee anthracnose caused by *Colletotrichum coffeanum*. *Journal of Agricultural Technology*, *12*(3), 623–635.

Stone, J. K., Bacon, C. W., & White, J. F., Jr. (2000). An overview of endophytic microbes: Endophytism defined. In *Microbial endophytes* (pp. 17–44). CRC Press.

Strobel, G. A. (1998). Endophytic microbes embody pharmaceutical potential. *ASM News*, *5*, 263–268.

Strobel, G. A. (2002). Rainforest endophytes and bioactive products. *Critical Reviews in Biotechnology*, *22*(4), 315–333.

Strobel, G., & Daisy, B. (2003). Bioprospecting for microbial endophytes and their natural products. *Microbiology and Molecular Biology Reviews*, *67*(4), 491–502.

Strobel, G., Daisy, B., Castillo, U., & Harper, J. (2004). Natural products from endophytic microorganisms. *Journal of Natural Products*, *67*(2), 257–268.

Suay, I., Arenal, F., Asensio, F. J., Basilio, A., Cabello, M. A., Díez, M. T., et al. (2000). Screening of basidiomycetes for antimicrobial activities. *Antonie Van Leeuwenhoek*, *78*(2), 129–140.

Sumarah, M. W., Puniani, E., Sørensen, D., Blackwell, B. A., & Miller, J. D. (2010). Secondary metabolites from anti-insect extracts of endophytic fungi isolated from *Picea rubens*. *Phytochemistry*, *71*(7), 760–765.

Sutjaritvorakul, T., Whalley, A. J. S., Sihanonth, P., & Roengsumran, S. (2011). Antimicrobial activity from endophytic fungi isolated from plant leaves in Dipterocarpous forest at Viengsa district Nan province, Thailand. *Journal of Agricultural Technology*, *7*(1), 115–121.

Talontsi, F. M., Dittrich, B., Schüffler, A., Sun, H., & Laatsch, H. (2013). Epicoccolides: Antimicrobial and antifungal polyketides from an endophytic fungus *Epicoccum* sp. associated with *Theobroma cacao*. *European Journal of Organic Chemistry*, *2013*(15), 3174–3180.

Tan, R. X., & Zou, W. X. (2001). Endophytes: A rich source of functional metabolites. *Natural Product Reports*, *18*(4), 448–459.

Taylor, D. L., Hollingsworth, T. N., McFarland, J. W., Lennon, N. J., Nusbaum, C., & Ruess, R. W. (2014). A first comprehensive census of fungi in soil reveals both hyperdiversity and fine-scale niche partitioning. *Ecological Monographs*, *84*(1), 3–20.

Tejesvi, M. V., Kini, K. R., Prakash, H. S., Subbiah, V., & Shetty, H. S. (2007). Genetic diversity and antifungal activity of species of *Pestalotiopsis* isolated as endophytes from medicinal plants. *Fungal Diversity*, *24*(3), 1–18.

Teles, H. L., Silva, G. H., Castro-Gamboa, I., da Silva Bolzani, V., Pereira, J. O., Costa-Neto, C. M., et al. (2005). Benzopyrans from *Curvularia* sp., an endophytic fungus associated with *Ocotea corymbosa* (Lauraceae). *Phytochemistry*, *66*(19), 2363–2367.

Teles, H. L., Sordi, R., Silva, G. H., Castro-Gamboa, I., da Silva Bolzani, V., Pfenning, L. H., et al. (2006). Aromatic compounds produced by *Periconia atropurpurea*, an endophytic fungus associated with *Xylopia aromatica*. *Phytochemistry*, *67*(24), 2686–2690.

Thompson, S. M., Tan, Y. P., Shivas, R. G., Neate, S. M., Morin, L., Bissett, A., et al. (2015). Green and brown bridges between weeds and crops reveal novel *Diaporthe* species in Australia. *Persoonia: Molecular Phylogeny and Evolution of Fungi*, *35*, 39.

Tian, J., Ban, X., Zeng, H., He, J., Chen, Y., & Wang, Y. (2012). The mechanism of antifungal action of essential oil from dill (*Anethum graveolens* L.) on *Aspergillus flavus*. *PLoS ONE*, *7*(1), e30147.

Tilton, F., Aneesh Nair, A., Maany Ramanan, M., & Devi, S. (2014). Bioassay guided characterisation of endomycophytes from *Barringtonia acutangula* L. leaves. *International Journal of Development Research*, *4*(2), 384–387.

Timmermans, K. (2001). Trips, CBD and traditional medicines: Concepts and questions. In *Report of an ASEAN workshop on the TRIPS agreement and traditional medicine, Jakarta, February 2001 (report)*. http://digicollection.org/hss/en/d/Jh2996e/#Jh2996e.

UNDP. (2016). *Bioprospecting, financing solutions for sustainable development*. http://www.undp.org/content/dam/sdfinance/doc/Bioprospecting%20_%20UNDP.pdf.

Walker, G. M., & White, N. A. (2017). Introduction to fungal physiology. In *Fungi: Biology and applications* (pp. 1–35). John Wiley & Sons Ltd.

Wang, Y., Xu, L., Ren, W., Zhao, D., Zhu, Y., & Wu, X. (2012). Bioactive metabolites from *Chaetomium globosum* L18, an endophytic fungus in the medicinal plant *Curcuma wenyujin*. *Phytomedicine*, *19*(3–4), 364–368.

Waqas, M., Khan, A. L., Kamran, M., Hamayun, M., Kang, S. M., Kim, Y. H., et al. (2012). Endophytic fungi produce gibberellins and indoleacetic acid and promotes host-plant growth during stress. *Molecules*, *17*(9), 10754–10773.

Weber, D., Sterner, O., Anke, T., Gorzalczancy, S., Martino, V., & Acevedo, C. (2004). Phomol, a new antiinflammatory metabolite from an endophyte of the medicinal plant *Erythrina cristagalli*. *The Journal of Antibiotics*, *57*(9), 559–563.

Wilson, D. (1995). Endophyte: The evolution of a term, and clarification of its use and definition. *Oikos, 73*, 274–276.

Wiyakrutta, S., Sriubolmas, N., Panphut, W., Thongon, N., Danwisetkanjana, K., Ruangrungsi, N., et al. (2004). Endophytic fungi with anti-microbial, anti-cancer and anti-malarial activities isolated from Thai medicinal plants. *World Journal of Microbiology and Biotechnology, 20*(3), 265–272.

Wu, X. Z., Cheng, A. X., Sun, L. M., & Lou, H. X. (2008). Effect of plagiochin E, an antifungal macrocyclic bis (bibenzyl), on cell wall chitin synthesis in *Candida albicans* 1. *Acta Pharmacologica Sinica, 29*(12), 1478–1485.

Wu, L., Han, T., Li, W., Jia, M., Xue, L., Rahman, K., et al. (2013). Geographic and tissue influences on endophytic fungal communities of *Taxus chinensis* var. *mairei* in China. *Current Microbiology, 66*(1), 40–48.

Yadav, S. K., & Mishra, G. C. (2013). Biodiversity management open avenues for bioprospecting. *International Journal of Agriculture and Food Science Technology, 4*(6), 635–642.

Yan, X. N., Sikora, R. A., & Zheng, J. W. (2011). Potential use of cucumber (*Cucumis sativus* L.) endophytic fungi as seed treatment agents against root-knot nematode *Meloidogyne incognita*. *Journal of Zhejiang University Science B, 12*(3), 219–225.

Yu, H., Zhang, L., Li, L., Zheng, C., Guo, L., Li, W., et al. (2010). Recent developments and future prospects of antimicrobial metabolites produced by endophytes. *Microbiological Research, 165*(6), 437–449.

Zang, W., Becker, D., & Cheng, Q. A. (2006). Mini review of recent WO. *Resent Patents on Anti Infective Drug Discovery, 1*, 1.225–230.

Zhang, C. L., Zheng, B. Q., Lao, J. P., Mao, L. J., Chen, S. Y., Kubicek, C. P., et al. (2008). Clavatol and patulin formation as the antagonistic principle of *Aspergillus clavatonanicus*, an endophytic fungus of *Taxus mairei*. *Applied Microbiology and Biotechnology, 78*(5), 833–840.

Zhao, J., Fu, Y., Luo, M., Zu, Y., Wang, W., Zhao, C., et al. (2012). Endophytic fungi from pigeon pea [*Cajanus cajan* (L.) Millsp.] produce antioxidant cajaninstilbene acid. *Journal of Agricultural and Food Chemistry, 60*(17), 4314–4319.

Zhao, J., Mou, Y., Shan, T., Li, Y., Zhou, L., Wang, M., et al. (2010). Antimicrobial metabolites from the endophytic fungus *Pichia guilliermondii* isolated from *Paris polyphylla* var. *yunnanensis*. *Molecules, 15*(11), 7961–7970.

Zhou, X., Wang, Z., Jiang, K., Wei, Y., Lin, J., Sun, X., et al. (2007). Screening of taxol-producing endophytic fungi from *Taxus chinensis* var. *mairei*. *Applied Biochemistry and Microbiology, 43*(4), 439–443.

Zhou, S. L., Yang, F., Lan, S. L., Xu, N., & Hong, Y. H. (2009). Huperzine A producing conditions from endophytic fungus in SHB *Huperzia serrata*. *Journal of Microbiology, 3*, 32–36.

Zou, W. X., Meng, J. C., Lu, H., Chen, G. X., Shi, G. X., Zhang, T. Y., et al. (2000). Metabolites of *Colletotrichum gloeosporioides*, an endophytic fungus in *Artemisia mongolica*. *Journal of Natural Products, 63*(11), 1529–1530.

Chapter 17

Bioprospecting appraisal of Himalayan pindrow fir for pharmacological applications

Rezwana Assad[a], Zafar Ahmad Reshi[a], Showkat Hamid Mir[a], Irfan Rashid[a], Yogesh Shouche[b], and Dhiraj Dhotre[b]
[a]*Department of Botany, University of Kashmir, Srinagar, Jammu and Kashmir, India,*
[b]*National Centre for Microbial Resource (NCMR), National Centre for Cell Science, Pune, Maharashtra, India*

1. Introduction

Bioprospecting, especially of plant products, has ensnared global attention ever since the beginning of human civilization. Plants possess enormous medicinal properties and have been used since centuries for therapeutic and pharmacological purposes, owing to their diverse phytochemical constitution. However, phytochemical composition and therapeutic effects of merely diminutive number of plant species have been investigated so far (Al Rashid, Kundu, Mandal, Wangchuk, & Mandal, 2020; Heinrich & Gibbons, 2001). Consequently, this area of "bioprospecting of pharmacologically active plant products" is yet amenable for further exploration.

Himalayan forests are dwelling to intriguing plants that possess diverse therapeutic effects, and scores of such plants have been commercialized globally. Gymnosperms, in particular conifers, are the dominant constituents of Himalayan forests (Dar & Dar, 2006). Among these conifers, firs were first documented as the genus *Abies* by Miller in 1754 (Farjon, 2001; Kim & Park, 2018). The genus *Abies* belonging to the family Pinaceae consists of 78 species that are widely distributed in the temperate regions of Asia, Africa, America, and Europe (Farjon, 2010; Plant List, 2020; Verma, 2017). This genus originated in North America and subsequently migrated to the rest of the world (Xiang, Xiang, Guo, & Zhang, 2009).

The present chapter provides an overview of morphological characteristics, distribution, phytochemical composition, and potential pharmacological applications of one of the most imperative fir species, "Himalayan pindrow fir."

Phytomedicine: A Treasure of Pharmacologically Active Products from Plants
https://doi.org/10.1016/B978-0-12-824109-7.00003-0

1.1 Botanical description and distribution of *Abies pindrow* (Royle ex D. Don) Royle

Abies pindrow (Royle ex D. Don) Royle, commonly known as Himalayan pindrow fir/West Himalayan silver fir/Pindrau/Talisapatra/Tosh/Tung/Drewar/Budul, is an economically important endemic conifer of Kashmir Himalayan temperate coniferous forests (Dar & Dar, 2006). The species name *pindrow* has been derived from one of its common name in Nepali "Pindrau," which refers to this conifer's native range "Pindrau hills." The accepted plant name of *A. pindrow* and its synonyms are mentioned in Table 1.

(a) Classification
   Kingdom: Plantae
      Division: Pinophyta
         Class: Pinopsida
            Order: Pinales
               Family: Pinaceae
                  Genus: *Abies*
                     Species: *Abies pindrow*
(b) Morphological characteristics and ecology
   A. pindrow is a tall evergreen tree growing up to 45–55 m, with a trunk diameter of up to 2–3 m, tapering gradually upward forming slender pyramidal contour. It has a cylindrical crown with level branches; upper branches are generally flat, whereas lower branches are faintly drooping with upward curved ends. The bark of young trees is smooth, glabrous, and silvery, while it is grayish brown, deeply and longitudinally fissured on mature trees. Shoots are monomorphic. Leaves are shiny dark green, needle-shaped, narrowly linear, flattened, spirally arranged, spreading (Fig. 1), acute, 1–7 cm × 0.1 mm, apices bifid, margins recurved, two grayish bands of stomata on either side

TABLE 1 Accepted plant name of *Abies pindrow* with its synonyms.

Accepted name	Synonyms
Abies pindrow (Royle ex D. Don) Royle	• *Abies chiloensis* Carriére • *Abies himalayensis* Lavallée • *Abies pindrow* var. *intermedia* A. Henry • *Abies pindrow* var. *pindrow* • *Abies webbiana* var. *pindrow* (Royle ex D. Don) Brandis • *Picea herbertiana* Madden • *Picea pindrow* (Royle ex D. Don) Loudon • *Pinus naphta* Antoine [Invalid] • *Pinus pindrow* Royle ex D. Don • *Pinus spectabilis* var. *pindrow* (Royle ex D. Don) Voss • *Taxus lambertiana* Wall. [Invalid]

Source: http://www.theplantlist.org/tpl1.1/record/kew-2610035 (Accessed on March 12, 2020).

FIG. 1 Morphology of *Abies pindrow.*

of shallow midrib (Fig. 2). Male cones are solitary, cylindrical, 1–2×0.3–0.6 cm, very short stalked (0.2–0.5 cm), reddish green, consisting of numerous, spiral, short-stalked microsporophylls. Female cones are solitary, cylindrical, 8–15×4–7 cm, short stalked (1–1.5 cm), dark purple when young and turn reddish-brown at maturity. Seeds are reddish brown, triangular shaped, 1–1.2 cm long, attached with a delicate wing (American Conifer Society, 2020; Burdi, Samejo, Bhanger, & Khan, 2007; Dar & Dar, 2006; Efloraofindia, 2020; Farjon, 2010; Kumar, 2016; Verma, 2017).

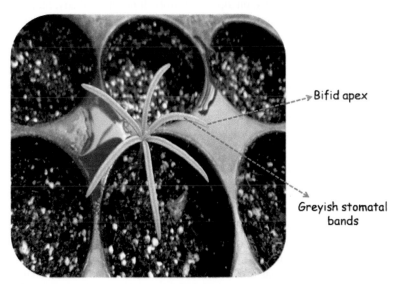

FIG. 2 Morphological characteristics of Himalayan pindrow fir seedling (bifid needle apex and two grayish bands of stomata on either side of midrib).

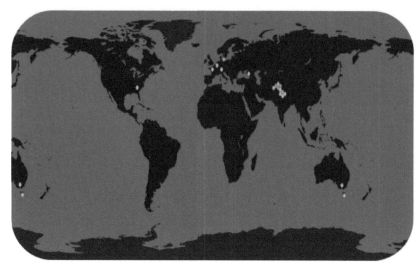

FIG. 3 Map showing global distribution of *Abies pindrow* (*yellow* classic hexagons). *(The map was adapted from Global Biodiversity Information Facility (GBIF) (Accessed on March 6, 2020).)*

(c) Distribution

A. pindrow is native to West Himalayan mountain range, from northeast Afghanistan through northern Pakistan and India (Himachal Pradesh, Jammu and Kashmir, and Uttar Pradesh) to Nepal.

Himalayan pindrow fir is a species of high mountains, occurring between 2000 m and 3300 masl, on alpine lithosols. It occupies shady, moist, and cool north-facing forest slopes. It grows in pure stands or in association with *Picea smithiana, Pinus wallichiana, Tsuga dumosa,* and *Cedrus deodara*; at lower elevations, broad-leaved trees, for example, *Quercus semecarpifolia, Q. dilatata, Juglans regia, Aesculus indica, Acer* spp., *Prunus* spp., and *Ulmus* spp., become more important, replacing the conifers below 1600 m. *A. pindrow* is listed as "least concern" species (Version 3.1) owing to its wide distribution (IUCN Red List, 2020; Xiang, Carter, & Rushforth, 2013) (Fig. 3).

1.2   Phytochemical constituents of Himalayan pindrow fir

Himalayan pindrow fir has been reported to be an imperative resource of scores of phytochemicals of pharmacological importance such as carbohydrates, fatty acids, fatty alcohols, flavonoids, glycosides, hydrocarbons, lignans, phenols, quinines, proteins, resins, saponins, steroids, tannins, terpenoids, and volatile oils (Burdi et al., 2007; Devrani, Rawat, Chandra, Prasad, & Bisht, 2017; Kumar, 2016; Kumar & Kumar, 2016, 2017; Padalia, Verma, Chauhan, Goswami, & Chanotiya, 2014; Samejo, Ndukwe, Burdi, Bhanger, & Khan, 2009; Samejo, Burdi, Bhanger, Talpur, & Khan, 2010; Singh, Pandey, Tripathi, & Pandey, 2001a; Singh, Pandey, Tripathi, & Pandey, 2001b; Singh & Pandey,

2015; Tiwari & Minocha, 1980; Tripathi, Jha, Singh, & Pandey, 2000; Tripathi, Jain, and Pandey, 1996a; Tripathi, Jain, Pandey, Ray, and Rücker, 1996b; Verma, 2017; Vidal, 2012; Willför et al., 2009). Several phytochemicals that have been isolated from Himalayan pindrow fir hitherto are demonstrated in Table 2.

2′,3′,4′:3,4-Pentahydroxy-chalcone-4′-(L-arabinofuranosyl-α-1 → 4-D-glucopyranoside-β), camphene, (E)-α-bisabolene, gallic acid, limonene, maltol, myrcene, okanin, pindrolactone, pinitol, piperitone, shikimic acid, terephthalic acid dimethyl ester (TADE), α-pinene, and β-pinene are some of the characteristic phytochemicals isolated from Himalayan pindrow fir (Fig. 4).

1.3 Pharmacological applications of Himalayan pindrow fir

Himalayan pindrow fir possesses scores of diverse pharmacological applications and has been named as "Taalisa" in Ayurvedic system of medication and "Zarnab" in Unani system (Majeed et al., 2013). The pharmacological applications of this plant have not been yet studied meticulously. Thus current appraisal is of great substance, since it provides comprehensive information regarding pharmacological applications of Himalayan pindrow fir vis-à-vis its diverse phytoconstituents, which will not only provide an impetus to plant bioprospecting studies in general but will also further assist scientists during advance research on this particular plant species.

Pharmacological applications of *A. pindrow* include analgesic, antiasthmatic, anticataract, antidiabetic, antidiarrheal, antihelminthic, antihypertensive, antiinflammatory, antimicrobial, antioxidant, antipyretic, antirheumatic, antispasmodic, antitumor, antitussive, antiulcerogenic, anxiolytic, and neuroprotective activities in conjunction with other nonpharmacological applications (Aggarwal et al., 2011; Al Rashid et al., 2020; Bharti et al., 2014; Dahanukar, Kulkarni, & Rege, 2000; Dubey, Saha, & Saraf, 2015; Gupta, Bhardwaj, & Gupta, 2011; Hussain et al., 2004; Kalim, Bhattacharyya, Banerjee, & Chattopadhyay, 2010; Kumar, 2016; Kumar & Kumar, 2015a, 2015b, 2017; Kumar et al., 2000; Kumar, Alok, et al., 2014a; Kumar, Jamwal, et al., 2014b; Majeed et al., 2013; Mukhtar, Goyal, & Kumar, 2018; Mushtaq, Chaudhry, & Rahman, 2015; Nikhat, Khan, & Ahmad, 2012; Patel, Kumar, Singh, & Kumar, 2014; Singh, Nath, Goel, & Bhattacharya, 1998; Singh, Bhattacharya, & Acharya, 2000; Singh et al., 2001a, 2001b; Singh & Pandey, 1997, 2015; Sinha, 2019; Verma, 2017; Vidal, 2012; Yang et al., 2008). Various prominent pharmacological values of Himalayan pindrow fir are represented in Fig. 5.

(a) Analgesic activity

Leaf extract of Himalayan pindrow fir has strong analgesic property and produces hypnotic effects in rats (Majeed et al., 2013; Singh et al., 1998; Singh et al., 2000; Singh et al., 2001a; Yang et al., 2008). Resin produced by this plant species has been found to alleviate headache (Chopra, Nayar, & Chopra, 1956; Khan, Ishfaq, & Ali, 1979). The potential mechanism of its analgesic activity involves its phytoconstituents, namely, flavonoids and terpenoids (Singh et al., 1998; Dahanukar et al., 2000).

TABLE 2 Phytochemical constituents of Himalayan pindrow fir.

S. no.	Class of compounds	Constituents	References
1.	Carbohydrates/hydrocarbons	1-Docosene 1-Octadecene 2,6,10,14-Tetramethylhexadecane Docosane Eicosane Heneicosane Heptadecane Nonadecane Octadecane Tetracosane Tricosane	Samejo et al. (2010) Kumar (2016) Devrani et al. (2017)
2.	Fatty acids	5,9-Octadecadienoic acid 14-Methylhexadecanoic acid 14-Methylpentadecanoic acid 16-Methylheptadecanoic acid 17-Methyloctadecanoic acid cis-9-Octadecenoic acid Cyclopentaneundecenoic acid Docosanoic acid n-Pentadecanoic acid n-Tetradecanoic acid Tetracosanoic acid	Burdi et al. (2007) Kumar and Kumar (2018)

3.	Flavonoids		
	Chalcones	2′,3′,4′:3,4-Pentahydroxy-chalcone-4′-(L-arabinofuranosyl-α-1 → 4-D-glucopyranoside-β)	Tiwari and Minocha (1980)
		Butein-4′-O-β-D-glucopyranoside	Yang, Li, Shen, and Zhang (2008)
		Okanin	
		Okanin-4′-O-β-D-glucopyranoside	
	Flavonoid glycosides	Afzelin	Vidal (2012)
		Kaempferol-3-O-α-L-(4″-E-p-coumaroyl)rhamnoside	
		Kaempferol-3-O-(2″, 4″-di-E-p-coumaroyl)-α-L-rhamnopyranoside	
		Quercitrin	
	Flavonone	8,3′4′-Trihydroxyflavonone-7-O-β-D-glucopyranoside	Tiwari and Minocha (1980)
4.	Phenols		
		3,4-Dihydroxybenzoic acid	Kumar (2016)
		4-Hydroxybenzoic acid	Kumar and Kumar (2017)
		Apigenin	Kumar (2016)
			Kumar and Kumar (2017)
		Dipalmitin	Kumar (2016)
		Gallic acid	Kumar (2016)
			Kumar and Kumar (2017)
		Hesperitin	Kumar (2016)
			Kumar and Kumar (2017)
		Lignans	Willför et al. (2009)
		Matairesinol	
		Pinoresinol	
		Unidentified lignan derivatives	

Continued

TABLE 2 Phytochemical constituents of Himalayan pindrow fir—cont'd

S. no.	Class of compounds	Constituents	References
		Maltol (3-hydroxy-2-methyl-4H-pyran-4-one)	Samejo et al. (2009) Kumar (2016) Kumar and Kumar (2017)
		n-Dodecan-1-ol	Devrani et al. (2017)
		Pinitol (3-O-methyl-D-chiro-inositol)	Tripathi et al. (2000) Singh et al. (2001a) Yang et al. (2008) Poongothai and Sripathi (2013) Kumar (2016) Kumar and Kumar (2016) Kumar and Kumar (2017)
		p-Coumaric acid	Kumar (2016) Kumar and Kumar (2017)
		Shikimic acid	Kumar (2016) Kumar and Kumar (2016, 2017)
5.	Proanthocyanidins	Catechin Epicatechin Epigallocatechin Gallocatechin	Willför et al. (2009)

6.	Steroids	Campesterol	Willför et al. (2009)
		Sitosterol	
		Sterols and triterpenyl alcohols	
7.	Terpenoids		
	Monoterpenoid	Terephthalic acid dimethyl ester (TADE)	Singh et al. (2001b)
			Singh and Pandey (2015)
			Kumar (2016)
	Triterpenoid	Pindrolactone [3α-hydroxylanosta-7,9(11),22E,24-tetraen-26,23-olide]	Tripathi, Jain et al. (1996a) and Tripathi, Jain, Pandey et al. (1996b)
			Yang et al. (2008)
8.	Volatile oils	L-Bornyl acetate	Khanum and Razack (2010)
		l-Codeine	Kumar, Singh, Jaiswal, Bhattacharya, and Acharya (2000)
		Dipentene	
		Limonene	
		T3-carene	
		α-Pinene	
	Aldehyde	Dodecanal	Devrani et al. (2017)
	Ether	Carvacryl methyl ether	Devrani et al. (2017)
	Terpenes	Abietadiene	Padalia et al. (2014)
		Abietatriene	
		Camphene hydrate	
		Camphene	
		(E)-Calamenene	
		(E)-β-Farnesene	
		Germacrene D	
		Humulene oxide II	
		Kaurene	

Continued

TABLE 2 Phytochemical constituents of Himalayan pindrow fir—cont'd

S. no.	Class of compounds	Constituents	References
		Limonene	
		Linalool	
		Myrcene	
		p-Cymene	
		p-Cymenene	
		Sabinene	
		Terpinolene	
		trans-Sabinene hydrate	
		(Z)-Calamenene	
		α-Calacorene	
		α-Humulene	
		α-Longipinene	
		α-Muurolene	
		α-Pinene	
		α-Selinene	
		α-Terpinene	
		α-Thujene	
		β-Caryophyllene	
		β-Elemene	
		β-Pinene	
		γ-Elemene	
		γ-Muurolene	
		γ-Terpinene	
		δ-2-Carene	
		δ-Cadinene	
		δ-Elemene	

	Compounds	Reference
	(+)-(R)-Limonene	Devrani et al. (2017)
	(+)-3-Carene	
	Abietadiene	
	Camphene hydrate	
	Camphene	
	(E)-α-Bisabolene	
	Germacrene D	
	Myrcene	
	Sabinene	
	Santene	
	Tricyclene	
	α-Humulene	
	α-Phellandrene	
	α-Pinene	
	β-Alaskene	
	β-Bisabolene	
	β-Pinene	
	γ-Bisabolene	
	γ-Terpinene	
Terpene alcohol	1-Epi-Cubenol	Padalia et al. (2014)
	6-Camphenol	
	10-Epi-γ-Eudesmol	
	13-Epi-Manool	
	Abienol	
	Borneol	
	cis-p-Menth-2-en-1-ol	
	(E)-Nerolidol	
	Epi-α-Cadinol	
	Isoborneol	
	Myrtenol	
	Selin-11-en-4-α-ol	
	Terpinen-4-ol	
	trans-p-Menth-2-en-1-ol	
	α-Cadinol	
	α-Terpineol	
	β-Eudesmol	
	γ-Eudesmol	

Continued

TABLE 2 Phytochemical constituents of Himalayan pindrow fir—cont'd

S. no.	Class of compounds	Constituents	References
		Abienol Cadin-4-en-10-ol cis-Piperitol Citronellol (E)-Nerolidol (E)-p-2-Menthen-1-ol Epicubenol Epi-α-Bisabolol Fenchyl alcohol Isoborneol Terpinen-4-ol T-Muurolol trans-Piperitol α-Terpineol β-Linalool	Devrani et al. (2017)
	Terpene esters	Bornyl acetate Linalyl acetate	Padalia et al. (2014)
		Bornyl acetate Citronellyl acetate Geranyl acetate Menthyl acetate Myrtenyl acetate trans-Piperitol acetate α-Terpinyl acetate	Devrani et al. (2017)
	Terpene ketone	Piperitone Undecan-2-one	Devrani et al. (2017)

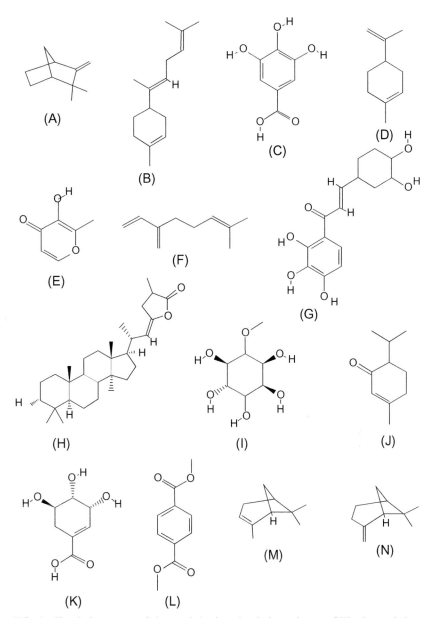

FIG. 4 Chemical structures of characteristic phytochemical constituents of Himalayan pindrow fir: (A) Camphene, (B) (*E*)-α-bisabolene, (C) gallic acid, (D) limonene, (E) maltol, (F) myrcene, (G) okanin, (H) pindrolactone, (I) pinitol, (J) piperitone, (K) shikimic acid, (L) terephthalic acid dimethyl ester, (M) α-pinene, and (N) β-pinene. *(Source: PubChem Database; Accessed on April 3, 2020).)*

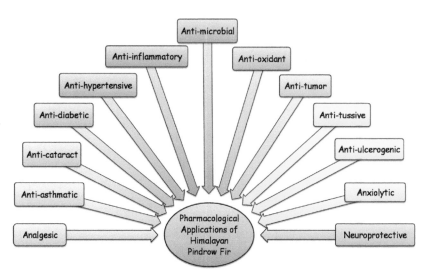

FIG. 5 Prominent pharmacological applications of Himalayan pindrow fir.

(b) Antiasthmatic activity

Decoction of desiccated leaves and bark of Himalayan pindrow fir is used for the treatment of various pulmonary and other respiratory ailments like asthma, chronic bronchitis, and tuberculosis due to its bronchodilator and bronchoprotective capability (Anonymous, 2004; Chopra et al., 1956; Gupta et al., 2011; Hussain et al., 2004; Kumar, 2016; Mushtaq et al., 2015; Nadkarni, 1927; Singh et al., 2000, 2001a; Singh & Pandey, 2015; Yang et al., 2008). Flavonoids and terpenoids found in this plant play a protective role against histamine-induced bronchospasm through their calcium channel blocking capability (Mushtaq et al., 2015; Singh et al., 2000). Singh et al. (2001b) reported that a monoterpenoid, terephthalic acid dimethyl ester (TADE), is the main phytoconstituent in the leaves of this fir species that inhibit histamine-induced bronchial asthma, owing to its antiinflammatory potential. TADE treatment possesses mast cell stabilizing ability and significantly reduces mast cell degranulation in mice, pointing toward its potential application in the treatment of severe respiratory ailments (Singh & Pandey, 2015).

(c) Anticataract activity

Oxidative stress incites cataract disease of eye lens leading to blurry vision. *A. pindrow* leaf extract has been used as a remedy for cataract disease owing to the antioxidant and antiinflammatory potential of its flavonoid and phenolic phytoconstituents (Dubey et al., 2015; Sinha, 2019). Dubey et al. (2015) reported that leaf extract of this fir species demonstrated anticataract potential in isolated goat lenses having hydrogen peroxide-induced cataract, either by delaying the onset of cataract or by preventing the cataract progression. Application of this leaf extract reduces oxidative stress and lens opacity by decreasing the concentration of malondialdehyde and glutathione (GSH) and simultaneously escalating the concentration of superoxide dismutase (Dubey et al., 2015; Sinha, 2019).

(d) Antidiabetic activity

A. pindrow leaf extract has antidiabetic or antihyperglycemic properties owing to its hypoglycemic activity (Gupta et al., 2011; Hussain et al., 2004; Kumar, 2016; Kumar, Alok, et al., 2014a; Kumar, Jamwal, et al., 2014b; Rahman & Zaman, 1989; Saravanamuttu & Sudarsanam, 2012; Singh et al., 2000; Sinha, 2019). Extracts of this plant demonstrated robust hypoglycemic effect in rabbits with experimentally induced diabetes through insulin secretagogue activity (stimulate insulin secretion) from INS-1 cells (Bhushan et al., 2010; Hussain et al., 2004; Kumar, Alok, et al., 2014a; Kumar, Jamwal, et al., 2014b; Manukumar, Shiva Kumar, Chandrasekhar, Raghava, & Umesha, 2016; Patel et al., 2012; Singh et al., 2000). In hypoinsulinemic STZ-diabetic mice the main active phytoconstituent D-pinitol (3-O-methyl-D-chiro-inositol) exerted insulin-mimetic effect and improved glycemic control (Manukumar et al., 2016; Saravanamuttu & Sudarsanam, 2012).

(e) Antidiarrheal activity

This plant can be also used for the treatment of diarrhea through its calcium channel obstruction capability (Mushtaq et al., 2015; Singh et al., 2000). Mechanism of calcium channel obstruction based antidiarrheal activity of *A. pindrow* has been explained in detail by Mushtaq et al. (2015).

(f) Antihelminthic activity

Bark extract of *A. pindrow* possess strong antihelminthic activity and expels internal parasitic worms (helminths) from the body (Anonymous, 2004; Bharti et al., 2014). Phytoconstituents like carbohydrates, glycosides, saponins, and tannins in this extract account for its antihelminthic activity (Bharti et al., 2014).

(g) Antihypertensive activity

Antihypertensive activity has been reported to be present in crude extract of Himalayan pindrow fir (Majeed et al., 2013; Singh et al., 1998, 2000; Yang et al., 2008). Leaf extract has been reported to cause hypotension in dogs (Al Rashid et al., 2020; Singh et al., 1998, 2000).

(h) Antiinflammatory activity

Antiinflammatory activity has been reported in Himalayan pindrow fir extract owing to the presence of flavonoids, glycoside, steroids, and terpenoids (Gupta et al., 2011; Kumar, 2016; Majeed et al., 2013; Singh et al., 1998, 2000, 2001a, 2001b; Singh & Pandey, 1997, 2015; Sinha, 2019; Yang et al., 2008). These multiple active constituents have prominent antiinflammatory effects on both chronic and immunological inflammations (Singh & Pandey, 1997). More specifically, antiinflammatory effect of this plant has been ascribed to its important leaf phytoconstituents such as pinitol, alpha-pinene, terephthalic acid dimethyl ester (TADE), and chalcones

(Aggarwal et al., 2011; Kim et al., 2015; Singh et al., 2001a, 2001b; Sinha, 2019). The leaf extracts have been reported to be beneficial in the treatment of bladder catarrh (inflammation of bladder), including other bladder diseases (Rahman et al., 1986; Sinha, 2019). Detailed mechanism of antiinflammatory effect of this plant has been discussed by Aggarwal et al. (2011).

(i) Antimicrobial activity

Himalayan pindrow fir leaves possess antiseptic activity and prevents the growth of infection-causing microbes (Fayaz, Jain, Bhat, & Kumar, 2019; Majeed et al., 2013; Verma, 2017). Moreover, resin of this plant facilitates prompt healing of cuts and wounds (Verma, 2017).

(j) Antioxidant activity

A. pindrow leaf extract possess antioxidant activity due to the presence of significant amount of flavonoids and phenols (Gupta et al., 2011; Kalim et al., 2010; Kumar, 2016; Kumar, Alok, et al., 2014a; Kumar, Jamwal, et al., 2014b; Verma, 2017). These phenols and flavonoids exhibit antioxidant property by preventing oxidative DNA damage and by scavenging, capturing, or deactivating the free radicals produced in the body, which otherwise may cause several diseases including aging (Kalim et al., 2010). This plant species acts as a potential source of natural antioxidants, which makes it ideal base for upcoming phytopharmaceuticals.

(k) Antipyretic activity

Leaf powder and fresh leaf juice of Himalayan pindrow fir are used as a medication for the treatment of fever (Gupta et al., 2011; Majeed et al., 2013; Rahman et al., 1986; Verma, 2017).

(l) Antirheumatic activity

Bark decoction of *A. pindrow* is used for the cure of rheumatism (Verma, 2017).

(m) Antispasmodic activity

Leaves of Himalayan pindrow fir have antispasmodic property (Burdi et al., 2007; Khan et al., 1979; Majeed et al., 2013; Sinha, 2019).

(n) Antitumor activity

Bioactive compounds like flavonoids isolated from Himalayan pindrow fir serve as a remedy for tumors (Anonymous, 2004; Majeed et al., 2013; Vidal, 2012).

(o) Antitussive activity

Bark extract and leaf powder of *A. pindrow* are used for the treatment of cough, throat infection, and hemoptysis (coughing up of blood) (Anonymous, 2004; Hussain et al., 2004; Majeed et al., 2013; Malik et al., 2019; Rahman & Zaman, 1989; Singh et al., 2000; Sinha, 2019).

(p) Antiulcerogenic activity

Himalayan pindrow fir extracts have been reported to be helpful in healing aspirin-induced ulcer in rats (Nikhat et al., 2012; Singh et al.,

1998, 2000, 2001a; Yang et al., 2008). This ulcer defensive feat is accredited to the steroids present in this particular plant species (Singh et al., 2000).

(q) Anxiolytic activity

Ethanolic extract of *A. pindrow* leaves showed significant anxiolytic activity on all paradigms of anxiety (like in open-field exploratory behavior, elevated zero maze, and elevated plus maze trials of anxiety) in rodents (Gilhotra & Dhingra, 2008; Khanum & Razack, 2010; Kumar, 2016; Kumar & Kumar, 2015a, 2015b, 2017; Kumar et al., 2000; Singh et al., 1998; Sinha, 2019). This plant's antianxiety activity is ascribed to the flavonoids and triterpenoids present in the aerial parts of this fir species (Gilhotra & Dhingra, 2008; Kumar, 2016; Kumar & Kumar, 2015a).

(r) Neuroprotective activity

Himalayan pindrow fir has been traditionally used as a remedy for several nervous disorders (Quattrocchi, 2012). Methanolic extract of Himalayan pindrow fir aerial parts demonstrated strong neuropharmacological activities like anticonvulsant, antidepressant, antistress, and hypnotic activities, attributed to its flavonoid phytoconstituents (Al Rashid et al., 2020; Kumar & Kumar, 2015a, 2015b; Sinha, 2019).

(s) Other pharmacological applications

Himalayan pindrow fir possesses **astringent** (constricts body tissues to reduce discharge of mucus and blood serum), **carminative** (prevents formation and accumulation of gas in the gastrointestinal tract), **diuretic** (promotes urine production), **expectorant** (promotes expulsion of mucus from the respiratory tract), and **stomachic** (aids digestion and promotes appetite) activities (Anonymous, 2004; Burdi et al., 2007; Chopra et al., 1956; Fayaz et al., 2019; Khan et al., 1979; Majeed et al., 2013; Sinha, 2019; Tiwari & Minocha, 1980).

A. pindrow leaf juice can function as a tonic during parturition (Burdi et al., 2007; Chopra et al., 1956; Khan et al., 1979; Rahman & Zaman, 1989; Sinha, 2019). Leaves of this plant serve as an important constituent of Ayurvedic oral contraceptive (Asolkar et al., 1992).

(t) Other nonpharmacological applications

Besides having existential value and diverse pharmacological applications, Himalayan pindrow fir has aesthetic value and is grown as an ornamental tree. Leaves and bark are used as fodder, and cones are used for decorative function. Wood of this plant is also used as fuel, as material for handicrafts, and as timber for construction purposes in manufacturing furniture, fruit containers, tea boxes, windows, doors, ceilings, house beams, floor boards, stairs, and bridges, attributable to its durable and microbe/insect-resistant nature (Fayaz et al., 2019; Kirtikar & Basu, 1993; Verma, 2017).

2. Conclusion and future prospects

Himalayan forests are dwelling to intriguing plants that possess diverse therapeutic effects, and scores of such plants have been commercialized globally. Gymnosperms, in particular conifers, are the dominant constituents of these forests. Among these conifers, the genus *Abies* consists of 78 species that are widely distributed in the temperate regions of world. This chapter provides an overview of morphological characteristics, distribution, phytochemical composition, and potential pharmacological applications of one of the most imperative fir species *Abies pindrow*, commonly known as "Himalayan pindrow fir."

Himalayan pindrow fir has been reported to be an imperative resource of scores of phytochemicals of pharmacological importance such as carbohydrates, fatty acids, fatty alcohols, flavonoids, glycosides, hydrocarbons, lignans, phenols, quinines, proteins, resins, saponins, steroids, tannins, terpenoids, and volatile oils. Some of the characteristic phytochemicals isolated from Himalayan pindrow fir include 2′,3′,4′:3,4-pentahydroxy-chalcone-4′-(L-arabinofuranosyl-α-1 → 4-D-glucopyranoside-β), camphene, (E)-α-bisabolene, gallic acid, limonene, maltol, myrcene, okanin, pindrolactone, pinitol, piperitone, shikimic acid, terephthalic acid dimethyl ester (TADE), α-pinene, and β-pinene.

Pharmacological applications of *A. pindrow* include analgesic, antiasthmatic, anticataract, antidiabetic, antidiarrheal, antihelminthic, antihypertensive, antiinflammatory, antimicrobial, antioxidant, antipyretic, antirheumatic, antispasmodic, antitumor, antitussive, antiulcerogenic, anxiolytic, and neuroprotective activities in conjunction with other nonpharmacological applications. The pharmacological applications of this plant have not been yet studied meticulously. Consequently, this area of "bioprospecting of pharmacologically active plant products" is yet amenable for further exploration. Current appraisal is of great substance, since it provides comprehensive information regarding pharmacological applications of Himalayan pindrow fir vis-à-vis its diverse phytoconstituents, which will not only provide an impetus to plant bioprospecting studies in general but will also further assist scientists during advance research on this particular plant species. Further studies must incorporate greater endeavors toward segregation, purification, identification, and bioprospection of active phytoconstituents from this fir species, along with their correlation to pharmacological demeanor and therapeutic efficacy.

Acknowledgments

The authors profoundly acknowledge G.B. Pant National Institute of Himalayan Environment and Sustainable Development (NMHS-IERP) for providing financial support under the grant number GBPI/IERP-NMHS/15-16/10/03. We also thank Head, Department of Botany, University of Kashmir, India, for providing the necessary laboratory facilities.

Conflict of interest

The authors declare that they have no conflict of interest.

References

Aggarwal, B. B., Prasad, S., Reuter, S., Kannappan, R., Yadav, V., Park, B., et al. (2011). Identification of novel anti-inflammatory agents from ayurvedic medicine for prevention of chronic diseases: "reverse pharmacology" and "bedside to bench" approach. *Current Drug Targets, 12,* 1595–1653.

Al Rashid, H., Kundu, A., Mandal, V., Wangchuk, P., & Mandal, S. C. (2020). Preclinical and clinical trials of Indian medicinal plants in disease control. In *Herbal medicine in India* (pp. 119–142). Singapore: Springer Nature.

American Conifer Society. (2020). https://conifersociety.org/conifers/abies-pindrow/. (Accessed 26 April 2020).

Anonymous. (2004). *Ayurvedic pharmacopoeia of India, part I. 1st. Vol. 4.* New Delhi: Govt of India.

Asolkar, L. V., Kakkar, R. R., & Chakre, O. (1992). *Second supplement to glossary of Indian medicinal plants with active principles. Part I (1965–1981)* (pp. 2–3). New Delhi: PID & CSIR.

Bharti, S. P., Kumar, H., Parashar, B., Deswal, S., Chahar, N., & Devgan, M. (2014). *In-Vitro* anthelmintic activity on aqueous and ethanol extracts of *Abies pindrow* bark. *Research Journal of Pharmacognosy and Phytochemistry, 6,* 112–114.

Bhushan, M. S., Rao, C. H. V., Ojha, S. K., Vijayakumar, M., & Verma, A. (2010). An analytical review of plants for anti diabetic activity with their phytoconstituent and mechanism of action. *International Journal of Pharmaceutical Sciences and Research, 1,* 29–46.

Burdi, D. K., Samejo, M. Q., Bhanger, M. I., & Khan, K. M. (2007). Fatty acid composition of *Abies pindrow* (West Himalayan fir). *Pakistan Journal of Pharmaceutical Sciences, 20,* 15–19.

Chopra, R. N., Nayar, S. L., & Chopra, I. C. (1956). *Glossary of Indian medicinal plants* (p. 212). New Dehli: Council of Scientific and Industrial Research (CSIR).

Dahanukar, S. A., Kulkarni, R. A., & Rege, N. N. (2000). Pharmacology of medicinal plants and natural products. *Indian Journal of Pharmacology, 32,* S81–S118.

Dar, A. R., & Dar, G. H. (2006). The wealth of Kashmir Himalaya–gymnosperms. *Asian Journal of Plant Sciences, 5,* 251–259.

Devrani, M. K., Rawat, K., Chandra, D., Prasad, K., & Bisht, G. (2017). Phytochemical studies and GC-MS analysis of the *Abies pindrow. World Journal of Pharmaceutical Research, 6,* 1639–1644.

Dubey, S., Saha, S., & Saraf, S. A. (2015). *In vitro* anti-cataract evaluation of standardised *Abies pindrow* leaf extract using isolated goat lenses. *Natural Product Research, 29,* 1145–1148.

Efloraofindia. (2020). https://sites.google.com/site/efloraofindia/species/m—z/p/pinaceae/abies/abies-pindrow. (Accessed 26 April 2020).

Farjon, A. (2001). *World checklist and bibliography of conifers.* Kew: Royal Botanic Gardens.

Farjon, A. (2010). *A handbook of the world's conifers. Vol. 2* (pp. 59–113). Netherlands: Brill Publishers.

Fayaz, M., Jain, A. K., Bhat, M. H., & Kumar, A. (2019). Ethnobotanical survey of Daksum forest range of Anantnag District, Jammu and Kashmir, India. *Journal of Herbs Spices & Medicinal Plants, 25,* 55–67.

Gilhotra, N., & Dhingra, D. (2008). A review on antianxiety plants. *Natural Product Radiance, 8,* 476–483.

Gupta, D., Bhardwaj, R., & Gupta, R. K. (2011). *In vitro* antioxidant activity of extracts from the leaves of *Abies pindrow* Royle. *African Journal of Traditional, Complementary, and Alternative Medicines*, *8*, 391–397.

Heinrich, M., & Gibbons, S. (2001). Ethnopharmacology in drug discovery: An analysis of its role and potential contribution. *The Journal of Pharmacy and Pharmacology*, *53*, 425–432.

Hussain, Z., Waheed, A., Qureshi, R. A., Burdi, D. K., Verspohl, E. J., Khan, N., et al. (2004). The effect of medicinal plants of Islamabad and Murree region of Pakistan on insulin secretion from INS-1 cells. *Phytotherapy Research*, *18*, 73–77.

IUCN Red List. (2020). http://www.iucnredlist.org. (Accessed 2 March 2020).

Kalim, M. D., Bhattacharyya, D., Banerjee, A., & Chattopadhyay, S. (2010). Oxidative DNA damage preventive activity and antioxidant potential of plants used in Unani system of medicine. *BMC Complementary and Alternative Medicine*, *10*, 77.

Khan, A. A., Ishfaq, M., & Ali, M. N. (1979). *Pharmacognostic studies of selected indigenous plants of Pakistan* (p. 75). Peshawar: Pakistan Forest Institute.

Khanum, F., & Razack, S. (2010). Anxiety–herbal treatment: A review. *Research and Reviews in Biomedicine and Biotechnology*, *1*, 71–89.

Kim, D. S., Lee, H. J., Jeon, Y. D., Han, Y. H., Kee, J. Y., Kim, H. J., et al. (2015). Alpha-pinene exhibits anti-inflammatory activity through the suppression of MAPKs and the NF-κB pathway in mouse peritoneal macrophages. *The American Journal of Chinese Medicine*, *43*, 731–742.

Kim, J., & Park, E. J. (2018). Chemical and biological properties of the genus Abies. In *Advances in plant phenolics: From chemistry to human health* (pp. 225–236). American Chemical Society.

Kirtikar, K. R., & Basu, B. D. (1993). *Indian medicinal plants* (Vols. 1–4). Delhi: Periodical Experts.

Kumar, D. (2016). *Phytochemical and anxiolytic evaluation of Abies Pindrow Royle and Calotropis Gigantea (L.) Dryand*. Doctoral thesis India: Punjabi University.

Kumar, M., Alok, S., Jain, S. K., & Verma, A. (2014a). Antidiabetic activity of plants with their phytoconstituents: A review. *International Journal of Pharmacognosy*, *1*, 9–22.

Kumar, D., Jamwal, A., Madaan, R., & Kumar, S. (2014b). Evaluation of antioxidant activity of selected Indian medicinal plants. *Journal of Fundamental Pharmaceutical Research*, *2*, 1–10.

Kumar, D., & Kumar, S. (2015a). Screening of antianxiety activity of *Abies pindrow* Royle aerial parts. *Indian Journal of Pharmaceutical Education and Research*, *49*, 66–70.

Kumar, D., & Kumar, S. (2015b). Neuropharmacological activities of *Abies pindrow* aerial parts in mice. *Journal of Pharmaceutical Technology, Research and Management*, *3*, 141–151.

Kumar, D., & Kumar, S. (2016). Quantitative determination of Shikimic acid and Pinitol in *Abies pindrow* aerial parts using TLC. *Indian Journal of Pharmaceutical Sciences*, *78*, 287–290.

Kumar, D., & Kumar, S. (2017). Isolation and characterization of bioactive phenolic compounds from *Abies pindrow* aerial parts. *Pharmaceutical Chemistry Journal*, *51*, 205–210.

Kumar, D., & Kumar, S. (2018). A complete monographic study on *Abies pindrow* Royle aerial parts. *Indian Journal of Pharmaceutical Sciences*, *79*, 1001–1007.

Kumar, V., Singh, R. K., Jaiswal, A. K., Bhattacharya, S. K., & Acharya, S. B. (2000). Anxiolytic activity of Indian *Abies pindrow* Royle leaves in rodents: An experimental study. *Indian Journal of Experimental Biology*, *38*, 343–346.

Majeed, H., Bokhari, T. Z., Sherwani, S. K., Younis, U., Shah, M. H. R., & Khaliq, B. (2013). An overview of biological, phytochemical, and pharmacological values of *Abies pindrow*. *Journal of Pharmacognosy and Phytochemistry*, *2*, 182–187.

Malik, K., Ahmad, M., Zafar, M., Sultana, S., Tariq, A., & Rashid, N. (2019). Medicinal plants used for treatment of prevalent diseases in northern Pakistan of Western Himalayas. In *Medicinal plants-use in prevention and treatment of diseases* IntechOpen.

Manukumar, H. M., Shiva Kumar, J., Chandrasekhar, B., Raghava, S., & Umesha, S. (2016). Evidences for diabetes and insulin mimetic activity of medicinal plants: Present status and future prospects. *Critical Reviews in Food Science and Nutrition*, *57*, 2712–2729.

Mukhtar, H. M., Goyal, R., & Kumar, H. (2018). Pharmacognostical standardization of *Abies pindrow* bark. *Asian Journal of Biochemical and Pharmaceutical Research*, *4*, 1–49.

Mushtaq, S., Chaudhry, M. A., & Rahman, H. M. A. (2015). Calcium channels blocked activity: Providing the basis for medicinal use of *Abies pindrow* in diarrhea and bronchitis. *Bangladesh Journal of Pharmacology*, *10*, 430–435.

Nadkarni, K. M. (1927). *The Indian materia medica* (p. 2113). Bombay: Popular Book Depo.

Nikhat, S., Khan, J. A., & Ahmad, G. (2012). Some experimentally proved herbs in peptic ulcer disease. *International Journal of Pharmaceutical Sciences and Research*, *3*, 2387–2392.

Padalia, R. C., Verma, R. S., Chauhan, A., Goswami, P., & Chanotiya, C. S. (2014). Chemical analysis of volatile oils from West Himalayan Pindrow Fir *Abies pindrow*. *Natural Product Communications*, *9*, 1181–1184.

Patel, K., Kumar, V., Singh, P. K., & Kumar, V. (2014). Phytochemistry and pharmacology of *Abies pindrow* (Royle ex D. Don) Royle. *Pharmagene*, *2*, 36–39.

Patel, D. K., Prasad, S. K., Kumar, R., & Hemalatha, S. (2012). An overview on antidiabetic medicinal plants having insulin mimetic property. *Asian Pacific Journal of Tropical Biomedicine*, *2*, 320–330.

Plant List. (2020). http://www.theplantlist.org. (Accessed 12 March 2020).

Poongothai, G., & Sripathi, S. K. (2013). A review on insulinomimetic pinitol from plants. *International Journal of Pharma and Bio Sciences*, *4*, 992–1009.

Quattrocchi, U. (2012). *CRC world dictionary of medicinal and poisonous plants: Common names, scientific names, eponyms, synonyms and etymology*. Boca Raton, FL: CRC Press.

Rahman, A. U., Said, H. M., & Ahmed, V. U. (1986). *Pakistan encyclopaedia planta medica* (pp. 19–20). Nazimabad, Karachi: Hamdard Foundation Press.

Rahman, A. U., & Zaman, K. (1989). Medicinal plants with hypoglycemic activity. *Ethnopharmacology*, *6*, 1–55.

Samejo, M. Q., Burdi, D. K., Bhanger, M. I., Talpur, F. N., & Khan, K. M. (2010). Identification of hydrocarbons from *Abies pindrow* leaves. *Chemistry of Natural Compounds*, *46*, 132–134.

Samejo, M. Q., Ndukwe, G. I., Burdi, D. K., Bhanger, M. I., & Khan, K. M. (2009). Isolation and crystal structure of maltol from *Abies pindrow*. *Journal of Medicinal Plant Research*, *3*, 55–60.

Saravanamuttu, S., & Sudarsanam, D. (2012). Antidiabetic plants and their active ingredients: A review. *International Journal of Pharmaceutical Sciences and Research*, *3*, 3639–3650.

Singh, R. K., Bhattacharya, S. K., & Acharya, S. B. (2000). Pharmacological activity of *Abies pindrow*. *Journal of Ethnopharmacology*, *73*, 47–51.

Singh, R. K., Nath, G., Goel, R. K., & Bhattacharya, S. K. (1998). Pharmacological actions of *Abies pindrow* Royle leaf. *Indian Journal of Experimental Biology*, *36*, 187–191.

Singh, R. K., & Pandey, B. L. (1997). Further study of antiinflammatory effects of *Abies pindrow*. *Phytotherapy Research*, *11*, 535–537.

Singh, R. K., & Pandey, B. L. (2015). Acute-toxicity and *in vitro* rat mast cell studies on terephthalic acid dimethyl ester (TADE) from *Abies pindrow* leaves. *International Journal of Research Studies in Biosciences*, *3*, 108–110.

Singh, R. K., Pandey, B. L., Tripathi, M., & Pandey, V. B. (2001a). Anti-inflammatory effect of (+)-pinitol. *Fitoterapia*, *72*, 168–170.

Singh, R. K., Pandey, B. L., Tripathi, M., & Pandey, V. B. (2001b). Anti-inflammatory effect of terephthalic acid dimethyl ester (TADE) from *Abies pindrow*. *Journal of Medicinal and Aromatic Plant Sciences*, *23*, 357–360.

Sinha, D. (2019). Ethnobotanical and pharmacological importance of Western Himalayan Fir *Abies pindrow* (Royle ex D. Don) Royle: A review. *Journal of Pharmaceutical Research International*, *31*, 1–14.

Tiwari, K. P., & Minocha, P. K. (1980). A chalcone glycoside from *Abies pindrow*. *Phytochemistry*, *19*, 2501–2503.

Tripathi, M., Jain, L., & Pandey, V. B. (1996a). Flavonoids of *Abies pindrow*. *Fitoterapia*, *67*, 477.

Tripathi, M., Jain, L., Pandey, V. B., Ray, A. B., & Rücker, G. (1996b). Pindrolactone, a lanostane derivative from the leaves of *Abies pindrow*. *Phytochemistry*, *43*, 853–855.

Tripathi, M., Jha, R. N., Singh, V. P., & Pandey, V. B. (2000). Chemical constituents of *Abies pindrow* leaves. *Indian Journal of Natural Products*, *16*, 30–31.

Verma, U. (2017). *Assessment of antimicrobial and antioxidant potential and phytochemical characterization of some gymnosperms of Kumaun Himalaya*. Doctoral thesis India: Kumaun University.

Vidal, C. O. A. S. (2012). *Phytochemicals as a source of novel drugs against prostate cancer-preparation of animal experiments and isolation and identification of flavonoid glycosides from Abies pindrow*. Master's thesis Norway: The University of Bergen.

Willför, S., Ali, M., Karonen, M., Reunanen, M., Arfan, M., & Harlamow, R. (2009). Extractives in bark of different conifer species growing in Pakistan. *Holzforschung*, *63*, 551–558.

Xiang, Q., Carter, G., & Rushforth, K. (2013). *Abies pindrow. The IUCN Red List of threatened species 2013: e.T42294A2970337*. https://doi.org/10.2305/IUCN.UK.2013-1.RLTS. T42294A2970337.en.

Xiang, Q. P., Xiang, Q. Y., Guo, Y. Y., & Zhang, X. C. (2009). Phylogeny of *Abies* (Pinaceae) inferred from nrITS sequence data. *Taxon*, *58*, 141–152.

Yang, X. W., Li, S. M., Shen, Y. H., & Zhang, W. D. (2008). Phytochemical and biological studies of *Abies* species. *Chemistry and Biodiversity*, *5*, 56–81.

Further reading

GBIF/Global Biodiversity Information Facility. (2020). https://www.gbif.org/species/2685683. (Accessed 6 March 2020).

PubChem Database. (2020). https://pubchem.ncbi.nlm.nih.gov. (Accessed 3 April 2020).

Chapter 18

Date palm (*Phoenix dactylifera* L.) secondary metabolites: Bioactivity and pharmaceutical potential

Heba I. Mohamed[a], Hossam S. El-Beltagi[b,c], S. Mohan Jain[d], and Jameel M. Al-Khayri[b]

[a]*Biological and Geological Science Department, Faculty of Education, Ain Shams University, Cairo, Egypt,* [b]*Agricultural Biotechnology Department, College of Agriculture and Food Sciences, King Faisal University, Al-Ahsa, Saudi Arabia,* [c]*Biochemistry Department, Faculty of Agriculture, Cairo University, Giza, Egypt,* [d]*Department of Agricultural Sciences, PL-27, University of Helsinki, Helsinki, Finland*

1. Introduction

It has become increasingly apparent over the years that nature has multiple medicinal plants with significant pharmacological activity including potential antioxidant properties (Ekor, 2014; El-Beltagi, Mohamed, Abdelazeem, Youssef, & Safwat, 2019; El-Beltagi, Mohamed, Elmelegy, Eldesoky, & Safwat, 2019; El-Beltagi, Mohamed, Safwat, Gamal, & Megahed, 2019; El-Beltagi, Mohamed, Safwat, Megahed, Gamal, 2018). The World Health Organization reports that as many as 80% of people still rely on conventional medicines (Hamed, Abd El-Mobdy, Kamel, Mohamed, & Bayoumi, 2019; Yadav, Kadam, Patel, & Patil, 2014).

Phoenix dactylifera is a valuable typical plant of Arecaceae family (Sirisena, Ng, & Ajlouni, 2015). There are 14 species in the genus Phoenix, including *P. dactylifera* grown in the Middle East for at least 6000 years (Copley et al., 2001). Its phytochemical analysis showed that fruits contain carotenoids, anthocyanins, procyanidins, phenolics, and flavonoids; sterols have free radical scavenging, antioxidants, antimicrobials, antimutagens, anti-inflammatory, antihyperlipidemic, hepatoprotective, gastroprotective, anticancer, nephroprotective, and immunostimulants (Baliga, Baliga, Kandathil, Bhat, & Vayalil, 2011; El-Beltagi, Aly, & El-Desouky, 2019; El-Far, Shaheen, Abdel-Daim, Al Jaouni, & Mousa, 2016).

Phytomedicine: A Treasure of Pharmacologically Active Products from Plants
https://doi.org/10.1016/B978-0-12-824109-7.00018-2

All products of *P. dactylifera* like fruits, leaves, pollen, seeds, and the syrup have beneficial applications for humans and animals. Previous research demonstrated immune and antioxidant stimulant activity of *P. dactylifera* seeds supplemented as a feed additive with broiler chickens (El-Far, Ahmed, & Shaheen, 2016). *P. dactylifera* products have been widely eaten as everyday food in Islamic countries for over 1400 years of history.

1.1 Botany, distribution, and biodiversity

1.1.1 Botanical description

Phoenix dactylifera is a palm grown for its sweet edible fruit and is graded as follows:

Kingdom: Plantae
Subkingdom: Tracheobionta
Superdivision: Spermatophyta
Division: Magnoliophyta
Class: Liliopsida
Subclass: Arecidae
Order: Arecales
Family: Arecaceae
Genus: *Phoenix* L.
Species: *Phoenix dactylifera* L.

Date palms may grow alone or form a clump of multiple stems from a single root system up to 25–30 m in height. Root system is highly evolved to support such elevated vertical growth and reaches very deep for water supply. It has a vascular network, which originates from a bulb at base of trunk. Primary roots are on average 4 m long and can reach up to 10 m deep in light soils. Primary roots allow length and diameter of the secondary roots to form shorter tertiary roots (Zaid & de Wet, 2002). It is derived from seeds but can also continue to evolve if the date palm develops with an offshoot or tissue-cultivated seedling. The date palm trunk or stripe is a dense, vertical, equal-diameter cylinder lined with fiber-closed leaf bases, an evolutionary mechanism for defending the trunk from herbivorous insects and animals, and water-reducing insulation (Uhl & Dransfield, 1987). A vascular tissue consisting of closely packed bundles of vascular material transfects water and nutrients. During the fascicular transformation, the stem develops vertically and laterally at terminal bud (phyllophor or phyllogen). Leaves at midrib, called fronds, are 4–6 m tall and 0.5 m broad, narrowing toward both ends of the plant, with spines pinning around the base on petiole and spiraling around the base (Al-Yahyai & Al-Kharusi, 2012).

The amount of leaves generated annually ranges from 10 to 26, and between 100 and 125 leaves can have a mature palm; photosynthesis requires 50% leaves (Uhl & Dransfield, 1987). Around 150 leaflets shape the rising leaf; the leaflets are about 30 cm long and about 2 cm thick. Upon their senescence, the leaves stay attached to the tree and need to be pruned periodically. At the base of any herb can be found vegetative, floral, or intermediate auxiliary buds. Throughout the tree's juvenile life, these buds form the so-called offshoots or suckers that may grow into an adult palm and bring forth fruit when they are mature.

Alternatively, the rosette terminal contains auxiliary date palm flowers and branched cluster type. Date palm is indeed a dioecious species; male and female flowers can be present in various palms (Al-Yahyai & Al-Kharusi, 2012), and female palms only may bear fruit when fertilized either naturally or artificially.

Small, white, new and old flowers are formed in strands attached to a rachis creating an inflorescence called a spadix or rise consisting of the main stem named a rachis and several spikeletal strands (normally 50–150 lateral branches). The immature inflorescence is surrounded by a greenish bract, called spathe. During anthesis, for pollination purposes, it turns brown and divides longitudinally revealing the whole inflorescence. The spathes of males are narrower and wider than those of females. In females, Spikelet has an inflorescence of as many as 8000 to 10,000 and more in males than a great number of little flowers.

At the end of rachis, male inflorescence is packed while at the end of rachis female cluster inflorescence divisions are less packed (Uhl & Dransfield, 1987). The sweet-scented male flower normally consists of six stamens, surrounded (three in each) by sepals and waxy petals. Rising stamen comprises two small, yellowish bags of pollen.

Female flower is around 3 to 4 mm in diameter, and the ovary is superior (hypogynous) with the simple stamens being pressed closely together and three carpels. For divergence between the tips only, the three sepals and three petals are joined together. The three sepals and three petals are joined to diverge only between the tips. They display more yellow color when the female flowers are opened, while male flowers show white dust produced once it shakes. Pollen sacs normally open after spathe bursts out in an hour or two. Pollen sacs usually open after the spathe bursts within an hour or two. Just one of the three ovaries will turn into a drupe (Al-Yahyai & Al-Kharusi, 2012). Palm pollination is mostly windborne, so higher yields are also common practice for natural pollination.

Every year, a mature and completely successful palm may bear 60–100 kg of fruit, by several cultivars yielding up to 180 kg (Huntrods, 2011). The dates are cylindrical fruits with a small, yellowish to reddish brown epicarp, with a sweet mesocarp of flesh. At each bearing spikelet, fruits are normally placed on a few tens of individual dates. Spikelets are clustered onto a central stem to form a bunch (Mater, 1991). Depending on cultivar and environmental factors, the number of bunches for a tree ranges from 5 to 30.

Date fruit is a thin, oblong, single-seeded product with terminal stigma, fleshy pericarp, and membranous endocarp (among flesh and seed). Color may range from yellow to black, and soft to dry consistency. Typically, a seed is oblong with a thin embryo and rough endosperm that consists of a cellulose cover inside the walls. The properties of fruit and seed differ widely according to variety, environmental factors, and field farming practices (Al-Yahyai & Al-Kharusi, 2012).

1.1.2 Distribution

There are more than 2000 varieties of date palms worldwide (Ait-Oubahou & Yahia, 1999). Worldwide, there are over 1500 varieties of date (Popenoe, 1973). Whereas, it is recorded, there are over 455 cultivars in Iraq and 350 in Oman (Vittoz, 1979). These cultivars are seed propagated. In the major date palm-producing countries, very few cultivars are cultivated extensively. Khadrawy, Hillawy, Zahidi, Shalabi, Maktoom, Sukari, Khustawy, and Sayer are generally grown in Iraq; they are grown commercially in Egypt, Duwaki, Hayani, Samani, Saidy, and Zaghlol; Bikraari, Deglet Noor, and Saidy in Libya, Bousthami, Bouskri, Boufgouss, Medjhool, and Jihel in Morocco; Deglet Beida, Deglet Nour, and Rhars in Algeria; Ftimi and Deglet Noor in Tunisia; Chichap, Shahaani, Halawi, Barhi, Bureim, and Shanker in India; Duwaiki, Khalas, Khasab, Anbara, Sukari, Kheneizy, Khudairi, and Ruzeis in Saudi Arabia; Kabkab, Shahani, and Sayer in Iran,; and Abdandan, Kalud, Karba, and Jowan Sor in Pakistan. Naghal, Oum Sila, Kamri, Fardh, and Mobsouli are the major cultivars in Oman (Vittoz, 1979). In the United States, commercial production is dominated by Medjhool (Paulsen, 2005), Hallawi, Khadrawi, Deglet Nour, and Zahidi (Hodel & Johnson, 2007).

1.1.3 Biodiversity

In warmer regions of all continents (excluding Antarctica), date palm can be found, however it varies in area and size. Sawaya clarified its geographical distribution via the value of latitude and altitude (Sawaya, 2000). The dates range from 10°N (Somalia) to 39°N (Elche/Spain or Turkmenistan) in the southern and northern hemisphere, while favorable regions vary from 24°N to 34°N in Morocco, Tunisia, Libya, Algeria, Egypt, Iraq, Pakistan, and Iran (Zohary & Hopf, 2000). Sea level is ideal for cultivation of date palms, by an altitude range of 1892 m from 392 m to 1500 m above. The main yield areas come from North Africa, Middle East, and Pakistan, while some low-scale yield originates from parts of South Europe and North America (Fig. 1).

1.2 Cultivation, harvest, and postharvest practices

1.2.1 Cultivation

Hababauk, Kimri, Khalal, Rutab, and Tamer are five stages of the year's premat-uration, ripening, and maturation (Al-Mssallem et al., 2013) (Fig. 2). Different external and internal variations in color, brightness, texture, and chemical composition are observed depending on the degree of maturity and maturity during the year's growth and development (Al-Shahib & Marshall, 2013).

1.2.2 Indexes on maturity and harvesting

Level of maturity at which fruit is harvested relies on expected consumption of both the cultivar and the fruit. The harvest time depends on the amount of sugar and moisture, the size of the fruit, and its texture. Dates are also selected for immediate sale while there is still a high degree of moisture, whereas the

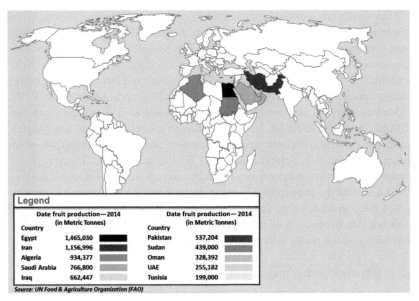

FIG. 1 World global distribution of the palm dates.

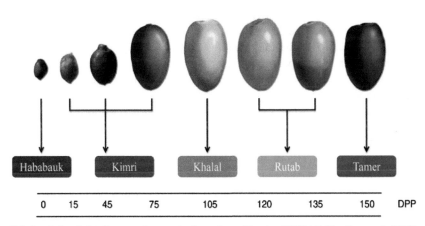

FIG. 2 A date fruit's five growth stages by days after pollination (DPP) (Al-Mssallem et al., 2013).

dates that are to be processed are placed on the palm to remove excess moisture for natural healing. Habababouk, Khalal, Khimri, Rutab, and Tamar are early stages of development. Many of the date varieties, abundant in sugars and poor in tannins, including Barhi (Berhi, Barhee), Samany, Zaghlol, and Hayani, are harvested at Khalal stage (mainly mature) when red or yellow (dependent on variety), but several customers consider them astringent (increased because of tannin levels). Artificial maturation of the dates of harvested cultivars are held until full maturity. The premature dates cannot be stored artificially and would thus be of low quality (Lobo, Yahia, & Kader, 2014).

Many dates are selected at the complete maturation stage of Rutab (soft and light-brown) and Tamar (soft and dark-brown, dry or semi-dry) while high in sugar, lower in humidity and tannins, and smoother than Khalal's stage dates. Deglet Noor dates must be harvested until texture and amber-to-cinnamon coloration are yield-to-pliable. On the perianth end, fruits collected with a reddish circle get a higher storage capacity than fruits left on palm until the circle is more mature (Rygg, 1975).

Various physical and chemical changes are measured as maturity and harvest indices (Yahia & Kader, 2011), including rises in total solids, total sugars, changes in cultivar color from green to yellow or orange or red or purple, major decreases in fruit size, dramatic decreases in temperature, rises in sugar drops, decreases in sucrose, and decreases in acidity and tannin loss. Ibrahim (Ibrahim, 1996) assessed the probability of using an increase in production of ethylene as a physiological maturity indicator at Hillawi dates, and results showed that an increase in production of ethylene started 7–10 days before fruit maturation (depending on the pollen parent used to pollinate female flowers).

1.2.3   Postharvest physiology of dates

Nonclimatic dates with comparatively low respiration rates: for Khalal stage dates <25 mL/kg/h and for Rutab and Tamar stage dates <5 mL CO_2/kg/h for $20°C$. At higher temperatures, the respiration rate and moisture content increases. The production of ethylene in dates is very small at $20°C$, for Khalal <0.5–1 kg/h and for Rutab and Tamar <0.1–1 kg/h (Yahia, 2004).

Serrano, Pretel, Botella, and Amoros (2001) investigated multiple maturing physicochemical properties and their interaction with ethylene in dates (cv. Negros), which were collected and graded in 16 maturing phases, ranging in color from yellow-green to dark brown. Fruit firmness reduced during various phases of maturation, while the maturation index was defined as a relationship between soluble solids and increased acidity; greater lack of fruit firmness was associated for the largest rises in α-galactosidase and polygalacturonase activity. A slight raise in ethylene synthesis was noticed in early stages of maturation, followed by raise in respiration rate; changes in color, total solid content, fruit weight, and acidity in the plant hormones were responsible for ethylene. With respect to the ethylene answer, there was no effect of revealing Khalal stage yellow Barhee dates to 100 ppm ethylene at $20°C$ and 85%–90% relative humidity for up to 48 h. Khalal stage dates are, however, more appropriate for maturing, reacting to the action of ethylene at higher temperatures. Ethylene exposure does not affect the dates of the Rutab and Tamar stage, but it can easily absorb the fragrance of other stored aromatic items. Therefore, dates should not be stored with onion, potato, garlic, or other aromatic products (Kader & Hussein, 2009).

1.3   Economic importance and commercial products

Date palm trees are cultivated primarily for its fruits. In the industrial and urban contexts, it has different uses as well. Any of these findings are probably imperative or close to date fruit itself (Weber, 2010). Date palm tree calls for much sunshine. Arranging is the most evident and common use of the business of date palm trees. It would give the area an impressive and tasteful atmosphere to see set avenues with date palm trees. This can tolerate high and low temperatures but should not fall below $-6.66°C$. Date seed oils are used as part of cleaners and skin care goods, which are abundant in a few unsaturated fats. Date palm has a high content of tocol (tocopherols and tocotrienols) relative to other oils, such as olive, etc. This suggests that the date palm has increased oxidative stability (Nehdi, Sbihi, Tan, Rashid, & Si, 2018). As an ingredient of oxalic acid, this is made equal by the chemical composition of date seed oil. Medical benefits of date seed include protective effects against early diabetic complications, to prevent harm to DNA, and healthy kidney and liver. Date seed can be burned, no matter its small size, and used as a viable charcoal. The date leaves long, thin, but durable stalk or spine that can withstand the heaviness of some people. They are used on roofs as rafters, networks, walls, cabin floors, and simple furnishing (Khan & Khan, 2016). Fruits of date palm are widely distributed and use rich sources of agents to investigate the prevention of sugar, fiber, and phenolic cancer. Because of their conceivable use, date fruits provide high sustenance raw materials from a wide variety of assortments at three stages of advancement. Due to large demand, date fruits are underused, and more engaged work is necessary to raise this yield's value. Thanks to its nutritional and financial interest, the fruits of the date have a high degree and the capacity to be used as food. There is a huge potential for growing nutritious goods using high-quality preventive fibers and phenolic cancer found in seeds and fruits (Ghnimi, Umer, Karim, & Kamal-Eldin, 2017).

2. Nutritional and health benefits of dates

2.1 Moisture content

Date fruits have a rise moisture content of up to 85% at kimri stage; however, it is reduced to 13.25% at tamer phase of Davee date cultivar. Average humidity percent of common date fruit varieties is 17.66%. Reduced date fruit moisture content makes it ideal as a staple food for energy. It can store more calories during the fasting period that can serve better. Date fruit dryness protects them from infestation due to contamination and lack of nutrients. The date fruit tamer stage is nice to store and use them fora healthy 1-year life span as an energy-rich food.

2.2 Carbohydrate content

The date fruits become underused due to the high demand for total carbohydrates, and more active research is required to maximize the yield benefit. Thanks to its nutritional and financial interest, the fruits of the date have a high degree and capacity for use as food. There is a tremendous potential for growing nutritious products using high-quality preventive fiber and phenolic cancer found in seeds, and fruits can be contrasted with dates with moisture/water content where a disparity in the total carbohydrate content is suggested by the maturation and drying phase of the date fruit. Only high amounts of carbohydrates in dried dates allow them to break the fasting that provides Kcal with high energy. Average total dried carbohydrates are 77.13 g/100 g and 308.52 kcal of energy, extracted from the most common cultivars and estimated as previously noted for most dates (Al-Farsi & Lee, 2008; Habib & Ibrahim, 2011; Ismail, Haffar, Baalbaki, Mechref, & Henry, 2006).

Glucose, mannose, fructose, cellulose, sucrose, and starch are included in the overall carbohydrate (Mohamed, Akladious, & El-Beltagi, 2018). Except for sucrose, other sugars in the dried fruit stage were reported to increase (Al-Farsi & Lee, 2008). Average of fructose dates were approximately 29.39 g/100 g and glucose to fructose proportion was nearly equal to 1 (Habib & Ibrahim, 2011), indicating larger fructose content in dried fruit dates. A higher percentage of sucrose was only present in fresh dates and was small in dried dates and not detectable (Al-Farsi & Lee, 2008). Although there are more dates for the total sugar available such as glucose and fructose, in date varieties, fructose is popular, which increases fruit sweetness and has a high satiety advantage with a lower intake date (Al-Farsi & Lee, 2008). Nevertheless, the calories can increase by eating dates or any dried fruit (Gardner, 2017). For traditional medicine, diabetes mellitus is treated with normal fructose-rich foods. Date fructose was linked to negative glycemic index (GI) correlation. The low GI index showed in Oman for three varieties (Ali, Al-Kindi, & Al-Said, 2009). Throughout postprandial evaluation, low GI value foods participate significantly to the reduction of the blood sugar content. Because it reduces the carbohydrate supply for digestion (Duke, 2001), Deglet Nour has been recorded to contain 38% of sucrose as low GI fruit among all varieties. Date fruits eaten in isocaloric amounts did not increase blood sugar levels for varieties dated Sukkari (Famuyiwa et al., 1992).

2.3 Protein

Due to the presence of different amino acid bases, the protein content is comparatively high in different date varieties (Al-Farsi & Lee, 2008). Contrary to other fruits like banana, peach, and apple, nitrogen amount and reduction of water can have a high protein content (Al-Harrasi et al., 2014). Protein content average increased from 2.0% to 6.4%, varying from the date fruit maturing stages (Al-Orf et al., 2012). Current analyzes found an average protein amount of about 2.61% (estimated from date varieties that were most discovered). Nitrogen within a range of 0.25–0.5 g/100 g was assessed. Nitrogen is an essential element in the production of amino acids and, therefore, at maturation levels, amino acids were found to rise and in dried dates were significantly higher at 2.14 g/100 g. Additionally, these amino acids play a crucial function in synthesis and metabolism of proteins required to mature the year's fruits. Dates of glutamic and aspartic acid have been reported to be rich (Al-Shahib & Marshall, 2003). Protein dates contains 23 high concentration amino acids compared with other fruits widely popular for the benefits of wellness. Although proteins from dates do not reach the RDI (roughly 0.84 g/kg/day), they serve to support other essential amino acids such as egg-protein (Bouaziz et al., 2008). Dates such as Dan mali, Fari, and Dagalla have recorded the greatest amount of alanine, glutamic acid, aspartic acid, and proline as essential protein building blocks for different metabolisms (Uba et al., 2015). These amino acids often act as precursors for growth, metabolism and normal function of proteins, receptors, hormones, gene expression, transport of nutrients, and immunity of a healthy human body (Wang et al., 2013; Wu, 2010).

2.4 Minerals

Dietary minerals play an integral part in many of the human body's principal functions. There are a variety of important minerals including macronutrients, micronutrient minerals, and trace elements that help prevent diabetes, obesity, high blood pressure, and other cardiovascular diseases, and certain dietary complement are used to complete mineral shortages due to these cases. Varieties of date palm function as one of those dietary foods to inhibit monitoring and treatment for the mineral deficiency disease. Calcium, sodium, phosphorus, magnesium, potassium, iron, zinc, manganese, molybdenum, cobalt, copper, selenium, aluminum, barium, arsenic, chromium, lead, cadmium, nickel, vanadium, and strontium are some of the minerals mentioned in the date fruits.

Although mineral supplements are available. Medication formulations, such as tablets, sachets, and injections have several side effects such as manifestation of pain and absorption problems causing headache, gastritis, anorexia, and nausea (El-Zoghbi, 1997). Date palm fruit was described as the wealthiest dietary source of vitamins and minerals (Vayalil, 2012).

The date varieties showed substantial improvements in both qualitative and quantitative mineral content, which had previously been noted due to genetic differences in harvest time and maturing phases (Amira et al., 2011). Farming parameters such as soil minerals, resources, and environmental factors are defined (Marzouk & Kassem, 2011). Some of main findings that can be listed are that, in general, all dried fruit mineral content (1.7%) as reported (Al-Shahib & Marshall, 2002) was comparatively smaller than any fresh date level. It was stated that the potassium content was high with low sodium constituent. This also contains higher amounts of iron, supplemented in many date varieties by other trace elements such as fluoride and selenium. In all rising date varieties, fluorides and selenium are not recorded (Al-Shahib & Marshall, 2013), but because selenium is present they have been reported to have high industrial value (Al-Farsi & Lee, 2008; Ghnimi et al., 2017). In several variations of date, certain minerals such as iodine and fluoride were not identified (Vayalil, 2012).

2.5 Vitamins

Date fruits result in at least 10 enzymes being produced as a medium source of vitamins where the vitamin A levels are lower. It is abundant in water-soluble vitamins and plays a major role in the human body's vital function. The liver is covered by vitamin C. Date fruit effectiveness in treating anemia was possible because iron was easily absorbed by vitamin C (Al-Mamary, Al-Habor, & Al-Zubairi, 2011).

Date pulp contains vitamins, such as folic acid, biotin, thiamine, riboflavin, and ascorbic acid, which are important for the body. Dates in B-complex vitamins like thiamine (B1), riboflavin (B2), niacin (B3), pantothenic acid (B5), pyridoxin (B6), folate (B9), and vitamin K are abundant (Al-Farsi & Lee, 2008). It should be noted that some vitamins (B3, B5, B6, and B9) are higher in dates than other popular fruits like apples, peaches, and berries. The level of niacin is very high, ranging from 1.27 to 1.61 mg/100 g. The quantitative study of moisture-soluble vitamins (B1, B2, B3, B5, B6, B9, B12) noticed a substantial difference in different cultivars and in date fruit production (Aslam, Khan, & Khan, 2011).

2.6 Dietary Fiber

The varieties of date palms excel as a rich source of dietary fibers. Commonly consumed date fruit tamer stage contains 3% of the dietary fiber, hitting 32% of the fiber with 100 g intake as suggested by RDA 25 g/day. Dietary fibers contain a limited amount of highly indigestible, insoluble polysaccharides, tannins, and lignins, both soluble and large components. Soluble polysaccharides are pectins and hydrocolloids that have been reduced at dry dates in dried fruits. Cellulose, hemicelluloses, and lignin are the insoluble polysaccharides. In two ways, indigestible dietary fibers help delay the ability to absorb antinutritional factors through the creation of matrix gel using soluble polysaccharides. The other approach is to use high holding capacity of water and oil to bulk the insoluble fecal mass of polysaccharides. Consequently, the maximum insoluble ratio of polysaccharides suggests an increase in fecal transit time. It promotes satiety benefit and contributes significantly to the metabolism of glucose by enhancing insulin sensitivity, development of gut hormones, and other metabolic and inflammatory factors associated with metabolic disorder and disease, such as resistance to lactose and bowel irritability. Fiber intake in general controls the absorption of glucose through maximal insulin secretion with reduced levels of Hb A1C in diabetes (Mohammad & Habibi, 2011). As a result, fruit is a potential factor in diabetes prevention and progression (Vayalil, 2012).

Soluble polysaccharides throughout dietary fiber of date palm types resist atherosclerosis by halting the mechanism of absorption of intestinal cholesterol in which stroke and hyperlipidemia are prevented and thereby avoiding conditions such as coronary heart disease (Abuelgassim, 2010; Lusis, 2002). The fiber in date fruit has hypocholesterolemic effects inhibiting hepatic cholesterol biosynthesis by raising cholesterol absorption, decreasing insulin secretion, and short chain fatty acids generated throughout dietary fiber creation. The date fiber also includes 0.5%–3.9% pectin.

Pectin-rich foods minimize heart disease-related metabolic risk factors by reducing blood cholesterol levels (Al-Shahib & Marshall, 2002). Although pectin levels decrease to 0.5% in the fruit tamer stage caused by large pectin esterase activity, they are considered superior to other fruits (El-Zoghbi, 1997). Moreover, dietary fiber minimize the production of toxic compounds as well, which lead to hepatotoxicity due to the low fecal transit time, which likewise prevents constipation through acting as a laxative (Al-Farsi & Lee, 2008). It decreases the risk of cancer in bladder, intestine, and colon (Marlett, McBurney, & Slavin, 2002). Two antineoplastic glucans (1–3)-β-D glucans with separate (1–6) branches of mono-, di-, and trisaccharide residues and (1–3)-β-D glucopyranosyl residues are primarily associated with this date fruit property (Ishurd et al., 2007; Ishurd & Kennedy, 2005). Equally, the doses of β-glucans in prostate cancer cell line PC-3 have been documented as having increased apoptotic activity through decreasing cell viability (Fullerton et al., 2000). Dietary fibers by fecal mass and gastrointestinal movement serve as active scavengers of exogenous and endogenous mutagens (Al-Qarawi, Ali, Al-Mougy, & Mousa, 2003). Date palm dietary fiber is used as an antimicrobial agent in addition to the previously discussed laxative and prevents diseases of diventricles (Al-Qarawi et al., 2003; Shafiei, Karimi, & Taherzadeh, 2010).

2.7 Fats and fatty acids

The amount of fat was very low in dried fruits and was slightly different in varieties. Fat amount of date fruit is found in two areas: the fruit date skin and the fruit seed. Drying helps dated fruit to rise in fat (Al-Farsi & Lee, 2008). The main function of fat in date fruits is to avoid nutrient leakage, oxidation, and humidity (Shafiei et al., 2010). Date flesh contains 0.2%–0.5% saponifiable oils (Al-Hooti, Sidhu, & Qabazard, 1998). Fatty acids have been described as being saturated and unsaturated in date fruits (Al-Shahib & Marshall, 2003). Saturated fatty acids, including palmitoleic, oleic, linoleic, and linolenic acids, include apric, lauric, myristic, palmitic, stearic, margaric, arachidic, henscicosanoic, behenic, and tricosanoic acids. Fruits of seedless date principally produce eight fatty acids, namely myristic, palmitic, stearic, margaric, arachidic, hencicosanoic, oleic, and linoleic. A key explanation for hypocholesterolic effect of the fruit dates is the low fat content of dated fruit (Al-Saif et al., 2007). Medjool consumption dates decreased VLDL-cholesterol by 8%–15% for 4 weeks (Rock et al., 2009).

2.8 Phytochemical compounds in dates

Phytochemicals are non-nutrient bioactive components that are primarily responsible for scavenging toxic radicals after oxidative stress by generating antioxidants, the main cause of most chronic diseases (Al-Harrasi et al., 2014). Fruit phytochemicals displayed high antioxidant capacities linked to lower incidence of degenerative diseases and lower mortality average in humans (Baliga et al., 2011; Wang et al., 2013). Phenolic acids, flavonoids, tannins, carotenoids, isoflavons, sterols, and lignans are some of date phytochemicals of the fruits that are considered bioactive. Table 1 includes the role of these phytochemicals and their levels in date fruits when it comes to disease prevention.

2.9 Phenolic acids

Phenolic acids are one of the most important classes of secondary metabolites and have in recent years been the subject of extensive research (Aly, Mansour,

TABLE 1 Phytochemicals and the role of date varieties (Gnanamangai et al., 2019).

Phytochemical	Quantity	Function	References
Carotenoid:			
β Carotene	3%–10%	Protects testicular functions and have gonadotropic and hepatoprotective activity	Boudries, Kefalas, and Hornero-Mendez (2007), Domitrovic, Jakovac, Grebic, Milin, and Radosevic-Stasic (2008), Eustache et al. (2009), Jana et al. (2008), Janbaz, Saeed, and Gilani (2005), Said, Banni, Kerkeni, Said, and Messaoudi (2010)
Lutein	89%–94%		
Minor Carotenoids	2%–8%		
Phenols:			
Cinnamic acids, acetylated flavonols, caffeoglshikimic acid hexoside, caffeoylsinapoyl monohexoside, dihexoside, hydroxyl benzoates, hydroxyl cinnamates	30,000ppm (3.0g/100g)	Less glucose absorbance, antioxidant activity, inhibits Angiotensin II converting enzymes and reduces hypertension and anticancer	Neori et al. (2013), Surh (2003)
Phenolic acids:			
Gallic acid, protocatechuic, phydroxyl benzoic, vanillic, caffeic, dactyliferic acid, ellagic acid, syringic, p-coumaric ferulic, o-coumaric	14.18–49.67mg/100g (bound)	Hepatoprotective	Vayalil, (2002), Al-Farsi and Lee (2008), Al-Farsi, Alasalvar, Morris, Baron, and Shahidi (2005), Janbaz et al. (2005), Domitrovic et al. (2008)
	6.1–14.8mg/100g (free)		
Tannins:			
Flavanoids, flavones, flavonols, flavanones, procynidines, proanthocyanidins, flavonoid glycosides (luteolin, quercetin, apigenin), Chrysoeriol, isohamnetic, anthocyanins	0.0162–5446g/kg	Hepatoprotective	Hong, Tomas-Barberan, Kader, and Mitchell (2006), Al-Hooti et al. (1998), Janbaz et al. (2005), Domitrovic et al. (2008)

Sterols:			
Phytosterols:			
Campesterol, β-sitosterol, lupenone, lupeol, 24-methylene cy-cloartanol, propylidene cholesterol	1.83%–2.57%	Hepatoprotective, inhibits cholesterol synthesis and causes hypocholestrolemia	John, Sorokin, and Thompson (2007), Liolios, Sotiroudis, and Chinou (2009)
Phytoestrogens:			
Isoflavones, lignans, genistein, daidzein		Anticancer and antidiabetic	Bhathena and Velasquez (2002)

Mohamed, & Abd-Elsalam, 2012; Aly, Mohamed, Mansour, & Omar, 2013). It involves a hydroxylated benzene ring that is directly or indirectly connected by one or more groups of carboxyl (Fig. 3). Mansouri, Embarek, Kokkalou, and Kefalas (2005) studied phenolic profile of Algerian date types and found that they include sinapic, p-coumarial, and ferulic acids; several cinnamic acid derivatives; and three isomers of 5-o-caffeoyl shikimic acid (Fig. 3). Reports showed the existence of both free (protocatechic, vanillic, ferulic, and syringic acids) and bonded phenolic acids (gallic, caffeic, protocatechic, ferulic, syringic, benzoic, p-coumaric, o-coumaric acids, and vanillic) in three forms (Al Farsi, Alasalvar, Morris, Baron, & Shahidi, 2005) (Fig. 3). Relative experiments with fresh and dried dates showed that drying results in a large increase in phenolic content, likely due to tannin degradation and ripening at higher enzyme temperatures (Al Farsi et al., 2005). The Mermella variety has also recently been found to be the lowest in Tunisian date cultivars, while Korkobbi cultivar recorded the highest phenolic content (Chaira et al., 2009). The phenolic content was stated to be nonstable in varieties (Karasawa, Uzuhashi, Hirota, & Otani,

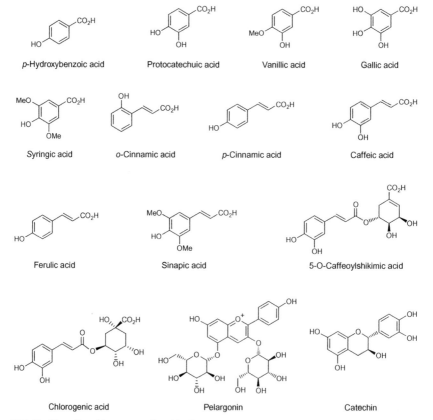

FIG. 3 Phenolic acid structures found in dates.

2011). This plays a significant function in production of antioxidants, ranging from 40% to 80% (Al-Harrasi et al., 2014). Findings on factors influencing total phenolics should be carried out to generate varieties with an accurate estimate of phenolics and antioxidant activity.

2.10   Sterols

Sterols or steroid alcohols at the 3-position of the A-ring are a subgroup of hydroxyl clustered steroids and are amphipathic lipids. Plant sterols have many safety benefits and are called phytosterols (Liolios et al., 2009). Date sterols were tested and cholesterol, stigmasterol, campesterol, isofucosterol, and β-sitosterolwere found (Kikuchi & Miki, 1978) (Fig. 4).

2.11   Carotenoids

Carotenoids are a type of fat-soluble natural pigments that give the plants their bright color. These are important provenance of vitamin A and, by serving as antioxidants, they protect cells against harmful effects of free radicals (Di Mascio, Murphy, & Sies, 1991). Research have established dates of carotenoids; β-carotene, neoxanthin, and lutein (Boudries et al., 2007) (Fig. 5). β-carotene content in the Algerian fresh date cultivars Deglet Noor, Tantebouchte, and Hamraya is 6.4, 3.3, and 2.5 µg/100 g, respectively, while lutein content was 156, 28, and 33.6 µg/100 g, respectively (Al-Farsi & Lee, 2008; Boudries et al., 2007). During transition from khalal to tamar, there is a small decrease in carotenoid levels and a slight increase in provitamin A levels in the Deglet-Noor range

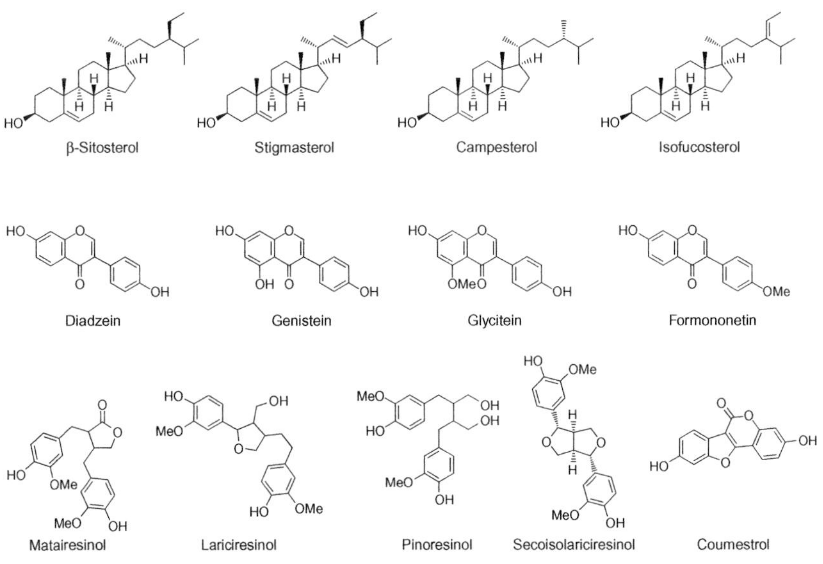

FIG. 4   Phytoestrol and phytoestrogen structures in dates.

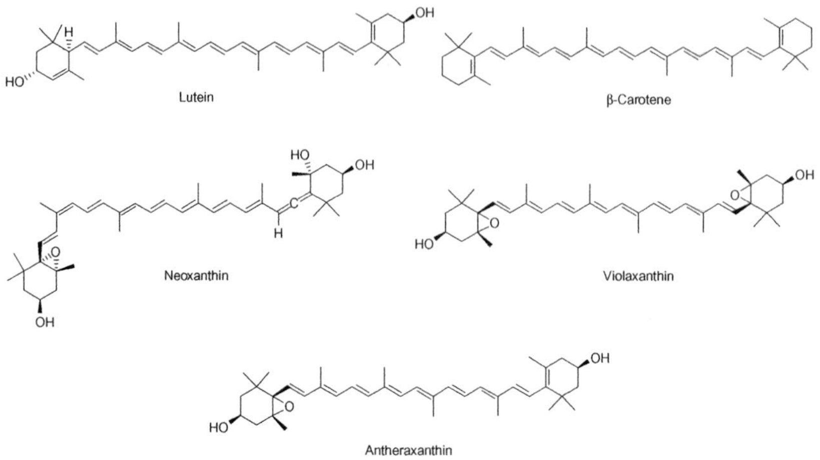

FIG. 5 Carotenoid structures found in dates.

during the maturation process, while their levels in Hamraya and Tantebougte decreased (Boudries et al., 2007). There is a substantial decrease in carotenoid levels during transformation from khalal to tamar and a slight increase in provitamin A levels in the Deglet-Noor range during the maturation cycle, while their levels in Hamraya and Tantebougte decreased (Al Farsi et al., 2005).

2.12 Procyanidins

Procyanidins are abundant tannins in seeds, fruits, berries, nuts, flowers, and barks and are principal precursors of red and blue-violet pigments (Fine, 2000). Hong et al. (2006) extracted Deglet Noor procyanidins using the method of extraction of acetone-water-acetic acid solvent from a variety of dates during Khalal maturity time. Chemical analysis shows that procyanidine existed as polymers, undecamers, and decamers with high molecular weight (Niazi et al., 2017) (Fig. 6).

2.13 Flavonoids

Health benefits of flavonoids is that they contain antioxidants and radical scavenging, reducing certain chronic diseases, avoiding certain cardiovascular disorders, and the processes of plant cancer (Tapas, Sakarkar, & Kakde, 2008). It is reported that flavonoid content of Deglet Noor cultivar during Khalal ripening phase found 13 luteoline, quercetin, and apigenin flavonoid glycosides (Hong et al., 2006) (Fig. 7). Also, it was noted that as mono-, di-, and triglycosylated conjugates, all types of methylated and sulfated luteoline and quercetin are present, whereas apigenin exists only as diglycoside. While apigenin existed as the C-glycoside, luteolin and quercetine formed predominantly O-glycosid associations. The date is also special because it is the only food that contains flavonoid sulphates (Hong et al., 2006). Among the common Tunisian dates, Chaira et al. (2009) recently recorded the highest flavonoid content of Korkobbi cultivar.

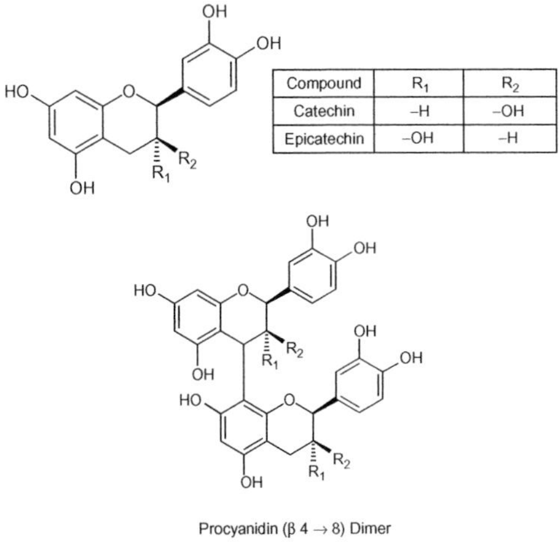

Compound	R₁	R₂
Catechin	–H	–OH
Epicatechin	–OH	–H

Procyanidin (β 4 → 8) Dimer

FIG. 6 Procyanidine structures present in dates.

Flavonoids were extensively studied in various stages and varieties, including in basic phenolic compounds (Chaira et al., 2009). Varieties of fruit, quercetin, apigenin, isoquercetrin, luteolin, and rutin were identified in another analysis of 11 various varieties of Saudi date (Hamad et al., 2015). Kaempferol, a natural flavanol, was present in Hallawi and Amari fruits, where it contained five flavanols and Hallawi was found to contain one large flavanol (Borochov-Neori et al., 2015). The overall flavonoid content decreased considerably from Khalal to Tamer level for seven several date fruit types they assessed (Lemine et al., 2014). Other work found that during the Khalal to Tamer maturity stages, the flavonoid content of four different Tunisian date fruit types decreased (Amira et al., 2011).

2.14   Anthocyanins

Anthocyanins (Fig. 8) are water-soluble vacuolar pigments found in blue, red, or purple plants. These are popular in numerous fruits, grains, vegetables of cereals, and flowers, and can improve health (Wang, Cao, & Prior, 1997). Previous studies found that highest concentrations of anthocyanins in the fresh date varieties examined were present in the Khasab varieties, followed by the Fard varieties and the Khalas varieties, and that there was a strong association between the quantities of anthocyanin and fruit color. Only fresh dates have anthocyanins indicating that sun-drying can harm them (Al-Farsi et al., 2005).

2.15   Other sources of secondary metabolites: Pits, pollen, and spathe

Typically, date fruit consists of fleshy pericarp, seed, or pit (Fig. 9). Seeds are known as having rising extractable added importance components, in addition to the date product. Various studies on the chemical composition of the dates

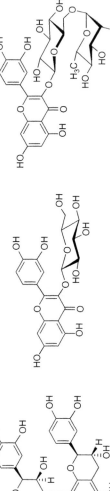

FIG. 7 Flavonoid structures found in dates.

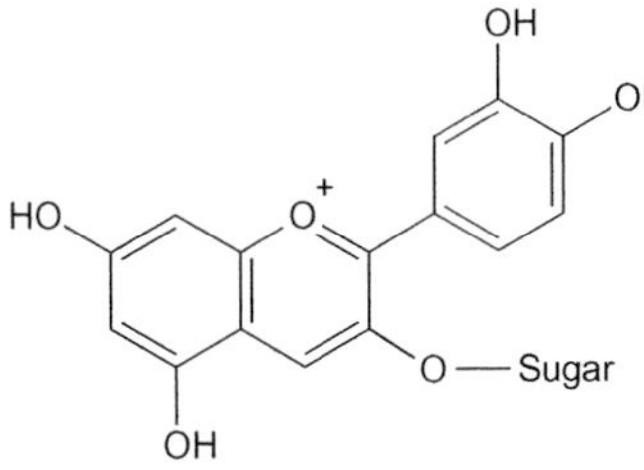

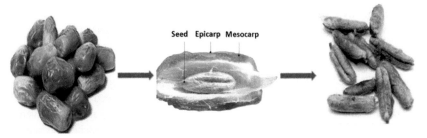

FIG. 8   Anthocyanin structures that are present in dates.

FIG. 9   Date palm and seed parts.

found that industry by-products are primarily wasted due to their technical and biological transformation. Therefore, greater proportion of date palm pits can be obtained from different dates or by using their discarded by-products (Hossain, Waly, Singh, Sequeira, & Rahman, 2014; Khalid, Ahmad, & Kaleem, 2017).

Date pits are often underused and labeled agricultural by-products of low value. During the last century, it has not discovered its true potential; it is now regarded as a good source of dietary components with wide application for drug formulation. For example, date pits are regarded as a perfect precursor of carbon formulation, which is the most common and efficient adsorbent used in various industries (Yaish & Kumar, 2015). Hence, it is effective in extracting and characterizing galactomannan, glucomannan, and heteroxylan from date seeds. Significant quantities of carbohydrates (62.90%), essential fibers (19.99%–39.66%), and lipids are also present in date pits (7.87%–9.76%) (Khalid, Ahmad, Masud, Asad, & Sandhu, 2016). Nevertheless, considerable attention has now been given to the oil or lipids extraction of date plants, due to their possible biological and industrial applications.

Several studies have shown a complex physiochemical composition and classification of fatty acids in pit oils of the year. Date seed oil recorded higher levels of oleic (47.66%) and lauric acids (17.39%) but also moderate to low levels of linoleic (10.54%), palmitic (10.20%), and myristic acids (10.06%) (Khalid, Khalid, Khan, Ahmed, & Ahmad, 2017).

Fruits and date pits are known to be filled with different nutritional attributes. Several studies established the nutritional structure of flesh and pit of the date palm. Proteins and fats, flesh, and date seeds were found to produce higher rates of sugar reduction compared with other amino acids, whereas the date boxes have higher protein, crude fiber, and crude fat proportions than the date flesh

(Khalid, Khalid, et al., 2017). Recent studies identified the near structure of date flesh and pits, and demonstrated existence of ash, moisture, glucose, galactose fructose, maltose, fats, and fibers in date flesh and pits. Date pits or seeds contain approximately 83.0% of carbohydrates in relation to current meat, including 42% of cellulose, 8% of hemicellulose, and 25% of total sugar and other constituents (Ahmed & Theydan, 2012). Similarly, in date seed endosperms, carbohydrates are considered to be essential storage reservoirs mainly in the form of β-D mannan. Many other parts of hemicellulose that are known in date pits consist primarily of water-soluble gluco- and galactomanans, and alkali-soluble heteroxylene. Alkali-soluble heteroxylene is usually composed of xylose (82%) and 4-O-methylglucuronic acid (17%) with low amounts of arabinoses and very small amounts of galactose, mannose, and glucose (Assirey, 2015).

2.15.1 Phenolic contents

The date pits are known to be perfect reservoirs of gallic acid components and antioxidants (Ahmad et al., 2012). Several studies have shown that seed oils are regarded as an abundant source of phenolic content when extracted with multiple solvents compared with various other edible oils. The Iranian date seeds matched the radical scavenging activity and highest antioxidant of several date types, which can be used for various pharmaceutical and commercial reasons (Suresh et al., 2013). There have been records of varying amounts of phenolic gallic acid, p-hydroxybenzoic acid, vanillic acid, caffeic acid, p-coumaric acid, ferulic acid, protocatechuic acid, o-coumaric acid, and m-coumaric acid. Therefore, hydroxybenzoic, protocatechic, and m-coumaric acids were phenolic components, which contributed significant date seed antioxidant ability (Waly, Al-Ghafri, Guizani, & Rahman, 2015).

2.15.2 Flavonoids

The most-rich phenolic compounds found in dated fruits are known to be flavonoids. Some of those polyphenols in date tissue have enormous health benefits, such as radical scavenging and antioxidant. These phenolic compounds are involved in reducing cardiovascular and chronic diseases due to their antioxidant ability, with a positive response to impaired cell proliferation. The fruit date is loaded with various flavonoids, mainly rutin, quercetin, and isoquercetin, and can be checked with HPLC (Kaleem, Ahmad, Khalid, & Azam, 2016). The important antioxidant flavonoids function the same for different types of the kernel or date palm pits. Nevertheless, the overall flavonoid content of Ajwa date palm is 2.79–4.35 mg/100 g while the date pit is 1.35–3.67 mg/100 g quercetin as the predominant flavonoid.

Likewise, due to differences in genetic conditions, laboratory circumstances, and hydration levels, the precise quantities of these compounds vary considerably between pits and date fruit flesh. Phytochemicals (phenolics and flavonoids) have a high antioxidant capacity in date fruit and have significant health benefits such as diabetes reduction, cardiovascular disorders, and cancer cell prevention. The therapeutic potential of flavonoids in the fight against microbial and viral infections is comparable to other flavonoids in fruit (Ahmed et al., 2015).

3.   Biotic factors influencing biosynthesis

3.1   Cultivar

The overall amount of bioactive compounds also varies according to *P. dactylifera* cultivars (Al-Turki, Shahba, & Stushnoff, 2010). For example, Khalas exhibited stronger antioxidant capacity, total phenolic acid, and carotenoid content than other cultivars (Ghnimi et al., 2017). Total phenolic content of certain types of Saudi *P. dactylifera* is usually significantly different (Hamad et al., 2015), whereas in some *P. dactylifera* cultivars, the concentration of total antioxidant compounds and antioxidant activity was strongly correlated (Awad, Al-Qurashi, & Mohamed, 2011).

Three main parts of the fruit date are the outer skin, middle flesh, and inner seed or stone (Shafiei et al., 2010). Each of the varieties and their production varies in size and form. These are usually small oval fruits with length of 3–7 cm, diameter of 2–7 cm, varying in stone size from 2 to 2.5 cm (Ateeq, Dutta Sunil, Varun, & Santosh, 2013). The fruit date color and softness depends on the degree of ripeness. Around 5000 date palm cultivars are known all over the world, but few are checked for their consistency. In Table 2, some of major varieties from several countries are listed.

3.2   Pollination

Date palm is a monocotyledonous dioecious tree, and fertilization occurs either naturally or artificially. One cultivar of pollens will fertilize another. Pollination,

TABLE 2 Some popular date-palm varieties in different countries.

Country	Famous variety	Reference
Pakistan	Aseel, Begum Jangi, Dhakki, Karabalian, Fasli, Muzawati, Halawi	Al-Shahib and Marshall (2013)
Iran	Khenizi, Sayer, Lasht, Kabkab, Shahabi, Majoul, Khazui, Zahedi	Ardekani, Khanavi, Hajimahmoodi, Jahangiri, and Hadjiakhoondi (2010)
Algeria	Deglet nour (semidry), Degla beida (dry dates)	Mansouri et al. (2005)
Tunisia	Alligh, Goundi, Ikhouat, Lagou, Touzerzaillet, Tranj	Borchani et al. (2010)
Saudi Arabia	Suqaey, Sofry, Ajwa, Safawy	Al-Shahib and Marshall (2013)
Egypt	Lobanah, Masery, Saidi	Al-Shahib and Marshall (2013)
Iraq	Shorcy, Tamriraq	Al-Shahib and Marshall (2013)

however, can have significant effects on the physical and chemical properties of resulting products. Among other things, it affects fruit variety, size, maturation period, seeds, and eating consistency, as well as the date's chemical constituency such as antioxidants, an effect known as metaxenia in plant sciences. The dates of pollen patents can have a major effect on total phenolics (Maryam, 2014). Total Hallawi cultivar phenols increased from 190 to 491 mg GAE/100 g in control and from 212 to 480 mg GAE/100 g in Khadhrawi cultivar to fertilize two separate cultivars using eight male pollen patents. Similar effect with ascorbic acid has also been found. Farag, Elsabagh, and ElAshry (2012) found that one of two groups of pollinators substantially increased the ascorbic acid and anthocyanin content but not tannins.

Pollination is one of the key preharvesting factors affecting the dates' quality (Al-Delami & Ali, 1969). For commercial planting (hand or mechanical pollination), female trees are artificially pollinated with male pollen. Pollinator selection is of utmost significance in date palm, as the form of pollen parent determines the fruit size, time of maturation, and chemical composition of fruit called metaxenia (Abbas, 1997).

3.3 Plant organ

The quantity of bioactive compounds in *P. dactylifera* is greatly different, depending on tissue type. For example, the highest phytosterol concentration was found in pollen grains with the lowest roots (El Hadrami, Daayf, & El Hadrami, 2011). Generally, *P. dactylifera* fruit contains high levels of bioactive compounds, and the roots contain the least amount (Biglari, AlKarkhi, & Easa, 2008).

3.4 Maturity stage

At a preliminary stage of fruit production, content of certain bioactive components, like phenolic acid, begins to increase and reach maximum concentration at Bisir level (Fig. 2) (El Arem, Saafi, Mechri, et al., 2012; Lemine et al., 2014). The overall carotenoid content at the Bisir level, for example, was highest compared with its Tamer level content (Boudries et al., 2007). Phenolic acid reduction was around 25% during the ripening period (Awad et al., 2011). However, the study of the relationship between the maturing stage of *P. dactylifera* fruits and antioxidant activity showed which stage Bisir has highest antioxidant activity relative to Rutab and Tamer stage (Allaith, 2008; Lemine et al., 2014). This may be due to hormonal homeostasis during fruit growth and production (Al-Alawi, Al-Mashiqri, Al-Nadabi, Al-Shihi, & Baqi, 2017).

Maturation of the date fruit is a long process, and the maximum postpollination stage of maturation takes about 7 months to achieve. They pass through five key stages of maturation according to the Arabic tradition and generally agreed terminology: Habbuk, Khalal, Rutab, Kimri, and Tamr (Al-Shahib & Marshall, 2013). Color, scale, texture, content of moisture, sugar content, and sweetness vary from each point to the next (Table 3); other than that, chemical composition is different also.

TABLE 3 Stages of maturation of the date palm (Mirza et al., 2019).

Stage	Color	Duration	Moisture content	Sugar content (g/100g)	Image
Hababouk	Green	4–5 weeks	–	–	
Kimri	Green	9–14 weeks	80%	6.2	
Khalal	Greenish, yellow, or red	6 weeks	65%	26.6	
Rutab	Yellow, brown	2–4 weeks	43%	45.2	
Tamr	Brown or black	4 weeks	24%	50.8	
Tamr (dry)	Yellowish brown	–	Hydrated	–	

4. Abiotic factors influencing biosynthesis

4.1 Sunlight

Postharvesting processes, such as storage temperature and sun drying in *P. dactylifera* fruits, can influence the total bioactive compound content (Fig. 10). There is a significant loss of carotenoid concentration during sun drying (Al-Farsi et al., 2005). This reduction may be due to enzymatic oxidation or carotenoid thermal degradation (Al-Farsi & Lee, 2008). Fall in the concentration of carotenoids ranged from 4% to 30% (Al-Farsi et al., 2005). Nonetheless, an important increase in total phenolic acids occurred during sun drying or cold storage. This rise could be due to tannin degradation during the sun-drying process due to thermal and enzymatic effects or due to enzymatic activity with cold storage (Al-Najada & Mohamed, 2014). Incidentally, several Saudi *P. dactylifera* fruit cultivars have experienced a significant rise in anthocyanin during cold storage at 4°C (Samad, Hashim, Simarani, & Yaacob, 2016).

4.2 Irrigation

In the antioxidant constituency of the three Saudi cultivars, the effect of sewage water irrigation found higher levels of total phenols and total flavonoids increased antioxidant activity through complex DPPH, ABTS, and phosphor-molybdenum testing, and increased production of glutathione-S-transferase, polyphenol oxidase, and peroxidase in water irrigation dates compared with municipal water irrigation dates (Abdulaal et al., 2017). Following the increased degree of these criteria was increased accumulation of heavy metals (Cr, Cu, Fe, Mn, Pb, and Zn) in

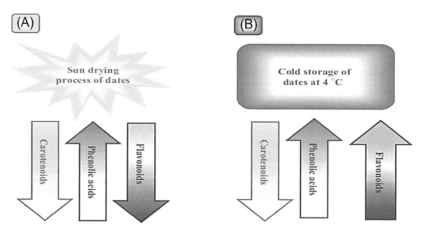

FIG. 10 The effect of date processing on the consistency of certain bioactive compounds, (A) the sun drying effect, and (B) cold storage. ⇑ = increase; ⇓ = decrease (Al-Mssallem, Alqurashi, & Al-Khayri, 2019).

years of irrigated wastewater. Three research cultivars showed different reactions to complete flavonoids and phenols. In Agwa and Safawi, the average phenol level increased over time by 28%–30%, while in Anbr it increased by just 8%. However, the rise in TF was very similar in the three cultivars (Agwa, 41%; Safawi, 50%; and Anbr, 50%), while levels of Fe, Zn, and Ni in treated sewage water were significantly higher than those of irrigated groundwater. Levels of Cu, Cd, Pb, and B were substantially higher in soil-irrigated fruits than in irrigated sewage water. These contradictory results can be due in part to the degree of care used for drinking water, i.e., secondary or tertiary care. There was no significant difference in concentration of multiple heavy metals between groundwater and nonirrigated, untreated sewage dates in our own (unpublished) findings with multiple date cultivars grown locally (Al-Busaidi, Shaharoona, Al-Yahyai, & Ahmed, 2015).

5. Pharmaceutical potentialities

5.1 Antioxidant

Studies of *P. dactylifera*'s antioxidant effect showed that reactive oxygen species (ROS) and reactive nitrogen species (RNS) decreased the body's antioxidant defense mechanism and caused cellular oxidant damage. In the Trolox equivalent antioxidant efficiency (TEAC) study, *P. dactylifera* showed antioxidant activity, 2,2′-azinobis (3-ethylbenzothiazoline-6-sulphonic acid) radical cation (ABTS+) test, and the ferric power reduction/antioxidant test (FRAP) method (Biglari et al., 2008). *P. dactylifera* also contains a high proportion of vitamins C, A, and E plus a high total phenolic content (Hasan et al., 2010; Samad et al., 2016). Such results from in vitro studies prompted researchers to investigate the antioxidant function of *P. dactylifera* extracts in vivo against various toxicants, such as carbon tetrachloride (CCl4), streptozotocin-induced isoproterenol, cadmium, and diabetes in rats (Abdelaziz & Ali, 2014; Abdelaziz, Ali, & Mostafa, 2015; Al-Yahya et al., 2016; El-Neweshy, El-Maddawy, & El-Sayed, 2013). The protective role of *P. dactylifera* can be associated with increased activity of antioxidant enzymes such as catalase (CAT), GPx, SOD, glutathione S-transferase (GST), glutathione reductase (GR), and a substantial reduction in malondialdehyde (MDA). To sum up, *P. dactylifera* can be said to be a healthy antioxidant food (Hoseinifar, Dadar, Khalili, Cerezuela, & Esteban, 2017; Sharifi, Bashtani, Naserian, & Farhangfar, 2017).

Bouhlali et al. (2017) investigated various date palm varieties and reported highly associated antioxidant activity with polyphenol and flavonoid content. Vinson, Zubik, Bose, Samman, and Proch (2005) identified the date fruits with the highest concentration of polyphenol between dried fruits. This can be because of their increased exposure to sunlight and extreme temperature. Antioxidants are defensive in nature that works against ROS, which play an important role in many diseases. The free radicals are neutralized by antioxidants, which render them harmless to other cells. Endogenous antioxidant production is inadequate for some cases to meet the needs, with exogenous supplementation helping to protect the body from the harmful effects of ROS.

The fruits, vitamins, carotenoids, tannins, polyphenols, etc. found in dates confer high free radical scavenging and antioxidant power on them. Fruit dates also increase antioxidant activity, such as catalase and superoxide dismutase (Al-Farsi & Lee, 2008). The date seeds do have antioxidant potential and can be used as a cheap natural source of antioxidants (Thouri et al., 2017).

For the first time, in vitro experiments with Vayalil (2002) showed that the date fruit aqueous extract was an effective scavenger of superoxide and hydroxyl radicals and prevented concentration-dependent lipid peroxidation and protein oxidation in the rat brain. Other investigators later checked those findings with combinations of different dates (Allaith, 2008; Mansouri et al., 2005). A study recently measured antioxidant activity on the Tunisian date, and the study showed the best radical lipoperoxyl scavenging behavior of the Korkobbi variety, while the Rotbi variety has been very effective in scavenging hydroxyl radicals (Chaira et al., 2009).

Korkobbi > Bouhattam > Baht = Smiti > Bekreri = Garn ghzal > Mermilla = Kenta > Nefzaoui = Rotbi has been recognized as the Trolox equivalent antioxidant value. The authors assume the greater antiradical effects are responsible for the highest amount of flavonoids in the Korkobbi variety (Chaira et al., 2009).

Also, animal studies have shown an improvement in the expression of the antioxidant enzyme gene in cardiac tissue by oral feeding of present p-coumaric acid (Yeh, Ching, & Yen, 2008). The presence of phenolic compounds, anthocyanins, flavonoid glycosides, and procyanidins caused the antioxidant activity observed in the dates (Allaith, 2008). Selenium, too, is believed to contribute to the antioxidant activity in dates. Several studies have shown that this important trace element mainly exercises its antioxidant function in the form of residues of selenocysteine, an integral component of the ROS-detoxifying selenoenzymes (GPx, thioredoxin reductases, and possibly selenoprotein P) (Steinbrenner & Sies, 1790). It is very clear, if taken as a whole, that the existence of different phenolic compounds and selenium could have been responsible for the free radical scavenging and antioxidant effects observed (Ferguson, Philpott, & Karunasinghe, 2004).

5.2 Anticancer

Antitumor activity has been demonstrated in fruit constituents of *P. dactylifera.* In vitro studies have been conducted to evaluate the anticancer activity of *P. dactylifera* extracts against various cancer cell lines such as human epithelial colorectal adenocarcinoma (Caco-2) and human melanoma derived cell line (IGR-39) (Chakroun et al., 2016; Eid et al., 2014). The mechanism by which the methanol extract of Ajwa date inhibited the development of human breast adenocarcinoma (MCF7) by up-regulating proapoptotic molecules, p53, Bcl-2-associated X protein (Bax), Fas, and Fas ligand (FasL) along with down-regulation of B-cell lymphoma 2 (Bcl-2) were investigated (Khan et al., 2016). Two of the key advantages of date fruit are the production of polyphenols, flavonoids, folic acid, etc.

and dates are used to determine the resistance and defense against cancer. Khan et al. (2017) investigated the effect of ajwa date fruit extract on hepatocellular carcinoma on the rat model and showed that date fruit extracts can restore the hepatic to normal conditions by restoring antioxidant enzymes, liver enzymes, cytokine balance, etc. A previous study (Ishurd & Kennedy, 2005) offered evidence of date fruit's role as an anticancer in mice. Lybian-dated fruit extract was capable of inhibiting female mice's development of sarcoma. Date fruits in rats have a strong protection against mammary cancer caused by DMBA (Al-Sayyed, Takruri, & Shomaf, 2014). At different stages of carcinogenesis, the date fruit exerts its effect on initiation, progress, and growth. By inhibiting proliferation of colon cancer cells, daily intake of date fruits will improve health of the colon and encourage growth of beneficial bacterial cells too. The digested date extract and the polyphenol extract significantly prevented human colorectal adenocarcinoma cells (Caco-2) from developing. This antiproperty of the date fruit is attributable to the presence of phenolics and flavonoids that play major roles in up-apoptotic molecules such as p53, caspases, and Bax or acting as down-apoptotic antimolecules, viz., Akt, Bcl-2, nuclear factor πB (NFÿB), etc. Methanol extract of 29 Saudi varieties displayed lung (NCI-H460), breast (MCF-7), gastric (AGS), prostate (DU-145 and LNCaP), and colon (HCT-116) in vitro cancer cell lines (Zhang, Aldosari, Vidyasagar, Shukla, & Nair, 2017).

5.3 Antidiabetic

Several studies have demonstrated the antidiabetic function of *P. dactylifera* in rat diabetes by elevating plasma insulin with normalization of alloxane-induced plasma glucose, triacylglycerol, and cholesterol (Mard, Jalalvand, Jafarinejad, Balochi, & Naseri, 2010). If this is due to an improvement in the growth of healthy β-cells or the regeneration of alloxane-injured cells, the real cause of rise in plasma insulin should be considered.

Date seeds often contain rise nutrients, rise carbon content, and high fatty acids. The effect of date seed extract on diabetic rats is caused by streptozotocin. Date seed extract repair kidney and liver function to alleviate oxidizing stress in diabetes-induced laboratory animals (Hasan & Mohieldein, 2016). The study specifically states that antidiabetic drugs are derived from the seeds of origin. Michael, Salib, and Eskander (2013) studied effects of glycoside diosmetin (isolated from the date fruit epicarp) on laboratory diabetic animals affected by alloxane. The results showed a substantial rise in serum glucose levels, liver function, antioxidant enzymes, and a large decrease in cholesterol and triglycerides. Insulin production and inhibiting glucose absorption can contribute to antidiabetic action. This behavior may be responsible for flavonoids, saponins, phenols, and hormones found in dates. Date phenolics are efficient inhibitors of alpha glycosidase and alpha amylase, resulting in decreased carbohydrate absorption and digestion, which can decrease hyperglycemic conditions (Ranilla, Kwon, Genovese, Lajolo, & Shetty, 2008). A study carried out by researchers on healthy volunteers found a low glycemic level for days of eating, either alone or in mixed meals with plain yogurt. Fruit-eating diabetes patients help control their lipid profile and glycemic index (Miller, Dunn, & Hashim, 2003).

5.4 Antinephrotoxicity

It has also been reported that *P. dactylifera* plays an essential role in the treatment and control of various xenobiotic nephrotoxicities. In a study carried out using a CCl4-encouraged toxicity model in rats, a dose-dependent hydroacetone extract (50 or 100 mg/kg/rat), intended from *P. dactylifera* seeds, was led to confer significant safety on the kidney (Ahmed et al., 2015). According to scientists, the extract's nephroprotective potential can be clarified by its ability to actually scavenge free radical produced during CCl4 metabolism, possibly even though extract produced large concentrations of proanthocyanidins, which had strong antioxidant activity (Ahmed et al., 2015). Another study by researchers showed that continuous treatment of aqueous extracts of *P. dactylifera* could restore function in the renal system (Hasan & Mohieldein, 2016). Throughout this study, it was found that diabetic rats had different degrees of renal dysfunction while treatment with aqueous *P. dactylifera* extracts increased progressive deterioration during renal dysfunction of rats being treated.

Aflatoxin B1 (AFB1) causes histopathological changes in kidney related to renal failure as specified by rising in plasma creatinine levels and urea. Intraperitoneal administration of AFB1 was responsible for toxicity. Treatment with aqueous *P. dactylifera* extract has produced a substantial decrease in plasma creatinine and urea levels for 2 weeks, as well as a major increase in kidney architecture. They suggested that the effect found could be due to antioxidant properties of the *P. dactylifera* extract (Al-Ghasham, Ata, El-Deep, Meki, & Shehada, 2008).

Ironically, the administration of either *P. dactylifera* flesh extract with food (50% w/w) or seed extract in drinking water (2:1 w/v) resulted in a substantial reversal of recorded kidney toxicity indices. Nephroprotection has been suggested to be dependent on melatonin, vitamin E, and ascorbic acid, which are abundantly present in the extract and can function synergistically to counteract the overwhelming effect produced by free radicals (Al-Qarawi, Abdel-Rahman, Mousa, Ali, & El-Mougy, 2008).

5.5 Antimicrobial

Various extracts and oils from *P. dactylifera* demonstrate strong antimicrobial activity. Others investigated the effect of *P. dactylifera* on streptolysin O hemolytic activity and stated that *P. dactylifera* fruit extract decreased Streptococcus pyrogens growth by 88.5% compared with control (Abuharfeil, El Sukhon, Msameh, & Sallal, 1999). In a disc diffusion test, it was measured that *P. dactylifera* aqueous fruit and ethanol extracts had strong antibacterial activity against *Salmonella enterica*, *Escherichia coli*, and *Bacillus subtilis*, and that *Enterococcus faecalis* and *Staphylococcus aureus* were moderately inhibited. This antibacterial activity is triggered by presence of esculetin, tannic acid, low gallic acid content, itaconic acid, and traces of ferulic acid found in extracts from *P. dactylifera* (El Sohaimy, Abdelwahab, Brennan, & Aboul-enein, 2015).

Mabroom, Ajwa, and Safawi and even Iranian *P. dactylifera* cultivar Mariami assessed for methanol and acetone extract antibacterial activity in another study. The methanol extract from Mabroom was most effective against *Staphylococcus aureus* than the methanol extract from Ajwa and Mariami. The lowest concentration of Mabrooom and Mariami methanol extracts capable of inhibiting *Bacillus cereus* growth was 500 mg/mL while Safawi and Ajwa methanol extracts that inhibit the growth of this bacterial species was at 400 mg/mL (Samad et al., 2016).

Finally, Boufgous, Bousthammi, Bouskri, Bousrdon, Jihl, and Majdoul were tested toward Gram-positive (*B. cereus*, *Bacillus subtilis*, and *S. aureus*) and Gram-negative bacteria (*E. coli*, *Salmonella abony*, *Pseudomonas aeruginosa*) using a disc diffusion method. Except for extracts from Majdoul and Bouskri, all fruit extracts tested with *P. dactylifera* exhibited antibacterial activity, but this antibacterial impact was lower than gentamicine. The Bousrdon and Jihl extracts were more effective inhibitors with minimum concentration (MIC) values of 2.5–10 mg/mL for all the bacterial strains examined (Bouhlali, Ramchoun, et al., 2017).

In the other hand, the more effective molecules attributed to antimicrobial activity in *P. dactylifera* are phenolic compounds, containing hydrogen peroxide that inhibits bacterial growth (Taleb, Maddocks, Morris, & Kanekanian, 2016a). Phenolic compounds, especially Gram-positive, are presumed to use redox active metals while dealing with bacteria (Taleb, Maddocks, Morris, & Kanekanian, 2016b). Samad et al. (2016) indicate that extracts of the date could be commonly used in the nutraceutical and pharmaceutical industries due to their antibacterial and antioxidant properties. The Ajwa date methanol extract from the fruit has shown antibacterial activity against *Bacillus cereus*, *Escherichia coli*, *Staphylococcus aureus*, and *Serratia marcescens*. *Fusarium oxysporum* antifungal operation has the date palm seed and pollen water extract (Bentrad, Gaceb-Terrak, Benmalek, & Rahmania, 2017). The origin of the polyphenols in date extracts is important for their antifungal activity (Boulenouar, Marouf, & Cheriti, 2011), whereas the existence of fatty acids such as palmitic, stearic, oleic, linoleic acids, and so on imparts their antibacterial potential (Bentrad, Gaceb-Terrak, & Rahmania, 2017). Extracts from Tunisian cultivar date had antimicrobial activity toward both Gram-positive and Gram-negative bacteria; in particular, strong activity against E.coli was exercised (Kchaou et al., 2016).

5.6 Antiviral

Extracts of crude acetone from the dates pit against Pseudomonas phage ATCC 14209-B1, using *Pseudomonas aeruginosa* ATCC 25668 as the host cell (Jassim & Naji, 2010). Pseudomonas phage ATCC 14209-B1 was inhibited by date-pit extract, and it eliminated bacterial lysis against *E. coli*. Antiviral activity of date pit was observed to be mediated by the binding phage with an inhibitory amount of at least 10 mg/mL (Jassim & Naji, 2010). Antiviral work should be extended in the midst of those findings to other viruses that are essential to humans.

5.7 Antifungal

The treatment with Barhi date extract of *Candida albicans* caused the cell wall to be bent, weakened, and partially collapsed. At high concentrations, leakage of cytoplasmic material, subsequent cell death, and dramatic damage in the form of cell lysis was observed (Shraideh, Abu-Elteen, & Sallal, 1998). Overall, these results suggest that, after the date extract, phytochemicals can have many effects on Candida, and further work should be carried out for therapeutic applications. In vitro experiments have shown that flavonoids demonstrate antifungal activity toward *C. albicans* as well as *C. krusei*, and may have been responsible for the reported antifungal effects and their inclusion in the extract (Belmir, Boucherit, Boucherit-Otmani, & Belhachemi, 2016; Ozçelik, Orhan, & Toker, 2006).

5.8 Antiinflammatory

P. dactylifera's anti-inflammatory effects could be demonstrated as a decrease in the amount of paw edema (Ali Haimoud, Allem, & Merouane, 2016), wound healing activity (Abdennabi et al., 2016), and cardioprotective activity. Studies summarize the anti-inflammatory properties of *P. dactylifera* in accordance with this evidence (Al-Yahya et al., 2016).

P. dactylifera aqueous and ethanolic undialyzed and dialyzed extracts have been effective in improving gastric ulceration levels and reducing the increased levels of histamine and gastrine caused by ethanol and decreased levels of mucin gastric. According to scientists, the anti-inflammatory ability of the extract can be explained by antioxidant activity of the extract (Al-Qarawi et al., 2008).

The phenolic and flavonoid content can inhibit the production of prostaglandin endoperoxide, ending inflammatory mediators such as prostaglandins and thromboxane (Zhang, Aldosari, Vidyasagar, Nair, & Nair, 2013). Researchers looked at the anti-inflammatory role of *P. dactylifera* fruits grown as methanolic extracts in Algeria. In Swiss albino mice, they recorded a substantial decrease in paw volume as a reaction to an oral dose of methanol extract in the carrageenan-induced acute paw oedema experimental model (Ali Haimoud et al., 2016).

In traditional medicine, *P. dactylifera* was used to treat inflammatory-related diseases (Yasin, El-Fawal, & Mousa, 2015). This was used alongside drugs such as paracetamol and ibuprofen as a pain reliever (Maryam et al., 2015; Sani et al., 2015). There is no detailed explanation for these findings, but *P. dactylifera* has been documented to have effective compounds that may interact with prostaglandin synthesis (Taleb et al., 2016b), inhibit development of inflammatory cytokines like Interleukin 6, Interleukin, Interleukin 10, tumor necrosis factor (TNF-α), and growth factor insulin-like (IGF-1) (Al-Yahya et al., 2016). Analgesic results from the presence of vitamin C and E (Maryam et al., 2015) and increased expression of growth factor (TGF-β) (Elberry et al., 2011). On the other hand, a potential mechanism for anti-inflammatory activity of *P. dactylifera* may be to potentiate the antioxidant system, as previously described.

Taleb et al. (2016b) also stated being closely linked to the anti-inflammatory role of the date with its secondary metabolites and antioxidants. Free radicals play an significant part in inflammatory response upregulation. Secondary metabolites, especially phenolics and flavonoids, act as suppressors of NF-KB, and act as anti-inflammatory factors.

The difference between different date varieties in anti-inflammatory activity depends in large part on the variations in their phenolic and flavonoid content (Bouhlali et al., 2017). Researchers studied different extracts of ajwa date fruits and documented the potent antioxidant and anti-inflammatory properties of bioactive compounds in them (Zhang et al., 2013). Enzymes such as cyclooxygenase, Cox-1, and Cox-2, the major inflammatory mediators, were inhibited by these compounds (Zhang et al., 2017). A model of chronic inflammation reported the effects of date methanolic fruit and water extracts in rat-induced adjuvant arthritis (Al-Okbi & Mohamed, 2012). Researchers assume the extracts will substantially reduce swelling in the foot while normalizing plasma contents of antioxidants. Algerian date fruit extract was able to reduce formalin-caused edema rates, homocysteine contents, and C-reactive protein levels in rats (Kehili, Zerizer, Beladjila, & Kabouche, 2016).

Studies have shown that, in the model of rat with adjuvant arthritis, methanol and aqueous extract from date fruit pulp as well as methanol extract from date seeds have anti-inflammatory benefit. The extracts boosted the plasma contents of antioxidants (β-carotene, vitamin A, C, and E) and lowered levels of lipid peroxide. The extract of pulp methanol was most active in reducing all the extracts from foot swelling, ESR, and plasma fibrinogenics. The extracts raised body weight and proved their importance by the food efficiency ratio. Acute toxicity tests on mice found methanol extract and pulp aqueous extract stable up to 6 g/kg, while LD50 was 6.75 g/kg for seed methanol extract (Doha & Al-Okbi, 2004). Countless studies have shown that date components of proanthocyanidin, flavonoids, polyphenols, and β-carotene have anti-inflammatory properties in various models of the test and could have therapeutic effects (Gescher, 2004; Robak & Gryglewski, 1996; Subarnas & Wagner, 2000; Uteshev et al., 2000).

5.9 Antihemolytic

Streptococcus pyogenes, which induces erythrocyte hemolysis, produces Streptococcus O (SLO) bacterial toxin. The date-fruit extract could neutralize SLO's hemolytic activity at low concentrations (Abuharfeil et al., 1999). Date fruit extract can also inhibit hemolytic activity in snake and scorpion venom (Sallal, Rim, Amr, & Disi, 1997). A study inferred that date-fruit extract showed substantial antihemolytic activity in AAPH (free radical generator)-mediated rabbit blood, positively linked to polyphenols and flavonoids (Bouhlali et al., 2016). Flavonoids and other polyphenols boost the stability of erythrocyte membrane, whereas antioxidants scavenge lipid peroxyl radicals that are generated by AAPH.

5.10 Antihyperglycemic

One of *P. dactylifera's* antihyperglycemic activities is suppression of either α-amylase orα-glucosidase in vitro, possibly slowing digestion and absorption of carbohydrates contributing to plasma glucose levels becoming normalized. This result was caused by the hydroalcoholic extract of *P. dactylifera* leaves Mabrouk, Chakroun, El Abed, Drira, Khemakhem, Bouaziz, Marrakchi, Makni, and Mejdoub and aqueous extract of *P. dactylifera* seeds (Chakroun et al., 2016; Khan et al., 2016). With each rodent, Hasan and Mohieldein (Hasan & Mohieldein, 2016) have also treated diabetic streptozotocin-induced rats with 10 mL aqueous extract of *P. dactylifera* seeds per day. The extract being tested substantially invert diabetic rats' high serum glucose, cholesterol, and triacylglycerol levels to near normal values.

5.11 Antihyperlipidemic

Hyperlipidemia is the main risk factor contributing to morbidity and mortality in Western culture for atherosclerosis, and scientists are seeking to find new approaches to protect against hyperlipidemia. Many pathways improve the characteristic 3-hydroxy-3-methylglutaryl-coenzyme A (HMG-CoA) that has an important role to play in organizing serum lipid profile. In addition, some herbal extracts stimulate lipid accumulation repression during adipogenesis or prevent lipid development, resulting in lower serum triacylglycerol, total cholesterol, and LDLC. An alloxan diabetic rats study was performed over 4 weeks on hypolipidemic effect of *P. dactylifera* leaves (DPL) along with flax seed extracts (FS) (Abuelgassim, 2010). Either FS or DPL extract resulted in a substantial 4th week decrease in FS and DPL levels of total serum cholesterol by 40% and 31%, respectively.

All extracts decreased serum LDL-C levels significantly in the 2nd week, although no major alterations were identified in serum high-density lipoprotein cholesterol (HDL-C) levels. Results propose FS and DPL extracts in diabetic rats may have hypolipidemic effect. To minimize diabetic complications, the experiment period should be prolonged to examine the effect of both extracts on glycation products like hemoglobin A1c (HbA1c) and advanced glycation end products. It will also be helpful to investigate the hypolipidemic role of *P. dactylifera*.

The most recent studies note the antihyperlipidemic effect of palm date. Hypolipidemic effects on the diabetic rats caused by streptozotocin (Hasan & Mohieldein, 2016) and hypercholesterolemic rats (Takaeidi et al., 2014) were reported in an experimental study of Saudi (Ajwa and Sukkary) and Iranian date palm seeds. Another study found that date fruit (Aseel) extract in male albino rats has significantly reduced cholesterol and triglyceride levels (Ahmed, 2016). The experimental work has therefore shown that date palm has the capability to reduce cholesterol levels and protect the heart.

5.12   Antiinfertility

5.12.1   Impact on male infertility

In experiments with male rats, date extracts provided defense against oxidative damage induced by dichloroacetic acid (El Arem et al., 2017). This recorded therapeutic potential of ADE can be attributed to different contributing factors, primarily due to composition of those fruits in minerals (iron, copper, calcium, zinc, magnesium, selenium, manganese, cobalt) and vitamins (A, B, C) (Baliga et al., 2011). The presence of polyphenolic compounds, flavonoids, which are highly protective toward ROS-mediated damage, often contributes to testicular defense (Garcia & Castillo, 2008). Indeed, existence of quercetin, an active bioflavonoid in date extracts, is one of the significant elements responsible for the testicular protective effect. Quercetine suppresses production of free radicals in cells and suppress lipid peroxidation. Antioxidant and antiapoptotic effects of quercetin have substantially prevented arsenic-caused testicular damage in rats (Baltaci et al., 2016). Microelements, which have affirmative effect on male fertility in date palm pollen, extract viz., estrone, sterols, and other chemicals. The suspension of the palm pollen extract also increases levels of plasma estradiol and testosterone (Bahmanpour et al., 2006).

5.12.2   Impact on female labor and lactation

Zangeneh, Moezi, and Amir (2009) found that the late consumption of date fruits during pregnancy improve rate of labor and increase pain tolerance. The fatty acids of date fruit help to save energy and reinforce the uterus muscles. Studies have shown that date fruits have an effect on oxytocin receptors, stimulate uterine muscles to react to oxytocins more easily, and prepare the uterus and cervix better for delivery (Al-Kuran, Al-Mehaisen, Bawadi, Beitawi, & Amarin, 2011). The study suggests that the intake of date fruits during the last 4 weeks of pregnancy decreases the need for induction, increases labor, and provides a favorable outcome in delivery. Others stated that fruit from the date impedes a postpartum hemorrhage (Khadem, Sharaphy, Latifnejad, Hammod, & Ibrahimzadeh, 2007). Other study reported on higher cervical stretching in females consuming date fruit. For this purpose, it is recommended that pregnant women eat date fruit, particularly during the last week of gestation (Kordi, Aghaei, Tara, Nemati, & Taghi, 2014). In addition, date fruit has minimized labor interference without influencing mother and child (Razali, Mohd Nahwari, Sulaiman, & Hassan, 2017). Prolactin synthesis is believed to be due to the existence of potassium, glycine, and threonine in date fruits. Date fruit is also suggested in feeding mothers as a galactogogue, due to oxytocin abundance (Tang, Shi, & Aleid, 2013).

5.13 Neuroprotective

Driven by oxidative stress, the effect of free radical development has been well established to demonstrate the pathogenesis of different neurological disorders like a cognitive deterioration in aging, Parkinson's disease, Alzheimer's disease, and vascular dementia (Uttara, Singh, Zamboni, & Mahajan, 2009). Thanks to their ability to prevent and depart free radicals, nutraceutical antioxidants from plant-derived foods are now recognized as an effective therapy toward solemn neuronal failure, thus providing neuroprotection. Nevertheless, *P. dactylifera* was found to be neuroprotective and antioxidant toward regular bilateral common carotid arteries (BCCA) (Pujari, Vyawahare, & Kagathara, 2011).

They reported that persistent occlusion of BCCA led to significant raise in lipid peroxidation, as demonstrated by elevated levels of malondialdehyde. A general decrease was also observed in endogenous antioxidants, including glutathione, GST, GPx, CAT, GR, and SOD. Researchers concluded that extracts of 100 and 300 mg/kg significantly showed improvements and identified the protective function of the extract in ischemic hypoperfusion. The observed neuroprotective effect can be accounted for by the extract's polyphenolic constituents lik eplant sterols and flavonoids, as well as their content of ascorbic acid (Pujari, Vyawahare, & Thakurdesai, 2014).

A related research to investigate the neuroprotective effect of aqueous *P. dactylifera* fruit extract in focal cerebral ischemia in rats suggested that a 250 mg/kg extract dose substantially precluded cerebral ischemia-induced neuronal harm. Rats were transported in different extract doses (125, 250, 500, and 1000 mg/kg). The extract was given once daily over 2 weeks; the extract was observed with the greatest beneficial effects at a dosage of 250 mg/kg. A lower beneficial effect at 500 mg/kg was detected, and a harmful effect at 1000 mg/kg was observed due to elevated levels of toxic antioxidants. They concluded that the 250 mg/kg extract could memorize the neurons from insults induced by ischemia-reperfusion (Majid, Marzieh, Shahriar, Zahed, & Pari, 2008). Therefore *P. dactylifera* has a great potential as antitoxic agent in treatment and management of hepatotoxicity, nephrotoxicity, and neurotoxicity.

Date-palm fruit is an excellent source of dietary fiber, abundant in total phenolics and natural antioxidants such as anthocyanins, caffeic acid, ferulic acid, and protocatechic acid. These are polyphenolic compounds and have neuroprotection. Known as Kharjura, Ayurvedic date palm is recommended for treating depression, anxiety, cognitive dysfunction, and nervous system disorders (Shanmugapriya & Patwardhan, 2012). Date fruit has demonstrated strong brain-protective activity in mice toward cerebral ischemia (Kalantaripour, Asadi-Shekaari, Basiri, & Najar, 2012). This process is responsible for the production of flavonoids, sterols, and ascorbic acid. Researchers examined the neuropharmacological and analgesic properties of date fruit in rats and said this effect can be due to the existence of polyphenols viz., catechin, epicatechin, and transferulic acid (Sheikh et al., 2016).

6. Metabolomics

Using downstream gene expression, plant metabolomics was applied to address the information gap between plant genotypes and phenotypes, representing different biological endpoints. Extensive knowledge of the composition of metabolites would enable the evaluation of useful genotypic or phenotypic variations in plant varieties and species (Arbona, Iglesias, Talón, & Gómez-Cadenas, 2009; Ku et al., 2010). Metabolomics has recently been successfully applied to determine variations in metabolite composition between different plant taxa, varieties, and/or introgressive lines. Using these techniques may help to establish rational breeding programs, especially if desirable nutritional features do not correlate with specific DNA markers (Harrigan, Martino-Catt, & Glenn, 2007). Many widely used metabolomic techniques combine mass spectrometry (MS) with separation techniques such as HPLC, GC, or capillary electrophoresis (Sumner & Hall, 2013).

Fruits of *P. dactylifera* are metabolically classified using MS analysis. The composition of the date palm seeds, cultivars, and postharvest methods varies considerably. Metabolic characteristics of 123 cultivars of *P. dactylifera* occurred at various stages of maturing. Phenotype classification is based on seed length, width, seed thickness, and weight (Stephan et al., 2018). Metabolomic methods have been employed to quantify source variation, fruit processing conditions, and cultivars. The maturation of the dates was the principal determinant of the *P. dactylifera* fruit metabolome (Diboun et al., 2015).

6.1 Analysis of secondary metabolites: Date palm fruit varieties

Together with multivariate data analysis, SPME-GCMS was used to profile volatile components from 13 date cultivars grown in Egypt to explore the composition of fruit aromas and explore possible future applications by the food industry. A total of 89 volatiles were recorded, with the volatile extracted from lipids and phenylpropanoid derivatives being the main components of date fruit aroma. Multivariate data analysis showed that 2,3-butanediol, hexanol, hexanal, and cinnamaldehyde were main contributors to the classification of the specific species. This research promotes more detailed map of volatile Egyptian date fruit with more distinct aroma among the Siwi and Sheshi varieties studied (Khalil, Fekry, & Farag, 2017). Siwi is named after an isolated oasis in western Egypt's desert and is thus less interbred with other cultivars. Crossing date varieties can be required to include varieties with minor variations in chemical composition like volatiles. Most cultivars meet market requirements, with only a few considered to be superior. In terms of fruit consistency, cheaper varieties make no sacrifices but only in terms of flesh size and pulp-to-fruit ratio (El Arem, Saafi, Flamini, et al., 2012). The connection between metabolomics and genomic data can provide a greater understanding about how crossing affects metabolic pathways that yield unpredictable fruits. It can also be presumed that the gene, which controls the morphometric properties of fruit, is isolated from those which regulate maturation and metabolism. It will provide insights into future crossing scheduling. Multidate genomes were sequenced recently. Comprehensive genome survey was performed of 70 date cultivars (Al-Mssallem et al., 2013), but varieties were categorized using morphometric date fruit characters (Rabei, Said, Rizk, & El Sharabasy, 2012) and with PCR-based molecular markers, e.g., ISSR or RAPD (Arabnezhad, Bahar, Mohammadi, & Latifian, 2012). Comparing this information with transcriptomic and metabolomic information would response questions concerning factors, which may influence the ripening and chemical composition of the fruit, even the phylogenetic relationship among various types in different countries.

7. Conclusion

Fresh or dried dates are eaten. Dried dates can be categorized as foods with intermediate humidity. Aside from being a great source of carbohydrates, dietary fibers, other basic vitamins, and minerals, a number of phytochemicals such as carotenoids, anthocyanins, phenolics, procyanidins, and flavonoids frequently prevail in dates. They also have high levels of insoluble fiber, which is essential for gastrointestinal safety. Furthermore, date boxes were a significant source of minerals, lipids, dietary fiber, and protein. Phytochemicals as well as their pharmacological effects often contribute to the nutritional and sensory properties of the dates. For decades, this fruit has been used in date-producing countries to treat a range of ailments in different traditional medicine systems. Over the past few years, researchers have interfaced to determine the health benefits of the dates and published multiple articles on the beneficial contribution of dates to the human diet. Similar to other vegetables and fruits called functional foods (e.g., carrots and grapes), fiber and antioxidants make the dates equally important. Ultimately, because of their broad range of nutritional and functional properties, dates can be considered as a safe food, which can play a critical nutritional and health role.

References

Abbas, M. F. (1997). Metaxenic effect in date palm (*Phoenix dactylifera* L.) cv. Hillawi in relation to levels of endogenous hormones. *Basrah Journal of Science, 15*, 29–36.

Abdelaziz, D. H., & Ali, S. A. (2014). The protective effect of *Phoenix dactylifera* L. seeds against CCl4-induced hepatotoxicity in rats. *Journal of Ethnopharmacology, 155*(1), 736–743.

Abdelaziz, D. H., Ali, S. A., & Mostafa, M. M. (2015). *Phoenix dactylifera* seeds ameliorate early diabetic complications in streptozotocin-induced diabetic rats. *Pharmaceutical Biology, 53*(6), 792–799.

Abdennabi, R., Bardaa, S., Mehdi, M., Rateb, M. E., Raab, A., Alenezi, F. N., et al. (2016). *Phoenix dactylifera* L. sap enhances wound healing in Wistar rats: Phytochemical and histological assessment. *International Journal of Biological Macromolecules, 88*, 443–450.

Abdulaal, W. H., Zeyadi, M., Baothman, O. A. S., Zamzami, M. A., Choudhry, H., Almulaiky, Y. Q., et al. (2017). Investigation of antioxidant and detoxifying capacities of some date cultivars (*Phoenix dactylifera* L.) irrigated with sewage water. *RSC Advances, 7*, 12953–12958. https://doi.org/10.1039/c6ra28760c.

Abuelgassim, A. O. (2010). Effect of flax seeds and date palm leaves extracts on serum concentrations of glucose and lipids in alloxan diabetic rats. *Pakistan Journal of Biological Sciences, 13*, 1141–1145.

Abuharfeil, N. M., El Sukhon, S., Msameh, Y., & Sallal, A. J. (1999). Effect of date fruits, *Phoenix dactyliferia* L., on the hemolytic activity of Streptolysin O. *Pharmaceutical Biology, 37*(5), 335–339.

Ahmad, T., Danish, M., Rafatullah, M., Ghazali, A., Sulaiman, O., Hashim, R., et al. (2012). The use of date palm as a potential adsorbent for wastewater treatment: A review. *Environmental Science and Pollution Research, 19*, 1464–1484.

Ahmed, S. (2016). Anti hyperlipidemic and hepatoprotective effects of native date fruit variety Aseel (*Phoenix dactylifera*). *Pakistan Journal of Pharmaceutical Sciences, 29*, 1945–1950.

Ahmed, A. F., Al-Qahtani, J. H., Al-Yousef, H. M., Al-Said, M. S., Ashour, A. E., Al-Sohaibani, M., et al. (2015). Proanthocyanidin-rich date seed extract protects against chemically induced hepatorenal toxicity. *Journal of Medicinal Food, 18*(3), 280–289.

Ahmed, M. J., & Theydan, S. K. (2012). Equilibrium isotherms, kinetics and thermodynamics studies of phenolic compounds adsorption on palm-tree fruit stones. *Ecotoxicology and Environmental Safety, 84*, 39–45.

Ait-Oubahou, A., & Yahia, E. M. (1999). Postharvest handling of dates. *Postharvest News and Information, 10*, 67N–74N.

Al Farsi, M., Alasalvar, C., Morris, A., Baron, M., & Shahidi, F. (2005). Compositional and sensory characteristics of three native sun-dried date (*Phoenix dactylifera* L.) varieties grown in Oman. *Journal of Agricultural and Food Chemistry, 53*, 7586–7591.

Al-Alawi, R. A., Al-Mashiqri, J. H., Al-Nadabi, J. S. M., Al-Shihi, B. I., & Baqi, Y. (2017). Date palm tree (*Phoenix dactylifera* L.): Natural products and therapeutic options. Front. *Plant Science, 8*, 845.

Al-Busaidi, A., Shaharoona, B., Al-Yahyai, R., & Ahmed, M. (2015). Heavy metal concentrations in soils and date palms irrigated by groundwater and treated wastewater. *Pakistan Journal of Agricultural Sciences, 52*(1), 129–134.

Al-Delami, K. S., & Ali, S. H. (1969). The effect of different date pollen on maturation and quality of Zahdi date fruit. *Journal of the American Society for Horticultural Science, 94*, 638–639.

Al-Farsi, M., Alasalvar, C., Morris, A., Baron, M., & Shahidi, F. (2005). Comparison of antioxidant activity, anthocyanins, carotenoids, and phenolics of three native fresh and sun-dried date (*Phoenix dactylifera* L.) varieties grown in Oman. *Journal of Agricultural and Food Chemistry, 53*, 7592–7599.

Al-Farsi, M. A., & Lee, C. Y. (2008). Nutritional and functional properties of dates: A review. *Critical Reviews in Food Science and Nutrition, 48*, 877–887.

Al-Ghasham, A., Ata, H. S., El-Deep, S., Meki, A. R., & Shehada, S. (2008). Study of protective effect of date and Nigella sativa on aflatoxin b(1) toxicity. *International Journal of Health Sciences (Qassim), 2*(2), 26–44.

Al-Harrasi, A., Rehman, N. U., Hussain, J., Khan, A. L., Al-Rawahi, A., Gilani, S. A., et al. (2014). Nutritional assessment and antioxidant analysis of 22 date palm (*Phoenix dactylifera*) varieties growing in Sultanate of Oman. *Asian Pacific Journal of Tropical Medicine, 7*(Suppl 1), S591–S598.

Al-Hooti, S., Sidhu, J. S., & Qabazard, H. (1998). Chemical composition of seeds of date fruit cultivars of United Arab Emirates. *Journal of Food Science and Technology, 35*, 44–46.

Ali, A., Al-Kindi, Y. S. M., & Al-Said, F. (2009). Chemical composition and glycemic index of three varieties of Omani dates. *International Journal of Food Sciences and Nutrition, 60*, 51–62.

Ali Haimoud, S., Allem, R., & Merouane, A. (2016). Antioxidant and anti-inflammatory properties of widely consumed date palm (*Phoenix Dactylifera* L.) fruit varieties in Algerian Oases. *Journal of Food Biochemistry, 40*(4), 463–471.

Al-Kuran, O., Al-Mehaisen, L., Bawadi, H., Beitawi, S., & Amarin, Z. (2011). The effect of late pregnancy consumption of date fruit on labour and delivery. *Journal of Obstetrics and Gynaecology, 31*, 29–31.

Allaith, A. A. (2008). Antioxidant activity of Bahraini date palm (*Phoenix dactylifera* L.) fruit of various cultivars. *International Journal of Food Science and Technology, 43*, 1033–1040.

Al-Mamary, M., Al-Habor, M., & Al-Zubairi, A. S. (2011). The in vitro antioxidant activity of different types of palm dates (*Phoenix dactylifera*) syrups. *Arabian Journal of Chemistry, 7*, 964–971.

Al-Mssallem, M. Q., Alqurashi, R. M., & Al-Khayri, J. M. (2019). Bioactive compounds of date palm (*Phoenix dactylifera* L.). In H. N. Murthy, & V. A. Bapat (Eds.), *Bioactive compounds in underutilized fruits and nuts, reference series in phytochemistry* (pp. 1–15).

Al-Mssallem, I. S., Hu, S., Zhang, X., Lin, Q., Liu, W., Tan, J., et al. (2013). Genome sequence of the date palm Phoenix dactylifera L. *Nature Communications, 4*, 2274.

Al-Najada, A. R., & Mohamed, S. A. (2014). Changes of antioxidant capacity and oxidoreductases of Saudi date cultivars (*Phoenix dactylifera* L.) during storage. *Scientia Horticulturae, 170*, 275–280.

Al-Okbi, S. Y., & Mohamed, D. A. (2012). Preparation and evaluation of functional foods in adjuvant arthritis. *Grasas y Aceites, 63*, 394–402.

Al-Orf, S. M., Mona, H. M., Norah, A., Al-Atwai, Al-Zaidi, H., Dehwah, A., et al. (2012). Review: Nutritional properties and benefits of the date fruits (*Phoenix dactylifera* L.). *Bulletin of the National Nutrition Institute of the Arab Republic of Egypt, 39*, 97–129.

Al-Qarawi, A. A., Abdel-Rahman, H., Mousa, H. M., Ali, B. H., & El-Mougy, S. A. (2008). Nephroprotective Action of *Phoenix dactylifera*. in gentami-cin-induced nephrotoxicity. *Pharmaceutical Biology, 46*(4), 227–230.

Al-Qarawi, A. A., Ali, B. H., Al-Mougy, S. A., & Mousa, H. M. (2003). Gastrointestinal transit in mice treated with various extracts of date (*Phoenix dactylifera* L.). *Food and Chemical Toxicology, 41*(1), 37–39.

Al-Saif, M. A., Khan, L. K., Alhamdan, A. A. H., Alorf, S. M., Harfi, S. H., Al-Othman, A. M., et al. (2007). Effect of dates and Gahwa (*Arabian coffee*) supplementation on lipids in Hypercholesterolemic hamsters. *International Journal of Pharmacology, 3*, 123–129.

Al-Sayyed, H. F., Takruri, H. R., & Shomaf, M. S. (2014). The effect of date palm fruit (*Phoenix dactylifera* L.) on the hormone 17-β-estradiol in 7,12-dimethylbenz (α) anthracene induced mammary cancer in rats. *Mediterranean Journal of Nutrition and Metabolism, 7*(1), 5–10.

Al-Shahib, W., & Marshall, R. J. (2002). Dietary fiber content of dates from 13 varieties of date palm Phoenix dactylifera L. *International Journal of Food Science and Technology, 37*(6), 719–721.

Al-Shahib, W., & Marshall, R. J. (2003). The fruit of the date palm: Its possible use as the best food for the future. *International Journal of Food Sciences and Nutrition, 54*, 247–259.

Al-Shahib, W., & Marshall, R. J. (2013). The fruit of the date palm: It's possible use as the best food for the future? *International Journal of Food Sciences and Nutrition, 54*, 247–259.

Al-Turki, S., Shahba, M. A., & Stushnoff, C. (2010). Diversity of antioxidant properties and phenolic content of date palm (*Phoenix dactylifera* L.) fruits as affected by cultivar and location. *Journal of Food, Agriculture and Environment, 8*, 253–260.

Aly, A. A., Mansour, M. T. M., Mohamed, H. I., & Abd-Elsalam, K. A. (2012). Examination of correlations between several biochemical components and powdery mildew resistance of flax cultivars. *Plant Pathology Journal, 28*(2), 149–155.

Aly, A. A., Mohamed, H. I., Mansour, M. T. M., & Omar, M. R. (2013). Suppression of powdery mildew on flax by foliar application of essential oils. *Journal of Phytopathology, 161*, 376–381.

Al-Yahya, M., Raish, M., AlSaid, M. S., Ahmad, A., Mothana, R. A., Al-Sohaibani, M., et al. (2016). 'Ajwa' dates (*Phoenix dactylifera* L.) extract ameliorates isoproterenol-induced cardiomyopathy through downregulation of oxidative, in-flammatory and apoptotic molecules in rodent model. *Phytomedicine, 23*(11), 1240–1248.

Al-Yahyai, R., & Al-Kharusi, L. (2012). Physical and chemical quality characteristics of freeze-estored dates. *International Journal of Agriculture and Biology, 14*, 97–100.

Amira, E. A., Guido, F., Behija, S. E., Manel, I., Nesrine, Z., Ali, F., et al. (2011). Chemical and aroma volatile compositions of date palm (*Phoenix dactylifera* L.) fruits at three maturation stages. *Food Chemistry, 127*, 1744–1754.

Arabnezhad, H., Bahar, M., Mohammadi, H. R., & Latifian, M. (2012). Development, characterization and use of microsatellite markers for germplasm analysis in date palm (*Phoenix dactylifera* L.). *Scientia Horticulturae, 134*, 150–156.

Arbona, V., Iglesias, D. J., Talón, M., & Gómez-Cadenas, A. (2009). Plant phenotype demarcation using nontargeted LC–MS and GC–MS metabolite profiling. *Journal of Agricultural and Food Chemistry, 57*(16), 7338–7347.

Ardekani, M. R. S., Khanavi, M., Hajimahmoodi, M., Jahangiri, M., & Hadjiakhoondi, A. (2010). Comparison of antioxidant activity and total phenol contents of some date seed varieties from Iran. *Iranian Journal of Pharmaceutical Research, 9*(2), 141–146.

Aslam, J., Khan, S. H., & Khan, S. A. (2011). Quantification of water soluble vitamins in six date palm (*Phoenix dactylifera* L.) cultivar's fruits growing in Dubai, United Arab Emirates, through high performance liquid chromatography. *Journal of Saudi Chemical Society, 17*, 9–16.

Assirey, E. A. R. (2015). Nutritional composition of fruit of 10 date palm (*Phoenix dactylifera* L.) cultivars grown in Saudi Arabia. *Journal of Taibah University for Science, 9*, 75–79.

Ateeq, A., Dutta Sunil, S., Varun, S. K., & Santosh, M. K. (2013). *Phoenix dactylifera* Linn. (Pind Kharjura): A review. *International Journal of Research in Ayurveda and Pharmacy, 4*(3), 447–451.

Awad, M. A., Al-Qurashi, A. D., & Mohamed, S. A. (2011). Antioxidant capacity, antioxidant compounds and antioxidant enzyme activities in five date cultivars during development and ripening. *Scientia Horticulturae, 129*, 688–693.

Bahmanpour, S., Talaei, T., Vojdani, Z., Panjehshahin, M. R., Poostpasand, A., Zareei, S., et al. (2006). Therapeutic effect of *Phoenix dactylifera* pollen on sperm parameters and reproductive system of adult male rats. *Iranian Journal of Medical Sciences, 31*, 8–12.

Baliga, M. S., Baliga, B. R. V., Kandathil, S. M., Bhat, H. P., & Vayalil, P. K. (2011). A review of the chemistry and pharmacology of the date fruits (*Phoenix dactylifera* L.). *Food Research International, 44*, 1812–1822.

Baltaci, B. B., Uygur, R., Caglar, V., Aktas, C., Aydin, M., & Ozen, O. A. (2016). Protective effects of quercetin against arsenic-induced testicular damage in rats. *Andrologia, 48*(10), 1202–1213.

Belmir, S., Boucherit, K., Boucherit-Otmani, Z., & Belhachemi, M. H. (2016). Effect of aqueous extract of date palm fruit (*Phoenix dactylifera* L.) on therapeutic index of amphotericin B. *Phytothérapie, 14*, 97–101.

Bentrad, N., Gaceb-Terrak, R., Benmalek, Y., & Rahmania, F. (2017). Studies on chemical composition and antimicrobial activities of bioactive molecules from date palm (*Phoenix dactylifera* L.) pollens and seeds. *African Journal of Traditional, Complementary, and Alternative Medicines*, *4*(3), 242–256.

Bentrad, N., Gaceb-Terrak, R., & Rahmania, F. (2017). Identification and evaluation of antibacterial agents present in lipophilic fractions isolated from sub-products of *Phoenix dactilyfera*. *Natural Product Research*, *31*(21), 2544–2548.

Bhathena, S. J., & Velasquez, M. T. (2002). Beneficial role of dietary phytoestrogens in obesity and diabetes. *The American Journal of Clinical Nutrition*, *76*, 1191–1201.

Biglari, F., AlKarkhi, A. F. M., & Easa, A. M. (2008). Antioxidant activity and phenolic content of various date palm (*Phoenix dactylifera* L.) fruits from Iran. *Food Chemistry*, *107*, 1636–1641.

Borchani, C., Besbes, S., Blecker, C., Masmoudi, M., Baati, R., & Attia, H. (2010). Chemical properties of 11 date cultivars and their corresponding fiber extracts. *African Journal of Biotechnology*, *9*(12), 4096–4105.

Borochov-Neori, H., Judeinstein, S., Greenberg, A., Volkova, N., Rosenblat, M., & Aviram, M. (2015). Antioxidant and antiatherogenic properties of phenolic acid and flavonol fractions of fruits of "Amari" and "Hallawi" date (*Phoenix dactylifera* L.) varieties. *Journal of Agricultural and Food Chemistry*, *63*, 3189–3195.

Bouaziz, M. A., Besbes, S., Blecker, C., Wathelet, B., Deroanne, C., & Attia, H. (2008). Protein and amino acid profiles of Tunisian Deglet Nour and Allig date palm fruit seeds. *The International Journal of Tropical and Subtropical Horticulture*, *63*, 37–43.

Boudries, H., Kefalas, P., & Hornero-Mendez, D. (2007). Carotenoid composition of Algerian date varieties (*Phoenix dactylifera*) at different edible maturation stages. *Food Chemistry*, *101*(4), 1372–1377.

Bouhlali, E. T., Alema, C., Ennassir, J., Benlyas, M., Mbark, A. N., & Zegzouti, Y. F. (2017). Phytochemical compositions and antioxidant capacity of three date (*Phoenix dactylifera* L.) seeds varieties grown in the South East Morocco. *Journal of the Saudi Society of Agricultural Sciences*, *16*(4), 350–357.

Bouhlali, E. T., Bammou, M., Sellam, K., Benlyas, M., Alem, C., & Filali-Zegzouti, Y. (2016). Evaluation of antioxidant, antihemolytic and antibacte-rial potential of six Moroccan date fruit (*Phoenix dactylifera* L.) varieties. *Journal of King Saud University—Science*, *28*(2), 136–142.

Bouhlali, E. T., Ramchoun, M., Alem, C., Ghafoor, K., Ennassir, J., & Zegzouti, Y. F. (2017). Functional composition and antioxidant activities of eight Moroccan date fruit varieties (Phoenix dactylifera L.). *Journal of the Saudi Society of Agricultural Sciences*, *16*(3), 257–264.

Boulenouar, N., Marouf, A., & Cheriti, A. (2011). Antifungal activity and phytochemical screening of extracts from *Phoenix dactylifera* L. cultivars. *Natural Product Research*, *25*(20), 1999–2002.

Chaira, N., Smaali, M. I., Martinez-Tomé, M., Mrabet, A., Murcia, M. A., & Ferchichi, A. (2009). Simple phenolic composition, flavonoid contents and antioxidant capacities in water–methanol extracts of Tunisian common date cultivars (*Phoenix dactylifera* L.). *International Journal of Food Sciences and Nutrition*, *60*, 316–329.

Chakroun, M., Khemakhem, B., Mabrouk, H. B., El Abed, H., Makni, M., Bouaziz, M., et al. (2016). Evaluation of anti-diabetic and anti-tumoral activities of bioactive compounds from *Phoenix dactylifera* L's leaf: In vitro and in vivo approach. *Biomedicine & Pharmacotherapy*, *84*, 415–422.

Copley, M. S., Rose, P. J., Clapham, A., Edwards, D. N., Horton, M. C., & Ever-shed, R. P. (2001). Detection of palm fruit lipids in archaeological pottery from Qasr Ibrim, Egyptian Nubia. *Proceedings of the Biological Sciences*, *268*(1467), 593–597.

Di Mascio, P., Murphy, M. E., & Sies, H. (1991). Antioxidant defense systems: The role of carotenoids, tocopherols, and thiols. *The American Journal of Clinical Nutrition, 53*, 194S–200S.

Diboun, I., Mathew, S., Al-Rayyashi, M., Elrayess, M., Torres, M., Halama, A., et al. (2015). Metabolomics of dates (*Phoenix dactylifera*) reveals a highly dynamic ripening process accounting for major variation in fruit composition. *BMC Plant Biology, 15*, 291.

Doha, M. A., & Al-Okbi, S. Y. (2004). In vivo evaluation of antioxidant and anti-inflammatory activity of different extracts of date fruits in adjuvant arthritis. *Polish Journal of Food and Nutrition Sciences, 13*, 397–402.

Domitrovic, R., Jakovac, H., Grebic, D., Milin, C., & Radosevic-Stasic, B. (2008). Dose- and time-dependent effects of luteolin on liver metallothioneins and metals in carbon tetrachloride-induced hepatotoxicity in mice. *Biological Trace Element Research, 126*, 176–185.

Duke, J. A. (2001). *Handbook of phytochemical constituents of GRAS herbs and other economic plants*. Boca Raton: CRC Press.

Eid, N., Enani, S., Walton, G., Corona, G., Costabile, A., Gibson, G., et al. (2014). The impact of date palm fruits and their component polyphenols, on gut microbial ecology, bacterial metabolites and colon cancer cell proliferation. *Journal of Nutritional Science, 3*, e46.

Ekor, M. (2014). The growing use of herbal medicines: Issues relating to adverse reactions and challenges in monitoring safety. *Frontiers in Pharmacology, 4*, 177.

El Arem, A., Lahouar, L., Saafi, E. B., Thouri, A., Ghrairi, F., Houas, Z., et al. (2017). Dichloroacetic acid-induced testicular toxicity in male rats and the protective effect of date fruit extract. *BMC Pharmacology and Toxicology, 18*, 17.

El Arem, A., Saafi, E. B., Flamini, G., Issaoui, M., Ferchichi, A., Hammami, M., et al. (2012). Volatile and nonvolatile chemical composition of some date fruits (*Phoenix dactylifera* L.) harvested at different stages of maturity. *International Journal of Food Science and Technology, 47*(3), 549–555.

El Arem, A., Saafi, E. B., Mechri, B., Lahouar, L., Issaoui, M., Hammami, M., et al. (2012). Effects of the ripening stage on phenolic profile, phytochemical composition and antioxidant activity of date palm fruit. *Journal of Agricultural and Food Chemistry, 60*, 10896–10902.

El Hadrami, A., Daayf, F., & El Hadrami, I. (2011). Secondary metabolites of date palm. In S. M. Jain, J. M. Al-Khayri, & D. V. Johnson (Eds.), *Date palm biotechnology* (pp. 653–674). Dordrecht: Springer.

El Sohaimy, S., Abdelwahab, A., Brennan, C., & Aboul-enein, A. (2015). Phenolic content, antioxidant and antimicrobial activities of Egyptian date palm (*Phoenix dactylifera* L.) fruits. *Australian Journal of Basic and Applied Sciences, 9*(1), 141–147.

El-Beltagi, H. S., Aly, A. A., & El-Desouky, W. (2019). Effect of gamma irradiation on some biochemical properties, antioxidant and antimicrobial activities of Sakouti and Bondoky dry dates fruits genotypes. *Journal of Radiation Research and Applied Science, 12*(1), 437–446.

El-Beltagi, H. S., Mohamed, H. I., Abdelazeem, A. S., Youssef, R., & Safwat, G. (2019). GC-MS analysis, antioxidant, antimicrobial and anticancer activities of extracts from *Ficus sycomorus* fruits and leaves. *Notulae Botanicae Horti Agrobotanici Cluj-Napoca, 47*(2), 493–505.

El-Beltagi, H. S., Mohamed, H. I., Elmelegy, A. A., Eldesoky, S. E., & Safwat, G. (2019). Phytochemical screening, antimicrobial, antioxidant, anticancer activities and nutritional values of cactus (*Opuntia Ficus Indicia*) pulp and peel. *Fresenius Environmental Bulletin, 28*(2A), 1534–1551.

El-Beltagi, H. S., Mohamed, H. I., Safwat, G., Gamal, M., & Megahed, B. M. H. (2019). Chemical composition and biological activity of *Physalis peruviana* L. *Gesunde Pflanzen, 71*, 113–122.

El-Beltagi, H. S., Mohamed, H. I., Safwat, G., Megahed, B. M. H., & Gamal, M. (2018). Evaluation of some chemical constituents, antioxidant, antibacterial and anticancer activities of *Beta vulgaris* L. root. *Fresenius Environmental Bulletin, 27*(9), 6369–6378.

Elberry, A. A., Mufti, S. T., Al-Maghrabi, J. A., Abdel-Sattar, E. A., Ashour, O. M., Ghareib, S. A., et al. (2011). Anti-inflammatory and antiprolifera-tive activities of date palm pollen (*Phoenix dactylifera*) on experi-mentally-induced atypical prostatic hyperplasia in rats. *Journal of Inflammation*, 8(1), 40.

El-Far, A. H., Ahmed, H. A., & Shaheen, H. M. (2016). Dietary supplementation of phoenix dactylifera seeds enhances performance, immune response, and antioxidant status in broilers. *Oxidative Medicine and Cellular Longevity*, 2016, 9.

El-Far, A. H., Shaheen, H. M., Abdel-Daim, M. M., Al Jaouni, S. K., & Mousa, S. A. (2016). Date palm (*Phoenix dactylifera*): Protection and remedy food. *Current Trends in Nutraceuticals*, 1(29), 1–10.

El-Neweshy, M. S., El-Maddawy, Z. K., & El-Sayed, Y. S. (2013). Therapeutic effects of date palm (*Phoenix dactylifera* L.) pollen extract on cadmium-induced testicular toxicity. *Andrologia*, 45(6), 369–378.

El-Zoghbi, M. (1997). Biochemical changes in some tropical fruits during ripening. *Food Chemistry*, 49, 33–37.

Eustache, F., Mondon, F., Canivenc-Lavier, M. C., Lesaffre, C., Fulla, Y., Berges, R., et al. (2009). Chronic dietary exposure to a low-dose mixture of genistein and vinclozolin modifies the reproductive axis, testis transcriptome, and fertility. *Environmental Health Perspectives*, 117, 1272–1279.

Famuyiwa, O. O., Elhazmi, M. A. F., Aljasser, S. J., Sulimani, R. A., Jayakumar, R. V., Alnuaim, A. A., et al. (1992). A comparison of acute glycemic and insulin-response to dates (*Phoenix dactylifera*) and oral dextrose in diabetic and nondiabetic subjects. *Saudi Medical Journal*, 13(5), 397–402.

Farag, K. M., Elsabagh, A. S., & ElAshry, H. A. (2012). Fruit characteristics of "Zaghloul" date palm in relation to Metaxenic influences of used pollinator. *American-Eurasian Journal of Agricultural & Environmental Sciences*, 12(7), 842–855.

Ferguson, L. R., Philpott, M., & Karunasinghe, N. (2004). Dietary cancer and prevention using antimutagens. *Toxicology*, 198, 147–159.

Fine, A. M. (2000). Oligomeric proanthocyanidin complexes: History, structure, and phytopharmaceutical applications. *Alternative Medicine Review: A Journal of Clinical Therapeutic*, 5, 144–151.

Fullerton, S. A., Samadi, A. A., Tortorelis, D. G., Choudhury, M. S., Mallouh, C., Tazaki, H., et al. (2000). Induction of apoptosis in human prostatic cancer cells with betaglucan (Maitake mushroom polysaccharide). *Molecular Urology*, 4, 7–13.

Garcia, O. B., & Castillo, J. (2008). Update on uses and properties of citrus flavonoids: New findings in anticancer, cardiovascular, and anti-inflammatory activity. *Journal of Agricultural and Food Chemistry*, 56, 6185–6205.

Gardner, E. (2017). Alternative sugars: Dates. *British Dental Journal*, 223(6), 393.

Gescher, A. (2004). Polyphenolic phytochemicals versus non-steroidal anti-inflammatory drugs: Which are better cancer chemopreventive agents? *Journal of Chemotherapy*, 16, 3–6.

Ghnimi, S., Umer, S., Karim, A., & Kamal-Eldin, A. (2017). Date fruit (*Phoenix dactylifera* L.): An underutilized food seeking industrial valorization. *NFS Journal*, 6, 1–10.

Gnanamangai, B. M., Saranya, S., Ponmurugan, P., Kavitha, S., Pitchaimuthu, S., & Divya, P. (2019). Analysis of antioxidants and nutritional assessment of date palm fruits. In M. Naushad, & E. Lichtfouse (Eds.), *Vol. 34. Sustainable agriculture reviews* (pp. 19–40). Cham: Springer.

Habib, H. M., & Ibrahim, W. H. (2011). Nutritional quality evaluation of eighteen date fruit varieties. *International Journal of Food Sciences and Nutrition*, 60(S1), 99–111.

Hamad, I., Abdelgawad, H., Jaouni, S. K., Zinta, G., Asard, H., Hassan, S. H., et al. (2015). Metabolic analysis of various date palm fruit (*Phoenix dactylifera* L.) cultivars from Saudi Arabia to assess their nutritional quality. *Molecules*, 20, 13620–13641.

Hamed, M. M., Abd El-Mobdy, M. A., Kamel, M. T., Mohamed, H. I., & Bayoumi, A. E. (2019). Phytochemical and biological activities of two asteraceae plants *Senecio vulgaris* and *Pluchea dioscoridis* L. *Pharmacologyonline*, *2*, 101–121.

Harrigan, G. G., Martino-Catt, S., & Glenn, K. C. (2007). Metabolomics, metabolic diversity and genertic variation in crops. *Metabolomics*, *3*, 259–272.

Hasan, N. S., Amom, Z. H., Nor, A. I., Mokhtarrud, N., Mohd Esa, N., & Azlan, A. (2010). Nutritional composition and in vitro evaluation of the antioxidant properties of various dates extracts (*Phoenix dactylifera* L.) from Libya. *Asian Journal of Clinical Nutrition*, *2*(4), 208–214.

Hasan, M., & Mohieldein, A. (2016). In vivo evaluation of anti diabetic, hypolipidemic, antioxidative activities of saudi date seed extract on streptozotocin induced diabetic rats. *Journal of Clinical and Diagnostic Research*, *10*(3), 6–12.

Hodel, D. R., & Johnson, D. V. (2007). *Dates, imported and American varieties of dates in the United States* (p. 3498). Oakland, CA: Univ Calif, Agri & Natural Resources. ANR Publication. 112 p.

Hong, Y. J., Tomas-Barberan, F. A., Kader, A. A., & Mitchell, A. E. (2006). The flavonoid glycosides and procyanidin composition of Deglet Noor dates (*Phoenix dactylifera*). *Journal of Agricultural and Food Chemistry*, *54*, 2405–2411.

Hoseinifar, S. H., Dadar, M., Khalili, M., Cerezuela, R., & Esteban, M.Á. (2017). Effect of dietary supplementation of palm fruit extracts on the tran-scriptomes of growth, antioxidant enzyme and immune-related genes in common carp (*Cyprinus carpio*) fingerlings. *Aquaculture Research*, *48*(7), 3684–3692.

Hossain, M. Z., Waly, M. I., Singh, V., Sequeira, V., & Rahman, M. S. (2014). Chemical composition of datepits and its potential for developing value-added product—A review. *Polish Journal of Food and Nutrition Sciences*, *64*, 215–226.

Huntrods, D. (2011). *Date profile. Agricultural marketing resource center bulletin* (p. 3). Washington, DC: USDA.

Ibrahim, M. A. (1996). *Effect of source of pollen on the physiology of fruit ripening of the fruit of date palm (Phoenix dactylifera L.) cv. Hillawi*. MSc Thesis (p. 2). Basrah, Iraq: Basrah University (in Arabic with English summary).

Ishurd, O., & Kennedy, J. F. (2005). The anticancer activity of polysaccharide prepared from Libyan dates (*Phoenix dactylifera* L.). *Carbohydrate Polymers*, *59*, 531–535.

Ishurd, O., Zgheela, F., Kermagia, A., Flefleaa, N., Elmabruka, M., Kennedy, J. F., et al. (2007). (1–3)-β D-glucans from Lybian dates (*Phoenix dactylifera* L.) and their anticancer activities. *Journal of Biological Sciences*, *7*(3), 554–557.

Ismail, B., Haffar, I., Baalbaki, R., Mechref, Y., & Henry, J. (2006). Physico-chemical characteristics and total quality of five date varieties grown in the United Arab Emirates. *International Journal of Food Science and Technology*, *41*, 919–926.

Jana, K., Samanta, P. K., Manna, I., Ghosh, P., Singh, N., Khetan, R. P., et al. (2008). Protective effect of sodium selenite and zinc sulfate on intensive swimming-induced testicular gamatogenic and steroidogenic disorders in mature male rats. *Applied Physiology, Nutrition, and Metabolism*, *33*, 903–914.

Janbaz, K. H., Saeed, S. A., & Gilani, A. H. (2005). Studies on the protective effects of caffeic acid and quercetin on chemical-induced hepatotoxicity in rodents. *Phytomedicine*, *11*, 424–430.

Jassim, S. A. A., & Naji, M. A. (2010). In vitro evaluation of the antiviral activity of an extract of date palm (*Phoenix dactylifera* L.) pits on a pseudomonas phage. *Evidence-based Complementary and Alternative Medicine*, *7*, 57–62.

John, S., Sorokin, A. V., & Thompson, P. D. (2007). Phytosterols and vascular disease. *Current Opinion in Lipidology*, *18*, 35–40.

Kader, A. A., & Hussein, A. M. (2009). *Harvesting and postharvest handling of dates* (p. 15). Aleppo, Syria: ICARDA.

Kalantaripour, T., Asadi-Shekaari, M., Basiri, M., & Najar, A. G. (2012). Cerebroprotective effect of date seed extract (*Phoenix dactylifera*) on focal cerebral ischemia in male rats. *Journal of Biological Sciences, 12,* 180–185.

Kaleem, M., Ahmad, A., Khalid, S., & Azam, M. T. (2016). HPLC condition optimization for identification of flavonoids from *Carissa opaca*. *Science International (Lahore), 28*(1), 343–348.

Karasawa, K., Uzuhashi, Y., Hirota, M., & Otani, H. (2011). A matured fruit extract of date palm tree (*Phoenix dactylifera* L.) stimulates the cellular immune system in mice. *Journal of Agricultural and Food Chemistry, 59,* 11287–11293. https://doi.org/10.1021/jf2029225.

Kchaou, W., Abbès, F., Mansour, R. B., Blecker, C., Attia, H., & Besbes, S. (2016). Phenolic profile, antibacterial and cytotoxic properties of second grade date extract from Tunisian cultivars (*Phoenix dactylifera* L.). *Food Chemistry, 194,* 1048–1055.

Kehili, H. E., Zerizer, S., Beladjila, K. A., & Kabouche, Z. (2016). Anti-inflammatory effect of Algerian date fruit (*Phoenix dactylifera*). *Food and Agricultural Immunology, 27,* 820–829.

Khadem, N., Sharaphy, A., Latifnejad, R., Hammod, N., & Ibrahimzadeh, S. (2007). Comparing the efficacy of dates and oxytocin in the management of postpartum hemorrhage. *Shiraz E-Medical Journal, 8*(2), 64–71.

Khalid, S., Ahmad, A., & Kaleem, M. (2017). Antioxidant activity and phenolic contents of Ajwa date and their effect on lipo-protein profile. *Functional Foods in Health and Disease, 7*(6), 396–410.

Khalid, S., Ahmad, A., Masud, T., Asad, M., & Sandhu, M. (2016). Nutritional assessment of Ajwa date flesh and pits in comparison to local varieties. *Journal of Animal and Plant Sciences, 26,* 1072–1080.

Khalid, S., Khalid, N., Khan, R. S., Ahmed, H., & Ahmad, A. (2017). A review on chemistry and pharmacology of Ajwa date fruit and pit. *Trends in Food Science and Technology, 63,* 60–69.

Khalil, M. N. A., Fekry, M. I., & Farag, M. A. (2017). Metabolome based volatiles profiling in 13 date palm fruit varieties from Egypt via SPME GC–MS and chemometrics. *Food Chemistry, 217,* 171–181.

Khan, F., Ahmed, F., Pushparaj, P. N., Abuzenadah, A., Kumosani, T., Barbour, E., et al. (2016). Date (*Phoenix dactylifera* L.) extract inhibits human breast adenocarcinoma (MCF7) cells in vitro by inducing apoptosis and cell cycle Ar-rest. *PLoS One, 11*(7), e0158963.

Khan, S. A., Al Kiyumi, A. R., Al Sheidi, M. S., Al Khusaibi, T. S., Al Shehhi, N. M., & Alam, T. (2016). In vitro inhibitory effects on α-glucosidase and α-amylase level and antioxidant potential of seeds of *Phoenix dactylifera* L. *Asian Pacific Journal of Tropical Biomedicine, 6*(4), 322–329.

Khan, H., & Khan, S. A. (2016). Date palm revisited. *Research Journal of Pharmaceutical, Biological and Chemical Sciences, 7*(3), 2010–2019.

Khan, F., Khan, T. J., Kalamegam, G., Pushparaj, P. N., Chaudhary, A., Abuzenadah, A., et al. (2017). Anticancer effects of Ajwa dates (*Phoenix dactylifera* L.) in diethylnitrosamine induced hepatocellular carcinoma in Wistar rats. *BMC Complementary and Alternative Medicine, 17,* 418.

Kikuchi, N., & Miki, T. (1978). The separation of date (Phoenix dactylifera) sterols by liquid chromatography. *Mikrochimica Acta, 69,* 89–96.

Kordi, M., Aghaei, M. F., Tara, F., Nemati, M., & Taghi, S. M. (2014). The effect of late pregnancy consumption of date fruit on cervical ripening in nulliparous women. *Journal of Midwifery & Women's Health, 2*(3), 150–156.

Ku, K. M., Choi, J. N., Kim, J., Kim, J. K., Yoo, L. G., Lee, S. J., et al. (2010). Metabolomics analysis reveals the compositional differences of shade grown tea (*Camellia sinensis* L.). *Journal of Agricultural and Food Chemistry, 58*(1), 418–426.

Lemine, M. F. M., Ahmed, M. M. V. O., Maoulainine, B. M. L., Zei, B. A., Samb, A., & Boukhary, O. A. O. (2014). Antioxidant activity of various Mauritanian date palm (*Phoenix dactylifera* L.) fruits at two edible ripening stages. *Food Science & Nutrition, 2*, 700–705.

Liolios, C. C., Sotiroudis, G. T., & Chinou, I. (2009). Fatty acids, sterols, phenols and antioxidant activity of *Phoenix theophrasti* fruits growing in Crete, Greece. *Plant Foods for Human Nutrition, 64*(1), 52–61.

Lobo, M. G., Yahia, E. M., & Kader, A. A. (2014). Biology and postharvest physiology of date fruit. In M. Siddiq, S. M. Aleid, & A. A. Kader (Eds.), *Dates: Postharvest science, processing technology and health benefits* (1st ed.). John Wiley & Sons, Ltd.

Lusis, A. J. (2002). Atherosclerosis. *Nature, 14*, 233–241.

Majid, A. S., Marzieh, P., Shahriar, D., Zahed, S. K., & Pari, K. T. (2008). Neuropro-tective effects of aqueous date fruit extract on focal cerebral ische-mia in rats. *Pakistan Journal of Medical Sciences, 24*(5), 661–665.

Mansouri, A., Embarek, G., Kokkalou, E., & Kefalas, P. (2005). Phenolic profile and antioxidant activity of the Algerian ripe date palm fruit (*Phoenix dactylifera*). *Food Chemistry, 89*(3), 411–420.

Mard, S. A., Jalalvand, K., Jafarinejad, M., Balochi, H., & Naseri, M. K. (2010). Evaluation of the antidiabetic and antilipaemic activities of the hy-droalcoholic extract of *Phoenix dactylifera* palm leaves and its fractions in alloxan-induced diabetic rats. *Malaysian Journal of Medical Sciences, 17*(4), 4–13.

Marlett, J. A., McBurney, M. I., & Slavin, J. L. (2002). Position of the American Dietetic Association: Health implications of dietary fiber. *Journal of the American Dietetic Association, 102*, 993–1000.

Maryam. (2014). *Studies on metaxenial effect and use of molecular approaches for detection of hybrids, genetic diversity and sex in date palm (Phoenix dactylifera L.)*. Thesis Faisalabad, Pakistan: Institute of Horticultural Sciences, Faculty of Agriculture, University of Agriculture.

Maryam, U. I., Simbak, N., Umar, A., Sani, I. H., Baig, A. A., Zin, T., et al. (2015). Anti-inflammatory and analgesic activities of aqueous extract date palm (*Phoenix dactylifera* L) fruit in rats. *International Journal of Novel Research in Healthcare and Nursing, 2*, 166–172.

Marzouk, H. A., & Kassem, H. A. (2011). Improving fruit quality, nutritional value and yield of Zaghloul dates by the application of organic and/or fertilizers. *Scientia Horticulturae, 127*, 249–254.

Mater, A. A. (1991). *Cultivation and production of date palms*. Iraq: Basrah University.

Michael, H. N., Salib, J. Y., & Eskander, E. F. (2013). Bioactivity of diosmetin glycosides isolated from the epicarp of date fruits, *Phoenix dactylifera*, on the biochemical profile of alloxan diabetic male rats. *Phytotherapy Research, 27*, 699–704.

Miller, C. J., Dunn, E. V., & Hashim, I. B. (2003). The glycaemic index of dates and date/yoghurt mixed meals. Are dates 'the candy that grows on trees'? *European Journal of Clinical Nutrition, 57*, 427–430.

Mirza, M. B., Syed, F. Q. I., Khan, F., Elkady, A. I., Al-Attar, A. M., & Hakeem, K. R. (2019). Ajwa Dates: A highly nutritive fruit with the impending therapeutic application. In M. Ozturk, & K. R. Hakeem (Eds.). *Plant and human health, Vol. 3*, 209–229.

Mohamed, H. I., Akladious, S. A., & El-Beltagi, H. S. (2018). Mitigation the harmful effect of salt stress on physiological, biochemical and anatomical traits by foliar spray with trehalose on wheat cultivars. *Fresenius Environmental Bulletin, 27*(10), 7054–7065.

Mohammad, B., & Habibi, N. (2011). Date seeds: A novel and inexpensive source of dietary fiber. In *Vol. 9. International conference on food engineering and biotechnology* (pp. 323–326).

Nehdi, I. A., Sbihi, H. M., Tan, C. P., Rashid, U., & Si, A.-R. (2018). Chemical composition of date palm (*Phoenix dactylifera* L.) seed oil from six Saudi Arabian cultivars. *Journal of Food Science, 83*(3), 624–630.

Neori, B. H., Judeinstein, S. A., Greenberg, N., Volkova, M., Rosenblat, M., & Aviram, M. (2013). Date (*Phoenix dactylifera* L.) fruit soluble phenolics composition and antiatherogenic properties in nine israeli varieties. *Journal of Agricultural and Food Chemistry, 61*(18), 4278–4286.

Niazi, S., Khan, I. M., Pasha, I., Rasheed, S., Ahmad, S., & Shoaib, M. (2017). Date palm: Composition, health claim and food applications. *International Journal of Public Health and Health Systems, 2*(1), 9–17.

Ozçelik, B., Orhan, I., & Toker, G. (2006). Antiviral and antimicrobial assessment of some selected flavonoids. *Zeitschrift fur Naturforschung. C, Journal of Biosciences, 61*, 632–638.

Paulsen, M. E. (2005). *The amazing story of the fabulous medjool date* (p. 152). Tualatin, OR: Marc Paulsen Press.

Popenoe, P. (1973). In H. Field (Ed.), *The date palm* (p. 139). Coconut Grove. Miami, FL: Field Research Projects.

Pujari, R. R., Vyawahare, N. S., & Kagathara, V. G. (2011). Evaluation of antioxidant and neuroprotective effect of date palm (*Phoenix dactylifera* L.) against bilateral common carotid artery occlusion in rats. *Indian Journal of Experimental Biology, 49*(8), 627–633.

Pujari, R. R., Vyawahare, N. S., & Thakurdesai, P. A. (2014). Neuroprotective and antioxidant role of *Phoenix dactylifera* in permanent bilateral common carotid occlusion in rats. *Journal of Acute Diseases, 3*(2), 104–114.

Rabei, S., Said, W. M., Rizk, R. M., & El Sharabasy, S. F. (2012). Morphometric taxonomy of date palm diversity growing in Egypt. *Egyptian Journal of Botany*, 175–189.

Ranilla, L. G., Kwon, Y. I., Genovese, M. I., Lajolo, F. M., & Shetty, K. (2008). Antidiabetes and antihypertension potential of commonly consumed carbohydrate sweeteners using in vitro models. *Journal of Medicinal Food, 11*, 337–348.

Razali, N., Mohd Nahwari, S. H., Sulaiman, S., & Hassan, J. (2017). Date fruit consumption at term: Effect on length of gestation, labour and delivery. *Journal of Obstetrics and Gynaecology, 37*(5), 595–600.

Robak, J., & Gryglewski, R. J. (1996). Bioactivity of flavonoids. *Polish Journal of Pharmacology, 48*, 555–564.

Rock, W., Rosenblat, M., Borochov-Neori, H., Volkova, N., Judeinstein, S., Elias, M., et al. (2009). Effects of date (Phoenix dactylifera L., Medjool or Hallawi variety) consumption by healthy subjects on serum glucose and lipid levels and on serum oxidative status: A pilot study. *Journal of Agricultural and Food Chemistry, 57*, 8010–8017.

Rygg, G. L. (1975). Date development, handling and packing in the United States. In *USDA agriculture handbook* (p. 56). Washington, DC: Agricultural Research Service. No. 482.

Said, L., Banni, M., Kerkeni, A., Said, K., & Messaoudi, I. (2010). Influence of combined treatment with zinc and selenium on cadmium induced testicular pathophysiology in rat. *Food and Chemical Toxicology, 48*, 2759–2765.

Sallal, A. K., Rim, J., Amr, Z. S., & Disi, A. M. (1997). Inhibition of haemolytic activity of snake and scorpion venom by date extract. *Biomedical Letters, 55*(217), 51–56.

Samad, M., Hashim, S., Simarani, K., & Yaacob, J. (2016). Antibacterial properties and effects of fruit chilling and extract storage on antioxidant activity, total phenolic and anthocyanin content of four date palm (*Phoenix dactylifera*) cultivars. *Molecules, 21*(4), 419.

Sani, I., Bakar, N., Rohin, M., Suleiman, I., Umar, M., & Mohamad, N. (2015). *Phoenix dactylifera* linn as a potential novel anti-oxidant in treating major opioid toxicity. *Journal of Applied Pharmaceutical Science, 5*, 167–172.

Sawaya, W. N. (2000). *Proposal for the establishment of a regional network for date-palm in the near East and North Africa. A draft for discussion.* FAO/RNE.

Serrano, M., Pretel, M. T., Botella, M. A., & Amoros, A. (2001). Physicochemical changes during date ripening related to ethylene production. *Food Science and Technology International, 7,* 31–36.

Shafiei, M., Karimi, K., & Taherzadeh, M. J. (2010). Palm date fibers: Analysis and enzymatic hydrolysis. *International Journal of Molecular Sciences, 11,* 4285–4296.

Shanmugapriya, M., & Patwardhan, K. (2012). Uses of date palm in Ayurveda. In A. Manickavasagan, M. M. Essa, & E. Sukumar (Eds.), *Dates: Production, processing, food, and medicinal values* (pp. 377–385). CRC Press: Boca Raton.

Sharifi, M., Bashtani, M., Naserian, A. A., & Farhangfar, H. (2017). The effect of increasing levels of date palm (*Phoenix dactylifera* L.) seed on the Performance, ruminal fermentation, antioxidant status and milk fat-ty acid profile of Saanen dairy goats. *Journal of Animal Physiology and Animal Nutrition, 101*(5), e332–e341.

Sheikh, B. Y., Zihad, S. M. N. K., Sifat, N., Uddin, S. J., Shilpi, J. A., Hamdi, O. A., et al. (2016). Comparative study of neuropharmacological, analgesic properties and phenolic profile of Ajwah, Safawy and Sukkari cultivars of date palm (*Phoenix dactylifera*). *Oriental Pharmacy and Experimental Medicine, 16*(3), 175–183.

Shraideh, Z. A., Abu-Elteen, K. H., & Sallal, A. K. J. (1998). Ultrastructural effects of date extract on Candida albicans. *Mycopathologia, 142,* 119–123.

Sirisena, S., Ng, K., & Ajlouni, S. (2015). The emerging australian date palm industry: Date fruit nutritional and bioactive compounds and valuable processing by-products. *Comprehensive Reviews in Food Science and Food Safety, 14*(6), 813–823.

Steinbrenner, H., & Sies, H. (1790). Protection against reactive oxygen species by selenoproteins. *Biochimica et Biophysica Acta, 2009,* 1478–1485.

Stephan, N., Halama, A., Mathew, S., Hayat, S., Bhagwat, A., Mathew, L. S., et al. (2018). A comprehensive metabolomic data set of date palm fruit. *Data in Brief, 18,* 1313–1321.

Subarnas, A., & Wagner, H. (2000). Analgesic and anti-inflammatory activity of the proanthocyanidin shellegueain A from Polypodium feei METT. *Phytomedicine, 7,* 401–405.

Sumner, L. W., & Hall, R. D. (2013). Metabolomics across the globe. *Metabolomics, 9*(1), 258–264.

Suresh, S., Guizani, N., Al-Ruzeiki, M., Al-Hadhrami, A., Al-Dohani, H., Al-Kindi, I., et al. (2013). Thermal characteristics, chemical composition and polyphenol contents of date-pits powder. *Journal of Food Engineering, 119,* 668–679.

Surh, Y. J. (2003). Cancer chemoprevention with dietary phytochemicals. *Nature Reviews. Cancer, 3*(10), 768–780.

Takaeidi, M. R., Jahangiri, A., Khodayar, M. J., Siahpoosh, A., Yaghooti, H., Rezaei, S., et al. (2014). The effect of date seed (*Phoenix dactylifera*) extract on paraoxonase and arylesterase activities in hypercholesterolemic rats. *Jundishapur Journal of Natural Pharmaceutical Products, 9*(1), 30–34.

Taleb, H., Maddocks, S. E., Morris, R. K., & Kanekanian, A. D. (2016a). The antibacterial activity of date syrup polyphenols against S. aureus and E. coli. *Frontiers in Microbiology, 7,* 198.

Taleb, H., Maddocks, S. E., Morris, R. K., & Kanekanian, A. D. (2016b). Chemical characterisation and the anti-inflammatory, anti-angiogenic and antibacterial properties of date fruit (*Phoenix dactylifera* L.). *Journal of Ethnopharmacology, 194,* 457–468.

Tang, Z. X., Shi, L. E., & Aleid, S. M. (2013). Date fruit: Chemical composition, nutritional and medicinal values, products. *Journal of the Science of Food and Agriculture, 93,* 2351–2361.

Tapas, A. R., Sakarkar, A. M., & Kakde, R. B. (2008). Flavonoids as nutraceuticals: A review. *Tropical Journal of Pharmaceutical Research, 7,* 1089–1099.

Thouri, A., Chahdoura, H., El Arem, A., Omri Hichri, A., Ben Hassin, R., & Achour, L. (2017). Effect of solvents extraction on phytochemical components and biological activities of Tunisian date seeds (var. Korkobbi and Arechti). *BMC Complementary and Alternative Medicine*, *17*, 248.

Uba, A., Abdullahi, M. I., Yusuf, A. J., Ibrahim, Z. Y. Y., Lawal, M., Nasir, I., et al. (2015). Mineral profile, proximate and amino acid composition of three dates varieties (*Phoenix dactylifera* L.). *Der Pharma Chemica*, *7*(5), 48–53.

Uhl, N. W., & Dransfield, J. (1987). *Genera palmarum: A classification of palms based on the work of Harold E. Moore, Jr, L.H. Baily* (pp. 214–217). Lawrence, KS: Hortorium and the International Palm Society.

Uteshev, D. B., Kostriukov, E. B., Karabinenko, A. A., Kovaleva, V. L., Makarova, O. V., & Storozhakov, G. I. (2000). The anti-inflammatory activity of intal and beta-carotene in a model of experimental granulomatous lung inflammation. *Patologicheskaia fiziologiia i èksperimental'naia terapiia*, *2*, 19–22.

Uttara, B., Singh, A., Zamboni, P., & Mahajan, R. (2009). Oxidative stress and neurodegenerative diseases: A review of upstream and downstream antioxidant therapeutic options. *Current Neuropharmacology*, *7*(1), 65–74.

Vayalil, P. K. (2002). Antioxidant and antimutagenic properties of aqueous extract of date fruit (*Phoenix dactylifera* L. Arecaceae). *Journal of Agricultural and Food Chemistry*, *50*, 610–617.

Vayalil, P. K. (2012). Date fruits (*Phoenix dactylifera* Linn): An emerging medicinal food. *Critical Reviews in Food Science and Nutrition*, *52*(3), 249–271.

Vinson, J. A., Zubik, L., Bose, P., Samman, N., & Proch, J. (2005). Dried fruits: Excellent in-vitro and in-vivo antioxidants. *Journal of the American College of Nutrition*, *24*, 44–50.

Vittoz, J. (1979). Le palmier-dattier en Oman [The date palm in Oman]. *Fruits*, *34*, 609–621.

Waly, M. I., Al-Ghafri, B. R., Guizani, N., & Rahman, M. S. (2015). Phytonutrient effects of date pit extract against azoxymethane-induced oxidative stress in the rat colon. *Asian Pacific Journal of Cancer Prevention*, *16*, 3473–3477.

Wang, H., Cao, G., & Prior, R. (1997). Oxygen radical absorbing capacity of anthocyanins. *Journal of Agricultural and Food Chemistry*, *45*, 304–309.

Wang, Z., Wu, Z., Dai, Y., Yang, J., Wang, W., & Wu, G. (2013). Glycine metabolism in animals and humans: Implications for nutrition and health. *Amino Acids*, *45*(3), 463–477.

Weber, R. W. (2010). On the cover—Date palm. *Annals of Allergy, Asthma & Immunology*, *105*(4), A4.

Wu, G. (2010). Functional amino acids in growth, reproduction, and health. *Advances in Nutrition*, *1*, 31–37.

Yadav, K., Kadam, P., Patel, J., & Patil, M. (2014). Strychnos potatorum: Phytochemical and pharmacological review. *Pharmacognosy Reviews*, *8*(15), 61–66.

Yahia, E. M. (2004). Date. In K. Gross, C. Y. Wang, & M. Saltveit (Eds.), *The commercial storage of fruits, vegetables and florist and nursery crops. Agriculture handbook 66* (p. 3). Beltsville, MD: USDA.

Yahia, E. M., & Kader, A. A. (2011). Date (*Phoenix dactylifera* L.). In E. M. Yahia (Ed.), *Postharvest biology and technology of tropical and subtropical fruits* (pp. 41–79). Cambridge, UK: Woodhead Publishing.

Yaish, M. W., & Kumar, P. P. (2015). Salt tolerance research in date palm tree (*Phoenix dactylifera* L.), past, present, and future perspectives. *Frontiers in Plant Science*, *6*, 348.

Yasin, B., El-Fawal, H., & Mousa, S. (2015). Date (*Phoenix dactylifera*) poly-phenolics and other bioactive compounds: A traditional Islamic remedy's potential in prevention of cell damage, cancer therapeutics and beyond. *International Journal of Molecular Sciences*, *16*(12), 30075–30090.

Yeh, C. T., Ching, L. C., & Yen, G. C. (2008). Inducing gene expression of cardiac antioxidant enzymes by dietary phenolic acids in rats. *The Journal of Nutritional Biochemistry, 20,* 163–171.

Zaid, A., & de Wet, P. F. (2002). Botanical and systematic description of the date palm. In A. Zaid, & E. J. Arias-Jiménez (Eds.), *Date palm cultivation* (p. 156). Rome, Italy: FAO Plant Production and Protection Paper.

Zangeneh, F., Moezi, L., & Amir, Z. A. (2009). The effect of palm date, fig and olive fruits regimen on weight, pain threshold and memory in mice. *Iranian Journal of Medicinal and Aromatic Plants, 25,* 149–158.

Zhang, C. R., Aldosari, S. A., Vidyasagar, P. S., Nair, K. M., & Nair, M. G. (2013). Antioxidant and anti-inflammatory assays confirm bioactive compounds in Ajwa Date fruit. *Journal of Agricultural and Food Chemistry, 61,* 5834–5840.

Zhang, C. R., Aldosari, S. A., Vidyasagar, P. S. P. V., Shukla, P., & Nair, M. G. (2017). Health-benefits of date fruits produced in Saudi Arabia based on in vitro antioxidant, anti-inflammatory and human tumor cell proliferation inhibitory assays. *Journal of the Saudi Society of Agricultural Sciences, 16*(3), 287–293.

Zohary, D., & Hopf, M. (2000). *Domestication of plants in the old world: the origin and spread of cultivated plants in West Asia, Europe, and the Nile Valley* (3rd ed.). Oxford: Oxford University Press.

Chapter 19

The inhibitory role of the metabolites of *Moringa oleifera* seeds in cancer cells by apoptosis and cell cycle arrest activation

Ismail Abiola Adebayo[a,b], Hasni Arsad[b], and Mohammed Razip Samian[c]

[a]*Microbiology and Immunology Department, Faculty of Biomedical Sciences, Kampala International University, Ishaka-Bushenyi, Uganda,* [b]*Integrative Medicine Cluster, Advanced Medical and Dental Institute, Universiti Sains Malaysia, Penang, Malaysia,* [c]*School of Biological Sciences, Universiti Sains Malaysia, Penang, Malaysia*

1. Introduction

Moringa oleifera is a tree found in several regions of the world from Africa, to Asia, to North America. The leaves, seed, bark, root, and all other part of the tree are useful for medicinal, nutritional, and economic benefits (Fahey, 2005). For these reasons, it is named a "wonderful" or "miracle tree." The traditional usage has led to the curiosity of the researchers to find out the phytochemical contents of the plant and their activities. Several research-based articles have reported that the plant's parts have anticancer, antiinflammatory, antiulcer, and antidiabetic activities, among others (Fahey, 2005). Our research group has worked extensively on the anticancer properties of the seeds and leaves of *Moringa oleifera* (Adebayo, Arsad, Kamal, & Samian, 2020; Adebayo, Arsad, & Samian, 2017, 2018a, 2018b). In this chapter, we concisely discussed the effects of chemical compounds found in the plant seeds on the cancer cell's apoptosis and cell cycle progression.

2. *Moringa oleifera* tree is miraculous

Moringa oleifera tree (Fig. 1), which is popularly named drumstick and horseradish tree (Parrotta, 1993; Roloff, Weisgerber, Lang, & Stimm, 2009), is considered to be a miracle tree by the folklore medicine practitioners in African

Phytomedicine: A Treasure of Pharmacologically Active Products from Plants
https://doi.org/10.1016/B978-0-12-824109-7.00008-X
533

FIG. 1   A tree of *Moringa oleifera* showing its branches with green leaves.

and Asian continents because of its several economic, health, and nutritional uses (Goyal, Agrawal, Goyal, & Mehta, 2007). Other common names of the tree are benzolive tree, ben oil tree, and moonga. The plant origin is traced to Himalayan foothills of India (Mahmood, Mugal, & Haq, 2010; Ramachandran, Peter, & Gopalakrishnan, 1980). It is a member of Moringaceae family, which has 14 species (Roloff et al., 2009), and it is much more cultivated and studied than other species of the family. It is also grown in several countries such as Malaysia, Thailand, Indonesia, and Pakistan and in West African and North American countries (Mahmood et al., 2010; Roloff et al., 2009).

Moringa oleifera is a 5–10 m tall perennial tree with poor-quality timber, and its evergreen leaves are either tripinnate or bipinnate (Dubey, Dora, Kumar, & Gulsan, 2013; Fahey, 2005; Goyal et al., 2007). The plant has pods that are long and contain seeds of the plant (Roloff et al., 2009). The hull encapsulated the seeds. The seeds are green when they are not matured. After maturation, they become dry, solid, sweet, white seeds (Negi, 1977; Roloff et al., 2009).

FIG. 2   *Moringa oleifera* ripe seeds. Capsulated seeds (left image) and decapsulated seeds (right image).

The tropics, with 25–30°C temperature range, is the most ideal for the cultivation of *Moringa oleifera*, especially in its sandy, less acidic, and less basic soils (Gopalakrishnan, Doriya, & Kumar, 2016). The seeds, flower, and leaves of *Moringa oleifera* are edible, and they are consumed in the form of powder or decoctions and are also eaten in raw forms or cooked to make curries. The ripped, white sweet seeds are peoples' favorite for consumption (Fahey, 2005). *Moringa oleifera* leaves, flowers, and seeds have medicinal phytochemicals and they are nutrient rich. The seeds (Fig. 2) have inherent coagulating power, and they are used to coagulate and remove dirt from drinking water in some African countries (Zaku, Emmanuel, Tukur, & Kabir, 2015).

3. Pharmacological activities of the seeds of *Moringa oleifera*

There are many scientific studies that have reported the pharmacological activities of the *Moringa oleifera* seeds because the seeds possess many important phytochemicals and metabolites. Some of these studies will be highlighted in this section. Ruttarattanamongkol and Petrasch (2015) showed that *Moringa oleifera* seed extract and oil have antimicrobial activity against foodborne microorganisms, *Staphylococcus aureus*. Lar, Ojile, Dashe, and Oluoma (2011) showed that the *Moringa oleifera* seed aqueous extract had inhibitory effects on diarrheal-causing microbes, *Escherichia coli* and *Shigella flexneri*. The plant seeds also have anticyanobacterial activity against *Microcystis aeruginosa* (Lürling & Beekman, 2010). The seed of *Moringa oleifera* was proven to have diuretic activity, and it helps in lowering the blood pressure (Anwar, Latif, Ashraf, & Gilani, 2007).

Similarly the antimicrobial effect of *Moringa oleifera* seed aqueous extract was also reported by Auwal et al. (2013). It was revealed that the extract possesses significant inhibitory activity against *Streptococcus pyogenes, Staphylococcus aureus, Bacillus subtilis, Pseudomonas aeruginosa, Corynebacterium pyogenes, Salmonella typhi, Klebsiella pneumoniae*, and *E. coli* (Auwal et al., 2013).

Minaiyan, Asghari, Taheri, Saeidi, and Nasr-esfahani (2014) reported the antiinflammatory property of *Moringa oleifera* seed hydroalcoholic (70% ethanol) extract and its chloroform fraction. It was found that both extracts were effective in treating experimental acute colitis that was induced in rat by acetic acid. The hydroalcoholic extract was found to be more effective. Another study on antiinflammatory effect of the seed ethanol extract on airway inflammation induced in guinea pigs using ovalbumin revealed that the seed extract improved the observed respiratory parameters that indicated inflammation (Mahajan & Mehta, 2007).

The antidiabetic activity of *Moringa oleifera* seeds was reported by Jaja-Chimedza et al. (2018). An extract of the seed that is rich in isothiocyanate that was prepared by an improved extraction method using water and ethanol solvent was found to improve the metabolic health of mice with obesity when they were fed with the extract-supplemented diet. There was significant decrease in body weight and adiposity after 12 weeks of feeding. The supplemented diet also improved the glucose tolerance, increased the antioxidant gene expression, and lowered the inflammatory gene expression in mice.

Hot water infusion prepared from *Moringa oleifera* seeds was showed to possess antispasmodic effect by inhibiting rat contraction that was induced by acetylcholine. The authors also reported that the extract has antiinflammatory activity as it was showed to inhibit carrageenan-induced edema in rats. It was also reported by the researchers that the extract has antidiuretic activity at a dose of 1000 mg/kg in rats (Caceres et al., 1992). The potential of *Moringa oleifera* seed phytochemicals to be antianaphylactic agent has been reported by Mahajan and Mehta (2007). It was found that the ethanol extract (0.001–0.1 g/kg) of the seeds completely inhibited the compound 48/80 induced anaphylactic shock in rat model. Also the extract inhibited the local cutaneous anaphylaxis in rats that was induced by 1 g/kg dose of anti-IgE antibody.

4. Antioxidant and anticancer activities of the seeds of *Moringa oleifera*

The antioxidant and anticancer activities of the seeds of *Moringa oleifera* are well reported in many research publications, and this section will discuss some of the reports. The antioxidant capacities and phenolic contents of *Moringa oleifera* seeds, pod, and leaves cultivated in three Thailand regions were evaluated (Wangcharoen & Gomolmanee, 2011). As estimated by Folin-Ciocalteu method using gallic acid standard, the *Moringa oleifera* seeds, pods, and leaf ethanolic and aqueous extracts have significant phenolic contents. The phenolic contents of the ethanol extracts ranged from 0.95 to 5.26 mg GAE/g, while the phenolic contents of the aqueous extracts were generally higher, which ranged from 1.20 to 9.69 mg GAE/g. The antioxidant activities of the extracts were evaluated using radical scavenging effect assays, which are FRAP, DPPH, and ABTS assays, and the results were expressed as ascorbic equivalence in milligram per weight of the dry sample in gram. The experimental analyses revealed the extracts have significant and relatively high antioxidant activities. The following are the result ranges for the assays used: FRAP (0.73–5.44 mg ascorbic acid/g sample), DPPH (0.51–4.43 mg ascorbic acid/g sample), and ABTS (0.56–10.10 mg ascorbic acid/g sample).

The antioxidant activity of ethanolic seed tissue extract was determined using DPPH. The results showed that 2.5 and 5 mg/mL of the seed tissue extract have 2.6% and 10.8% inhibitory percentages, respectively. The IC_{50} concentration that scavenged the DPPH radical was estimated to be 2.92 mg/g DPPH (Santos, Argolo, Paiva, & Coelho, 2012).

An optimized method for extraction of flavonoid of the seeds *Moringa oleifera* was developed by Mahmud (2017) by determining the effect of certain environmental factors such as pH, temperature, and extraction time on the optimum extraction of flavonoids. The flavonoid extracted with the optimized method has an antioxidant capacity of 5 Fe^{+2}/mg sample as estimated by FRAP assay.

The phenolic content and DPPH and ABTS antioxidant activities of 70% methanol extract of defatted seed cake of *Moringa oleifera* seeds that were harvested from three regions of Egypt (Asyut, Ismailia, and Monufia governorates) were determined by Barakat and Ghazal (2016). The results revealed that the seed methanol extracts from the regions have total phenolic contents that ranged from 16.94 ± 0.12 mg gallic acid equivalence (GAE)/g dry weight sample to 18.456 ± 0.19 mg GAE/g dry weight sample. The DPPH estimated antioxidant capacities of the seed extracts ranged from 0.17 ± 0.04 µmol Trolox equivalent (TE)/g dry weight of sample to 0.28 ± 0.05 µmol TE/g dry weight of sample. The ABTS estimated antioxidant capacities of the extracts ranged from 4.19 ± 1.12 µmol TE/g dry weight sample to 6.29 ± 1.25 µmol TE/g dry weight sample.

The oil extracted from the seeds of *Moringa oleifera* has also been reported to have antioxidant capacities by many researchers, and some of them are mentioned later. Bhatnagar and Krishna (2013) extracted oil from Jaffna variety of the seeds of *Moringa oleifera* using hexane and Soxhlet extraction method. DPPH antioxidant test for the determination of the radical scavenging effect of the oil gave 35.5 mg/mL IC_{50} concentration of the oil.

Ogbunugafor et al. (2011) harvested *Moringa oleifera* seeds grown in Nigeria and extracted its oil using *n*-hexane solvent. The researchers determined the oil total antioxidant capacity using phosphomolybdenum assay method and DPPH free radical scavenging activity assay. Their findings revealed that the oil has antioxidant capacity of 37.94 ± 0.02 mg ascorbic acid/g sample. The authors also showed that the oil has dose-dependent percentage radical scavenging effect.

In addition, *Moringa oleifera* seed oil extracted using petroleum ether and Soxhlet extraction method was evaluated for its antioxidant activity using DPPH assay. The results revealed that the oil has a concentration-dependent antioxidant capacity. The minimum concentration used was $10 \mu g/mL$ and had 65% inhibition, while the maximum concentration used ($500 \mu g/mL$) had 84% inhibition (Ojiako & Okeke, 2013).

Purwal, Pathak, and Jain (2010) investigated the inhibitory ability of *Moringa oleifera* leaves and fruits against melanoma skin cancer using in vivo model. Cold maceration method with 100% and 75% (*v/v*) methanol solvents was used to extract phytochemicals from *Moringa oleifera* leaves and fruits. Then, artificial skin tumor was induced by injecting mice with viable cells of B16F10 cancerous cells. The mice that were grouped into different groups were orally fed with the extracts for 15 days. The findings revealed the leaves and fruit extracts decreased and lowered the rate of tumor growth in the mice. Therefore it was suggested that the plant leaves and fruits could help prolong the cancer patients' life span.

Bharali, Tabassum, and Azad (2003) reported that *Moringa oleifera* seed ethanol extract has chemopreventive effect against chemical carcinogenesis in vivo. The researchers observed the chemomodulatory effect of the extract on skin papillomagenesis that was induced by 17,12-dimethylbenz(*a*)anthracene (DMBA) in mice. Their results showed that there was a substantial reduction in the population/number of mice with papillomas and the average number of papillomas in papilloma-bearing mice when they were fed with the seed extract.

Elsayed, Sharaf-eldin, and Wadaan (2015) investigated the cytotoxicity of *Moringa oleifera* seed essential oil on HeLa, HepG2, MCF7, and Caco-2 cancer cells. The cancer cells were exposed to varying doses of the oil for 24 h after which MTT assay was used for determination of the cell viability. It was revealed that the oil inhibited the proliferation and growth of MCF7, HepG2, and HeLa cells at IC_{50} values of 226.1, 751.9, and 442.8 $\mu g/mL$, respectively. At 1 mg/mL concentration, the oil decreased the viability of Caco-2 cells to $50.28 \pm 5.86\%$.

5. Apoptosis and its pathways

Apoptosis is a well-regulated and energy-dependent cellular process that is vital for normal homoeostasis. It is a programmed cell death whereby the malfunctioning and deficient cells induced self-death; hence, this is the reason the apoptotic process is the hallmark of cancer because cancer cells are cells with abnormal growth rate (Karna & Yang, 2009). Apoptotic cells have specific morphological characteristics at different stages; early apoptotic cells have nuclear fragmentation, condensed chromatins, retraction of pseudopods, and pyknosis (Kroemer et al., 2005). The late apoptotic cells possess characteristics such as internal membrane-bound phosphatidylserine exposure to the outer cell surface, cellular shrinkage, membrane blebbing, and apoptotic body formation (De bruin & Medema, 2008).

Apoptosis has two main pathways, which are the intrinsic and the extrinsic pathways. The extrinsic pathway is activated when death ligands such as TRAIL bind with their specific death receptor (such as TNFR1, fas, DR4, and DR5) at the outer membrane of the cell surface (Fig. 3). The binding triggers the receptor trimerization and clustering of the receptors' cytosolic death domains, which results in the interaction of the receptor with the adaptor molecules such as FADD at the cytoplasm. The binding of the death receptor, ligand, and adaptor molecule forms a complex called the death-inducing signaling complex (DISC) (De bruin & Medema, 2008; Nagata, 1997). The example of a DISC in Fig. 3 is the complex formed by CD95L, CD95, and FADD.

Once the DISC is formed, the DISC binds and interacts with initiator caspases 8 and 10 to activate them by cleavage as a result of induced proximity (Bao & Shi, 2007; Boatright et al., 2003) (Fig. 3). It should be noted that the activation of the caspases can be hindered and prevented by FLICE-inhibitory protein (FLIP). The activated initiator caspases execute downstream effector caspase 3 activation, which in turn causes cell death by apoptosis (Adebayo, Balogun, & Arsad, 2017). The intrinsic pathway begins when the internal stimuli (like oxidative stress, DNA damage, growth factor shortage, and high cytosolic Ca^{2+} concentration) trigger its activation (De bruin & Medema, 2008).

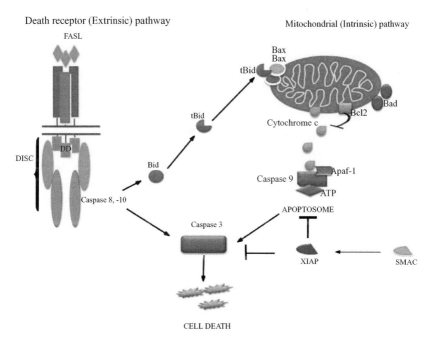

FIG. 3 The intrinsic and extrinsic pathways of apoptosis. *(Reproduced from Kalimuthu, S., Se-Kwon, K. (2013). Cell survival and apoptosis signaling as therapeutic target for cancer: Marine bioactive compounds. International Journal of Molecular Sciences, 14(2), 2334–2354.)*

The activation causes the membrane of mitochondria to be more permeable, which results in the migration of proapoptotic proteins SMAC/DIABLO and cytochrome c to the cell cytoplasm (De bruin & Medema, 2008; Johnstone, Ruefli, & Lowe, 2002). SMAC/DIABLO binds and interacts with inhibitors of apoptosis (IAPs) to prevent the IAPs from having interactions with effector caspases 3 and 7 to halt the apoptotic action of the caspases (Du, Fang, Li, Li, & Wang, 2000; Verhagen et al., 2000). The release of the proapoptotic proteins from the mitochondria is regulated by Bcl2 family proteins, which include apoptosis-inducing and apoptosis-inhibiting proteins.

Bax, Bcl-2, Bad, Bcl-xl, and Bak are examples of the Bcl-2 family proteins. The apoptosis-inducing proteins trigger the cytochrome c transportation/migration from the mitochondria to the cell cytoplasm, while the apoptosis-inhibiting proteins antagonize the migration to inhibit apoptotic cell death (Reed, 1997). In general the resultant effect of the Bcl2 apoptosis-inducing and apoptosis-inhibiting family proteins will determine whether the cytochrome c will be released from the mitochondria or not.

The released cytochrome c interacts and binds with apoptotic protease activating factor 1 (Apaf 1) and procaspase 9 to form a complex called apoptosome. Then, procaspase 9 is activated by cleavage within apoptosome by induced proximity (Pop, Timmer, Sperandio, & Salvesen, 2006; Stennicke et al., 1999). Activated caspase 9 activates procaspase 3/7 by cleavage to give caspase 3/7. The activated caspase 3/7 executes apoptotic cell death by breaking down and cleavage of proteins that are essential for the survival of the cells (Hengartner, 2000) (Fig. 3). Another process that is involved in the apoptosis activation is the fragmentation of DNA that is caused by the dissociation of caspase-activated DNase (CAD) from inhibitor of CAD (ICAD) that was induced by caspase 3. Activation of apoptosis is also caused when apoptosis-inducing factor (AIF) is released by mitochondria, which also caused DNA fragmentation. NF-κB, p53, and endoplasmic reticulum pathways are other pathways that are involved in the apoptosis process (Reed, 2000).

The disruption of the apoptosis process in cells leads to cancer whereby cancerous cells grow indefinitely without regulation and they become insensitive to death signals (Wong, 2011). The disruption occurs by certain mechanisms such as genetic mutation of p53 (Vikhanskaya, Lee, Mazzoletti, Broggini, & Sabapathy, 2007), inactivity of caspases (Devarajan et al., 2002; Fong et al., 2006; Shen et al., 2010), dysregulation of Bcl2 family proteins (Fulda, Meyer, & Debatin, 2002; Raffo et al., 1995), and inactivity of the signaling of death receptor signaling (Wong, 2011). Therefore phytochemical extracts that can induce cancer cell apoptosis house potential effective anticancer compounds.

6. Ability of compounds in *Moringa oleifera* seeds to activate cancer cell apoptosis

The ability of *Moringa oleifera* to activate cancer cell apoptosis heavily relies on its phytochemical constituents' antioxidant activities, which are mainly natural phenolics (Lu, Ou, & Lu, 2013; Rodríguez, Estrela, & Ortega, 2013; Rushworth & Micheau, 2009). These antioxidant chemical compounds reduce free radicals with their free unbonded electron to stabilize free reactive oxygen species to avert oxidative stress (Dai & Mumper, 2010). Oxidative stress is implicated in several diseases such as Alzheimer, Huntington, Parkinson, and cancer that are degenerative in nature (Valadez-Vega et al., 2013).

Moringa oleifera has vast amount of phenol types or varieties. The phenolic and flavonoid compounds that have been found in the plant seeds and have effects on apoptosis of cancer cells are ellagic acid, gallic acids, cinnamic acid, catechin, kaempferol, caffeic acid, protocatechuic acid, quercetin, ferulic acid, and p-coumaric acid (Singh et al., 2009; Singh, Negi, & Radha, 2013).

Ellagic acid induced the cleavage and activation of Bax, caspase 9, and caspase 3 while downregulating the antiapoptotic genes Bcl-x_L in cancer cells (Bisen, Bundela, & Sharma, 2012). Ellagic acid was found to inhibit lymph node carcinoma of the prostate (LNCaP) cell proliferation and growth by inducing apoptosis as it downregulated the expression of antiapoptotic proteins, heme oxygenase-1 (HO-1), human antigen R (HUR), and silent information regulator 1 (SIRT1) (Vanella et al., 2013). Ellagic acid has antiproliferative activity on U251 human glioblastoma cells as it triggered apoptosis in the cells by downregulating survivin, B-cell lymphoma 2 (Bcl-2), and X-linked inhibitor of apoptosis protein (XIAP) and upregulating caspase 3 and Bax in the cancer cells (Wang et al., 2016).

Gallic acid is one of the most widespread phenols that are abundant in plants. It has been found to exert growth inhibition of many cancer cells by activating apoptosis in the cancer cells. Sourani, Pourgheysari, Beshkar, Shirzad, and Shirzad (2016) investigated the inhibitory action of gallic acid on C121 lymphoblastic leukemia cell line. It was revealed that gallic acid hinders the growth of the cancer cells and induced apoptosis in the cells. However, the action mechanism of gallic acid on the cells was not determined.

Liu, Li, Yu, and Niu (2012) showed that gallic acid selectively induced apoptosis in MIA PaCa-2 pancreatic cancer cells but had no cytotoxic effect on nontumor hepatocyte HL-7702 cells. It was further revealed that the apoptosis was triggered by activation of Bax, caspase 9, and caspase 3. An in vitro research that was conducted by Ohno et al. (1999) revealed gallic acid significantly inhibited the proliferation of A549 and SBC-3/cisplatin-resistant H460 (CDDP) cells, epidermal bladder cell-1 (EBC-1), and SBC-3 lung cancer cells by triggering apoptosis in them via oxidative process and caspase activation. Gallic acid activated the ROS-dependent mitochondrial pathway to induce apoptosis in H446 lung cancer cells. It upregulated DIABLO, Apaf-1, p53, and bax expressions and downregulated XIAP protein expression to activate the apoptotic pathway (Wang et al., 2016).

Cinnamic acid is a phenolic compound, which is also called beta-phenylacrylic acid. It was found to impede the proliferation of CNE2 human nasopharyngeal carcinoma cells in time- and concentration-dependent manner.

Cinnamic acid (2 mmol/L) significantly triggered and activated the cancer cell apoptosis. Western blotting analyses revealed it downregulated the expression of Bcl-2 and cyclin D1 antiapoptotic proteins and upregulated the expression of bax, fas-L, p53, caspase 3, and KCF6 proapoptotic proteins. Therefore the authors concluded that cinnamic acid has potent antitumor effect (Qi et al., 2016).

Cinnamic acid induced apoptosis and disruption of cytoskeleton in HT-144 melanoma cells. It was observed HT-144 cells exposed to 2 mM cinnamic acid for incubation period of 24 h showed apoptotic properties by activating caspase 3 (De Oliveira Niero & Machado-Santelli, 2013).

Catechin had been found to possess antiproliferative activity on cancer cells by induction of apoptosis in several ways:

(1) Activation of p53/caspase-dependent pathway of apoptosis. It was found that catechin scavenged intracellular free radicals of breast cancer cells, thereby inducing apoptosis by regulation of caspase 8, caspase 3, and p53 proteins (Alshatwi, 2010).

(2) Triggering of cell cycle arrest in T47D breast cancer cells at G2/M phase, which led to the regulation of cyclin B1 and cyclin A proteins and eventually apoptosis (Deguchi, Fujii, Nakagawa, Koga, & Shirouzu, 2002).

(3) Suppression of antiapoptotic proteins of breast cancer cells, which are Bcl-2, survivin, and Bcl-x$_L$, and also upregulation of proapoptotic proteins Bcl-2 family in MDA-MB-231 and MCF-7 cells to activate apoptosis (Jo et al., 2007).

(4) Suppression of fas protein in MCF7 cells to activate apoptosis (Xiang et al., 2016).

(5) Induction of s phase arrest in T47D cells by regulating nitric oxide/nitric oxide synthase (NO/NOS) system to trigger apoptosis (Nifli, Kampa, Alexaki, Notas, & Castanas, 2005).

(6) Activation of the migration of cytosolic Ca^{2+} to endoplasmic reticulum to regulate Bcl-2 proteins of MCF7 cells to induce apoptosis (Hsu & Liou, 2011; Xiang et al., 2016).

Caffeic acid is a natural phenol that is in abundance in coffee, olive oil, vegetables, and fruits. It possesses antioxidant, antiproliferative, and antiinflammatory activities. It has been reported that Henrietta Lacks (HeLa) and MF-180 human cervical cancer cells treated with caffeic acid exhibited morphological properties of apoptotic cells (Kanimozhi & Prasad, 2015).

An extensive research was performed by Yin, Lin, Wu, Tsao, and Hsu (2009) to investigate the apoptotic effects of protocatechuic acid in lung (A549 cells), prostate (LNCaP cells), cervix (HeLa cells), liver (HepG2 cells), and breast (MCF7 cells) cancer cells, and it was found that the phenol significantly decreased the viability of the cancer cells and increased the level of caspase 3 activities in the cells. The authors concluded that the observed decrease in cell viability was caused by apoptosis activation because it corresponded to the observed increase in caspase 3 activity.

In another study, it was revealed protocatechuic acid decreased the growth of A549, Calu-6, and H3255 nonsmall lung cancer cells and elevated the expression of Bax and cleaved caspase 3 proapoptotic proteins. It also downregulated antiapoptotic protein Bcl-2 expression (Tsao, Hsia, & Yin, 2014).

Quercetin is a nonsynthetic, naturally occurring phenolic compound that is housed by many plants including *Moringa oleifera* seeds (Leone et al., 2016). Hashemzaei et al. (2017) investigated the antiproliferative effect of quercetin on LNCaP cells, colon tumor 26 (CT-26) cells, PC3 prostate cancer cells, Chinese hamster ovary (CHO) cancer cells, MCF7 breast cancer cells, MOLT-4T-lymphoblastic leukemia cancer cells, Raji lymphoid cancer cells, and U266B1 human myeloma cancer cells. The results revealed that quercetin inhibited the proliferation of these cancer cells by triggering apoptosis as tested by annexin A5 (annexin V)/propidium iodide (PI) staining. However, the molecular mechanisms of action through which the apoptosis was activated by quercetin in the cancer cells were not investigated.

In a related work, Zhou et al. (2017) also reported that quercetin has anticancer activity on HepG2 cancer cells in vitro and in vivo. The study revealed that HepG2 cell growth was suppressed by activation of apoptosis triggered by quercetin. Deng, Song, Zhou, Yuan, and Zheng (2013) observed that the activation of MCF7 cell apoptosis triggered by quercetin was by downregulation of survivin expression. Similarly, quercetin repressed the proliferation of SKOV-3 ovarian cancer cells by suppression of survivin (Ren, Deng, Ai, Yuan, & Song, 2015). Moreover, quercetin caused regression of tumor in mice by inducing cell cycle arrest and activating intrinsic mitochondrial-mediated apoptotic pathway (Srivastava et al., 2016).

Kaempferol is a major flavonoid of plants that is abundantly available in most plants including in *Moringa oleifera* seeds (Singh et al., 2009). It upregulated the expressions of Bad and Bax genes and downregulated expression of Bcl-2 and Bcl-x$_L$ genes of A549 lung cancer cells to induce apoptosis through intrinsic pathway (Nguyen et al., 2003). It was documented by Bestwick, Milne, and Duthie (2007) and Kang et al. (2009) that kaempferol triggered activity of caspase 3 in oral cavity and human leukemia HL-60 cancer cells to induce apoptosis. Kaempferol also has the potential to activate apoptosis through extrinsic death receptor pathway in osteoclasts and osteoblasts by downregulating receptor activator of NF-κB ligand (RANKL) and tumor necrosis factor-α (TNF-α) (Pang et al., 2006).

Kaempferol enhanced the apoptotic activity of cisplatin on ovarian cancer cell line 3 (OVCAR-3) to cause the cell death by downregulating myelocytomatosis viral oncogene homolog (cMyc). It was observed that kaempferol/cisplatin combined treatment of OVCAR-3 cancer cells activated apoptosis and decreased cMyc mRNA levels in the cells (Luo, Daddysman, Rankin, Jiang, & Chen, 2010).

Similarly, Dang et al. (2015) reported that kaempferol suppressed bladder cancer cell growth and triggered apoptosis activation in vitro. The finding was validated in in vivo experimental studies using mouse models. The apoptotic effect of kaempferol in the study was attributed to its ability to significantly downregulated the mesenchymal-epithelial transition factor (c-Met)/p38 signaling pathway. Kaempferol induced apoptosis in QBC939 and HCCC9810 cholangiocarcinoma cancer cells by upregulating the expressions of proapoptotic proteins caspase 9, caspase 3, caspase 8, poly (ADP-ribose) polymerase (PARP), first apoptosis signal (fas), and bax while downregulating bcl-2 antiapoptotic protein expression (Qi et al., 2016).

p-Coumaric acid inhibited HCT15 colon cancer cell growth and triggered apoptosis in the cells. Morphological property of apoptotic cells, namely, membrane blebbing, was detected in HCT15 cells treated with p-coumaric acid (Jaganathan, Supriyanto, & Mandal, 2013). Kolahi, Tabandeh, Saremy, Hosseini, and Hashemitabar (2016) related that p-coumaric acid had antiproliferative activity on MCF7 cancer cells and the apoptosis of the cancer cells was induced by 300 mM of p-coumaric acid after 24 h treatment. p-Coumaric acid had cytotoxic effect on neuroblastoma 2a (N-2a) cells by activation of apoptosis via caspase 8 dependent p53-mediated pathway. The pathway was activated by p-coumaric acid by elevation of the intracellular ROS of the cancer cells (Shailasree, Venkataramana, Niranjana, & Prakash, 2015).

Ferulic acid is a phenolic compound that was first extracted and purified from *Ferula foetida* in the year 1886, but it has since been found in many other plants (Sgarbossa, Giacomazza, & Di Carlo, 2015). Zhang et al. (2016) published the effect of ferulic acid on the growth and apoptosis of MDA-MB-231 estrogen negative breast cancer cells. It was found that the phenolic compound significantly inhibited the cancer cell proliferation and annexin V/PI staining showed that ferulic acid activated apoptosis induction in MDA-MB-231 cells. The extent of the apoptotic inducing activity of the ferulic acid in the cancer cells was directly proportional to its concentration. In the same study, it was revealed that ferulic acid activated the cell apoptosis via caspase 3 dependent pathway.

In another study, Abaza, Afzal, Al-attiyah, and Guleri (2016) published that ferulic acid isolated from *Tamarix aucheriana* had cytotoxic activity against colorectal cancer cells and induced time- and dose-dependent apoptosis in the cells. The study revealed that ferulic acid inhibited NF-κB DNA binding activity, increased the expression of proapoptotic genes (p53, p19, p18, p21, p27, and p57), and reduced the expression of antiapoptotic genes (cyclin-dependent kinases 1, 2, 4, and 6).

7.   Fatty acid contents of *Moringa oleifera* seeds

Moringa oleifera seed oil is highly fortified with monounsaturated, polyunsaturated, and saturated fatty acids, and many authors have extracted the fatty acids using various methods and identified them usually using gas-liquid chromatography (Anwar & Rashid, 2007; Ogunsina et al., 2014; Tsaknis, Lalas, Gergis, Dourtoglou, & Spiliotis, 1999). For instance, Ogunsina et al.

(2014) extracted oil from *Moringa oleifera* seeds that originated from India using two methods, which are cold pressing method and solvent (hexane) extraction method.

The oil was also extracted from the seeds grown in Kenya using solvent extraction methods (*n*-hexane and an equal mixture of chloroform/methanol methods) and cold pressing method (Tsaknis et al., 1999). Anwar and Rashid (2007) also extracted oil from *Moringa oleifera* seeds that originated from Pakistan, and the extracted oil was characterized to identify its fatty acid content.

Myristic acid, stearic acid, palmitic acid, behenic acid, and arachidic acid are the monounsaturated fatty acids identified in the oil (Ogunsina et al., 2014). The commercial names given to the oil from the seeds are Ben and Behen oil, and this is because behenic acid is the dominant saturated fatty acid in the oil (Leone et al., 2016; Tsaknis et al., 1999). A low content of margaric acid was also found in the oil (Barakat & Ghazal, 2016). The monounsaturated fatty acid contents of the seeds oil are palmitoleic acid, gadoleic acid, oleic acid, and heptadecenoic acid (Anwar & Rashid, 2007; Leone et al., 2016; Ogunsina et al., 2014; Tsaknis et al., 1999). Oleic acid is the predominant fatty acid in the seed oil, and it is estimated to be 73.57% of the total fatty acid content of the seeds (Leone et al., 2016). The polyunsaturated fatty acids that have been found in the oil of the seeds are linoleic and linolenic acids (Barakat & Ghazal, 2016).

Some fatty acids that are identified in the seeds of *Moringa oleifera* and have also been reported to activate cancer cell apoptosis induction are stearic acid, palmitic acid, oleic acid, and caprylic acid. Oleic acid triggered apoptosis in tongue squamous cell carcinoma (TSCC) by upregulating the expression of p53 and cleaved caspase 3 and downregulating the expression of Bcl-2 (Jiang et al., 2017). Stearic acid inhibited the growth and invasiveness of breast cancer cells by enhancing and raising caspase 3 activity to induce apoptosis (Evans, Cowey, Siegal, & Hardy, 2009).

Palmitic acid was found to hinder the growth of MDA-MB-231 human breast cancer cells by cytochrome *c* release from mitochondria to the cell cytosol to activate mitochondrial-mediated intrinsic apoptosis (Hardy et al., 2003). A combined treatment of palmitic acid and epalrestat caused enhanced inhibitory effect against cholangiocarcinoma RBE cell proliferation, which was a result of activation of apoptosis induction in the carcinoma cells (Qining, Wenmin, Rongping, Yulan, & Songqing, 2015).

Caprylic acid inhibited the proliferation of human skin carcinoma (A-431), colorectal carcinoma (HCT-116), and mammary gland adenocarcinoma (MDA-MB-231) cells. It downregulated cyclin-dependent kinase 4 (CDK4), cell division cycle 28 (CDC28) protein kinase regulatory subunit 1B (CKS1B), cyclin-dependent kinase 2 (CDK2), cyclin A1, and cyclin A2 to trigger cell cycle arrest in the cancer cells. It also increased the activity of caspase 8 proapoptotic protein, suggesting that it induced apoptosis in the cancer cells (Narayanan, Baskaran, Amalaradjou, & Venkitanarayanan, 2015).

8. Induction of cell cycle arrest by *Moringa oleifera* seed bioactive compounds

There are several phenolic compounds and fatty acids that have been detected in the seeds of *Moringa oleifera* and have been reported to be cell cycle arrest inducers in many cancer cells. This section briefly discusses the effect of bioactive compounds of *Moringa oleifera* seeds on cancer cell cycle progression.

8.1 Cell cycle progression events

The cell cycle is a series of events that regularly take place in cells that is necessary for cell replication. Cell cycle occurs in four major phases, which are gap 1 (G1), synthesis (S), gap 2 (G2), and mitosis (M) phases. G1 phase marks the completion of cytokinesis of a former cell division and the start of a new S phase. It is the phase at which the fate of the cell is determined as to whether it will enter a resting phase G0 or it will continue with the cell cycle process and eventually divide (Visconti, Della monica, & Grieco, 2016). At G1 phase, cells proliferate in readiness for DNA replication. At initial state of G1 phase, availability of growth factors to the cells is necessary for the cell to successfully progress to next phase (S phase) of cell cycle (Wenzel & Singh, 2018). Active replication of DNA and replication of chromosome to produce a pair of sister chromatids take place at S phase. G2 phase is the gap that bridges the S phase ending with the beginning of mitosis at M phase (Wenzel & Singh, 2018). During G2 phase, cells synthesize molecules such as proteins and cytokinesis. At the start of the phase, mitosis occurs in five different phases (viz., prophase, prometaphase, metaphase, anaphase, and telophase) (Baserga, 1962). Cytokinesis signified the completion of the processes at M phase of cell cycle after which genetically similar daughter cells are reproduced (Wenzel & Singh, 2018).

8.2 Cell cycle checkpoints

The cell cycle checkpoints exist to maintain adequate regulation of cell cycle process that will lead to ideal and proper identical daughter cell production. When there is irregularity or abnormality in the cell growth, cell cycle process is temporarily stopped until the fate of the cell is determined. If the damage is repaired, the cells die by one or more of the death processes such as apoptosis or they undergo senescence (Visconti et al., 2016). The cell cycle checkpoints are G1/S, S, and G2/M checkpoints. To activate G1/S checkpoint, DNA double-strand breaks trigger ataxia telangiectasia mutated (ATM), which in turns activated checkpoint kinase 2 (Chk2) by phosphorylation. Chk2 interacts and inhibits cell cycle division 25 homolog A (Cdc25A) protein, which is a phosphatase that regulates the activities of cyclin A/cyclin-dependent kinase (Cdk) 2 and cyclin E/Cdk2 complexes, thereby hindering the cells from entering the S phase (Falck, Mailand, Syljuåsen, Bartek, & Lukas, 2001). p53 protein plays significant role at G1/S phase checkpoint when activated by phosphorylation that is induced by ATM. The phosphorylated p53 has increased stability and less affinity for MDM2 protein (inhibitor of p53) (Visconti et al., 2016). The stabilized p53 triggers activation of p21, which binds and inhibits cyclin A/cdk2 and cyclin E/cdk2 complexes and DNA repair proteins, thus triggering arrest of cell cycle (Harper, Adami, Wei, Keyomarsi, & Elledge, 1993).

The S phase checkpoint becomes active when DNA damage occurs at S phase, which is a result of static, nonprogressive nucleotide repair/excision process and/or static replication forks (Cimprich, 2007; Errico & Costanzo, 2012). The damage sensitizes ataxia telangiectasia and Rad3-related (ATR) kinase to induce the degradation of checkpoint kinase 1 (chk1) and Cdc25A proteasome, which inhibit the progression of cell cycle at S phase (Xiao et al., 2003). G2/M checkpoint is triggered by ATR and chk1, which prevents cells that have been damaged from progression into mitosis during cell cycle process (Visconti et al., 2016).

8.3 Effects of *Moringa oleifera* seed bioactive compounds on cancer cell cycle arrest

Ellagic acid downregulated CDK2 expression to induce G1 phase arrest in bladder cancer cells (T24 cells) (Li et al., 2005). The compound also downregulated cyclin A and B1 expressions to trigger S phase cell cycle arrest in Caco-2 colorectal cancer cells (Larrosa, Tomás-barberán, & Espín, 2006). Ellagic acid induced G1 phase cell cycle arrest in PA-1 and epidermal stem 2 (ES-2) ovarian cancer cells (Chung et al., 2013). Vanella et al. (2013) also discovered that ellagic acid triggered G0/G1 phase arrest in prostate cancer cells (LNCaP).

Quercetin triggered S phase cell cycle arrest in breast cancer cells (T47D and Ehrlich ascites carcinoma (EAC)) and leukemic cancer cells (Nalm6, K562, and CEM) (Srivastava et al., 2016). G1 phase cell cycle arrest of SKOV-3 was induced by quercetin (Ren et al., 2015). Kaempferol induced cell cycle arrest of renal cell carcinoma (RCC) by regulating biomarker proteins such as chk1, CDK2, p21, and p35 (Song et al., 2014).

Caprylic acid downregulated CDK4, CKSIb, CDK2, cyclin A1, and cyclin A2 to induce cell cycle arrest in skin cancer (A-431), colorectal cancer (HCT-116), and breast cancer (MDA-MB-231) cells (Narayanan et al., 2015). Oleic acid was proven to induce G1 cell cycle arrest by downregulating the cyclin D1 expression in tongue squamous cell carcinoma (TSCC) (Jiang et al., 2017).

9. Conclusion

In conclusion, based on the existing academic reports, the seeds of *Moringa oleifera* are rich in phytochemicals and metabolites that are capable of activating apoptosis and inducing cell cycle arrest in cancer cells. The activities result in cancer cell death; hence the seeds have promising potent anticancer agents that need to be further explored especially in clinical research.

References

Abaza, M. S. I., Afzal, M., Al-attiyah, R., & Guleri, R. (2016). Cytotoxic and chemo-sensitizing effects of ferulic acid, from *Tamarix aucheriana*, on colorectal cancer cells: insights into mechanisms of therapeutic efficacy. *The FASEB Journal, 30*(Suppl. 1), 1193.7.

Adebayo, I. A., Arsad, H., Kamal, N. N. S. B. N. M., & Samian, M. R. (2020). The hexane fraction of the *Moringa oleifera* Lam seed extract induces apoptosis, causes cell cycle arrest, and modulates expression of HSP60, NPM, PGK1, RCN1, and PDIA1 in MCF7 cells. *South African Journal of Botany, 129*, 379–387. https://www.sciencedirect.com/science/article/abs/pii/S0254629919305241.

Adebayo, I. A., Arsad, H., & Samian, M. R. (2017). Antiproliferative effect on breast cancer (MCF7) of *Moringa oleifera* seed extracts. *African Journal of Traditional, Complementary, and Alternative Medicines, 14*(2), 282–287.

Adebayo, I. A., Arsad, H., & Samian, M. R. (2018a). Methyl elaidate: A major compound of potential anticancer extract of *Moringa oleifera* seeds binds with bax and MDM2 (p53 inhibitor) *in silico. Pharmacognosy Magazine, 14*(59), 554–557.

Adebayo, I. A., Arsad, H., & Samian, M. R. (2018b). Total phenolics, total flavonoids, antioxidant capacities, and volatile compounds gas-chromatography-mass spectrometry profiling of *Moringa oleifera* ripe seed polar fractions. *Pharmacognosy Magazine, 14*(54), 191–194.

Adebayo, I. A., Balogun, W. G., & Arsad, H. (2017). Moringa oleifera: An apoptosis inducer in cancer cells. *Tropical Journal of Pharmaceutical Research, 16*(9), 2289–2296.

Alshatwi, A. A. (2010). Catechin hydrate suppresses MCF-7 proliferation through TP53/Caspase-mediated apoptosis. *Journal of Experimental & Clinical Cancer Research, 29*(1), 167.

Anwar, F., Latif, S., Ashraf, M., & Gilani, A. H. (2007). *Moringa oleifera*: A food plant with multiple medicinal uses. *Phytotherapy Research, 21*(1), 17–25.

Anwar, F., & Rashid, U. (2007). Physicochemical characteristics of *Moringa oleifera* seeds and seed oil from a wild provenance of Pakistan. *Pakistan Journal of Botany, 39*(5), 1443–1453.

Auwal, M. S., Tijjani, A. N., Sadiq, M. A., Saka, S., Mairiga, I. A., Shuaibu, A., et al. (2013). Antimicrobial and haematological activity of *Moringa oleifera* aqueous seed extract in wistar albino rats. *Sokoto Journal of Veterinary Science, 11*(1), 28–37.

Bao, Q., & Shi, Y. (2007). Apoptosome: A platform for the activation of initiator caspases. *Cell Death & Differentiation, 14*(1), 56–65. https://doi.org/10.1038/sj.cdd.4402028.

Barakat, H., & Ghazal, G. A. (2016). Physicochemical properties of *Moringa oleifera* seeds and their edible oil cultivated at different regions in Egypt. *Food and Nutrition Sciences, 7*(6), 472–484.

Baserga, R. (1962). A study of nucleic acid synthesis in ascites tumor cells by two-emulsion autoradiography. *The Journal of Cell Biology, 12*(3), 633–637.

Bestwick, C. S., Milne, L., & Duthie, S. J. (2007). Kaempferol induced inhibition of HL-60 cell growth results from a heterogeneous response, dominated by cell cycle alterations. *Chemico-Biological Interactions, 170*(2), 76–85.

Bharali, R., Tabassum, J., & Azad, M. R. H. (2003). Chemomodulatory effect of *Moringa oleifera*, Lam, on hepatic carcinogen metabolising enzymes, antioxidant parameters and skin papillomagenesis in mice. *Asian Pacific Journal of Cancer Prevention, 4*(2), 131–139.

Bhatnagar, A. S., & Krishna, A. G. G. (2013). Natural antioxidants of Jaffna variety of *Moringa oleifera* seed oil of Indian origin as compared to other vegetable oils. *Grasas y Aceites, 64*(5), 537–545.

Bisen, P. S., Bundela, S. S., & Sharma, A. (2012). Ellagic acid—Chemopreventive role in oral cancer. *Journal of Cancer Science and Therapy, 4*(2), 23–30.

Boatright, K. M., Renatus, M., Scott, F. L., Sperandio, S., Shin, H., Pedersen, I. M., et al. (2003). A unified model for apical caspase activation. *Molecular Cell*, *11*(2), 529–541.

Caceres, A., Saravia, A., Rizzo, S., Zabala, L., De leon, E., & Nave, F. (1992). Pharmacologic properties of *Moringa oleifera*. 2: Screening for antispasmodic, antiinflammatory and diuretic activity. *Journal of Ethnopharmacology*, *36*(3), 233–237.

Chung, Y. C., Lu, L. C., Tsai, M. H., Chen, Y. J., Chen, Y. Y., Yao, S. P., et al. (2013). The inhibitory effect of ellagic acid on cell growth of ovarian carcinoma cells. *Evidence-based Complementary and Alternative Medicine*, *2013*, 306705.

Cimprich, K. A. (2007). Probing ATR activation with model DNA templates. *Cell Cycle*, *6*(19), 2348–2354.

Dai, J., & Mumper, R. J. (2010). Plant phenolics: Extraction, analysis and their antioxidant and anticancer properties. *Molecules*, *15*(10), 7313–7352.

Dang, Q., Song, W., Xu, D., Ma, Y., Li, F., Zeng, J., et al. (2015). Kaempferol suppresses bladder cancer tumor growth by inhibiting cell proliferation and inducing apoptosis. *Molecular Carcinogenesis*, *54*(9), 831–840.

De bruin, E. C., & Medema, J. P. (2008). Apoptosis and non-apoptotic deaths in cancer development and treatment response. *Cancer Treatment Reviews*, *34*(8), 737–749.

De Oliveira Niero, E. L., & Machado-Santelli, G. M. (2013). Cinnamic acid induces apoptotic cell death and cytoskeleton disruption in human melanoma cells. *Journal of Experimental & Clinical Cancer Research*, *32*, 31.

Deguchi, H., Fujii, T., Nakagawa, S., Koga, T., & Shirouzu, K. (2002). Analysis of cell growth inhibitory effects of catechin through MAPK in human breast cancer cell line T47D. *International Journal of Oncology*, *21*(6), 1301–1305.

Deng, X. H., Song, H. Y., Zhou, Y. F., Yuan, G. Y., & Zheng, F. J. (2013). Effects of quercetin on the proliferation of breast cancer cells and expression of survivin in vitro. *Experimental and Therapeutic Medicine*, *6*(5), 1155–1158.

Devarajan, E., Sahin, A. A., Chen, J. S., Krishnamurthy, R. R., Aggarwal, N., Brun, A. M., et al. (2002). Down-regulation of caspase 3 in breast cancer: A possible mechanism for chemoresistance. *Oncogene*, *21*(57), 8843–8851.

Du, C., Fang, M., Li, Y., Li, L., & Wang, X. (2000). Smac, a mitochondrial protein that promotes cytochrome c–dependent caspase activation by eliminating IAP inhibition. *Cell*, *102*(1), 33–42.

Dubey, D. K., Dora, J., Kumar, A., & Gulsan, R. K. (2013). A multipurpose tree—*Moringa oleifera*. *International Journal of Pharmaceutical and Chemical Sciences*, *2*, 415–423.

Elsayed, E. A., Sharaf-eldin, M. A., & Wadaan, M. (2015). In vitro evaluation of cytotoxic activities of essential oil from *Moringa oleifera* seeds on HeLa, HepG2, MCF-7, CACO-2 and L929 cell lines. *Asian Pacific Journal of Cancer Prevention*, *16*(11), 4671–4675.

Errico, A., & Costanzo, V. (2012). Mechanisms of replication fork protection: A safeguard for genome stability. *Critical Reviews in Biochemistry and Molecular Biology*, *47*(3), 222–235.

Evans, L. M., Cowey, S. L., Siegal, G. P., & Hardy, R. W. (2009). Stearate preferentially induces apoptosis in human breast cancer cells. *Nutrition and Cancer*, *61*(5), 746–753.

Fahey, J. W. (2005). *Moringa oleifera*: A review of the medical evidence for its nutritional, therapeutic, and prophylactic properties. Part 1. *Trees for Life Journal*, *1*, 1–15.

Falck, J., Mailand, N., Syljuåsen, R. G., Bartek, J., & Lukas, J. (2001). The ATM–Chk2–Cdc25A checkpoint pathway guards against radioresistant DNA synthesis. *Nature*, *410*(6830), 842–847.

Fong, P., Xue, W., Ngan, H., Chiu, P., Chan, K., Tsao, S., et al. (2006). Caspase activity is down-regulated in choriocarcinoma: A cDNA array differential expression study. *Journal of Clinical Pathology*, *59*(2), 179–183.

Fulda, S., Meyer, E., & Debatin, K. M. (2002). Inhibition of TRAIL-induced apoptosis by Bcl-2 overexpression. *Oncogene, 21*, 2283–2294.

Gopalakrishnan, L., Doriya, K., & Kumar, D. S. (2016). *Moringa oleifera*: A review on nutritive importance and its medicinal application. *Food Science and Human Wellness, 5*(2), 49–56.

Goyal, B. R., Agrawal, B. B., Goyal, R. K., & Mehta, A. A. (2007). Phyto-pharmacology of *Moringa oleifera* Lam.: An overview. *Natural Product Radiance, 6*(4), 347–353.

Hardy, S., El-assad, W., Przybytkowski, E., Joly, E., Prentki, M., & Langelier, Y. (2003). Saturated fatty acids induced apoptosis in MDA-MB-231 breast cancer cells: A role for cardiolipin. *Journal of Biological Chemistry, 278*(34), 31861–31870.

Harper, J. W., Adami, G. R., Wei, N., Keyomarsi, K., & Elledge, S. J. (1993). The p21 Cdk-interacting protein Cip1 is a potent inhibitor of G1 cyclin-dependent kinases. *Cell, 75*(4), 805–816.

Hashemzaei, M., Delarami far, A., Yari, A., Heravi, R. E., Tabrizian, K., Taghdisi, S. M., et al. (2017). Anticancer and apoptosis-inducing effects of quercetin in vitro and in vivo. *Oncology Reports, 38*(2), 819–828.

Hengartner, M. O. (2000). The biochemistry of apoptosis. *Nature, 407*(6805), 770–776.

Hsu, Y. C., & Liou, Y. M. (2011). The anti-cancer effects of (−)-epigalocathine-3-gallate on the signaling pathways associated with membrane receptors in MCF-7 cells. *Journal of Cellular Physiology, 226*(10), 2721–2730.

Jaganathan, S. K., Supriyanto, E., & Mandal, M. (2013). Events associated with apoptotic effect of p-coumaric acid in HCT-15 colon cancer cells. *World Journal of Gastroenterology, 19*(43), 7726.

Jaja-Chimedza, A., Zhang, L., Wolff, K., Graf, B. L., Kuhn, P., Moskal, K., et al. (2018). A dietary isothiocyanate-enriched moringa (*Moringa oleifera*) seed extract improves glucose tolerance in a high-fat-diet mouse model and modulates the gut microbiome. *Journal of Functional Foods, 47*, 376–385. https://doi.org/10.1016/j.jff.2018.05.056.

Jiang, L., Wang, W., He, Q., Wu, Y., Lu, Z., Sun, J., et al. (2017). Oleic acid induces apoptosis and autophagy in the treatment of Tongue Squamous cell carcinomas. *Scientific Reports, 7*, 11277.

Jo, E. H., Lee, S. J., Ahn, N. S., Park, J. S., Hwang, J. W., Kim, S. H., et al. (2007). Induction of apoptosis in MCF-7 and MDA-MB-231 breast cancer cells by oligonol is mediated by Bcl-2 family regulation and MEK/ERK signaling. *European Journal of Cancer Prevention, 16*(4), 342–347.

Johnstone, R. W., Ruefli, A. A., & Lowe, S. W. (2002). Apoptosis: A link between cancer genetics and chemotherapy. *Cell, 108*(2), 153–164.

Kang, G. Y., Lee, E. R., Kim, J. H., Jung, J. W., Lim, J., Kim, S. K., et al. (2009). Downregulation of PLK-1 expression in kaempferol-induced apoptosis of MCF-7 cells. *European Journal of Pharmacology, 611*(1–3), 17–21.

Kanimozhi, G., & Prasad, N. (2015). Anticancer effect of caffeic acid on human cervical cancer cells. In *Coffee in health and disease prevention* Elsevier.

Karna, P., & Yang, L. (2009). Apoptotic signaling pathway and resistance to apoptosis in breast cancer stem cells. In *Apoptosis in carcinogenesis and chemotherapy* Springer.

Kolahi, M., Tabandeh, M. R., Saremy, S., Hosseini, S. A., & Hashemitabar, M. (2016). The study of apoptotic effect of p-coumaric acid on breast cancer cells MCF-7. *Journal of Shahid Sadoughi University of Medical Sciences and Health Services, 24*(3), 211–221.

Kroemer, G., El-deiry, W., Golstein, P., Peter, M., Vaux, D., Vandenabeele, P., et al. (2005). Classification of cell death: Recommendations of the nomenclature committee on cell death. *Cell Death and Differentiation, 12*, 1463–1467.

Lar, P., Ojile, E., Dashe, E., & Oluoma, J. (2011). Antibacterial activity on *Moringa oleifera* seed extracts on some Gram negative bacterial isolates. *African Journal of Natural Sciences, 14*, 57–62.

Larrosa, M., Tomás-barberán, F. A., & Espín, J. C. (2006). The dietary hydrolysable tannin punica-lagin releases ellagic acid that induces apoptosis in human colon adenocarcinoma Caco-2 cells by using the mitochondrial pathway. *The Journal of Nutritional Biochemistry, 17*(9), 611–625.

Leone, A., Spada, A., Battezzati, A., Schiraldi, A., Aristil, J., & Bertoli, S. (2016). *Moringa oleifera* seeds and oil: Characteristics and uses for human health. *International Journal of Molecular Sciences, 17*(12), 2141.

Li, T. M., Chen, G. W., Su, C. C., Lin, J. G., Yeh, C. C., Cheng, K. C., et al. (2005). Ellagic acid induced p53/p21 expression, G1 arrest and apoptosis in human bladder cancer T24 cells. *Anti-cancer Research, 25*(2A), 971–979.

Liu, Z., Li, D., Yu, L., & Niu, F. (2012). Gallic acid as a cancer-selective agent induces apoptosis in pancreatic cancer cells. *Chemotherapy, 58*(3), 185–194.

Lu, L. Y., Ou, N., & Lu, Q. B. (2013). Antioxidant induces DNA damage, cell death and mutagenic-ity in human lung and skin normal cells. *Scientific Reports, 3*, 3169.

Luo, H., Daddysman, M. K., Rankin, G. O., Jiang, B. H., & Chen, Y. C. (2010). Kaempferol en-hances cisplatin's effect on ovarian cancer cells through promoting apoptosis caused by down regulation of cMyc. *Cancer Cell International, 10*, 16.

Lürling, M., & Beekman, W. (2010). Anti-cyanobacterial activity of *Moringa oleifera* seeds. *Journal of Applied Phycology, 22*(4), 503–510.

Mahajan, S. G., & Mehta, A. A. (2007). Inhibitory action of ethanolic extract of the seeds of *Moringa oleifera* Lam. on systemic and local anaphylaxis. *Journal of Immunotoxicology, 4*(4), 287–294.

Mahmood, K. T., Mugal, T., & Haq, I. U. (2010). *Moringa oleifera*: A natural gift—A review. *Journal of Pharmaceutical Sciences and Research, 2*(11), 775–781.

Mahmud, S. A. (2017). Optimal extraction of flavonoids from the *Moringa oleifera* seeds extract and study of its antioxidant activity. *Journal of Chemical and Pharmaceutical Research, 9*(4), 326–330.

Minaiyan, M., Asghari, G., Taheri, D., Saeidi, M., & Nasr-esfahani, S. (2014). Anti-inflammatory effect of *Moringa oleifera* Lam. seeds on acetic acid-induced acute colitis in rats. *Avicenna Journal of Phytomedicine, 4*(2), 127–136.

Nagata, S. (1997). Apoptosis by death factor. *Cell, 88*(3), 355–365.

Narayanan, A., Baskaran, S. A., Amalaradjou, M. A. R., & Venkitanarayanan, K. (2015). Anticar-cinogenic properties of medium chain fatty acids on human colorectal, skin and breast cancer cells in vitro. *International Journal of Molecular Sciences, 16*(3), 5014–5027.

Negi, S. (1977). Fodder trees in Himachal Pradesh. *Indian Forester, 103*(9), 616–622.

Nguyen, T., Tran, E., Ong, C., Lee, S., Do, P., Huynh, T., et al. (2003). Kaempferol-induced growth inhibition and apoptosis in A549 lung cancer cells is mediated by activation of MEK-MAPK. *Journal of Cellular Physiology, 197*(1), 110–121.

Nifli, A. P., Kampa, M., Alexaki, V. I., Notas, G., & Castanas, E. (2005). Polyphenol interac-tion with the T47D human breast cancer cell line. *Journal of Dairy Research, 72*(Suppl. 1), 44–50.

Ogbunugafor, H. A., Eneh, F. U., Ozumba, A. N., Igwo-ezikpe, M. N., Okpuzor, J., Igwilo, I. O., et al. (2011). Physico-chemical and antioxidant properties of *Moringa oleifera* seed oil. *Pakistan Journal of Nutrition, 10*(5), 409–414.

Ogunsina, B. S., Indira, T. N., Bhatnagar, A. S., Rahda, C., Debnath, S., & Gopala krishna, A. G. (2014). Quality characteristics and stability of *Moringa oleifera* seed oil of Indian origin. *Journal of Food Science and Technology, 51*(3), 503–510.

Ohno, Y., Fukuda, K., Takemura, G., Toyota, M., Watanabe, M., Yasuda, N., et al. (1999). Induction of apoptosis by gallic acid in lung cancer cells. *Anti-Cancer Drugs, 10*(9), 845–851.

Ojiako, E. N., & Okeke, C. C. (2013). Determination of antioxidant of *Moringa oleifera* seed oil and its use in the production of a body cream. *Asian Journal of Plant Science and Research*, *3*(3), 1–4.

Pang, J. L., Ricupero, D. A., Huang, S., Fatma, N., Singh, D. P., Romero, J. R., et al. (2006). Differential activity of kaempferol and quercetin in attenuating tumor necrosis factor receptor family signaling in bone cells. *Biochemical Pharmacology*, *71*(6), 818–826.

Parrotta, J. A. (1993). *Moringa oleifera Lam: Resedá, Horseradish Tree, Moringaceae, Horseradish-tree family*. International Institute of Tropical Forestry, US Department of Agriculture, Forest Service.

Pop, C., Timmer, J., Sperandio, S., & Salvesen, G. S. (2006). The apoptosome activates caspase-9 by dimerization. *Molecular Cell*, *22*(2), 269–275.

Purwal, L., Pathak, A., & Jain, U. (2010). *In vivo* anticancer activity of the leaves and fruits of *Moringa oleifera* on mouse melanoma. *Pharmacologyonline*, *1*, 655–665.

Qi, G., Chen, J., Shi, C., Wang, Y., Mi, S., Shao, W., et al. (2016). Cinnamic acid (cinn) induces apoptosis and proliferation in human nasopharyngeal carcinoma cells. *Cellular Physiology and Biochemistry*, *40*(3), 589–596.

Qining, J., Wenmin, Y., Rongping, Z., Yulan, L., & Songqing, H. (2015). Apoptotic effect of combination of epalrestat and palmitic acid in cholangiocarcinoma RBE cells. *Journal of Nutrition & Food Sciences*, *5*(6), 1.

Raffo, A. J., Perlman, H., Chen, M. W., Day, M. L., Streitman, J. S., & Buttyan, R. (1995). Overexpression of bcl-2 protects prostate cancer cells from apoptosis *in vitro* and confers resistance to androgen depletion *in vivo*. *Cancer Research*, *55*(19), 4438–4445.

Ramachandran, C., Peter, K., & Gopalakrishnan, P. (1980). Drumstick (*Moringa oleifera*): A multipurpose Indian vegetable. *Economic Botany*, *34*(3), 276–283.

Reed, J. C. (1997). Bcl-2 family proteins: Regulators of apoptosis and chemoresistance in hematologic malignancies. *Seminars in Hematology*, *34*(4 Suppl. 5), 9–19.

Reed, J. C. (2000). Mechanisms of apoptosis. *The American Journal of Pathology*, *157*(5), 1415–1430.

Ren, M. X., Deng, X. H., Ai, F., Yuan, G. Y., & Song, H. Y. (2015). Effect of quercetin on the proliferation of the human ovarian cancer cell line SKOV-3 *in vitro*. *Experimental and Therapeutic Medicine*, *10*(2), 579–583.

Rodríguez, M., Estrela, J., & Ortega, Á. (2013). Natural polyphenols and apoptosis induction in cancer therapy. *Journal of Carcinogenesis and Mutagenesis*, *6*, 1–10.

Roloff, A., Weisgerber, H., Lang, U., & Stimm, B. (2009). *Moringa oleifera LAM., 1785*. Sea 10 WILEY-VCH.

Rushworth, S. A., & Micheau, O. (2009). Molecular crosstalk between TRAIL and natural antioxidants in the treatment of cancer. *British Journal of Pharmacology*, *157*(7), 1186–1188.

Ruttarattanamongkol, K., & Petrasch, A. (2015). Antimicrobial activities of *Moringa oleifera* seed and seed oil residue and oxidative stability of its cold pressed oil compared with extra virgin olive oil. *Songklanakarin Journal of Science and Technology*, *37*(5), 587–594.

Santos, A. F. S., Argolo, A. C. C., Paiva, P. M. G., & Coelho, L. C. B. B. (2012). Antioxidant activity of *Moringa oleifera* tissue extracts. *Phytotherapy Research*, *26*(9), 1366–1370.

Sgarbossa, A., Giacomazza, D., & Di Carlo, M. (2015). Ferulic acid: A hope for Alzheimer's disease therapy from plants. *Nutrients*, *7*(7), 5764–5782.

Shailasree, S., Venkataramana, M., Niranjana, S., & Prakash, H. (2015). Cytotoxic effect of p-coumaric acid on neuroblastoma, N2a cell via generation of reactive oxygen species leading to dysfunction of mitochondria inducing apoptosis and autophagy. *Molecular Neurobiology*, *51*(1), 119–130.

Shen, X. G., Wang, C., Li, Y., Wang, L., Zhou, B., Xu, B., et al. (2010). Downregulation of caspase-9 is a frequent event in patients with stage II colorectal cancer and correlates with poor clinical outcome. *Colorectal Disease, 12*(12), 1213–1218.

Singh, R. S. G., Negi, P. S., & Radha, C. (2013). Phenolic composition, antioxidant and antimicrobial activities of free and bound phenolic extracts of *Moringa oleifera* seed flour. *Journal of Functional Foods, 5*(4), 1883–1891.

Singh, B. N., Singh, B. R., Singh, R. L., Prakash, D., Dhakarey, R., Upadhyay, G., et al. (2009). Oxidative DNA damage protective activity, antioxidant and anti-quorum sensing potentials of *Moringa oleifera. Food and Chemical Toxicology, 47*(6), 1109–1116.

Song, W., Dang, Q., Xu, D., Chen, Y., Zhu, G., Wu, K., et al. (2014). Kaempferol induces cell cycle arrest and apoptosis in renal cell carcinoma through EGFR/p38 signaling. *Oncology Reports, 31*(3), 1350–1356.

Sourani, Z., Pourgheysari, B., Beshkar, P., Shirzad, H., & Shirzad, M. (2016). Gallic acid inhibits proliferation and induces apoptosis in lymphoblastic leukemia cell line (C121). *Iranian Journal of Medical Sciences, 41*(6), 525–530.

Srivastava, S., Somasagara, R. R., Hegde, M., Nishana, M., Tadi, S. K., Srivastava, M., et al. (2016). Quercetin, a natural flavonoid interacts with DNA, arrests cell cycle and causes tumor regression by activating mitochondrial pathway of apoptosis. *Scientific Reports, 6*, 24049.

Stennicke, H. R., Deveraux, Q. L., Humke, E. W., Reed, J. C., Dixit, V. M., & Salvesen, G. S. (1999). Caspase-9 can be activated without proteolytic processing. *Journal of Biological Chemistry, 274*(13), 8359–8362.

Tsaknis, J., Lalas, S., Gergis, V., Dourtoglou, V., & Spiliotis, V. (1999). Characterization of *Moringa oleifera* variety mbololo seed oil of Kenya. *Journal of Agricultural and Food Chemistry, 47*(11), 4495–4499.

Tsao, S. M., Hsia, T. C., & Yin, M. C. (2014). Protocatechuic acid inhibits lung cancer cells by modulating FAK, MAPK, and NF-κ B pathways. *Nutrition and Cancer, 66*(8), 1331–1341.

Valadez-vega, C., Delgado-olivares, L., Morales-gonzález, J., García, E. A., Ibarra, J. R. V., Moreno, E. R., et al. (2013). The role of natural antioxidants in cancer disease. In J. A. Morales-Gonzalez (Ed.), *Oxidative stress and chronic degenerative diseases-a role for antioxidants* (pp. 391–418). Rijeka, Croatia: InTech.

Vanella, L., Di giacomo, C., Acquaviva, R., Barbagallo, I., Cardile, V., Kim, D. H., et al. (2013). Apoptotic markers in a prostate cancer cell line: Effect of ellagic acid. *Oncology Reports, 30*(6), 2804–2810.

Verhagen, A. M., Ekert, P. G., Pakusch, M., Silke, J., Connolly, L. M., Reid, G. E., et al. (2000). Identification of DIABLO, a mammalian protein that promotes apoptosis by binding to and antagonizing IAP proteins. *Cell, 102*(1), 43–53.

Vikhanskaya, F., Lee, M. K., Mazzoletti, M., Broggini, M., & Sabapathy, K. (2007). Cancer-derived p53 mutants suppress p53-target gene expression—Potential mechanism for gain of function of mutant p53. *Nucleic Acids Research, 35*(6), 2093–2104.

Visconti, R., Della monica, R., & Grieco, D. (2016). Cell cycle checkpoint in cancer: A therapeutically targetable double-edged sword. *Journal of Experimental & Clinical Cancer Research, 35*, 153.

Wang, D., Chen, Q., Liu, B., Li, Y., Tan, Y., & Yang, B. (2016). Ellagic acid inhibits proliferation and induces apoptosis in human glioblastoma cells. *Acta Cirúrgica Brasileira, 31*(2), 143–149.

Wang, R., Ma, L., Weng, D., Yao, J., Liu, X., & Jin, F. (2016). Gallic acid induces apoptosis and enhances the anticancer effects of cisplatin in human small cell lung cancer H446 cell line via the ROS-dependent mitochondrial apoptotic pathway. *Oncology Reports, 35*(5), 3075–3083.

Wangcharoen, W., & Gomolmanee, S. (2011). Antioxidant capacity and total phenolic content of *Moringa oleifera* grown in Chiang Mai, Thailand. *Thai Journal of Agricultural Science, 44*(5), 118–124.

Wenzel, E. S., & Singh, A. T. (2018). Cell-cycle checkpoints and aneuploidy on the path to cancer. *In Vivo, 32*(1), 1–5.

Wong, R. S. (2011). Apoptosis in cancer: From pathogenesis to treatment. *Journal of Experimental & Clinical Cancer Research, 30*(1), 87.

Xiang, L. P., Wang, A., Ye, J. H., Zheng, X. Q., Polito, C. A., Lu, J. L., et al. (2016). Suppressive effects of tea catechins on breast cancer. *Nutrients, 8*(8), 458.

Xiao, Z., Chen, Z., Gunasekera, A. H., Sowin, T. J., Rosenberg, S. H., Fesik, S., et al. (2003). Chk1 mediates S and G2 arrests through Cdc25A degradation in response to DNA-damaging agents. *Journal of Biological Chemistry, 278*(24), 21767–21773.

Yin, M. C., Lin, C. C., Wu, H. C., Tsao, S. M., & Hsu, C. K. (2009). Apoptotic effects of protocatechuic acid in human breast, lung, liver, cervix, and prostate cancer cells: Potential mechanisms of action. *Journal of Agricultural and Food Chemistry, 57*(14), 6468–6473.

Zaku, S., Emmanuel, S., Tukur, A., & Kabir, A. (2015). *Moringa oleifera*: An underutilized tree in Nigeria with amazing versatility: A review. *African Journal of Food Science, 9*(9), 456–461.

Zhang, X., Lin, D., Jiang, R., Li, H., Wan, J., & Li, H. (2016). Ferulic acid exerts antitumor activity and inhibits metastasis in breast cancer cells by regulating epithelial to mesenchymal transition. *Oncology Reports, 36*(1), 271–278.

Zhou, J., Fang, L., Liao, J., Li, L., Yao, W., Xiong, Z., et al. (2017). Investigation of the anti-cancer effect of quercetin on HepG2 cells in vivo. *PLoS One, 12*(3), e0172838.

Chapter 20

The medicinal properties of *Olax subscorpioidea*

Tariq Oluwakunmi Agbabiaka[a,b] and Ismail Abiola Adebayo[c]

[a]*Department of Microbiology, Faculty of Life Sciences, University of Ilorin, Ilorin, Nigeria,*
[b]*Microbiology Unit, Department of Science Laboratory Technology, School of Science and Technology, Federal Polytechnic, Damaturu, Nigeria,* [c]*Microbiology and Immunology Department, Faculty of Biomedical Sciences, Kampala International University, Ishaka-Bushenyi, Uganda*

1. Introduction

Before orthodox medicine and antibiotics, folk medicine—largely reliant on medicinal plants (Thomford et al., 2018)—was the mainstay of healthcare world over (Abe & Ohtani, 2013; Cunningham, Shanley, & Laird, 2012). Concerns, however, about safety relegated phytotherapy to a complementary and alternative status (Bruschi, Sugni, Moretti, Signorini, & Fico, 2019). Attitudes and perception, in recent times, appear to be changing. This is chiefly fueled by increasing awareness and concern about drugs, which are mostly synthetic (Moyo, Ndhlala, Finnie, & Van Staden, 2010; Nergard et al., 2015); cost; access (Nergard et al., 2015; Tait et al., 2013); and inefficacy and drug resistance (Fokunang et al., 2011; Kigen, Ronoh, Kipkore, & Rotich, 2013), and this is true for both developed and developing countries (Nergard et al., 2015; Qiu, 2007; Tait et al., 2013). This paradigm shift, spurred by the "WHO Traditional Medicine Strategy 2014–2023" (Oyebode, Kandala, Chilton, & Lilford, 2016; Sen & Chakraborty, 2017), is best exemplified in the formulation of national policies recognizing the status of traditional medicine (TM), also referred to as complementary and alternative medicine (CAM), and integrating it into healthcare delivery.

Chinese medicine (Cunningham et al., 2012), in spite of the controversies around it (Qiu, 2007), is popular and so sophisticated that a pharmacological network that establishes a nexus between the systems and the molecular basis of traditional Chinese medicine (Shao & Zhang, 2013) exists. India has a Department of AYUSH, which governs traditional Indian medicine (Sen & Chakraborty, 2017; Sen, Chakraborty, & De, 2011). The Department of Traditional Medicine serves a similar role in Mali (Nergard et al., 2015); across

Phytomedicine: A Treasure of Pharmacologically Active Products from Plants
https://doi.org/10.1016/B978-0-12-824109-7.00019-4

555

Africa, at least 36 countries possess national policies recognizing TM with Nigeria, South Africa, and Ghana going further by establishing and funding research in TM (James, Wardle, Steel, & Adams, 2018); TRAMIL (Traditional Medicines in the Islands), in the Caribbean Islands, promotes the safe use of medicinal plants in primary healthcare with a similar structure available in Asia, Western Pacific (Bhattarai, Chaudhary, Quave, & Taylor, 2010; Park & Canaway, 2019), and Latin America (Cunningham et al., 2012; Hartmann, 2016; Payyappallimana, 2010). In the West, herb-based drugs and supplements have been approved in the United Kingdom (Dzobo et al., 2018; Thomford et al., 2018; Zhokhova et al., 2019), the United States (Katz & Baltz, 2016; Li & Lou, 2018; Patridge, Gareiss, Kinch, & Hoyer, 2016; Thomford et al., 2018), Germany and France (Dzobo et al., 2018; Thomford et al., 2018), and elsewhere (Grazina, Amaral, & Mafra, 2020). The use of medicinal plants is, interestingly, not limited to chemotherapy, but has uses even in regenerative medicine (Dzobo et al., 2018; Thomford et al., 2018).

This movement is additionally predicated on the fact that plants, under the nomenclature of natural products, were, and still are, the original source of most drugs. These plants serve as the template for discovery and development of modern medicines (Atanasov et al., 2015; Ehrenworth & Peralta-Yahya, 2017; Gali-Muhtasib, Hmadi, Kareh, Tohme, & Darwiche, 2015; Lazari et al., 2017; Leonti & Verpoorte, 2017; Thomford et al., 2015, 2018; Veeresham, 2012; Zhang, Zhang, Pan, & Gong, 2016) and account for about 11% of drugs regarded by WHO as basic and essential (Veeresham, 2012). Consequently, studies (Albuquerque et al., 2014; Anyanwu, Nisar-ur, Onyeneke, & Rauf, 2015; Mothana, Al-Musayeib, Al-Ajmi, Cos, & Maes, 2014; Mukungu, Abuga, Okalebo, Ingwela, & Mwangi, 2016; Ullah et al., 2016) are documenting hits and leads of medicinal plants with diverse therapeutic uses. One such plant is *Olax subscorpioidea* (Adeoluwa, Aderibigbe, & Olonode, 2014; Ovais et al., 2018; Ukwe, Michael, & Johnny, 2010).

2. Distribution, description and traditional uses of *Olax subscorpioidea*

O. subscorpioidea Oliv., a dicot, is a member of the family Olacaceae (Adegbite et al., 2015; Gbadamosi, Raji, Oyagbemi, & Omobowale, 2017; Popoola et al., 2016) that exists as a tree or shrub (Adegbite et al., 2015), widely distributed in Nigeria, the Democratic Republic of Congo, Senegal (Gbadamosi et al., 2017), Côte d'Ivoire (Koné, Vargas, & Keiser, 2012), and other African countries (Adeoluwa, Aderibigbe, Agboola, Olonode, & Ben-Azu, 2019). Dzoyem et al. (2014), domiciled in Cameroon, reported its in vitro and in vivo activity against *Candida albicans*, thus suggesting that the plant is present in Cameroon. Other species appear to be present in South Africa (Mavundza, Maharaj, Finnie, Kabera, & Van Staden, 2011) and tropical India (Mujeeb et al., 2020).

The genus *Olax* is paraphyletic on the basis of the DNA sequences of the nucleus and chloroplast with around 40 species distributed in the tropics of the Old World, south of Madagascar, South Africa, and Australia's temperate areas (Wanntorp & De Craene, 2009). Significantly, however, the Global Biodiversity Information Facility (GBIF, 2008; Robertson et al., 2014) georeferencing tool (https://www.gbif.org/species/7397039) (Fig. 1) identifies only locations in West and Central Africa for *O. subscorpioidea*. Currently, no evidence of their presence in the New World exists, suggesting evolution and adaptation biased toward largely tropical regions of the world.

O. subscorpioidea has many local names in many regions of Africa. It is called *Ifon* in Yoruba (Adebayo, Adegbite, Olugbuyiro, Famodu, & Odenigbo, 2014; Adegbite et al., 2015; Gbadamosi et al., 2017); *Aziza* (Adebayo et al., 2014; Gbadamosi et al., 2017), *Osaja*, *Atu-ogili*, and *Igbulu* (Odoma, Zezi, Danjuma, Ahmed, & Magaji, 2017) in Igbo; *Ocheja* in Igala (Odoma et al., 2017); *Ukpakon* in Edo (Odoma et al., 2017); *Gwano kurmi* (Gbadamosi et al., 2017) and *Gwaanon raafii* (Odoma et al., 2017) in Hausa; and *Mtungapwez* in Swahili (Adebayo et al., 2014). It grows up to 10 m in height (Adebayo et al., 2014) with leafy branches bearing decurrent lines. The distichous leaves, often 2.5 to approximately 4 in. long, are broad, leathery to touch, variable in shape from oval to lanceolate with the base being cuneate or round. They bear 0.5–0.75 in. long racemes and white auxiliary flowers (Oliver & Thiselton-Dyer, 1868).

The availability of *O. subscorpioidea* led to its utilization in traditional therapeutics. Traditionally, *O. subscorpioidea* is used as an antimicrobial (Gbadamosi et al., 2017; Gottardi, Bukvicki, Prasad, & Tyagi, 2016; Ishola, Akinyede, Lawal, Popoola, & Lawal, 2015; Konan et al., 2015; Koné et al., 2012;

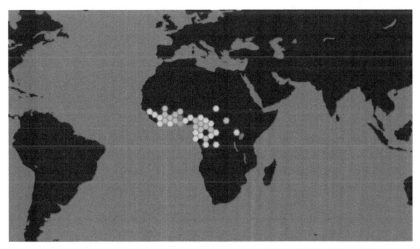

FIG. 1 Geolocation of *Olax subscorpioidea* distribution (GBIF, 2008) (Global Biodiversity Information Facility).

Ovais et al., 2018), anticancer, and antiinflammatory (Adeoluwa, Aderibigbe, & Agu, 2016; Adeoluwa et al., 2014; Attiq, Jalil, & Husain, 2017; Ezeani et al., 2019; Gbadamosi et al., 2017; Kazeem, Ayeleso, & Mukwevho, 2015; Kuete & Efferth, 2015; Ogbole, Segun, & Adeniji, 2017; Oloyede, Okpuzor, Omidiji, & Odeigah, 2011; Ovais et al., 2018; Oyedapo & Famurewa, 1995; Popoola et al., 2016, 2019; Taur & Patil, 2011; Ukwe et al., 2010) agent, as analgesic (Odoma et al., 2017), and even as an aphrodisiac (Adebayo et al., 2014; Ishola et al., 2015)! It is also used in the management of diabetes (Ayoola, Adebajo, Obuotor, Oladapo, & Fleischer, 2017; Gbadamosi et al., 2017; Kazeem et al., 2015; Olabanji et al., 2014; Saliu & Olabiyi, 2017), ulcer (Adeoluwa et al., 2016; Ishola et al., 2015; Ukwe et al., 2010), and neurological conditions (Adeoluwa, Aderibigbe, & Bakre, 2015; Adeoluwa et al., 2014, 2016; Ishola et al., 2015; Mo, Nwacheta, Aliegbere, & Akpan, 2017; Ovais et al., 2018; Saliu & Olabiyi, 2017). Decotions are made from the leaves, stem, and/or roots using both polar and nonpolar solvents (Adeoluwa et al., 2014, 2015, 2016; Odoma et al., 2017; Popoola et al., 2016; Ukwe et al., 2010) individually or in comination with other medicinal plants. It is obvious that the use of *O. subscorpioidea* in traditional medicine spans the full spectrum from infectious to physiological diseases that is indicative of its potency and year-round availability because it exists as a shrub or tree, which are perennial.

3. Anticancer properties of *Olax subscorpioidea*

The use of *O. subscorpioidea*, in traditional medicine, was folkloric, based on traditions and perhaps placebo effect. The lack of scientific basis for the use of *O. subscorpioidea* in traditional medicine has led to some studies aimed at determining the justification or otherwise of the use of *O. subscorpioidea* while evaluating its overall safety.

Adegbite et al. (2015) revealed that methanolic leaf extract of *O. subscorpioidea* can inhibit cancer cells at a dose-dependent rate. Other studies (Gbadamosi et al., 2017; Konan et al., 2015; Kuete et al., 2011; Oyedapo & Famurewa, 1995; Womeni, Djikeng, Tiencheu, & Linder, 2013) have reported similar results on the anticancer activities of *O. subscorpioidea*. A list of some available studies is presented in Table 1.

The cutoff point in these studies, however, exceeds the threshold of $\leq 30\,\mu g/mL$ (except in Kuete et al. (2011)) set by the American National Cancer Institute (Pinto, Seca, & Silva, 2017), suggesting the need to optimize dosing when they are considered as viable clinical chemotherapy candidates.

Since alteration of cellular processes such as oxidative stress, sustenance of angiogenesis, and apoptosis evasion has been linked to initiation and progression of tumors (Greenwell & Rahman, 2015; Monteiro et al., 2014; Pinto et al., 2017; Shaikh, Pund, Dawane, & Iliyas, 2014; Wen, Luo, Huang, Liao, & Yang, 2019), most anticancer agents work by preventing and/or interrupting these alterations. The alterations majorly are those involving cell signaling and

TABLE 1 Anticancer activity of *Olax subscorpioidea* extracts.

Part	Extract	In vitro and in vivo	Type of cancer	LD$_{50}$/IC$_{50}$	References
Root	Ethanolic	In vivo	Edema	400 mg/kg	Popoola et al. (2016)
Root	Ethanolic	In vitro and in vivo	ND	400 mg/kg	Popoola et al. (2019)
Leaf	Methanolic	In vivo	Liver	112 µg/mL	Adegbite et al. (2015)
Leaf	Ethanolic	In vivo	Liver	100 mg/kg	Konan et al. (2015)
Root	Saline	In vitro	ND	ND	Oyedapo and Famurewa (1995)
Seed	Methanolic	In vivo	Pancreatic cancer cell line (MiaPaCa-2), leukemia CCRF-CEM cells, CEM ADR500	20 µg/mL	Kuete et al. (2011)

ND, not determined.

apoptotic pathways (Adebayo et al., 2019; Adebayo, Balogun, & Arsad, 2017; Adeoluwa et al., 2014; Choudhari, Mandave, Deshpande, Ranjekar, & Prakash, 2020; Gunathilake, Ranaweera, & Rupasinghe, 2018; Hafezi, Hemmati, Abbaszadeh, Valizadeh, & Makvandi, 2020; Maleki et al., 2019; Miao & Xiang, 2020; Pinto et al., 2017; Rai, Jogee, Agarkar, & Santos, 2016; Zheng, Shen, Wang, Lu, & Ho, 2016). The anticancer activity of medicinal plants generally is thus premised on stimulating cell death even though the impact of nonapoptotic pathways such as senescence, autophagy, and mitotic catastrophe is now being considered (Gali-Muhtasib et al., 2015; Pinto et al., 2017; Rai et al., 2016). Same is true for *O. subscorpioidea*.

Adegbite et al. (2015) showed that methanolic leaf extracts of *O. subscorpioidea* induce intrinsic apoptosis by initiating permeability transition of the mitochondrial membrane. This is one of the effects of free radicals (Daniel, Adeoye, Ojowu, & Olorunsogo, 2018; Ovais et al., 2018) including reactive oxygen species. The disruption of the mitochondrial membrane integrity will lead to ATP seepage into the cytoplasm, subsequent hydrolysis, and ultimately cell death as a result of energy loss. Significantly, apoptosis-inducing oxidative stress has been linked to mitochondrial lipid peroxidation (Konan et al., 2015) since mitochondria, the site of cellular respiration, are reactive oxygen species' major source (Iyanda-Joel, Ajetunmobi, Chinedu, Iweala, & Adegbite, 2019), which conveniently explains the interplay of signaling pathways and apoptosis to elicit anticancer effect.

To interrupt cell signaling pathways, the plant depends on its ability to competitively inhibit different categories of kinases and COX enzymes, thus preventing cell communications that initiate and promote cancer. This inhibition is known to lead to cell cycle inhibition and proapoptotic and antiproliferative effects (Pinto et al., 2017; Popoola et al., 2016; Tariq et al., 2017).

Other mechanisms that have been reported are DNA-damaging, antioxidative, and antimitotic effects (Popoola et al., 2019). The effects of these activities are hinged on the significance of oxidative stress in the initiation and progression of tumors (Gangwar et al., 2017; Konan et al., 2015). Consequently, if the carcinogenic trigger is prevented, initiation is avoided, while progression is similarly prevented postinitiation, if administered postinitiation. A few studies have also suggested the ability of anticancer plants to upregulate tumor suppressing genes and downregulate epidermal growth factor receptor (EGFR) and transcriptional factors, thus reversing the epigenetic initiation of carcinogenesis (Greenwell & Rahman, 2015; Rai et al., 2016; Tariq et al., 2017; Xu et al., 2010).

These anticancer activities are dependent on the bioactive phytochemicals present in *O. subscorpioidea*. Cardiac glycosides, deoxysugar, tannins, phlobatannins, saponins, terpenoids, flavonoids, carbohydrates, anthocyanins, cardenolides, alkaloids, and reducing sugars (Adegbite et al., 2015; Ishola et al., 2015; Kazeem et al., 2015; Odoma et al., 2017; Popoola et al., 2016) have all been reported though variation in constituents exist, which has been linked to difference in geographical location (Dzoyem et al., 2014; Shah et al.,

2011). These constituents include both primary and secondary metabolites, thus explaining the diversity of usage *O. subscorpioidea* enjoys. Significant among these constituents are the secondary metabolites, such as terpenoids, alkaloids, and flavonoids, that serve purposes of cell signaling, protection, and defense (Griesser et al., 2015). Flavonoids, tannins, and anthocyanins, which are all phenolics, are known for regulating microRNAs and autophagy (Gali-Muhtasib et al., 2015); their high cell arrest ability, free radical scavenging, and cytotoxic effect on cancer cells (Gali-Muhtasib et al., 2015; Greenwell & Rahman, 2015) have been linked to their structure (Tariq et al., 2017) though the specific agent mediating this activity has not been isolated (Pinto et al., 2017) and differentiated from the general phytochemical classes. Other *Olax* species have enjoyed attention elucidating different agents they possess such as the four novel flavonoids (kaempferol 3-*O*-[β-D-arabinopyranosyl-(1–4)-α-L-rhamnopyranoside]-7-*O*-α-L-rhamnopyranoside, kaempferol 3-*O*-[α-D-apiofuranosyl-(1–2)-α-L-arabinofuranoside]-7-*O*-α-L-rhamnopyranoside, kaempferol 3-*O*-α-L-rhamnopyranoside, kaempferol 3-*O*-[β-D-glucopyranosyl-(1–2)-α-L-arabinofuranoside]-7-*O*-α-L-rhamnopyranoside) isolated from *O. mannii* Oliv. (Ovais et al., 2018). This is the current gap in the studies of phytochemical components of *O. subscorpioidea* that necessitate the active agents be isolated and characterized and their effects (and interactions) be independently studied and dosage, solubility, and bioavailability improved. The potential active agent for *O. subscorpioidea* is putatively suggested to be santalbic acid (Cantrell et al., 2003; Kuete, Karaosmanoğlu, & Sivas, 2017). Thus the anticancer effects these agents exert can be synergistically or independently investigated.

Since majority of the phytochemicals present in *O. subscorpioidea* and responsible for its bioactive properties are phenolics and terpenes, it's vital to discuss, briefly, their biosynthesis.

Phenolics, produced in response to biotic and abiotic factors (Zhang & Tsao, 2016), are arguably the most widespread (Cheynier, Comte, Davies, Lattanzio, & Martens, 2013) and abundant secondary metabolites (Soto-Vaca, Gutierrez, Losso, Xu, & Finley, 2012) with more than 8000 already identified (Crozier, Jaganath, & Clifford, 2009; Martins, Barros, Henriques, Silva, & Ferreira, 2015; Tsao, 2010) that are majorly produced by only plants and bryophytes, suggesting that phenolics production evolved with plants colonizing terrestrial habitats (Cheynier et al., 2013). In spite of the diversity, structurally, phenolics possess aromatic rings bearing a single (phenol) or multiple (polyphenol and tannins) hydroxyl group and other functional groups like ethers, esters, glycosides, methyl backbone (Martins et al., 2015). Now "plant phenolics" refer to secondary metabolites derived from phenylpropanoid/shikimate (where the phenylpropanoid precursors are derived) or acetate-malonate pathways (Cheynier et al., 2013), and the enzyme complexes regulating synthesis are present in the endoplasmic reticulum membranes (cytosolic phase) (Chouhan, Sharma, Zha, Guleria, & Koffas, 2017).

Biosynthesis (Fig. 2) begins with deamination of phenylalanine by phenyl-alanine ammonia lyase (PAL) to yield cinnamic acid (Chouhan et al., 2017) to give the C6–C3 structural backbone (Tsao, 2010). This product is then oxidized by cinnamate-4-hydroxylase (C4H) to form 4-coumaric acid, which is converted to 4-coumaroyl-CoA by 4-coumarate-CoA ligase (4CL). The next step involves condensation of malonyl-CoA and CoA ester in a ratio of three to one as catalyzed by chalcone synthase (CHS) to give the naringenin chalcone (Chouhan et al., 2017; Tsao, 2010). Subsequent downstream modifications of chalcone by enzymes like acyltransferases, glycosyltransferases, methyltransferases, isomerases, oxidoreductases, and hydroxylases lead to the formation of the various categories of phenolic compounds (Cheynier et al., 2013; Chouhan et al., 2017). The major classes are lignins (consisting of lignins of softwoods, hardwoods, and grasses); xanthones with subclasses such as xanthone glycosides, simple oxygenated xanthones, prenylated and related xanthones, and xanthonolignoids and related structures; flavonoids with three major subclasses of (a) isoflavonoids (3-benzopyrans) containing isoflavan, isoflavone, isoflavanone, isoflav-3-ene, isoflavanol, rotenoid, coumestane, 3-arylcoumarin, coumaronochromene, coumaronochromone, and pterocarpa; (b) neoflavonoids (4-benzopyrans), which

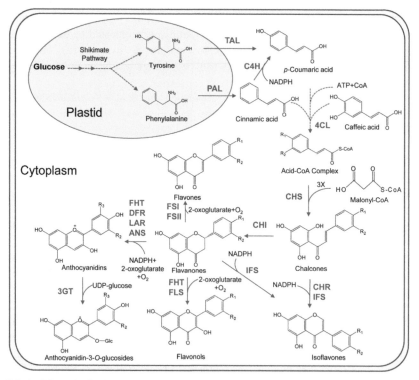

FIG. 2 Biosynthesis of phenolics (Chouhan et al., 2017).

include 4-arylcoumarins, neoflavenes, and 3,4-dihydro-4-arylcoumarins; and (c) flavonoids (2-phenylbenzopyrans) consisting of flavan, flavanone, flavone, flavonol, dihydroflavonol, flavan-3-ol, flavan-4-ol, and flavan-3,4-diol. Other classes are lignans; coumarins, which include umbelliferone, esculetin, and scopoletin; hydroxybenzoic acids including acids like protocatechuic acid, vanillic acid, gentisic acid, salicylic acid, ρ-hydroxybenzoic acid, and gallic acid; and lastly hydroxycinnamic acids such as ρ-coumaric acid, ferulic acid, caffeic acid, and sinapic acid (Martins et al., 2015). It should be noted that this is a generalized scheme not specific to *O. subscorpioidea* as no study has charted the biosynthetic path yet.

Terpenes, classified on the basis of the amount of isoprene (C5) components and the isopentenyl diphosphate (IPP) and dimethylallyl diphosphate (DMADP) precursors (Alvarez, 2014; Pazouki & Niinemets, 2016; Singh & Sharma, 2015) they contain, include hemiterpenes (C1), monoterpenes (C10), sesquiterpenes (C15), diterpenes (C20), triterpenes (C30), tetraterpenes (C40), and polyterpenes (greater than 80-carbon units) (Alvarez, 2014; Martins et al., 2015), which are derived from the methylerythritol phosphate (in the plastids) or mevalonate (in the cytosol) pathways (Alvarez, 2014; Pazouki & Niinemets, 2016; Yang et al., 2012). At least 60,000 (Pazouki & Niinemets, 2016) have been identified, making them an extremely large group (Gershenzon & Kreis, 2018; Kabera, Semana, Mussa, & He, 2014; Singh & Sharma, 2015).

To generate the IPP precursor in the cytosol, two molecules of acetyl-CoA are condensed to form 3-hydroxy-3-methylglutaryl-CoA, which is reduced by 3-hydroxy-3-methylglutaryl-CoA reductase to yield mevalonic acid via the mevalonic acid (MVA) pathway. MVA is then phosphorylated and decarboxylated to yield isopentenyl diphosphate. In the plastid, on the other hand, pyruvic acid and glyceraldehyde-3-phosphate (GAP) are condensed by 1-deoxy-d-xylulose 5-phosphate synthase (DXS) to give 1-deoxy-d-xylulose 5-phosphate (DOXP). The next step involves the reduction of the product by 1-deoxy-d-xylulose 5-phosphate reductoisomerase (DXR) to give 2-*C*-methyl-d-erythritol 4-phosphate (MEP), which is followed by its conjugation, as catalyzed by 2-*C*-methyl-d-erythritol 4-phosphate cytidylyltransferase (MCT), with 4-cytidine 5-phosphate (CMP) to produce 4-cytidine 5-phospho-2C-methyl erythritol (CDP-ME). Subsequently, 2-*C*-methyl erythritol 2,4-cyclodiphosphate synthase (MDS) converts CDP-ME to 2-C-methyl erythritol 2,4-cyclodiphosphate (CDP-ME2P), which in turn is converted to hydroxymethylbutenyl 4-diphosphate (HMBPP) by hydroxymethylbutenyl 4-diphosphate synthase (HDS). HMBPP finally branches to yield IPP and DMAP (Nagegowda, 2010; Singh & Sharma, 2015).

The IPP and DMAP generated are condensed by geranyl prenyltransferases (GDS) to give prenyl diphosphates (GDP, C10), precursor of monoterpenes; farnesyl diphosphate synthase (FDS), on the other hand, produce farnesyl diphosphate (FDP, C15), a sesquiterpene and triterpene precursor. Other transferases like geranylgeranyl diphosphate synthase (GGDS) and squalene synthase yield geranylgeranyl diphosphate (GGDP, C20) (diterpene and tetraterpene precursor) and squalene (C30). Downstream modifications of these products form the large family of enzymes referred to as terpene synthases (TPSs) to form the diverse categories of terpenes that exist (Nagegowda, 2010; Pazouki & Niinemets, 2016; Singh & Sharma, 2015). A schematic representation of the biosynthesis route is provided in Fig. 3.

In spite of these lofty traits, the major concern of orthodox medicine with traditional medicine—toxicity—remains. The main claim is that phytotherapy fails the selective toxicity criterion (Tariq et al., 2017). This concern is justified since the diversity of bioactive phytochemicals, under certain conditions or at high doses, can be injurious to consumers (Tariq et al., 2017). Adeoluwa et al. (2015) reported an LD_{50} of 300 mg/kg i.p. body weight for ethanolic extracts of *O. subscorpioidea*. Konan et al. (2015), on the other hand, reported a higher dose of 1000 mg/kg for ethanolic extracts also. This significant difference can be attributed to a difference in plant source that has been shown to influence phytochemical constituents (Shah et al., 2011), which invariably will affect the concentration of extract yield. Another important point in this is that the LD_{50} is 1000 and 3333.33% higher than the American National Cancer Institute recommended dose. Consequently, there is an imperative need to sort current dosing and toxicity challenges. Conversely, it may lead to the review of the current threshold.

4. Antimicrobial properties of *Olax subscorpioidea*

One of the claims of traditional medicine is that medicinal plants are broad spectrum, thus applicable to treatment of both infectious and noninfectious diseases.

O. subscorpioidea, traditionally, is used in the treatment of infectious diseases (Adebayo et al., 2014; Fankam, Kuete, Voukeng, Kuiate, & Pages, 2011; Ibukunoluwa, Olusi, & Dada, 2015; Kazeem et al., 2015; Kuete & Efferth, 2015; Orabueze, Amudalat, & Usman, 2016). This has led to a number of studies investigating the scientific justification.

Ayandele and Adebiyi (2007) reported the antimicrobial activity of both ethanolic and aqueous extracts of *O. subscorpioidea* stem against clinical and environmental isolates of bacteria and fungi, namely, *Pseudomonas aeruginosa*, *Staphylococcus aureus*, *Salmonella* sp., *Escherichia coli*, *Proteus vulgaris*, *Aspergillus tamarii*, *A. niger*, *Fusarium oxysporum*, and *Saccharomyces cerevisiae* using the disk diffusion assay. The ethanolic extract displayed activity against all isolates with the minimum inhibitory concentration varying between 5 and 45 mg/mL unlike aqueous extract that showed limited activity against bacteria and no activity against fungal isolates. Relative to a positive control of gentamycin (10 μg) and aqueous extract, the ethanolic extract generally possesses better activity. This could be because the ethanolic extract contained more fractions of phytochemicals (tannins, alkaloids, glycosides, flavonoids, and steroids) compared with the aqueous extract (tannins, glycosides, and saponins). This difference in extract constituents has been tied to the solubility

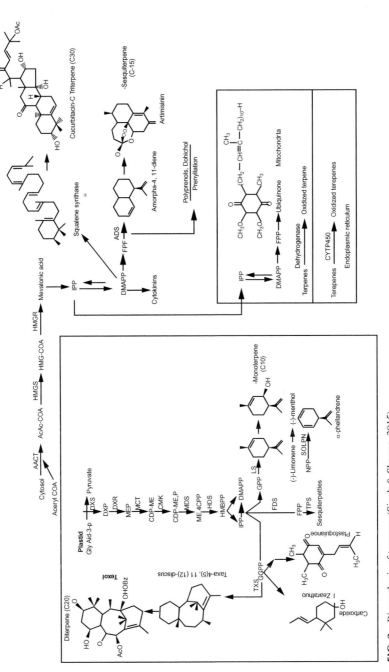

FIG. 3 Biosynthesis of terpenes (Singh & Sharma, 2015).

of the phytochemical groups in the solvents, chemical compositions and structures, amount of polymerization (Felhi et al., 2017), extract polarity, and solvent to solute mixing ratio (Iloki-Assanga et al., 2015). Cantrell et al. (2003) suggest that an inverse relationship exists between organic and aqueous extracts; thus, when the organic extract is active, aqueous extracts tend to be inactive. Consequently, it is logical that improved extraction capacity will mediate enhanced concerted activity by the phytochemicals.

Dzoyem et al. (2014) in a thorough investigation demonstrated the in vitro and in vivo activities of *O. subscorpioidea* and 21 other species against different opportunistically pathogenic yeasts, namely, *C. albicans* ATCC 9002, *C. lusitaniae* ATCC 200950, *C. tropicalis* ATCC 750, *C. parapsilosis* ATCC 22019, *C. krusei* ATCC 6258 (American Type Culture Collection), *Cryptococcus neoformans* IP95026 (Institut Pasteur France), *C. glabrata*, and *C. guilliermondii* (Centre Pasteur Yaoundé-Cameroun), with *O. subscorpioidea* being the most potent of the 22 extracts. In this study the solvent mixture of methanol and dichloromethane (3:1 *v/v*) was used for extraction. Broth microdilution assay revealed that the crude extract had a minimum inhibitory concentration (MIC) of 0.097 mg/mL and 0.048 mg/mL for *C. albicans* and *C. tropicalis*, respectively, and a corresponding minimum fungicidal concentration (MFC) of 0.19 mg/mL for both organisms. This led to an in vivo investigation of dosage in rats with experimentally induced systemic candidiasis. Results showed that oral administration of the extract at 2 g/kg of body weight led to reduction in fungal count by 24 h and to undetectable levels by 72 h. This result suggests the promising application of *O. subscorpioidea* in combating sepsis either as a prophylaxis or chemotherapeutic agent.

A similar study reported an MIC of 125 mg/mL against clinical isolates of *Klebsiella pneumoniae, E. coli, P. aeruginosa, Salmonella typhi*, and *S. aureus* and the molds *A. niger* and *Penicillium* sp. for ethanolic extracts of *O. subscorpioidea* leaves. Relative to chloramphenicol (0.2 mg/mL) and fluconazole (0.5 mg/mL) positive controls and *Ocimum sanctum* leaves' ethanolic extract, *O. subscorpioidea* extract had higher inhibitory activity with zones of inhibition ranging from 13 ± 0.50 and 16 ± 1.00 at the MIC (Wisdom et al., 2016).

Badawe et al. (2018) reported activity of methanolic extract of *O. subscorpioidea* seeds against clinical isolates of *S. aureus* with multiple drug resistant phenotype. An MIC and minimum bactericidal concentration (MBC) of 1.024 mg/mL was recorded and improved activity when modulated with some conventional antibiotic. As promising as this result is, it begs the question of adverse effect of drug interaction. This concern—that their activities are antagonistic—stems from the lack of certainty in the outcome of interaction (Shi & Klotz, 2012) even though the significance of current clinical reports has varied from probable and limited to adverse (Izzo, 2012). These interactions are modulated by drug transporters like cytochrome P450 (CYP450) and p-glycoprotein (P-gp) (Bahramsoltani, Rahimi, & Farzaei, 2017; Izzo, 2012; Oga, Sekine, Shitara, & Horie, 2016).

In another study (Fankam et al., 2011), methanolic extract of *O. subscorpioidea* was used to challenge clinical and reference strains of the enteric organisms *P. aeruginosa, Providencia stuartii, E. coli, K. pneumoniae, Enterobacter aerogenes*, and *Enterobacter cloacae* in the presence of a bacterial efflux pump inhibitor, phenylalanine-arginine *β*-naphthylamide (PAßN). MIC ranged between 0.512 and 1.024 mg/mL. This study suggests the possibility of organisms acquiring resistance to *O. subscorpioidea* extracts via at least efflux; thus, it may be imperative to consider incorporation of efflux inhibitors if *O. subscorpioidea* ever becomes a template for conventional drug development.

Evidence of activity against the parasites *Schistosoma mansoni, Echinostoma caproni, Heligmosomoides bakeri, Ancylostoma ceylanicum*, and *Trichuris muris* has been reported (Koné et al., 2012). In vitro minimum lethal concentration (MLC) ranged from 40 to 2,000 mg/mL. Oral administration of NRMI mice with 800 mg/kg of body weight only resulted in a marginal 15.5% in *E. caproni* population as opposed to a higher fraction of 60.2% reduction in *S. mansoni* population from a 400 mg/kg single oral dose administration. A less promising result from Orabueze et al. (2016) showed that among *Streptococcus* spp., *P. aeruginosa, A. fumigatus, Lactobacillus acidophilus, C. albicans, and S. aureus* of oral origin, only *A. fumigatus* was susceptible to *O. subscorpioidea* in vitro at an MIC of 51.2 mg/mL. A summary of these studies is presented in Table 2.

These data should be interpreted with caution because a phytochemical is currently defined as an antimicrobial if its MIC ranges from 0.1 to 1 mg/mL (Fankam et al., 2011). On that basis, only a handful will meet the criteria.

Despite the broad spectrum of activity against bacteria, fungi, and nematodes, we did not find any report of antiviral activity for *O. subscorpioidea*. This could be because of the lack of positive results or the difficulty associated with carrying out antiviral assays (in vitro and in vivo) in Africa, where *O. subscorpioidea* is widely available. Similarly a limited number of organic solvents that are available have been explored; solvents across the polarity spectrum should be utilized. Consequently, these current gaps need to be filled.

Another avenue to be genuinely explored is the synthesis of nanoparticles from *O. subscorpioidea*. Reports abound of impressive inhibitory effect of green-synthesized nanoparticles (Zhang, Liu, Shen, & Gurunathan, 2016). Mukherjee et al. (2014) reported the synthesis of a green-synthesized silver nanoparticles that are biocompatible and possess antibacterial, anticancer, and bioimaging capacity from *O. scandens*, which belongs to the same genus as *O. subscorpioidea*.

Significantly, none of the studies reviewed in the previous summary investigated the mechanism of action of the extracts, thus signifying another chasm worthy of filling. Generally, however, medicinal plants mediate their antimicrobial activity via secondary metabolites to effect lysis; protein denaturation; cell membrane and wall destruction; inhibition and disruption of substrate, homeostasis, cell-to-cell communication, and metabolism (Bacha et al., 2016; Martins et al., 2015; Sánchez, García, & Heredia, 2010; Silva & Fernandes

TABLE 2 Antimicrobial activity of *Olax subscorpioidea*.

Part	Extract	In vitro and in vivo	Test isolates	MIC/MBF/MFC/ MLC/LD	References
Stem	Ethanolic and aqueous	In vitro	Bacteria and fungi (mold)	5–45 mg/mL	Ayandele and Adebiyi (2007)
Not specified	Methanol-dichloromethane (3:1 v/v)	In vitro and in vivo	Fungi (yeast)	MIC: 0.097 mg/mL; MFC: 0.19 mg/mL; LD: 2 g/kg	Dzoyem et al. (2014)
Leaf	Ethanolic	In vitro	Bacteria and fungi	125 mg/mL	Wisdom et al. (2016)
Seed	Methanolic	In vitro	Bacterium	1.024 mg/mL	Badawe et al. (2018)
Fruit	Methanolic	In vitro	Bacteria	0.256–1.024 mg/mL	Fankam et al. (2011)
Root	Ethanolic	In vitro and in vivo	Parasites	40–2000 mg/mL	Koné et al. (2012)
Stem bark	Methanolic	In vitro	Bacteria and fungi	51.2 mg/mL	Orabueze et al. (2016)

Júnior, 2010; Tamokou, Mbaveng, & Kuete, 2017); immunostimulation; and cytotoxicity (Mwitari, Ayeka, Ondicho, Matu, & Bii, 2013). These activities are largely due to alkaloids, phenolics, and, to a lesser extent, terpenoids (Tamokou et al., 2017), which are abundant in *O. subscorpioidea* extracts.

5. Other medicinal properties of *Olax subscorpioidea*

5.1 Antiulcer, analgesic, and neurological activity

Reports of anticancer and analgesic effect of *O. subscorpioidea* extracts have been made. Ukwe et al. (2010) reported that methanolic extracts of *O. subscorpioidea* roots possessed ulcer inhibitory effect in a dose-dependent manner when administered orally in rats with experimentally induced ulcer at an acute toxicity LD_{50} of 2154 mg/mL. The authors suggest that the inhibitory effect observed is due to the phytochemical constituents' ability to attack free radicals, thus preventing/repairing lipid peroxidation and mucosal injury coupled with an ability to effect antibiotic effect on *Helicobacter pylori*. They attribute these to the colloidal behavior of terpenoids and saponins in solution, which enable it to form a protective coat on the mucosa with alkaloids stimulating surface repair.

Intraperitoneal administration of Swiss male albino mice with ethanolic extract of *O. subscorpioidea* leaves caused significant inhibition of novelty-induced behaviors (NIB) such as grooming, rearing, and locomotor activity (which are taken as a measure of sedation) and increased sleep time at a dose-dependent rate. The extract, however, displayed only moderate anticonvulsive effect at 50 mg/kg just as no anxiolytic/anxiogenic effect was observed. It was insinuated that the inhibitory mechanism of action is dependent on the ability of the phytochemical constituent(s) to inhibit substrates for the α_2-adrenergic receptor while depressant action is achieved by central nervous system hyperpolarization using the major neurotransmitter (gamma-aminobutyric acid [GABA]) or neighboring benzodiazepine receptor (Adeoluwa et al., 2016).

Antinociceptive effect of *O. subscorpioidea* was evaluated using formalin-induced nociception, tail immersion model, and mouse writhing test in male and female albino rats and Swiss albino mice. Extract obtained from *O. subscorpioidea* leaves using cornstarch water reduced writhes by 68.30% at a peak oral administration dose of 50 mg/kg (comparable with 67.70% for ibuprofen control) and at 200 mg/kg (oral administration); first- and second-phase duration of hand licking reduced by 54.60% and 70.50%, respectively; and a dose-dependent reaction latency for tail immersion was recorded. The mode of action, attributed to saponins present in the extract, was identified as due to antiinflammatory effect and the interaction of extract with potassium ATP channels, dopamine D_2, and serotonin as demonstrated by reversal of antinociceptive effect when glibenclamide (ATP-sensitive potassium channel inhibitor), sulpiride (dopamine D2 analogue), and metergoline (serotonin antagonist) reversed observed effects (Ishola et al., 2015).

Ominously, Ishola et al. (2015) and Adeoluwa et al. (2016) disagree on the mechanism *O. subscorpioidea* employs to achieve its neuromodulatory effects. Ishola et al. (2015) rule out the involvement of α_1- and α_2-adrenoceptors (noradrenergic system) that Adeoluwa et al. (2016) suggests is responsible! We, however, are more predisposed to support the position of Ishola et al. (2015) as their study investigated mechanism with controls even though they admit that the activities of serotonin receptors are dependent on several factors and sometimes contrasting and complex. The presence of controls that reversed observed effect, however, makes their position compelling over mere opposing suggestion by Adeoluwa et al. (2016) that was not investigated. Determining, conclusively, the exact mechanism of action would require thoroughly researched corroborating reports from either groups or third-party groups. While we could have easily tied differences to phytochemical constituents, Adeoluwa et al. (2016), in their exquisite report, did not perform phytochemical screening, which would have formed a basis of comparison for the presence of phlobatannins, polyphenols, flavonoids, steroids, anthraquinones, saponins, cardiac glycoside, and tannins reported by Ishola et al. (2015).

Adeoluwa et al. (2019) reported that butanol fraction of *O. subscorpioidea* leaves has antidepressant ability in Swiss mice. Decreased immobility (taken as a depression cue) was recorded at 10 mg/kg for forced swimming test (FST) and 5 and 10 mg/kg for tail suspension test (TST). This observation highlights antidepressant effect the extract had. The study identified serotonergic, adrenergic, and dopaminergic mechanism due to the effect of flavonoids (morin, quercetin, and rutin) and caffeic acid content of the extract. This result corroborates the report of Ishola et al. (2015) over Adeoluwa et al. (2016) discussed in the preceding paragraph.

These reports justify the use of *O. subscorpioidea* in the treatment and management of neurological conditions in traditional medicine. The results of a study by Adeoluwa et al. (2014) reiterate this when results demonstrated potent antinociceptive ability by central and peripheral effect.

5.2 Hypolipidemic effects

In view of the traditional use of *O. subscorpioidea* in weight management schemes, Gbadamosi et al. (2017) investigated its hypolipidemic effect in Wistar rats. Ethanolic extract of the leaves caused 18.36% reduction in total cholesterol (TC), 52.46% reduction in total triglyceride (TG), marginal 6.0% in low-density lipoproteins (LDL), and an increase of 13.92% in high-density lipoproteins (HDL) at 200 mg/kg relative to mice experimentally induced with hyperlipidemia after 2 weeks of administration. Improved effect was observed at 400 mg/kg with the same pattern of increase in HDL and reduction in TC, TG, and LDL. This result appears to justify short-term antihyperlipidemic usage. Ogbonnia, Mbaka, Igbokwe, Emordi, and Ota (2017) reported that Katoka mixtures, a herbal mixture, with 4% *O. subscorpioidea* did not elicit acute toxicity in male and female Wistar rats, but LD_{50} was evaluated to be 9.0 g/kg body weight. Liver tissues showed vacuolation, necroinflammation, and edematous initiation on a dose-dependent pattern; partial shrinkage of renal corpuscles; marginal enlargement of convoluted tubules' luminal diameter; and mild inflammation at high doses in the kidney; no adverse effect was observed in the heart and testicular tissues. This result reiterates the need for cautious use of herbal mixture.

5.3 Postprandial hyperglycemia

O. subscorpioidea is also used in the management of diabetes. Its activity is mediated by inhibition of the major glucose regulation enzymes, α-glucosidase and α-amylase, in the hexane fraction of *O. subscorpioidea* leaves orally administered at an IC_{50} 0.10 mg/mL and 0.72 mg/mL, respectively. α-Glucosidase and α-amylase were noncompetitively and competitively inhibited by extract (Kazeem et al., 2015). Methanolic extract of *O. nana* has also been reported to possess this activity (Ovais et al., 2018).

6. Conclusion

O. subscorpioidea is widely distributed in West and Central Africa. This availability has enabled its widespread usage in traditional medicine where it is used in the management and treatment of both infectious and noninfectious disease. This usage has spurred many studies investigating the pharmacological basis of application.

Poignantly the leaves and roots are the choice parts with ethanol and methanol being the major solvents used, which highlights the gap yet to be filled in the use of other solvents across the polarity spectrum. Majority of the reports affirm the folkloric usage with effects often dose dependent and generally safe. It should, however, be mentioned that while anticancer studies abound and substantial antibacterial and antifungal reports exist, a major gap is yet to be serviced in antiviral studies.

Similarly, there is a necessity to incorporate rigorously tested mechanism of action and identify and characterize the active principle(s) mediating these activities. In silico assays, with ability to save resources and count down significantly on turnover time, have yet to made inroads in studies involving *O. subscorpioidea* with in vitro and in vivo studies still dominating the scene. The vast experience and resources of pharmaceutical companies can come to bear here. This is an area scientists and pharmas can pull resources together to improve success in development and discovery of novel drugs.

While most reports contain toxicity studies that largely declare *O. subscorpioidea* safe, it is necessary to harmonize and develop a universal standard to eliminate the probability of subjectivity influencing interpretation.

References

Abe, R., & Ohtani, K. (2013). An ethnobotanical study of medicinal plants and traditional therapies on Batan Island, the Philippines. *Journal of Ethnopharmacology, 145*(2), 554–565.

Adebayo, H., Adegbite, O. S., Olugbuyiro, J. A., Famodu, O. O., & Odenigbo, K. B. (2014). Toxicological evaluation of extract of Olax subscorpioidea on albino Wistar rats. *African Journal of Pharmacy and Pharmacology, 8*(21), 570–578.

Adebayo, I. A., Balogun, W. G., & Arsad, H. (2017). Moringa oleifera: An apoptosis inducer in cancer cells. *Tropical Journal of Pharmaceutical Research, 16*(9), 2289–2296.

Adebayo, I. A., Gagman, H. A., Balogun, W. G., Adam, M. A. A., Abas, R., Hakeem, K. R., et al. (2019). Detarium microcarpum, Guiera senegalensis, and Cassia siamea induce apoptosis and cell cycle arrest and inhibit metastasis on MCF7 breast cancer cells. *Evidence-based Complementary and Alternative Medicine: Ecam, 2019*, 6104574. https://doi.org/10.1155/2019/6104574.

Adegbite, O. S., Akinsanya, Y. I., Kukoyi, A. J., Iyanda-Joel, W. O., Daniel, O. O., & Adebayo, A. H. (2015). Induction of rat hepatic mitochondrial membrane permeability transition pore opening by leaf extract of Olax subscorpioidea. *Pharmacognosy Research, 7*(Suppl. 1), S63–S68. https://doi.org/10.4103/0974-8490.157998.

Adeoluwa, A. O., Aderibigbe, O. A., Agboola, I. O., Olonode, T. E., & Ben-Azu, B. (2019). Butanol fraction of Olax subscorpioidea produces antidepressant effect: Evidence for the involvement of monoaminergic neurotransmission. *Drug Research, 69*(01), 53–60.

Adeoluwa, O. A., Aderibigbe, A. O., & Agu, G. O. (2016). Pharmacological evaluation of central nervous system effects of ethanol leaf extract of Olax subscorpioidea in experimental animals. *Drug Research, 66*(04), 203–210. https://doi.org/10.1055/s-0035-1564137.

Adeoluwa, O. A., Aderibigbe, A. O., & Bakre, A. G. (2015). Evaluation of antidepressant-like effect of Olax subscorpioidea Oliv. (Olacaceae) extract in mice. *Drug Research, 65*(06), 306–311. https://doi.org/10.1055/s-0034-1382010.

Adeoluwa, O. A., Aderibigbe, A. O., & Olonode, E. T. (2014). Antinociceptive property of Olax subscorpioidea Oliv (Olacaceae) extract in mice. *Journal of Ethnopharmacology, 156*, 353–357. https://doi.org/10.1016/j.jep.2014.08.040.

Albuquerque, U. P., Medeiros, P., Ramos, M. A., Ferreira Júnior, W. S., Nascimento, A. L. B., Torres, W., et al. (2014). Are ethnopharmacological surveys useful for the discovery and development of drugs from medicinal plants? *Revista Brasileira de Farmacognosia, 24*(2), 110–115.

Alvarez, M. A. (2014). *Plant biotechnology for health: From secondary metabolites to molecular farming.* Springer.

Anyanwu, G. O., Nisar-ur, R., Onyeneke, C. E., & Rauf, K. (2015). Medicinal plants of the genus Anthocleista—A review of their ethnobotany, phytochemistry and pharmacology. *Journal of Ethnopharmacology, 175*, 648–667. https://doi.org/10.1016/j.jep.2015.09.032.

Atanasov, A. G., Waltenberger, B., Pferschy-Wenzig, E.-M., Linder, T., Wawrosch, C., Uhrin, P., et al. (2015). Discovery and resupply of pharmacologically active plant-derived natural products: A review. *Biotechnology Advances, 33*(8), 1582–1614. https://doi.org/10.1016/j.biotechadv.2015.08.001.

Attiq, A., Jalil, J., & Husain, K. (2017). Annonaceae: Breaking the wall of inflammation. *Frontiers in Pharmacology, 8*, 752. https://doi.org/10.3389/fphar.2017.00752.

Ayandele, A., & Adebiyi, A. (2007). The phytochemical analysis and antimicrobial screening of extracts of Olax subscorpioidea. *African Journal of Biotechnology, 6*(7), 868–870.

Ayoola, M., Adebajo, A., Obuotor, E., Oladapo, T., & Fleischer, T. (2017). Anti-hyperglycaemic and anti-oxidant activities of five Nigerian antidiabetic plants. *Journal of Science and Technology (Ghana), 37*(2), 71–84.

Bacha, K., Tariku, Y., Gebreyesus, F., Zerihun, S., Mohammed, A., Weiland-Bräuer, N., et al. (2016). Antimicrobial and anti-Quorum sensing activities of selected medicinal plants of Ethiopia: Implication for development of potent antimicrobial agents. *BMC Microbiology, 16*(1), 139. https://doi.org/10.1186/s12866-016-0765-9.

Badawe, G., Fankam, A. G., Nayim, P., Wamba, B. E., Mbaveng, A. T., & Kuete, V. (2018). Anti-staphylococcal activity and antibiotic-modulating effect of Olax subscorpioidea, Piper guineense, Scorodophloeus zenkeri, Fagara leprieurii, and Monodora myristica against resistant phenotypes. *Investigational Medicinal Chemistry and Pharmacology, 1*(2), 17.

Bahramsoltani, R., Rahimi, R., & Farzaei, M. H. (2017). Pharmacokinetic interactions of curcuminoids with conventional drugs: A review. *Journal of Ethnopharmacology, 209*, 1–12. https://doi.org/10.1016/j.jep.2017.07.022.

Bhattarai, S., Chaudhary, R. P., Quave, C. L., & Taylor, R. S. L. (2010). The use of medicinal plants in the trans-Himalayan arid zone of Mustang district, Nepal. *Journal of Ethnobiology and Ethnomedicine, 6*(1), 14. https://doi.org/10.1186/1746-4269-6-14.

Bruschi, P., Sugni, M., Moretti, A., Signorini, M. A., & Fico, G. (2019). Children's versus adult's knowledge of medicinal plants: An ethnobotanical study in Tremezzina (Como, Lombardy, Italy). *Revista Brasileira de Farmacognosia, 29*(5), 644–655. https://doi.org/10.1016/j.bjp.2019.04.009.

Cantrell, C. L., Berhow, M. A., Phillips, B. S., Duval, S. M., Weisleder, D., & Vaughn, S. F. (2003). Bioactive crude plant seed extracts from the NCAUR oilseed repository. *Phytomedicine, 10*(4), 325–333. https://doi.org/10.1078/094471103322004820.

Cheynier, V., Comte, G., Davies, K. M., Lattanzio, V., & Martens, S. (2013). Plant phenolics: Recent advances on their biosynthesis, genetics, and ecophysiology. *Plant Physiology and Biochemistry, 72*, 1–20.

Choudhari, A. S., Mandave, P. C., Deshpande, M., Ranjekar, P., & Prakash, O. (2020). Phytochemicals in cancer treatment: From preclinical studies to clinical practice. *Frontiers in Pharmacology, 10*, 1614. https://doi.org/10.3389/fphar.2019.01614.

Chouhan, S., Sharma, K., Zha, J., Guleria, S., & Koffas, M. A. G. (2017). Recent advances in the recombinant biosynthesis of polyphenols. *Frontiers in Microbiology, 8*(2259). https://doi.org/10.3389/fmicb.2017.02259.

Crozier, A., Jaganath, I. B., & Clifford, M. N. (2009). Dietary phenolics: Chemistry, bioavailability and effects on health. *Natural Product Reports, 26*(8), 1001–1043.

Cunningham, A. B., Shanley, P., & Laird, S. (2012). Health, habitats and medicinal plant use. In *Human health and forests* (pp. 57–84). Routledge.

Daniel, O. O., Adeoye, A. O., Ojowu, J., & Olorunsogo, O. O. (2018). Inhibition of liver mitochondrial membrane permeability transition pore opening by quercetin and vitamin E in streptozotocin-induced diabetic rats. *Biochemical and Biophysical Research Communications, 504*(2), 460–469. https://doi.org/10.1016/j.bbrc.2018.08.114.

Dzobo, K., Thomford, N. E., Senthebane, D. A., Shipanga, H., Rowe, A., Dandara, C., et al. (2018). Advances in regenerative medicine and tissue engineering: Innovation and transformation of medicine. *Stem Cells International, 2018*, 2495848. https://doi.org/10.1155/2018/2495848.

Dzoyem, J. P., Tchuenguem, R. T., Kuiate, J. R., Teke, G. N., Kechia, F. A., & Kuete, V. (2014). In vitro and in vivo antifungal activities of selected Cameroonian dietary spices. *BMC Complementary and Alternative Medicine, 14*, 58. https://doi.org/10.1186/1472-6882-14-58.

Ehrenworth, A. M., & Peralta-Yahya, P. (2017). Accelerating the semisynthesis of alkaloid-based drugs through metabolic engineering. *Nature Chemical Biology, 13*(3), 249–258. https://doi.org/10.1038/nchembio.2308.

Ezeani, N. N., Ibiam, U. A., Orji, O. U., Igwenyi, I. O., Aloke, C., Alum, E., et al. (2019). Effects of aqueous and ethanol root extracts of Olax subscorpioidea on inflammatory parameters in complete Freund's adjuvant-collagen type II induced arthritic albino rats. *Pharmacognosy Journal, 11*(1), 16–25.

Fankam, A. G., Kuete, V., Voukeng, I. K., Kuiate, J. R., & Pages, J.-M. (2011). Antibacterial activities of selected Cameroonian spices and their synergistic effects with antibiotics against multidrug-resistant phenotypes. *BMC Complementary and Alternative Medicine, 11*(1), 104. https://doi.org/10.1186/1472-6882-11-104.

Felhi, S., Daoud, A., Hajlaoui, H., Mnafgui, K., Gharsallah, N., & Kadri, A. (2017). Solvent extraction effects on phytochemical constituents profiles, antioxidant and antimicrobial activities and functional group analysis of Ecballium elaterium seeds and peels fruits. *Food Science and Technology, 37*(3), 483–492.

Fokunang, C., Ndikum, V., Tabi, O., Jiofack, R., Ngameni, B., Guedje, N., et al. (2011). Traditional medicine: Past, present and future research and development prospects and integration in the national health system of Cameroon. *African Journal of Traditional, Complementary, and Alternative Medicines, 8*(3), 284.

Gali-Muhtasib, H., Hmadi, R., Kareh, M., Tohme, R., & Darwiche, N. (2015). Cell death mechanisms of plant-derived anticancer drugs: Beyond apoptosis. *Apoptosis, 20*(12), 1531–1562.

Gangwar, R., Meena, A. S., Shukla, P. K., Nagaraja, A. S., Dorniak, P. L., Pallikuth, S., et al. (2017). Calcium-mediated oxidative stress: A common mechanism in tight junction disruption by different types of cellular stress. *The Biochemical Journal, 474*(5), 731–749. https://doi.org/10.1042/BCJ20160679.

Gbadamosi, I. T., Raji, L. A., Oyagbemi, A. A., & Omobowale, T. O. (2017). Hypolipidemic effects of Olax subscorpioidea Oliv. root extract in experimental rat model. *African Journal of Biomedical Research, 20*(3), 293–299.

GBIF. (2008). *Training manual 1: Digitisation of natural history collections data.* Retrieved from https://www.gbif.org/species/7397039. Accessed 30.03.20.

Gershenzon, J., & Kreis, W. (2018). Biochemistry of terpenoids: Monoterpenes, sesquiterpenes, diterpenes, sterols, cardiac glycosides and steroid saponins. In J. A. Roberts (Ed.), *Annual plant reviews online* (pp. 218–294). Wiley.

Gottardi, D., Bukvicki, D., Prasad, S., & Tyagi, A. K. (2016). Beneficial effects of spices in food preservation and safety. *Frontiers in Microbiology, 7*, 1394. https://doi.org/10.3389/fmicb.2016.01394.

Grazina, L., Amaral, J. S., & Mafra, I. (2020). Botanical origin authentication of dietary supplements by DNA-based approaches. *Comprehensive Reviews in Food Science and Food Safety.* https://doi.org/10.1111/1541-4337.12551.

Greenwell, M., & Rahman, P. K. S. M. (2015). Medicinal plants: Their use in anticancer treatment. *International Journal of Pharmaceutical Sciences and Research, 6*(10), 4103–4112. https://doi.org/10.13040/IJPSR.0975-8232.6(10).4103-12.

Griesser, M., Weingart, G., Schoedl-Hummel, K., Neumann, N., Becker, M., Varmuza, K., et al. (2015). Severe drought stress is affecting selected primary metabolites, polyphenols, and volatile metabolites in grapevine leaves (Vitis vinifera cv. Pinot noir). *Plant Physiology and Biochemistry, 88*, 17–26. https://doi.org/10.1016/j.plaphy.2015.01.004.

Gunathilake, K. D. P. P., Ranaweera, K. K. D. S., & Rupasinghe, H. P. V. (2018). In vitro antiinflammatory properties of selected green leafy vegetables. *Biomedicine, 6*(4), 107. https://doi.org/10.3390/biomedicines6040107.

Hafezi, K., Hemmati, A. A., Abbaszadeh, H., Valizadeh, A., & Makvandi, M. (2020). Anticancer activity and molecular mechanisms of α-conidendrin, a polyphenolic compound present in Taxus yunnanensis, on human breast cancer cell lines. *Phytotherapy Research*, 1–12. https://doi.org/10.1002/ptr.6613.

Hartmann, C. (2016). Postneoliberal public health care reforms: Neoliberalism, social medicine, and persistent health inequalities in Latin America. *American Journal of Public Health, 106*(12), 2145–2151. https://doi.org/10.2105/AJPH.2016.303470.

Ibukunoluwa, M. R., Olusi, T. A., & Dada, E. O. (2015). Assessment of chemical compositions of three antimalarial plants from Akure, Southwestern Nigeria: A preliminary study. *African Journal of Plant Science, 9*(8), 313–319.

Iloki-Assanga, S. B., Lewis-Luján, L. M., Lara-Espinoza, C. L., Gil-Salido, A. A., Fernandez-Angulo, D., Rubio-Pino, J. L., et al. (2015). Solvent effects on phytochemical constituent profiles and antioxidant activities, using four different extraction formulations for analysis of Bucida buceras L. and Phoradendron californicum. *BMC Research Notes, 8*(1), 396. https://doi.org/10.1186/s13104-015-1388-1.

Ishola, I. O., Akinyede, A., Lawal, S. M., Popoola, T. D., & Lawal, A. M. (2015). Antinociceptive and anti-inflammatory effects of Olax subscorpioidea Oliv. (Olacaceae) leaf extract in rodents: Possible mechanisms of antinociceptive action. *West African Journal of Pharmacy, 26*(1), 99–112.

Iyanda-Joel, W. O., Ajetunmobi, O. B., Chinedu, S. N., Iweala, E. E. J., & Adegbite, O. S. (2019). Phytochemical, antioxidant and mitochondrial permeability transition studies on fruit-skin ethanol extract of Annona muricata. *Journal of Toxicology, 2019*, 7607031. https://doi.org/10.1155/2019/7607031.

Izzo, A. A. (2012). Interactions between herbs and conventional drugs: Overview of the clinical data. *Medical Principles and Practice, 21*(5), 404–428.

James, P. B., Wardle, J., Steel, A., & Adams, J. (2018). Traditional, complementary and alternative medicine use in Sub-Saharan Africa: A systematic review. *BMJ Global Health, 3*(5). https://doi.org/10.1136/bmjgh-2018-000895.

Kabera, J. N., Semana, E., Mussa, A. R., & He, X. (2014). Plant secondary metabolites: Biosynthesis, classification, function and pharmacological properties. *The Journal of Pharmacy and Pharmacology, 2*, 377–392.

Katz, L., & Baltz, R. H. (2016). Natural product discovery: Past, present, and future. *Journal of Industrial Microbiology & Biotechnology, 43*(2–3), 155–176.

Kazeem, M. I., Ayeleso, A. O., & Mukwevho, E. (2015). Olax subscorpioidea Oliv. Leaf alleviates postprandial hyperglycaemia by inhibition of α-amylase and α-glucosidase. *International Journal of Pharmacology, 11*(5), 484–489.

Kigen, G. K., Ronoh, H. K., Kipkore, W. K., & Rotich, J. K. (2013). Current trends of traditional herbal medicine practice in Kenya: A review. *African Journal of Pharmacology and Therapeutics, 2*(1), 32–37.

Konan, K., Justin, N.d. K., Lydie, B., Souleymane, M., Francis, Y. A., & David, N.g. J. (2015). Hepatoprotective and in vivo antioxidant activity of Olax subscorpioidea Oliv. (Olacaceae) and Distemonathus benthamianus Baill. (Caesalpiniaceae). *Pharmacognosy Magazine, 11*(41), 111–116. https://doi.org/10.4103/0973-1296.149723.

Koné, W. M., Vargas, M., & Keiser, J. (2012). Anthelmintic activity of medicinal plants used in Côte d'Ivoire for treating parasitic diseases. *Parasitology Research, 110*(6), 2351–2362.

Kuete, V., & Efferth, T. (2015). African flora has the potential to fight multidrug resistance of cancer. *BioMed Research International, 2015*, 914813. https://doi.org/10.1155/2015/914813.

Kuete, V., Karaosmanoğlu, O., & Sivas, H. (2017). Chapter 10—Anticancer activities of African medicinal spices and vegetables. In V. Kuete (Ed.), *Medicinal spices and vegetables from Africa* (pp. 271–297). Academic Press.

Kuete, V., Krusche, B., Youns, M., Voukeng, I., Fankam, A. G., Tankeo, S., et al. (2011). Cytotoxicity of some Cameroonian spices and selected medicinal plant extracts. *Journal of Ethnopharmacology, 134*(3), 803–812. https://doi.org/10.1016/j.jep.2011.01.035.

Lazari, D., Alexiou, G. A., Markopoulos, G. S., Vartholomatos, E., Hodaj, E., Chousidis, I., et al. (2017). N-(p-coumaroyl) serotonin inhibits glioblastoma cells growth through triggering S-phase arrest and apoptosis. *Journal of Neuro-Oncology, 132*(3), 373–381. https://doi.org/10.1007/s11060-017-2382-3.

Leonti, M., & Verpoorte, R. (2017). Traditional Mediterranean and European herbal medicines. *Journal of Ethnopharmacology, 199*, 161–167. https://doi.org/10.1016/j.jep.2017.01.052.

Li, G., & Lou, H.-X. (2018). Strategies to diversify natural products for drug discovery. *Medicinal Research Reviews, 38*(4), 1255–1294. https://doi.org/10.1002/med.21474.

Maleki, M., Aidy, A., Karimi, E., Shahbazi, S., Safarian, N., & Abbasi, N. (2019). Synthesis of a copolymer carrier for anticancer drug luteolin for targeting human breast cancer cells. *Journal of Traditional Chinese Medicine, 39*(4), 474–481.

Martins, N., Barros, L., Henriques, M., Silva, S., & Ferreira, I. C. F. R. (2015). Activity of phenolic compounds from plant origin against Candida species. *Industrial Crops and Products, 74*, 648–670. https://doi.org/10.1016/j.indcrop.2015.05.067.

Mavundza, E. J., Maharaj, R., Finnie, J. F., Kabera, G., & Van Staden, J. (2011). An ethnobotanical survey of mosquito repellent plants in uMkhanyakude district, KwaZulu-Natal province, South Africa. *Journal of Ethnopharmacology, 137*(3), 1516–1520. https://doi.org/10.1016/j.jep.2011.08.040.

Miao, M., & Xiang, L. (2020). Chapter three—Pharmacological action and potential targets of chlorogenic acid. In G. Du (Ed.), *Vol. 87. Advances in pharmacology* (pp. 71–88). Academic Press.

Mo, O. L., Nwacheta, C., Aliegbere, A., & Akpan, M. (2017). Assessment of antibacterial activities crude leaf extracts of selected medicinal plants from Ezza North Ebonyi state Nigeria against Staphylococcus aureus, Klebsiella pneumoniae, Pseudomonas aeruginosa. Escherichia coli, and Streptococcus mutans. *International Journal of Pharmaceutical Science Invention, 6*(9), 13–18.

Monteiro, L.d. S., Bastos, K. X., Barbosa-Filho, J. M., de Athayde-Filho, P. F., Diniz, M. D. F. F. M., & Sobral, M. V. (2014). Medicinal plants and other living organisms with antitumor potential against lung cancer. *Evidence-based Complementary and Alternative Medicine: Ecam, 2014*, 604152. https://doi.org/10.1155/2014/604152.

Mothana, R. A., Al-Musayeib, N. M., Al-Ajmi, M. F., Cos, P., & Maes, L. (2014). Evaluation of the in vitro antiplasmodial, antileishmanial, and antitrypanosomal activity of medicinal plants used in Saudi and Yemeni traditional medicine. *Evidence-based Complementary and Alternative Medicine: Ecam, 2014*, 905639. https://doi.org/10.1155/2014/905639.

Moyo, M., Ndhlala, A. R., Finnie, J. F., & Van Staden, J. (2010). Phenolic composition, antioxidant and acetylcholinesterase inhibitory activities of Sclerocarya birrea and Harpephyllum caffrum (Anacardiaceae) extracts. *Food Chemistry, 123*(1), 69–76.

Mujeeb, A. A., Khan, N. A., Jamal, F., Badre Alam, K. F., Saeed, H., Kazmi, S., et al. (2020). Olax scandens mediated biogenic synthesis of Ag-Cu nanocomposites: Potential against inhibition of drug-resistant microbes. *Frontiers in Chemistry, 8*, 103. https://doi.org/10.3389/fchem.2020.00103.

Mukherjee, S., Chowdhury, D., Kotcherlakota, R., Patra, S., Vinothkumar, B., Bhadra, M. P., et al. (2014). Potential theranostics application of bio-synthesized silver nanoparticles (4-in-1 system). *Theranostics, 4*(3), 316–335. https://doi.org/10.7150/thno.7819.

Mukungu, N., Abuga, K., Okalebo, F., Ingwela, R., & Mwangi, J. (2016). Medicinal plants used for management of malaria among the Luhya community of Kakamega East sub-County, Kenya. *Journal of Ethnopharmacology, 194*, 98–107. https://doi.org/10.1016/j.jep.2016.08.050.

Mwitari, P. G., Ayeka, P. A., Ondicho, J., Matu, E. N., & Bii, C. C. (2013). Antimicrobial activity and probable mechanisms of action of medicinal plants of Kenya: Withania somnifera, Warbugia ugandensis, Prunus africana and Plectrunthus barbatus. *PLoS One, 8*(6), e65619. https://doi.org/10.1371/journal.pone.0065619.

Nagegowda, D. A. (2010). Plant volatile terpenoid metabolism: Biosynthetic genes, transcriptional regulation and subcellular compartmentation. *FEBS Letters, 584*(14), 2965–2973.

Nergard, C. S., Ho, T. P. T., Diallo, D., Ballo, N., Paulsen, B. S., & Nordeng, H. (2015). Attitudes and use of medicinal plants during pregnancy among women at health care centers in three regions of Mali, West-Africa. *Journal of Ethnobiology and Ethnomedicine, 11*(1), 73. https://doi.org/10.1186/s13002-015-0057-8.

Odoma, S., Zezi, A. U., Danjuma, N. M., Ahmed, A., & Magaji, M. G. (2017). Elucidation of the possible mechanism of analgesic actions of butanol leaf fraction of Olax subscorpioidea Oliv. *Journal of Ethnopharmacology, 199*, 323–327.

Oga, E. F., Sekine, S., Shitara, Y., & Horie, T. (2016). Pharmacokinetic herb-drug interactions: Insight into mechanisms and consequences. *European Journal of Drug Metabolism and Pharmacokinetics, 41*(2), 93–108.

Ogbole, O. O., Segun, P. A., & Adeniji, A. J. (2017). In vitro cytotoxic activity of medicinal plants from Nigeria ethnomedicine on Rhabdomyosarcoma cancer cell line and HPLC analysis of active extracts. *BMC Complementary and Alternative Medicine, 17*(1), 494. https://doi.org/10.1186/s12906-017-2005-8.

Ogbonnia, S., Mbaka, G., Igbokwe, H., Emordi, J., & Ota, D. (2017). Toxicological, histopathological and purity evaluation of polyherbal drug katoka mixtures in rodents. *Journal of Advances in Medicine and Medical Research*, 1–16.

Olabanji, S., Adebajo, A., Omobuwajo, O., Ceccato, D., Buoso, M., & Moschini, G. (2014). PIXE analysis of some Nigerian anti-diabetic medicinal plants (II). *Nuclear Instruments and Methods in Physics Research Section B: Beam Interactions with Materials and Atoms, 318*, 187–190.

Oliver, D., & Thiselton-Dyer, W. T. (1868). *Flora of tropical Africa. Vol. 1*. L. Reeve and Company.

Oloyede, A., Okpuzor, J., Omidiji, O., & Odeigah, P. (2011). Evaluation of sub-chronic oral toxicity of joloo: A traditional medicinal decoction. *Pharmaceutical Biology, 49*(9), 936–941.

Orabueze, I., Amudalat, A., & Usman, A. (2016). Antimicrobial value of Olax subscorpioidea and Bridelia ferruginea on micro-organism isolates of dental infection. *Journal of Pharmacognosy and Phytochemistry, 5*(5), 398–406.

Ovais, M., Ayaz, M., Khalil, A. T., Shah, S. A., Jan, M. S., Raza, A., et al. (2018). HPLC-DAD finger printing, antioxidant, cholinesterase, and α-glucosidase inhibitory potentials of a novel plant Olax nana. *BMC Complementary and Alternative Medicine, 18*(1), 1. https://doi.org/10.1186/s12906-017-2057-9.

Oyebode, O., Kandala, N.-B., Chilton, P. J., & Lilford, R. J. (2016). Use of traditional medicine in middle-income countries: A WHO-SAGE study. *Health Policy and Planning, 31*(8), 984–991. https://doi.org/10.1093/heapol/czw022.

Oyedapo, O. O., & Famurewa, A. J. (1995). Antiprotease and membrane stabilizing activities of extracts of Fagara zanthoxyloides, Olax subscorpioides and Tetrapleura tetraptera. *International Journal of Pharmacognosy, 33*(1), 65–69. https://doi.org/10.3109/13880209509088150.

Park, Y. L., & Canaway, R. (2019). Integrating traditional and complementary medicine with national healthcare systems for universal health coverage in Asia and the Western Pacific. *Health Systems & Reform, 5*(1), 24–31.

Patridge, E., Gareiss, P., Kinch, M. S., & Hoyer, D. (2016). An analysis of FDA-approved drugs: Natural products and their derivatives. *Drug Discovery Today, 21*(2), 204–207. https://doi.org/10.1016/j.drudis.2015.01.009.

Payyappallimana, U. (2010). Role of traditional medicine in primary health care: An overview of perspectives and challenging. *Yokohama Journal of Social Sciences, 14*(6), 57–77. Retrieved from: https://ci.nii.ac.jp/naid/110009587206/en/.

Pazouki, L., & Niinemets, Ü. (2016). Multi-substrate terpene synthases: Their occurrence and physiological significance. *Frontiers in Plant Science, 7*(1019). https://doi.org/10.3389/fpls.2016.01019.

Pinto, D. C. G. A., Seca, A. M. L., & Silva, A. M. S. (2017). Insight approaches of medicinal plants for the discovery of anticancer drugs. In M. S. Akhtar, & M. K. Swamy (Eds.), *Vol. 3. Anticancer plants: Clinical trials and nanotechnology* (pp. 105–151). Singapore: Springer.

Popoola, T. D., Awodele, O., Babawale, F., Oguns, O., Onabanjo, O., Ibanga, I., et al. (2019). Antioxidative, antimitotic, and DNA-damaging activities of Garcinia kola stem bark, Uvaria chamae root, and Olax subscorpioidea root used in the ethnotherapy of cancers. *Journal of Basic and Clinical Physiology and Pharmacology.* https://doi.org/10.1515/jbcpp-2019-0073.

Popoola, T. D., Awodele, O., Omisanya, A., Obi, N., Umezinwa, C., & Fatokun, A. A. (2016). Three indigenous plants used in anti-cancer remedies, Garcinia kola Heckel (stem bark), Uvaria chamae P. Beauv. (root) and Olax subscorpioidea Oliv. (root) show analgesic and anti-inflammatory activities in animal models. *Journal of Ethnopharmacology, 194,* 440–449. https://doi.org/10.1016/j.jep.2016.09.046.

Qiu, J. (2007). A culture in the balance. *Nature, 448*(7150), 126–128. https://doi.org/10.1038/448126a.

Rai, M., Jogee, P. S., Agarkar, G., & Santos, C. A.d. (2016). Anticancer activities of Withania somnifera: Current research, formulations, and future perspectives. *Pharmaceutical Biology, 54*(2), 189–197.

Robertson, T., Döring, M., Guralnick, R., Bloom, D., Wieczorek, J., Braak, K., et al. (2014). The GBIF integrated publishing toolkit: Facilitating the efficient publishing of biodiversity data on the internet. *PLoS One, 9*(8), e102623. https://doi.org/10.1371/journal.pone.0102623.

Saliu, J. A., & Olabiyi, A. A. (2017). Aqueous extract of Securidaca longipendunculata Oliv. and Olax subscropioidea inhibits key enzymes (acetylcholinesterase and butyrylcholinesterase) linked with Alzheimer's disease in vitro. *Pharmaceutical Biology, 55*(1), 252–257. https://doi.org/10.1080/13880209.2016.1258426.

Sánchez, E., García, S., & Heredia, N. (2010). Extracts of edible and medicinal plants damage membranes of Vibrio cholerae. *Applied and Environmental Microbiology, 76*(20), 6888–6894.

Sen, S., & Chakraborty, R. (2017). Revival, modernization and integration of Indian traditional herbal medicine in clinical practice: Importance, challenges and future. *Journal of Traditional and Complementary Medicine, 7*(2), 234–244. https://doi.org/10.1016/j.jtcme.2016.05.006.

Sen, S., Chakraborty, R., & De, B. (2011). Challenges and opportunities in the advancement of herbal medicine: India's position and role in a global context. *Journal of Herbal Medicine, 1*(3), 67–75. https://doi.org/10.1016/j.hermed.2011.11.001.

Shah, G., Shri, R., Panchal, V., Sharma, N., Singh, B., & Mann, A. S. (2011). Scientific basis for the therapeutic use of Cymbopogon citratus, stapf (Lemon grass). *Journal of Advanced Pharmaceutical Technology & Research, 2*(1), 3–8. https://doi.org/10.4103/2231-4040.79796.

Shaikh, R., Pund, M., Dawane, A., & Iliyas, S. (2014). Evaluation of anticancer, antioxidant, and possible anti-inflammatory properties of selected medicinal plants used in Indian traditional medication. *Journal of Traditional and Complementary Medicine, 4*(4), 253–257. https://doi.org/10.4103/2225-4110.128904.

Shao, L., & Zhang, B. (2013). Traditional Chinese medicine network pharmacology: Theory, methodology and application. *Chinese Journal of Natural Medicines, 11*(2), 110–120.

Shi, S., & Klotz, U. (2012). Drug interactions with herbal medicines. *Clinical Pharmacokinetics, 51*(2), 77–104.

Silva, N., & Fernandes Júnior, A. (2010). Biological properties of medicinal plants: A review of their antimicrobial activity. *Journal of Venomous Animals and Toxins Including Tropical Diseases, 16*(3), 402–413.

Singh, B., & Sharma, R. A. (2015). Plant terpenes: Defense responses, phylogenetic analysis, regulation and clinical applications. *3 Biotech, 5*(2), 129–151. https://doi.org/10.1007/s13205-014-0220-2.

Soto-Vaca, A., Gutierrez, A., Losso, J. N., Xu, Z., & Finley, J. W. (2012). Evolution of phenolic compounds from color and flavor problems to health benefits. *Journal of Agricultural and Food Chemistry, 60*(27), 6658–6677.

Tait, E. M., Laditka, J. N., Laditka, S. B., Nies, M. A., Racine, E. F., & Tsulukidze, M. M. (2013). Reasons why older Americans use complementary and alternative medicine: Costly or ineffective conventional medicine and recommendations from health care providers, family, and friends. *Educational Gerontology, 39*(9), 684–700. https://doi.org/10.1080/03601277.2012.734160.

Tamokou, J. D. D., Mbaveng, A. T., & Kuete, V. (2017). Chapter 8—Antimicrobial activities of African medicinal spices and vegetables. In V. Kuete (Ed.), *Medicinal spices and vegetables from Africa* (pp. 207–237). Academic Press.

Tariq, A., Sadia, S., Pan, K., Ullah, I., Mussarat, S., Sun, F., et al. (2017). A systematic review on ethnomedicines of anti-cancer plants. *Phytotherapy Research, 31*(2), 202–264. https://doi.org/10.1002/ptr.5751.

Taur, D. J., & Patil, R. Y. (2011). Some medicinal plants with antiasthmatic potential: A current status. *Asian Pacific Journal of Tropical Biomedicine, 1*(5), 413–418. https://doi.org/10.1016/S2221-1691(11)60091-9.

Thomford, N. E., Dzobo, K., Chopera, D., Wonkam, A., Skelton, M., Blackhurst, D., et al. (2015). Pharmacogenomics implications of using herbal medicinal plants on African populations in health transition. *Pharmaceuticals, 8*(3), 637–663. Retrieved from: https://www.mdpi.com/1424-8247/8/3/637.

Thomford, N. E., Senthebane, D. A., Rowe, A., Munro, D., Seele, P., Maroyi, A., et al. (2018). Natural products for drug discovery in the 21st century: Innovations for novel drug discovery. *International Journal of Molecular Sciences, 19*(6), 1578. Retrieved from: https://www.mdpi.com/1422-0067/19/6/1578.

Tsao, R. (2010). Chemistry and biochemistry of dietary polyphenols. *Nutrients, 2*(12), 1231–1246.

Ukwe, V. C., Michael, U. C., & Johnny, M. U. (2010). Evaluation of the antiulcer activity of Olax subscorpioidea Oliv. roots in rats. *Asian Pacific Journal of Tropical Medicine, 3*(1), 13–16. https://doi.org/10.1016/S1995-7645(10)60022-3.

Ullah, N., Nadhman, A., Siddiq, S., Mehwish, S., Islam, A., Jafri, L., et al. (2016). Plants as Anti-leishmanial agents: Current scenario. *Phytotherapy Research, 30*(12), 1905–1925. https://doi.org/10.1002/ptr.5710.

Veeresham, C. (2012). Natural products derived from plants as a source of drugs. *Journal of Advanced Pharmaceutical Technology & Research, 3*(4), 200–201. https://doi.org/10.4103/2231-4040.104709.

Wanntorp, L., & De Craene, L. P. (2009). Perianth evolution in the sandalwood order Santalales. *American Journal of Botany, 96*(7), 1361–1371. https://doi.org/10.3732/ajb.0800236.

Wen, Q., Luo, K., Huang, H., Liao, W., & Yang, H. (2019). Xanthoxyletin inhibits proliferation of human oral squamous carcinoma cells and induces apoptosis, autophagy, and cell cycle arrest by modulation of the MEK/ERK signaling pathway. *Medical Science Monitor: International Medical Journal of Experimental and Clinical Research, 25*, 8025–8033. https://doi.org/10.12659/MSM.911697.

Wisdom, N. N., Bassey, E. E., Jelani, F. B., Ishaku, G. A., Uwem, U. M., & Joseph, S. C. (2016). Biochemical studies of Ocimum sanctum and Olax subscorpioidea leaf extracts. *Journal of Pharmaceutical Research International*, 1–9.

Womeni, H. M., Djikeng, F. T., Tiencheu, B., & Linder, M. (2013). Antioxidant potential of methanolic extracts and powders of some Cameroonian spices during accelerated storage of soybean oil. *Advances in Biological Chemistry, 3*(3), 304–313.

Xu, J., Timares, L., Heilpern, C., Weng, Z., Li, C., Xu, H., et al. (2010). Targeting wild-type and mutant p53 with small molecule CP-31398 blocks the growth of rhabdomyosarcoma by inducing reactive oxygen species-dependent apoptosis. *Cancer Research, 70*(16), 6566–6576. https://doi.org/10.1158/0008-5472.CAN-10-0942.

Yang, C. Q., Fang, X., Wu, X. M., Mao, Y. B., Wang, L. J., & Chen, X. Y. (2012). Transcriptional regulation of plant secondary metabolism F. *Journal of Integrative Plant Biology, 54*(10), 703–712.

Zhang, C., Zhang, G., Pan, J., & Gong, D. (2016). Galangin competitively inhibits xanthine oxidase by a ping-pong mechanism. *Food Research International, 89*, 152–160. https://doi.org/10.1016/j.foodres.2016.07.021.

Zhang, H., & Tsao, R. (2016). Dietary polyphenols, oxidative stress and antioxidant and antiinflammatory effects. *Current Opinion in Food Science, 8*, 33–42.

Zhang, X. F., Liu, Z. G., Shen, W., & Gurunathan, S. (2016). Silver nanoparticles: Synthesis, characterization, properties, applications, and therapeutic approaches. *International Journal of Molecular Sciences, 17*(9), 1534. https://doi.org/10.3390/ijms17091534.

Zheng, Y. M., Shen, J. Z., Wang, Y., Lu, A. X., & Ho, W. S. (2016). Anti-oxidant and anti-cancer activities of Angelica dahurica extract via induction of apoptosis in colon cancer cells. *Phytomedicine, 23*(11), 1267–1274.

Zhokhova, E. V., Rodionov, A. V., Povydysh, M. N., Goncharov, M. Y., Protasova, Y. A., & Yakovlev, G. P. (2019). Current state and prospects of DNA barcoding and DNA fingerprinting in the analysis of the quality of plant raw materials and plant-derived drugs. *Biology Bulletin Reviews, 9*(4), 301–314. https://doi.org/10.1134/S2079086419040030.

Chapter 21

Phytotherapeutic agents for neurodegenerative disorders: A neuropharmacological review

Andleeb Khan[a], Sadaf Jahan[b], Saeed Alshahrani[a], Bader Mohammed Alshehri[b], Aga Syed Sameer[c], Azher Arafah[d], Ajaz Ahmad[d], and Muneeb U. Rehman[d]

[a]Department of Pharmacology and Toxicology, College of Pharmacy, Jazan University, Jazan, Saudi Arabia, [b]Medical Laboratories Department, College of Applied Medical Sciences, Majmaah University, Al Majma'ah, Kingdom of Saudi Arabia, [c]Department of Basic Medical Sciences and Quality Unit, College of Medicine, King Saud Bin Abdulaziz University for Health Sciences, King Abdullah International Medical Research Centre (KAIMRC), Jeddah, Kingdom of Saudi Arabia, [d]Department of Clinical Pharmacy, College of Pharmacy, King Saud University, Riyadh, Saudi Arabia

1. Introduction

Neurodegenerative disease (ND) is a superordinate phrase describing various conditions that affect nerve cells and the nervous system. They are complex, fatal, disabling sicknesses that result in gradual neuronal loss in both the central nervous system (CNS) and peripheral nervous system by destruction of neuronal networks. Many of these diseases are genetic with a few caused by medical conditions like stroke while others are due to toxins or chemicals in the environment. NDs can cause problems related to movement (ataxias) or mental functioning (dementia) and can lead to death, having profound social and economic implications (Wynford-Thomas & Robertson, 2017). Alzheimer's disease (AD) alone represents 60%–70% of dementia cases all over the world (Qiu, Kivipelto, & von Strauss, 2009). Other NDs include cognitive and behavioral disorders, Parkinson's disease (PD), amyotrophic lateral sclerosis (ALS), Huntington's disease (HD), spinocerebellar ataxia (SCA), spinal muscular atrophy (SMA), cerebral ischemia, etc. Although several medications are clinically available for the treatment of these diseases, most of them have symptomatic relief (Chen & Pan, 2014). Still there is always a chance for better drugs in this field having multifactorial applications (Mizuno, 2014).

Phytomedicine: A Treasure of Pharmacologically Active Products from Plants
https://doi.org/10.1016/B978-0-12-824109-7.00012-1
581

The main aim of neuroprotection is to prevent neuronal loss, prompt neuronal network regeneration, and alleviate brain dysfunction. Phytochemicals are complex compounds with multiple-target efficacy found mainly in plants (Harvey & Cree, 2010). Phytochemicals have disease-modifying ability by acting as antioxidants (resveratrol, hesperidin), antiinflammatory agents (cineole, thymoquinine), inhibitors of GABA$_A$ receptors (diterpenes and cyclodepsipeptides), inhibitors of MAO-B (selegiline rasagiline), and bioenergetic agents (coenzyme Q), but clinical studies have not fully proven their ability to prevent disease (Schapira & Olanow, 2004). The large number of pharmacological or biological activities of the phytochemicals have made them appropriate candidates for the treatment of NDs (Balunas & Kinghorn, 2005; Harvey, Clark, Mackay, & Johnston, 2010; Kimura, 2006). Combating disease symptoms by phytochemicals and herbal nutraceuticals is a hot-button issue, as several studies using phytochemicals to prevent NDs have been published (Venkatesan, Ji, & Kim, 2015).

The main objective of this chapter is to compile most of the recent advances in the field of neuroprotection by herbal products on various NDs and to elicit their potential mechanism of actions with emphasis on nano formulation of the phytochemicals.

2. Neurodegenerative diseases and their progression

As stated before, NDs are disorders that are circumscribed by the loss of either cognitive or muscular functions. AD, PD, HD, ALS, trauma, and ischemic stroke are under same umbrella of neurological disorders. The pathological hallmarks of these diseases are mainly neuronal loss and protein misfolding and aggregation leading to different signs and symptoms like dementia and motor disorders (Sampson, Blanchard, Jones, Tookman, & King, 2009) (Fig. 1).

2.1 Alzheimer's disease

AD is known as the most common cause of irreversible cognitive impairment (Reitz & Mayeux, 2014). It is a multifactorial disease where various genetic and environmental factors work together (Iqbal & Grundke-Iqbal, 2010). The main neuropathological features of AD comprises of neuronal loss, saline plaque, and neurofibrillary tangles (Macdonald, Rockwood, Martin, & Darvesh, 2014). Various mechanisms work together for the progression of the disorder, which includes amyloid deposition, tau hyperphosphorylation and aggregation, loss of cholinergic system, oxidative stress, inflammation, and apoptosis (Webber et al., 2005).

To date, no wonder drug has come to market to reverse the symptoms altogether. The only approved drugs for AD are acetyl cholinesterase inhibitors (galantamine, rivastigmine, and doneprezil), which have the ability to delay symptoms and not to totally treat them (Kumar, Singh, & Ekavali, 2015). As

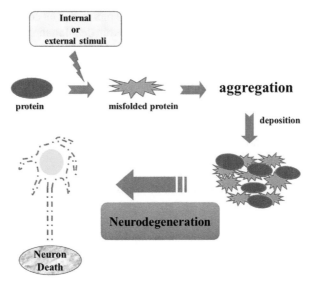

FIG. 1 The diagram represents the effect of internal or external stimuli leading to conformational changes in proteins causing misfolding and aggregation in several neurodegenerative disorders. These aggregate deposits inside and in between the neurons lead to disruption in cell-to-cell communication ultimately causing neuronal death.

AD is multifactorial, it will be impossible for one drug to treat symptoms in all directions. Researchers are finding efficient drugs and targets to combat the progression of the disease (Kumar et al., 2015). There are reports that herbal products and nutraceuticals have the ability to act as a multitarget approach as suggested in traditional system of medicines (Hosseinkhani, Sahragard, Namdari, & Zarshenas, 2017).

2.2 Parkinson's disease

PD is the second most prevalent ND of the elderly that is characterized by dopaminergic neuronal loss mainly in substantia nigra pars compacta region of the brain and the striatum. The nigrostriatal pathway is disrupted in the case of PD patients leading to loss of muscular control. The main sign and symptoms of PD patients are tremors, rigidity, bradykinesia, muscular impairment, and difficulty in walking (Leverenz et al., 2009). The pathological findings are the presence of Lewy Bodies (aggregation of α-synuclein) with impairment in the DNA mutations in the brain of PD patients (Zecca et al., 1996). The treatment of PD mainly rotates around enhanced activities of dopaminergic neurons or cholinergic effects inhibition (Albin, 2006). There is no cure for PD, but drugs available in the market increases the life expectancy with relief in the symptoms (Albin, 2006). Researchers are in continuous process of finding a new drug that could have the ability to treat the disease. Several scientists have reported plant compounds to be effective in the treatment of PD. This is discussed in detail in the next segment.

2.3 Huntington's disease

HD is a type of inherited and familial disease in which there is progressive loss of brain and muscle functions. This is a genetic disorder in which there is programmed degeneration of neurons within different regions of the brain. The main symptoms found are uncontrolled movements, emotional disorders, and loss of mental ability. In normal people, there are 6–35 CAG repeats in exon 1 of the Huntingtin (HTT) gene on chromosome number 4 where as in an affected person, this repeat is more than 36 times present (Szlachcic, Switonski, Krzyzosiak, Figlerowicz, & Figiel, 2015). As a result of this, there is an accumulation of HTT protein in the neuron. This protein has a long polyglutamine region that causes neuronal death and disruption of cell-to-cell signaling causing degeneration of neuronal network in the brain. Early symptoms include concentration problems, short-term memory, tumbling, lack of focus, clumsiness, and depression (Manoharan et al., 2016). Late symptoms include weight loss, feeding problems, difficulty in speech, uncontrolled face movements, itching, etc. There is a lot of research going on in the field of HD prevention and treatment. Many experimental models have been experimentally validated to test the efficacy of different drugs. There is no cure for it, and available drugs only give symptomatic relief. A promising treatment could be in the efficacy of different natural compounds against HD.

2.4 Amyotrophic lateral sclerosis

ALS, also called motor neuron disease (MND) or Lou Gehrig's disease, is another ND in which both upper and lower motor neurons are degenerated causing motor symptoms. The initial symptoms varies between individuals, some exhibiting spinal-onset disease with muscle weakness of limbs and others present with bulbar-onset disease in which they have dysarthria and dysphagia (Hardiman et al., 2017). The cause of this disease is unknown to date, but a few cases are linked to familial history with gene mutations as a major contributor (Al-Chalabi, van den Berg, & Veldink, 2017). There are reports that 50% of the cases of ALS have motor dysfunctions along with cognitive impairment and memory loss (Elamin et al., 2013; Phukan et al., 2012). Other symptoms present in these patients are spasticity, sialorrhea, pain and muscular cramps, deep venous thrombosis, mood alterations, and respiratory insufficiency to name a few. The treatment of ALS is basically symptomatic with only two FDA-approved drugs: Riluzole (Rilutek) and Edaravone (Radicava). These drugs can only increase the life expectancy by 3 to 6 months. This is a life-threatening disorder like other NDs. Much of the efforts of physicians and other health workers are to make the quality of life (QOL) better for ALS patients.

Much of the research work all over the world contributes several preclinical data, but their translation to human studies is still questionable. Moreover, there are several reported clinical trials that failed in humans, which may be due to poor understanding of the disease or due to heterogeneity of the disease. Several studies on natural plants and their active compounds are in progress to combat the dreadful symptoms of this disease. The molecular mechanisms and potential therapeutic role of the plant compounds has to be studied and implemented.

2.5 Stroke/cerebral ischemia

Stroke is the third major cause of death in brain disorders worldwide (Hicks & Jolkkonen, 2009). There is transient or permanent reduction of the blood flow to particular area in the brain in the case of an ischemic stroke. The main region affected in case of ischemia is cornu ammonis 1 (CA1) of the hippocampus (Kirino, 1982; Pulsinelli, Brierley, & Plum, 1982). There is death of pyramidal neurons of CA1 region, delayed to 4 to 5 days of ischemia, hence the name "delayed neuronal death" (DND). Many major pathological processes work together including excitotoxicity, oxidative stress, apoptosis, inflammation, and gliosis of astrocytes and microglia in CA1 region after ischemia/reperfusion (Ozbal et al., 2008; Park et al., 2017; Yousuf et al., 2009). All these changes lead to mitochondrial dysfunction leading to a lack of ATP and subsequently death of the neurons.

There are few effective neuroprotective agents available for the prevention of stroke, but still there is need of some drugs having multidisciplinary action causing prevention of various symptoms of the disease. From ancient medicine, it is proven that there are many plants and nutraceuticals that have the capability of preventing various diseases including stroke. Most mechanisms of action of natural products with minerals, flavonoids, and polyphenols have antioxidant and antiinflammatory activities. The use of natural compounds in ischemia patients is discussed in detail in the coming section.

3. Phytochemicals overview and mechanism of actions

Neurodegenerative diseases (NDs) are generally described as the loss of neurons. There are numerous compounds being tested for the treatment of NDs. Various compounds have been tested against NDs, but unfortunately beside having curing properties, they have side effects on the body as well (Velmurugan, Rathinasamy, Lohanathan, Thiyagarajan, & Weng, 2018). There are lots of factors that are responsible for NDs like chemical exposure, lifestyle, low levels of nerve growth factors (NGF, BDNF, etc.), and aging. The prominent responsible factor for NDs is aging. As per the literature and surveys, NDs are increasing day by day all over the world just because of the large proportion of aged population (Johnson, 2015). During the aging process, all physiological and metabolic processes are slowed down, which also hampers the function of the brain, many growth factors (NGF, BDNF, etc.), and damages the neurons that leads to the NDs. The NDs finally convert into life-threatening disorders. Early detection of ND-related symptoms can awaken hope in the patients with promising treatment and therefore, with low difficulty, we can overcome these health disorders (Schulz & Deuschl, 2015).

For treatment purposes, when talking about commercial neurological medicines, we must mention their side effects (Longin, Teich, Koelfen, & König, 2002; Park & Kwon, 2008). To deal with aging people with minimal side effects, it is recommended to focus toward natural therapy, which can be taken by aging people without any side effects (Albarracin et al., 2012; Geun Kim & Sook Oh, 2012). Having potential activities, viz., antioxidative, antiinflammatory, and neuroprotective, the polyphenols from vegetables and fruits have been explored for a long time. Therefore, herbal medicines are finding attraction of researchers to make polyphenols as potential and promising therapy against many disorders (Azam et al., 2019; Jabir, Khan, & Tabrez, 2018; Solanki, Parihar, & Parihar, 2016). Daily uptake of vegetables, fruits, and herbs should be recommended for a healthy lifestyle. It is well known that fruits, herbs, and vegetables are the best source of phytochemicals (Davinelli et al., 2016). The phytochemicals are those semisynthetic precursors of any medicinal plant's parts (like fruits, leaves, stems, flowers, etc.) that contain the therapeutic values, which are known as phytochemicals (Patra, 2012). Being the natural precursors of plants, the phytochemicals play a role in antimicrobial and pest control activity (Barbieri et al., 2017; Omojate Godstime, Enwa Felix, Jewo Augustina, & Eze Christopher, 2014). Apart from the beneficiary effects, phytochemicals are considered as an attractive source to treat many disorders. Because of its power to boost immunity and antioxidative activity, in ancient times, a fruits and vegetables-based diet was the essential part of food component for a healthy lifestyle. Such diet gives strength to fight against many diseases like cancer, tumor, neurodegeneration, diabetes, etc. (Jahan et al., 2017). Phytochemicals (Phyto + chemicals = natural chemicals obtained from plants) are obtained naturally, and they are found to be effective against many disorders including inflammatory diseases and oxidative stress-induced damage (Gupta & Sharma, 2017; Shal, Ding, Ali, Kim, & Khan, 2018; Wang, Song, Chen, & Leng, 2018; Zhang et al., 2013). Our body has free radicals after many biochemical processes, but at the same time it has pathways for scavenging of free radicals. When the body is not capable balancing the free radicals, it results in oxidative stress, mitochondrial dysfunction, and inflammation (Huang, Zhang, & Chen, 2016; Lushchak, 2014). In the case of NDs, oxidative stress, mitochondrial dysfunctions, and inflammation are significant reasons for inducing NDs. There are many easily available phytochemicals that we use on a daily basis, viz., resveratrol, quercetin, curcumin, and many more. Polyphenolic compounds play a specific role to strengthen the immune system and are also helpful in decreasing platelet aggregation and regulate hormone imbalance. Researchers have discussed the importance of various phytochemicals as well including olive oil, avocado, kiwi, grapes, and many more. It is stated that there was a significant reduction in overall mortality (13%) in PD and AD patients by consuming an olive oil-containing Mediterranean diet, which is a significant improvement in health status (Kahkeshani, Saeidnia, & Abdollahi, 2015). Accumulative evidence is available to prove that phyto-chemicals can act as nutraceuticals, which can alleviate the condition of NDs, but the mechanism of action of these phytochemicals is not known (Zhang et al., 2015). Another study reported that, by their antioxidative and radical scavenging capacity, phytochemicals can be a promising therapeutic agent. Although, mechanistic activity needs to be explored more. Many hypothesis are documented that suggest their activity by controlling the enzyme metabo-lism, gene expression and by regulating many tiny cascade molecules that are initiated in response to healing as well as stress condition (Paredes-Gonzalez et al., 2015). One more hypothesis stated the role of phytochemicals in terms of key lock theory. It is stated that phytochemicals behave like a ligand and bind to specific receptors present on the cells and then further indirectly participate in a cascade mechanism, which would result in potential therapeutic activity (Krajka-Kuźniak, Paluszczak, Szaefer, & Baer-Dubowska, 2015). Resveratrol (RV) is a phytochemical found in grapes. RV is found to be neuroprotective by inducing the signaling pathway via PKA-mediated induction of GSK3β, β catenin, CREB, and ERK1/2 (Jahan, Singh, et al., 2018). This study was carried out on human stem cells derived from umbilical cord blood to see the neuroprotective effect of RV. It was reported that there was a positive effect of RV over the survival, proliferation, and neuronal differentiation in human um-bilical cord blood-derived mesenchymal stem cells (hCBMSCs) (Jahan, Singh, et al., 2018). Here we are going to discuss phytochemicals based on their activ-ity to overcome NDs.

3.1   Antioxidant and antiinflammatory agent

The most common factors responsible for the NDs are protein aggregates, free radicals, and neuroinflammation. Our brain is very complex, therefore, oxygen consumption capacity of the brain is high, but the response against oxidative stress is comparatively low (Lopresti, Maker, Hood, & Drummond, 2014) and then oxidative stress results in neuronal cell injury in various pathological states of the brain, including NDs. The brain only has 2% of total body weight but its requirement of O_2 is high in comparison to the rest of the body (20% of all oxygen available) (Russell et al., 2012). So much high demand of oxygen makes the brain a most vulnerable organ for oxidative damage. To overcome the problem, phytochemicals come into the picture due to their therapeutic strategy. Antioxidants also make the immune system able to deal with the diseased state. Oxidative stress can be overcome by the CNS by natural precursors like bioactive compounds, vitamins and various antioxidative enzymes, and redox sensitive protein transcriptional factors (Miller, Wallace, & Walker, 2012). Polyphenols could cope up with such problems through their antioxidative activity. The most important dietary phytochemicals (such as polyphenols, quinones, flavonoids, catechins, coumarins, terpenoids) and the smaller molecules like ascorbic acid (vitamin C), alpha-tocopherol (vitamin E), beta-carotene vitamin-E, and supplements are available (Menendez et al., 2013). The mode of action needs more research. Still, phytochemicals offer hope against many NDs like PD, AD, HD, ALS, and ischemic and hemorrhagic stroke.

Ginkgo biloba is used as a memory booster agent. It works on the abnormality associated with memory loss due to the blood circulation. This phytochemical increases the oxygen supply to the brain cells thus helping against oxidative stress. Phytoconstituents include flavonoids, steroids (sitosterol and stigmasterol) and organic acids (ascorbic, benzoic shikimic, and vanillic acid), terpenoids, bilobalide, and ginkgolides (Kumar & Khanum, 2012).

Curcumin is a very well-known spice and is found in almost every Indian kitchen. It is known as an antiseptic, antibiotic, and having the property of antioxidative. Curcumin is also used for the treatment purpose of biliary disorders, hepatic disorders, and diabetes as well. Curcumin is also used against NDs, specifically for AD by inhibiting the amyloid plaques through the downregulation of NF-κβ. In one study, curcumin was found to have high affinity against plaques among 214 compounds. Curcumin also carried out its curative activity via CREB and BDNF levels (Motaghinejad, Motevalian, Fatima, Faraji, & Mozaffari, 2017). TRK and PI3K cascade molecules are regulated by curcumin-induced therapeutic strategy, and it is also associated with the upregulation of BDNF expression. This study was accompanied in a PD model. Curcumin is also helpful in the reduction of TNF-α and caspase levels and thus stimulates the upregulation of BDNF level. For increasing the bioavailability of curcumin, research was carried out for curcumin nanoparticles, and the administration of these nanoparticles relieved the cognitive impairment condition by upregulation of BDNF level. Curcumin provides a promising and cheap therapeutic strategy against many NDs. Now, research is focusing on the increment in the bioavailability of curcumin to make it able to cross the BBB. So, it can show the therapeutic effect in more prominent way. Amyloid plaques reduction study is done in vitro as well in vivo. Curcumin also prevented the maturation of amyloid-β precursor protein (APP) in mouse neurons, and the associated mechanism is completed by the inhibition of NF-κB (Eftekharzadeh et al., 2012). Curcumin was also found effective against other NDs like PD by destabilization of α-synuclein protein (Singh et al., 2013).

Resveratrol is a polyphenolic compound. It is found in red wine and grapes. It has many properties like antioxidative, and antiinflammatory, neuroprotective, and anticancerous. In the rat model, a study was carried out that showed that many cognitive defects were reversible by using resveratrol by hampering the TNF-α and IL-1β levels and higher BDNF levels in the hippocampus region (Szűcs, 2015). Resveratrol also helped in soothing of the inflammation effect by increasing IL-10 levels. IL-10 enhances the anti-inflammatory effect by hampering NF-κβ levels and TNF-α level via leading ERK1–2/CREB signaling pathways. Here, resveratrol protects the neurons via upregulation of GDNF and BDNF levels. When NDs occur, the glial cells start to release the cytokines, which have inflammatory activity Apart from that, many neurotoxic molecules are released. So here, resveratrol acts on the glial cells and suppresses their activity (Cianciulli et al., 2015). In another study, when streptozotocin is used as a toxic compound, then resveratrol was found to be promising to reverse the effect. Like curcumin, resveratrol also promotes the destabilization of amyloid β plaques. In one recent study, it was observed that 6-OHDA-induced toxicity, which was also reversed by the administration of resveratrol via inducing SIRT-1, i.e., sirtinol. Here, resveratrol inhibits the deacetylation of substrate of SIRT 1, i.e., p53 and PGC-1α (Menzies, Singh, Saleem, & Hood, 2013). In a PD model having rotenone-induced toxicity, the effects were reversed by using resveratrol via activation of autophagy pathway, i.e., AMPK-SIRT1. Phenolic acids are classified into benzoic acid derivatives and cinnamic acid derivatives. It is a major constituent of orange, tomato, carrot, and sweet corn (Manuja, Sachdeva, Jain, & Chaudhary, 2013; Mathew, Abraham, & Zakaria, 2015) (Fig. 2).

Sulforaphane (isothiocyanato-4-(methylsulfinyl)-butane) (SF). Among phytochemicals, SF also shows neuroprotective activity. SF is able to target and modulate different kinds of pathways for neuronal protection.

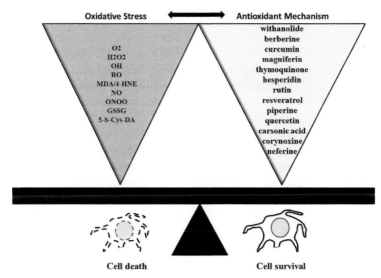

FIG. 2   An illustration of oxidative stress and antioxidant mechanism of different phytochemicals determining the fate of the neuron. This figure depicts that there will be cell survival if the antioxidant load is more than the oxidative stress in the cell, which leads to cell death.

3.2 Mitochondrial function restoring agent

In the eukaryotic cells, mitochondria are the organelles that are completely responsible for energy production and biogenesis. The organelle also supports iron homeostasis, synthesis of protein and differentiation. Mitochondrial dysfunction leads to an imbalance between metabolic activities and impairment of neuronal activity (Andreux, Houtkooper, & Auwerx, 2013). The proper mitochondrial function is maintained by numerous pathways (Andreux et al., 2013; Li et al., 2013). There are many phytochemicals present in nature that promote the modulation in mitochondrial structure and function. We have discussed many phytochemicals having more than one kind of application but collectively leading toward neuroprotection. Resveratrol has an antioxidative property and thus modulates the function and structure of mitochondria. One study was carried out in human umbilical cord blood-derived stem cells. Monocrotophos was used as the neurotoxic compound. Resveratrol just reversed the mitochondrial damaging effect via inducing the PI3K cascade molecule (Jahan, Kumar, et al., 2018). With ongoing research and explored features of the mitochondria, it is very easy now to dig out the therapeutic targets. This chapter focuses on the main pathways responsible for maintaining and/or restoring proper mitochondrial function, like mitochondrial biogenesis and metabolic flexibility; mitochondrial dynamics, including fusion and fission; and mitochondrial quality control through proteostasis and mitophagy (Youle & Van Der Bliek, 2012).

3.3 Enhancers of mitochondrial biogenesis

Mitochondria is known as the powerhouse of the cell. Any unwanted hindrance results in deformed function and structure. Mitochondria is an energetic cell organelle and facilitates many functions like mitogenesis, differentiation, and regulation of many metabolic and biochemical reactions. Mitogenesis is a process of formation of new mitochondria from the existing one (Palikaras, Lionaki, & Tavernarakis, 2015). There are many factors that stimulate the mitochondrial biogenesis. These factors are exercise, cell cycle, caloric restriction, etc. Free radicals play an important role to damage and imbalance the mitochondria and associated components (electron transport chain, mitochondrial DNA). This impairment further results in NDs, viz., AD and PD. Furthermore, mitogenesis is also associated with the formation of the inner mitochondrial membrane (IMM), outer mitochondrial membrane (OMM), mitochondrial deoxy carbo nucleic acid (mtDNA)-encoded protein, genesis and gist of nucleus-encoded protein, and replication of mitochondrial DNA (mtDNA) (O'Neill, Holloway, & Steinberg, 2013). The association between mitochondrial dysfunction and neurodegeneration is confirmed by in vitro studies, genetic and toxin animal models, post-mortem human brain tissue, and human genetic studies (Wang et al., 2014). Among the therapeutic strategy, there are lots of antioxidants like mitochondria-targeted antioxidants such as MitoQ and antioxidants such as

CoQ10 treatments that are protective against mitochondrial damage and have shown efficacy in animals (Walters, Porter Jr, & Brookes, 2012). Some signaling cascade molecules like NRF2, PGC1α, and NAD+, etc. phytochemicals stimulate mitochondrial biogenesis to show a neuroprotective effect. It is documented that resveratrol and epigallocatechin gallate (EGCG) increase the level of 5-AMP-activated protein kinase (AMPK) and Sirt-1, thus promoting biogenesis of mitochondria by inducing PGC-1α and showing neuroprotective property. Wogonin, genistein, quercetin, hydroxytryrosol, daidzein, and wogonin promote mitochondrial biogenesis (Wood dos Santos et al., 2018). It was reported that quercetin shows an increment in the number of muscle mtDNA by enhancing the physical activity in adult males (Somerville, Bringans, & Braakhuis, 2017).

3.4 Antiapoptotic agents

Neuroprotective therapy is also associated with apoptotic molecules and mitochondria. In the series of neurodegeneration, lots of changes occur in which oxidative stress is one from all series. The oxidative stress affects the mitochondria and leads to damage via apoptosis. In the concerned chapter, we are discussing the polyphenolic compounds that diminished the mitochondrial damage via anti apoptotic activity. Vitamin E (antioxidant), MitoQ (mitochondria targeted antioxidants), and coenzyme Q10, creatinine are known as mitochondria-activating compounds. The neuroprotective effects are supported by using the in vitro as well as in vivo study (Naoi, Wu, Shamoto-Nagai, & Maruyama, 2019). Numerous natural compounds have been studied for digging out the neuroprotective effects against oxidative stress and mitochondrial damage. Inhibitors of type B monoamine oxidase (MAO-B), saelegiline and rasagiline are the examples of studied polyphenolic compounds. Molecular pathways associated with them are also studied (Inaba-Hasegawa, Akao, Maruyama, & Naoi, 2012). Polyphenols alter the expression in cancer cells as well as mitochondrial associated genes (Wcislo, 2014). Polyphenols also affect mitochondrial structure and function via by harmonizing the mitophagy, transport, division, and amalgamation. Amphipathic properties and redox potential of polyphenols are also key markers for imparting a role in the protection activity. Synthesis of mitochondriotropic phytochemical derivatives are playing roles as anticancerous, antioxidant, antiapoptosis, and antiinflammatory activities via increased expression of protective genes like Bcl-2 (Naoi, Wu, et al., 2019). Black tea extract, resveratrol, quercetin, rosmarinic acid, and astaxanthin inhibited membrane permeabilization and safeguarded cells from programmed cell death caused by 1-methyl-4-phenyl-1,2,3,6-tetrahydropyridine, i.e., MPTP and formed aggregation of plaques and inadequate blood supply. The phytochemicals are also associated with formation and prevention of mitochondrial membrane pores in in vitro model of programmed cell death, which is initiated by the PK11195. It is a well-known ligand of TSPO, i.e., outer membrane (Wu, Shamoto-Nagai,

Maruyama, Osawa, & Naoi, 2017). Sesamolin and ferulic acid's lipophilic derivative, named astaxanthin, hampered the pore formation at IMM. The pores are formed by the cyclophilin-D and nucleotide translocator (ANT). Curcumin was also found helpful in inhibiting mitochondrial pore formation to overcome oxidative stress. Ferulic derivatives also overpowered the mitochondrial permeability transition pore, as well as inhibition of efflux of cytochrome *c* (Cyt c), by inhibiting the OMM pore opening. Sesamin was also found effective against oxidative stress by promoting the mitochondrial permeability transition pore (mPTP) formation and apoptosis (Naoi, Wu, Shamoto-Nagai, & Maruyama, 2016). GSK3 thus inhibited the opening of pore (Suh, Kim, Kim, Chung, & Song, 2013). Many more phytochemicals like bailarein, quercetin, EGCG, and naringenin stimulate the cascade molecules for programmed cell death via alteration in expression of Bax, BCl2.

3.5 Agonists of neurotrophic factor inducing neuroprotection

Neurotrophic factors (NTFs) are a family of biomolecules made up of small proteins (peptides). These factors are responsible for the survival, growth, and differentiation of both developing and mature neurons. In this chapter, we already have discussed the factors that lead toward the compromised neurons or NDs. From all those factors, the reduction of NTF is the foremost reason for the emergence of NDs. There are many reasons for reduced NTFs level, viz., lifestyle, aging, chemical exposure, etc. Nowadays, knowingly or unknowingly we encounter many chemicals whether they are present in edible products or in industry. The repeated chemical exposure result in NDs (Tanner, Goldman, Ross, & Grate, 2014). Apart from that, some neurotransmitters present in the cells also behave in a negative manner. Examples for that is glutamate and aspartate. Glutamate is the major excitatory neurotransmitter found in the brain. Excess extracellular glutamate may turn into excitotoxicity. Glutamate binds and activates both ligand-gated ion channels (ionotropic glutamate receptors) and a class of G-protein-coupled receptors (metabotropic glutamate receptors). The glutamate concentration is found in very low quantity, i.e., in micromolar range. This glutamate-induced excitotoxicity is a chief cause of neuronal cell death in the human brain, especially in perinatal brain injury. Antagonists of the ionotropic *N*-methyl-D-aspartic acid (NMDA) glutamate receptor are known as vigorous neuroprotective agents, which were studied in several animal models with perinatal brain lesions (Burd, Welling, Kannan, & Johnston, 2016). NMDA receptors have shown success in the survival, growth, proliferation, and differentiation of neurons but not as much as glutamate does in normal concentration. From this fact, it can be said that, by blocking the glutamate receptors, specific neurodevelopmental stages can be hindered and turned into the hampered brain development. Another agonist, dexmedetomidine shows neuroprotective effects against neonatal glutamate-induced injury through an increase in the expression of brain-derived neurotrophic factor (BDNF). It was demonstrated that astrocytes, but not neurons, are involved in dexmedetomidine-induced BDNF expression, and that astrocyte expression of BDNF and dexmedetomidine neuroprotective effects are both linked to the ERK1/2 pathway. Oleuropein, resveratrol, curcumin, and rosmarinic instigated the expression of GDNF and BDNF and exhibited the neuroprotective effect by following the PI3K/Akt pathways or by ERK/cAMP response element-binding protein (CREB) in primarily and clonally cultured neurons and astrocytes. Catalpol, naringen, and harpagoside escalated the expression of GDNF in animal models of PD. It is also noticed that the flavonoids are also helpful in the upregulation of BDNF in in vivo and in vitro (Zuo et al., 2016). On the other hand, non-flavonoids (like curcumin, resveratrol, etc.) show neuroprotective effects by increasing the expression of GDNF (Naoi, Shamoto-Nagai, & Maruyama, 2019). Polyphenols also promote neurogenesis via stimulating the expression of TRK increase (Jahan, Singh, et al., 2018). Olive polyphenols escalated the expression of Tyrosine kinase B and A (TrkB and TrkA) in the hippocampus region but not in other parts like the frontal cortex and striatum, and shows antidepressant-like effects in animal models, i.e., in vivo model of depression. EGCG prevents the pathophysiological condition via working on the BDNF, Tyrosine kinase B and A (TrkB and TrkA) expression in in vivo, study for AD model. Deoxygedunin and nobiletin extracted from citrus fruits show an increased expression of TrkB and BDNF in mice model and thus promote neurogenesis in hippocampus region in rat model that has depression.

3.6 Neuromodulator of cell signaling pathway (neuroprotection)

Polyphenols are well documented for having therapeutic potential against NDs. The protection has a specific mechanism, and this mechanism is regulated by cascade molecules. The interaction between the cascade molecules are associated with cell survival, growth differentiation, and cell death. We will discuss the involvement of signaling molecules that play specific roles on cell regulation. Polyphenols are associated in the cell-survival signaling pathways including MAPKs (mitogen-activated protein kinase), Protein Kinase (PKC), and PI3K/Akt. PKC is kindred in the activation by Ca^{2+} and phospholipids, whcih arbitrates the normal cell function. Genistein and EGCG escalate the PKCα and PKCε and fortify the neurons from apoptosis instigated by neurotoxins (Das, Ramani, & Suraju, 2016). The expression of prosurvival genes including antioxidant enzymes are arbitrated by MAPK signal pathways. MAPKs are organized into extracellular signal-regulated kinase (ERK), c-Jun N-terminal kinase (JNK), and p38. Extracellular signal-regulated kinase (ERK1/2) regulates cAMP response element-binding protein (CREB) and upregulates Bcl-2 and Bcl-xL expression. JNK regulates the transcription-dependent apoptotic signals, whereas p38 is associated with cell death, cell cycle, senescence, and carcinogenesis. Catechins, genistein, and quercetin upregulate ERK/CREB pathway, myricetin downregulates p38, and JNK promotes cell survival. PI3K/ Akt pathway activates transcription factors and increases levels of neurotrophic. In the brain, neurons interact with glia, and glia-derived NTFs, cytokines, ROS/ RNS, and Ca^{2+} activate signal transduction and gene expression in neurons to maintain the function and structure. In one study, it is shown that resveratrol has worked as neuronal differentiation factor in mesenchymal stem cells (MSCs) by leading the PKA-GSK3β and β-Catenin signaling pathway. Resveratrol also worked on the improvement of the expression of fusion-mediating optic atrophy protein (Opa1), mitofisin-1 and -2 (Mfn1, Mfn2), fission-regulating dynamin-related protein 1 (Drp1), and mitochondrial fission protein 1 (Fis1), and protected PC12 cells against rotenone-induced cytotoxicity (Gao et al., 2015). Many phytochemicals like naringenin, hesperetin, and bilobalide de-rived from *Ginkgo biloba* leaf have been reported to activate the PI3K/Akt and PKC by binding to the ATP-binding site on enzymes and receptors thus avoid-ing programmed cell death of neurons and boosting the evocation. Quercetin, luteolin, amentoflavone, and resveratrol also have shown effects in the same direction. Many flavonoids stimulate the upstream molecules like MAPK-kinase-kinase, downregulate JunN-terminal kinase expression, and slowdown the programmed cell death caused by free radicals. Resveratrol, curcumin, and *Salvia miltiorrhiza*-derived salvianolic acid initiate the PI3K/Akt pathways and provide protection to the neurons. The effects of those phytochemicals are confirmed by both in vivo and in vitro studies. Resveratrol inhibited the mito-chondrial dysfunction and programmed cell death via disabling the GSK-3β by phosphorylated form of Akt. The impairment of mitochondrial dysfunction was induced by the toxicant MPP+. One more phytochemical named hesper-etin saved the hippocampus neuronal activity by upregulation of Akt and down regulation of proapoptotic proteins, such as apoptosis signal-regulating kinase 1 (ASK1), caspade-3, 9 and Bad. Forkhead box O3 (FOXO3), which is a tran-scriptional factor for longevity and inhibitors of NF-κB (IκB), is phosphorylated by the Akt and furthermore, Akt activates NF-κB to increase the expression of Bcl-2 protein family and pro survival IAP (inhibitor of apoptosis protein). Through in vivo and in vitro study, it was revealed that many signaling cascades like ERK-CREB-BDNF or PI3K/AKT are stimulated by the phytochemicals like caffeic acid, EGCG, hesperetin, curcumin, and resveratrol. Thus, they are imparting their role as therapeutic agents for the protection of neurons, revers-ing the memory deficit condition and depression-associated symptoms (Bawari et al., 2019).

Keep1/Nrf2/antioxidant response element (ARE) pathway is also evocated in the response of the phytochemicals via regulation of antioxidative enzymes. PKC, a member of a serine-threonine kinase superfamily, is activated by poly-phenols and regulates cell growth, development, cell survival, cell differentia-tion, programmed cell death, and learning and behavioral brain activity. EGCG activates the PKCε and PKCα in the striatum of MPTP-treated mice. PKCγ a signaling molecule is also activated by the resveratrol and EGCG and makes more new neuronal interconnections in the hippocampus region of a rat model, i.e., in vivo model. PKCα and PI3K/Akt signal pathways activates the Keap1/Nrf2/ARE pathway and participates in the neuroprotection, antioxidative activity of polyphenolic compounds (Lu, Ji, Jiang, & You, 2016).

3.7 Non aggregator of misfolded proteins

In the pathogenesis of NDs, the misfolding of protein is emerging as a very pivotal element. Consequently, to develop natural therapy, the identification of the envoy needs to be studied. The natural therapy should have the ability to hamper and relapse the protein aggregation and protein misfolding as well (Popovic, Vucic, & Dikic, 2014). Thus, the neuroprotective property of the polyphenols makes them significant agents against different amyloidosis diseases by following the different molecular cascade mechanisms. Polyphenolic compounds are capable of binding with amyloids and show significant effect to inhibit the misfolding of proteins, i.e., amyloid plaques. Natural polyphenols have been documented with having the power to hamper the self-assembly of several proteins that are linked with the amyloid diseases (Ngoungoure, Schluesener, Moundipa, & Schluesener, 2015). The presence of phenolic rings is the most common structural feature of medically significant polyphenols. The hydroxyl groups present in the structure can have important roles in the aggregation and fibrillation prevention. The OH group present in the phenolic molecules set out as an antagonistic agent for cyclic reactions, which fabricate reactive oxygen species (ROS) as well as reactive nitrogen species (RNS) (Hardy et al., 2018). The polyphenols produce further free radicals when they interconnect with prime free radicals. It happens when another radical encounters the OH group and the π electrons of the benzene ring. Hence, nearly long-lived radicals are produced and stabilized by delocalization and are enough of an obstruction of radical-mediated oxidation processes. Again, the metal ion-chelating potential of the antioxidants contributes greatly to their free radical-scavenging quality by having metal chelating quality. The polyphenolic structures also have the ability to link with the proteins via their H-bonding and benzenoid moieties through the phenolic OH group. When we talk about protein in NDs, it is well known that α-Synuclein (α-syn) is a presynaptic protein that regulates the release of neurotransmitters from synaptic vesicles in the brain. The aggregates, also having Lewy bodies, are is characteristic features of both familial forms as well as sporadic form in PD (Wood-Kaczmar, Gandhi, & Wood, 2006). The commercial drugs provide only symptomatic treatment because they are having gradually low effect despite taking the drugs on regular basis. The connection between etiopathogenesis, α-syn, and progression of PD is already documented in the literature. Studies indicate that α-syn is an important therapeutic target for phytochemicals to inhibit α-syn aggregation, oligomerization, and fibrillation. Therefore, phytochemical-induced therapy is considered a promising therapy without any noted side effects (Wang et al., 2014). The αSyn oligomer formation and assembly into aggregates is obstructed by many phytochemicals like kaempferol, theaflavins, curcumin, nordihydroguaiaretic acid, rosmarinic acid, and ferulic acid. The mature αSyn fibrils were converted by EGCG into nontoxic, smaller, amorphous aggregates. Curcumin was reported to bind to preformed toxic oligomer and fibrils and alter the hydrophobic surface into more soluble species.

One more polyphenolic compound, named morin, is mostly obtained from fruits and red wine as well. The morin is similar to quercetin (flavonoid) by structure. It has neuroprotective activity specifically for AD (Lemkul & Bevan, 2010). Polyphenol morin alters the tertiary and quaternary structure of Aβ but not the secondary structure, although it does not bind directly to the monomeric and dimeric Aβ peptides. Morin binds to the hydrophobic residues of the peptide and inhibits the higher degree of polymerization and also affects the tertiary structure of the amyloid peptide. Mutant HTT aggregates are also inhibited by morin via acting through protein-trafficking pathways and thus destabilizing the HTT amyloid. Fisetin (flavonoid) and its derivatives are also investigated against AD for having antiamyloidogenic activity. Ferulic acid, tannic acid (a polyphenol), polymethoxyflavones (tangeretin,nobiletin, sinensetin), galangin and genistein also have shown inhibitory activity on β-secretase and also play significant roles in improving the behavioral impairment of AD via in vivo study. Furthermore, resveratrol, luteolin, icariin, rutin, quercetin, curcumin, and 7,8-dihydroxyflavine also have roles in β-secretase activity. Curcumin, oleuropein, and genistein promoted APP cleavage by α-secretase into N-terminal product (which is nontoxic) named as soluble APPα and C-terminal product. Rosmarinic acid, curcumin and myricetin generate soluble Aβ oligomer and profibrillar species and were found helpful in decreasing the toxic fibrils as well as oligomers. EGCG is studied for the interaction with the amino acids 14–24 and 27–37 of insoluble Aβ, for breaking the β-sheet motif, and thus it prevents fibril formation and converts large and mature αSyn fibrils and Aβ fibrils into smaller, amorphous, and nontoxic protein aggregates (Singh, Dutta, & Modi, 2017). Myricetin and luteolin, bound to the hydrophobic region of the amyloid pentamer, inhibited Aβ aggregation. Resveratrol, EGCG, curcumin, caffeic acid, and altenusin restrained the hyperphosphorylation of Tau protein. Disparity in phosphatases and kinases resulted in the hyperphosphorylation of tau, which turns in the synthesis of neurofibrillary tangles. In the case of AD, the oligomer is documented as most toxic (Sengupta, Nilson, & Kayed, 2016).

4. Specific neuroprotective phytochemicals and their mode of action on neurodegenerative disorders

"Herbal medicines" is the term given for herbal preparations and phytochemicals from plant fruits and vegetables for the prevention of various NDs (Bagli et al., 2004; Rahman & Chung, 2010). The main mechanism of action by which these phytochemicals work is their antioxidant activities. However, they have been reported to modulate multiple signaling pathways by acting on various enzymes and their products (Bagli et al., 2004; Harvey & Cree, 2010). The immune system of the body is induced and hormone systems are regulated by the consumption of these phytochemicals (Farooqui, 2013). Several corroborative evidences stated that phytochemicals acting as nutraceuticals have the ability to alleviate the conditions of NDs, but the exact mechanism of action is not known. In the previous section, we discussed the plausible mechanism of actions of phytochemicals in detail.

4.1 Evidence of effect of phytochemicals in memory, cognition, and Alzheimer's disease

The effect of various phytochemicals in the restoration of cognitive skills and memory functions has been known for decades. Recently extensive research has been conducted that supports their potential in vitro, in vivo, and in clinical studies. Almond is a natural memory enhancer, as the oleic acid found in it acts as an NTF in neurons leading to the growth of dendrites and axons (Polo-Hernández et al., 2014). It has been reported to significantly reduce the activity of propyl endopeptidase found in high concentration in AD brains (Park, Jang, Lee, Hahn, & Paik, 2006). From *Withania somnifera*, active constituents called withanolide A, withanoside IV, and withanoside VI have been shown to increase the neurotic outgrowth in human neuroblastoma SHSY5Y cell line (Kuboyama et al., 2002). Berberine from cortex of Coptis chinensis has been found to prevent the pathology of AD in a multidirectional manner. It has reduced the extracellular Aβ production in H4 neuroglioma cells and HEK293 cells (Asai et al., 2007; Zhu et al., 2011). Moreover, neuroinflammation induced by Aβ was also reduced significantly by the pretreatment of berberine in primary, BV2 microglia cells and SK-N-SH neuroblastoma cells suggesting its protection against AD (Jia et al., 2012; Xu, Zhang, Yang, & Yu, 2013). Likewise, a polyphenol curcumin from rhizome of *Curcuma longa* has antioxidant, antiinflammatory, antiapoptotic, antiamyloid, anti-tau, neuroprotective activities (Kim, Kim, & Yang, 2014). In addition, it binds with main aggregating protein in AD, amyloid beta, and prevents their aggregation (Reinke & Gestwicki, 2007). Acetylcholine is the responsible neurotransmitter for memory and cognition, and magniferin from *Canscora decussata* has elevated its level (Urbain et al., 2008). Oxidative stress and mitochondrial damages are also linked with AD pathogenesis (Takuma, Yan, Stern, & Yamada, 2005), which were significantly ameliorated by Thymoquinone, a bioactive component of black seed oil, in Aβ-treated PC12 cells (Khan, Raza, et al., 2012; Khan, Vaibhav, et al., 2012). In accordance with the neuroprotective effects of phytochemicals, another bioflavonoid, hesperidin, found abundantly in oranges and lemons, have also been reported to protect against Aβ-treated cells and ICV-STZ-induced mouse model of AD (Javed et al., 2015). Similar results have also been found in bacosides from Bacopa monnieeri (Singh & Dhawan, 1997), Ginsenoside Rg1 from Panax notoginseng (Kim et al., 2014;

Shukia, Khanna, & Godhwani, 1987; Urbain et al., 2008), Puerarin (Shukitt-Hale et al., 2005, 2015), Silibinin (Trouillas et al., 2008; Yin et al., 2011), resveratrol (Jhang, Park, Kim, & Chong, 2017), hyperforin (Huang et al., 2017), and capsaicin (Xu et al., 2017).

4.2 Evidence of effect of phytochemicals in Parkinson's disease and parkinsonism

PD is a social, psychological, and economic burden as 1% of the aged population all over the world suffer from this disease. Parkinsonism is a clinical syndrome associated with the symptoms of PD. Extensive research with in vitro and in vivo models have explored the potential of natural compounds against PD and parkinsonism. The main neurotoxic factors responsible for this disease are mitochondrial dysfunctions leading to oxidative stress, inflammatory cytokines, apoptosis, and dopamine depletion (Khan et al., 2010; Khan, Ishrat, Ahmad, et al., 2010; Khan, Raza, et al., 2012; Khan, Vaibhav, et al., 2012; Shrivastava et al., 2013).

Several plant components have been reported to have free radical scavenging properties including rutin, resveratrol, piperine, and quercetin (Khan, Ahmad, et al., 2010; Khan, Ishrat, et al., 2010; Khan, Raza, et al., 2012; Khan, Vaibhav, et al., 2012; Sharma et al., 2016; Shrivastava et al., 2013). Neuroprotective effect of carsonic acid against 6-hydroxydopamine-induced Parkinson's was studied by Wu et al. (2015). Another study suggested that hesperidin ameliorated 6-OHDA-induced oxidative stress in aged mice (Antunes, Goes, Boeira, Prigol, & Jesse, 2014). In another research, it was reported that sesame seed oil has antioxidant efficacy against 6-hydroxydopamine-induced neurotoxicity in mice (Ahmad et al., 2012). They all concluded that the antioxidant potential of these plants is responsible for the neuroprotective effect in different models of PD. Naringenin is known to increase the dopamine levels and tyrosine hydroxylase positive cells in 6-OHDA-induced PD model (Zbarsky et al., 2005).

Another pathway that is activated in PD is autophagy-lysosome pathway (ALP) (Gan-Or, Dion, & Rouleau, 2015; Pan, Kondo, Le, & Jankovic, 2008). The neurodegeneration in PD by impairment of ALP has both genetic and environmental factors (Gan-Or et al., 2015; Pan et al., 2008). Some small molecules from plant origin have shown a promising strategy for disease-modifying treatment of PD. Corynoxine B and corynoxine from Chinese herb *Uncaria rhynchophylla* have been shown to act as neuronal autophagy inducers (Chen et al., 2014; Song et al., 2014). It was also reported that Corynoxine B promotes clearance of α-synuclein in neuronal cells and in Drosophila. Other nutraceuticals exhibiting the same property are conophylline, curcumin, resveratrol, celastrol, and kaempferol (Ferretta et al., 2014; Huang et al., 2013; Sasazawa, Sato, Umezawa, & Simizu, 2015; Song et al., 2016).

4.3   Evidence of effect of phytochemicals in Huntington's disease

This is a genetic ND and can be found at an early age too. In cerebral cortex and striatum, the pathological changes elicit the development of dyskinesia and cognitive impairment leading to early deaths. Men and women are equally effected in their 40s or 50s with 50% chances of children inheriting the disease from HD-affected parents (Manoharan et al., 2016). The exact cause of death in HD is unknown like any other NDs, but one of the pathological hallmarks could be oxidative stress. The brain requires a large amount of oxygen as energy hence is more prone to oxidative stress (Walker, 2007). To scientifically validate the potential of natural plants against HD, researchers have utilized several HD models. The active compound Neferine, isolated from aquatic plant Lotus (*Nelumbo nucifera*), is reported to remove the neurotoxic protein aggregates by autophagy in PC12 cells treated with mutant HTT (Wong, Wu, Wang, Liu, & Law, 2015). Other studies are focused on targeting the apoptotic proteins, caspase 1 and 3 which are linked to the progression of NDs by natural compounds like rosmarinic acid, curcumin, luteolin, and huperzine A as potential caspase inhibitors (Khan et al., 2015).

Ginkgo biloba has down regulated striatal glyceraldehyde-3-phosphate dehydrogenase and unregulated *Bcl-xl* expression levels, which is supported by histopathological studies suggesting the neuroprotection in HD (Mahdy, Tadros, Mohamed, Karim, & Khalifa, 2011). Another common plant *Withania somnifera* has also been reported to act on HD by GABAergic system, which has a major role in HD (Kumar & Kumar, 2009b). *Centella asiatica* (CA) also exhibits protection against neuronal damage in HD induced by oxidative stress and mitochondrial dysfunction (Shinomol & Muralidhara, 2008). Several other flavonoids namely celastrol, trehalose, lycopene, and sesamol exhibits protection against the pathogenesis of HD (Cleren, Calingasan, Chen, & Beal, 2005; Kumar, Kalonia, & Kumar, 2010; Kumar & Kumar, 2009a; Sarkar, Davies, Huang, Tunnacliffe, & Rubinsztein, 2007).

4.4   Evidence of effect of phytochemicals in amyotrophic lateral sclerosis

Complementary and alternative medicines (CAM) is a nonmainstream treatment used together with conventional medicine. CAM has gained a lot of attention for the treatment of ALS especially acupuncture, homeopathy, and naturopathy (Wasner, Klier, & Borasio, 2001). Ginseng comprises of various substances including carbohydrates, proteins, vitamins, fatty acids and ginsenosides. It has shown protection against ALS by attenuation of motor neurons in the lumber spinal cord of hSOD1 animals (Cai & Yang, 2016). There is strong evidence of mutation of Cu/Zn superoxide dismutase (SOD1) gene in ALS patients. The presence of excessive ROS production in the brain along with deteriorative anti oxidative system leads to neuronal cell death in ALS. In a transgenic SOD1-G93A mice model of ALS, madecassoside from *Centella asiatica* was found to increase the survival time of mice suggesting the protection of motor neurons degeneration (Liu, Kobayashi, Li, et al., 2006). Epigallocatechin gallate (EGCG) is the main active constituent of green tea. It has a strong antioxidant property protecting neurodegeneration including ALS (Koh et al., 2006; Mandel, Weinreb, Amit, & Youdim, 2004). Garlic is known to have a strong antioxidant effect due to the presence of Allicin, an organic sulfide. Diallyl trisulfide (DATS), the most active monomer in allicin, has various biological activities. It has activated the phase II detoxifying enzyme called heme oxygenase-1 (HO-1) responsible for enhancing antioxidant activity and down-regulated the glial fibrillary acidic protein (GFAP) in the SOD1-G93A model of ALS (Guo et al., 2011). Kiaei et al. (2005) demonstrated that celastrol administration in SOD1-G93A model of ALS delayed the onset of the disease, survival time of mice was increased, and the motor functions were improved significantly. Many other components of plant origin namely curcumin, resveratrol, obovatol, wogonin, and paeonol have been suggested as nutraceuticals for the prevention of symptoms of ALS by their antiinflammatory properties (Ock et al., 2010; Sikora, Scapagnini, & Barbagallo, 2010).

4.5 Evidence of effect of phytochemicals in cerebral ischemia

Stroke is the third leading cause of death worldwide according to the World Health Organization. Maximum cases of stroke are ischemic in nature due to the blockage of blood flow and oxygen in the cerebral arteries. The pathological mechanisms linked to ischemia are excitotoxicity, inflammation, oxidative stress, and apoptosis (Yousuf et al., 2009). Although the pathophysiology of ischemic brain damage has been thoroughly studied, the attempts to develop new therapeutic options are always open. From the aspect of drug discovery, validation of the target is the most critical step to find effective therapeutic agents. As the main pathophysiology is oxidative stress, natural compounds exhibiting antioxidative properties have shown remarkable success in the prevention of ischemia. Khan et al. (2009) used rutin, a bioflavonoid, against the neuronal damage induced by transient focal cerebral ischemia in rats. In another study, he used sesamin from sesame seed oil against the damage caused by middle cerebral artery occlusion (MCAO) (Khan, Ahmad, et al., 2010; Khan, Ishrat, et al., 2010). Quercetin has offered significant protection against ischemia/reperfusion model by depleting lipid peroxidation, caspase 3 activity, p53 activation, and elevation of internal antioxidative system (Ahmad et al., 2011). Silymarin and hesperidin have also protected the neurons from oxidative stress-associated damage in focal cerebral ischemia (Raza, Khan, Ahmad, et al., 2011; Raza, Khan, Ashafaq, et al., 2011).

The transcription factor, hypoxia inducible factor-1 (HIF-1), is activated in the case of hypoxic environment (Semenza & Wang, 1992). It acts as a key regulator of oxygen homeostasis in the cell. Agents that have the ability to activate HIF-1 may prevent ischemia/reperfusion injuries. Dibenzoylmethane found in licorice has been reported to maintain HIF-1 protein in various cell lines (Mabjeesh et al., 2003). Another flavonoid, quercetin has shown the activation of HIF-1 in HeLa cells (Wilson & Poellinger, 2002).

5.   Nano formulations from active phytochemicals having efficacy in neurodegenerative disorders

There has been extensive literature regarding the therapeutic effects of different phytochemicals on NDs. Still their application in clinical studies are not resolved fully due to many reasons. One of the main reasons is the inability of maximum molecules to cross the BBB (Begley, 2004). Thus to obtain therapeutic levels, high doses are required, which can cause serious adverse effects (Mathias & Hussain, 2010). BBB is extremely dynamic in structure and function maintained collectively by its constituents. It has been known that only highly lipophilic molecules with less than 400 Da can cross the BBB while hydrophilic molecules crossing is highly limited. In fact, the most limiting factor in drug development for brain disease is BBB, which makes the neurotherapeutics a challenge.

In the past few years, this challenge was somewhat sufficed by the introduction of nanotechnology (Herrán et al., 2013, 2014; Wong, Wu, & Bendayan, 2012). There are different kinds of nano formulations targeting different models of NDs (Table 1). Nanosized bioactive compounds vary from 10 to 1000 nm in size, enhances bioavailability, increases safety, and decreases toxicity (Cheng et al., 2013; Fathi, Mozafari, & Mohebbi, 2012; Frozza et al., 2013; Kumar et al., 2013). The delivery method of the nano formulation used in ND should be carefully designed (Tsai et al., 2011; Wu et al., 2014). The nanoparticle designed should be in such a way that it targets the bioactive compound to the brain with least toxicity (Kumar et al., 2013). Several methods have been developed such as liposomes, nanomicelles, exosomes, niosomes, dendrimers, inorganic and organic nanoparticles, polymeric nanoparticles, hybrid nanoparticles, and protein nanoparticles. Several nanophytobioactive compounds delivery has shown better bioavailability as compared to the crude bioactive compound (Neves, Lúcio, Martins, Lima, & Reis, 2013; Pandita, Kumar, Poonia, & Lather, 2014). A few of the bioactive compounds having neuroprotective activity are discussed in this section.

5.1   Nanocurcumin

Curcumin in its natural form is highly hydrophobic in nature, as a water-insoluble compound from *Curcuma longa*. Curcumin is known to have multiple benefits on the body and acts as antimicrobial, antioxidant, antiinflammatory, antiaging, anti-Alzheimer, and anti-Parkinson (Abrahams, Haylett, Johnson,

TABLE 1 Few examples of nanotechnology and nanocarriers for plant component delivery system for neurodegenerative disorders.

Nanoparticles type	Bioactive compound	Target disease	Model studied	Refs.
Liposomes	curcumin	Parkinson's disease	MPTP-induced PD mouse model	Zhang et al. (2018)
Nanomicelles	resveratrol	Alzheimer's disease	Aβ toxicated PC12 cells	Lu et al. (2009)
Solid lipid NP (SLNP)	curcumin	Cerebral ischemia	Cerebral ischemia (BCCAO model) in rats	Kakkar, Muppu, Chopra, and Kaur (2013)
Exosomes	curcumin	Neurodegenerative disease	Endothelial junction proteins	Kalani and Chaturvedi (2017)
Nanosomes	quercetin	Parkinson's disease	6-hydroxydopamine (6-OHDA) injection in the Substantia Nigra (SN) in a model of PD in rats	Díaz, Vaamonde, and Dajas (2015)
poly(lactic-co-glycolic acid) (PLGA)	ginsenoside Rg3	Alzheimer's disease	In-vitro transwell model of the BBB	Aalinkeel et al. (2018)
	quercetin	Alzheimer's disease	Aβ toxicated SH-SY5Y cel line	Sun et al. (2016)
Cerium oxide nanoparticles	none	Amyotrophic lateral sclerosis	Mouse model of ALS	DeCoteau et al. (2016)

Carr, & Bardien, 2019; Cheng et al., 2015; Maiti & Dunbar, 2018; Schiborr et al., 2014). Many methods of incorporation of curcumin in different nanoparticles are available that increases its bioavailability in systemic circulation (Cheng et al., 2015; Fan et al., 2018; Kidd, 2009; Maiti, Mukherjee, Gantait, Saha, & Mukherjee, 2007). Salehi et al. (2020) also reported that various nanoformulations of curcumin have the ability to increase the bioavailability of curcumin to the brain curing various NDs including AD. Another report suggests that curcumin loaded solid lipid nanoparticles (SLNS) have the ability to cross the BBB treating ischemia (Kakkar et al., 2013). Curcumin exosomes have the potential to restore junction proteins and endothelial cells proliferation (Abdolahi et al., 2017). Microspheres and microcapsules have been reported to increase the curcumin's stability and bioavailability (Paolino et al., 2016). Liposomal curcumin is another type of nanoparticle that showed protection against migraine headaches induced in mice in combination with sumatriptan (Bulboacă et al., 2018). Several other curcumin nanoformulations include curcumin polymeric micelles, curcumin microemulsions, curcumin nanogels, curcumin SLNs, and curcumin polymers nanoparticles that have efficacy in several disorders (Lin et al., 2014; Vaz et al., 2017; Zhao et al., 2012). Most recently, curcumin encapsulated exosomes have been a promising target. Kalani & Chaturvedi reported that curcumin encapsulated exosomes were effective in rat models of AD (Kalani & Chaturvedi, 2017). Magnetic curcumin nanoparticles under external magnetic fields can be effective against several NDs including AD, PD, and ALS (Huo, Zhang, Jin, Li, & Zhang, 2019; Maiti & Dunbar, 2018).

5.2 Nanoginsenosides

Ginseng (*Panax ginseng*) has been used for thousands of years as a medicinal herb in Asia and China. The active components found in ginseng are ginsenosides (Rhim, Kim, Lee, Oh, & Nah, 2002) with ginsenoside Rg3 as the main pharmacologically active ingredient (Aalinkeel et al., 2018). The extract has shown efficacy against the cognitive defects in AD patients (Heo et al., 2011). Research suggests that ginsenoside Rg3 has an important role in memory and cognition in treatment of AD (Yang et al., 2009). It has antiinflammatory action by inhibiting the release of proinflammatory cytokines by blocking NF-κB signaling pathways and protein kinases in different neurological disorders (Joo et al., 2008). Aalinkeel et al. (2018) evaluated the efficacy of poly(lactic-*co*-glycolic acid) (PLGA) nanoformulations of ginsenoside Rg3 in AD pathogenesis and found that nano formulation had better bioavailability as compared to the mother ginsenoside. They also reported that these nanoparticles decrease the mitochondrial DNA suggesting a decrease in oxidative stress. In an in vitro study, H_2O_2-induced oxidative stress was decreased by ginsenoside Rg1 in PD model (Liu, Kou, & Yu, 2011). Another study reveals that toxicity induced by iron and 6-hydroxydopamine-induced cell death was restored significantly by ginsenoside Rg1 (Ge, Chen, Xie, & Wong, 2010; Xu, Jiang, Wang, & Xie, 2010). Ganesan, Ko, Kim, and Choi (2015) reported that nanoformulation of ginsenoside Rg1 and Rb1 have increased bioavailability in the brain. Another study reports that in L929 cells the protective effect of ginseng crude extract rich in ginsenosides nanoliposomes against H_2O_2-induced oxidative stress in L929 cells (Tsai, Li, Yin, Yu, & Wen, 2012). Wu et al. (2009) concluded that ginsenosides Rb1, Rg5 and Rc can serve as medicine for HD treatment. Another study indicates the potentiality of few ginsenosides against superoxide dismutase (SOD1) and TAR DNA Binding Protein (TARDBP) targets in ALS (Merchant, Rao, Bhat, et al., 2015). In the case of cerebral ischemia, ginsenoside Rg1 downregulates expression of protease-activated receptor 1 (PAR-1) (Xie, Li, Wang, Zheng, & Wang, 2015) and in the brain injury model protects BBB disruption (Chen et al., 2015; Zhou et al., 2014). Zhang et al. (2012) reported that panax notoginsenoside loaded in nanoliposomal vesicles offer protection against cerebral ischemia in rats.

5.3   Nanoquercetin

Quercetin is a flavonoid in many fruits and vegetables and exhibits strong antioxidant and antiinflammatory properties (Oliveira et al., 2016; Shi et al., 2009). Quercetin is likely to affect various pathways leading to amyloid beta neurotoxicity (Ohta et al., 2011). It can cross the BBB (Youdim, Shukitt-Hale, & Joseph, 2004), but its delivery by oral route is limited. It is less soluble in water hence it shows poor bioavailability in rats (approx. 17%) and in humans (2%), which leads to sub therapeutic levels (Gao et al., 2009). To overcome this problem, Quercetin was loaded in nano formulations, increasing its bioavailability to the brain. Sun et al. (2016) designed PLGA-quercetin nanoparticle having therapeutic effect on AD-induced transgenic mice. Another study reported the efficacy of nanoencapsulated quercetin by oral route in mouse model of AD (Moreno et al., 2017). In the case of PD, nanosomes of quercetin were used to protect the brain (Díaz et al., 2015); quercetin nanocrystal has protected toxicity of 6-OHDA model of PD (Ghaffari, Hajizadeh Moghaddam, & Zare, 2018). Ghosh, Sarkar, Mandal, and Das (2013) reported that quercetin nanoparticles have the ability to provide absolute enzymatic antioxidant status with antiinflammatory and antiapoptotic effects in ischemic rats. They also reported that oral treatment with nanoquercetin have the potential to protect CA1 and CA3 regions of hippocampus in ischemia reperfusion in aged and young rats (Ghosh et al., 2013).

5.4   Nanoresveratrol

Resveratrol (Res) (3,5,4′-trihydroxy-*trans*-stilbene) is a nonflavonoid polyphenol found in various fruits exhibiting numerous activities including neuroprotective effects (Bastianetto, Menard, & Quirion, 2015; Oomen et al., 2009). Although the research related to Res are limited to date and long-term clinical effects are missing, this plant product's studies have presented it as a promising alternative treatment for many neurological disorders. This has been shown to prevent AD, PD, ischemic stroke, and HD (Rege, Geetha, Griffin, Broderick, & Babu, 2014) mainly as it enters the blood stream as glucuronide conjugates and has the ability to pass the BBB (Baur & Sinclair, 2006). Despite these mentioned advantages of resveratrol, its bioavailability is still puzzling. One reason could be its poor solubility in water, which is reported to be less than 1 mg/mL (López-Nicolás & García-Carmona, 2008). Hence the studies could not give promising results in in vivo models and clinical studies. Various nanoparticles were designed, and the drug was loaded into it to combat the previous problems (Amiot et al., 2013; Amri, Chaumeil, Sfar, & Charrueau, 2012; Frozza et al., 2010). Frozza et al. (2010) tested the nanocapsules loaded with Res to be effective against Amyloid beta-exposed rats. In another study, Res-loaded micelle was developed and compared with free res in PC 12 cells exposed to amyloid beta. The free Res was toxic to the cells decreasing the cell viability while loaded Res was nontoxic and prevented the amyloid beta toxicity in cells via reduction in oxidative stress and apoptotic activity (Lu et al., 2009). In the case of PD, Res-loaded nanoparticles also showed better prevention than free resveratrol in MPTP-induced parkinsonism in C57BL/6 mice (da Rocha Lindner et al., 2015). Palle and Neerati (2018) studied the prevention of rotenone-induced Parkinson's rats by RES encapsulated into NPs. Res was loaded in Human Serum Albumin as nanoparticles (RES-HAS-NPs) and were tested against the transient middle cerebral artery occlusion rats. It was found in the study that RES-HAS-NPs was able to improve neuronal outcomes by reducing oxidative stress and apoptosis in neurons (Xu et al., 2018).

6. Conclusion

Aging is the main risk factor of various neurological disorders namely AD, PD, HD, ALS, and age-related cerebral ischemia. The current drugs available gives only symptomatic treatments, and there is always a need for a multifactorial approach in age-related neurological disorders. Nature has given us the opportunity to do experiments with components of variable complexity and heterogeneity, which is impossible to find in a synthetic formulation. This chapter discussed some of the most commonly used plant products for preventing NDs. The nanoformulations from these bio compounds have also been shown to be promising approaches in the line of treatment. The potential benefits of these phytochemicals have been studied, but more extensive in vivo and clinical studies have to be established further to assess the long-term efficacy of these phytochemicals as therapy against various age-related neurological diseases.

References

Aalinkeel, R., Kutscher, H. L., Singh, A., Cwiklinski, K., Khechen, N., Schwartz, S. A., et al. (2018). Neuroprotective effects of a biodegradable poly(lactic-co-glycolic acid)-ginsenoside Rg3 nanoformulation: A potential nanotherapy for Alzheimer's disease? *Journal of Drug Targeting*, *26*(2), 182–193.

Abdolahi, M., Tafakhori, A., Togha, M., Okhovat, A. A., Siassi, F., Eshraghian, M. R., et al. (2017). The synergistic effects of ω-3 fatty acids and nano-curcumin supplementation on tumor necrosis factor (TNF)-α gene expression and serum level in migraine patients. *Immunogenetics*, *69*(6), 371–378.

Abrahams, S., Haylett, W. L., Johnson, G., Carr, J. A., & Bardien, S. (2019). Antioxidant effects of curcumin in models of neurodegeneration, aging, oxidative and nitrosative stress: A review. *Neuroscience*, *406*, 1–21.

Ahmad, S., Khan, M. B., Hoda, M. N., Bhatia, K., Haque, R., Fazili, I. S., et al. (2012). Neuroprotective effect of sesame seed oil in 6-hydroxydopamine induced neurotoxicity in mice model: Cellular, biochemical and neurochemical evidence. *Neurochemical Research*, *37*(3), 516–526.

Ahmad, A., Khan, M. M., Hoda, M. N., Raza, S. S., Khan, M. B., Javed, H., et al. (2011). Quercetin protects against oxidative stress associated damages in a rat model of transient focal cerebral ischemia and reperfusion. *Neurochemical Research*, *36*(8), 1360–1371.

Albarracin, S. L., Stab, B., Casas, Z., Sutachan, J. J., Samudio, I., Gonzalez, J., et al. (2012). Effects of natural antioxidants in neurodegenerative disease. *Nutritional Neuroscience*, *15*(1), 1–9.

Albin, R. L. (2006). Parkinson's disease: Background, diagnosis, and initial management. *Clinics in Geriatric Medicine*, *22*(4), 735–751.

Al-Chalabi, A., van den Berg, L. H., & Veldink, J. (2017). Gene discovery in amyotrophic lateral sclerosis: Implications for clinical management. *Nature Reviews Neurology*, *13*(2), 96–104.

Amiot, M. J., Romier, B., Dao, T. M., Fanciullino, R., Ciccolini, J., Burcelin, R., et al. (2013). Optimization of trans-resveratrol bioavailability for human therapy. *Biochimie*, *95*(6), 1233–1238.

Amri, A., Chaumeil, J. C., Sfar, S., & Charrueau, C. (2012). Administration of resveratrol: What formulation solutions to bioavailability limitations? *Journal of Controlled Release: Official Journal of the Controlled Release Society*, *158*(2), 182–193.

Andreux, P. A., Houtkooper, R. H., & Auwerx, J. (2013). Pharmacological approaches to restore mitochondrial function. *Nature Reviews Drug Discovery*, *12*(6), 465–483.

Antunes, M. S., Goes, A. T., Boeira, S. P., Prigol, M., & Jesse, C. R. (2014). Protective effect of hesperidin in a model of Parkinson's disease induced by 6-hydroxydopamine in aged mice. *Nutrition (Burbank, Los Angeles County, Calif.)*, *30*(11–12), 1415–1422.

Asai, M., Iwata, N., Yoshikawa, A., Aizaki, Y., Ishiura, S., Saido, T. C., et al. (2007). Berberine alters the processing of Alzheimer's amyloid precursor protein to decrease Abeta secretion. *Biochemical and Biophysical Research Communications*, *352*(2), 498–502.

Azam, S., Jakaria, M., Kim, I.-S., Kim, J., Haque, M. E., & Choi, D.-K. (2019). Regulation of Toll-Like Receptor (TLR) signaling pathway by polyphenols in the treatment of age-linked neurodegenerative diseases: Focus on TLR4 signaling. *Frontiers in Immunology*, *10*, 1000.

Bagli, E., Stefaniotou, M., Morbidelli, L., Ziche, M., Psillas, K., Murphy, C., et al. (2004). Luteolin inhibits vascular endothelial growth factor-induced angiogenesis; inhibition of endothelial cell survival and proliferation by targeting phosphatidylinositol 3′-kinase activity. *Cancer Research*, *64*(21), 7936–7946.

Balunas, M. J., & Kinghorn, A. D. (2005). Drug discovery from medicinal plants. *Life Sciences*, *78*(5), 431–441.

Barbieri, R., Coppo, E., Marchese, A., Daglia, M., Sobarzo-Sánchez, E., Nabavi, S. F., et al. (2017). Phytochemicals for human disease: An update on plant-derived compounds antibacterial activity. *Microbiological Research, 196*, 44–68.

Bastianetto, S., Menard, C., & Quirion, R. (2015). Neuroprotective action of resveratrol. *Biochimica et Biophysica Acta, 1852*(6), 1195–1201.

Baur, J. A., & Sinclair, D. A. (2006). Therapeutic potential of resveratrol: The in vivo evidence. *Nature Reviews. Drug Discovery, 5*(6), 493–506.

Bawari, S., Tewari, D., Argüelles, S., Sah, A. N., Nabavi, S. F., Xu, S., et al. (2019). Targeting BDNF signaling by natural products: Novel synaptic repair therapeutics for neurodegeneration and behavior disorders. *Pharmacological Research, 148*, 104458.

Begley, D. J. (2004). Delivery of therapeutic agents to the central nervous system: The problems and the possibilities. *Pharmacology & Therapeutics, 104*, 29–45.

Bulboacă, A. E., Bolboacă, S. D., Stănescu, I. C., Sfrângeu, C. A., Porfire, A., Tefas, L., et al. (2018). The effect of intravenous administration of liposomal curcumin in addition to sumatriptan treatment in an experimental migraine model in rats. *International Journal of Nanomedicine, 13*, 3093–3103.

Burd, I., Welling, J., Kannan, G., & Johnston, M. V. (2016). Excitotoxicity as a common mechanism for fetal neuronal injury with hypoxia and intrauterine inflammation. In *Vol. 76. Advances in pharmacology* (pp. 85–101). Elsevier.

Cai, M., & Yang, E. J. (2016). Ginsenoside re attenuates neuroinflammation in a symptomatic ALS animal model. *The American Journal of Chinese Medicine, 44*(2), 401–413.

Chen, W., Guo, Y., Yang, W., Zheng, P., Zeng, J., & Tong, W. (2015). Protective effect of ginsenoside Rb1 on integrity of blood-brain barrier following cerebral ischemia. *Experimental Brain Research, 233*(10), 2823–2831.

Chen, X., & Pan, W. (2014). The treatment strategies for neurodegenerative diseases by integrative medicine. *Integrative Medicine International, 1*, 223–225.

Chen, L. L., Song, J. X., Lu, J. H., Yuan, Z. W., Liu, L. F., Durairajan, S. S., et al. (2014). Corynoxine, a natural autophagy enhancer, promotes the clearance of alpha-synuclein via Akt/mTOR pathway. *Journal of Neuroimmune Pharmacology: The Official Journal of the Society on NeuroImmune Pharmacology, 9*(3), 380–387.

Cheng, K. K., Chan, P. S., Fan, S., Kwan, S. M., Yeung, K. L., Wáng, Y. X., et al. (2015). Curcumin-conjugated magnetic nanoparticles for detecting amyloid plaques in Alzheimer's disease mice using magnetic resonance imaging (MRI). *Biomaterials, 44*, 155–172.

Cheng, K. K., Yeung, C. F., Ho, S. W., Chow, S. F., Chow, A. H., & Baum, L. (2013). Highly stabilized curcumin nanoparticles tested in an in vitro blood-brain barrier model and in Alzheimer's disease Tg2576 mice. *The AAPS Journal, 15*(2), 324–336.

Cianciulli, A., Dragone, T., Calvello, R., Porro, C., Trotta, T., Lofrumento, D. D., et al. (2015). IL-10 plays a pivotal role in anti-inflammatory effects of resveratrol in activated microglia cells. *International Immunopharmacology, 24*(2), 369–376.

Cleren, C., Calingasan, N. Y., Chen, J., & Beal, M. F. (2005). Celastrol protects against MPTP- and 3-nitropropionic acid-induced neurotoxicity. *Journal of Neurochemistry, 94*(4), 995–1004.

da Rocha Lindner, G., Bonfanti Santos, D., Colle, D., Gasnhar Moreira, E. L., Daniel Prediger, R., Farina, M., et al. (2015). Improved neuroprotective effects of resveratrol-loaded polysorbate 80-coated poly(lactide) nanoparticles in MPTPinduced Parkinsonism. *Nanomedicine, 10*, 1127–1138.

Das, J., Ramani, R., & Suraju, M. O. (2016). Polyphenol compounds and PKC signaling. *Biochimica et Biophysica Acta (BBA) - General Subjects, 1860*(10), 2107–2121.

Davinelli, S., Maes, M., Corbi, G., Zarrelli, A., Willcox, D. C., & Scapagnini, G. (2016). Dietary phytochemicals and neuro-inflammaging: From mechanistic insights to translational challenges. *Immunity & Ageing, 13*(1), 16.

DeCoteau, W., Heckman, K. L., Estevez, A. Y., Reed, K. J., Costanzo, W., Sandford, D., et al. (2016). Cerium oxide nanoparticles with antioxidant properties ameliorate strength and prolong life in mouse model of amyotrophic lateral sclerosis. *Nanomedicine: Nanotechnology, Biology and Medicine, 12*(8), 2311–2320.

Díaz, M., Vaamonde, L., & Dajas, F. (2015). Assessment of the protective capacity of nanosomes of quercetin in an experimental model of parkinson's disease in the rat. *General Medicine, 3*(5), 207.

Eftekharzadeh, B., Ramin, M., Khodagholi, F., Moradi, S., Tabrizian, K., Sharif, R., et al. (2012). Inhibition of PKA attenuates memory deficits induced by β-amyloid (1–42), and decreases oxidative stress and NF-κB transcription factors. *Behavioural Brain Research, 226*(1), 301–308.

Elamin, M., Bede, P., Byrne, S., Jordan, N., Gallagher, L., Wynne, B., et al. (2013). Cognitive changes predict functional decline in ALS: A population-based longitudinal study. *Neurology, 80*(17), 1590–1597.

Fan, C., Song, Q., Wang, P., Li, Y., Yang, M., Liu, B., et al. (2018). Curcumin protects against chronic stress-induced dysregulation of neuroplasticity and depression-like behaviors via suppressing IL-1β pathway in rats. *Neuroscience, 392*, 92–106.

Farooqui, A. A. (2013). Effect of lifestyle, aging, and phytochemicals on the onset of neurological disorders. In *Phytochemicals, signal transduction, and neurological disorders*. New York, NY: Springer.

Fathi, M., Mozafari, M. R., & Mohebbi, M. (2012). Nanoencapsulation of food ingredients using lipid based delivery systems. *Trends in Food Science and Technology, 23*(1), 13–27.

Ferretta, A., Gaballo, A., Tanzarella, P., Piccoli, C., Capitanio, N., Nico, B., et al. (2014). Effect of resveratrol on mitochondrial function: Implications in parkin-associated familiar Parkinson's disease. *Biochimica et Biophysica Acta, 1842*(7), 902–915.

Frozza, R. L., Bernardi, A., Hoppe, J. B., Meneghetti, A. B., Battastini, A. M., Pohlmann, A. R., et al. (2013). Lipid-core nanocapsules improve the effects of resveratrol against Abeta-induced neuroinflammation. *Journal of Biomedical Nanotechnology, 9*(12), 2086–2104.

Frozza, R. L., Bernardi, A., Paese, K., Hoppe, J. B., da Silva, T., Battastini, A. M., et al. (2010). Characterization of trans-resveratrol-loaded lipid-core nanocapsules and tissue distribution studies in rats. *Journal of Biomedical Nanotechnology, 6*(6), 694–703.

Ganesan, P., Ko, H. M., Kim, I. S., & Choi, D. K. (2015). Recent trends in the development of nanophytobioactive compounds and delivery systems for their possible role in reducing oxidative stress in Parkinson's disease models. *International Journal of Nanomedicine, 10*, 6757–6772.

Gan-Or, Z., Dion, P. A., & Rouleau, G. A. (2015). Genetic perspective on the role of the autophagy-lysosome pathway in Parkinson disease. *Autophagy, 11*(9), 1443–1457.

Gao, B., Chang, C., Zhou, J., Zhao, T., Wang, C., Li, C., et al. (2015). Pycnogenol protects against rotenone-induced neurotoxicity in PC12 cells through regulating NF-κB-iNOS signaling pathway. *DNA and Cell Biology, 34*(10), 643–649.

Gao, Y., Wang, Y., Ma, Y., Yu, A., Cai, F., Shao, W., et al. (2009). Formulation optimization and in situ absorption in rat intestinal tract of quercetin-loaded microemulsion. *Colloids and Surfaces. B, Biointerfaces, 71*(2), 306–314.

Ge, K. L., Chen, W. F., Xie, J. X., & Wong, M. S. (2010). Ginsenoside Rg1 protects against 6-OHDA-induced toxicity in MES23.5 cells via Akt and ERK signaling pathways. *Journal of Ethnopharmacology, 127*(1), 118–123.

Geun Kim, H., & Sook Oh, M. (2012). Herbal medicines for the prevention and treatment of Alzheimer's disease. *Current Pharmaceutical Design, 18*(1), 57–75.

Ghaffari, F., Hajizadeh Moghaddam, A., & Zare, M. (2018). Neuroprotective effect of quercetin nanocrystal in a 6-hydroxydopamine model of Parkinson disease: Biochemical and behavioral evidence. *Basic and Clinical Neuroscience, 9*(5), 317–324.

Ghosh, A., Sarkar, S., Mandal, A. K., & Das, N. (2013). Neuroprotective role of nanoencapsulated quercetin in combating ischemia-reperfusion induced neuronal damage in young and aged rats. *PLoS One, 8*(4), e57735.

Guo, Y., Zhang, K., Wang, Q., Li, Z., Yin, Y., Xu, Q., et al. (2011). Neuroprotective effects of diallyl trisulfide in SOD1-G93A transgenic mouse model of amyotrophic lateral sclerosis. *Brain Research, 1374*, 110–115.

Gupta, V., & Sharma, B. (2017). Role of phytochemicals in neurotrophins mediated regulation of Alzheimer's disease. *International Journal of Complementary & Alternative Medicine, 7*(4), 00231.

Hardiman, O., Al-Chalabi, A., Chio, A., Corr, E. M., Logroscino, G., Robberecht, W., et al. (2017). Amyotrophic lateral sclerosis. *Nature Reviews. Disease Primers, 3*, 17071.

Hardy, M., Zielonka, J., Karoui, H., Sikora, A., Michalski, R., Podsiadły, R., et al. (2018). Detection and characterization of reactive oxygen and nitrogen species in biological systems by monitoring species-specific products. *Antioxidants & Redox Signaling, 28*(15), 1416–1432.

Harvey, A. L., Clark, R. L., Mackay, S. P., & Johnston, B. F. (2010). Current strategies for drug discovery through natural products. *Expert Opinion on Drug Discovery, 5*(6), 559–568.

Harvey, A. L., & Cree, I. A. (2010). High-throughput screening of natural products for cancer therapy. *Planta Medica, 76*, 1080–1086.

Heo, J. H., Lee, S. T., Oh, M. J., Park, H. J., Shim, J. Y., Chu, K., et al. (2011). Improvement of cognitive deficit in Alzheimer's disease patients by long term treatment with korean red ginseng. *Journal of Ginseng Research, 35*(4), 457–461.

Herrán, E., Requejo, C., Ruiz-Ortega, J. A., Aristieta, A., Igartua, M., Bengoetxea, H., et al. (2014). Increased antiparkinson efficacy of the combined administration of VEGF- and GDNF-loaded nanospheres in a partial lesion model of Parkinson's disease. *International Journal of Nanomedicine, 9*, 2677–2687.

Herrán, E., Ruiz-Ortega, J. A., Aristieta, A., Igartua, M., Requejo, C., Lafuente, J. V., et al. (2013). In vivo administration of VEGF- and GDNF-releasing biodegradable polymeric microspheres in a severe lesion model of Parkinson's disease. *European Journal of Pharmaceutics and Biopharmaceutics, 85*, 1183–1190.

Hicks, A., & Jolkkonen, J. (2009). Challenges and possibilities of intravascular cell therapy in stroke. *Acta Neurobiologiae Experimentalis, 69*(1), 1–11.

Hosseinkhani, A., Sahragard, A., Namdari, A., & Zarshenas, M. M. (2017). Botanical sources for Alzheimer's: A review on reports from traditional Persian medicine. *American Journal of Alzheimer's Disease and Other Dementias, 32*(7), 429–437.

Huang, W., Cheng, P., Yu, K., Han, Y., Song, M., & Li, Y. (2017). Hyperforin attenuates aluminum-induced Aβ production and Tau phosphorylation via regulating Akt/GSK-3β signaling pathway in PC12 cells. *Biomedicine & Pharmacotherapy = Biomedecine & Pharmacotherapie, 96*, 1–6.

Huang, W. W., Tsai, S. C., Peng, S. F., Lin, M. W., Chiang, J. H., Chiu, Y. J., et al. (2013). Kaempferol induces autophagy through AMPK and AKT signaling molecules and causes G2/M arrest via downregulation of CDK1/cyclin B in SK-HEP-1 human hepatic cancer cells. *International Journal of Oncology, 42*(6), 2069–2077.

Huang, W. J., Zhang, X., & Chen, W. W. (2016). Role of oxidative stress in Alzheimer's disease. *Biomedical Reports, 4*(5), 519–522.

Huo, X., Zhang, Y., Jin, X., Li, Y., & Zhang, L. (2019). A novel synthesis of selenium nanoparticles encapsulated PLGA nanospheres with curcumin molecules for the inhibition of amyloid β aggregation in Alzheimer's disease. *Journal of Photochemistry and Photobiology. B, Biology*, *190*, 98–102.

Inaba-Hasegawa, K., Akao, Y., Maruyama, W., & Naoi, M. (2012). Type A monoamine oxidase is associated with induction of neuroprotective Bcl-2 by rasagiline, an inhibitor of type B monoamine oxidase. *Journal of Neural Transmission*, *119*(4), 405–414.

Iqbal, K., & Grundke-Iqbal, I. (2010). Alzheimer's disease, a multifactorial disorder seeking multitherapies. *Alzheimer's & Dementia : The Journal of the Alzheimer's Association*, *6*(5), 420–424.

Jabir, N. R., Khan, F. R., & Tabrez, S. (2018). Cholinesterase targeting by polyphenols: A therapeutic approach for the treatment of Alzheimer's disease. *CNS Neuroscience & Therapeutics*, *24*(9), 753–762.

Jahan, S., Kumar, D., Chaturvedi, S., Rashid, M., Wahajuddin, M., AKhan, Y., et al. (2017). Therapeutic targeting of NLRP3 inflammasomes by natural products and pharmaceuticals: A novel mechanistic approach for inflammatory diseases. *Current Medicinal Chemistry*, *24*(16), 1645–1670.

Jahan, S., Kumar, D., Singh, S., Kumar, V., Srivastava, A., Pandey, A., et al. (2018). Resveratrol prevents the cellular damages induced by monocrotophos via PI3K signaling pathway in human cord blood mesenchymal stem cells. *Molecular Neurobiology*, *55*(11), 8278–8292.

Jahan, S., Singh, S., Srivastava, A., Kumar, V., Kumar, D., Pandey, A., et al. (2018). PKA-GSK3β and β-catenin signaling play a critical role in trans-resveratrol mediated neuronal differentiation in human cord blood stem cells. *Molecular Neurobiology*, *55*(4), 2828–2839.

Javed, H., Vaibhav, K., Ahmed, M. E., Khan, A., Tabassum, R., Islam, F., et al. (2015). Effect of hesperidin on neurobehavioral, neuroinflammation, oxidative stress and lipid alteration in intracerebroventricular streptozotocin induced cognitive impairment in mice. *Journal of the Neurological Sciences*, *348*(1–2), 51–59.

Jhang, K. A., Park, J. S., Kim, H. S., & Chong, Y. H. (2017). Resveratrol ameliorates tau hyperphosphorylation at Ser396 site and oxidative damage in rat hippocampal slices exposed to vanadate: Implication of ERK1/2 and GSK-3β signaling cascades. *Journal of Agricultural and Food Chemistry*, *65*(44), 9626–9634.

Jia, L., Liu, J., Song, Z., Pan, X., Chen, L., Cui, X., et al. (2012). Berberine suppresses amyloidbeta-induced inflammatory response in microglia by inhibiting nuclear factor-kappaB and mitogen-activated protein kinase signalling pathways. *The Journal of Pharmacy and Pharmacology*, *64*(10), 1510–1521.

Johnson, I. P. (2015). Age-related neurodegenerative disease research needs aging models. *Frontiers in Aging Neuroscience*, *7*, 168.

Joo, S. S., Yoo, Y. M., Ahn, B. W., Nam, S. Y., Kim, Y. B., Hwang, K. W., et al. (2008). Prevention of inflammation-mediated neurotoxicity by Rg3 and its role in microglial activation. *Biological & Pharmaceutical Bulletin*, *31*, 1392–1396.

Kahkeshani, N., Saeidnia, S., & Abdollahi, M. (2015). Role of antioxidants and phytochemicals on acrylamide mitigation from food and reducing its toxicity. *Journal of Food Science and Technology*, *52*(6), 3169–3186.

Kakkar, V., Muppu, S. K., Chopra, K., & Kaur, I. P. (2013). Curcumin loaded solid lipid nanoparticles: An efficient formulation approach for cerebral ischemic reperfusion injury in rats. *European Journal of Pharmaceutics and Biopharmaceutics*, *85*(3 Pt A), 339–345.

Kalani, A., & Chaturvedi, P. (2017). Curcumin-primed and curcumin-loaded exosomes: Potential neural therapy. *Neural Regeneration Research*, *12*(2), 205–206.

Khan, S., Ahmad, K., Alshammari, E. M., Adnan, M., Baig, M. H., Lohani, M., et al. (2015). Implication of caspase-3 as a common therapeutic target for multineurodegenerative disorders and its inhibition using nonpeptidyl natural compounds. *BioMed Research International, 2015*, 379817.

Khan, M. M., Ahmad, A., Ishrat, T., Khan, M. B., Hoda, M. N., Khuwaja, G., et al. (2010). Resveratrol attenuates 6-hydroxydopamine-induced oxidative damage and dopamine depletion in rat model of Parkinson's disease. *Brain Research, 1328*, 139–151.

Khan, M. M., Ahmad, A., Ishrat, T., Khuwaja, G., Srivastawa, P., Khan, M. B., et al. (2009). Rutin protects the neural damage induced by transient focal ischemia in rats. *Brain Research, 1292*, 123–135.

Khan, M. M., Ishrat, T., Ahmad, A., et al. (2010). Sesamin attenuates behavioral, biochemical and histological alterations induced by reversible middle cerebral artery occlusion in the rats. *Chemico-Biological Interactions, 183*(1), 255–263.

Khan, M. M., Raza, S. S., Javed, H., Ahmad, A., Khan, A., Islam, F., et al. (2012). Rutin protects dopaminergic neurons from oxidative stress in an animal model of Parkinson's disease. *Neurotoxicity Research, 22*(1), 1–15.

Khan, A., Vaibhav, K., Javed, H., Khan, M. M., Tabassum, R., Ahmed, M. E., et al. (2012). Attenuation of Aβ-induced neurotoxicity by thymoquinone via inhibition of mitochondrial dysfunction and oxidative stress. *Molecular and Cellular Biochemistry, 369*(1–2), 55–65.

Kiaei, M., Kipiani, K., Petri, S., Chen, J., Calingasan, N. Y., & Beal, M. F. (2005). Celastrol blocks neuronal cell death and extends life in transgenic mouse model of amyotrophic lateral sclerosis. *Neuro-Degenerative Diseases, 2*(5), 246–254.

Kidd, P. M. (2009). Bioavailability and activity of phytosome complexes from botanical polyphenols: The silymarin, curcumin, green tea, and grape seed extracts. *Alternative Medicine Review: A Journal of Clinical Therapeutic, 14*(3), 226–246.

Kim, M. H., Kim, S. H., & Yang, W. M. (2014). Mechanisms of action of phytochemicals from medicinal herbs in the treatment of Alzheimer's disease. *Planta Medica, 80*(15), 1249–1258.

Kimura, I. (2006). Medical benefits of using natural compounds and their derivatives having multiple pharmacological actions. *Yakugaku zasshi: Journal of the Pharmaceutical Society of Japan, 126*(3), 133–143.

Kirino, T. (1982). Delayed neuronal death in the gerbil hippocampus following ischemia. *Brain Research, 239*(1), 57–69.

Koh, S. H., Lee, S. M., Kim, H. Y., Lee, K. Y., Lee, Y. J., Kim, H. T., et al. (2006). The effect of epigallocatechin gallate on suppressing disease progression of ALS model mice. *Neuroscience Letters, 395*(2), 103–107.

Krajka-Kuźniak, V., Paluszczak, J., Szaefer, H., & Baer-Dubowska, W. (2015). The activation of the Nrf2/ARE pathway in HepG2 hepatoma cells by phytochemicals and subsequent modulation of phase II and antioxidant enzyme expression. *Journal of Physiology and Biochemistry, 71*(2), 227–238.

Kuboyama, T., Tohda, C., Zhao, J., Nakamura, N., Hattori, M., & Komatsu, K. (2002). Axon- or dendrite-predominant outgrowth induced by constituents from Ashwagandha. *Neuroreport, 13*(14), 1715–1720.

Kumar, A., Chen, F., Mozhi, A., Zhang, X., Zhao, Y., Xue, X., et al. (2013). Innovative pharmaceutical development based on unique properties of nanoscale delivery formulation. *Nanoscale, 5*(18), 8307–8325.

Kumar, P., Kalonia, H., & Kumar, A. (2010). Protective effect of sesamol against 3-nitropropionic acid-induced cognitive dysfunction and altered glutathione redox balance in rats. *Basic & Clinical Pharmacology & Toxicology, 107*(1), 577–582.

Kumar, G. P., & Khanum, F. (2012). Neuroprotective potential of phytochemicals. *Pharmacognosy Reviews, 6*(12), 81.

Kumar, P., & Kumar, A. (2009a). Effect of lycopene and epigallocatechin-3-gallate against 3-nitropropionic acid induced cognitive dysfunction and glutathione depletion in rat: A novel nitric oxide mechanism. *Food and Chemical Toxicology: An International Journal Published for the British Industrial Biological Research Association, 47*(10), 2522–2530.

Kumar, P., & Kumar, A. (2009b). Possible neuroprotective effect of *Withania somnifera* root extract against 3-nitropropionic acid-induced behavioral, biochemical, and mitochondrial dysfunction in an animal model of Huntington's disease. *Journal of Medicinal Food, 12*(3), 591–600.

Kumar, A., Singh, A., & Ekavali. (2015). A review on Alzheimer's disease pathophysiology and its management: An update. *Pharmacological Reports, 67*(2), 195–203.

Lemkul, J. A., & Bevan, D. R. (2010). Destabilizing Alzheimer's Aβ42 protofibrils with morin: Mechanistic insights from molecular dynamics simulations. *Biochemistry, 49*(18), 3935–3946.

Leverenz, J. B., Quinn, J. F., Zabetian, C., Zhang, J., Montine, K. S., & Montine, T. J. (2009). Cognitive impairment and dementia in patients with Parkinson disease. *Current Topics in Medicinal Chemistry, 9*(10), 903–912.

Li, Y.-g., Zhu, W., Tao, J.-p., Xin, P., Liu, M.-y., Li, J.-b., et al. (2013). Resveratrol protects cardiomyocytes from oxidative stress through SIRT1 and mitochondrial biogenesis signaling pathways. *Biochemical and Biophysical Research Communications, 438*(2), 270–276.

Lin, C. C., Lin, H. Y., Chi, M. H., Shen, C. M., Chen, H. W., Yang, W. J., et al. (2014). Preparation of curcumin microemulsions with food-grade soybean oil/lecithin and their cytotoxicity on the HepG2 cell line. *Food Chemistry, 154*, 282–290.

Liu, Y. J., Kobayashi, Y., Li, S. H., et al. (2006). Therapeutic potential of madecassoside in transgenic mice of amyotrophic lateral sclerosis. *Chinese Traditional and Herbal Drugs, 37*, 718–720.

Liu, Q., Kou, J. P., & Yu, B. Y. (2011). Ginsenoside Rg1 protects against hydrogen peroxide-induced cell death in PC12 cells via inhibiting NF-κB activation. *Neurochemistry International, 58*(1), 119–125.

Longin, E., Teich, M., Koelfen, W., & König, S. (2002). Topiramate enhances the risk of valproate-associated side effects in three children. *Epilepsia, 43*(4), 451–454.

López-Nicolás, J. M., & García-Carmona, F. (2008). Aggregation state and pKa values of (E)-resveratrol as determined by fluorescence spectroscopy and UV-visible absorption. *Journal of Agricultural and Food Chemistry, 56*(17), 7600–7605.

Lopresti, A. L., Maker, G. L., Hood, S. D., & Drummond, P. D. (2014). A review of peripheral biomarkers in major depression: The potential of inflammatory and oxidative stress biomarkers. *Progress in Neuro-Psychopharmacology and Biological Psychiatry, 48*, 102–111.

Lu, M. C., Ji, J. A., Jiang, Z. Y., & You, Q. D. (2016). The Keap1–Nrf2–ARE pathway as a potential preventive and therapeutic target: An update. *Medicinal Research Reviews, 36*(5), 924–963.

Lu, X., Ji, C., Xu, H., Li, X., Ding, H., Ye, M., et al. (2009). Resveratrol-loaded polymeric micelles protect cells from Abeta-induced oxidative stress. *International Journal of Pharmaceutics, 375*, 89–96.

Lushchak, V. I. (2014). Free radicals, reactive oxygen species, oxidative stress and its classification. *Chemico-Biological Interactions, 224*, 164–175.

Mabjeesh, N. J., Willard, M. T., Harris, W. B., Sun, H. Y., Wang, R., Zhong, H., et al. (2003). Dibenzoylmethane, a natural dietary compound, induces HIF-1 alpha and increases expression of VEGF. *Biochemical and Biophysical Research Communications, 303*(1), 279–286.

Macdonald, I. R., Rockwood, K., Martin, E., & Darvesh, S. (2014). Cholinesterase inhibition in Alzheimer's disease: Is specificity the answer? *Journal of Alzheimer's Disease, 42*(2), 379–384.

Mahdy, H. M., Tadros, M. G., Mohamed, M. R., Karim, A. M., & Khalifa, A. E. (2011). The effect of *Ginkgo biloba* extract on 3-nitropropionic acid-induced neurotoxicity in rats. *Neurochemistry International, 59*(6), 770–778.

Maiti, P., & Dunbar, G. L. (2018). Use of curcumin, a natural polyphenol for targeting molecular pathways in treating age-related neurodegenerative diseases. *International Journal of Molecular Sciences*, *19*(6), 1637.

Maiti, K., Mukherjee, K., Gantait, A., Saha, B. P., & Mukherjee, P. K. (2007). Curcumin-phospholipid complex: Preparation, therapeutic evaluation and pharmacokinetic study in rats. *International Journal of Pharmaceutics*, *330*(1–2), 155–163.

Mandel, S., Weinreb, O., Amit, T., & Youdim, M. B. (2004). Cell signaling pathways in the neuroprotective actions of the green tea polyphenol (−)-epigallocatechin-3-gallate: Implications for neurodegenerative diseases. *Journal of Neurochemistry*, *88*(6), 1555–1569.

Manoharan, S., Guillemin, G. J., Abiramasundari, R. S., Essa, M. M., Akbar, M., & Akbar, M. D. (2016). The role of reactive oxygen species in the pathogenesis of Alzheimer's disease, Parkinson's disease, and Huntington's disease: A mini review. *Oxidative Medicine and Cellular Longevity*, *2016*, 8590578.

Manuja, R., Sachdeva, S., Jain, A., & Chaudhary, J. (2013). A comprehensive review on biological activities of p-hydroxy benzoic acid and its derivatives. *International Journal of Pharmaceutical Sciences Review and Research*, *22*(2), 109–115.

Mathew, S., Abraham, T. E., & Zakaria, Z. A. (2015). Reactivity of phenolic compounds towards free radicals under in vitro conditions. *Journal of Food Science and Technology*, *52*(9), 5790–5798.

Mathias, N. R., & Hussain, M. A. (2010). Non-invasive systemic drug delivery: Developability considerations for alternate routes of administration. *Journal of Pharmaceutical Sciences*, *99*, 1–20.

Menendez, J. A., Joven, J., Aragonès, G., Barrajón-Catalán, E., Beltrán-Debón, R., Borrás-Linares, I., et al. (2013). Xenohormetic and anti-aging activity of secoiridoid polyphenols present in extra virgin olive oil: A new family of gerosuppressant agents. *Cell Cycle*, *12*(4), 555–578.

Menzies, K. J., Singh, K., Saleem, A., & Hood, D. A. (2013). Sirtuin 1-mediated effects of exercise and resveratrol on mitochondrial biogenesis. *Journal of Biological Chemistry*, *288*(10), 6968–6979.

Merchant, N., Rao, P., Bhat, S., et al. (2015). Synthesis of ginsenoside nanoparticles and insilico docking of ginsenosides to sod1 and tardbp targets in amyotrophic lateral sclerosis (Als). *International Journal of ChemTech Research*, *7*(2), 790–794.

Miller, S. L., Wallace, E. M., & Walker, D. W. (2012). Antioxidant therapies: A potential role in perinatal medicine. *Neuroendocrinology*, *96*(1), 13–23.

Mizuno, Y. (2014). Recent research progress in and future perspective on treatment of Parkinson's disease. *Integrative Medicine International*, *1*, 67–79.

Moreno, L., Puerta, E., Suárez-Santiago, J. E., Santos-Magalhães, N. S., Ramirez, M. J., & Irache, J. M. (2017). Effect of the oral administration of nanoencapsulated quercetin on a mouse model of Alzheimer's disease. *International Journal of Pharmaceutics*, *517*(1–2), 50–57.

Motaghinejad, M., Motevalian, M., Fatima, S., Faraji, F., & Mozaffari, S. (2017). The neuroprotective effect of curcumin against nicotine-induced neurotoxicity is mediated by CREB–BDNF signaling pathway. *Neurochemical Research*, *42*(10), 2921–2932.

Naoi, M., Shamoto-Nagai, M., & Maruyama, W. (2019). Neuroprotection of multifunctional phytochemicals as novel therapeutic strategy for neurodegenerative disorders: Antiapoptotic and antiamyloidogenic activities by modulation of cellular signal pathways. *Future Neurology*, *14*(1), FNL9.

Naoi, M., Wu, Y., Shamoto-Nagai, M., & Maruyama, W. (2016). Bioactive dietary compounds regulate mitochondrial apoptosis signaling in ambivalent way to function as neuroprotective or antitumor agents. In *Vol. 51. Studies in natural products chemistry* (pp. 201–222). Elsevier.

Naoi, M., Wu, Y., Shamoto-Nagai, M., & Maruyama, W. (2019). Mitochondria in neuroprotection by phytochemicals: Bioactive polyphenols modulate mitochondrial apoptosis system, function and structure. *International Journal of Molecular Sciences, 20*(10), 2451.

Neves, A. R., Lúcio, M., Martins, S., Lima, J. L., & Reis, S. (2013). Novel resveratrol nanodelivery systems based on lipid nanoparticles to enhance its oral bioavailability. *International Journal of Nanomedicine, 8,* 177–187.

Ngoungoure, V. L. N., Schluesener, J., Moundipa, P. F., & Schluesener, H. (2015). Natural polyphenols binding to amyloid: A broad class of compounds to treat different human amyloid diseases. *Molecular Nutrition & Food Research, 59*(1), 8–20.

O'Neill, H. M., Holloway, G. P., & Steinberg, G. R. (2013). AMPK regulation of fatty acid metabolism and mitochondrial biogenesis: Implications for obesity. *Molecular and Cellular Endocrinology, 366*(2), 135–151.

Ock, J., Han, H. S., Hong, S. H., Lee, S. Y., Han, Y. M., Kwon, B. M., et al. (2010). Obovatol attenuates microglia-mediated neuroinflammation by modulating redox regulation. *British Journal of Pharmacology, 159*(8), 1646–1662.

Ohta, K., Mizuno, A., Li, S., Itoh, M., Ueda, M., Ohta, E., et al. (2011). Endoplasmic reticulum stress enhances γ-secretase activity. *Biochemical and Biophysical Research Communications, 416*(3–4), 362–366.

Oliveira, M. R., Nabavi, S. M., Braidy, N., Setzer, W. N., Ahmed, T., & Nabavi, S. F. (2016). Quercetin and the mitochondria: A mechanistic view. *Biotechnology Advances, 34,* 532–549.

Omojate Godstime, C., Enwa Felix, O., Jewo Augustina, O., & Eze Christopher, O. (2014). Mechanisms of antimicrobial actions of phytochemicals against enteric pathogens—A review. *Research Journal of Pharmaceutical, Biological and Chemical Sciences, 2*(2), 77–85.

Oomen, C. A., Farkas, E., Roman, V., van der Beek, E. M., Luiten, P. G., & Meerlo, P. (2009). Resveratrol preserves cerebrovascular density and cognitive function in aging mice. *Frontiers in Aging Neuroscience, 1,* 4.

Ozbal, S., Erbil, G., Koçdor, H., Tuğyan, K., Pekçetin, C., & Ozoğul, C. (2008). The effects of selenium against cerebral ischemia-reperfusion injury in rats. *Neuroscience Letters, 438*(3), 265–269.

Palikaras, K., Lionaki, E., & Tavernarakis, N. (2015). Coupling mitogenesis and mitophagy for longevity. *Autophagy, 11*(8), 1428–1430.

Palle, S., & Neerati, P. (2018). Improved neuroprotective effect of resveratrol nanoparticles as evinced by abrogation of rotenone-induced behavioral deficits and oxidative and mitochondrial dysfunctions in rat model of Parkinson's disease. *Naunyn-Schmiedeberg's Archives of Pharmacology, 391*(4), 445–453.

Pan, T., Kondo, S., Le, W., & Jankovic, J. (2008). The role of autophagy-lysosome pathway in neurodegeneration associated with Parkinson's disease. *Brain: A Journal of Neurology, 131*(Pt 8), 1969–1978.

Pandita, D., Kumar, S., Poonia, N., & Lather, V. (2014). Solid lipid nanoparticles enhance oral bioavailability of resveratrol, a natural polyphenol. *Food Research International, 62,* 1165–1174.

Paolino, D., Vero, A., Cosco, D., Pecora, T. M., Cianciolo, S., Fresta, M., et al. (2016). Improvement of oral bioavailability of curcumin upon microencapsulation with methacrylic copolymers. *Frontiers in Pharmacology, 7,* 485.

Paredes-Gonzalez, X., Fuentes, F., Jeffery, S., Saw, C. L. L., Shu, L., Su, Z. Y., et al. (2015). Induction of NRF2-mediated gene expression by dietary phytochemical flavones apigenin and luteolin. *Biopharmaceutics & Drug Disposition, 36*(7), 440–451.

Park, Y. S., Jang, H. J., Lee, K. H., Hahn, T. R., & Paik, Y. S. (2006). Prolyl endopeptidase inhibitory activity of unsaturated fatty acids. *Journal of Agricultural and Food Chemistry, 54*(4), 1238–1242.

Park, S.-P., & Kwon, S.-H. (2008). Cognitive effects of antiepileptic drugs. *Journal of Clinical Neurology, 4*(3), 99–106.

Park, J. H., Lee, T. K., Ahn, J. H., Shin, B. N., Cho, J. H., Kim, I. H., et al. (2017). Pre-treated *Populus tomentiglandulosa* extract inhibits neuronal loss and alleviates gliosis in the gerbil hippocampal CA1 area induced by transient global cerebral ischemia. *Anatomy & Cell Biology, 50*(4), 284–292.

Patra, A. K. (2012). *Dietary phytochemicals and microbes.* Springer Science & Business Media.

Phukan, J., Elamin, M., Bede, P., Jordan, N., Gallagher, L., Byrne, S., et al. (2012). The syndrome of cognitive impairment in amyotrophic lateral sclerosis: A population-based study. *Journal of Neurology, Neurosurgery, and Psychiatry, 83*(1), 102–108.

Polo-Hernández, E., Tello, V., Arroyo, A. A., Domínguez-Prieto, M., de Castro, F., Tabernero, A., et al. (2014). Oleic acid synthesized by stearoyl-CoA desaturase (SCD-1) in the lateral periventricular zone of the developing rat brain mediates neuronal growth, migration and the arrangement of prospective synapses. *Brain Research, 1570*, 13–25.

Popovic, D., Vucic, D., & Dikic, I. (2014). Ubiquitination in disease pathogenesis and treatment. *Nature Medicine, 20*(11), 1242.

Pulsinelli, W. A., Brierley, J. B., & Plum, F. (1982). Temporal profile of neuronal damage in a model of transient forebrain ischemia. *Annals of Neurology, 11*(5), 491–498.

Qiu, C., Kivipelto, M., & von Strauss, E. (2009). Epidemiology of Alzheimer's disease: Occurrence, determinants, and strategies toward intervention. *Dialogues in Clinical Neuroscience, 11*(2), 111–128.

Rahman, I., & Chung, S. (2010). Dietary polyphenols, deacetylases and chromatin remodeling in inflammation. *Journal of Nutrigenetics and Nutrigenomics, 3*(4–6), 220–230.

Raza, S. S., Khan, M. M., Ahmad, A., Ashafaq, M., Khuwaja, G., Tabassum, R., et al. (2011). Hesperidin ameliorates functional and histological outcome and reduces neuroinflammation in experimental stroke. *Brain Research, 1420*, 93–105.

Raza, S. S., Khan, M. M., Ashafaq, M., Ahmad, A., Khuwaja, G., Khan, A., et al. (2011). Silymarin protects neurons from oxidative stress associated damages in focal cerebral ischemia: A behavioral, biochemical and immunohistological study in Wistar rats. *Journal of the Neurological Sciences, 309*(1–2), 45–54.

Rege, S. D., Geetha, T., Griffin, G. D., Broderick, T. L., & Babu, J. R. (2014). Neuroprotective effects of resveratrol in Alzheimer disease pathology. *Frontiers in Aging Neuroscience, 6*, 218.

Reinke, A. A., & Gestwicki, J. E. (2007). Structure-activity relationships of amyloid beta-aggregation inhibitors based on curcumin: Influence of linker length and flexibility. *Chemical Biology & Drug Design, 70*(3), 206–215.

Reitz, C., & Mayeux, R. (2014). Alzheimer disease: Epidemiology, diagnostic criteria, risk factors and biomarkers. *Biochemical Pharmacology, 88*(4), 640–651.

Rhim, H., Kim, H., Lee, D. Y., Oh, T. H., & Nah, S. Y. (2002). Ginseng and ginsenoside Rg3, a newly identified active ingredient of ginseng, modulate Ca2+ channel currents in rat sensory neurons. *European Journal of Pharmacology, 436*(3), 151–158.

Russell, D. M., Garry, E. M., Taberner, A. J., Barrett, C. J., Paton, J. F., Budgett, D. M., et al. (2012). A fully implantable telemetry system for the chronic monitoring of brain tissue oxygen in freely moving rats. *Journal of Neuroscience Methods, 204*(2), 242–248.

Salehi, B., Calina, D., Docea, A. O., Koirala, N., Aryal, S., Lombardo, D., et al. (2020). Curcumin's nanomedicine formulations for therapeutic application in neurological diseases. *Journal of Clinical Medicine, 9*(2), 430.

Sampson, E. L., Blanchard, M. R., Jones, L., Tookman, A., & King, M. (2009). Dementia in the acute hospital: Prospective cohort study of prevalence and mortality. *The British Journal of Psychiatry: the Journal of Mental Science, 195*(1), 61–66.

Sarkar, S., Davies, J. E., Huang, Z., Tunnacliffe, A., & Rubinsztein, D. C. (2007). Trehalose, a novel mTOR-independent autophagy enhancer, accelerates the clearance of mutant huntingtin and alpha-synuclein. *The Journal of Biological Chemistry*, *282*(8), 5641–5652.

Sasazawa, Y., Sato, N., Umezawa, K., & Simizu, S. (2015). Conophylline protects cells in cellular models of neurodegenerative diseases by inducing mammalian target of rapamycin (mTOR)-independent autophagy. *The Journal of Biological Chemistry*, *290*(10), 6168–6178.

Schapira, A. H., & Olanow, C. W. (2004). Neuroprotection in Parkinson disease: Mysteries, myths, and misconceptions. *JAMA*, *291*(3), 358–364.

Schiborr, C., Kocher, A., Behnam, D., Jandasek, J., Toelstede, S., & Frank, J. (2014). The oral bio-availability of curcumin from micronized powder and liquid micelles is significantly increased in healthy humans and differs between sexes. *Molecular Nutrition & Food Research*, *58*(3), 516–527.

Schulz, J., & Deuschl, G. (2015). Influence of lifestyle on neurodegenerative diseases. *Der Nerve-narzt*, *86*(8), 954–959.

Semenza, G. L., & Wang, G. L. (1992). A nuclear factor induced by hypoxia via de novo protein synthesis binds to the human erythropoietin gene enhancer at a site required for transcriptional activation. *Molecular and Cellular Biology*, *12*(12), 5447–5454.

Sengupta, U., Nilson, A. N., & Kayed, R. (2016). The role of amyloid-β oligomers in toxicity, propagation, and immunotherapy. *eBioMedicine*, *6*, 42–49.

Shal, B., Ding, W., Ali, H., Kim, Y. S., & Khan, S. (2018). Anti-neuroinflammatory potential of natural products in attenuation of Alzheimer's disease. *Frontiers in Pharmacology*, *9*, 548.

Sharma, D. R., Wani, W. Y., Sunkaria, A., Kandimalla, R. J., Sharma, R. K., Verma, D., et al. (2016). Quercetin attenuates neuronal death against aluminum-induced neurodegeneration in the rat hippocampus. *Neuroscience*, *324*, 163–176.

Shi, C., Zhao, L., Zhu, B., Li, Q., Yew, D. T., Yao, Z., et al. (2009). Protective effects of *Ginkgo biloba* extract (EGb761) and its constituents quercetin and ginkgolide B against beta-amyloid peptide-induced toxicity in SH-SY5Y cells. *Chemico-Biological Interactions*, *181*(1), 115–123.

Shinomol, G. K., & Muralidhara. (2008). Prophylactic neuroprotective property of *Centella asiatica* against 3-nitropropionic acid induced oxidative stress and mitochondrial dysfunctions in brain regions of prepubertal mice. *Neurotoxicology*, *29*(6), 948–957.

Shrivastava, P., Vaibhav, K., Tabassum, R., Khan, A., Ishrat, T., Khan, M. M., et al. (2013). Anti-apoptotic and anti-inflammatory effect of Piperine on 6-OHDA induced Parkinson's rat model. *The Journal of Nutritional Biochemistry*, *24*(4), 680–687.

Shukia, B., Khanna, N. K., & Godhwani, J. L. (1987). Effect of Brahmi Rasayan on the central nervous system. *Journal of Ethnopharmacology*, *21*(1), 65–74.

Shukitt-Hale, B., Bielinski, D. F., Lau, F. C., Willis, L. M., Carey, A. N., & Joseph, J. A. (2015). The beneficial effects of berries on cognition, motor behaviour and neuronal function in ageing. *The British Journal of Nutrition*, *114*(10), 1542–1549.

Shukitt-Hale, B., Galli, R. L., Meterko, V., Carey, A., Bielinski, D. F., McGhie, T., et al. (2005). Dietary supplementation with fruit polyphenolics ameliorates age-related deficits in behavior and neuronal markers of inflammation and oxidative stress. *Age (Dordrecht, Netherlands)*, *27*(1), 49–57.

Sikora, E., Scapagnini, G., & Barbagallo, M. (2010). Curcumin, inflammation, ageing and age-related diseases. *Immunity & Ageing*, *7*(1), 1.

Singh, H., & Dhawan, B. (1997). Neuropsychopharmacological effects of the Ayurvedic nootropic *Bacopa monniera* Linn.(Brahmi). *Indian Journal of Pharmacology*, *29*(5), 359.

Singh, S. K., Dutta, A., & Modi, G. (2017). α-Synuclein aggregation modulation: An emerging approach for the treatment of Parkinson's disease. *Future Medicinal Chemistry*, *9*(10), 1039–1053.

Singh, P. K., Kotia, V., Ghosh, D., Mohite, G. M., Kumar, A., & Maji, S. K. (2013). Curcumin modulates α-synuclein aggregation and toxicity. *ACS Chemical Neuroscience, 4*(3), 393–407.

Solanki, I., Parihar, P., & Parihar, M. S. (2016). Neurodegenerative diseases: From available treatments to prospective herbal therapy. *Neurochemistry International, 95,* 100–108.

Somerville, V., Bringans, C., & Braakhuis, A. (2017). Polyphenols and performance: A systematic review and meta-analysis. *Sports Medicine, 47*(8), 1589–1599.

Song, J. X., Lu, J. H., Liu, L. F., Chen, L. L., Durairajan, S. S., Yue, Z., et al. (2014). HMGB1 is involved in autophagy inhibition caused by SNCA/α-synuclein overexpression: A process modulated by the natural autophagy inducer corynoxine B. *Autophagy, 10*(1), 144–154.

Song, J. X., Sun, Y. R., Peluso, I., Zeng, Y., Yu, X., Lu, J. H., et al. (2016). A novel curcumin analog binds to and activates TFEB in vitro and in vivo independent of MTOR inhibition. *Autophagy, 12*(8), 1372–1389.

Suh, D. H., Kim, M.-K., Kim, H. S., Chung, H. H., & Song, Y. S. (2013). Mitochondrial permeability transition pore as a selective target for anti-cancer therapy. *Frontiers in Oncology, 3,* 41.

Sun, D., Li, N., Zhang, W., Zhao, Z., Mou, Z., Huang, D., et al. (2016). Design of PLGA-functionalized quercetin nanoparticles for potential use in Alzheimer's disease. *Colloids and Surfaces. B, Biointerfaces, 148,* 116–129.

Szlachcic, W. J., Switonski, P. M., Krzyzosiak, W. J., Figlerowicz, M., & Figiel, M. (2015). Huntington disease iPSCs show early molecular changes in intracellular signaling, the expression of oxidative stress proteins and the p53 pathway. *Disease Models & Mechanisms, 8*(9), 1047–1057.

Szűcs, L. (2015). *The role of resveratrol treatment and regular physical activity on sirtuins in brain of rats artificially selected for intrinsic aerobic running capacity.* Testnevelési Egyetem.

Takuma, K., Yan, S. S., Stern, D. M., & Yamada, K. (2005). Mitochondrial dysfunction, endoplasmic reticulum stress, and apoptosis in Alzheimer's disease. *Journal of Pharmacological Sciences, 97*(3), 312–316.

Tanner, C. M., Goldman, S. M., Ross, G. W., & Grate, S. J. (2014). The disease intersection of susceptibility and exposure: Chemical exposures and neurodegenerative disease risk. *Alzheimer's & Dementia, 10,* S213–S225.

Trouillas, P., Marsal, P., Svobodová, A., Vostálová, J., Gazák, R., Hrbác, J., et al. (2008). Mechanism of the antioxidant action of silybin and 2,3-dehydrosilybin flavonolignans: A joint experimental and theoretical study. *The Journal of Physical Chemistry. A, 112*(5), 1054–1063.

Tsai, Y. M., Jan, W. C., Chien, C. F., Lee, W. C., Lin, L. C., & Tsai, T. H. (2011). Optimised nanoformulation on the bioavailability of hydrophobic polyphenol, curcumin, in freely-moving rats. *Food Chemistry, 127*(3), 918–925.

Tsai, W. C., Li, W. C., Yin, H. Y., Yu, M. C., & Wen, H. W. (2012). Constructing liposomal nanovesicles of ginseng extract against hydrogen peroxide-induced oxidative damage to L929 cells. *Food Chemistry, 132*(2), 744–751.

Urbain, A., Marston, A., Grilo, L. S., Bravo, J., Purev, O., Purevsuren, B., et al. (2008). Xanthones from *Gentianella amarella ssp. acuta* with acetylcholinesterase and monoamine oxidase inhibitory activities. *Journal of Natural Products, 71*(5), 895–897.

Vaz, G. R., Hädrich, G., Bidone, J., Rodrigues, J. L., Falkembach, M. C., Putaux, J. L., et al. (2017). Development of nasal lipid nanocarriers containing curcumin for brain targeting. *Journal of Alzheimer's Disease, 59*(3), 961–974.

Velmurugan, B. K., Rathinasamy, B., Lohanathan, B. P., Thiyagarajan, V., & Weng, C.-F. (2018). Neuroprotective role of phytochemicals. *Molecules, 23*(10), 2485.

Venkatesan, R., Ji, E., & Kim, S. Y. (2015). Phytochemicals that regulate neurodegenerative disease by targeting neurotrophins: A comprehensive review. *BioMed Research International, 2015,* 814068.

Walker, F. O. (2007). Huntington's disease. *Lancet (London, England), 369*(9557), 218–228.

Walters, A. M., Porter, G. A., Jr., & Brookes, P. S. (2012). Mitochondria as a drug target in ischemic heart disease and cardiomyopathy. *Circulation Research, 111*(9), 1222–1236.

Wang, L., Das, U., Scott, D. A., Tang, Y., McLean, P. J., & Roy, S. (2014). α-synuclein multimers cluster synaptic vesicles and attenuate recycling. *Current Biology, 24*(19), 2319–2326.

Wang, J., Song, Y., Chen, Z., & Leng, S. X. (2018). Connection between systemic inflammation and neuroinflammation underlies neuroprotective mechanism of several phytochemicals in neuro-degenerative diseases. *Oxidative Medicine and Cellular Longevity, 2018.*

Wang, X., Wang, W., Li, L., Perry, G., Lee, H.-G., & Zhu, X. (2014). Oxidative stress and mitochon-drial dysfunction in Alzheimer's disease. *Biochimica et Biophysica Acta (BBA) - Molecular Basis of Disease, 1842*(8), 1240–1247.

Wasner, M., Klier, H., & Borasio, G. D. (2001). The use of alternative medicine by patients with amyotrophic lateral sclerosis. *Journal of the Neurological Sciences, 191*(1–2), 151–154.

Wcislo, G. (2014). Resveratrol inhibitory effects against a malignant tumor: A molecular introduc-tory review. In *Polyphenols in human health and disease* (pp. 1269–1281). Elsevier.

Webber, K. M., Raina, A. K., Marlatt, M. W., Zhu, X., Prat, M. I., Morelli, L., et al. (2005). The cell cycle in Alzheimer disease: A unique target for neuropharmacology. *Mechanisms of Ageing and Development, 126*(10), 1019–1025.

Wilson, W. J., & Poellinger, L. (2002). The dietary flavonoid quercetin modulates HIF-1 alpha activity in endothelial cells. *Biochemical and Biophysical Research Communications, 293*(1), 446–450.

Wong, H. L., Wu, X. Y., & Bendayan, R. (2012). Nanotechnological advances for the delivery of CNS therapeutics. *Advanced Drug Delivery Reviews, 64*, 686–700.

Wong, V. K., Wu, A. G., Wang, J. R., Liu, L., & Law, B. Y. (2015). Neferine attenuates the protein level and toxicity of mutant huntingtin in PC-12 cells via induction of autophagy. *Molecules (Basel, Switzerland), 20*(3), 3496–3514.

Wood dos Santos, T., Cristina Pereira, Q., Teixeira, L., Gambero, A., A Villena, J., & Lima Ribeiro, M. (2018). Effects of polyphenols on thermogenesis and mitochondrial biogenesis. *Interna-tional Journal of Molecular Sciences, 19*(9), 2757.

Wood-Kaczmar, A., Gandhi, S., & Wood, N. (2006). Understanding the molecular causes of Parkin-son's disease. *Trends in Molecular Medicine, 12*(11), 521–528.

Wu, J., Jeong, H. K., Bulin, S. E., Kwon, S. W., Park, J. H., & Bezprozvanny, I. (2009). Ginsen-osides protect striatal neurons in a cellular model of Huntington's disease. *Journal of Neurosci-ence Research, 87*(8), 1904–1912.

Wu, W., Lee, S. Y., Wu, X., Tyler, J. Y., Wang, H., Ouyang, Z., et al. (2014). Neuroprotective ferulic acid (FA)-glycol chitosan (GC) nanoparticles for functional restoration of traumatically injured spinal cord. *Biomaterials, 35*(7), 2355–2364.

Wu, Y., Shamoto-Nagai, M., Maruyama, W., Osawa, T., & Naoi, M. (2017). Phytochemicals pre-vent mitochondrial membrane permeabilization and protect SH-SY5Y cells against apoptosis induced by PK11195, a ligand for outer membrane translocator protein. *Journal of Neural Transmission, 124*(1), 89–98.

Wu, C. R., Tsai, C. W., Chang, S. W., Lin, C. Y., Huang, L. C., & Tsai, C. W. (2015). Carnosic acid protects against 6-hydroxydopamine-induced neurotoxicity in in vivo and in vitro model of Parkinson's dis-ease: Involvement of antioxidative enzymes induction. *Chemico-Biological Interactions, 225*, 40–46.

Wynford-Thomas, R., & Robertson, N. P. (2017). The economic burden of chronic neurological disease. *Journal of Neurology, 264*(11), 2345–2347.

Xie, C. L., Li, J. H., Wang, W. W., Zheng, G. Q., & Wang, L. X. (2015). Neuroprotective effect of ginsenoside-Rg1 on cerebral ischemia/reperfusion injury in rats by downregulating protease-activated receptor-1 expression. *Life Sciences, 121*, 145–151.

Xu, H., Hua, Y., Zhong, J., Li, X., Xu, W., Cai, Y., et al. (2018). Resveratrol delivery by albumin nanoparticles improved neurological function and neuronal damage in transient middle cerebral artery occlusion rats. *Frontiers in Pharmacology, 9*, 1403.

Xu, H., Jiang, H., Wang, J., & Xie, J. (2010). Rg1 protects iron-induced neurotoxicity through antioxidant and iron regulatory proteins in 6-OHDA-treated MES23.5 cells. *Journal of Cellular Biochemistry, 111*(6), 1537–1545.

Xu, W., Liu, J., Ma, D., Yuan, G., Lu, Y., & Yang, Y. (2017). Capsaicin reduces Alzheimer-associated tau changes in the hippocampus of type 2 diabetes rats. *PLoS One, 12*(2), e0172477.

Xu, J., Zhang, H., Yang, F., & Yu, J. X. (2013). Intervention effect of berberine on expressions of TNF-alpha and receptor type I in Abeta25-35-induced inflammatory reaction in SH-SY5Y cell lines. *China Journal of Chinese Materia Medica, 38*(9), 1327–1330.

Yang, L., Hao, J., Zhang, J., Xia, W., Dong, X., Hu, X., et al. (2009). Ginsenoside Rg3 promotes beta-amyloid peptide degradation by enhancing gene expression of neprilysin. *The Journal of Pharmacy and Pharmacology, 61*, 375–380.

Yin, F., Liu, J., Ji, X., Wang, Y., Zidichouski, J., & Zhang, J. (2011). Silibinin: A novel inhibitor of Aβ aggregation. *Neurochemistry International, 58*(3), 399–403.

Youdim, K. A., Shukitt-Hale, B., & Joseph, J. A. (2004). Flavonoids and the brain: Interactions at the blood-brain barrier and their physiological effects on the central nervous system. *Free Radical Biology & Medicine, 37*, 1683–1693.

Youle, R. J., & Van Der Bliek, A. M. (2012). Mitochondrial fission, fusion, and stress. *Science, 337*(6098), 1062–1065.

Yousuf, S., Atif, F., Ahmad, M., Hoda, N., Ishrat, T., Khan, B., et al. (2009). Resveratrol exerts its neuroprotective effect by modulating mitochondrial dysfunctions and associated cell death during cerebral ischemia. *Brain Research, 1250*, 242–253.

Zbarsky, V., Datla, K. P., Parkar, S., Rai, D. K., Aruoma, O. I., & Dexter, D. T. (2005). Neuroprotective properties of the natural phenolic antioxidants curcumin and naringenin but not quercetin and fisetin in a 6-OHDA model of Parkinson's disease. *Free Radical Research, 39*(10), 1119–1125.

Zecca, L., Shima, T., Stroppolo, A., Goj, C., Battiston, G. A., Gerbasi, R., et al. (1996). Interaction of neuromelanin and iron in substantia nigra and other areas of human brain. *Neuroscience, 73*(2), 407–415.

Zhang, Y.-J., Gan, R.-Y., Li, S., Zhou, Y., Li, A.-N., Xu, D.-P., et al. (2015). Antioxidant phytochemicals for the prevention and treatment of chronic diseases. *Molecules, 20*(12), 21138–21156.

Zhang, J., Han, X., Li, X., Luo, Y., Zhao, H., Yang, M., et al. (2012). Core-shell hybrid liposomal vesicles loaded with panax notoginsenoside: Preparation, characterization and protective effects on global cerebral ischemia/reperfusion injury and acute myocardial ischemia in rats. *International Journal of Nanomedicine, 7*, 4299–4310.

Zhang, F., Wang, H., Wu, Q., Lu, Y., Nie, J., Xie, X., et al. (2013). Resveratrol protects cortical neurons against microglia-mediated neuroinflammation. *Phytotherapy Research, 27*(3), 344–349.

Zhang, N., Yan, F., Liang, X., Wu, M., Shen, Y., Chen, M., et al. (2018). Localized delivery of curcumin into brain with polysorbate 80-modified cerasomes by ultrasound-targeted microbubble destruction for improved Parkinson's disease therapy. *Theranostics, 8*(8), 2264–2277.

Zhao, L., Du, J., Duan, Y., Zang, Y., Zhang, H., Yang, C., et al. (2012). Curcumin loaded mixed micelles composed of Pluronic P123 and F68: Preparation, optimization and in vitro characterization. *Colloids and Surfaces. B, Biointerfaces, 97*, 101–108.

Zhou, Y., Li, H. Q., Lu, L., Fu, D. L., Liu, A. J., Li, J. H., et al. (2014). Ginsenoside Rg1 provides neuroprotection against blood brain barrier disruption and neurological injury in a rat model of cerebral ischemia/reperfusion through downregulation of aquaporin 4 expression. *Phytomedicine: International Journal of Phytotherapy and Phytopharmacology, 21*(7), 998–1003.

Phytomedicine: A treasure of pharmacologically active products from plants

Zhu, F., Wu, F., Ma, Y., Liu, G., Li, Z., Sun, Y., et al. (2011). Decrease in the production of β-amyloid by berberine inhibition of the expression of β-secretase in HEK293 cells. *BMC Neuroscience, 12*, 125.

Zuo, D., Lin, L., Liu, Y., Wang, C., Xu, J., Sun, F., et al. (2016). Baicalin attenuates ketamine-induced neurotoxicity in the developing rats: Involvement of PI3K/Akt and CREB/BDNF/Bcl-2 pathways. *Neurotoxicity Research, 30*(2), 159–172.

Chapter 22

Hepatotoxicity: Its physiological pathways and control measures using phyto-polyphenols

Rajesh Kumar[a], Raksha Rani[a], Sanjay Kumar Narang[b], Seema Rai[c], and Younis Ahmad Hajam[a]

[a]*Department of Biosciences, Division Zoology, School of Basic and Applied Sciences, Career Point University, Hamirpur, Himachal Pradesh, India,* [b]*S.V.G. College Ghumarwin, Bilaspur, Himachal Pradesh, India,* [c]*Department of Zoology, Guru Ghasidas Vishwavidyalaya, Bilaspur, Chhattisgarh, India*

1. Introduction

Liver is one of the major organ involved in various metabolic pathways, especially energy production inside the body of animals. It stores glucose in the form of glycogen and also utilizes carbohydrates for synthesis of cholesterol. Metabolically, it is the most active organ and plays major role in regulating metabolic processes, namely, (i) metabolizing and detoxifying toxins, (ii) releasing glucose from glycogen and maintaining glycemic levels during fasting through glycogenolysis and gluconeogenesis, and (iii) regulating acetyl-CoA for the synthesis of fatty acids and cholesterol as it absorbs fat (triglycerides) and cholesterol from the diet.

Liver plays a major role in the conversion of extra fatty acids into ketone bodies, which provides energy to body parts during fasting/starvation. Thus it contributes to maintaining metabolic homeostasis inside the body of animals (Chang, 2014). It gets affected by high doses of paracetamol; thioacetamide (TAA); hepatitis A, B and C; viruses; carbon tetrachloride (CCl4); and certain chemotherapeutic agents (Saleem et al., 2010). Liver is the primary site for the synthesis of various proteins like albumins and also carries fatty acids, amino acids, vitamins and drugs. Therefore it helps in maintaining osmotic pressures for coagulation of blood and production of immunoglobulins to be used in defense system of the body. It also synthesizes and degrades heme prosthetic groups in mitochondrial cytochromes, protoporphyrin IX in hemoglobins, and microsomal cytochrome P450 (Chang, 2014). The filtration of complete blood is carried out by the liver first, and then it is passed to various body parts. The

Phytomedicine: A Treasure of Pharmacologically Active Products from Plants
https://doi.org/10.1016/B978-0-12-824109-7.00007-8

liver also synthesizes many proteins and associated components for supporting immune system (Dey, Saha, & Sen, 2013).

No other body organ can compensate all these important functions; that's why liver is the most important body organ. Out of the total population, about 10% is affected worldwide with liver ailments including alcoholic steatosis, hepatitis, fibrosis, hepatocellular carcinoma, and liver cirrhosis. Nowadays, liver ailments are becoming major public health concern causing severe morbidity and mortality (Zhang et al., 2013). Liver cells may get damaged by alcohol-based drinks, various metabolites, and other xenobiotics. All these components release AST and ALT into the bloodstream. These cause hepatitis, jaundice, increased bilirubin, increased clotting time, and liver encephalopathy, which are considered main symptoms of liver disorders (Chang, 2014). There are very few medicines composed of corticosteroids, and immunosuppressive agents that cause several serious adverse effects are the only drugs available for the treatment of liver ailments. The researchers are still struggling to develop safe medication against liver injuries.

2. Liver as main metabolic organ

Liver is one of the important organ that maintains biochemical equilibrium inside the animal body, especially lipid and glucose. The first and foremost function of the liver is detoxification caused by drugs. Glycolysis that leads to glucose oxidation and energy production takes place in the liver. Glycogen is synthesized with the help of excess glucose, and energy is released through glycogenolysis. These activities take place inside the liver. The production of bile salts also takes place inside the liver with the help of cholesterol. Free fatty acids are released after hydrolysis of triglycerides inside adipose tissues. These are transported to the liver for beta-oxidation. Metabolic homeostasis is maintained by the liver, which involves pathways for synthesis of glucose and lipid. Liver toxification or imbalance in metabolic homeostasis may lead to chronic liver diseases, fatty liver disease, FLD, insulin resistance, diabetes, and obesity (Chang, 2014) (Fig. 1).

3. Hepatotoxic effects of some plant extracts/metabolites

Hepatotoxicity refers to liver damage or injury caused by some chemical formulations. These formulations are generally called hepatotoxins. Apart from this, overdoses of certain drugs may have adverse effects on the liver. Many developed countries have banned number of drugs due to their adverse implications on the liver (Singh, Kumar, Rana, & Sharma, 2012). Nowadays, liver diseases and disorders have become worldwide problem as there are no any reliable protective drugs available in allopathy (Gurusamy, Kokilavani, Arumugasamy, & Sowmia, 2009). It has now become fifth most vulnerable problem leading to death around the globe (Williams, 2006).

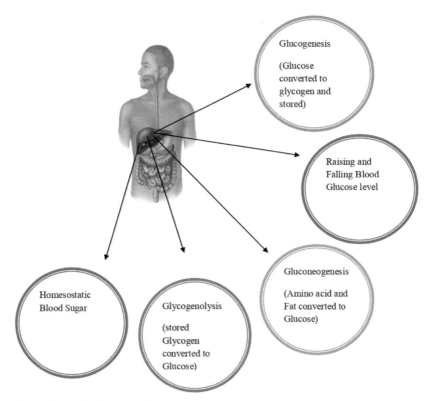

FIG. 1 Metabolic functions of liver.

3.1   *Lantana camara* Linn.

Lantana camara L., commonly known as "Phulnoo," is a noxious angiosperm weed distributed in tropical and subtropical regions worldwide that belongs to the class Lamiales and the family Verbenaceae (Garcia et al., 2010).

Chemical constituent(s): Lantadenes A and B are the main components of *L. camara* that are responsible for causing liver toxicity (Singh, Bhat, & Singh, 2003; Singh, Sharma, Dawra, Kanwar, & Mahato, 1999).

3.1.1   *Hepatotoxicity by Lantadenes A and B: Possible causes*

The lantana toxins (lantadene A and lantadene B) after assimilation are carried to the liver.

How lantadenes A and B get assimilated is still unknown. These toxins however, resemble cholesterol, and its cholesterol absorption is facilitated by esterification with an enzyme called cholesterol esterase. Lantana toxins are used to cause injury using bile canalicular membrane as primary site. In a study on guinea pig, it was reported that lantadenes could not be detected in any of part, that is, liver, gallbladder, or blood and urine samples. It was reported that lantadenes may be detected in the lower part of gastrointestinal tract and in fecal matter. Photosensitization is caused by lantana poisoning during intrahepatic cholestasis (Sharma, Makkar, & Dawra, 1983; Sharma, Makkar, Dawra, & Negi, 1981; Sharma, Makkar, Pal, & Negi, 1980; Singh, Bhat, & Sharma, 2011)

3.2 Cycas revoluta

Cycas revoluta, commonly known as sago palm (bonsai plant), is an extremely poisonous gymnosperm that belongs to the family Cycadaceae and the order Cycadales. It can grow in both temperate and tropical areas.

Chemical constituent(s): The extract of *C. revoluta* leaves is composed of alkaloids, steroids, and tannins and sugars. Seeds contain cycasin, which is highly toxic along with beta-methylamino-L-alanine, a neurotoxic amino acid.

3.2.1 Cycasin as potent hepatotoxic agent

Cycas revoluta is poisonous to almost all types of animals. All parts of this plant are toxic; however, seeds contain maximum cycasin component, a potent toxin. Chemically, it is methylazoxymethanol-β-D-glucoside (Morgan & Hoffmann, 1983). The cycasin may cause gastrointestinal exasperation, which may lead to liver damage, if consumed in higher doses (Bigoniya, Shukla, & Singh, 2010; Nishida, Kobayashi, & Nagahama, 1955; Nishida, Kobayashi, & Nagahama, 1956; Zarchini, Hashemabadi, Kaviani, Fallahabadi, & Negahdar, 2011) (Fig. 2).

3.3 Blighia sapida

Blighia sapida is an evergreen tree that is native to West Africa and belongs to the family Sapindaceae. It is commonly called Ackee. Jamaica has declared it as its national food. All the plant parts have been reported to be toxic except fruits that are luscious, and many dishes can be prepared using fruits as main ingredient.

Chemical constituent(s): *B. sapida* has two main toxins (soapberry): hypoglycins A and B. *major* concentrations of these toxic components are present in inedible parts of fruit.

3.3.1 Hepatotoxic effect of hypoglycins A and B

The inedible portions of fruits contain hypoglycins A and B. Hypoglycin A is available in seeds and in aril part, whereas hypoglycin B can be extracted from seeds only. Methylene cyclopropyl acetic acid is produced from hypoglycin that inhibits majority of enzymes that are involved in the dissociation of acyl-CoA. Hypoglycin inhibits β-oxidation of fatty acids by binding to coenzyme A, carnitine, and carnitine acyltransferases I and II. Inhibition of β-oxidation causes

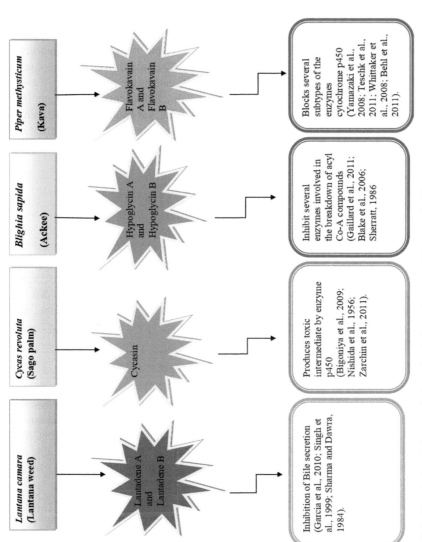

FIG. 2 Description of plant metabolites, chemical constituents, and hepatotoxic effects.

hypoglycemia due to depletion of glucose leading to Jamaican vomiting sickness (Blake, Bennink, & Jackson, 2006; Gaillard et al., 2011; Sherratt, 1986).

3.4 *Piper methysticum* (Kava)

Piper methysticum is native to the Pacific Island, commonly called kava, that belongs to the order Piperales and the family Piperaceae. Some of plant parts like roots are of immense medicinal importance, but intake of plant extract causes adverse health risks including liver injury. The kava thrives well in loose, well-drained soils where plenty of air and water reaches the roots.

Chemical constituent(s): Eighteen bioactive components are present in the Kava; however, only few of them like kavain, yangonin, dihydrokavain, and methysticin are mainly accountable for major pharmacological activity. Flavokavain A, flavokavain B, and pipermethystine are the toxic alkaloids present in the plant that cause liver toxicity and other associated ailments.

3.4.1 Role of Flavokavain B in causing Liver toxicity

Flavokavain B that is found in the rhizome of plant contributes to causing toxic effects. It has also been reported that that some kava pyrones are capable of blocking several cytochrome P450 subtypes. Pipermethystine is also reported to be potent source of liver toxicity in humans. It is also supposed to increase cytochrome P450 activity and is responsible for the activation of potent toxins such as aflatoxins and others that cause carcinogenic conditions (Behl et al., 2011; Lüde et al., 2008; Teschke, Genthner, & Wolff, 2009; Teschke, Qiu, Xuan, & Lebot, 2011; Whittaker et al., 2008; Yamasaki et al., 2008) and may cause liver injury in certain cases.

4. Detoxification in liver

The pathways used by liver for detoxification are categorized into two phases, that is, phase I and phase II. Phase I covers P450 enzymes, and phase II covers UDP-glucuronosyltransferases, amino acid transferases, glutathione-*S*-transferases, *N*-acetyltransferases, and methyltransferases.

4.1 Phase I enzymes

4.1.1 Cytochrome P450 enzymes

Phase I is addition of oxygen that creates reactive site on the surface of toxic compound, and phase II is described as conjugation or adding water-soluble group to the reactive site (Danielson, 2002; Ullrich, 1979). Cytochrome P450 is the microsomal membrane-bound proteins present in the liver. However, sometimes, it may be found in enterocytes and kidneys, and lungs and sometimes even in the brain. This may be accountable for the oxidation and reduction of various substrates (Danielson, 2002; Paine, 1981). P450 enzymes execute explicit tasks by adding a functional group (Danielson, 2002). Ability of any substance to metabolize 90% of drugs in use is solely based on genetic expression of these enzymes (Chen, Zhang, Wang, & Wei, 2011). Variations in the expression of individual enzymes are the result of CYP450 genes' genetic polymorphism.

4.1.2 CYP1A enzymes

CYP1A is associated with the metabolization of procarcinogen and hormones. It has immense role in carcinogenic bioactivation of polycyclic aromatic hydrocarbons (James, Sacco, & Faux, 2008); Ma & Lu, 2007). Low activity of CYP1A2 may lead to higher risk of testicular cancer (Vistisen et al., 2004). Excessive activity of CYP1A enzymes may increase the effects of procarcinogens (Božina, Bradamante, & Lovrić, 2009). These may be implicated in the production of estrogen metabolites (Tsuchiya, Nakajima, & Yokoi, 2005). CYP1 activity is altered by various food and phytonutrients. Cruciferous vegetables might act as inducers of CYP1A1 and 1A2 in humans (Hakooz & Hamdan, 2007; (Horn, Reichert, Bliss, & Malejka-Giganti, 2002); Tsuchiya et al., 2005). On the other hand, berries and their constituent components may reduce the activity of CYP1A1 (Aiyer & Gupta, 2010). Cruciferous vegetables are known for their stronger induction of CYP1A (Horn et al., 2002; (Lord, Bongiovanni, & Bralley, 2002); Walters et al., 2004). There are various foods that may induce or inhibit CYP1 enzymes. Curcumin (0.1%) in diet may induce Cyp1A1 (Bansal et al., 2014); however, 1% turmeric in diet was found to be inhibiting the activity (Thapliyal & Maru, 2001). Black tea (54 mL/day) has been reported to induce CYP1A1 and CYP1A2 (Yao et al., 2014). 100 mg/kg soybean in diet significantly upregulates CYP1A1 activity (Bogacz et al., 2014), whereas 1 g/kg black soybean extract and 200 mg daidzein twice a day inhibit CYP1A1 (Peng & Zhou, 2003; Zhang et al., 2013). More in-depth work is required to validate the effect of doses and their impacts on humans.

4.1.3 CYP2A-E enzymes

CYP2 enzymes are mainly associated with drug metabolism, hormones, and xenobiotics. These are used to metabolize various endogenous compounds like glycerol, ketones, and fatty acids (Danielson, 2002; Nissar et al., 2013). CYP2D polymorphisms have also been reported to contribute in Parkinson's disease (Danielson, 2002). There is strong clinical evidence that quercetin and broccoli induce CYP2A6 (Chen et al., 2009; Hakooz & Hamdan, 2007). It has been reported that chicory is used to induce CYP2A enzymes among animals (Rasmussen, Brunius, Zamaratskaia, & Ekstrand, 2012), whereas CYP2B activity is upregulated by rosemary and garlic (Li et al., 2006). CYP2D6 inhibition has been reported using resveratrol and garden cress (Chow et al., 2010). CYP2E1 enzymes also play role in various diseases. The dietary substances like green tea, black tea, and other vegetables of crucifer group have been reported to inhibit CYP2 enzyme activity. CYP2E1 is used to regulate metabolism related to bioactivating procarcinogenic nitrosamines and aflatoxin B1 (Danielson, 2002; Sheweita, 2000). It also triggers the production of free radicals (Danielson, 2002). CYP2E1 polymorphisms have also been reported to involve in coronary diseases and gastric cancer (González, Sala, & Capellá, 2002). Oxidative stress generated by CYP2E1 may cause impairment in insulin activity by suppressing GLUT4 expression (Armoni, Harel, Ramdas, & Karnieli, 2014). In humans, garlic and watercress have been reported to inhibit CYP2E1 enzymes. It has also been revealed that CYP2E1 activity is downregulated by various substances present in green tea, ellagic acid, black tea, chrysin, and dandelion (Celik et al., 2013; Lieber et al., 2007; Nissar et al., 2013; Park, Kweon, & Choi, 2002; Tahir & Sultana, 2011; Yao et al., 2014).

4.1.4 CYP3A enzymes

CYP3A enzymes are family of enzymes that are tissue specific (Danielson, 2002). Dietary materials like garlic, fish oil, and Rooibos tea are used to activate CYP3A and its isoform (Chen et al., 2003; Davenport & Wargovich, 2005; Wu et al., 2002). Foods like green tea, quercetin, and black tea causes inhibition of CYP3A (Misaka et al., 2013; Park et al., 2009). CYP3A4 enzymes have big role in pharma drugs; therefore a lot of research work is going on in this area (Liu, Tawa, & Wallqvist, 2013). Tanaka et al. (2013) have reported grapefruit juice as potent inhibitor of CYP3A (Tanaka et al., 2013). High doses of resveratrol and garden cress also used to inhibit these enzymes in humans (Al-Jenoobi et al., 2014; Chow et al., 2010). Curcumin may regulate 3A4 activity. Davenport and Wargovich (2005) have observed anticarcinogenic effects using lower doses of garlic organosulfur compounds, whereas allyl sulfides when administered in higher doses (200 mg/kg) lead to minor liver toxicity (Davenport & Wargovich, 2005). It has been observed that sulforaphanes inhibit the activity, while as indole-3-carbinol increases the activity; however, both are components of cruciferous vegetables (Yoxall et al., 2005).

4.1.5 CYP4 enzymes

The CYP4 family of enzymes plays trifling role in drug metabolism; therefore very less is known about these enzymes. The hypolipidemic drugs are known to induce these enzymes (Danielson, 2002). Bioactivation of carcinogenic and pneumotoxic compounds and MCT metabolism is regulated by CYP4B1 isoform (Baer & Rettie, 2006). Polymorphisms and overexpression of this isoform may cause bladder cancer (Danielson, 2002) and colitis (Ye et al., 2009). Caffeic acid has been reported to promote CYP4B1 activity that is supposed to cause decreased inflammation that further strengthens the bond between colitis and CYP4B1 activity (Ye et al., 2009). Another report suggests enhanced activity of CYP4A1 using green tea (Bu-Abbas, Clifford, Walker, & Ioannides, 1994; Strassburg et al., 2000). More significant research work is required to be carried out to validate the influence of food on activity of this family of enzymes.

4.2 Phase II enzymes

4.2.1 Conjugation enzymes

Conjugation is shifting of hydrophilic compounds through enzymes like glutathione by glutathione transferases, glucuronic acid via glucuronyl transferases, amino acids with amino acid transferases, and methyl group through N- and O-methyltransferases (Xu, Li, & Kong, 2005). The function of these enzymes is affected by genetic polymorphisms and has prospective insinuation on variety of cancers (Ginsberg et al., 2010). Intonation of these enzymes through food-based compounds may prove beneficial in the people having genetic polymorphisms due to altered enzyme activity. It is also advantageous for those patients that are having toxic load caused by pollution in environment or imbalance in hormones. It has been recommended by James et al. (2008) that pollutants in environment may be eliminated by either glucuronidation or sulfonation carried out with the help of bioactive compounds (James et al., 2008).

4.2.2 UDP-glucuronosyltransferases

UDP-Glucuronosyltransferases involve proteins that are crucial for elimination of toxins present in fecal matter and urine and bilirubin (Chang, 2014).

The main function of these enzymes is to break bondage between glucuronic acid and UDP-glucuronic acid. This process is called glucuronidation (Rowland, Miners, & Mackenzie, 2013). It takes place mainly in the liver but may occur at some other locations like the small intestine (Rowland et al., 2013). Bilirubin is released into the intestine after its conjugation with UGT1A1 in hepatocytes (Wells et al., 2004). This family of enzymes is of utmost importance as 40%–70% administered in humans undergoes glucuronidation (Wells et al., 2004).

It has been clinically proved that UGT enzymes are induced by various bioactive components and food ingredients, mainly citrus and vegetables of the Cruciferae family ((Chow et al., 2010); Saracino et al., 2009; Walters et al., 2004). Animal studies suggest the potential for foods like curcumin, green tea, black tea, and soybean to enhance UGT activity (Debersac, Vernevaut, Amiot, Suschetet, & Siess, 2001; Gradelet, Leclerc, Siess, & Astorg, 1996; Maliakal & Wanwimolruk, 2001; Marnewick, Joubert, Swart, van der Westhuizen, & Gelderblom, 2003; Van der Logt, Roelofs, Nagengast, & Peters, 2003). It has been reported by various workers that the effect of enzymes varies as per gender and genotype (Chang, 2014; Navarro et al., 2009; Saracino et al., 2009). The observations may be vague; however, results in some of the dietary studies may not have a potential effect on UGT enzyme when compared with that of a diet not composed of fruits and vegetables (Chang, 2014). Inhibition of UGT enzyme inhibition activity proves a tool for hormonal modulation and the risk of prostate cancer (Jenkinson, Petroczi, & Naughton, 2013). A study carried out by (Jenkinson et al., 2013) revealed that foods like green tea and black tea inhibit UGT activity; still, in-depth research is required to validate the clinical effects.

4.2.3 Sulfotransferases

Sulfotransferases are the enzymes accountable for transferring the sulfuryl group to hydroxyl or amine groups. It is called sulfation and takes place in the liver and other organs (James & Ambadapadi, 2013). However, technically, sulfation is called sulfonation. Environment and chemicals may disturb thyroid or reproductive hormone level due to decrease in activity of enzymes (Kodama & Negishi, 2013; Wang & James, 2006).

4.2.4 Glutathione S-transferases

Glutathione S-transferase is a category of enzymes that accelerate the attachment of glutathione group to a biotransformed metabolite. These enzymes are produced through gene transcription and xenobiotic-responsive element (Hayes & Pulford, 1995). The dietary components and available nutrients have been reported to increase the synthesis of vitamin B6, glutathione, and magnesium (Galluzzi et al., 2012; (Howard, Davies, & Hunnisett, 1994)). The ingredients of turmeric and milk thistle and folic acid are used to wash out GSH in humans (Ansar, Mazloom, Kazemi, & Hejazi, 2011); Child et al., 2004; Kalpravidh et al., 2010; Lucena et al., 2002). The glutathione may be regulated through dietary supplements as reported by El Morsy and Kamel (2015) that vegetables or products related to the Cruciferae family may have glutathione shielding effects.

4.2.5 Amino acid transferases

It has been reported that protein-enriched diets are essential for detoxifying procedure to provide suitable substrate to amino acid transferases. Also, different amino acids like glycine may be used to attach the molecule for the purpose of excretion.

4.2.6 N-Acetyltransferases (NAT)

N-Acetyltransferases regulates the shifting of acetyl group that helps in the conversion of aromatic amines to amides and hydrazines to hydrazides. This mechanism supports the person who takes medications like hydralazine and isoniazid (Jancova, Anzenbacher, & Anzenbacherova, 2010). The genetic polymorphism in N-acetyltransferase enzymes may be involved in liver toxicity in some of the patients (Makarova, 2008). It has been reported by Chen et al. (2009) that administration of quercetin (500 mg once a day) increases N-acetyltransferase activity. More in-depth study is required to establish the relation between diet-based nutrients and N-acetyltransferase activity.

4.2.7 Methyltransferases

This class of enzymes is less significant among phase II enzymes in the medical community.

The methionine group from S-adenosylmethionine (SAM) is the main conjugating donor of methyltransferase (Kohalmy & Vrzal, 2011). Catechol-O-methyltransferase (COMT) is the major enzyme among methyltransferases that is essential for hormonal detoxication (Yager, 2015). Diet has profound effect on this family of enzymes as revealed that diet rich in sucrose may inhibit activity of methylation enzymes (Busserolles, Zimowska, Rock, Rayssiguier, & Mazur, 2002).

5. Paracetamol-induced liver injury

Paracetamol, which is chemically known as acetaminophen (APAP), is a secure and effective drug used in the treatment of fever, pain, and inflammation. It may be used by the people of all age groups even without doctor's prescription. Overdose of this medicine may result in acute liver failure. Toxicity of the liver induced by acetaminophen may be controlled by using clinically approved drug called *N*-acetylcysteine that regulates GSH level in the liver. It is now well-established fact that plant extracts/metabolites or herbal products, which are in use for the treatment of liver ailments, produce significant results. It has also been revealed that plant metabolites/phytochemicals alleviate cell death in localized areas and provide safety against acetaminophen-mediated liver diseases by limiting oxidative stress, regulating defense mechanism, etc. CYP450 is a heme-containing monooxygenase that helps in acetaminophen metabolism and plays significant role in therapeutics of metabolic modulation (Gonzalez, 2005).

6. Attenuation of hepatotoxicity using natural phytochemicals

A lot of research work has been carried out using plant products/metabolites. (Rao, Manjunath, Bhagawati, and Thippeswamy (2014)) have reported liver protective properties in chemical-induced hepatotoxicity using natural products. These plant products and drugs formulated with the help of natural plant-based ingredients have proven their abilities for safeguarding liver and associated diseases by regulating oxidative stress and decreasing localized cell mortality (Mitchell, Jollow, Potter, Gillette, & Brodie, 1973). A summary of these phytochemicals has been illustrated in Table 1.

The table depicts the role of phytochemicals against acetaminophen-induced hepatotoxicity.

6.1 Methoxypsoralen

Methoxypsoralen is a plant-based compound extracted especially from citrus fruits. Its chemical name is 4-methoxy-7H-furo[3,2-*g*]chromen-7-one as shown in Fig. 3.

Methoxypsoralen is well known for its therapeutic use in the treatment of vitiligo and psoriasis. It is used to improve liver condition by restoring hepatic enzymes and decreasing penetration of inflammatory cells (Liu et al., 2013).

TABLE 1 Summary of plant phytochemicals and their source and therapeutics involved.

Sr. no.	Phytochemicals	Source of phytochemicals	Therapeutic potential	Parameters	References
1.	Methoxypsoralen (5-MOP)	Obtained from essential oil of bergamot and citrus fruit including grapefruits	Improves liver function by decreasing necrosis and decreasing infiltration of inflammatory cells	APAP ↑ ALT, AST and LDH and 5-MOP ↓ ALT, AST and LDH levels, MDA level and ↑ GSH/GSSG ratio	Lee and Wu (2005)
2.	Ajoene	Isolated from processed garlic	Inhibits thiol and GSH depletion and restore the liver enzymes	APAP ↑ GPT level and Ajoene ↓ GPT	Hattori, Yamada, Nishikawa, Fukuda, and Fujino (2001); Karmen, Wróblewski, and LaDue (1955)
3.	Curcumin (CMN)	Obtained from the rhizomes of *Curcuma longa*	Attenuated liver toxicity by improving antioxidant and restoring liver enzymes	CMN ↑ Bcl-2/Bax ratio; thus ↓ APAP-induced apoptosis	Kheradpezhouh et al. (2010); Yousef, Omar, El-Guendi, and Abdelmegid (2010); Li et al. (2013)
4.	Arjunolic acid (AA)	Obtained from the bark of *Terminalia arjuna*	Prevents GSH depletion and APAP metabolite formation	AA ↓ APAP-induced CYP2E1 activity, ↓ JNK phosphorylation, Bcl-2 and Bcl-xL	Ghosh, Das, Manna, and Sil (2010)

#	Compound	Source	Effect	Mechanism	Reference
5.	Berberine	Obtained from *Berberis aristata*	Reduces mortality, restores liver enzymes, and inhibits the inflammasome components	BBR ↑ SOD, GPX, CAT, and GSSH and ↓ MDA	Vivoli et al. (2016); Almani, Qureshi, Shaikh, Uqaili, and Khoharo (2017)
6.	Bixin	Obtained from the seeds of *Bixa orellana*	Prevents DNA damage and lipid peroxidation	Bixin ↓ leukocyte infiltration and TNF-α levels in the liver and inhibition of ERK signaling pathway	Rao et al. (2014); Lee, Yu, Liao, Chou, and Liu (2019)
7.	Caffeic acid	Obtained from fruits, coffee, and honey	↓ Expression of growth arrest and DNA damage-inducible protein	↓ Myeloperoxidase (MPO) ↑Nrf2, HO-1 and NAD NQO1 and ↓ expression of Keap1	Pang et al. (2016)

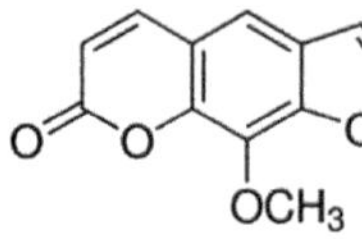

FIG. 3 Chemical structure of methoxypsoralen.

The influence of 5-methoxypsoralen was studied in liver cell injuries and reported to be a potent hepatoprotective agent (Acharya & Lau-Cam, 2010). Like all other drugs, 5-methoxypsoralen in higher doses may cause liver injury. As 5-methoxypsoralen decreases serum level in dose-dependent manner, it may prove helpful in severe acetaminophen-induced liver toxicity. 5-Methoxypsoralen is also reported to improve stress induced by paracetamol by decreasing MDA and increasing GSH/GSSG ratio. Along with antioxidative property, 5-methoxypsoralen also has biological activity that inhibits CYPs as experimented in both human and mouse (Lee & Wu, 2005). Liver toxicity as induced by acetaminophen may be treated by 5-methoxypsoralen by inhibiting CYPs that do not have adverse effect on the liver when administered in balanced doses.

6.2 Ajoene

Ajoene is extracted from garlic and is an allyl sulfur compound. The chemical name of ajoene is [(E)-1-(prop-2-enyldisulfanyl)-3-prop-2-enylsulfinylprop-1-ene)] (Fig. 4) (Kaschula et al., 2012). Ajoene inhibits thiol and GSH depletion from the liver and also refurbishes the enzymes in the liver in which toxicity is induced by APAP (Hattori et al., 2001).

The APAP treatment (300 mg/kg) induces hepatic injury in mice. When the liver cells get damaged, glutamic pyruvic transaminase (GPT) is released from the damaged cells and enters the serum, resulting in the serum GPT level being increased. The GPT activity in the serum may be calculated as per (Karmen et al.'s (1955) method using a GPT-UV test kit. The ajoene-pretreated group showed reduction of serum GPT activity (dose dependent), and it was found that ajoene at 50 and 100 mg/kg was most effective in treating APAP-induced

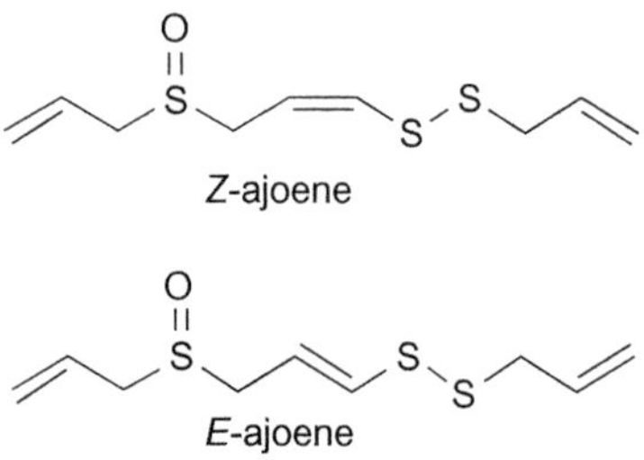

FIG. 4 Chemical structure of ajoene.

liver toxicity in mice model (Hattori et al., 2001). Overdoses of APAP may lead to glucuronidation saturation by forming N-acetyl-p-benzoquinone imine (NAPQI) that further lowers hepatic GSH level (Potter & Hinson, 1986). To elucidate the mechanism for liver protection by ajoene, GSH can be calculated as per Griffith (Griffith, 1980).

Administration of ajoene (0, 20, 50, and 100 mg/kg) to APAP (300 mg/kg; 4 different groups) induced hepatic injury in rats, GSH level was found increased in dose-dependent manner (Jollow et al., 1973). The group treated with a high dosage (100 mg/kg) maintained its hepatic GSH level at 80% compared with that of control group. These results show that the pretreatment with ajoene suppressed the decrease in hepatic GSH level. Total hepatic protein thiol content in the liver has been estimated just before and after 6 hours of APAP administration by following the method of Sedlak and Lindsay (1968). It was observed that hepatic protein thiol content in the APAP-treated group decreases to 82% in comparsion to normal group after 6 h, whereas the ajoene pretreatment (50 mg/kg) maintained the hepatic protein thiol content at the normal level 6 h after the APAP administration. These results revealed that pretreatment with ajoene might suppress the decrease in the hepatic protein and maintain it at the normal level dose-dependently. It is obvious from these results that ajoene proves to be strong against acetaminophen-induced liver toxicity and its properties get interpreted by regulating hepatic GSH level and protein thiol content (Hattori et al., 2001).

6.3 Curcumin

Curcumin is the major bioactive component present in *Curcuma longa*, which is generally known as turmeric. It is yellow pigment and has immense importance as a part of diet, food additive, and other cosmetic preparations. Its chemical name is (1E,6E)-1,7-bis(4-hydroxy-3-methoxyphenyl)hepta-1,6-diene-3,5-dione (Fig. 5). The potential use of curcumin in liver toxicity, inflammation, oxidative stress, and cell death has been proved and reported by various researchers. In a study carried out by Donatus and Vermeulen (1990), the medicinal properties of curcumin were proved in rat liver cells. LDH and GSH were not found to be depleted when curcumin was administered in lower doses; however, hepatoprotection was observed in case of higher doses. The protective effects were also reported in rats (in vivo) where curcumin restored liver enzymes and improved antioxidants and salvage pathway to recover liver and kidney-associated toxicity (Kheradpezhouh et al., 2010); (Yousef et al., 2010). Another study proved curcumin a potential agent in maintaining genome stability, decreasing cell mortality, and improving level of oxidative stress in the liver. Curcumin also works against apoptosis, cytokine inflammation, and DNA fragmentation and restores liver enzymes (Bulku et al., 2012). Similar type of hepatoprotective properties of curcumin has been reported by (Somanawat, Thong-Ngam, and Klaikeaw (2013) and Li et al. (2013) in rat models.

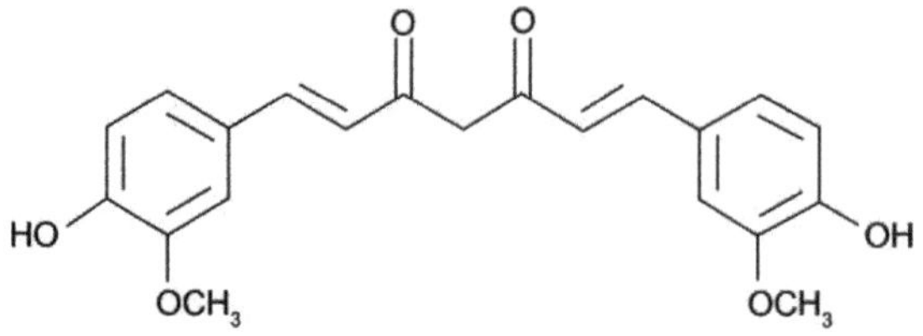

FIG. 5 Chemical structure of curcumin.

Curcumin is also used as a rescue therapy against hepatic injuries likewise it significantly decreases APAP-induced hepatotoxicity when administered for 2 h after APAP overdose. N-Acetyl-p-benzoquinone imine is the compound generated as a result of APAP-induced hepatic toxicity. At low doses of curcumin, NAPQI got easily detoxified by GSH. At higher doses, N-acetyl-p-benzoquinone imine causes reduction in glutathione and binds to proteins, which ultimately leads to lipid peroxidation (Hinson, Roberts, & James, 2010; Yin, Oltvai, & Korsmeyer, 1994). Malondialdehyde has been reported to be crucial lipid peroxidation indicator (Lee, Wu, Chen, & Chung, 2003). It has been observed that curcumin may could decrease MDA induced by APAP, thereby acting as potent agent in lipid peroxidation. Lee et al. (2003) suggested that organisms have capability to protect body cells from oxidative stress with the help of various enzymes like superoxide dismutase and catalase. Superoxide dismutase acts as the first line of defense against free radicals by converting superoxide anion into hydrogen peroxide, and then it is converted to oxygen and water by glutathione peroxidase (Wu et al., 2010). It has been reported that curcumin may maintain SOD activity, namely, CCL_4 induced liver toxicity and inflammation (Naik, Thakare, & Patil, 2011). It has been reported that APAP induces apoptosis in the liver cell apoptosis due to its hepatic toxic nature (Hu & Colletti, 2010; (Kass, Macanas-Pirard, Lee, & Hinton, 2003). It is not only seen in primary hepatocytes (Sharma, Singh, Pandey, & Dhawan, 2012) but also in livers of mice that are treated with high doses of APAP (Hu et al., 2010). A study also reveals that after 16 h of APAP administration, hepatocytic apoptosis significantly increases in the liver of mice and is decreased by curcumin pretreatment. Bcl-2 protein is an antiapoptotic factor and inhibits apoptosis of cells by preventing depolarization of mitochondrial membrane. Being a member of the Bcl-2 protein family, Bax is used to inactivate it (Rajan et al., 2012). Li et al. (2013) have reported that curcumin pretreatment downregulates the mRNA expression of Bax and upregulates the mRNA expression of Bcl-2 compared with APAP-treated group, which means curcumin increases Bcl-2/Bax ratio, thus reducing APAP-induced apoptosis.

It has been reported that curcumin shows strong antiapoptotic effect via inhibiting TGF-β as inducer of caspase-3 mediated apoptosis in kidney and lung tissues (Awad & El-Sharif, 2011). The mechanisms by which curcumin regulates cell apoptosis in APAP-induced liver toxicity should further be investigated.

The protection of curcumin may lower lipid peroxidation and oxidative stress. Li et al. (2013) reported that curcumin restores Bcl-2/Bax ratio, thereby reducing APAP-induced hepatocyte apoptosis.

6.4   Arjunolic acid

Arjunolic acid extracted from bark of *Terminalia arjuna* plant, chemically it is 2,3,23-trihydroxyolean-12-en-28-oic acid (Fig. 6). It has been reported to be hepatoprotective in chemical-induced liver injury and hepatocarcinogenesis (Manna, Sinha, Pal, & Sil, 2007). Ghosh et al. (2010) have reported that arjunolic acid prevents reduction of GSH from the liver. Arjunolic acid also regulates APAP metabolite formation by inhibiting CYP450 and also helps in metabolic activation (Ghosh et al., 2010). Further, it inhibits the indulgence of mitochondrial membrane potential and cytochrome C discharge and decrease in JNK activation and mitochondrial permeabilization.

Pretreatment with arjunolic acid (AA) significantly reduces glutathione depletion induced by APAP and inhibits the production of APAP metabolites, as evidenced by decrease in plasma levels of APAP metabolites. This may be because of inhibition of the bioactivation of APAP by AA that causes a decrease in NAPQI formation.

It has been reported that different mechanisms might be involved in APAP-induced hepatotoxicity (Dahlin, Miwa, Lu, & Nelson, 1984), suggesting that this toxicity is mediated by cytochrome P450 metabolism of APAP to NAPQI, which covalently binds to a number of critical proteins leading to their inactivation, especially after GSH depletion. CYP2E1 is known to be most active cytochrome P450 in catalysis of APAP metabolism to hepatotoxic NAPQI (Bessems, De Groot, Baede, Te Koppele, & Vermeulen, 1989). (Lee, Buters, Pineau, Fernandez-Salguero, and Gonzalez (1996)) reported CYP2E1 to be most important in APAP metabolism. To validate the CYP2E1 relation with potent protective effects of AA on liver toxicity, expression of the former was analyzed by western blotting. The results revealed that AA reduces APAP-induced CYP2E1 expression. These results were consistent with that of decrease in CYP2E1 activity. This shows that hepatoprotective effects of arjunolic acid in the pretreated groups are related to CYP2E1 suppression that further reduces NAPQI formation

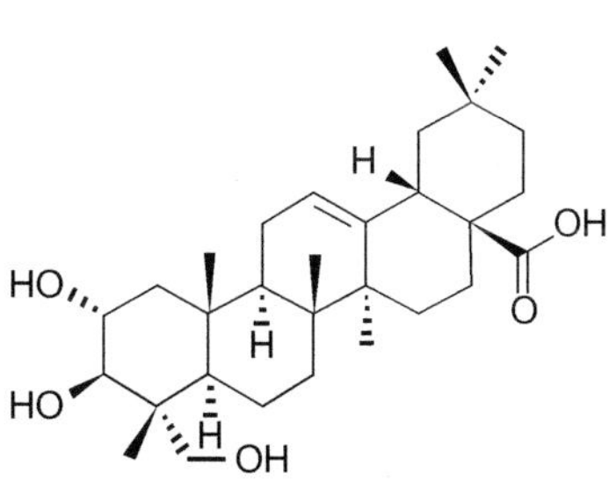

FIG. 6   Chemical structure of arjunolic acid.

associated with APAP liver toxicity and protects the liver against that pathophysiology. Oxidative stress plays important role in initiating lipid peroxidation and protein oxidation, which damage hepatocytes and promote liver toxicity during APAP-induced hepatotoxicity (Jaeschke, Knight, & Bajt, 2003).

APAP treatment increases the production of reactive oxygen species in hepatocytes and production of nitric oxide in the liver. This occurs as AA itself has a radical scavenging property being evidenced from its hydroxyl, superoxide, and nitric oxide radical quenching ability in a cell. It is therefore probable that beneficial effects of AA may result from the detoxification of reactive oxygen species and nitric oxide produced during APAP intoxication. The reports suggest that reactive oxygen species not only damage molecules but also act as important signaling molecules that activate JNK in APAP-induced liver toxicity (Hanawa et al., 2008). Ghosh et al. (2010) revealed that when arjunolic acid was administered 4 h after APAP exposure, it attenuates the expression of p-JNK (Ghosh et al., 2010). JNK activation also acts as vital regulator in mitochondrial permeabilization (Latchoumycandane, 2007). JNK phosphorylates Bcl-2 and Bcl-xL of the Bcl-2 protein family, thereby blocking their antiapoptotic functions (Deng et al., 2001). APAP stimulates phosphorylation of Bcl-2, and AA prevents phosphorylation, thereby favoring relative abundance of antiapoptotic proteins versus their proapoptotic partners (Ghosh et al., 2010).

It has also been reported that JNK phosphorylates the proapoptotic protein Bax (Gunawan et al., 2006), as it is the balance between proapoptotic and antiapoptotic factors. This determines the fate of a cell; this imbalance can lead to cell demise (Ghosh et al., 2010) and provide strong evidence that arjunolic acid suppresses APAP-induced phosphorylation of JNK and, hence, of Bcl-2 and Bcl-xL. This prevents the mitochondrial permeabilization and release of dead proteins from mitochondria. This causes loss of ATP production and ultimately protects cells from necrotic death. Therefore arjunolic acid may be explored as a potent protective agent against mitochondria-mediated toxicity. Ghosh et al. (2010) have suggested that APAP-induced hepatic damage may be due to (i) JNK-independent event and (ii) JNK-dependent event. Both of these are required during necrosis. Arjunolic acid treatment may provide an insight for the protection against severe liver ailments due to APAP overdose (Ghosh et al., 2010).

6.5 Berberine

Berberine has been extracted from plants like *Berberis aristata*. Its chemical name is 5,6-dihydro-9,10-dimethoxybenzo(g)-1,3-benzodioxolo(5,6-*a*) quinolizinium sulfate (Fig. 7) and has strong hepatoprotective properties. It has been proved that berberine shows hepatoprotective activity against APAP-induced liver toxicity (Janbaz & Gilani, 2000). Its effect was demonstrated in various experimental models, namely, APAP-induced liver toxicity and hepatitis. Berberine has also been found to reduce mortality, restore liver enzymes, and

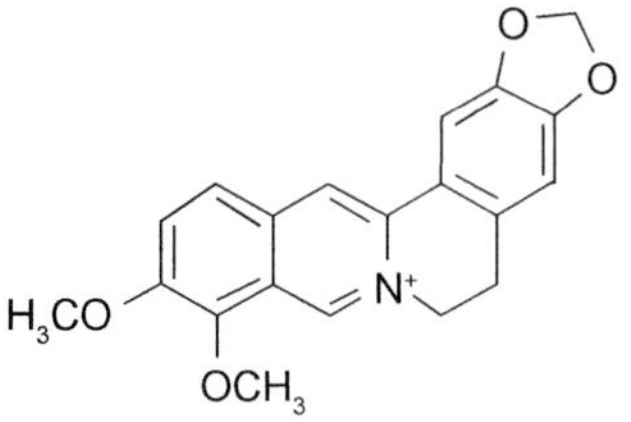

FIG. 7 Chemical structure of berberine.

inhibit the inflammasome components. Inflammasome activation in APAP-induced liver toxicity is supposed to be a main agent in activation of immune cells, damage in liver cells, and cell mortality. Therefore, inhibits the activation of P2X7, which is a purinergic receptor and mediates inflammasome activation by berberine may prove a new approach (Vivoli et al., 2016).

Berberine has been evaluated for its free radical scavenging activity in comparison with a standard drug; *N*-acetylcysteine (NAC). The liver injury due to acetaminophen leads to increase in the levels of various biochemical parameters such as AST, ALT, PT, ALP, and *g*-GT in the experimental group. Acetaminophen-treated group showed a relentless increase in liver aminotransferases. The NAC- and BBR-treated groups showed amelioration of cytoplasmic and mitochondrial enzymes. The BBR-treated animals showed more significant amelioration in comparison with NAC-treated group. In the case of liver cell injury, the cytoplasmic and mitochondrial enzymes leak through the cell membrane (Rajesh & Latha, 2004) and increase in the blood (Abraham, 2005). Treatment with NAC and BBR showed decline in ALT, AST, ALP, *g*-GT, and PT. This might be because of free radical scavenging activity of BBR and cell membrane stabilization. Previous studies (Abraham, 2005; Rabiul et al., 2011) also support the same mechanisms. It has been reported that due to cholestasis more and more alkaline phosphatase (ALP) is secreted from biliary canaliculi, hence the level of ALP increase in circulatory serum (Rabiul et al., 2011). Free radical scavenging activity of BBR and NAC was evaluated as markers of oxidation and peroxidation by (Almani et al., 2017). The BBR treated rats expressed statistically different increase in serum and tissue SOD, GPX, CAT, and GSSH and a decrease in MDA. Antioxidant and antilipid peroxidant effect of BBR was superior to the NAC (*N*-acetylcysteine). It has been reported that the increase in MDA indicates failure of cellular antioxidant enzymes (Shao, Chu, Lu, & Kang, 2008). The BBR- and NAC-treated rats showed a significant decrease in MDA; this suggests they possess antioxidant and antiperoxidant potential. The BBR prevents hepatocyte cell injury by scavenging free radicals through an increase in cellular antioxidants—the SOD, GPX, CAT, and GSSH. The berberine mitigates the oxidative injury by raising the SOD, GPX, CAT, and GSSH, which is visibly evident in the present study (Almani et al., 2017). The study carried out by Almani et al. (2017) revealed significant liver protective effect of berberine against acetaminophen-induced liver injury. It also shows antioxidant and antiperoxidant potential. Berberine increases the antioxidant activity of enzymes, namely, glutathione peroxidase, superoxide dismutase, reduced glutathione, and catalase.

6.6 Bixin

Bixin is used as food additive and in cosmetics. It is found in seed of *Bixa orellana* plant constitutes bioactive compound having antioxidant and antiinflammatory activitiesthat protects the body from any harm, particularly the liver. Rao et al. (2014) have demonstrated that solid lipid nanoparticles of this plant demonstrate continuous discharge with the help of first-order diffusion kinetics. They also revealed that this property of liver protection is due to localization of bixin nanoparticles in liver tissues as studied in rat models (Fig. 8).

B. orellana also possess protective activities against CCl_4 and ethanol-induced hepatotoxicity. In fact, contrary to flavonoids, some alkaloids could possess liver toxic effects as previously reported (Júnior et al., 2005). The extract of this plant has profound effect on oxidative stress that shows that MDA (malondialdehyde), SOD (superoxide dismutase), CAT (catalase), and GSH (glutathione) levels were significantly affected by APAP treatment. This occurs as a result of liver cell obliteration or may be due to variation in membrane permeability, which indicates hepatocellular destruction caused by APAP.

However, treatment with the bark infusion attenuates these changes in liver, suggesting its antioxidant property. The observations seem to be similar with antioxidant effects with the bark extract of *B. orellana* in the rat liver (Bell et al., 2012). On the basis of these information, it was suggested that this activity may happen through glutathione reductase activation and glutathione peroxidase inhibition or lipid peroxidation inhibition as demonstrated by decreased MDA concentrations and increased GSH values. The mechanisms of APAP hepatotoxicity are well described, and the role of inflammation in initiating and propagating liver injury was highlighted by evidence (Sabir & Rocha, 2008). In fact, Kupffer cells generate proinflammatory and chemotactic cytokines, leading to the infiltration and activation of neutrophils in damaged liver tissues and, therefore, to aggravation of liver injury (Hinson et al., 2010). Insertion of the bark infusion (250 mg/kg) cause significant reduction in liver inflammation as revealed by decrease in leukocyte infiltration and TNF-α levels in the liver tissue of treated mice. The decrease of this cytokine suggests that bark infusion would interfere with the signaling pathways involved in the inflammatory

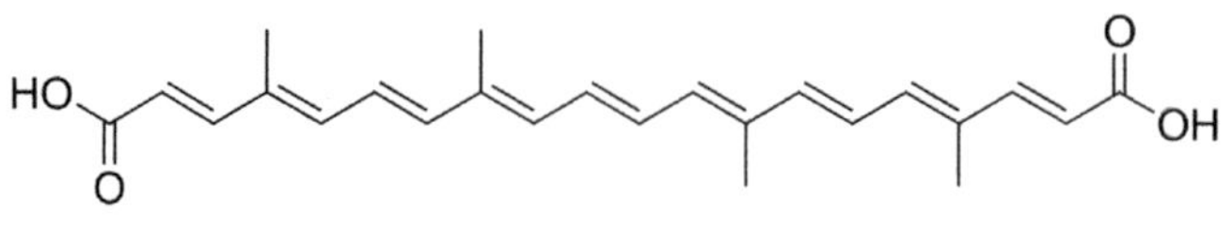

FIG. 8 Chemical structure of bixin.

process. Previous studies on ethanol-induced hepatotoxicity with this plant reported such antiinflammatory activity with leaf extracts (Lopez, Sumalapao, & Villarante, 2017). Furthermore the nonsignificant decrease was observed at 100 mg/kg dose, which can be demonstrated by the evidence that the active molecules are insufficient enough to inhibit the inflammatory process. These results are in accordance with that of Lee et al. (2019), demonstrated that the shielding effects of escin on APAP-induced liver toxicity are mediated by the inhibition of ERK signaling pathway (Lee et al., 2019). Therefore the bark's infusion of *B. orellana* has been shown in this study as an extract with antiinflammatory and antioxidant potential in APAP-induced liver damage. Extract from bark of *B. orellana* is more active than leaves and root extracts. Moreover the bark's infusion of *B. orellana* was a more potent hepatoprotective extract than maceration and decoction, acting mainly on subacute APAP liver injury through antiinflammatory and antioxidant mechanisms. In addition, the activity was comparable with that of silymarin. This work paved the way through the isolation of active ingredients from bark infusion, which will be advantageous to produce new bioactive compounds with greater activity against liver damage.

6.7 Caffeic acid

Caffeic acid is a plant-based polyphenolic compound extracted from coffee plant and honey. Its chemical name is trans-3,4-dihydroxycinnamic acid (Fig. 9). It improves APAP-induced hepatotoxicity by reducing myeloperoxidase (MPO) activity. Further, it enhances the expression of antioxidants, namely, Nrf2, and reduces expression of Keap1, thereby preventing the binding of Keap1 to Nrf2 and thus activating Nrf2 in hepatocytes. In vitro study showed minimal effect on the enzymatic activity of CYP3A4 and CYP2E1 (Pang et al., 2016). Pang et al. (2016) reported that caffeic acid downregulates the expression of mRNA in the liver. Moreover, it also inhibit the extracellular regulatory protein kinase (ERK1/2) signaling cascade activation. Overall, it was concluded that caffeic acid attributes to APAP-induced liver injury detoxification by inhibiting ERK1/2-mediated Egr1 transcriptional activation.

Assessment of ALT and AST level in serum and histological obervations reveals that 30 mg/kg of CA provides the better protection against APAP-induced liver injury in mice (Pang et al., 2016) reported that CA in lower doses 10 mg/kg, does not decrease activites of ALT/AST in APAP-induced liver toxicity. Caffeic acid in lower dose also regulates GSH level in liver and increases level of ROS

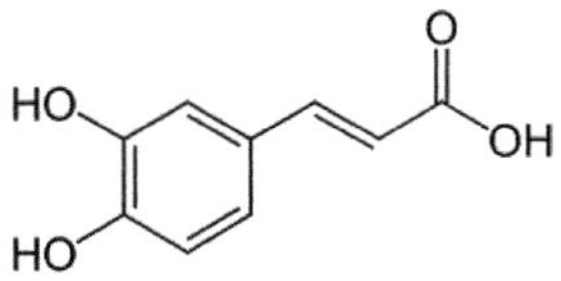

FIG. 9 Chemical structure of caffeic acid.

induced by APAP in mice. Thus CA (10 mg/kg) improves APAP-induced liver toxicity (Pang et al., 2016). These results coincides with the findings of Janbaz, Saeed, and Gilani (2004), reported that CA at 10 and 30 mg/kg had no significant effects on serum ALT/AST activities when CA was administered to mice for seven consecutive days. Also, CA reduces the activity of myeloperoxidase in APAP-induced liver toxicity. Enhanced liver myeloperoxidase activity indicates the occurrence of liver inflammation (Ramaiah & Jaeschke, 2007). These results revealed the inhibition of CA on APAP-induced hepatic inflammation.

GSH is important for the detoxification of APAP when conjugated with NAPQI. Accumulation of NAPQI depletes cellular GSH, thereby genereateing oxidative stress-induced liver toxicity (Jaeschke, Williams, & McGill, 2012; James, Mayeux, & Hinson, 2003). (Pang et al. (2016)) reported the reduction of liver GSH in mice after APAP treatment; however, this was upturned by CA (10 and 30 mg/kg). In the meantime, CA (10 and 30 mg/kg) also increased liver GSH level as when CA is given orally to mice, it may contribute to its antioxidant capacity. CA (10 and 30 mg/kg) reduces the APAP-induced formation of liver ROS in mice in vivo, and it also abrogated the formation of cellular ROS in L-02 and HepG2 cells induced by APAP. CA (10 and 30 mg/kg) alone had no obvious effect on the formation of liver ROS in mice. Gum and Cho (2013) have reported that Nrf2, which is an antioxidative transcription factor, plays an important role in protecting against APAP-induced liver toxicity. It was found that Nrf2 knockout mice were more vulnerable to APAP-induced liver toxicity (Enomoto et al., 2001). Various reports have demonstrated that natural products like quercetin, salvianolic acid B, sauchinone, ginsenoside Rg3, and oleanolic acid may prevent APAP-induced liver toxicity by inducing the transcriptional activation of Nrf2 (Ji, Sheng, Zheng, Shi, & Wang, 2015; Kay et al., 2011; Lin et al., 2015; Reisman, Aleksunes, & Klaassen, 2009).

It has also been reported that when cells were incubated with APAP for 36 h, expression of Nrf2 in the nucleus increases due to CA pretreatment. Secondly, luciferase reporter assay also suggested that CA-induced Nrf2 transcriptional activation takes place when hepatocytes get incubated with APAP for 36 h. Thirdly; Nrf2 siRNA significantly reduces the protection of CA against APAP-induced liver toxicity (Pang et al., 2016). It confirms role of Nrf2 in regulating the protection of CA against APAP-induced hepatotoxicity. Pang et al. (2016) also showed that APAP weakly induces Nrf2 activation when hepatocytes were incubated with APAP for 4, 8, and 12 h. These results are in line with Copple et al. (2008) who reported that the APAP metabolic product NAPQI directly activates Nrf2. However, CA decreased APAP-induced toxicity and then increased Nrf2 activation after liver cells were incubated with APAP for 4 and 8 h. It might be due to the decreased production of NAPQI as further results showed that CA weakly inhibits CYP2E1 and CYP3A4 enzyme activity. Nrf2 regulates the expression of antioxidative genes, namely, HO-1, NQO1, GCLC, and GCLM, when bounded to antioxidant-related elements (ARE) (Kaspar, Niture, & Jaiswal, 2009). CA reverses the APAP-induced decrease of mRNA and protein expression of NQO1 and HO-1. In addition, ZnPP and Dim reduce the protection of CA GCLM. Although APAP weakly enhanced the mRNA expression of GCLC, it had no significant effect on the protein expression of GCLC. CA is not supposed to increase APAP-induced increased mRNA expression of GCLC. An inhibitor for GCL and BSO is composed of GCLC and GCLM (Yeh, 1980); it also has no any effect on CA-induced protection against APAP-induced liver toxicity. Therefore, it has no role in CA-induced protection against APAP-induced hepatotoxicity. GCL is critical for the de novo biosynthesis of cellular GSH (Griffith, 1982), an important component for APAP detoxification.

Keap1 is an inhibitor of Nrf2 that acts as an adapter for Cul3/Rbx1-mediated degradation of Nrf2 (Kaspar et al., 2009). It was established that CA reduces the expression of Keap1 protein in the presence of APAP and may play role in the activation of Nrf2 induced by CA.

As an outcome of molecular docking, it has been suggested that CA may interact with Keap1 and occupy the Nrf2 binding site in the protein and thus lead to the dissociation of Keap1 from Nrf2. In the end, it induces the transcriptional activation of Nrf2 (Pang et al., 2016). It is reported by Laine, Auriola, Pasanen, and Juvonen (2009) and Raucy, Lasker, Lieber, and Black (1989) that CYP2E1, CYP3A4, and CYP1A2 are the main metabolic enzymes for converting APAP into NAPQI, which then depletes cellular GSH and causes oxidative stress-induced injury (Laine et al., 2009; Raucy et al., 1989). Recently, it was observed and established that CA-induced protection against APAP-induced liver toxicity takes place due to inhibition on CYP2E1, CYP3A4, and CYP1A2. It was also revealed that CA inhibits enzymatic activity of CYP2E1 and CYP3A4 weakly in vitro, but does not affect enzymatic activity of CYP1A2.

The study concluded that CA-induced inhibition on CYP2E1 and CYP3A4, therefore confirms the protective role of CA against APAP-induced liver toxicity (Pang et al., 2016). CA also reverses NAPQI-induced hepatotoxicity in liver cells. This study also indicates that CA may directly inhibit the hepatotoxicity induced by APAP's toxic metabolite NAPQI. CA-induced the weak inhibition on CYP2E1 and CYP3A4 is not essential for protection (Pang et al., 2016). A study conducted by Pang et al. (2016) reveals that CA induces Nrf2 activation by diminishing the expression of its inhibitor protein Keap1 and blocking the binding of Nrf2 with Keap1. It may lead to the increased expression of downstream antioxidative enzymes including NQO1 and HO-1, therefore preventing APAP-induced liver oxidative injury.

References

Abraham, P. (2005). Oxidative stress in paracetamol-induced pathogenesis:(I) Renal damage. *Indian Journal of Biochemistry & Biophysics*, *42*(1), 59–62.

Acharya, M., & Lau-Cam, C. A. (2010). Comparison of the protective actions of N-acetylcysteine, hypotaurine and taurine against acetaminophen-induced hepatotoxicity in the rat. *Journal of Biomedical Science*, *17*(S1), S35.

Aiyer, H. S., & Gupta, R. C. (2010). Berries and ellagic acid prevent estrogen-induced mammary tumorigenesis by modulating enzymes of estrogen metabolism. *Cancer Prevention Research*, *3*(6), 727–737.

Al-Jenoobi, F. I., Al-Thukair, A. A., Alam, M. A., Abbas, F. A., Al-Mohizea, A. M., Alkharfy, K. M., et al. (2014). Effect of garden cress seeds powder and its alcoholic extract on the metabolic activity of CYP2D6 and CYP3A4. *Evidence Based Complementary and Alternative Medicine*, *2014*, 634592.

Almani, S. A., Qureshi, F., Shaikh, T. Z., Uqaili, A. A., & Khoharo, H. K. (2017). Free radical scavenging activity of Berberine in acetaminophen induced liver injury. *International Journal of Surgery and Medicine*, *3*(1), 27–36.

Ansar, H., Mazloom, Z., Kazemi, F., & Hejazi, N. (2011). Effect of alpha-lipoic acid on blood glucose, insulin resistance and glutathione peroxidase of type 2 diabetic patients. *Saudi Medical Journal*, *32*(6), 584–588.

Armoni, M., Harel, C., Ramdas, M., & Karnieli, E. (2014). CYP2E1 impairs GLUT4 gene expression and function: NRF2 as a possible mediator. *Hormone and Metabolic Research*, *46*(07), 477–483.

Awad, A. S., & El-Sharif, A. A. (2011). Curcumin immune-mediated and anti-apoptotic mechanisms protect against renal ischemia/reperfusion and distant organ induced injuries. *International Immunopharmacology*, *11*(8), 992–996.

Baer, B. R., & Rettie, A. E. (2006). CYP4B1: An enigmatic P450 at the interface between xenobiotic and endobiotic metabolism. *Drug Metabolism Reviews*, *38*(3), 451–476.

Bansal, S. S., Kausar, H., Vadhanam, M. V., Ravoori, S., Pan, J., Rai, S. N., et al. (2014). Curcumin implants, not curcumin diet, inhibit estrogen-induced mammary carcinogenesis in ACI rats. *Cancer Prevention Research*, *7*(4), 456–465.

Behl, M., Nyska, A., Chhabra, R. S., Travlos, G. S., Fomby, L. M., Sparrow, B. R., et al. (2011). Liver toxicity and carcinogenicity in F344/N rats and B6C3F1 mice exposed to Kava Kava. *Food and Chemical Toxicology*, *49*(11), 2820–2829.

Bell, G. A. S., Shamna, R., Sangeetha, B., & Sasikumar, J. M. (2012). In vivo antioxidant activity of bark extract of Bixa orellana L. against acetaminophen–induced oxidative stress. *Asian Pacific Journal of Tropical Biomedicine*, *2*(2), S700–S705.

Bessems, J. G. M., De Groot, M. J., Baede, E. J., Te Koppele, J. M., & Vermeulen, N. P. E. (1989). Hydrogen atom abstraction of 3, 5-disubstituted analogues of paracetamol by horseradish peroxidase and cytochrome P450. *Xenobiotica*, *28*(9), 855–875.

Bigoniya, P., Shukla, A., & Singh, C. S. (2010). Evaluation of hepatic microsomal enzyme functional integrity on picroliv pretreatment against CCl4 induced hepatotoxicity. *International Journal of Pharmacology*, *6*(3), 200–207.

Blake, O. A., Bennink, M. R., & Jackson, J. C. (2006). Ackee (*Blighia sapida*) hypoglycin A toxicity: Dose response assessment in laboratory rats. *Food and Chemical Toxicology*, *44*(2), 207–213.

Bogacz, A., Bartkowiak-Wieczrek, J., Mikołajczak, P.Ł., Rakowska Mrozikiewicz, B., Grześkowiak, E., Wolski, H., et al. (2014). The influence of soybean extract on the expression level of selected drug transporters, transcription factors and cytochrome P450 genes encoding phase I drug-metabolizing enzymes. *Ginekologia Polska*, *85*(5), 348–353.

Božina, N., Bradamante, V., & Lovrić, M. (2009). Genetic polymorphism of metabolic enzymes P450 (CYP) as a susceptibility factor for drug response, toxicity, and cancer risk. *Arhiv za Higijenu Rada i Toksikologiju*, *60*(2), 217–242.

Bu-Abbas, A., Clifford, M. N., Walker, R., & Ioannides, C. (1994). Selective induction of rat hepatic CYP1 and CYP4 proteins and of peroxisomal proliferation by green tea. *Carcinogenesis*, *15*(1), 2575–2579.

Bulku, E., Stohs, J. S., Cicero, L., Brooks, T., Halley, H., & Ray, S. D. (2012). Curcumin exposure modulates multiple pro-apoptotic and anti-apoptotic signaling pathways to antagonize acetaminophen-induced toxicity. *Current Neurovascular Research, 9*(1), 58–71.

Busserolles, J., Zimowska, W., Rock, E., Rayssiguier, Y., & Mazur, A. (2002). Rats fed a high sucrose diet have altered heart antioxidant enzyme activity and gene expression. *Life Sciences, 71*(11), 1303–1312.

Celik, G., Semiz, A., Karakurt, S., Arslan, S., Adali, O., & Sen, A. (2013). A comparative study for the evaluation of two doses of ellagic acid on hepatic drug metabolizing and antioxidant enzymes in the rat. *BioMed Research International, 2013.*

Chang, J. (2014). Liver physiology: Metabolism and detoxification. In *Pathobiologies of human diseases* (pp. 1770–1782). https://doi.org/10.1016/B978-0-12-386456-7.04202-7.

Chen, H. W., Tsai, C. W., Yang, J. J., Liu, C. T., Kuo, W. W., & Lii, C. K. (2003). The combined effects of garlic oil and fish oil on the hepatic antioxidant and drug-metabolizing enzymes of rats. *British Journal of Nutrition, 89*(2), 189–200.

Chen, Y., Xiao, P., Ou-Yang, D. S., Fan, L., Guo, D., Wang, Y. N., et al. (2009). Simultaneous action of the flavonoid quercetin on cytochrome P450 (CYP) 1A2, CYP2A6, N-acetyltransferase and xanthine oxidase activity in healthy volunteers. *Clinical and Experimental Pharmacology and Physiology, 36*(8), 828–833.

Chen, Q., Zhang, T., Wang, J. F., & Wei, D. Q. (2011). Advances in human cytochrome p450 and personalized medicine. *Current Drug Metabolism, 12*(5), 436–444.

Child, D. F., Hudson, P. R., Jones, H., Davies, G. K., De, P., Mukherjee, S., et al. (2004). The effect of oral folic acid on glutathione, glycaemia and lipids in Type 2 diabetes. *Diabetes, Nutrition & Metabolism, 17*(2), 95–102.

Chow, H. S., Garland, L. L., Hsu, C. H., Vining, D. R., Chew, W. M., Miller, J. A., et al. (2010). Resveratrol modulates drug- and carcinogen-metabolizing enzymes in a healthy volunteer study. *Cancer Prevention Research, 3*(9), 1168–1175.

Copple, I. M., Goldring, C. E., Jenkins, R. E., Chia, A. J., Randle, L. E., Hayes, J. D., et al. (2008). The hepatotoxic metabolite of acetaminophen directly activates the Keap1-Nrf2 cell defense system. *Hepatology, 48*(4), 1292–1301.

Dahlin, D. C., Miwa, G. T., Lu, A. Y., & Nelson, S. D. (1984). N-acetyl-p-benzoquinone imine: A cytochrome P-450-mediated oxidation product of acetaminophen. *Proceedings of the National Academy of Sciences of the United States of America, 81*(5), 1327–1331.

Danielson, P.Á. (2002). The cytochrome P450 superfamily: Biochemistry, evolution and drug metabolism in humans. *Current Drug Metabolism, 3*(6), 561–597.

Davenport, D. M., & Wargovich, M. J. (2005). Modulation of cytochrome P450 enzymes by organosulfur compounds from garlic. *Food and Chemical Toxicology, 43*(12), 1753–1762.

Debersac, P., Vernevaut, M. F., Amiot, M. J., Suschetet, M., & Siess, M. H. (2001). Effects of a water-soluble extract of rosemary and its purified component rosmarinic acid on xenobiotic-metabolizing enzymes in rat liver. *Food and Chemical Toxicology, 39*(2), 109–117.

Deng, X., Xiao, L., Lang, W., Gao, F., Ruvolo, P., & May, W. S. (2001). Novel role for JNK as a stress-activated Bcl2 kinase. *Journal of Biological Chemistry, 276*(26), 23681–23688.

Dey, P., Saha, M. R., & Sen, A. (2013). An overview on drug-induced hepatotoxicity. *Asian Journal of Pharmaceutical and Clinical Research, 6*(4), 1–4.

Donatus, I. A., & Vermeulen, N. P. (1990). Cytotoxic and cytoprotective activities of curcumin: Effects on paracetamol-induced cytotoxicity, lipid peroxidation and glutathione depletion in rat hepatocytes. *Biochemical Pharmacology, 39*(12), 1869–1875.

El Morsy, E. M., & Kamel, R. (2015). Protective effect of artichoke leaf extract against paracetamol-induced hepatotoxicity in rats. *Pharmaceutical Biology, 53*(2), 167–173.

Enomoto, A., Itoh, K., Nagayoshi, E., Haruta, J., Kimura, T., O'Connor, T., et al. (2001). High sensitivity of Nrf2 knockout mice to acetaminophen hepatotoxicity associated with decreased expression of ARE-regulated drug metabolizing enzymes and antioxidant genes. *Toxicological Sciences, 59*(1), 169–177.

Gaillard, Y., Carlier, J., Berscht, M., Mazoyer, C., Bevalot, F., Guitton, J., et al. (2011). Fatal intoxication due to ackee (*Blighia sapida*) in Suriname and French Guyana. GC-MS detection and quantification of hypoglycin-A. *Forensic Science International, 206*(1–3), e103–e107.

Galluzzi, L., Vitale, I., Senovilla, L., Olaussen, K. A., Pinna, G., Eisenberg, T., et al. (2012). Prognostic impact of vitamin B6 metabolism in lung cancer. *Cell Reports, 2*(2), 257–269.

Garcia, A. F., Medeiros, H. C., Maioli, M. A., Lima, M. C., Rocha, B. A., da Costa, F. B., et al. (2010). Comparative effects of lantadene A and its reduced metabolite on mitochondrial bioenergetics. *Toxicon, 55*(7), 1331–1337.

Ghosh, J., Das, J., Manna, P., & Sil, P. C. (2010). Acetaminophen induced renal injury via oxidative stress and TNF-α production: Therapeutic potential of arjunolic acid. *Toxicology, 268*(1–2), 8–18.

Ginsberg, G., Guyton, K., Johns, D., Schimek, J., Angle, K., & Sonawane, B. (2010). Genetic polymorphism in metabolism and host defense enzymes: Implications for human health risk assessment. *Critical Reviews in Toxicology, 40*(7), 575–619.

Gonzalez, F. J. (2005). Role of cytochromes P450 in chemical toxicity and oxidative stress: Studies with CYP2E1. *Mutation Research: Fundamental and Molecular Mechanisms of Mutagenesis, 569*(1–2), 101–110.

González, C. A., Sala, N., & Capellá, G. (2002). Genetic susceptibility and gastric cancer risk. *International Journal of Cancer, 100*(3), 249–260.

Gradelet, S., Leclerc, J., Siess, M. H., & Astorg, P. O. (1996). β-Apo-8'-carotenal, but not β-carotene, is a strong inducer of liver cytochromes P4501A1 and 1A2 in rat. *Xenobiotica, 26*(9), 909–919.

Griffith, O. W. (1980). Determination of glutathione and glutathione disulfide using glutathione reductase and 2-vinylpyridine. *Analytical Biochemistry, 106*(1), 207–212.

Griffith, O. W. (1982). Mechanism of action, metabolism, and toxicity of buthionine sulfoximine and its higher homologs, potent inhibitors of glutathione synthesis. *Journal of Biological Chemistry, 257*(22), 13704–13712.

Gum, S. I., & Cho, M. K. (2013). Recent updates on acetaminophen hepatotoxicity: The role of nrf2 in hepatoprotection. *Toxicological Research, 29*(3), 165–172.

Gunawan, B. K., Liu, Z. X., Han, D., Hanawa, N., Gaarde, W. A., & Kaplowitz, N. (2006). c-Jun N-terminal kinase plays a major role in murine acetaminophen hepatotoxicity. *Gastroenterology, 131*(1), 165–178.

Gurusamy, K., Kokilavani, R., Arumugasamy, K., & Sowmia, C. (2009). Protective effect of ethanolic extract of polyherbal formulation on carbon tetrachloride induced liver injury. *Ancient Science of Life, 28*(3), 6.

Hakooz, N., & Hamdan, I. (2007). Effects of dietary broccoli on human in vivo caffeine metabolism: A pilot study on a group of Jordanian volunteers. *Current Drug Metabolism, 8*(1), 9–15.

Hanawa, N., Shinohara, M., Saberi, B., Gaarde, W. A., Han, D., & Kaplowitz, N. (2008). Role of JNK translocation to mitochondria leading to inhibition of mitochondria bioenergetics in acetaminophen-induced liver injury. *Journal of Biological Chemistry, 283*(20), 13565–13577.

Hattori, A., Yamada, N., Nishikawa, T., Fukuda, H., & Fujino, T. (2001). Protective effect of ajoene on acetaminophen-induced hepatic injury in mice. *Bioscience, Biotechnology, and Biochemistry, 65*(11), 2555–2557.

Hayes, J. D., & Pulford, D. J. (1995). The glutathione S-transferase supergene family: Regulation of GST and the contribution of the isoenzymes to cancer chemoprotection and drug resistance part I. *Critical Reviews in Biochemistry and Molecular Biology, 30*(6), 445–520.

Hinson, J. A., Roberts, D. W., & James, L. P. (2010). Mechanisms of acetaminophen-induced liver necrosis. In *Adverse drug reactions* (pp. 369–405). Berlin, Heidelberg: Springer.

Horn, T. L., Reichert, M. A., Bliss, R. L., & Malejka-Giganti, D. (2002). Modulations of P450 mRNA in liver and mammary gland and P450 activities and metabolism of estrogen in liver by treatment of rats with indole-3-carbinol. *Biochemical Pharmacology, 64*(3), 393–404.

Howard, J. M. C., Davies, S., & Hunnisett, A. (1994). Red cell magnesium and glutathione peroxidase in infertile women—Effects of oral supplementation with magnesium and selenium. *Magnesium Research, 7*(1), 49–57.

Hu, B., & Colletti, L. M. (2010). CXC receptor-2 knockout genotype increases X-linked inhibitor of apoptosis protein and protects mice from acetaminophen hepatotoxicity. *Hepatology, 52*(2), 691–702.

Hu, J., Yan, D., Gao, J., Xu, C., Yuan, Y., Zhu, R., et al. (2010). rhIL-1Ra reduces hepatocellular apoptosis in mice with acetaminophen-induced acute liver failure. *Laboratory Investigation, 90*(12), 1737–1746.

Jaeschke, H., Knight, T. R., & Bajt, M. L. (2003). The role of oxidant stress and reactive nitrogen species in acetaminophen hepatotoxicity. *Toxicology Letters, 144*(3), 279–288.

Jaeschke, H., Williams, C. D., & McGill, M. R. (2012). Caveats of using acetaminophen hepatotoxicity models for natural product testing. *Toxicology Letters, 215*(1), 40–41.

James, M. O., & Ambadapadi, S. (2013). Interactions of cytosolic sulfotransferases with xenobiotics. *Drug Metabolism Reviews, 45*(4), 401–414.

James, L. P., Mayeux, P. R., & Hinson, J. A. (2003). Acetaminophen-induced hepatotoxicity. *Drug Metabolism and Disposition, 31*(12), 1499–1506.

James, M. O., Sacco, J. C., & Faux, L. R. (2008). Effects of food natural products on the biotransformation of PCBs. *Environmental Toxicology and Pharmacology, 25*(2), 211–217.

Janbaz, K. H., & Gilani, A. H. (2000). Studies on preventive and curative effects of berberine on chemical-induced hepatotoxicity in rodents. *Fitoterapia, 71*(1), 25–33.

Janbaz, K. H., Saeed, S. A., & Gilani, A. H. (2004). Studies on the protective effects of caffeic acid and quercetin on chemical induced hepatotoxicity in rodents. *Phytomedicine, 11*(5), 424–430.

Jancova, P., Anzenbacher, P., & Anzenbacherova, E. (2010). Phase II drug metabolizing enzymes. *Biomedical Papers of the Medical Faculty of the University Palacky, Olomouc, Czech Republic, 154*(2), 103–116.

Jenkinson, C., Petroczi, A, & Naughton, D. P. (2013). Effects of dietary components on testosterone metabolism via UDP-glucuronosyltrahsferease. *Frontiers in Endocrinology, 4*, 80.

Ji, L. L., Sheng, Y. C., Zheng, Z. Y., Shi, L., & Wang, Z. T. (2015). The involvement of p62-Keap1-Nrf2 antioxidative signaling pathway and JNK in the protection of natural flavonoid quercetin against hepatotoxicity. *Free Radical Biology and Medicine, 85*, 12–23.

Jollow, D. J., Mitchell, J. R., Potter, W. Z., Davis, D. C., Gillette, J. R., & Brodie, B. B. (1973). Acetaminophen-induced hepatic necrosis. II. Role of covalent binding in vivo. *Journal of Pharmacology and Experimental Therapeutics, 187*(1), 195–202.

Júnior, A. C., Asad, L. M., Oliveira, E. B., Kovary, K., Asad, N. R., & Felzenszwalb, I. (2005). Antigenotoxic and antimutagenic potential of an annatto pigment (norbixin) against oxidative stress. *Genetics and Molecular Research, 4*(1), 94–99.

Kalpravidh, R. W., Siritanaratkul, N., Insain, P., Charoensakdi, R., Panichkul, N., Hatairaktham, S., et al. (2010). Improvement in oxidative stress and antioxidant parameters in β-thalassemia/Hb E patients treated with curcuminoids. *Clinical Biochemistry, 43*(4-5), 424–429.

Karmen, A., Wróblewski, F., & LaDue, J. S. (1955). Transaminase activity in human blood. *The Journal of Clinical Investigation, 34*(1), 126–133.

Kaschula, C. H., Hunter, R., Stellenboom, N., Caira, M. R., Winks, S., Ogunleye, T., et al. (2012). Structure–activity studies on the anti-proliferation activity of ajoene analogues in WHCO1 oesophageal cancer cells. *European Journal of Medicinal Chemistry, 50*, 236–254.

Kaspar, J. W., Niture, S. K., & Jaiswal, A. K. (2009). Nrf2: INrf2 (Keap1) signaling in oxidative stress. *Free Radical Biology and Medicine, 47*(9), 1304–1309.

Kass, G. E., Macanas-Pirard, P., Lee, P. C., & Hinton, R. H. (2003). The role of apoptosis in acetaminophen-induced injury. *Annals of the New York Academy of Sciences, 1010*(1), 557–559.

Kay, H. Y., Kim, Y. W., Ryu, D. H., Sung, S. H., Hwang, S. J., & Kim, S. G. (2011). Nrf2-mediated liver protection by sauchinone, an antioxidant lignan, from acetaminophen toxicity through the PKCδ-GSK3β pathway. *British Journal of Pharmacology, 163*(8), 1653–1665.

Kheradpezhouh, E., Panjehshahin, M. R., Miri, R., Javidnia, K., Noorafshan, A., Monabati, A., et al. (2010). Curcumin protects rats against acetaminophen-induced hepatorenal damages and shows synergistic activity with N-acetyl cysteine. *European Journal of Pharmacology, 628*(1–3), 274–281.

Kodama, S., & Negishi, M. (2013). Sulfotransferase genes: Regulation by nuclear receptors in response to xeno/endo-biotics. *Drug Metabolism Reviews, 45*(4), 441–449.

Kohalmy, K., & Vrzal, R. (2011). Regulation of phase II biotransformation enzymes by steroid hormones. *Current Drug Metabolism, 12*(2), 104–123.

Laine, J. E., Auriola, S., Pasanen, M., & Juvonen, R. O. (2009). Acetaminophen bioactivation by human cytochrome P450 enzymes and animal microsomes. *Xenobiotica, 39*(1), 11–21.

Latchoumycandane, C. (2007). Mitochondrial protection by the JNK inhibitor leflunomide rescues mice from acetaminophen-induced liver injury. *Hepatology, 45*(2), 412–421.

Lee, S. S., Buters, J. T., Pineau, T., Fernandez-Salguero, P., & Gonzalez, F. J. (1996). Role of CYP2E1 in the hepatotoxicity of acetaminophen. *Journal of Biological Chemistry, 271*(20), 12063–12067.

Lee, Y. M., & Wu, T. H. (2005). Effects of 5-methoxypsoralen (5-MOP) on arylamine N-acetyltransferase activity in the stomach and colon of rats and human stomach and colon tumor cell lines. *In vivo, 19*(6), 1061–1069.

Lee, Y. M., Wu, T. H., Chen, S. F., & Chung, J. G. (2003). Effect of 5-methoxypsoralen (5-MOP) on cell apoptosis and cell cycle in human hepatocellular carcinoma cell line. *Toxicology In Vitro, 17*(3), 279–287.

Lee, H. C., Yu, H. P., Liao, C. C., Chou, A. H., & Liu, F. C. (2019). Escin protects against acetaminophen-induced liver injury in mice via attenuating inflammatory response and inhibiting ERK signaling pathway. *American Journal of Translational Research, 11*(8), 5170.

Li, G., Chen, J. B., Wang, C., Xu, Z., Nie, H., Qin, X. Y., et al. (2013). Curcumin protects against acetaminophen-induced apoptosis in hepatic injury. *World journal of gastroenterology: WJG, 19*(42), 7440.

Li, Y., Xie, M., Yang, J., Yang, D., Deng, R., Wan, Y., et al. (2006). The expression of antiapoptotic protein survivin is transcriptionally upregulated by DEC1 primarily through multiple sp1 binding sites in the proximal promoter. *Oncogene, 25*(23), 3296–3306.

Lieber, C. S., Cao, Q., DeCarli, L. M., Leo, M. A., Mak, K. M., Ponomarenko, A., et al. (2007). Role of medium-chain triglycerides in the alcohol-mediated cytochrome P450 2E1 induction of mitochondria. *Alcoholism: Clinical and Experimental Research, 31*(10), 1660–1668.

Lin, M., Zhai, X., Wang, G., Tian, X., Gao, D., Shi, L., et al. (2015). Salvianolic acid B protects against acetaminophen hepatotoxicity by inducing Nrf2 and phase II detoxification gene expression via activation of the PI3K and PKC signaling pathways. *Journal of Pharmacological Sciences, 127*(2), 203–210.

Liu, J., Tawa, G. J., & Wallqvist, A. (2013). Identifying cytochrome P450 functional networks and their allosteric regulatory elements. *PLoS One, 8*(12), e81980.

Lopez, C. P., Sumalapao, D. E. P., & Villarante, N. R. (2017). Hepatoprotective activity of aqueous and ethanolic *Bixa orellana* L. leaf extracts against carbon tetrachloride-induced hepatotoxicity. *National Journal of Physiology, Pharmacy and Pharmacology, 7*(9), 972–976.

Lord, R. S., Bongiovanni, B., & Bralley, J. A. (2002). Estrogen metabolism and the diet-cancer connection: Rationale for assessing the ratio of urinary hydroxylated estrogen metabolites. *Alternative Medicine Review, 7*(2), 112–129.

Lucena, M. I., Andrade, R. J., De la Cruz, J. P., Rodriguez-Mendizabal, M., Blanco, E., & de la Cuesta, F. S. (2002). Effects of silymarin MZ-80 on oxidative stress in patients with alcoholic cirrhosis. *International Journal of Clinical Pharmacology and Therapeutics, 40*(1), 2–8.

Lüde, S., Török, M., Dieterle, S., Jäggi, R., Büter, K. B., & Krähenbühl, S. (2008). Hepatocellular toxicity of kava leaf and root extracts. *Phytomedicine, 15*(1–2), 120–131.

Ma, Q., & Lu, A. Y. (2007). CYP1A induction and human risk assessment: An evolving tale of in vitro and in vivo studies. *Drug Metabolism and Disposition, 35*(7), 1009–1016.

Makarova, S. I. (2008). Human N-acetyltransferases and drug-induced hepatotoxicity. *Current Drug Metabolism, 9*(6), 538–545.

Maliakal, P. P., & Wanwimolruk, S. (2001). Effect of herbal teas on hepatic drug metabolizing enzymes in rats. *Journal of Pharmacy and Pharmacology, 53*(10), 1323–1329.

Manna, P., Sinha, M., Pal, P., & Sil, P. C. (2007). Arjunolic acid, a triterpenoid saponin, ameliorates arsenic-induced cyto-toxicity in hepatocytes. *Chemico-Biological Interactions, 170*(3), 187–200.

Marnewick, J. L., Joubert, E., Swart, P., van der Westhuizen, F., & Gelderblom, W. C. (2003). Modulation of hepatic drug metabolizing enzymes and oxidative status by rooibos (Aspalathus linearis) and honeybush (Cyclopia intermedia), green and black (Camellia sinensis) teas in rats. *Journal of Agricultural and Food Chemistry, 51*(27), 8113–8119.

Misaka, S., Kawabe, K., Onoue, S., Werba, J. P., Giroli, M., Watanabe, H., et al. (2013). Green tea extract affects the cytochrome P450 3A activity and pharmacokinetics of simvastatin in rats. *Drug Metabolism and Pharmacokinetics, 28*(6), 514–518.

Mitchell, J. R., Jollow, D. J., Potter, W. Z., Gillette, J. R., & Brodie, B. B. (1973). Acetaminophen-induced hepatic necrosis. IV. Protective role of glutathione. *Journal of Pharmacology and Experimental Therapeutics, 187*(1), 211–217.

Morgan, R. W., & Hoffmann, G. R. (1983). Cycasin and its mutagenic metabolites. *Mutation Research/Reviews in Genetic Toxicology, 114*(1), 19–58.

Naik, R. S., Thakare, V. N., & Patil, S. R. (2011). Protective effect of curcumin on experimentally induced inflammation, hepatotoxicity and cardiotoxicity in rats: Evidence of its antioxidant property. *Experimental and Toxicologic Pathology, 63*(5), 419–431.

Navarro, S. L., Peterson, S., Chen, C., Makar, K. W., Schwarz, Y., King, I. B., et al. (2009). Cruciferous vegetable feeding alters UGT1A1 activity: Diet-and genotype-dependent changes in serum bilirubin in a controlled feeding trial. *Cancer Prevention Research, 2*(4), 345–352.

Nishida, K., Kobayashi, A., & Nagahama, T. (1955). 12. Studies on cycasin, a new toxic glycoside, of *Cycas revoluta Thunb*: Part 1. Isolation and the structure of cycasin. *Journal of Agricultural Chemical Society of Japan, 19*(1), 77–84.

Nishida, K., Kobayashi, A., & Nagahama, T. (1956). Studies on cycasin, a new toxic glycoside, of *Cycas revoluta Thunb*: Part V. Quantitative determination of cycasin in cycad seeds. *Journal of Agricultural Chemical Society of Japan, 20*(2), 74–76.

Nissar, A. U., Farrukh, M. R., Kaiser, P. J., Rafiq, R. A., Afnan, Q., Bhushan, S., et al. (2013). Effect of N-acetyl cysteine (NAC), an organosulfur compound from Allium plants, on experimentally induced hepatic prefibrogenic events in Wistar rat. *Phytomedicine, 20*(10), 828–833.

Paine, A. J. (1981). Hepatic cytochrome P-450. *Essays in Biochemistry, 17*, 85.

Pang, C., Zheng, Z., Shi, L., Sheng, Y., Wei, H., Wang, Z., et al. (2016). Caffeic acid prevents acetaminophen-induced liver injury by activating the Keap1-Nrf2 antioxidative defense system. *Free Radical Biology and Medicine, 91*, 236–246.

Park, D., Jeon, J. H., Shin, S., Joo, S. S., Kang, D. H., Moon, S. H., et al. (2009). Green tea extract increases cyclophosphamide-induced teratogenesis by modulating the expression of cytochrome P-450 mRNA. *Reproductive Toxicology, 27*(1), 79–84.

Park, K., Kweon, S. H., & Choi, H. M. (2002). Anticarcinogenic effect and modification of cytochrome P450 2E1 by dietary garlic powder in diethylnitrosamine-initiated rat hepatocarcinogenesis. *BMB Reports, 35*(6), 615–622.

Peng, W. X., & Zhou, H. H. (2003). Effect of daidzein on CYP1A2 activity and pharmacokinetics of theophylline in healthy volunteers. *European Journal of Clinical Pharmacology, 59*(3), 237–241.

Potter, D. W., & Hinson, J. A. (1986). Reactions of N-acetyl-p-benzoquinone imine with reduced glutathione, acetaminophen, and NADPH. *Molecular Pharmacology, 30*(1), 33–41.

Rabiul, H., Subhasish, M., Sinha, S., Roy, M. G., Sinha, D., & Gupta, S. (2011). Hepatoprotective activity of Clerodendron inerme against paracetamol induced hepatic injury in rats for pharmaceutical product. *International Journal of Drug Development and Research, 3*(1), 118–126.

Rajan, D., Wu, R., Shah, K. G., Jacob, A., Coppa, G. F., & Wang, P. (2012). Human ghrelin protects animals from renal ischemia-reperfusion injury through the vagus nerve. *Surgery, 151*(1), 37–47.

Rajesh, M. G., & Latha, M. S. (2004). Preliminary evaluation of the antihepatotoxic activity of Kamilari, a polyherbal formulation. *Journal of Ethnopharmacology, 91*(1), 99–104.

Ramaiah, S. K., & Jaeschke, H. (2007). Role of neutrophils in the pathogenesis of acute inflammatory liver injury. *Toxicologic Pathology, 35*(6), 757–766.

Rao, M. P., Manjunath, K., Bhagawati, S. T., & Thippeswamy, B. S. (2014). Bixin loaded solid lipid nanoparticles for enhanced hepatoprotection preparation, characterisation and in vivo evaluation. *International Journal of Pharmaceutics, 473*(1–2), 485–492.

Rasmussen, M. K., Brunius, C., Zamaratskaia, G., & Ekstrand, B. (2012). Feeding dried chicory root to pigs decrease androstenone accumulation in fat by increasing hepatic 3β hydroxysteroid dehydrogenase expression. *The Journal of Steroid Biochemistry and Molecular Biology, 130*(1–3), 90–95.

Raucy, J. L., Lasker, J. M., Lieber, C. S., & Black, M. (1989). Acetaminophen activation by human liver cytochromes P450IIE1 and P450IA2. *Archives of Biochemistry and Biophysics, 271*(1), 270–283.

Reisman, S. A., Aleksunes, L. M., & Klaassen, C. D. (2009). Oleanolic acid activates Nrf2 and protects from acetaminophen hepatotoxicity via Nrf2-dependent and Nrf2-independent processes. *Biochemical Pharmacology, 77*(7), 1273–1282.

Rowland, A., Miners, J. O., & Mackenzie, P. I. (2013). The UDP-glucuronosyl transferases: Their role in drug metabolism and detoxification. *The International Journal of Biochemistry & Cell Biology, 45*(6), 1121–1132.

Sabir, S. M., & Rocha, J. B. T. (2008). Antioxidant and hepatoprotective activity of aqueous extract of Solanum fastigiatum (false "Jurubeba") against paracetamol-induced liver damage in mice. *Journal of Ethnopharmacology, 120*(2), 226–232.

Saleem, T. M., Chetty, C. M., Ramkanth, S. V. S. T., Rajan, V. S. T., Kumar, K. M., & Gauthaman, K. (2010). Hepatoprotective herbs—A review. *International Journal of Research in Pharmaceutical Sciences, 1*(1), 1–5.

Saracino, M. R., Bigler, J., Schwarz, Y., Chang, J. L., Li, S., Li, L., et al. (2009). Citrus fruit intake is associated with lower serum bilirubin concentration among women with the UGT1A1* 28 polymorphism. *The Journal of Nutrition, 139*(3), 555–560.

Sedlak, J., & Lindsay, R. H. (1968). Estimation of total, protein-bound, and nonprotein sulfhydryl groups in tissue with Ellman's reagent. *Analytical Biochemistry, 25*, 192–205.

Shao, H. B., Chu, L. Y., Lu, Z. H., & Kang, C. M. (2008). Primary antioxidant free radical scavenging and redox signaling pathways in higher plant cells. *International Journal of Biological Sciences, 4*(1), 8.

Sharma, O. P., Makkar, H. P. S., & Dawra, R. K. (1983). Effect of lantana toxicity on lysosomal and cytosol enzymes in guinea pig liver. *Toxicology Letters, 16*(1–2), 41–45.

Sharma, O. P., Makkar, H. P. S., Dawra, R. K., & Negi, S. S. (1981). A review of the toxicity of *Lantana camara (Linn)* in animals. *Clinical Toxicology, 18*(9), 1077–1094.

Sharma, O. P., Makkar, H. P. S., Pal, R. N., & Negi, S. S. (1980). Lantadene a content and toxicity of the lantana plant (*Lantana camara, Linn.*) to guinea pigs. *Toxicon, 18*(4), 485–488.

Sharma, V., Singh, P., Pandey, A. K., & Dhawan, A. (2012). Induction of oxidative stress, DNA damage and apoptosis in mouse liver after sub-acute oral exposure to zinc oxide nanoparticles. *Mutation Research, Genetic Toxicology and Environmental Mutagenesis, 745*(1–2), 84–91.

Sherratt, H. S. A. (1986). Hypoglycin, the famous toxin of the unripe Jamaican ackee fruit. *Trends in Pharmacological Sciences, 7*, 186–191.

Sheweita, S. A. (2000). Drug-metabolizing enzymes mechanisms and functions. *Current Drug Metabolism, 1*(2), 107–132.

Singh, A., Bhat, T. K., & Sharma, O. P. (2011). Clinical biochemistry of hepatotoxicity. *Journal of Clinical Toxicology, S4*, 001. https://doi.org/10.4172/2161-0495.S4-001. J Clinic Toxicol Clinical Pharmacology: Research & Trials ISSN: 2161-0495 JCT, an open access journal. *vitro Systems.*

Singh, B., Bhat, T. K., & Singh, B. (2003). Potential therapeutic applications of some anti-nutritional plant secondary metabolites. *Journal of Agricultural and Food Chemistry, 51*(19), 5579–5597.

Singh, A., Sharma, O. P., Dawra, R. K., Kanwar, S. S., & Mahato, S. B. (1999). Biotransformation of lantadene A (22β-angeloyloxy-3-oxoolean-12-en-28-oic acid), the pentacyclic triterpenoid, by Alcaligenes faecalis. *Biodegradation, 10*(5), 373–381.

Singh, R., Kumar, S., Rana, A. C., & Sharma, N. (2012). Different models of hepatotoxicity and related liver diseases: A review. *International Research Journal of Pharmacy, 3*(7), 86–95.

Somanawat, K., Thong-Ngam, D., & Klaikeaw, N. (2013). Curcumin attenuated paracetamol overdose induced hepatitis. *World journal of gastroenterology: WJG, 19*(12), 1962.

Strassburg, C. P., Kneip, S., Topp, J., Obermayer-Straub, P., Barut, A., Tukey, R. H., et al. (2000). Polymorphic gene regulation and interindividual variation of UDP-glucuronosyltransferase activity in human small intestine. *Journal of Biological Chemistry, 275*(46), 36164–36171.

Tahir, M., & Sultana, S. (2011). Chrysin modulates ethanol metabolism in Wistar rats: A promising role against organ toxicities. *Alcohol and Alcoholism, 46*(4), 383–392.

Tanaka, S., Uchida, S., Miyakawa, S., Inui, N., Takeuchi, K., Watanabe, H., et al. (2013). Comparison of inhibitory duration of grapefruit juice on organic anion-transporting polypeptide and cytochrome P450 3A4. *Biological and Pharmaceutical Bulletin, 36*(12), 1936–1941.

Teschke, R., Genthner, A., & Wolff, A. (2009). Kava hepatotoxicity: Comparison of aqueous, ethanolic, acetonic kava extracts and kava-herbs mixtures. *Journal of Ethnopharmacology, 123*(3), 378–384.

Teschke, R., Qiu, S. X., Xuan, T. D., & Lebot, V. (2011). Kava and kava hepatotoxicity: Requirements for novel experimental, ethnobotanical and clinical studies based on a review of the evidence. *Phytotherapy Research, 25*(9), 1263–1274.

Thapliyal, R., & Maru, G. B. (2001). Inhibition of cytochrome P450 isozymes by curcumins in vitro and in vivo. *Food and Chemical Toxicology*, *39*(6), 541–547.

Tsuchiya, Y., Nakajima, M., & Yokoi, T. (2005). Cytochrome P450-mediated metabolism of estrogens and its regulation in human. *Cancer Letters*, *227*(2), 115–124.

Ullrich, V. (1979). Cytochrome P450 and biological hydroxylation reactions. In *Biochemistry* (pp. 67–104). Berlin, Heidelberg: Springer.

Van der Logt, E. M. J., Roelofs, H. M. J., Nagengast, F. M., & Peters, W. H. M. (2003). Induction of rat hepatic and intestinal UDP-glucuronosyltransferases by naturally occurring dietary anticarcinogens. *Carcinogenesis*, *24*(10), 1651–1656.

Vistisen, K., Loft, S., Olsen, J. H., Vallentin, S., Ottesen, S., Hirsch, F. R., et al. (2004). Low CYP1A2 activity associated with testicular cancer. *Carcinogenesis*, *25*(6), 923–929.

Vivoli, E., Cappon, A., Milani, S., Piombanti, B., Provenzano, A., Novo, E., et al. (2016). NLRP3 inflammasome as a target of berberine in experimental murine liver injury: Interference with P2X7 signalling. *Clinical Science*, *130*(20), 1793–1806.

Walters, D. G., Young, P. J., Agus, C., Knize, M. G., Boobis, A. R., Gooderham, N. J., et al. (2004). Cruciferous vegetable consumption alters the metabolism of the dietary carcinogen 2-amino-1-methyl-6-phenylimidazo[4,5-b]pyridine (PhIP) in humans. *Carcinogenesis*, *25*(9), 1659–1669.

Wang, L. Q., & James, M. O. (2006). Inhibition of sulfotransferases by xenobiotics. *Current Drug Metabolism*, *7*(1), 83–104.

Wells, P. G., Mackenzie, P. I., Chowdhury, J. R., Guillemette, C., Gregory, P. A., Ishii, Y., et al. (2004). Glucuronidation and the UDP-glucuronosyltransferases in health and disease. *Drug Metabolism and Disposition*, *32*(3), 281–290.

Whittaker, P., Clarke, J. J., San, R. H., Betz, J. M., Seifried, H. E., de Jager, L. S., et al. (2008). Evaluation of commercial kava extracts and kavalactone standards for mutagenicity and toxicity using the mammalian cell gene mutation assay in L5178Y mouse lymphoma cells. *Food and Chemical Toxicology*, *46*(1), 168–174.

Williams, R. (2006). Global challenges in liver disease. *Hepatology*, *44*(3), 521–526.

Wu, Y. L., Jiang, Y. Z., Jin, X. J., Lian, L. H., Piao, J. Y., Wan, Y., et al. (2010). Acanthoic acid, a diterpene in Acanthopanax koreanum, protects acetaminophen-induced hepatic toxicity in mice. *Phytomedicine*, *17*(6), 475–479.

Wu, C. C., Sheen, L. Y., Chen, H. W., Kuo, W. W., Tsai, S. J., & Lii, C. K. (2002). Differential effects of garlic oil and its three major organosulfur components on the hepatic detoxification system in rats. *Journal of Agricultural and Food Chemistry*, *50*(2), 378–383.

Xu, C., Li, C. Y. T., & Kong, A. N. T. (2005). Induction of phase I, II and III drug metabolism/transport by xenobiotics. *Archives of Pharmacal Research*, *28*(3), 249.

Yager, J. D. (2015). Mechanisms of estrogen carcinogenesis: The role of E2/E1-quinone metabolites suggests new approaches to preventive intervention—A review. *Steroids*, *99*, 56–60.

Yamasaki, I., Yamada, M., Uotsu, N., Teramoto, S., Takayanagi, R., Yamada, Y., et al. (2008). Inhibitory effects of kale ingestion on metabolism by cytochrome P450 enzymes in rats. *Biomedical Research*, *33*(4), 235–242.

Yao, H. T., Hsu, Y. R., Lii, C. K., Lin, A. H., Chang, K. H., & Yang, H. T. (2014). Effect of commercially available green and black tea beverages on drug-metabolizing enzymes and oxidative stress in Wistar rats. *Food and Chemical Toxicology*, *70*, 120–127.

Ye, Z., Liu, Z., Henderson, A., Lee, K., Hostetter, J., Wannemuehler, M., et al. (2009). Increased CYP4B1 mRNA is associated with the inhibition of dextran sulfate sodium-induced colitis by caffeic acid in mice. *Experimental Biology and Medicine*, *234*(6), 605–616.

Yeh, H. J. (1980). Ultrastructure of continuously cultured adult human liver cell. *Acta Bio-logiac Experimentalis Sinica, 13*, 361–364.

Yin, X. M., Oltvai, Z. N., & Korsmeyer, S. J. (1994). BH1 and BH2 domains of Bcl-2 are required for inhibition of apoptosis and heterodimerization with Bax. *Nature, 369*(6478), 321–323.

Yousef, M. I., Omar, S. A., El-Guendi, M. I., & Abdelmegid, L. A. (2010). Potential protective effects of quercetin and curcumin on paracetamol-induced histological changes, oxidative stress, impaired liver and kidney functions and haematotoxicity in rat. *Food and Chemical Toxicology, 48*(11), 3246–32261.

Yoxall, V., Kentish, P., Coldham, N., Kuhnert, N., Sauer, M. J., & Ioannides, C. (2005). Modulation of hepatic cytochromes P450 and phase II enzymes by dietary doses of sulforaphane in rats: Implications for its chemopreventive activity. *International Journal of Cancer, 117*(3), 356–362.

Zarchini, M., Hashemabadi, D., Kaviani, B., Fallahabadi, P. R., & Negahdar, N. (2011). Improved germination conditions in *Cycas revoluta L.* by using sulfuric acid and hot water. *Plant Omics Journal, 4*(7), 350–353.

Zhang, T., Jiang, S., He, C., Kimura, Y., Yamashita, Y., & Ashida, H. (2013). Black soybean seed coat polyphenols prevent B (a) P-induced DNA damage through modulating drug-metabolizing enzymes in HepG2 cells and ICR mice. *Mutation Research, Genetic Toxicology and Environmental Mutagenesis, 752*(1–2), 34–41.

Chapter 23

Nanosized delivery systems for plant-derived therapeutic compounds and their synthetic derivative for cancer therapy

Henna Amin[a], Andleeb Khan[b], Hafiz A. Makeen[c], Hina Rashid[b], Insha Amin[d], Mubashir Hussain Masoodi[a], Rehan Khan[e], Azher Arafah[f], and Muneeb U. Rehman[f]

[a]*Department of Pharmaceutical Sciences, University of Kashmir, Srinagar, Jammu and Kashmir, India,* [b]*Department of Pharmacology and Toxicology, College of Pharmacy, Jazan University, Jazan, Saudi Arabia,* [c]*Department of Clinical Pharmacy, College of Pharmacy, Jazan University, Jazan, Saudi Arabia,* [d]*Division of Veterinary Biochemistry, Faculty of Veterinary Sciences and Animal Husbandry, Sheri Kashmir University of Agricultural Science and Technology (SKUAST-K), Srinagar, Jammu and Kashmir, India,* [e]*Institute of Nano Science and Technology, Mohali, Punjab, India,* [f]*Department of Clinical Pharmacy, College of Pharmacy, King Saud University, Riyadh, Saudi Arabia*

1. Introduction

Cancer, a multifactorial disease, is recognized as one of the primary reasons of mortality around the globe. In 2019 1,762,450 cancer cases with death rates of 606,880 were reported only in the United States (Siegel, Miller, & Jemal, 2019). Despite the various advancements in the field of modern medicine, cancer treatment remains a challenging task. Treatment of cancer usually involves chemotherapy, radiotherapy, or surgery, out of which the most common method is chemotherapy (Liu, 2009). Cancer chemoprevention by definition means making use of bioactive or chemically synthesized compounds for inhibition and prevention of the cancer relapse (Hong & Sporn, 1997). The synthetic drugs used as cancer therapeutics have number of side effects and are costly, which increase the socioeconomic burden. The traditional system of medicines involves the use of compounds of natural origin. The rationale for the use of phytometabolites over synthetic drugs is their safety profile, ease of availability, and nontoxic nature. These metabolites are also known as "phytochemicals," which has got its origin from the Greek word "phyto," which refers to plant. Thus these

Phytomedicine: A Treasure of Pharmacologically Active Products from Plants
https://doi.org/10.1016/B978-0-12-824109-7.00020-0
655

are the active plant constituents found usually in fruits, cereals, herbs, and legumes. These are usually secondary metabolites produced by an organism as defense against predators or as an adaptive measure to various stressful conditions. These are not generally essential for the survival of the plant (Dewick, 2002). On the basis of the chemical nature, phytometabolites are divided into polyphenolic compounds, terpenes, and thiols. The polyphenol group consists of flavonoids, phenolic acids, and other nonflavonoid polyphenols such as tannin, curcumin, and lignans. Among these the phenolic compounds have proved their worth as potent antioxidant, antineoplastic, antimicrobic, hypoglycemic, and neuroprotective agent (Oz & Kafkas, 2017). Phytochemicals inhibit carcinogenesis by various underlying mechanisms such as disruption in the cancer signaling pathways, antiproliferative effect, prevention of angiogenesis, and additive effect with various chemotherapeutics. These bioactive constituents have played a vital role in the treatment of infections and various cancer types (Ojima, 2008). With the advancements in high-throughput screening and combinatorial chemistry, there was an increase in the screening of natural compounds as molecular leads for drug developmental processes and in the synthesis of new chemical entities (Nussbaum, Brands, Hinzen, Weigand, & Habich, 2006). The use of the natural products and their semisynthetic derivatives as cancer chemotherapeutics started from the late 1930s to early 1940s. In the recent era, about 7000 medicinal plants are used in the modern medicine based on the use in the folklore medicine (Eddouks, Chattopadhyay, Feo, & Chi-shing Cho, 2012). These are used as such or provide templates for development of analogues and prodrugs (Grothaus, Cragg, & Newman, 2010). Taxol, an anticancer drug originally isolated from *Taxus brevifolia* (western yew) that was approved for clinical use by FDA in the year 1992, has gained utmost importance in breast cancer treatment.

Out of 174 anticancer drugs that have been approved for cancer treatment, 53% are natural product-based drugs or their semisynthetic derivatives (Amaral, Dos Santos, Andrade, Severino, & Carvalho, 2019). Phytoconstituents being used as anticancer drugs are paclitaxel obtained from *Pacific yew*, vincristine from *Vinca rosea*, and camptothecin from *Camptotheca acuminate* Decne (Saklani & Kutty, 2008).

2. Phytochemicals and their synthetic analogues with anticancer potential

Paclitaxel (Taxol), a diterpene alkaloid obtained from *T. brevifolia* (Pacific yew), is mainly used as a chemotherapeutic in various carcinomas such as breast cancer, bronchial cancer, Kaposi sarcoma, bladder cancer, and endometrial and cervical cancer. The action is exerted by cell cycle arrest and mitotic inhibition by binding to tubulin (Ferlini et al., 2003). Taxol for medicinal use is now produced synthetically as it was obtained in minute quantities from the natural source. In addition to this, 10-deacetylbaccatin III (baccatin III) and other structurally related compounds that occur naturally in *T. brevifolia* are used as templates for taxol synthesis (Nicolaou et al., 1994). Docetaxel is the most potent synthetic analogue of paclitaxel first synthesized in laboratory by Bissery and associates with certain advantages over the original drug candidate such as increased bioavailability, better in vivo activity profile, and potent antineoplastic activity (Bissery, Gubnard, Gubritte-Voegelein, & Lavelle, 1991). The synthetic derivative of paclitaxel known as cabazitaxel is used as a combination therapy along with prednisone in the treatment of metastatic prostate carcinoma (Villanueva, Bazan, Kim, et al., 2011).

Podophyllotoxin is an active constituent obtained from the roots and rhizomes of *Podophyllum peltatum* L. or mayapple and *Podophyllum hexandrum* (Indian podophyllum). The constituent is a resin with structural similarities to those of aryltetralin lactone lignans, medicinally being used as an antineoplastic, antiviral, and anthelmintic agent. The compound prevents cancer progression by reversible binding to tubulin, which causes a disorientation of the microtubular assembly and inhibits mitosis (Guerram, Jiang, & Zhang, 2012). Its anticancer profile was first investigated by Hartwell and Shear in the year 1940. The compound despite being a potent antitumor agent was not approved for human use as high drug dose was required to elicit pharmacological action, which resulted in various side effects, the most common being gastrointestinal effects. It was in the year 1960 that group of Stahelin and Sandoz evaluated a panel of about 50 semisynthetic analogues of this compound in vivo. Two compounds, etoposide and teniposide, were found to have marked antineoplastic activity with reduced side effects (Loike, 1984). Etoposide is used as a chemotherapeutic in ovarian cancer, small cell lung cancer, lymphomas, brain tumor, and acute leukemia in combination with various other anticancer drugs (Van Maanen, Retel, De Vries, et al., 1988). Teniposide is used as medication in childhood acute lymphoblastic leukemia, brain tumors such as glioma, and Hodgkin's lymphomas.

Camptothecin is a bioactive compound obtained from *Camptotheca acuminate*, which belongs to the class of quinoline-type alkaloid and is used as a cytotoxic agent (Wall et al., 1966). The antitumor activity of the compound is due to the presence of lactone ring that attaches DNA topoisomerase I cleavage complex and causes DNA inhibition and apoptotic death in cancer cells. The limiting factors in the use of this phytoconstituent as anticancer agent include its insolubility and instability at the physiological pH. Although a number of camptothecin derivatives have been synthesized chemically, only two of them are clinically approved, which include irinotecan and topotecan. Irinotecan is used clinically in the remission of colorectal cancer, whereas topotecan is used for nonsmall cell lung cancer and ovarian and cervical carcinomas (Li, Zu, Shi, & Yao, 2006).

3. Problems encountered in use phytoconstituents as anticancer agents

Phytoconstituents are promising drug candidates but pose certain drawbacks that limited their use as therapeutics.

The nature consists of reservoir of phytoconstituents with various medicinal properties. Sometimes a single constituent possess different pharmacological activities. The choice of the right constituent for certain diseases is difficult from a plethora of compounds as it requires extensive screening; furthermore the extraction and isolation of the constituents is a cumbersome process.

Phytometabolites have limited clinical applications as the in vitro results are not well reflected when tested in in vivo models. This may be due the instability, inappropriate solubility, reduced lipophilicity, extensive hepatic first-pass metabolism, lack of target affinity, and drug decomposition in the gastric milieu resulting in reduced bioavailability (Cragg, Grothaus, & Newman, 2009). Phytoconstituents have high molecular masses, different types of chemical functionalities, and steric hindrance, which pose a challenging task for development of pharmaceutical formulations. These compounds have huge skeletons that are at times difficult to modify (Siddiqui, Iram, Siddqui, & Sahu, 2014). All these drawbacks in the clinical use of natural compounds urge to look for alternate options to attain the targeted and sustained delivery of these active constituents.

Various active metabolites such as epigallocatechin-3-gallate (EGCG), resveratrol, and curcumin have proved their worth as antitumor drugs in the treatment of gliomas, but due to the blood-brain barrier, inappropriate concentrations reach the tumor site, thus limiting their use. All these loopholes in the use of phytoconstituents demand the need for development of alternate means to achieve targeted delivery to the tumor sites (Pistollato et al., 2015).

4. Nanotechnology and phytomedicine

Phytometabolites despite having anticancer properties do not have many clinical applications in cancer treatment because of the solubility issues, in vivo instability, poor tissue specificity, decreased drug concentration at tumor site, and associated toxicities. In chemotherapy, pharmacologically active cancer drugs reach the tumor tissue in less specific concentrations with dose-limiting toxicity.

Nanotechnology, which refers to the use of nanoparticles, has gained much importance in the medical field for the targeted drug delivery, gene delivery, biosensing, artificial implants, etc. (Wagner, Dullaart, Bock, & Zweck, 2006). The delivery of anticancer agents with the help of these nanocarriers has helped to overcome various challenges in phytomedicine such as addressing the solubility, stability, and bioavailability issues with them. Among various classes of nanomaterials used, the most common products used are liposomal drug and drug polymer conjugate delivery systems. Ringsdorf in 1975 first gave an idea regarding the use of a polymer as anticancer drug delivery agent. He proposed that the pharmacological properties of the system could be altered by chemical and physical modifications in polymer material (Ringsdorf, 1975). Certain solubilizing agents can be used to overcome solubility issues. "Phytonanotechnology" deals with green synthesis of metal nanoparticles via use of phytoconstituents and their proper tailoring and targeted delivery to site of action (Table 1).

TABLE 1 Phytoconstituent-loaded nanodelivery systems with anticancer activity.

Phytoconstituent	Nanoparticles used	Cancer type	Inferences	References
Curcumin	Solid lipid nanoparticles	SKBR3 breast cancer	IC50 18.78 μM of SLNS-Cur, ↑antiproliferative effect, ↑cellular uptake	Wang et al. (2018)
Curcumin	Solid lipid nanoparticles	A549 lung cancer cell lines	Inhibition of lung cancer cell growth, ↑apoptosis, ↑volume of distribution	Wang, Zhang, Peng, et al. (2013)
Resveratrol	Solid lipid nanoparticles	MDA-MB-231 breast cancer cell lines	↓Cell proliferation	Siemann & Creasy (1992) and Wang et al. (2017)
EGCG	Polymeric nanoparticles	Prostate cancer	↑Antiproliferative effect, ↑apoptotic cell death and cell cycle modification	Sanna et al. (2017)
Coumarins	Polymeric nanoparticle	Human melanoma	↑Cellular uptake profile	Khuda-Bukhsh, Bhattacharyya, Paul, and Boujedaini (2010)
Apigenin	Polylactic-co-glycolic acid (PLGA)	Skin carcinoma	Inhibiting recurrence of chromosomal abnormalities; adjustment of certain apoptotic markers, thus preventing carcinogenesis	Das, Das, Samadder, Paul, and Khuda-Bukhsh (2013)
Resveratrol	Polyethylene glycol-polylactic acid (PEGPLA)	U87 glioma and mouse xenograft model of C6 glioma	↑Cancer cell death, ↑drug conc. ↓tumor mass	Guo et al. (2013)

Continued

TABLE 1 Phytoconstituent-loaded nanodelivery systems with anticancer activity—cont'd

Phytoconstituent	Nanoparticles used	Cancer type	Inferences	References
Hesperetin	Liposomes	H441 and MDA-MB-231 cancer cells	↑Drug stability and sustainability in the serum	Wolfram et al. (2016)
EGCG	Liposomes	Basal cell carcinoma	↑Conc. at the tumor sites, ↑death of tumor cells, ↑20 times drug conc.	Fang, Hung, Hwang, and Huang (2005) and Fang, Lee, Shen, and Huang (2006)
Quercetin	Liposome	MCF-7 breast cancer cell lines	↑Cellular uptake	Sun, Nie, Pan, et al. (2014)
Curcumin	Gold nanoparticles with hyaluronic acid conjugates	HeLa cells, glioma cells, and Caco 2 cells	↑Cellular uptake and localization of the particles by cells	Manju and Sreenivasan (2012)
EGCG	Gold nanoparticles	MBT-2 bladder cancer cell lines	Selective cytotoxicity, ↓ in tumor size	Hsieh et al. (2011)
EGCG	Gold nanoparticles	B16F10 murine melanoma cells	4.91-fold ↑in apoptosis, ↓tumor volume, antiproliferative effect ↑1.66 times	Chen et al. (2014)
Quercetin	Nanomicelles	A549 lung cancer xenograft murine model	↑Aqueous conc., ↑cellular penetration	Tan, Liu, Chang, et al. (2012)
Embelin	PEG micelles	PC3 xenograft murine model	↑Anticancer activity	Lu et al. (2013)

5. Phytoconstituents loaded nanoparticles with anticancer activity

5.1 Solid lipid nanoparticles

Solid lipid nanoparticles are nanocarriers produced either from synthetic or naturally occurring polymers. These are for both the systemic and topical drug delivery (Aditya et al., 2014).

Curcumin is a phytoconstituent isolated from the roots and rhizomes of *Curcuma longa*. It has well-established anticancer properties, but the constituent has not been so far used as drug as there is a distinctive behavior of the compound in in vitro and in vivo antineoplastic studies that might be due to certain physiochemical differences such as solubility and stability resulting in poor bioavailability. Curcumin prevents cancer progression by affecting the hallmarks of tumor development such as induces cell cycle arrest, antiproliferative effect, prevents colony formation of cancer cells, apoptosis in various gastrointestinal cancers including colorectal and pancreatic carcinomas (Gali-Muhtasib, Hamadi, Kareh, et al., 2015).

Curcumin is encapsulated into solid lipid nanoparticles (SLNs) to overcome the limitations in the use of curcumin as anticancer drug. The Cur-SLNs were evaluated in SKBR3 cancer cells by sulforhodamine B: IC50 28.42 and 18.78 μM for free Cur and Cur-SLNs. The nanocarrier-loaded curcumin demonstrated an increase in antiproliferative effect. Drug concentration and cellular uptake were increased at the tumor site, which in turn enhanced the anticancer effect of curcumin Wang et al., 2018).

Resveratrol is a naturally occurring polyphenol form of stilbene found in grapevine, raspberry, and mulberry, commonly developed in response to certain stress forms, such as sunlight, UV absorption, injury, and heavy metal presence. It is a bioactive compound as it possesses inherent antiproliferative, antioxidant, and antiinflammatory activity (Siemann & Creasy, 1992). Many in vitro and in vivo studies have concluded the antineoplastic effect in ovarian, breast, and prostate carcinoma (Vergara et al., 2017). The compound still faces challenges in human use as it has limited bioavailability, susceptibility to enzymatic degradation, and reduced target specificity.

Wang et al. carried a study using solid lipid nanoparticles as drug carriers of resveratrol against resistant MDA-MB-231 cancer cell lines, which resulted in an increase in antiproliferative activity of resveratrol. The nanoparticle-loaded Res-SLNs decreased the growth of MDA-MB-231 cancer cells by disrupting the invasion and migration of cancer cells. Thus solid lipid nanoparticles offer a greater advantage as vehicles for the delivery of resveratrol in cancer cells (Wang et al., 2017).

An investigation by Wang et al. led to the findings that solid lipid nanoparticles of curcumin when injected intraperitoneally in nude mice xenografted with A549 lung cancer cells resulted increase in vivo drug stability, inhibition of lung cancer cell growth, increased apoptosis, and increased volume of distribution resulting in increased anticancer activity of curcumin (Wang et al., 2013).

5.2 Polymeric nanoparticles

Polymeric NPs or microparticles are used as drug delivery systems in which the drug substances are loaded either by encapsulation or by covalent bonding. Structurally, these are built up of monomers generally present in body and are nontoxic. These nanocarriers have different characteristics, such as ecological, biostable, water soluble, ease of structural modifications, which have increased their ability as nanodelivery systems (Kumari, Yadav, & Yadav, 2010).

Sanna et al. formulated the polymeric nanoparticles of EGCG with three different nanoparticle carriers and evaluated them against mouse xenograft model of prostate cancer. The study led to the findings that the encapsulated EGCG augmented the anticancer effect of EGCG against the tested cancer cell line as compared with that of free EGCG, which was contributed to an increase in apoptotic cell death and cell cycle modification. The encapsulated EGCG has tumor specificity, increased drug concentration at site of action, and reduced associated toxicities, thus becoming a promising carrier for clinical utility (Sanna et al., 2017).

Coumarins that belong to the chemical class of benzo-α-pyrones have various biological activities such as antimicrobial, antineoplastic, antiinflammatory, and CNS stimulant. The compound has a good potential to be used as antitumor drug. Polymeric nanoparticles loaded with coumarin showed better bioavailability and increased cellular uptake profile as compared with that of free coumarin in human melanoma cells (Khuda-Bukhsh et al., 2010).

5.3 Polylactic-co-glycolic acid

Polylactic-*co*-glycolic acid is a biocompatible polymer used as nanovehicle for drug delivery in case of various medicinal formulations. This is usually produced from PLGA and other related polymers such as poly(lactic acid) and poly(glycolic acid) (Pandey, Jain, & Chakraborty, 2015).

Cui et al. developed T7-modified magnetic PLGA nanoparticulate system prepared with coencapsulation of the hydrophobic magnetic nanoparticles (MNP/T7-PLGA NPs), loaded with combination of paclitaxel and curcumin; in vitro cytotoxicity effects were examined in U87 cells; and IC50 for encapsulated system of PTX + CUR was reported to be 3.06 μg/mL with the coadministration of PTX + CUR. In vivo administration in murine model with orthotopic glioma showed increase in the bioavailability and targeted delivery (Cui et al., 2016).

Apigenin belonging to the chemical class of flavonoids occurs most commonly in the members of Asteraceae family, such as *Artemisia*, *Achillea*, and *Matricaria*. The compound usually possesses antioxidant, antiinflammatory, neuroprotective, and antitumor activity. Despite being a potent constituent, it is not used in therapeutics due to its low aqueous solubility and high permeability, which limit its use (Salehi et al., 2019).

Das et al. (2013) formulated apigenin-loaded PLGA nanoparticles and tested them against skin carcinoma both in vitro and in vivo. The nanoparticle-loaded apigenin offered the advantages of being tissue compatible, inhibiting recurrence of chromosomal abnormalities, and adjustment of certain apoptotic markers, thus preventing carcinogenesis (Das et al., 2013).

Guo et al. investigated the effect of transferrin (Tf)-modified polyethylene glycol-polylactic acid (PEGPLA) loaded resveratrol (Tf-PEG-PLA-Res) in vitro in U87 glioma and C6 glioma cells and in vivo in mouse xenograft model of C6 glioma The results revealed that the encapsulated resveratrol increased the cell death in the tested cell lines. In vivo administration into the murine model of C6 glioma demonstrated an increased drug concentration in the tumor site and reduction in the tumor mass, thus increasing the survival rate in the rat model (Guo et al., 2013).

5.4 Liposomes

Liposomes have structurally a mono- or multilayered vesicle wherein hydrophilic substances are entrapped in core and lipophilic substances adhered to the membrane of liposomes. The membrane of liposomes is usually made up of phospholipids. These targeted drug delivery vehicles usually increase the solubility, stability, and bioavailability of the enclosed drug substances (Wang, Langer, & Farokhzad, 2012; Wang, Wang, et al., 2012).

Hesperetin is a flavanone present abundantly in citrus fruits such as oranges and grapefruit. It has antiinflammatory, antioxidant, antiallergic, and antitumor properties (Kawaii, Tomono, Katase, Ogawa, & Yano, 1999). The anticancer effect of this flavone is due to inhibitory effect on cellular proliferation, apoptosis, angiogenesis in tumor outgrowths, and downregulation of various pathways (Aranganathan & Nalini, 2013). Use of hesperetin as drug candidate is limited because of its hydrophobic behavior.

Wolfram et al. developed liposome-loaded hesperetin as drug delivery system against H441 and MDA-MB-231 cancer cells. The nanocarriers provided an advantage in comparison with free hesperetin as it increased the drug stability and sustainability in the serum, thus potentiating its anticancer activity (Wolfram et al., 2016).

Fang et al. carried out in vivo anticancer study in female nude mice BCC (basal cell carcinoma) using liposomes as carriers for EGCG and other catechins. The drug was administered both topically and at intratumor sites. Intratumor injection resulted in the increased concentration of EGCG at the tumor sites (Fang et al., 2005). Another investigation done by the similar body prompted the conclusions that when EGCG-containing liposomes were administered into basal cell carcinoma (BCCs) resulting in increased death of tumor cells, the drug concentration was increased 20 times in comparison with unencapsulated drug (Fang et al., 2006).

Quercetin is a flavonoid that occurs naturally in fruits and vegetables. It is a natural antioxidant that plays a role in free radical scavenging. Its antioxidant property contributes to its anticancer, antihypertensive, antiviral, antiplatelet aggregatory, and hypolipidemic properties (Aguirre, Arias, Macarulla, Gracia, & Portillo, 2011). It has been found to be an effective therapy in the treatment of bronchial, breast, endometrial, gastric, and colon cancers. The shortcomings in the use of quercetin as a therapeutic is its hydrophobic nature, poor absorption, and reduced target specificity.

Formulation of liposomal drug delivery for quercetin against MCF-7 breast cancer cell lines resulted in increment in cellular uptake by the cancer cells, thus resulting in increased activity profile of the encapsulated drug (Sun et al., 2014).

5.5 Gold nanoparticles

Gold nanoparticles have indispensable position in the field of medicine due to their potent antimicrobial and antineoplastic effect (Tomar & Garg, 2013). Gold nanoparticles used as carriers for various active pharmacological components help achieve continuous and targeted delivery of specific drugs (Geetha et al., 2013).

Manju and Sreenivasan carried out a study using gold as a vehicle for water-soluble conjugates of curcumin and hyaluronic acid tested against a panel of cancer cell lines (HeLa cells, glioma cells, and Caco 2 cells), resulting in increased cellular uptake and localization of the particles by cells. HA-Cur@AuNPs exhibited more cytotoxicity compared with free Cur. The gold nanoparticles upon conjugation with folate-PEG improved cellular targeting. These encapsulated Cur particles showed increase in activity due to increase in tissue drug concentrations and availability at the tumor site (Manju & Sreenivasan, 2012).

Hsieh et al. studied the in vitro and in vivo anticancer activity of EGCG gold-coated nanoparticles against MBT-2 bladder cancer cell lines. Au-coated nanoparticles of EGCG exhibited increased activity profiles with selective cytotoxicity toward cancer cells without affecting the normal surrounding cells. In vivo studies in murine models showed reduction in tumor size as compared with free EGCG (Hsieh et al., 2011).

Chen et al. formulated gold nanoparticles of EGCG and evaluated in vitro and in vivo against melanoma. In vitro studies revealed that the nanoencapsulated particles displayed 4.91-fold increase in apoptosis in B16F10 murine melanoma cells compared with nonencapsulated EGCG. This nanocarrier likewise had biostability, inciting less damage to erythrocytes. In vivo outcomes revealed that the intratumoral injection of EGCG NPs reduced the tumor volume of a mouse melanoma model in contrast to the control used. The antiproliferative effect was amplified 1.66 times compared with free EGCG (Chen et al., 2014).

5.6 Silver nanoparticles

Silver nanoparticles have an immense importance in the field of nanotechnology. The importance is attributed due to the properties of silver particles such as its biocompatibility, physical and chemical stability, nontoxic behavior, and being economical and easily available. The formulation of biogenic silver nanoparticles has revolutionized the use of silver as drug delivery candidate (Zhang, Liu, Shen, & Gurunathan, 2016).

Prabhu et al. synthesized biogenic nanoparticles of Ag from the methanol extracts of *Vitex negundo* L. The nanoparticles resulted in 50% inhibition in malignancy of human colon cancer cell lines (HCT 15) when administered at dose of 20 µg/mL. The potentiation of the anticancer effect on HeLa cell lines upon dose increment has also been documented. The effects shown usually depend on the shape and size of Ag nanoparticle synthesized (Prabhu, Arulvasu, Babu, Manikandan, & Srinivasan, 2013).

5.7 Micelles

Micelles are nanocarriers with certain properties such as better solubility, tissue penetration, target selectivity, and in vivo stability (Keskin & Tezcaner, 2017).

Tan et al. loaded quercetin in nanomicelles and administered them in murine xenograft model of A549 lung cancer. Increase in the aqueous solubility of quercetin (by 110-fold to 3 mg/mL) by using solubilizing agents in the nanomicelles and increase in penetration of the micelles through epithelial tight junctions synergized the anticancer drug effect (Tan et al., 2012).

Embelin is a phytoconstituent obtained from the fruits of *Embelia ribes* Burm. The compound has medicinal importance due to its anticancer, hypoglycemic, antioxidant, hepatoprotective, and neuropharmacological effects. It has shown potent results in breast, colon, and prostate cancers. The compound has retarded aqueous solubility and as such cannot be formulated into a drug candidate (Kundap, Bhuvanendran, Kumari, Othman, & Shaikh, 2017).

Lu et al. evaluated the effect of embelin conjugated with PEG micelles. The anticancer activity of embelin was retained within these micelles and was found to have a synergistic effect with paclitaxel against various tested cancer cell lines. The results showed an increase in the drug concentrations in the tumor microenvironment of PC3 xenograft murine model. Delivery of paclitaxel via PEG_{5K}-embelin$_2$ micelles prompt better taxol activity in mouse models of breast and prostate malignancies (Lu et al., 2013).

5.8 Dendrimers

Dendrimers are globular, monodisperse, well-defined, three-dimensional moieties, the shape and size of which can be easily modified in accordance with the compound to be loaded. The structure represents number of dendritic branches (hydrophobic and hydrophilic moieties) rising out of a focal center dendrimer. These delivery vehicles are mostly suited for gene therapy, imaging, drug delivery for example in anticancer, antiinflammatory drugs (Mignani et al., 2018).

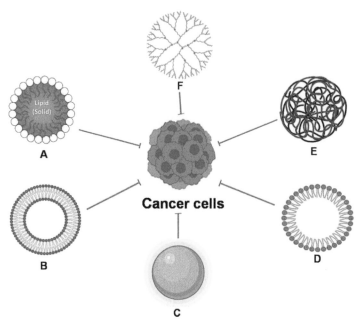

FIG. 1 Schematic representation of commonly used phytochemical nanoparticles for the treatment of cancer. (A) Solid lipid nanoparticles. (B) Liposomes. (C) Gold nanoparticle. (D) Micelle. (E) Polymeric nanoparticle. (F) Dendrimers.

Ooya et al. formulated polyglycerol dendrimers (PGDs) loaded with paclitaxel, which increased the aqueous solubility of the drug, which is otherwise the most common problem encountered in the use of free form of paclitaxel (Ooya, Lee, & Park, 2004) (Fig. 1).

6. Safety and toxicity of the developed nanoparticles

Nanotechnology has been an efficient tool for combating cancer, and there is no danger in the use of nanoparticles. As a science, nanotechnology is at its early age, but nanoparticles are as old as our Earth. In volcanic ash and sea spray to smoke and soot from fire, all contain nanoparticles. With the development of new technology, nanoparticles are continuously being tested. The nanosize of the particle and its reactivity and unique properties have always raised questions about the safety of nanoparticles for the environment, health, and safety (EHS).

Phytochemical-loaded nanoparticles have been extensively used to prevent various cancers. It has been reported that epigallocatechin gallate (EGCG), found in abundance in green tea, has chemopreventive potential, which acts by antiangiogenesis, apoptosis, and cell cycle arrest (Cao & Cao, 1999; Du, Zhang, Wen, et al., 2012; Lambert & Yang, 2003). The use of EGCG is limited in cancer prevention due to various factors like instability in water and physiological fluids (Hong et al., 2002), less bioavailability (Lee et al., 2002), quick degradation by liver enzymes, and low target specificity to tumors. EGCG nanoparticles could overcome these limitations and can be used as a therapeutic against cancer cells. Target ligands were incorporated on the surface of EGCG nanoparticles to target prostate cancer as suggested by Sanna et al. (2011). Another study made micellar nanocomplexes by using EGCG derivatives, which delivered herceptin to breast cancer cells. These nanocomplexes lowered cancer cell viability in vitro and inhibited tumor growth in vivo (Chung et al., 2014). Similarly, curcumin-loaded nanoparticles have increased their aqueous solubility, bioavailability, target delivery, and sustained-release property (Mimeault & Batra, 2011). Many forms of nanoparticles of curcumin like liposomes, SLNs, and PGLA nanoparticles have reported to be promising drug of choice in the treatment of various cancers (Nair et al., 2010; Mimeault & Batra, 2011; Yallapu, Jaggi, & Chauhan, 2012). Quercetin nanoparticles are also a good strategy for cancer prevention and treatment as suggested by many reports (Rezaei-Sadabady, Eidi, Zarghami, & Barzegar, 2016; Wang, Langer, & Farokhzad, 2012; Wang, Wang, et al., 2012). The efficacy of nanoquercetin to decrease the cell viability of A549 lung cancer cells was high as compared with free quercetin in vitro (Narayanan, Nargi, Randolph, & Narayanan, 2009). Nanotechnology can overpower the limitations of quercetin and can enhance the anticancer effects of quercetin. Similarly, nanoresveratrol and nanogenistein have also shown better results against cancers and tumors as compared with their parent compound (Guo et al., 2010; Phan, Walters, Brownlow, & Elbayoumi, 2013; Vergaro, Lvov, & Leporatti, 2012). Nanoparticles due to their target delivery cause the release of the phytochemical at the site, which hence increases their bioavailability and does not harm the other organs or healthy tissue.

To ensure that nanotechnology risks are extensively evaluated, the National Cancer Institute (NCI) in concert with the National Institute of Standards and Technology (NIST) and the US Food and Drug Administration (FDA) has established the Nanotechnology Characterization Laboratory (NCL) to perform preclinical efficacy and toxicity testing of nanoparticles in vitro and in animal models. The potential health risks associated with the manufacture and use of nanomaterials must be carefully studied, whether actual or observed. This will make head way in our understanding of nanotechnology and its benefits for cancer research.

The nanoparticle use in the cancer treatment can cause unnecessary toxicity by the interaction of NPs with biological entities (Wang, Santos, Evdokiou, & Losic, 2015). Several studies have reported the pernicious properties of nanoparticles due to the toxicity linked to them (Coradeghini et al., 2013; Ji et al., 2012). It is well known that the nanoparticle toxicity is due to its size, concentration, and type (Namdari, Eatemadi, Soleimaninejad, & Hammed, 2017). By the decomposition of the nanoparticle, the toxic substance release is the main cause of toxicity depending upon the material used to manufacture the nanoparticle

(Pelaz et al., 2013). Even the same nontoxic nanoparticle can behave as toxic if slight changes in surface coating, size, or charge are incorporated. Many variables have to be kept in mind while formulating a nanoparticle like material used, size, coating material, surface charge, agglomeration, and aggregation. Another issue is the complex behavior of correlation between in vitro and in vivo use of nanoparticle as they behave differently in different environment. The toxicity of nanoformulations also depends upon the route of administration and site of action. NPs have the ability to cross the organelle membranes and gets deposited into major organelles like mitochondria and nucleus leading to cytotoxicity. Gold nanoparticles are considered safe as its core is nontoxic and inert; still in one experiment, several gold particles with different capping agents have shown cytotoxicity against leukemia cell line (Connor, Mwamuka, Gole, Murphy, & Wyatt, 2005). The cytotoxicity depends on type of toxicity assay, cell line, and other properties (Patra, Banerjee, Chaudhuri, Lahiri, & Dasgupta, 2007). Silver nanoparticles are also used widely in several cancer therapies. As compared with other nanoparticles, silver nanoparticles have increased the reactive oxygen species (ROS) generation and lactate dehydrogenase (LDH) release in the cells (Hussain, Hess, Gearhart, Geiss, & Schlager, 2005). Polymeric and biodegradable NPs have been used to target the drugs or phytochemicals to the cancer cells and have shown very less toxicity. They are hydrolyzed to biocompatible metabolites. Recently a study reported that surface coating of these polymeric nanoparticles induced the toxicity toward human-like macrophages (Grabowski et al., 2015).

7. Clinical uses

The end-state vision of phytochemical-encapsulated nanoparticles is their use from laboratories to the clinics. Scientists are trying hard to achieve this goal, but there are several limitations and challenges to this approach. The significant challenges include maintenance of size of the particle, reproducibility from batch to batch, and scarcity of knowledge of interaction of nanoparticle with living system (Ryan & Brayden, 2014). Still few examples are available, which are under clinical trials and can be used as therapeutics against cancer cells.

Several drugs targeting isomerase are frequently used in cancer prevention. One of the examples is irinotecan, a semisynthetic analogue of camptothecin from *Camptotheca acuminata* tree that triggers apoptosis in cells by targeting topoisomerase I (Gitiaux et al., 2018). MM-398 is a liposome conjugated with irinotecan that is approved by FDA for the treatment of metastatic pancreatic cancer (Inman, 2015). Another target for chemoprevention could be tubulin assembly inhibitors. Paclitaxel and vincristine sulfate are examples of this category that inhibits the proliferation of cancer cells by blocking mitosis via inhibition of tubulin formation (Ngan, Bellman, Hill, Wilson, & Jordan, 2001; Ngan et al., 2000). Nanoparticle conjugated with paclitaxel increases treatment efficacy and has shown effect on the metastatic stage of breast cancers along with shRNA

(Gilmore et al., 2013). The combined use of chemotherapy with phytochemical nanoparticles has shown optimistic results against cancers. Breast, lung, and pancreatic cancers have been cured by an FDA-approved albumin-bound paclitaxel nanoparticle, Abraxane (Tran, DeGiovanni, Piel, & Rai, 2017). Another polymeric micelle conjugate with paclitaxel, Genexol-PM, has been approved by Korea for treatment of breast cancer (Tran et al., 2017). Marqibo, a vincristine sulfate conjugated with liposomes, is FDA-approved drug against acute lymphoblastic leukemia (Tran et al., 2017) (Fig. 2).

A study entitled "A Clinical Trial to Study the Effects of Nanoparticle Based Paclitaxel Drug, Which Does Not Contain the Solvent Cremophor, in Advanced Breast Cancer" is under phase 1 trial with identifier number NCT00915369 (ClinicalTrials.gov, NCT00915369, 2010). Another study titled "Neoadjuvant Chemotherapy of Nanoparticle Albumin-bound Paclitaxel Plus Carboplatin/ Cisplatin in Stage II B and IIIA Non-small Cell Lung Cancer Patients" is going on in phase 2 trial (ClinicalTrials.gov, NCT02016209, 2013). A study under clinical trials in phase 2/3 is undergoing to check the combination of capecitabine and paclitaxel (albumin-stabilized nanoparticle formulation) in treating women undergoing surgery for stage II or stage III breast cancer (ClinicalTrials.gov, NCT00397761, 2009).

Though many nanoencapsulated phytochemicals are developed against various cancers, very few can be therapeutically used for cancer cell targeting. The criteria for the manufacturing, testing, and safety evaluation of nanoparticles have to be checked and regulated. The interaction of nanoparticles with biological system will always influence the clearance rate and the target of nanoformulation. Extensive research has to be conducted to realize the widespread use of nanoparticles at clinical levels.

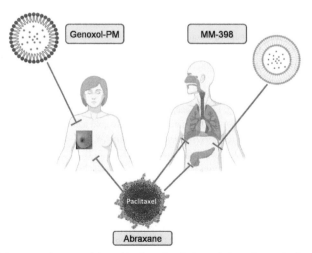

FIG. 2 FDA-approved nanoparticles with phytochemicals or their analogues in the treatment of various cancers.

8. Conclusion

This chapter emphasized on the use of medicinal plants on cancer therapy with special highlights on the green synthesis of nanoparticles. More than 65% of anticancer drugs are derived from natural resources. Chemical synthesis of synthetic analogues of phytochemicals have given a boom in the development of drugs for cancer treatment. The limitations of low solubility, poor bioavailability, and healthy organ toxicity have been overcome by encapsulating these phytochemicals in various types of nanoformulations. By coating the surface of the nanoparticles, they can be delivered to specific targets (tumors or cancers) without affecting the healthy organs. A lot of preclinical research has shown promising results in the use of green nanoparticles as anticancer drugs. Several clinical studies have approved the use of phytochemical-loaded nanoparticle along with conventional chemotherapy for use of various metastatic cancers, while others are under clinical trials (Pezzani et al., 2019). Still the use of nanoparticles in oncology from bench to the bedside requires efficient formulation targeting strategies and acquiescing to international standard for biocompatibility and toxicity.

References

Aditya, N. P., Macedo, A. S., Doktorovova, S., Souto, E. B., Kim, S. H., Chang, P. S., et al. (2014). Development and evaluation of lipid nanocarriers for quercetin delivery: A comparative study of solid lipid nanoparticles (SLN), nanostructured lipid carriers (NLC), and lipid nanoemulsions (LNE). *LWT—Food Science and Technology, 59*, 115–121.

Aguirre, L., Arias, N., Macarulla, M. T., Gracia, A., & Portillo, M. P. (2011). Beneficial effects of quercetin on obesity and diabetes. *The Open Nutraceuticals Journal, 4*, 189–198.

Amaral, R. G., Dos Santos, S. A., Andrade, L. N., Severino, P., & Carvalho, A. A. (2019). Natural products as treatment against cancer: A historical and current vision. *Clinical Oncology, 4*, 1562.

Aranganathan, S., & Nalini, N. (2013). Antiproliferative efficacy of hesperetin (citrus flavonoid) in 1,2-dimethylhydrazine-induced colon cancer. *Phytotherapy Research, 27*(7), 999–1005.

Bissery, M. C., Gubnard, D., Gubritte-Voegelein, F., & Lavelle, F. (1991). Experimental antitumor activity of taxotere (RP 56978, NSC 628503), a taxol analogue. *Cancer Research, 51*(18), 4845–4852.

Cao, Y., & Cao, R. (1999). Angiogenesis inhibited by drinking tea. *Nature, 398*(6726), 381.

Chen, C. C., Hsieh, D. S., Huang, K. J., Chan, Y. L., Hong, P. D., Yeh, M. K., et al. (2014). Improving anticancer efficacy of (−)-epigallocatechin-3-gallate gold nanoparticles in murine B16F10 melanoma cells. *Drug Design, Development and Therapy, 8*, 459–473.

Chung, J. E., Tan, S., Gao, S. J., Yongvongsoontorn, N., Kim, S. H., Lee, J. H., et al. (2014). Self-assembled micellar nanocomplexes comprising green tea catechin derivatives and protein drugs for cancer therapy. *Nature Nanotechnology, 9*(11), 907–912.

ClinicalTrials.gov. (2009). *[Internet]. Identifier NCT00397761, Capecitabine and paclitaxel (Albumin-stabilized nanoparticle formulation) in treating women undergoing surgery for stage II or stage III breast cancer*. Available from: https://clinicaltrials.gov/ct2/show/NCT00397761?cond=Capecitabine+and+Paclitaxel+%28Albumin-Stabilized+Nanoparticle+Formulation%29+in+Treating+Women+Undergoing+Surgery+for+Stage+II+or+Stage+III+Breast+Cancer&draw=2&rank=1.

ClinicalTrials.gov. (2010). *[Internet]. Identifier NCT00915369, A clinical trial to study the effects of nanoparticle based paclitaxel drug, which does not contain the solvent cremophor, in advanced breast cancer*. Available from: https://clinicaltrials.gov/ct2/show/NCT00915369?cond=NCT0 0915369&draw=2&rank=1.

ClinicalTrials.gov. (2013). *[Internet]. Identifier NCT02016209, Neoadjuvant chemotherapy of nanoparticle albumin-bound paclitaxel in lung cancer*. Available from: https://clinicaltrials. gov/ct2/show/NCT02016209?cond=NCT02016209&draw=2&rank=1.

Connor, E. E., Mwamuka, J., Gole, A., Murphy, C. J., & Wyatt, M. D. (2005). Gold nanoparticles are taken up by human cells but do not cause acute cytotoxicity. *Small, 1*(3), 325–327.

Coradeghini, R., Gioria, S., García, C. P., Nativo, P., Franchini, F., Gilliland, D., et al. (2013). Size-dependent toxicity and cell interaction mechanisms of gold nanoparticles on mouse fibroblasts. *Toxicology Letters, 217*(3), 205–216.

Cragg, G. M., Grothaus, P. G., & Newman, D. J. (2009). Impact of natural products on developing new anti-cancer agents. *Chemical Reviews, 109*(7), 3012–3043.

Cui, Y., Zhang, M., Zeng, F., Jin, H., Xu, Q., & Huang, Y. (2016). Dual-targeting magnetic PLGA nanoparticles for codelivery of paclitaxel and curcumin for brain tumor therapy. *ACS Applied Materials and Interfaces, 8*(47), 32159–32169.

Das, S., Das, J., Samadder, A., Paul, A., & Khuda-Bukhsh, A. R. (2013). Efficacy of PLGA-loaded apigenin nanoparticles in Benzo[a]pyrene and ultraviolet-B induced skin cancer of mice: Mitochondria mediated apoptotic signalling cascades. *Food and Chemical Toxicology, 62*, 670–680.

Dewick, P. M. (2002). *Medicinal natural products: A biosynthetic approach* (2nd ed., p. 520). West Sussex: John Wiley and Son.

Du, G. J., Zhang, Z., Wen, X. D., et al. (2012). Epigallocatechin Gallate (EGCG) is the most effective cancer chemopreventive polyphenol in green tea. *Nutrients, 4*, 1679–1691.

Eddouks, M., Chattopadhyay, D., Feo, V. D., & Chi-shing Cho, W. (2012). Medicinal plants in the prevention and treatment of chronic disease. *Evidence Based Complementary and Alternative Medicine, 2012*.

Fang, J. Y., Hung, C. F., Hwang, T. L., & Huang, Y. L. (2005). Physicochemical characteristics and *in vivo* deposition of liposome-encapsulated tea catechins by topical and intratumor administrations. *Journal of Drug Targeting, 13*(1), 19–27.

Fang, J. Y., Lee, W. R., Shen, S. C., & Huang, Y. L. (2006). Effect of liposome encapsulation of tea catechins on their accumulation in basal cell carcinomas. *Journal of Dermatological Science, 42*(2), 101–109.

Ferlini, C., Raspaglio, G., Mozzetti, S., Distefano, M., Filippetti, F., Martinelli, E., et al. (2003). Bcl-2 down-regulation is a novel mechanism of paclitaxel resistance. *Molecular Pharmacology, 64*, 51–58.

Gali-Muhtasib, H., Hamadi, R., Kareh, M., et al. (2015). Cell death and mechanisms of plant derived anticancer drugs: Beyond apoptosis. *Apoptosis, 20*(12), 1531–1562.

Geetha, R., Ashokkumar, T., Tamilselvan, S., Govindaraju, K., Sadiq, M., & Singaravelu, G. (2013). Green synthesis of gold nanoparticles and their anticancer activity. *Cancer Nanotechnology, 4*, 91–98.

Gilmore, D., Schulz, M., Liu, R., Zubris, K. A., Padera, R. F., Catalano, P. J., et al. (2013). Cytoreductive surgery and intraoperative administration of paclitaxel-loaded expansile nanoparticles delay tumor recurrence in ovarian carcinoma. *Annals of Surgical Oncology, 20*(5), 1684–1693.

Gitiaux, C., Kaminska, A., Boddaert, N., Barcia, G., Guéden, S., The Tich, S. N., et al. (2018). PLA2G6-associated neurodegeneration: Lessons from neurophysiological findings. *European Journal of Paediatric Neurology, 22*(5), 854–861.

Grabowski, N., Hillaireau, H., Vergnaud, J., Tsapis, N., Pallardy, M., Kerdine-Römer, S., et al. (2015). Surface coating mediates the toxicity of polymeric nanoparticles towards human-like macrophages. *International Journal of Pharmaceutics*, *482*(1–2), 75–83.

Grothaus, P. G., Cragg, G. M., & Newman, D. J. (2010). Plant natural products in anticancer drug discovery. *Current Organic Chemistry*, *14*, 1781–1791.

Guerram, M., Jiang, Z. Z., & Zhang, L. Y. (2012). Podophyllotoxin, a medicinal agent of plant origin: Past, present and future. *Chinese Journal of Natural Medicines*, *10*, 161–169.

Guo, W., Li, A., Jia, Z., Yuan, Y., Dai, H., & Li, H. (2013). Transferrin modified PEG-PLA-resveratrol conjugates: In vitro and in vivo studies for glioma. *European Journal of Pharmacology*, *718*(1–3), 41–47.

Guo, L., Peng, Y., Yao, J., Sui, L., Gu, A., & Wang, J. (2010). Anticancer activity and molecular mechanism of resveratrol-bovine serum albumin nanoparticles on subcutaneously implanted human primary ovarian carcinoma cells in nude mice. *Cancer Biotherapy & Radiopharmaceuticals*, *25*(4), 471–477.

Hong, J., Lu, H., Meng, X., Ryu, J. H., Hara, Y., & Yang, C. S. (2002). Stability, cellular uptake, biotransformation, and efflux of tea polyphenol (−)-epigallocatechin-3-gallate in HT-29 human colon adenocarcinoma cells. *Cancer Research*, *62*(24), 7241–7246.

Hong, W. K., & Sporn, M. B. (1997). Recent advances in chemoprevention of cancer. *Science*, *278*(5340), 1073–1077.

Hsieh, D. S., Wang, H., Tan, S. W., Huang, Y. H., Tsai, C. Y., Yeh, M. K., et al. (2011). The treatment of bladder cancer in a mouse model by epigallocatechin-3-gallate-gold nanoparticles. *Biomaterials*, *32*(30), 7633–7640.

Hussain, S. M., Hess, K. L., Gearhart, J. M., Geiss, K. T., & Schlager, J. J. (2005). In vitro toxicity of nanoparticles in BRL 3A rat liver cells. *Toxicology in Vitro: An International Journal Published in Association With BIBRA*, *19*(7), 975–983.

Inman, S. F. (2015). *Approves second-line MM-398 regimen for metastatic pancreatic cancer*. OncLive. http://www.onclive.com/web-exclusives/fda-approves-mm-398-regimen-for-metastatic-pancreatic-cancer.

Ji, Z., Wang, X., Zhang, H., Lin, S., Meng, H., Sun, B., et al. (2012). Designed synthesis of CeO2 nanorods and nanowires for studying toxicological effects of high aspect ratio nanomaterials. *ACS Nano*, *6*(6), 5366–5380.

Kawaii, S., Tomono, Y., Katase, E., Ogawa, K., & Yano, M. (1999). Quantification of flavonoid constituents in citrus fruits. *Journal of Agricultural and Food Chemistry*, *47*(9), 3565–3571.

Keskin, D., & Tezcaner, A. (2017). Micelles as delivery system for cancer treatment. *Current Pharmaceutical Design*, *23*(35), 5230–5241.

Khuda-Bukhsh, A. R., Bhattacharyya, S. S., Paul, S., & Boujedaini, N. (2010). Polymeric nanoparticle encapsulation of a naturally occurring plant scopoletin and its effects on human melanoma cell A375. *Zhong Xi Yi Jie He Xue Bao*, *8*(9), 853–862.

Kumari, A., Yadav, S. K., & Yadav, S. C. (2010). Biodegradable polymeric nanoparticles based drug delivery systems. *Colloids and Surfaces. B, Biointerfaces*, *75*(1), 1–18.

Kundap, U. P., Bhuvanendran, S., Kumari, Y., Othman, I., & Shaikh, M. F. (2017). Plant derived phytocompound, Emblin in CNS disorder: A systematic review. *Frontiers in Pharmacology*, *8*, 76.

Lambert, J. D., & Yang, C. S. (2003). Mechanisms of cancer prevention by tea constituents. *Journal of Nutrition*, *133*(10), 3262S–3267S.

Lee, M. J., Maliakal, P., Chen, L., Meng, X., Bondoc, F. Y., Prabhu, S., et al. (2002). Pharmacokinetics of tea catechins after ingestion of green tea and (−)-epigallocatechin-3-gallate by humans: Formation of different metabolites and individual variability. *Cancer Epidemiology,*

Biomarkers & Prevention: A Publication of the American Association for Cancer Research, cosponsored by the American Society of Preventive Oncology, 11(10 Pt 1), 1025–1032.

Li, Q. Y., Zu, Y. G., Shi, R. Z., & Yao, L. P. (2006). Review camptothecin: Current perspectives. *Current Medicinal Chemistry, 13*(17), 2021–2039.

Liu, F. S. (2009). Mechanisms of chemotherapeutic drug resistance in cancer therapy—A quick review. *Taiwanese Journal of Obstetrics & Gynecology, 48*(3), 239–244.

Loike, J. D. (1984). Podophyllotoxin and VP-16-213: A review of their divergent mechanisms of action. *Trends in Pharmacological Sciences, 5*, 30–33.

Lu, J., Huang, Y., Zhao, W., Marquez, R. T., Meng, X., Li, J., et al. (2013). PEG-derivatized embelin as a nanomicellar carrier for delivery of paclitaxel to breast and prostate cancers. *Biomaterials, 34*(5), 1591–1600.

Manju, S., & Sreenivasan, K. (2012). Gold nanoparticles generated and stabilised by water soluble curcumin-polymer conjugate: Blood compatibility evaluation and targeted drug delivery onto cancer cells. *Journal of Colloid and Interface Science, 368*(1), 144–151.

Mignani, S., Rodrigues, J., Tomas, H., Zablocka, M., Shi, X., Caminade, A. M., et al. (2018). Dendrimers in combination with natural products and analogues as anti-cancer. *Chemical Society Reviews, 47*, 514–532.

Mimeault, M., & Batra, S. K. (2011). Potential applications of curcumin and its novel synthetic analogs and nanotechnology-based formulations in cancer prevention and therapy. *Chinese Medicine, 6*, 31.

Nair, H. B., Sung, B., Yadav, V. R., Kannappan, R., Chaturvedi, M. M., & Aggarwal, B. B. (2010). Delivery of antiinflammatory nutraceuticals by nanoparticles for the prevention and treatment of cancer. *Biochemical Pharmacology, 80*(12), 1833–1843.

Namdari, M., Eatemadi, A., Soleimaninejad, M., & Hammed, A. T. (2017). A brief review on the application of nanoparticle enclosed herbal medicine for the treatment of infective endocarditis. *Biomedicine & Pharmacotherapy, 87*, 321–331.

Narayanan, N. K., Nargi, D., Randolph, C., & Narayanan, B. A. (2009). Liposome encapsulation of curcumin and resveratrol in combination reduces prostate cancer incidence in PTEN knockout mice. *International Journal of Cancer, 125*(1), 1–8.

Ngan, V. K., Bellman, K., Hill, B. T., Wilson, L., & Jordan, M. A. (2001). Mechanism of mitotic block and inhibition of cell proliferation by the semisynthetic Vinca alkaloids vinorelbine and its newer derivative vinflunine. *Molecular Pharmacology, 60*(1), 225–232.

Ngan, V. K., Bellman, K., Panda, D., Hill, B. T., Jordan, M. A., & Wilson, L. (2000). Novel actions of the antitumor drugs vinflunine and vinorelbine on microtubules. *Cancer Research, 60*(18), 5045–5051.

Nicolaou, K. C., Yang, Z., Liu, J. J., Ueno, H., Nantermet, P. G., Guy, R. K., et al. (1994). Total synthesis of taxol. *Nature, 367*(6464), 630–634.

Nussbaum, F. V., Brands, M., Hinzen, B., Weigand, S., & Habich, D. (2006). Antibacterial natural products in medicinal chemistry—Exodus or revival. *Angewandte Chemie, International Edition, 45*(31), 5072–5129.

Ojima, I. (2008). Modern natural products chemistry and drug discovery. *Journal of Medicinal Chemistry, 51*(9), 2587–2588.

Ooya, T., Lee, J., & Park, K. (2004). Hydrotropic dendrimers of generations 4 and 5: Synthesis, characterization and hydrotropic solubilization of paclitaxel. *Bioconjugate Chemistry, 15*(6), 1221–1229.

Oz, A. T., & Kafkas, E. (2017). Phytochemicals in fruits and vegetables. In *Super food and functional food—An overview of their processing and utilization* (pp. 175–184). InTech.

Pandey, A., Jain, D. S., & Chakraborty, S. (2015). Poly lactic-co-glycolic acid (PLGA) copolymer and its pharmaceutical application. In *Vol. 2. Handbook of polymer pharmaceutical technology* (pp. 151–172). Scrivener Publishing LLC.

Patra, H. K., Banerjee, S., Chaudhuri, U., Lahiri, P., & Dasgupta, A. K. (2007). Cell selective response to gold nanoparticles. *Nanomedicine: Nanotechnology, Biology, and Medicine, 3*(2), 111–119.

Pelaz, B., Charron, G., Pfeiffer, C., Zhao, Y., de la Fuente, J. M., Liang, X. J., et al. (2013). Interfacing engineered nanoparticles with biological systems: Anticipating adverse nano-bio interactions. *Small, 9*(9–10), 1573–1584.

Pezzani, R., Salehi, B., Vitalini, S., Iriti, M., Zuñiga, F. A., Sharifi-Rad, J., et al. (2019). Synergistic effects of plant derivatives and conventional chemotherapeutic agents: An update on the cancer perspective. *Medicina (Kaunas, Lithuania), 55*(4), 110.

Phan, V., Walters, J., Brownlow, B., & Elbayoumi, T. (2013). Enhanced cytotoxicity of optimized liposomal genistein via specific induction of apoptosis in breast, ovarian and prostate carcinomas. *Journal of Drug Targeting, 21*(10), 1001–1011.

Pistollato, F., Bremer-Hoffmann, S., Basso, G., Sumalla Cano, S., Elio, I., Vergara, M., et al. (2015). Targeting glioblastoma with the use of phytocompounds and nanoparticles. *Targeted Oncology, 11*(1), 1–16.

Prabhu, D., Arulvasu, C., Babu, G., Manikandan, R., & Srinivasan, P. (2013). Biologically synthesized green silver nanoparticles from leaf extract of *Vitex negundo* L. induce growth-inhibitory effect on human colon cancer cell line HCT15. *Process Biochemistry, 48*(2), 317–324.

Rezaei-Sadabady, R., Eidi, A., Zarghami, N., & Barzegar, A. (2016). Intracellular ROS protection efficiency and free radical-scavenging activity of quercetin and quercetin-encapsulated liposomes. *Artificial Cells, Nanomedicine, and Biotechnology, 44*(1), 128–134.

Ringsdorf, H. (1975). Structure and properties of pharmacologically active polymers. *Journal of Polymer Science, Polymer Symposia, 51*(1), 135–153.

Ryan, S. M., & Brayden, D. J. (2014). Progress in the delivery of nanoparticle constructs: Towards clinical translation. *Current Opinion in Pharmacology, 18*, 120–128.

Saklani, A., & Kutty, S. K. (2008). Plant-derived compounds in clinical trials. *Drug Discovery Today, 13*(3–4), 161–171.

Salehi, B., Venditti, A., Sharifi-Rad, M., Kręgiel, D., Sharifi-Rad, J., et al. (2019). The therapeutic potential of apigenin. *International Journal of Molecular Sciences, 20*(6), 1305.

Sanna, V., Pintus, G., Roggio, A. M., Punzoni, S., Posadino, A. M., Arca, A., et al. (2011). Targeted biocompatible nanoparticles for the delivery of (−)-epigallocatechin 3-gallate to prostate cancer cells. *Journal of Medicinal Chemistry, 54*(5), 1321–1332.

Sanna, V., Singh, C. K., Jashari, R., Adhami, V. M., Chamcheu, J. C., Rady, I., et al. (2017). Targeted nanoparticles encapsulating (−)-epigallocatechin-3-gallate for prostate cancer prevention and therapy. *Scientific Reports, 7*, 41573.

Siddiqui, A. A., Iram, F., Siddqui, S., & Sahu, K. (2014). Role of natural products in drug discovery process. *International Journal of Drug Development and Research, 6*, 172–204.

Siegel, R. L., Miller, K. D., & Jemal, A. (2019). Cancer statistics, 2019. *CA: A Cancer Journal for Clinicians, 69*(1), 7–34.

Siemann, E. H., & Creasy, L. L. (1992). Concentration of the phytoalexin resveratrol in wine. *American Journal of Enology and Viticulture, 43*, 49–52.

Sun, M., Nie, S., Pan, X., et al. (2014). Quercetin-nanostructured lipid carriers: Characteristics and anti-breast cancer activities *in vitro. Colloids and Surfaces. B, Biointerfaces, 113*, 15–24.

Tan, B. J., Liu, Y., Chang, K. L., et al. (2012). Perorally active nanomicellar formulation of quercetin in the treatment of lung cancer. *International Journal of Nanomedicine, 7*, 651–661.

Tomar, A., & Garg, G. (2013). Short review on application of gold nanoparticles. *Global Journal of Pharmacology*, 7(1), 34–38.

Tran, S., DeGiovanni, P. J., Piel, B., & Rai, P. (2017). Cancer nanomedicine: A review of recent success in drug delivery. *Clinical and Translational Medicine*, 6(1), 44.

Van Maanen, J. M., Retel, J., De Vries, J., et al. (1988). Mechanism of action of antitumor drug etoposide: A review. *Journal of National Cancer Institute*, 80(19), 1526–1533.

Vergara, D., De Domenico, S., Tinelli, A., Stanca, E., Del Mercato, L. L., Giudetti, A. M., et al. (2017). Anticancer effects of novel resveratrol analogues on human ovarian cancer cells. *Molecular Biosystems*, 13(6), 1131–1141.

Vergaro, V., Lvov, Y. M., & Leporatti, S. (2012). Halloysite clay nanotubes for resveratrol delivery to cancer cells. *Macromolecular Bioscience*, 12(9), 1265–1271.

Villanueva, C., Bazan, F., Kim, S., et al. (2011). Cabazitaxel: A novel microtubule inhibitor. *Drugs*, 71(10), 1251–1258.

Wagner, V., Dullaart, A., Bock, A. K., & Zweck, A. (2006). The emerging nanomedicine landscape. *Nature Biotechnology*, 24(10), 1211–1217.

Wall, M. E., Wani, M. C., Cooke, C. E., Palmer, K. H., McPhail, A. T., & Sim, G. A. (1966). Plant anti-tumor agents. The isolation and structure of camptothecin, a novel alkaloidal leukemia and tumor inhibitor from *Camptotheca acuminate*. *Journal of the American Chemical Society*, 88(16), 3888–3890.

Wang, W., Chen, T., Xu, H., Ren, B., Cheng, X., Qi, R., et al. (2018). Curcumin-loaded solid lipid nanoparticles enhanced anticancer efficiency in breast cancer. *Molecules*, 23(7), 1578.

Wang, A. Z., Langer, R., & Farokhzad, O. C. (2012). Nanoparticle delivery of cancer drugs. *Annual Review of Medicine*, 63, 185–198.

Wang, Y., Santos, A., Evdokiou, A., & Losic, D. (2015). An overview of nanotoxicity and nanomedicine research: Principles, progress and implications for cancer therapy. *Journal of Materials Chemistry B*, 3(36), 7153–7172.

Wang, G., Wang, J. J., Yang, G. Y., Du, S. M., Zeng, N., Li, D. S., et al. (2012). Effects of quercetin nanoliposomes on C6 glioma cells through induction of type III programmed cell death. *International Journal of Nanomedicine*, 7, 271–280.

Wang, W., Zhang, L., Chen, T., Guo, W., Bao, X., Wang, D., et al. (2017). Anticancer effects of resveratrol-loaded solid lipid nanoparticles on human breast cancer cells. *Molecules*, 22(11), 1814.

Wang, P., Zhang, L., Peng, H., et al. (2013). The formulation and delivery of curcumin with solid lipid nanoparticles for the treatment of non-small cell lung cancer both *in vitro* and *in vivo*. *Materials Science and Engineering: C*, 33(8), 4802–4808.

Wolfram, J., Scott, B., Boom, K., Shen, J., Borsoi, C., Suri, K., et al. (2016). Hesperetin liposomes for cancer therapy. *Current Drug Delivery*, 13(5), 711–719.

Yallapu, M. M., Jaggi, M., & Chauhan, S. C. (2012). Curcumin nanoformulations: A future nanomedicine for cancer. *Drug Discovery Today*, 17(1–2), 71–80.

Zhang, X. F., Liu, Z. G., Shen, W., & Gurunathan, S. (2016). Silver nanoparticles: Synthesis, characterization, properties, applications and therapeutic approaches. *International Journal of Molecular Sciences*, 17(9), 1534.

Chapter 24

Potential antioxidative response of bioactive products from *Ganoderma lucidum* and *Podophyllum hexandrum*

Saima Hamid

Centre of Research for Development/Department of Environmental Science, University of Kashmir, Srinagar, India

1. Introduction

There are certain reactions that occur inside the cells of living organisms, like humans, in which free radicals are generated, including oxygen- and nitrogen-free radical species, which have free electron pairs with the capacity to degrade DNA. This ultimately leads to protein and lipid oxidation (Fang, Yang, & Wu, 2002; Li, Chen, et al., 2015; Li, Tan, et al., 2015; Peng et al., 2014). However there are some mechanisms in the human body that can help reduce the generation of such free radicals inside the cells, which can nullify the formation of such to suppress the effects on macromolecules, which are known as antioxidants. In the current era, human population is more exposed to higher lifestyle stress, alcohol, pollutants, radiation, and other xenobiotic compounds like carbon tetrachloride (CCl_4), which leads to the formation of molecules like reactive oxygen species ($O\bullet$) and reactive nitrogen species ($N\bullet$). These can cause in long-term diseases by altering the natural mechanism balance between oxidation and antioxidation in the cells of the human body (Wang et al., 2016; Zhou, Zheng, Li, Xu, et al., 2016; Zhou, Zheng, Li, Zhou, et al., 2016). Various studies have demonstrated that tissue organs in the human body, like the heart, lung, brain, blood, liver, and kidney, leads into the production of reactive molecules upon exposure to CCl_4. Due to reaction inside the body, this CCl_4 is converted to a more toxic form, i.e., trichloromethyl radical ($CCl_3\bullet$) reacts oxygen into the trichloromethyl peroxyl radical ($CCl_3O_2\bullet$), which is a highly reactive compound (Baiano & Nobile, 2015; Cai, Luo, Sun, & Corke, 2004). Hence, to achieve the balance to reduce or suppress the levels of free radicals, an increase of the production of reducing agents needs to be taken via oral intake of antioxidant

Phytomedicine: A Treasure of Pharmacologically Active Products from Plants.
https://doi.org/10.1016/B978-0-12-824109-7.00013-3

molecules that neutralizes these free electron species inside the cells, thus preventing degenerative damage of macromolecules (Baiano & Nobile, 2015). It is necessary to take such foods that are rich in antioxidants like mushrooms, fruits, traditional medicinal herbs, flowers, spices, etc. and various by-products such as the extracts of natural varieties that are available in the market that can be taken exogenously (Deng et al., 2012; Shan, Cai, Sun, & Corke, 2005). A group of secondary metabolites in medicinal plants are a good source of naturally available antioxidants, which includes flavonoids, polyphenols (phenolic acids, anthocyanins, lignans, and stilbenes), and vitamin C and E (Jenab et al., 2006). Polyphenols and carotenoids have a wide range of biological properties by acting as antiaging, antioxidative, anti-inflammatory, anticancer, and antimicrobial agents (Manach, Scalbert, Morand, Remesy, & Jimenez, 2004). The food and drug industries with the aid of efficient technologies and, more importantly, different evaluation assays are used to extract the natural antioxidants from medicinal plants due to their tremendous health benefits. Several assessment methods of antioxidant activity have been developed to rank food items that can be used to prioritize the list of medicinal plants rich in antioxidants for consumption; such methods include ferric ion reducing antioxidant power (FRAP) assay, trolox equivalence antioxidant capacity (TEAC) assay, inhibiting the oxidation of low-density lipoprotein (LDL) assay, and oxygen radical absorbance capacity (ORAC) assay (Zheng et al., 2016; Zhou, Zheng, Li, Xu, et al., 2016; Zhou, Zheng, Li, Zhou, et al., 2016).

2. Bioactive compounds of *Podophyllum hexandrum* and *Ganoderma lucidum*

Since the ancient times, medicinal plants has been used to fulfill nutritional needs as well as to cure various health ailments in many parts of the world like Asia, Africa, China, and South America. Sixty percent of the population is still relying on traditional knowledge and use of these medicinal properties to cure their illnesses. The World Health organization (WHO) has also recognized that there is a huge potential for drug discovery from unexplored flora. Modern organic synthetic chemistry serves as a backbone for the pharmaceutical industries for the preparation of important novel drug discoveries on the basis of ethnopharmacological studies (da Bolzani, Valli, Pivatto, & Viegas, 2012; Duraipandiyan, Ayyanar, & Ignacimuthu, 2006). Therefore, medicinal plants contain a number of bioactive molecules that have enormous health benefits, like acting as antioxidants, which are found in mushrooms like Ganoderma lucideum, fruits, vegetables, and whole grains (Gil-Chavez et al., 2013). Various important medicinal plants like Podophyllum hexandrum and medicinal mushrooms contain a heterogeneous class of compounds (polyphenolic compounds, carotenoids, tocopherols, phytosterols, and polysaacharides), which exhibit huge health benefits like anticancer, antioxidative, antimicrobial, and anti-inflammatory. To date, the number of bioactive compounds with significant pharmacological properties have been reported like podophyllotoxone, epipodophyllotoxin, 4-methylpodophyllotoxin, aryltetra-hydronaphthalene lignans, flavonoids such as quercetin, quercetin-3-glycoside, 4-demethylpodophyllotoxin glycoside, podophyllotoxinglycoside, kaempferol, and kaempferol-3-glucoside, as all these exhibit diverse biological properties (Santana, Schell, & Williams, 1992; Viana, Oliveira, & Silva, 1991). Likewise, due to the presence of more than 400 bioactive compounds in Ganoderma lucidum, it is called the "immortal mushroom" because of the presence of such many compounds (triterpenoids, polysaccharides, nucleotides, sterols, steroids, fatty acids, and proteins/peptides) that are treatments against various diseases like cancer, arthritis, HIV/AIDS, herperax, diabetes, tumor, fungal infections, and arterial sclerosis (Krogh, 1993; Montaldo, Figoli, & Zanette, 1990). Also, these bioactive molecules possess antiaging and antioxidative properties and enhance immunity to fight against various health ailments.

3. Antioxidative property of *Ganoderma lucidum*

Among the edible mushrooms, *G. lucidum* is rich in antioxidants that may help prevent long-term diseases like cancer, diabetes, etc. Antioxidants perform various functions in the cells by protecting the cell organelles and macromolecules like DNA and prevent disruption of cellular mechanisms (Benzie & Wachtel-Galor, 2009). Various studies have been carried out to check the in vitro antioxidative activities of *G. Lucidum* due to presence of triterpenoids and particular polysaccharides (Wu & Wang, 2009). Galo, Szeto, Tomlinson, & Benzie, (2004) observed increased plasma total antioxidant activity due to absorption of antioxidant compounds after ingestion in humans. In the animal model studies, where mutagenic testing animals were used to study the role of polysaccharide peptide and protein-bound polysaccharide (PBP) to check antioxidant activity that resembles endogenous antioxidant superoxide dismutase (SOD), the presence of polysaccharides prevents oxidative damage (Ooi & Liu, 2007). Lee et al. (2001) concluded that, in vitro studies using agarose gel electrophoretic techniques along with the other reagents like metal-catalyzed fenton, UV radiation did not affect the molecular structure of DNA by inducing an OH• effect, hence antioxidative molecules of *G. lucidum* act as protective molecules. Hydrogen peroxide (H_2O_2) was induced to damage DNA, then a treatment of hot water extracts was given to visualize the effects on Raji cells as various concentrations have been used to check the structure of DNA of human lymphocytes Shi, James, Benzie, & Buswell, 2002. But at low concentrations ($< 0.001\%$ w/v), extracts were effective rather than at higher concentrations ($> 0.01\%$ w/v) (Wachtel-Galor, Choi, & Benzie, 2005). Hence upon comparison to the various extracts of *G. lucidum*, it has been reported that methanol averts kidney damage (induced by the anticancer drug cisplatin) through restoration of the renal antioxidant defense system and ethanolic extracts resulted in cell death by disruption of DNA molecules due to high levels of hydrogen peroxide (Ajith & Janardhanan, 2007). Whereas during an increment of intracellular reactive oxygen species

(ROS)-producing effect of doxorubicin (DOX) in hela cells due to presence of terpenes in *G. Lucidum*, the synergetic effects were subdued by scavenging molecules (Yuen & Gohel, 2008). Several studies were conducted to study the role of antioxidants and immunomodulatory effects in the mice model studies, which were diabetic and mutated, but to date no direct concentration has been confirmed. Meanwhile, enzymatic and nonenzymatic are in inverse reciprocate to lipid peroxidation (Jia et al., 2009) (Fig. 1).

The macro-mushroom *G. lucidum* is known to contain important metabolites like alkaloids, triterpenoids, steroids, nucleotides, polysaccharides, and lactones with immense biological properties Sanodiya, Thakur, Baghel, Prasad, & Bisen, 2009. Keliang et al. (2010) concluded that polysaccharides found in the *G. Lucidum* have been isolated from the submerged culture, hence valued as a Chinese medicinal fruit. Several types of fermentations like extracellular polysaccharides (EPS) and intracellular polysaccharides (IPS) are being used to produce polysaccharides from *G. lucidum,* and even truffle fermentation as per (Liu, Li, Li, & Tang, 2008) is used to make mycelia (Thulasi, Pillai, Nair, & Janardhanan, 2010). It has been evaluated that the production of polysaccharides and mycelial biomass are directly linked to available nutritional and environmental conditions (Vijayabaskar, Babinastarlin, Shankar, Sivakumar, & Anandapandian, 2011). High performance liquid chromatographic technique was used to characterize the composition of Ganoderma lucidum polysaccharide (GLP) by using boiling water extraction methods. It has been found that GLP consists of five monosaccharides and disaccharides, including glucose, fructose, xylose, sucrose, and maltose. Different concentrations of GLP doses

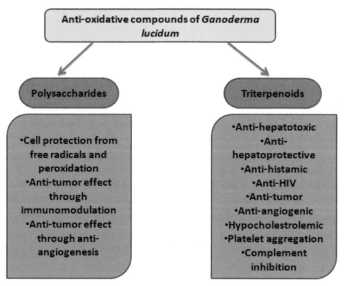

FIG. 1 Pharmacological properties of antioxidative compounds of *Ganoderma lucidum.*

were given to the diabetic rats, i.e., 100, 200, and 300 mg/kg to study the effects of prepared extracts over the lipid peroxidation metabolism and on the enzymes of antioxidants. Thus it has been concluded that, in the vivo studies, extracts showed reduction of oxidative stress and inhibition of apoptosis. Free radical scavenging activity of different mushrooms were compared by using methanolic and ethanolic extracts followed by DPPH assay, and the highest radical activity was found for the methanolic extracts [$IC_{50} = 3.82 \pm 0.04$ μg/mL] compared with the ethanolic extracts [$IC_{50} = 7.03 \pm 0.07$ μg/mL], even though the mushroom variety of Turkey showed antioxidative activity due to presence of phenolic compounds (Yuvalı Celık, Onbaslı, Altınsoy, & Allı, 2014). Jia et al. (2004) concluded that, besides the antioxidative property, GLP plays an immense role to inhibit the lipid peroxidation via metal chelation and free radical scavenging in the metabolic functions of the body. Scavenging activity was shown by GLP in a deoxyribose system with an IC_{50} value of 25 mg/mL for hydroxyl radicals and GLP effectively dose-dependent dose-based superoxide radical anion. In many biological systems, the amount of malondialdehyde was used as the oxidation index. GLP has demonstrated important antioxidant activity in homogeneous rat liver tissue and membrane peroxidation systems of mitochondrial membranes. GLP has also prevented autohemolysis of rat red blood cells in a dose-dependent manner. Various studies have been conducted to investigate the activity of ethanolic extract of G. lucidum for the hepatoprotective activity in different mice models. The key triterpenoids in GLE were analyzed using HPLC combined with a detector for the photodiode and ionizing mass spectrometry. Histopathology and biochemical study of serum enzymes examined the effect of GLE on hepatoprotection. To classify differentially expressed liver proteoma in the mice, we have used the relative and absolute quantitation (iTRAQ) tag in conjunction with tandem mass spectrometry. More than 4000 differentially expressed proteins were used; 40 proteins were used for further bioinformatic experiments with the most significantly modified proteins. Western blotting was tested for expression levels of cytochrome P450 2E1 and alcohol dehydrogenase 1, proteins closely associated with these processes. Tritrenoids, the main GLE components, preserved liver lesions caused by alcohol through lipid peroxidation inhibitors, increased antioxidant activity, and apoptotic cell death suppression (Chao & Lin, 2018). Due to bioactivities, particularly the antioxidant activity (GLP) has attracted much attention as one of water-solution polysaccharides. Deproteinization is a crucial step in the polysaccharide treatment process. For crude GLP therapy, three conventional deproteinization methods were evaluated in this study including neutral protease process, TCA precipitation, and salting $CaCl_2$. Protein removal (71.50%-87.36%) was possible, and a loss of polysaccharide (8.35%–11.39%) was observed. Structure analysis showed that this form ($P > 0.05$) of deproteinization had no substantial effect on polysaccharide molecular weight; however, the losses in glycoside bond (1.14%–64.05%) were variably large. In addition, in vitro hydroxyl radical reducing power, 2, 2-diphenylo 1-picryl-hydrazyl (DPPH) free radical testing, and ferric-reducing antioxidant power testing were calculated in antioxidants. Enzymolysis-related distilled GLP achieved the highest retention of antioxidants at a rate of 47.40% with a deproteinization rate of 74.03% and a loss of polysaccharides ratio of 11.39%, respectively (Xia, Li, & Yee, 2019). Various studies have been carried out to produce the extracellular polysaccharide (EPS) by using various novel strains of yeast-like fungus *Aureobasidium pullulans* near about 40.1 g/L and productivity of 12.5 g/L per day by using biogas reactor in fermentation tanks (Ravella et al., 2010).

4. Antioxidative properties of *Podophyllum hexandrum*

Among the treasures of medicinal plant wealth, a perennial plant *Podophyllum hexandrum,* also known by the name Himalayan Mayapple, has been reported to be used as an intestinal purgative and emetic, a cure for contaminated and necrotic wounds, and a tumor growth inhibitor over the ages and in modern times. The plant's rhizome contains a resin, generally and commercially known as Indian podophyllum resin, which can be processed to extract a neurotoxin called podophylotoxin or podophyllin. The key lignan in the resin is podophyllotoxin, and it is a medium-dimerized substance. A PDA-approved anticancer drug, the starting material for etoposide (Vepeside) is podophyllotoxin, which is used to treat testicular and lung cancer by inhibiting the replication of carcinogenic cells. In the treatment of leukemia; lung and testicular tumors; and dermatologic conditions such as warts, acne, and psoriasis, podophyllotoxin is used as a precursor in the semisynthetic topoisomerase inhibitors. Mengfei, Lanlan, Delong, Tiantian, and Wei (2012) carried out the biochemical and antioxidative activities of rhizome extracts by following the FRAP and DPPH assay as highest antioxidative capacity was observed for the following bioactive compounds: Podophyllotoxin deoxy (PODD), Podophyllotoxin (POD), and Polyneuridine (Fig. 2). Here, the free radical scavenging potential of these compounds were compared with ascorbic acid standard [60.78 ± 1.22 and $1267.5 \pm 30.24 \mu M$ ($P < 0.01$)]; the IC_{50} values were highest for the POD ($2923.98 \pm 21.89 \mu M$). Moreover, other components like BADE (1,2-Benzenedicarboxylic acid, diisooctyl ester), β-Sitosterol, and PADE (Phthalic acid, diisobutyl ester) also exhibited antioxidative potential. Yokozawa et al. (1998) concluded that, under the in vivo conditions, various antioxidative assays including DPPH scavenging assay, hydrogen peroxide assay, hydroxyl assay, and superoxide assay had been carried out over the ethyl acetate extract of the Rhizome of *P. hexandrum* to check the glutathione levels among albino rats. It has been reported that methanolic extract inhibits the lipid peroxidation and kidney and lung tissue protection against CCl_4-induced oxidative stress in the hydrogen peroxide method of oxidation due to suppression of activities of aspartate aminotransferase (AST), alanine aminotransferase (ALT), and lactate dehydrogenase activity (LDH), and enhances antioxidant enzyme activity (Duh, Tu, & Yen, 1999; Gordon, 1990; Yokozawa et al., 1998). Combined stress of radiation (250 Gy)

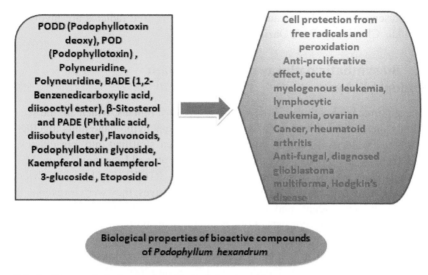

FIG. 2   Pharmocological properties of *Podophylum hexandrum.*

and iron/ascorbate at a concentration of 2000 µg/mL were given to mice models to measure the reducing the power of various fractionated solvents [(n-hexane (HE), chloroform (CE), alcohol (AE), hydro-alcohol (HA), and water (WE)]. Among all these extracts, chloroform showed the maximum antioxidative activity, and characterization has been performed by LC-MS technique to determine polyphenolic compounds (mg% of quercetin) to reveal their role in the inhibition of LPO activity as found in EA extracts, which was 93.05% of higher percentage in homogenated liver (Chawla et al., 2005). The usage of trypan blue exclusion checking, dichlorofluorescein diacetate, and DNA fragmentation assay was investigated in cell death, ROS, and apoptosis. After nitroblue tetrazolium, 2-deoxyribose degradation (DNA), and plasmid stimulation experiments, respectively, superoxide anions and hydroxyl radicals and DNA harm was predicted. Pre-irradiation of REC-2001 to peritoneal macrophages inside the radiation control community of 25–200 µg/mL greatly decreases the exposure caused by the formation of ROS, DNA disruption, apoptosis, and cell deaths suggesting a potential exposure defense category. Plasmid DNA tests found that REC-2001 is able to further improve antioxidative ability through disruption of 20 single- and double-strand Gy-mediated splits. Nonetheless, dose-based dose-dependent REC-2001 fragmentation, which suggests cytotoxic composition, caused cell death, ROS, and DNA. REC-2001 generated significant quantities of hydroxyl radicals and superoxide anions in the presence of 100 µM copper sulfate suggesting that it was possible to act as a pro-oxidant in the presence of metal ions. The anion superoxide production has been shown to be responsive to metal chelators such as EDTA and DFR. The findings suggest that REC-2001 can be responsible for cytotoxic and radioprotective effects as pro-oxidants in the presence of metal ions and antioxidants where free radicals are involved (Shukla, Chaudhary, Puri, Qazi, & Sharma, 2006).

5. Biosynthetic pathway of an antioxidant Podophyllotoxin

The shikimic acid pathway is the known to produce lignans, which are the secondary metabolites found in *P. hexandrum*. Two phenylpropane units together make one molecule of lignin, which can be further stratified into four classes, i.e., Trimers, Neolignans, Oxyneolignans, and combined Lignanoids (Luo, Hu, Kong, & Yang, 2014). Since the Podophyllotoxin (PTOX) ($C_{22}H_{22}O_8$) has been found to be an exclusive lignin as well as its derivatives, they were investigated to be potent antioxidants and antitumor agents (Liu et al., 2015). Cyclolignans in the group of lignans consists of two phenylpropane units present between two β-β′ positions and sided by C–C bonds. PTOX is formed by the aryltetralin of cyclolignans (Gordaliza, García, Del Corral, Castro, & Gomez-Zurita, 2004). According to Ramos, Luh, Goodrich, and Bronson (2001), Podwyssotzki was the first who isolated PTOX from *Podophyllum peltatum* in North America, commonly called American mandrake or May apple, and then the same compound was isolated in India from *P. emodi*, which is commonly known by the name Indian podophyllum. It is especially noteworthy that PTOX is the most abundant lignan in podophyllin, a resin formed by Podophyllum genus species and some Cupressaceae family endophytic microorganisms (Kusari, Hertweck, & Spiteller, 2012). To date, the biosynthetic pathway for PTOX is not fully understood, but few attempts were made just to find some major steps that are responsible for the production of podophyllotoxin. As per Petersen and Alfermann (2001), a few illustrations have been made in some plants like Podophyllum and Forsythia, Linum to find the major biosynthetic pathway for the formation of podophyllotoxin. As shown in Fig. 3, some key enzymes are involved in the formation of PTOX, namely ferulic acid and methylenedioxy-substituted cinnamic acid (Ardalani, Amir, & Majid, 2017). Methoxypodophyllotoxin was used as the principal cyclolignan in an in vitro experiment of *Linum flavum* L., but this compound was not observed when the cells were cultivated with PTOX. The outcome of this study showed that high kinetic constant affinity reflects PTOX, and PTOX will inhibit 6-methoxypodophyllotoxin development (Berim, Ebel, Schneider, & Petersen, 2008). PTOX biosynthesis evaluation is a promising alternative that can help find novel sources with sustainable PTOX production.

6. Biosynthetic pathway of Polysaccharide from *Ganoderma lucidum*

Polysaccharides (PS) found in the fungi are concerned with multiple cellular functions like cell signaling, protein glycosylation, building units of cell wall, and host-pathogen interactions (Ruthes, Smiderle, & Iacomini, 2016).

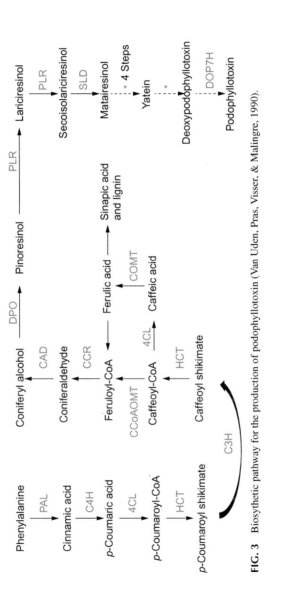

FIG. 3 Biosythetic pathway for the production of podophyllotoxin (Van Uden, Pras, Visser, & Malingre, 1990).

Membrane-level intracellular nucleotide sugars are involved to create key cell wall structures or to shape a gel-like matrix on the hyphal surface that will serve as a base for the formation of polysaccharides. Gel-like PS may be excreted more in the growth media in submerged fermentation and are referred to as extracellular polysaccharides (EPS) (Wu et al., 2014). Curing proof of gene clone and genetic transformation continues to be identified in the biosynthesis process of the PS mushroom because of the insufficient knowledge and understanding of associated enzymes and their process functions. However, a condensed biosynthetic pathway for mushroom PS was developed on the basis of the identification by numerous researchers of intermediate compounds and synthesis-related enzyme activities (Li, Chen, et al., 2015; Li, Tan, et al., 2015; Peng et al., 2016). For the study of the biosynthetic process, methods such as gene expression (Xu et al., 2015) and RNAi-mediated cell silencing (Li, Chen, et al., 2015; Li, Tan, et al., 2015) have also been added. In this model, nucleotide sugar precursor's biosynthetic mechanism in mushroom PS is close to that of plants (Caffall & Mohnen, 2009), offering a framework for study of other metabolic pathways of mushroom PS. It released the genomes of *Antrodia cinnamomea* (Lu et al., 2014), *Ganoderma lucidum* (Chen et al., 2011), Lentinula edodes (Chen et al., 2016), and *Lignosus rhinocerotis* (Yap et al., 2014). Such genomic studies offer insights into the genetic basis of recorded medicinal properties of PS and a forum to better identify putative bioactive PS and pathway enzymes (Fig. 4).

7. Conclusion

Successful fields of clinical research involve the discovery of bioactive substances. There are promising ways to reduce the likelihood of multiple illnesses, like chronic disorders and cardiovascular disease, by choosing bioactive compounds. Recent studies have demonstrated that coronary disease is an inflammation condition and is thus ideal for therapy by anti-inflammatory molecules. Researchers are demonstrating that harmful effects of oxidants on atherogenesis often increase the risk of cardioprotective benefits from antioxidants. A study offers a description of work methods to explore bioactive compounds' biological and safety consequences. To improve our knowledge of the physiology of bioactive compounds, a variety of various scientific methods must be taken into consideration of the numbers of bioactive compounds and the range of possible biological consequences. Various extraction methods need reconsideration and modification to produce an exogenous supply of available antioxidants to gain maximum benefits from the natural sources like medicinal plants, fruits, mushrooms, and other rich sources. Future attempts to use biotechnology to modify/fortify foods and food components to enhance consumer safety may rely on the discovery of novel health effects of bioactive compounds.

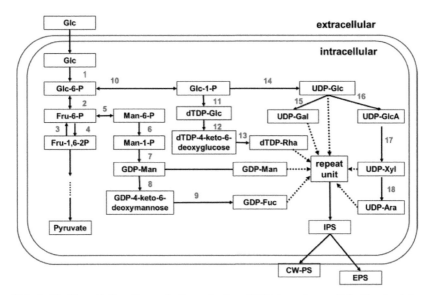

FIG. 4 The biosynthetic pathways for the production of ESPs in mushroom *G. lucidum* involving the following steps along with enzymes: 1. Glucokinase (GK); 2. Phosphoglucose isomerase (PGI); 3. Fructose-1,6-biphosphatase (FBPase); 4. Phosphofructokinase (FPK); 5. Phosphomannose isomerase (PMI); 6. Phosphomannose mutase (PMM); 7. GDP-Man pyrophosphorylase (GMP); 8. GDP-Man dehydratase (GMD); 9. GDP-4-keto-6-deoxymannose epimerase/reductase (GMER); 10. Phosphoglucose mutase (PGM); 11. dTDP-glucose pyrophosphorlase (dTDP-GPP); 12 and 13. dTDP-Rha synthase (dTRS); 14. UDP-Glc pyrophosphorylase (UGP); 15. UDP-Gal-4-epimerase (UGE); 16. UDP-Glc dehydrogenase (UGDG). 17. UDP-glucuronate decarboxylase/UDP-Xyl synthase (UXS); 18. UDP-Xyl-4-epimerase (UXE).*Source: Wang, Q., Wang, F., Xu, Z., Ding, Z. (2017). Bioactive mushroom polysaccharides: A review on monosaccharide composition, biosynthesis and regulation. Molecules 22(6), 955.*

References

Ajith, T. A., & Janardhanan, K. K. (2007). Indian medicinal mushrooms as a source of antioxidant and antitumor agents. *Journal of Clinical Biochemistry and Nutrition, 40*(3), 157–162. https://doi.org/10.3164/jcbn.40.157.

Ardalani, H., Amir, A., & Majid, G. M. (2017). Podophyllotoxin: A novel potential natural anticancer agent. *Avicenna Journal of Phytomedicine, 7*(4), 285–294.

Baiano, A., & Nobile, M. A. (2015). Antioxidant compounds from vegetable matrices: Biosynthesis, occurrence, and extraction systems. *Critical Reviews in Food Science and Nutrition, 56*, 2053–2068.

Benzie, I. F. F., & Wachtel-Galor, S. (2009). Biomarkers of long-term vegetarian diets. *Advances in Clinical Chemistry, 47*, 169–220.

Berim, A., Ebel, R., Schneider, B., & Petersen, M. (2008). UDP-glucose:(6-methoxy) podophyllotoxin 7-O-glucosyltransferase from suspension cultures of *Linum nodiflorum*. *Phytochemistry, 69*, 374–381.

Caffall, K. H., & Mohnen, D. (2009). The structure, function, and biosynthesis of plant cell wall pectic polysaccharides. *Carbohydrate Research, 344*, 1879–1900.

Cai, Y. Z., Luo, Q., Sun, M., & Corke, H. (2004). Antioxidant activity and phenolic compounds of 112 traditional Chinese medicinal plants associated with anticancer. *Life Sciences*, *74*, 2157–2184.

Chao, L. Z., & Lin, Z. B. (2018). Regulation on maturation and function of dendritic cells by *Ganoderma lucidum* polysaccharides. *Immunology Letters*, *83*, 163–169.

Chawla, R., Arora, R., Kumar, R., Sharma, A., Prasad, J., Singh, S., et al. (2005). Antioxidant activity of fractionated extracts of rhizomes of high-altitude Podophyllum hexandrum: Role in radiation protection. *Molecular and Cellular Biochemistry*, *273*(1-2), 193–208.

Chen, L. F., Gong, Y. H., Cai, Y. L., Liu, W., Zhou, Y., Xiao, Y., et al. (2016). Genome sequence of the edible cultivated mushroom *Lentinula edodes* (shiitake) reveals insights into lignocellulose degradation. *PLoS One*, *11*.

Chen, S. L., Xu, J., Liu, C., Zhu, Y. J., Nelson, D. R., Zhou, S. G., et al. (2011). Genome sequence of the model medicinal mushroom *Ganoderma lucidum*. *Nature Communications*, *3*, 913.

da Bolzani, V. S., Valli, M., Pivatto, M., & Viegas, C. (2012). Natural products from Brazilian biodiversity as a source of new models for medicinal chemistry. *Pure and Applied Chemistry*, *84*, 1837–1846.

Deng, G. F., Shen, C., Xu, X. R., Kuang, R. D., Guo, Y. J., Zeng, L. S., et al. (2012). Potential of fruit wastes as natural resources of bioactive compounds. *International Journal of Molecular Sciences*, *13*, 8308–8323.

Duh, P. D., Tu, Y. Y., & Yen, G. C. (1999). Antioxidant activity of water extract of Harng Jyur (*Chrysanthemum morifolium* Ramat). *Lebensmittel-Wissenschaft und Technologie*, *32*(5), 269–277.

Duraipandiyan, V., Ayyanar, M., & Ignacimuthu, S. (2006). Antimicrobial activity of some ethnomedicinal plants used by Palyar tribe from Tamil Nadu, India. *BMC Complementary and Alternative Medicine*, *6*, 35–41.

Fang, Y. Z., Yang, S., & Wu, G. (2002). Free radicals, antioxidants, and nutrition. *Nutrition*, *18*, 872–879.

Galo, S., Szeto, T., Tomlinson, B., & Benzie, B. (2004). *Ganoderma lucidum* ('Lingzhi'); acute and short-term biomarker response to supplementation. *International Journal of Food Sciences and Nutrition*, *55*(1), 75–83. https://doi.org/10.1080/09637480310001642510.

Gil-Chavez, G. J., Villa, J. A., Ayala-Zavala, F., Heredia, J. B., Sepulveda, D., Yahia, E. M., et al. (2013). Technologies for extraction and production of bioactive compounds to be used as nutraceuticals and food ingredients: An overview. *Comprehensive Reviews in Food Science and Food Safety*, *12*(1), 523.

Gordaliza, M., Garcia, P., Del Corral, J. M., Castro, M., & Gomez-Zurita, M. (2004). Podophyllotoxin: Distribution, sources, applications and new cytotoxic derivatives. *Toxicon*, *44*, 441–459.

Gordon, M. H. (1990). The mechanism of the antioxidant action in vitro. In B. J. E. Hudson (Ed.), *Food antioxidants* (pp. 1–18).

Jenab, M., Riboli, E., Ferrari, P., Sabate, J., Slimani, N., Norat, T., et al. (2006). Plasma and dietary vitamin C levels and risk of gastric cancer in the European Prospective Investigation into Cancer and Nutrition (EPIC-EURGAST). *Carcinogenesis*, *27*, 2250–2257.

Jia, J., Zhang, X., Hu, Y., (Eds.), et al. (2009). Evaluation of in vivo antioxidant activities of *Ganoderma lucidum* poly-saccharides in STZ-diabetic rats. *Food Chemistry*, *115*, 32–36.

Keliang, Y., Wen, Z., Hongbo, Y., Hongxun, W., & Xiaoyu, Z. (2010). New polysaccharides from *G. lucidum* and *L. barbarum*. *Food Technology and Biotechnology*, *48*(1), 94–101.

Krogh, C. M. E. (1993). In K. CME (Ed.), *Vol. 28. Compendium of pharmaceuticals and specialties* (pp. 1307–1308). Ottawa: Canadian Pharmaceutical Association.

Kusari, S., Hertweck, C., & Spiteller, M. (2012). Chemical ecology of endophytic fungi: Origins of secondary metabolites. *Chemistry & Biology*, *19*(7), 792–798. https://doi.org/10.1016/j.chembiol.2012.06.004.

Lee, J.-M., Kwon, H., Jeong, H., (Eds.), et al. (2001). Inhibition of lipid peroxidation and oxidative DNA damage by *Ganoderma lucidum*. *Phytotherapy Research*, *15*, 245–249.

Li, M. J., Chen, T. X., Gao, T., Miao, Z. G., Jiang, A. L., Shi, L., et al. (2015). UDP-glucose pyrophosphorylase influences polysaccharide synthesis, cell wall components, and hyphal branching in *Ganoderma lucidum* via regulation of the balance between glucose-1-phosphate and UDP-glucose. *Fungal Genetics and Biology*, *82*, 251–263.

Li, S., Tan, H. Y., Wang, N., Zhang, Z. J., Lao, L., Wong, C. W., et al. (2015). The role of oxidative stress and antioxidants in liver diseases. *International Journal of Molecular Sciences*, *16*, 26087–26124.

Liu, R. S., Li, H. M., Li, D. S., & Tang, Y. J. (2008). Response surface modeling the significance of nitrogen source on the submerged cultivation of Chinesetruffle *Tuber sinense*. *Process Biochemistry*, *43*, 868–876.

Liu, J., Shimizu, K., Konishi, F., (Eds.), et al. (2015). Anti-androgenic activities of the triterpenoids fraction of *Ganoderma lucidum*. *Food Chemistry*, *100*, 1691–1696.

Lu, M. Y., Fan, W. L., Wang, W. F., Chen, T., Tang, Y. C., Chu, F. H., et al. (2014). Genomic and transcriptomic analyses of the medicinal fungus *Antrodia cinnamomea* for its metabolite biosynthesis and sexual development. *Proceedings of the National Academy of Sciences of the United States of America*, *111*, 4743–4752.

Luo, J., Hu, Y., Kong, W., & Yang, M. (2014). Evaluation and structure-activity relationship analysis of a new series of arylnaphthalene lignans as potential anti-tumor agents. *PLoS One*, *9*, 93516.

Manach, C., Scalbert, A., Morand, C., Remesy, C., & Jimenez, L. (2004). Polyphenols: Food sources and bioavailability. *The American Journal of Clinical Nutrition*, *79*, 727–747.

Mengfei, L., Lanlan, Z., Delong, Y., Tiantian, L., & Wei, L. (2012). Biochemical composition and antioxidant capacity of extracts from Podophyllum hexandrumrhizome. *BMC Complementary and Alternative Medicine*, *12*, 263.

Montaldo, P. G., Figoli, F., & Zanette, M. I. (1990). Pharmacokinetics of intra pleural versus intravenous etoposide (VP-16) and teniposide (VM-26) in patient with malignant pleural effusion. *Oncology*, *47*, 55–61.

Ooi & Liu, 2007Ooi, V. E., & Liu, F. (2007). Immunomodulation and anti-cancer activity of polysaccharide-protein complexes. *Current Medicinal Chemistry*, *7*, 715–729.

Peng, L., Li, J., Liu, Y., Xu, Z. H., Wu, J. Y., Ding, Z. Y., et al. (2016). Effects of mixed carbon sources on galactose and mannose content of exopolysaccharides and related enzyme activities in *Ganoderma lucidum*. *RSC Advances*, *6*, 39284–39291.

Peng, C., Wang, X., Chen, J., Jiao, R., Wang, L., Li, Y. M., et al. (2014). Biology of ageing and role of dietary antioxidants. *BioMed Research International*, 831841.

Petersen, M. A., & Alfermann, W. (2001). Production of cytotoxic lignans by plant cell cultures. *Applied Microbiology and Biotechnology*, *55*(2), 135–142. https://doi.org/10.1007/s002530000510.

Ramos, D. M., Luh, B.-Y., Goodrich, J., & Bronson, J. J. (2001). Anti-MRSA cephems. Part 2: C-7 cinnamic acid derivatives. *Bioorganic & Medicinal Chemistry*, *11*, 265–279.

Ravella, S. R., James, S., Bond, C. J., Roberts, I. N., Cross, K., Retter, A., et al. (2010). *Cryptococcus shivajii* sp. nov.: A novel basidiomycetous yeast isolated from biogas reactor. *Current Microbiology*, *60*, 12–16. https://doi.org/10.1007/s00284-009-9493-9.

Ruthes, A. C., Smiderle, F. R., & Iacomini, M. (2016). Mushroom heteropolysaccharides: A review on their sources, structure and biological effects. *Carbohydrate Polymers*, *136*, 358–375.

Sanodiya, B. S., Thakur, G. S., Baghel, R. K., Prasad, G. B., & Bisen, P. S. (2009). Ganoderma lucidum: A potent pharmacological macrofungus. *Current Pharmaceutical Biotechnology*, *10*(8), 717–742.

Santana, V. M., Schell, M. J., & Williams, R. (1992). Escalating sequential high dose carboplatin and etoposide with autologous marrow support in children with relapsed solid tumours. *Bone Marrow Transplantation, 10*, 457–462.

Shan, B., Cai, Y. Z., Sun, M., & Corke, H. (2005). Antioxidant capacity of 26 spice extracts and characterization of their phenolic constituents. *Journal of Agricultural and Food Chemistry, 53*, 7749–7759.

Shi, Y. L., James, A. E., Benzie, I. F., & Buswell, J. A. (2002). Mushroom-derived preparations in the prevention of H_2O_2-induced oxidative damage to cellular DNA. Teratogenesis, Carcinogenesis. *and Mutagenesis, 22*, 103–111.

Shukla, S. K., Chaudhary, P., Puri, S. C., Qazi, G. N., & Sharma, R. K. (2006). Cytotoxic and radioprotective effects of Podophyllum hexandrum. *Environmental Toxicology and Pharmacology, 22*(1), 113–120.

Thulasi, G., Pillai, C. K. K., Nair, C., & Janardhanan, K. K. (2010). Enhancement of repair of radiation induced DNA strand breaks in human cells by Ganoderma mushroom polysaccharides. *Food Chemistry, 119*, 1040–1043.

Van Uden, W., Pras, N., Visser, J. F., & Malingre, T. M. (1990). Detection and identification of podophyllotoxin produced by cell cultures derived from *Podophyllum hexandrumroyle*. *Plant Cell Reports, 8*, 165–168.

Viana, M. B., Oliveira, M., & Silva, C. M. (1991). Etoposide in the treatment of six children with langerhans cell histiocytosis (histiocytosis X). *Medical and Pediatric Oncology, 19*, 289–294.

Vijayabaskar, P., Babinastarlin, S., Shankar, T., Sivakumar, T., & Anandapandian, K. T. K. (2011). Quantification and characterization of Exopolysaccharides from *Bacillus subtilis* (MTCC 121). *Advances in Biological Research, 5*(2), 71–76.

Wachtel-Galor, S., Choi, S. W., & Benzie, I. F. F. (2005). Effect of *Ganoderma lucidum* on human DNA is dose dependent and mediated by hydrogen peroxide. *Redox Report, 10*(3), 145–149.

Wang, F., Li, Y., Zhang, Y. J., Zhou, Y., Li, S., & Li, H. B. (2016). Natural products for the prevention and treatment of hangover and alcohol use disorder. *Molecules, 21*, 64.

Wu, Y., & Wang, D. (2009). A new class of natural glycopeptides with sugar moiety-dependent antioxidant activities derived from *Ganoderma lucidum* fruiting bodies. *Journal of Proteome Research, 8*, 436–442.

Wu, Z. W., Yang, Z. J., Dan, G., Fan, J. L., Dai, Z. Q., Wang, X. Q., et al. (2014). Influences of carbon sources on the biomass, production and compositions of exopolysaccharides from *Paecilomyces hepiali* HN1. *Biomass and Bioenergy, 67*, 260–269.

Xia, Y. Z., Li, S. Z., Yee, A., (Eds.), et al. (2019). *Ganoderma lucidum* inhibits tumour cell proliferation and induces tumour cell death. *Enzyme and Microbial Technology, 40*, 177–185.

Xu, J. W., Ji, S. L., Li, H. J., Zhou, J. S., Duan, Y. Q., Dang, L. Z., et al. (2015). Increased polysaccharide production and biosynthetic gene expressions in a submerged culture of *Ganoderma lucidum* by the overexpression of the homologous α-phosphoglucomutase gene. *Bioprocess and Biosystems Engineering, 38*, 399–405.

Yap, H. Y. Y., Chooi, Y. H., Firdausraih, M., Fung, S. Y., Ng, S. T., Tan, C. S., et al. (2014). The genome of the tiger milk mushroom, *Lignosus rhinocerotis*, provides insights into the genetic basis of its medicinal properties. *BMC Genomics, 15*, 635. https://doi.org/10.1186/1471-2164-15-635.

Yokozawa, T. C., Chen, P., Dong, E., Tanaka, T., Nonaka, G. I., & Nishioka, I. (1998). Study on the inhibitory effect of tannins and flavonoids against the 1,1-diphenyl-2-picrylhydrazyl radical. *Biochemical Pharmacology, 56*(2), 213–222.

Yuen, J. W., & Gohel, M. D. (2008). The dual roles of Ganoderma antioxidants on urothelial cell DNA under carcinogenic attack. *Journal of Ethnopharmacology, 118*, 324–330.

Yuvalı Celık, G., Onbaslı, D., Altınsoy, B., & Allı, H. (2014). In vitro antimicrobial and antioxidant properties of *Ganoderma lucidum* extracts grown in Turkey. *European Journal of Medicinal Plants*, *4*(6), 709–722. https://doi.org/10.9734/EJMP/2014/8546.

Zheng, J., Zhou, Y., Li, Y., Xu, D. P., Li, S., & Li, H. B. (2016). Spices for prevention and treatment of cancers. *Nutrients*, *8*, 495.

Zhou, Y., Zheng, J., Li, Y., Xu, D. P., Li, S., Chen, Y. M., et al. (2016). Natural polyphenols for prevention and treatment of cancer. *Nutrients*, *8*, 515.

Zhou, Y., Zheng, J., Li, S., Zhou, T., Zhang, P., & Li, H. B. (2016). Alcoholic beverage consumption and chronic diseases. *International Journal of Environmental Research and Public Health*, *13*, 522.

Chapter 25

Phytomedicine and the COVID-19 pandemic

Muhammad Irfan Sohail[a], Ayesha Siddiqui[b], Natasha Erum[c], and Muhammad Kamran[d]

[a]*Institute of Soil and Environmental Sciences, University of Agriculture, Faisalabad, Pakistan,* [b]*Department of Botany, University of Agriculture, Faisalabad, Pakistan,* [c]*Department of Biochemistry, University of Agriculture, Faisalabad, Pakistan,* [d]*Faculty of Agriculture, Gomal University, Dera Ismail Khan, Pakistan*

1. Introduction

Recent epidemic outbreak of novel human coronavirus baptized "SARS-CoV-2" (similar to previous virial strains MERS-CoV and SARS-CoV) inducted in China, and no single country is an exception from COVID-19 disease (Antonelli, Donelli, Maggini, & Firenzuoli, 2020; Li & De Clercq, 2020). Till date, no vaccination and cure against SARS-CoV-2 exist, while the current therapeutic options are limited (Li & De Clercq, 2020). With the continuously expanding terrible pandemic of COVID-19 infection, globally increasing numbers of COVID-19 patients and contacts progressively jump to heavy burdens beyond capabilities. Developing and poor countries are under severe humanity disasters; even the economies of developed nations have collapsed. As worldwide health authorities and hospitals are currently overloaded, home treatment of contacts and mild COVID-19 cases have become a fact necessitating some evidence-based suggested measures in addition to some health advices (Antonelli et al., 2020; El Sayed et al., 2020).

Coronaviruses cause acute respiratory and CNS diseases in many animals including humans (Kim et al., 2008). These were first reported by Tyrrell and Bynoe in 1966 who isolated and cultivated these viruses from patients with common cold symptoms. They are a family of enveloped viruses having a large, positive-sense single-stranded 27–32 kb genomic RNA and a helical nucleocapsid (Kim et al., 2010). The genomic RNA encodes seven to eight genes. Their genome (RNA) is packed together with nucleocapsid (N) protein and three envelope proteins, namely, E (envelope), M (membrane), and S (spike). The term "coronaviruses" was used for these viruses, which was based on their spherical morphology with solar corona-like surface projections (corona in Latin means

Phytomedicine: A Treasure of Pharmacologically Active Products from Plants
https://doi.org/10.1016/B978-0-12-824109-7.00005-4

crown) (Kim et al., 2008; Rathinavel, Palanisamy, Palanisamy, Subramanian, & Thangaswamy, 2020; Velavan & Meyer, 2020). Coronaviruses are animal viruses and belong to Coronaviridae family. These are further divided into alpha, beta, gamma, and delta subfamilies. The alpha and beta are provenanced from mammals, especially from bats. Among seven human-infecting subtypes, betacoronaviruses are responsible for severe fatalities (Velavan & Meyer, 2020; Yang et al., 2010). SARS-CoV-2 appears to be mutated form of SARS-CoV. It is a newly isolated coronavirus and belongs to B lineage of betacoronaviruses. It relates closely to SARS-CoV of Chinese chrysanthemum-headed bat origin (Benvenuto et al., 2020; Sun, Lu, Xu, Sun, & Pan, 2020).

The phytotherapy-based research and phytotherapic medicines can virtually probe and invent potential cures and remedies for viral infections like SARS, MERS, and CoV. Yet the horizon of phytomedicines needed to be explored. Globally, vast genera of pant and herb have been illustrated with clinical trials and have showed potential to cure based on their immune system boosting capacity against coronavirus diseases. Phytomedicines with strong preventive and immunity boosting and therapeutic effects with various mechanisms have been studied well in China. Cinchona bark which is pharmacological derivatives of chloroquine has already showed promising results in clinical trials. Mainly the laboratory studies have provided evidences for phytomedicines and plant extracts (Liu, Zhang, He, & Li, 2012). From various clinical studies of databases, Liu et al. (2012) concluded the methodology of various clinical trials remained questionable, yet the phytomedicines demonstrated promising potential to boost the human immune system, overall patient health, and lessening of COVID-19 and SARS symptoms. Therefore, in this chapter, a brief description of the potential phytomedicines and latest research in phytotherapy against coronavirus have been illustrated to provide researchers in the field with some vital hints to be used for planning future studies. An ideal anti-COVID-19 medication must be safe conferring three basic therapeutic effects, that is, enhancing the antiviral immunity, tissue protection/repair, and exerting potent antiviral effects.

2. The origin of COVID-19 pandemic

After its emergence in the end of 2019, the outbreak of COVID-19 soon became pandemic of great concern. The disease emerged as unexplained pneumonia cases in Wuhan, China, in December 2019 and declared as PHEIC (Public Health Emergency of International Concern) by WHO in January 2020. This is the sixth public health emergency of international concern after H1N1, polio, Ebola in West Africa, and Zika and Ebola in the Democratic Republic of Congo that occurred in 2009, 2014, 2014, 2016, and 2019, respectively. The disease was formally named as SARS-CoV-2 (severe acute respiratory syndrome coronavirus 2) by the International Committee on Taxonomy of Viruses. Now in middle of 2020, SARS-CoV-2 has been declared a clinical threat to general public worldwide (Lai, Shih, Ko, Tang, & Hsueh, 2020; Sun et al., 2020; Velavan & Meyer, 2020).

According to WHO report, SARS-CoV-2 was detected in samples collected from Huanan Seafood Market in China. Recent studies based on evolutionary analyses suggest that it is more likely that SARS-CoV-2 was novel and was independently introduced from animals to humans (Chen, Xiong, Bao, & Shi, 2020). The presence of bats and live animals in Huanan Seafood Market and findings based on genomic analyses further strengthen this idea that the virus might have transmitted through bat droppings as contamination (Velavan & Meyer, 2020). It is yet to be confirmed if a specific animal species is carrier of this virus. According to one study, SARS-CoV-2 was found to be a chimeric virus between bat and an unknown origin (Li et al., 2020). This study supports the theory that the transmission chain of SARS-CoV-2 started from bats to humans whereas human-to-human transmission more likely occurs through respiratory tract droplets (Lai et al., 2020; Velavan & Meyer, 2020). Comparative studies of SARS-CoV-2 genes with other animals suggest that snakes are most likely wildlife repository for this novel virus (Sun et al., 2020).

3. Pathogenesis of SARS-CoV-2

Recent reports show that COVID-19 may progress as both symptomatic and asymptomatic infections. The initial clinical sign in former case is pneumonia. The common symptoms include upper nasal tract infections such as high fever, headaches, nasal congestion, coughs, sore throat, fatigue, muscular pain, vomiting, and diarrhea. Among these symptoms, high fever ($\sim 88\%$) and coughs ($\sim 67\%$) are most common, whereas diarrhea (3.7%) and vomiting (5%) are rare signs. On progression of disease the patients were reported to develop dyspnea, hemoptysis, lymphocytopenia, leukopenia, myocarditis, and increased levels of aspartate aminotransferase and inflammation markers, that is, proinflammatory cytokines and C-reactive proteins. Abnormalities can be seen in computed chest tomography (CT) images of patients as ground glass-like and patchy consolidation areas in infected patients' bilateral lungs (Sun et al., 2020; Velavan & Meyer, 2020). The mean incubation period of 5 days was observed for SARS-CoV-2 before onset of disease. In the case of symptomatic infections, clinical signs usually appear in a period of less than a week followed by appearance of pneumonia in second or third week of infection (Guan et al., 2020; Li et al., 2020).

Receptor recognition by viruses is the first and essential step of viral infections of host cells. Knowledge about the receptor recognition mechanisms of coronaviruses is critical for understanding of their pathogenesis and for developing novel prevention or cure for the disease caused by them. It has been known that the receptor-binding domain (RBD) of coronaviruses recognizes a variety of host protein receptors including angiotensin-converting enzyme 2 (ACE2), aminopeptidase N (APN) (also known as CD13), dipeptidyl peptidase 4 (DPP4), carcinoembryonic antigen-related cell adhesion molecule 1 (CEACAM1), and cellular serine protease TMPRSS2. SARS-CoV β-coronavirus indeed utilizes

ACE2 and TMPRSS2 to enter into the host cells. In the light of the fact that the novel SARS-CoV-2 genome has high similarity with SARS-CoV β-coronavirus and that the most amino acid residues essential for ACE2 binding by SARS-CoV RBD were conserved in SARS-CoV-2 RBD, it was logical to predict that ACE2 and TMPRSS2 could also facilitate the entry of novel SARS-CoV-2 into the host cells. This prediction was indeed found true, and this so far has been a breakthrough to understand, develop, and establish options for prevention and treatment of COVID-19.

The fatality of COVID-19 is associated with damage to alveolar cells, which triggers a series of systemic reactions resulting in patient's death. SARS-CoV-2 infects lungs through entering alveolar epithelial cells by receptor-mediated endocytosis. Here, ACE2 (angiotensin-converting enzyme 2) serves as entry receptor for the virus. This is same as for other SARS coronaviruses (Ho, Wu, Chen, Li, & Hsiang, 2007; Zhou et al., 2020). In the case of SARS-CoV, SARS-CoV spike (S) protein binds to cellular receptors and mediates fusion of host and viral membranes. This protein also carries virus-neutralizing epitopes that neutralize antibodies in host. Pathogenesis of SARS-CoV is highly affected by gene mutations of this protein (Ho et al., 2007). ACE2 is expressed in type I and II alveolar epithelial cells in normal human lungs being more expressed (83%) in type II cells. Studies show that ACE2 is expressed more in men as compared with women, whereas in terms of ethnicity Asians have higher expression levels of ACE2 in their alveolar epithelial cells. Hence, Asian males appear more susceptible to SARS-CoV-2 infection. Binding of SARS-CoV-2 to ACE2 receptor triggers more expression of ACE2, which causes damage to alveoli, and this ultimately leads to death if not treated (Sun et al., 2020; Zhao et al., 2020). There is 10–20 times stronger receptor-binding ability for SARS-CoV-2 than SARS-CoV (Wrapp et al., 2020).

4.   Options for the treatment of Covid-19 infection: Is there room for phytomedicine?

4.1   The need for proper diagnosis

The first line of control of SARS-CoV2 infection and decisive factor in the initiation of course of its treatment is the proper diagnosis, particularly distinction from general cold infections. Generally, sputum examination and other diagnostic tests are conducted to confirm signs of early infections (Chhikara, Rathi, Singh, & Poonam, 2020). Fig. 1 enlists the up till date reliable laboratory tests for the diagnosis of SARS-CoV-2 infections.

4.2   Current and probable options for the treatment of COVID-19

The infection caused by SARS-CoV-2 is a novel and deadly respiratory condition; the existing antiviral drugs and vaccines have only been partially

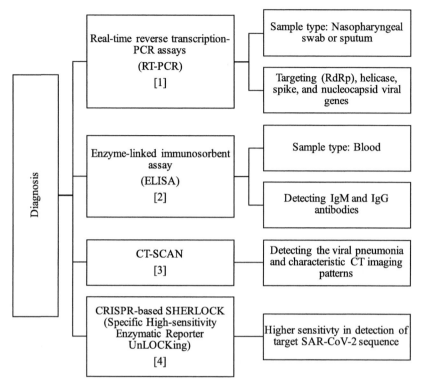

FIG. 1 Types and principles of diagnostic tests applicable for the detection of COVID-19. [1] Ai et al. (2020) and Chan et al. (2020); [2] Liu et al. (2020); [3] Shi et al. (2020); [4] Zhang, Abudayyeh, and Gootenberg (2020).

successful, and till date, no specified therapeutics have been elucidated. The only urgent management strategies include oxygen therapy, fluid management with conservative protocol in intensive care units, and the use of broad-spectrum antibiotics for treating secondary microbial infections (Huang et al., 2020). The World Health Organization (WHO) has made an urgent call for the development of vaccine, diagnostic tests for asymptomatic, and drugs for the cure. Currently the prevention of the outbreak is the utmost priority for protection (Cucinotta & Vanelli, 2020). Based on current information on mechanism and disease cycle from similar kind of viral infections, various therapeutic targets were identified to develop effective treatments against this novel virus.

4.2.1   Entry inhibitors

The SARS-CoV-2 virus infection begins when its crown-like projections made up of glycoprotein form a complex with the angiotensin-converting enzyme 2 receptors of the host cells from the respiratory system. A variety of heterocyclic molecules with the ability to make complexes might result in the development of entry inhibitor drugs (Yuan et al., 2017).

4.2.2 Replication inhibitors

Coronavirus genomes encode a protein RNA-dependent RNA polymerase (RdRp) that aids in replication of viral RNA using host machinery. RNA polymerase inhibitor–based drug such as remdesivir and DNA synthesis inhibitors such as lamivudine and tenofovir disoproxil have potential to inhibit its replication in infected cells. Protease inhibitors that interrupt protein translation such as lopinavir/ritonavir have also showed promising results (Sheahan et al., 2020). The protiens which have no structural frames such as 3CLpro (chymotrypsin-like protease) and PLP (papain-like protease) are crucial to complete coronavirus replication in the host cell. So, inhibitors of 3Clpro such as flavonoids and cinanserin (Chen et al., 2005; Jo, Kim, Shin, & Kim, 2020) and inhibitors of PLP such as diarylheptanoids (Park et al., 2012) also have prospective applications in the treatment of COVID-19 outbreak.

4.2.3 Heterocyclic drugs

Antimalarial drugs such as chloroquine inhibit glycosylation in many viruses, leading to alteration of newly synthesized proteins. These drugs have been proven effective against infections caused by SARS-CoV-19 in many studies. Other such drugs proved effective include galidesivir, garunavir, and umifenovir. Neuraminidase inhibitors (oseltamivir), vinylsulfone, and protease inhibitors have also showed potential anticoronaviral activity. Nanodrug delivery systems have been used in formulation with newly developed drugs to improve efficacy in targeting the drug (Liu, Liu, et al., 2020; Liu, Morse, Lalonde, & Xu, 2020).

4.2.4 Biological therapeutics: Antibodies and plasma therapy

Biological therapeutics using antibodies is another possibility for the cure of this infectious disease. Immune targeting of epitope of human monoclonal antibody (CR3022) and B- and T-cell epitopes of spike and nucleocapsid proteins of SARS-CoV-2 can be explored for the protection against SARS-CoV-2 (Shanmugaraj, Siriwattananon, Wangkanont, & Phoolcharoen, 2020). Moreover, to conduct clinical trials for plasma therapy, various convalescent patients donated the plasma against SARS-CoV-2. It exhibited positive results for the treatment of acute and severe SARS-CoV-2 infections (Chen et al., 2020).

4.2.5 Vaccines

It is imperative to design a vaccine for reducing SARS-CoV-2 infectious severity and viral transmissions, thereby controlling the outbreak. To treat SARS-CoV and MERS-CoV, in the recent years, several vaccination attempts following weakened virus, inactivated virus, recombinant DNA, vectors, protein vaccines, and subunit vaccines were tested in animals (Graham, Donaldson, & Baric, 2013). Such attempts to treat SARS-CoV-2 are also in progress, but it requires several months to years for the development of such cure.

4.2.6   Herbal drugs

Various herbal formulations including traditional Chinese medicine, traditional Indian ayurvedic medicine, and other herbal medicines have been used to mitigate symptoms associated with the outbreak. No clinical trials yet confirm the efficacy of these herbal extracts; however, phytochemicals have shown promising results based on their mode of actions inside the human body (Zhang, 2020).

5.   Prospective phytomedicines for COVID-19 and their possible mode of action

The emergence of infectious diseases caused by novel viral strains that are resistant to common antiviral drugs is a major worldwide issue. Interestingly, herbal medicines, also known as phytomedicines derived from traditional Chinese, Japanese, Indian, and European herbal medicine systems, are promising candidates for the discovery and development of novel antiviral drugs (Reichling & Schnitzler, 2011). Therefore, in the recent years, a huge number of experiments confirming the antiviral efficacies of medicinal plant extracts and secondary metabolites (i.e., such as flavonoids, naphthodianthrones, and anthraquinones) have been conducted (Abad, Guerra, Bermejo, Irurzun, & Carrasco, 2000; El-Toumy et al., 2018; Jabborova, Davranov, & Egamberdieva, 2019; Mateeva et al., 2017; Moshawih, Cheema, Ahmad, Zakaria, & Hakim, 2017; Simões et al., 1999; Sokmen et al., 2005; Vijayan, Raghu, Ashok, Dhanaraj, & Suresh, 2004). Particularly, in the last two decades, a number of medicinal plant extracts and/or related physiologically active ingredients have been reported to exhibit antiviral activities. Some of the phytochemicals proven effective against the viruses or symptoms related to what shown by COVID-19 along with mode of action are discussed in this section.

Chinese traditional medicine (TCM) is highly recommended by the government of China for the eradication of SARS-CoV-2 (Yang, Islam, Wang, Li, & Chen, 2020). It was reported that following medicinal plants and their derived formulations have been used in 23 provinces of China and proved effective for the treatment of COVID-19. These include *Agastache rugosa*, *Astragalus membranaceus*, *Radix platycodonis*, *Atractylodis Rhizoma*, *Cyrtomium fortunei*, *Lonicerae Japonicae*, *Glycyrrhiza uralensis*, *Fructus Forsythiae*, *Saposhnikoviae divaricata*, and *Rhizoma Atractylodis* (Luo et al., 2020). Although most of the treatments were found to lack proper statistical designs, effectiveness of these trials could be questioned. However, some TCM formulations and their possible mode of actions against novel coronavirus are listed in Table 1.

In addition to these formulations, many herbal extracts have been proposed as supplement to treat symptoms of COVID-19. For example, *Tinospora cordifolia* extract having immunomodulatory effect against human immunodeficiency virus is effective to treat related symptoms (Kalikar et al., 2008). Similarly, herbal extracts of *Anthemis hyalina*, *Nigella sativa*, and *Citrus sinensis* decreased the coronavirus replication and downregulated TRP genes that maybe involved in

TABLE 1 Phytochemicals identified and extracted from Chinese medicinal herbs and their mode of action against treatment of COVID-19.

Phytochemicals	Chinese herbs	Mode of action	Reference
Phenolic compounds in plant extract	*Isatis indigotica*	Inhibition of the cleavage activity of SARS-3CLpro enzyme	Lin et al. (2005)
Phenolic compounds in plant extract	*Houttuynia cordata*	Inhibition of viral RNA-dependent RNA polymerase activity (RdRp)	Lau et al. (2008)
Isobavachalcone, herbacetin, helichrysetin, quercetin, 3-β-D-glucoside	Multiple herbs	Inhibition of cleavage activity of MERS-3CLpro enzyme	Jo, Kim, Kim, Shin, and Kim (2019)
Glycyrrhizin	*Glycyrrhizae radix*	Inhibition of viral attachment and penetration	Wolkerstorfer, Kurz, Bachhofner, and Szolar (2009)
Flavonoids such as rhoifolin, pectolinarin, epigallocatechin gallate, gallocatechin gallate, quercetin, and herbacetin	*Litchi chinensis* and *Rheum palmatum* (Chinese rhubarb)	Inhibition of SARS-3CLpro activity	Xu, Xie, Hao, Jiang, and Wei (2011) and Jo et al. (2020)
Baicalin	*Scutellaria baicalensis*	Inhibition of angiotensin-converting enzyme (ACE)	Deng, Aluko, Jin, Zhang, and Yuan (2012)

the survival of coronavirus in epithelial cells in a study conducted by Ulasli et al. (2014). Likewise, medicinal plants such as *Heteromorpha* spp. and *Scrophularia scorodonia* possess various phytochemicals, for example, saiko-saponins, a derivative of triterpene-oleanane found abundantly across many angiosperm families (Li et al., 2018). It is reported to possess medicinal functions such as modulation of immune function, antiinflammation, antihepatoma, and antimicrobial effects; therefore it has been shown to be active against measles, herpes simplex, influenza, varicella zoster, and human immunodeficiency viruses and related symptoms (Bermejo et al., 2002; Chiang, Ng, Liu, Shieh, & Lin, 2003). A study conducted by Cheng, Ng, Chiang, and Lin (2006) indicated that saikosaponin B_2 has potent antiviral property against infection caused by human coronavirus 229E and possible mode of action includes inhibitory effect on attachment, penetration, and replication of the novel coronavirus. Similarly, *Zingiber officinale*-derived phytochemical 6-gingerol showed promising anti-coronaviral properties due to its high binding affinity against multiple SARS-CoV-2 targets, namely, RNA-binding protein, proteases, and spike proteins (Rathinavel et al., 2020).

6. Plant secondary metabolites and antiviral drugs

Plants produce a vast array of organic compounds that are differentially scattered among numerous taxa of the plant kingdom and are not primarily concerned with growth and development, known as plant secondary metabolites (Tiwari & Rana, 2015). These natural plant constituents or phytochemicals attribute characteristic medicinal properties to the plants. Extensive biological investigations have revealed a broad spectrum of pharmacological and physiological activities such as antiinflammatory, antioxidant, and anticancerous that led to its use in the formulation of promising drugs for the treatment of different diseases (Zhang et al., 2015). Many of these bioactive compounds are known to inhibit coronaviruses including MERS-CoV, SARS-CoV-1, and SARS-CoV-2 based on their potential to destroy proteases of coronavirus structural proteins and polymerases essential for its replication machinery (Khaerunnisa, Kurniawan, Awaluddin, Suhartati, & Soetjipto, 2020). Some of anticoronaviral activities and their respective mode of actions are discussed under this section. Plant secondary metabolites are mainly classified as alkaloids, phenolic compounds, and terpenoids.

6.1 Alkaloids

Alkaloids are a class of naturally occurring plant secondary metabolites primarily containing amine-type structure with basic nitrogen atom. Due to unique structural properties, many of alkaloids can be referred to as DNA intercalators such as berberine, emetine, and sanguinarine. The DNA intercalators can inhibit the transcription, replication, and translation of genetic material, in addition to its ability to stabilize the structure. Therefore these alkaloids have the potential to inhibit coronavirus replication and development within the host cell (Velu, Palanichamy, & Rajan, 2018). Many quinoline and isoquinoline alkaloids such as quinine, skimmianine, dictamine, cinchonine, and β-carboline have proved effective in treatment of many viruses including SARS-CoV-1. Chloroquine, a derivative of alkaloid quinine, has been clinically proved effective in treating SARS-CoV-2 infections (Wink, 2020). Moreover, colchicine, an alkaloid derived from the seeds of *Colchicum autumnale*, is under clinical trial for its strong antiinflammatory property in treatment of COVID-19 patients due to inhibition of NLRP3 inflammasomes and reduced activation of interleukin (Deftereos et al., 2020). Therefore it is assumed that medicinal plants producing alkaloids may be promising candidates in developing the treatment for COVID-19.

6.2 Terpenoids

Terpenoids such as monoterpenes and sesquiterpenes, the derivatives of isopentenyl diphosphate (IPP), are the most abundant class of plant secondary metabolites (Ashour, Wink, & Gershenzon, 2018). These contain pharmacological properties that play important ecological roles such as plant protection against insect predators, herbivores, and microbial pathogens; therefore they are important candidates for medicine and biotechnology (Cheng et al., 2007). Numerous studies confirmed that terpenoids can inhibit the protease activity of viruses by interfering with related amino acids. Intriguingly, terpenoid-based drugs such as thymoquinone, forskolin, ginkgolide A, menthol, salvinorin A, citral, noscapine, bilobalide, and beta-selinene have been identified as potent inhibitors of viral proteases by binding with aspartate, asparagine, and phenylalanine amino acid sites (Shaghaghi, 2020). Similarly, an important eucalyptus oil component known as eucalyptol (1,8 cineole) has been proved effective against main-protease (Mpro) during in silico studies; therefore, it may be regarded as a potent candidate for the treatment of COVID-19 among many other related terpenoids (Sharma & Kaur, 2020). In addition to that, it has been reported that various diterpenoids, triterpenoids, and essential oils having salicylaldehyde and trans-myrtanol derived from a wide range of medicinal plants are promising candidates as fumigates for the protection against COVID-19 (Nikhat & Fazil, 2020).

6.3 Polyphenols and flavonoids

Polyphenolic compounds such as phenolic acids, stilbenes, and lignans are the important natural substances found in many plants, and due to strong antioxidant potential, they have several pharmacological applications (Ma et al., 2011; Yáñez et al., 2013). Flavonoids are the major class of polyphenols that are widely found in vegetables and fruits. They have been used in numerous formulations for the treatment of diseases such as atherosclerosis, Alzheimer, and cancer owing to the biological properties such as antiinflammatory, antioxidant, anticancer, and antimutagenic (Leyva-López, Gutierrez-Grijalva, Ambriz-Perez, & Heredia, 2016; Rengasamy et al., 2019). Some of the flavonoids are reported to possess antiviral ability. In particular, apigenin, luteolin, quercetin, kaempferol, daidzein, amentoflavone, epigallocatechin, puerarin, and epigallocatechin gallate were reported to inhibit the proteolytic activity of chymotrypsin-like protease (3CLpro) of SARS-CoV (Jo et al., 2020; Nguyen et al., 2012; Ryu et al., 2010). Moreover, other polyphenols including both flavonoids and nonflavonoids have been reported to target spike (S) glycoproteins of SARS-CoV-2, which are, therefore, proven effective in destroying the structure of novel coronavirus. It was reported that polyphenols such as curcumin, kaempferol, and pterostilbene specifically target and bind to S1 domain of these proteins. Other polyphenolic compounds, particularly quercetin, apigenin, luteolin, genistein, resveratrol, isorhamnetin, and fisetin, were reported to interact and target the S2 domain of SARS-CoV-2 spike proteins (Pandey et al., 2020). The angiotensin-converting enzyme (ACE) receptors in humans are the potential receptors for SARS-CoV-2. Therefore research into finding the potential binders and inhibitors to the ACE receptors can open up new insights in COVID-19 therapy. Polyphenols, specially caffeic acid, flavonoids, chrysin, myricetin, rutin, hesperetin, pinocembrin, galangin, luteolin, and phenethyl ester, were reported to possess inhibition properties against ACE receptors and therefore are promising phytochemicals for the treatment of SARS-CoV-2 infection based on in silico and molecular docking studies (Güler, Tatar, Yildiz, Belduz, & Kolayli, 2020). Furthermore, many in-vitro and in-vivo studies will be helpful in the development of drugs based on these phytochemicals, which are more effective, safer, and inexpensive than other alternatives.

7. Conclusion

Since 1940, around 400 new infectious pathogens have been identified, and the production of vaccines has proven successful in the past, but in the case of COVID-19, no cure exists yet. COVID-19 infection has presented a grave threat to the global health and economy. It has claimed more than 258,000 deaths worldwide till date. Globally, researchers are in race to find a cure. Strict epidemiological measures have been implemented initially like the largest quarantine in history. However, there are attempts to find a chemical drug; several clinical trials have confirmed that medicinal plant extracts in combination with other drugs have shown promising results. Plant secondary metabolites or phytochemicals are products of plant metabolism that are involved in defense mechanisms against insect and pathogenic attack. Numerous bioactive compounds, broadly classified as alkaloids, polyphenols, and terpenoids, have been proved effective against novel coronaviruses in numerous in silico, in vivo, in vitro, and clinical trials carried out all over the world, that is, chloroquine derived from plant has already shown promising results. Primary mode of action for phytochemicals includes the entry inhibition of these viruses by binding with the specific receptor sites in targeted cells and/or halting the replication process of these viruses by destroying viral polymerases and proteases essential to perform important task in viral replication. Phytomedicines have also proven to boost up immunity against novel coronavirus. However, clinical trials confirming the effectivity of any drug on COVID-19 are absent. Yet the phytomedicines with immunity boosting properties seem potential candidates. Further investigations are needed to identify and test all possible targets.

References

Abad, M. J., Guerra, J. A., Bermejo, P., Irurzun, A., & Carrasco, L. (2000). Search for antiviral activity in higher plant extracts. *Phytotherapy Research, 14*(8), 604–607.

Ai, T., Yang, Z., Hou, H., Zhan, C., Chen, C., Lv, W., et al. (2020). Correlation of chest CT and RT-PCR testing in coronavirus disease 2019 (COVID-19) in China: A report of 1014 cases. *Radiology, 296*, E32–E40. https://doi.org/10.1148/radiol.2020200642.

Antonelli, M., Donelli, D., Maggini, V., & Firenzuoli, F. (2020). Phytotherapic compounds against coronaviruses: Possible streams for future research. *Phytotherapy Research, 34*(7), 1469–1470. https://doi.org/10.1002/ptr.6712.

Ashour, M., Wink, M., & Gershenzon, J. (2018). Biochemistry of terpenoids: Monoterpenes, sesquiterpenes and diterpenes. *Annual Plant Reviews Online*, 258–303.

Benvenuto, D., Giovanetti, M., Ciccozzi, A., Spoto, S., Angeletti, S., & Ciccozzi, M. (2020). The 2019-new coronavirus epidemic: Evidence for virus evolution. *Journal of Medical Virology, 92*(4), 455–459.

Bermejo, P., Abad, M. J., Díaz, A. M., Fernández, L., De Santos, J., Sanchez, S., et al. (2002). Antiviral activity of seven iridoids, three saikosaponins and one phenylpropanoid glycoside extracted from Bupleurum rigidum and Scrophularia scorodonia. *Planta Medica, 68*(02), 106–110.

Chan, J. F. W., Yip, C. C. Y., To, K. K. W., Tang, T. H. C., Wong, S. C. Y., Leung, K. H., et al. (2020). Improved molecular diagnosis of COVID-19 by the novel, highly sensitive and specific COVID-19-RdRp/Hel real-time reverse transcription-PCR assay validated in vitro and with clinical specimens. *Journal of Clinical Microbiology, 58*(5).

Chen, L., Gui, C., Luo, X., Yang, Q., Günther, S., Scandella, E., et al. (2005). Cinanserin is an inhibitor of the 3C-like proteinase of severe acute respiratory syndrome coronavirus and strongly reduces virus replication in vitro. *Journal of Virology, 79*(11), 7095–7103.

Chen, L., Xiong, J., Bao, L., & Shi, Y. (2020). Convalescent plasma as a potential therapy for COVID-19. *The Lancet Infectious Diseases, 20*(4), 398–400.

Cheng, A. X., Lou, Y. G., Mao, Y. B., Lu, S., Wang, L. J., & Chen, X. Y. (2007). Plant terpenoids: Biosynthesis and ecological functions. *Journal of Integrative Plant Biology, 49*(2), 179–186.

Cheng, P. W., Ng, L. T., Chiang, L. C., & Lin, C. C. (2006). Antiviral effects of saikosaponins on human coronavirus 229E in vitro. *Clinical and Experimental Pharmacology and Physiology, 33*(7), 612–616.

Chhikara, B. S., Rathi, B., Singh, J., & Poonam, F. N. U. (2020). Corona virus SARS-CoV-2 disease COVID-19: Infection, prevention and clinical advances of the prospective chemical drug therapeutics. *Chemical Biology Letters, 7*(1), 63–72.

Chiang, L. C., Ng, L. T., Liu, L. T., Shieh, D. E., & Lin, C. C. (2003). Cytotoxicity and anti-hepatitis B virus activities of saikosaponins from Bupleurum species. *Planta Medica, 69*(08), 705–709.

Cucinotta, D., & Vanelli, M. (2020). WHO declares COVID-19 a pandemic. *Acta Bio-Medica: Atenei Parmensis, 91*(1), 157–160.

Deftereos, S., Giannopoulos, G., Vrachatis, D. A., Siasos, G., Giotaki, S. G., Cleman, M., et al. (2020). Colchicine as a potent anti-inflammatory treatment in COVID-19: Can we teach an old dog new tricks? *European Heart Journal-Cardiovascular Pharmacotherapy*. https://doi.org/10.1093/ehjcvp/pvaa033.

Deng, Y. F., Aluko, R. E., Jin, Q., Zhang, Y., & Yuan, L. J. (2012). Inhibitory activities of baicalin against renin and angiotensin-converting enzyme. *Pharmaceutical Biology, 50*(4), 401–406.

El Sayed, S. M., Almaramhy, H. H., Aljehani, Y. T., Okashah, A. M., El-Anzi, M. E., AlHarbi, M. B., et al. (2020). The evidence-based TaibUVID nutritional treatment for minimizing COVID-19 fatalities and morbidity and eradicating COVID-19 pandemic: A novel approach for better outcomes (a treatment protocol). *American Journal of Public Health Research, 8*(2), 54–60.

El-Toumy, S. A., Salib, J. Y., El-Kashak, W. A., Marty, C., Bedoux, G., & Bourgougnon, N. (2018). Antiviral effect of polyphenol rich plant extracts on herpes simplex virus type 1. *Food Science and Human Wellness, 7*(1), 91–101.

Graham, R. L., Donaldson, E. F., & Baric, R. S. (2013). A decade after SARS: Strategies for controlling emerging coronaviruses. *Nature Reviews Microbiology, 11*(12), 836–848.

Guan, W. J., Ni, Z. Y., Hu, Y., Liang, W. H., Ou, C. Q., He, J. X., ... Du, B. (2020). Clinical characteristics of 2019 novel coronavirus infection in China. *New England Journal of Medicine, 382*, 1708–1720. https://doi.org/10.1056/NEJMoa2002032.

Güler, H. I., Tatar, G., Yildiz, O., Belduz, A. O., & Kolayli, S. (2020). An investigation of ethanolic propolis extracts: Their potential inhibitor properties against ACE-II receptors for COVID-19 treatment by molecular docking study. *ScienceOpen Preprints.* https://doi.org/10.14293/S2199-1006.1.SOR-.PP5BWN4.v1.

Ho, T. Y., Wu, S. L., Chen, J. C., Li, C. C., & Hsiang, C. Y. (2007). Emodin blocks the SARS coronavirus spike protein and angiotensin-converting enzyme 2 interaction. *Antiviral Research, 74*(2), 92–101.

Huang, C., Wang, Y., Li, X., Ren, L., Zhao, J., Hu, Y., et al. (2020). Clinical features of patients infected with 2019 novel coronavirus in Wuhan, China. *The Lancet, 395*(10223), 497–506.

Jabborova, D., Davranov, K., & Egamberdieva, D. (2019). Antibacterial, antifungal, and antiviral properties of medical plants. In *Medically important plant biomes: Source of secondary metabolites* (pp. 51–65). Singapore: Springer.

Jo, S., Kim, H., Kim, S., Shin, D. H., & Kim, M. S. (2019). Characteristics of flavonoids as potent MERS-CoV 3C-like protease inhibitors. *Chemical Biology & Drug Design, 94*(6), 2023–2030.

Jo, S., Kim, S., Shin, D. H., & Kim, M. S. (2020). Inhibition of SARS-CoV 3CL protease by flavonoids. *Journal of Enzyme Inhibition and Medicinal Chemistry, 35*(1), 145–151.

Kalikar, M. V., Thawani, V. R., Varadpande, U. K., Sontakke, S. D., Singh, R. P., & Khiyani, R. K. (2008). Immunomodulatory effect of Tinospora cordifolia extract in human immuno-deficiency virus positive patients. *Indian Journal of Pharmacology, 40*(3), 107.

Khaerunnisa, S., Kurniawan, H., Awaluddin, R., Suhartati, S., & Soetjipto, S. (2020). Potential inhibitor of COVID-19 main protease (Mpro) from several medicinal plant compounds by molecular docking study. *Preprints, 1–14.* https://doi.org/10.20944/preprints202003.0226.v1.

Kim, H. Y., Eo, E. Y., Park, H., Kim, Y. C., Park, S., Shin, H. J., & Kim, K. (2010). Medicinal herbal extracts of *Sophorae radix, Acanthopanacis cortex, Sanguisorbae radix* and *Torilis fructus* inhibit coronavirus replication in vitro. *Antiviral Therapy, 15*(5), 697–709.

Kim, H. Y., Shin, H. S., Park, H., Kim, Y. C., Yun, Y. G., Park, S., ... Kim, K. (2008). In vitro inhibition of coronavirus replications by the traditionally used medicinal herbal extracts, *Cimicifuga rhizoma, Meliae cortex, Coptidis rhizoma*, and *Phellodendron cortex. Journal of Clinical Virology, 41*(2), 122–128.

Lai, C. C., Shih, T. P., Ko, W. C., Tang, H. J., & Hsueh, P. R. (2020). Severe acute respiratory syndrome coronavirus 2 (SARS-CoV-2) and corona virus disease-2019 (COVID-19): The epidemic and the challenges. *International Journal of Antimicrobial Agents, 55*(3), 105924.

Lau, K. M., Lee, K. M., Koon, C. M., Cheung, C. S. F., Lau, C. P., Ho, H. M., et al. (2008). Immunomodulatory and anti-SARS activities of Houttuynia cordata. *Journal of Ethnopharmacology, 118*(1), 79–85.

Leyva-López, N., Gutierrez-Grijalva, E. P., Ambriz-Perez, D. L., & Heredia, J. B. (2016). Flavonoids as cytokine modulators: A possible therapy for inflammation-related diseases. *International Journal of Molecular Sciences, 17*(6), 921.

Li, G., & De Clercq, E. (2020). Therapeutic options for the 2019 novel coronavirus (2019-nCoV). *Nature Reviews Drug Discovery, 19*, 149–150.

Li, Q., Guan, X., Wu, P., Wang, X., Zhou, L., Tong, Y., ... Xing, X. (2020). Early transmission dynamics in Wuhan, China, of novel coronavirus-infected pneumonia. *New England Journal of Medicine, 382*(13), 1199–1207. https://doi.org/10.1056/NEJMoa2001316.

Li, X. Q., Song, Y. N., Wang, S. J., Rahman, K., Zhu, J. Y., & Zhang, H. (2018). Saikosaponins: A review of pharmacological effects. *Journal of Asian Natural Products Research, 20*(5), 399–411.

Lin, C. W., Tsai, F. J., Tsai, C. H., Lai, C. C., Wan, L., Ho, T. Y., et al. (2005). Anti-SARS corona-virus 3C-like protease effects of Isatis indigotica root and plant-derived phenolic compounds. *Antiviral Research, 68*(1), 36–42.

Liu, W., Liu, L., Kou, G., Zheng, Y., Ding, Y., Ni, W., et al. (2020). Evaluation of nucleocapsid and spike protein-based ELISAs for detecting antibodies against SARS-CoV-2. *Journal of Clinical Microbiology, 58*(6), 1–7. https://doi.org/10.1128/JCM.00461-20.

Liu, W., Morse, J. S., Lalonde, T., & Xu, S. (2020). Learning from the past: Possible urgent prevention and treatment options for severe acute respiratory infections caused by 2019-nCoV. *Chembiochem, 21*(5), 730–738. https://doi.org/10.1002/cbic.202000047.

Liu, X., Zhang, M., He, L., & Li, Y. (2012). Chinese herbs combined with Western medicine for severe acute respiratory syndrome (SARS). *Cochrane Database of Systematic Reviews, 10*, CD004882. https://doi.org/10.1002/14651858.CD004882.pub3.

Luo, H., Tang, Q. L., Shang, Y. X., Liang, S. B., Yang, M., Robinson, N., et al. (2020). Can Chinese medicine be used for prevention of corona virus disease 2019 (COVID-19)? A review of historical classics, research evidence and current prevention programs. *Chinese Journal of Integrative Medicine*, 1–8.

Ma, X., Wu, H., Liu, L., Yao, Q., Wang, S., Zhan, R., et al. (2011). Polyphenolic compounds and antioxidant properties in mango fruits. *Scientia Horticulturae, 129*(1), 102–107.

Mateeva, N., Eyunni, S. V., Redda, K. K., Ononuju, U., Hansberry, T. D., II, Aikens, C., et al. (2017). Functional evaluation of synthetic flavonoids and chalcones for potential antiviral and anticancer properties. *Bioorganic & Medicinal Chemistry Letters, 27*(11), 2350–2356.

Moshawih, S., Cheema, M. S., Ahmad, Z., Zakaria, Z. A., & Hakim, M. N. (2017). A comprehensive review on *Cosmos caudatus* (Ulam raja): Pharmacology, ethnopharmacology, and phytochemistry. *International Research Journal of Educational and Sciences, 1*(1), 14–31.

Nguyen, T. T. H., Woo, H. J., Kang, H. K., Kim, Y. M., Kim, D. W., Ahn, S. A., et al. (2012). Flavonoid-mediated inhibition of SARS coronavirus 3C-like protease expressed in Pichia pastoris. *Biotechnology Letters, 34*(5), 831–838.

Nikhat, S., & Fazil, M. (2020). Overview of Covid-19; its prevention and management in the light of Unani medicine. *Science of the Total Environment*, 138859.

Pandey, P., Rane, J. S., Chatterjee, A., Kumar, A., Khan, R., Prakash, A., & Ray, S. (2020). Targeting SARS-CoV-2 spike protein of COVID-19 with naturally occurring phytochemicals: An in silico study for drug development. *Journal of Biomolecular Structure and Dynamics*, 1–11.

Park, J. Y., Jeong, H. J., Kim, J. H., Kim, Y. M., Park, S. J., Kim, D., et al. (2012). Diarylheptanoids from *Alnus japonica* inhibit papain-like protease of severe acute respiratory syndrome coronavirus. *Biological and Pharmaceutical Bulletin, 35*(11), 2036–2042. https://doi.org/10.1248/bpb.b12-00623.

Rathinavel, T., Palanisamy, M., Palanisamy, S., Subramanian, A., Thangaswamy, S., et al. (2020). Phytochemical 6-Gingerol – A promising drug of choice for COVID-19. *International Journal of Advanced Science and Engineering, 6*(4), 1482–1489. https://doi.org/10.29294/IJASE.6.4.2020.1482-1489.

Reichling, J., & Schnitzler, P. (2011). Antiviral effects of essential oils used traditionally in Phytomedicine. In G. Bagetta, M. Cosentino, M. T. Corasaniti, & S. Sakurada (Eds.), *Herbal medicines: Development and validation of plant-derived medicines for human health* (p. 317). Taylor & Francis (Chapter 18).

Rengasamy, K. R., Khan, H., Gowrishankar, S., Lagoa, R. J., Mahomoodally, F. M., Khan, Z., et al. (2019). The role of flavonoids in autoimmune diseases: Therapeutic updates. *Pharmacology & Therapeutics, 194*, 107–131.

Ryu, Y. B., Jeong, H. J., Kim, J. H., Kim, Y. M., Park, J. Y., Kim, D., et al. (2010). Biflavonoids from *Torreya nucifera* displaying SARS-CoV 3CLpro inhibition. *Bioorganic & Medicinal Chemistry*, *18*(22), 7940–7947.

Shaghaghi, N. (2020). Molecular docking study of novel COVID-19 protease with low risk terpenoides compounds of plants. *ChemRxiv*. 10.

Shanmugaraj, B., Siriwattananon, K., Wangkanont, K., & Phoolcharoen, W. (2020). Perspectives on monoclonal antibody therapy as potential therapeutic intervention for coronavirus disease-19 (COVID-19). *Asian Pacific Journal of Allergy and Immunology*, *38*(1), 10–18.

Sharma, A. D., & Kaur, I. J. (2020). Eucalyptol (1,8 cineole) from eucalyptus essential oil a potential inhibitor of COVID 19 corona virus infection by molecular docking studies. *Preprints*, 1–8. https://doi.org/10.20944/preprints202003.0455.v1.

Sheahan, T. P., Sims, A. C., Leist, S. R., Schäfer, A., Won, J., Brown, A. J., et al. (2020). Comparative therapeutic efficacy of remdesivir and combination lopinavir, ritonavir, and interferon beta against MERS-CoV. *Nature Communications*, *11*(1), 1–14.

Shi, H., Han, X., Jiang, N., Cao, Y., Alwalid, O., Gu, J., et al. (2020). Radiological findings from 81 patients with COVID-19 pneumonia in Wuhan, China: A descriptive study. *The Lancet Infectious Diseases*, *20*(4), 425–434. https://doi.org/10.1016/S1473-3099(20)30086-4.

Simões, C. M. O., Falkenberg, M., Mentz, L. A., Schenkel, E. P., Amoros, M., & Girre, L. (1999). Antiviral activity of south Brazilian medicinal plant extracts. *Phytomedicine*, *6*(3), 205–214.

Sokmen, M., Angelova, M., Krumova, E., Pashova, S., Ivancheva, S., Sokmen, A., et al. (2005). In vitro antioxidant activity of polyphenol extracts with antiviral properties from Geranium sanguineum L. *Life Sciences*, *76*(25), 2981–2993.

Sun, P., Lu, X., Xu, C., Sun, W., & Pan, B. (2020). Understanding of COVID-19 based on current evidence. *Journal of Medical Virology*, *92*(6), 548–551.

Tiwari, R., & Rana, C. S. (2015). Plant secondary metabolites: A review. *International Journal of Engineering Research and General Science*, *3*(5), 661–670.

Ulasli, M., Gurses, S. A., Bayraktar, R., Yumrutas, O., Oztuzcu, S., Igci, M., et al. (2014). The effects of *Nigella sativa* (ns), Anthemis hyalina (Ah) and *Citrus sinensis* (Cs) extracts on the replication of coronavirus and the expression of TRP genes family. *Molecular Biology Reports*, *41*(3), 1703–1711.

Velavan, T. P., & Meyer, C. G. (2020). The COVID-19 epidemic. *Tropical Medicine & International Health*, *25*(3), 278.

Velu, G., Palanichamy, V., & Rajan, A. P. (2018). Phytochemical and pharmacological importance of plant secondary metabolites in modern medicine. In *Bioorganic phase in natural food: An overview* (pp. 135–156). Cham: Springer.

Verma, S. (2020). In search of feasible interventions for the prevention and cure of novel Coronavirus disease 2019. *OSF Preprints*, 1–13. https://doi.org/10.31219/osf.io/q6tsc.

Vijayan, P., Raghu, C., Ashok, G., Dhanaraj, S. A., & Suresh, B. (2004). Antiviral activity of medicinal plants of Nilgiris. *Indian Journal of Medical Research*, *120*, 24–29.

Wink, M. (2020). Potential of DNA intercalating alkaloids and other plant secondary metabolites against SARS-CoV-2 causing COVID-19. *Diversity*, *12*(5), 175.

Wolkerstorfer, A., Kurz, H., Bachhofner, N., & Szolar, O. H. (2009). Glycyrrhizin inhibits influenza a virus uptake into the cell. *Antiviral Research*, *83*(2), 171–178.

Wrapp, D., Wang, N., Corbett, K. S., Goldsmith, J. A., Hsieh, C. L., Abiona, O., … McLellan, J. S. (2020). Cryo-EM structure of the 2019-nCoV spike in the prefusion conformation. *Science*, *367*(6483), 1260–1263.

Xu, X., Xie, H., Hao, J., Jiang, Y., & Wei, X. (2011). Flavonoid glycosides from the seeds of Litchi chinensis. *Journal of Agricultural and Food Chemistry*, *59*(4), 1205–1209.

Yáñez, J. A., Remsberg, C. M., Takemoto, J. K., Vega-Villa, K. R., Andrews, P. K., Sayre, C. L., et al. (2013). Polyphenols and flavonoids: An overview. In N. M. Davies, & J. A. Yañez (Eds.), *Flavonoid pharmacokinetics. Methods of analysis, preclinical and clinical pharmacokinetics, safety and toxicology* (pp. 1–71). Hoboken, NJ: John Wiley & Sons, Inc.

Yang, C. W., Lee, Y. Z., Kang, I. J., Barnard, D. L., Jan, J. T., Lin, D., … Lee, S. J. (2010). Identification of phenanthroindolizines and phenanthroquinolizidines as novel potent anti-coronaviral agents for porcine enteropathogenic coronavirus transmissible gastroenteritis virus and human severe acute respiratory syndrome coronavirus. *Antiviral Research, 88*(2), 160–168.

Yang, Y., Islam, M. S., Wang, J., Li, Y., & Chen, X. (2020). Traditional Chinese medicine in the treatment of patients infected with 2019-new coronavirus (SARS-CoV-2): A review and perspective. *International Journal of Biological Sciences, 16*(10), 1708.

Yuan, Y., Cao, D., Zhang, Y., Ma, J., Qi, J., Wang, Q., … Zhang, X. (2017). Cryo-EM structures of MERS-CoV and SARS-CoV spike glycoproteins reveal the dynamic receptor binding domains. *Nature Communications, 8*, 15092.

Zhang, F., Abudayyeh, O. O., & Gootenberg, J. S. (2020). A protocol for detection of COVID-19 using CRISPR diagnostics. In *A protocol for detection of COVID-19 using CRISPR diagnostics* (p. 8).

Zhang, K. (2020). Is traditional Chinese medicine useful in the treatment of COVID-19? *The American Journal of Emergency Medicine.* https://doi.org/10.1016/j.ajem.2020.03.046. In press.

Zhang, Y. J., Gan, R. Y., Li, S., Zhou, Y., Li, A. N., Xu, D. P., et al. (2015). Antioxidant phytochemicals for the prevention and treatment of chronic diseases. *Molecules, 20*(12), 21138–21156.

Zhao, Y., Zhao, Z., Wang, Y., Zhou, Y., Ma, Y., Zuo, W., et al. (2020). Single-cell RNA expression profiling of ACE2, the receptor of SARS-CoV-2. *American Journal of Respiratory and Critical Care Medicine, 202*(5), 756–759. https://doi.org/10.1164/rccm.202001-0179LE.

Zhou, P., Yang, X. L., Wang, X. G., Hu, B., Zhang, L., Zhang, W., … Chen, H. D. (2020). A pneumonia outbreak associated with a new coronavirus of probable bat origin. *nature, 579*(7798), 270–273.

Chapter 26

Phytopharmaceutical marketing: A case study of USPs used for phytomedicine promotion

Sheikh Basharul Islam, Mushtaq Ahmad Darzi, and Suhail Ahmad Bhat
Department of Management Studies, University of Kashmir, Srinagar, Jammu and Kashmir, India

1. Introduction

Phytomedicines as a part of folk or traditional medicines consist of ancient knowledge systems related to treatment by herbs. About 4 billion people, which is 80% of global population, depend on phytomedicines as primary healthcare (Srivastava, Srivastava, Pandey, Khanna, & Pant, 2019). Developed over several generations within various cultures before the advent of synthetic medicine, it has been established in the form of several popular traditional medicine systems like Ayurveda, Unani, traditional Chinese medicine, and Kampo.

The demand of phytomedicine has been witnessing a continuous rise in both industrially developed and developing nations. In global market of herbal medicine, Europe entertains the largest share in the pie followed by Asia (Sen & Chakraborty, 2015). However, China and India are crowned as the major exporters of traditional medicines. The reason is the availability of vast treasure of plant species and related traditional knowledge in these countries (Lange, 2006). The mixture of one or more plants in the form of standardized preparations is used across globe for treatment of various diseases like diarrhea, headache, diabetes, burns, rashes, allergies, cold, menopause, and depression. The standard preparations are generally marketed in solid, viscous, or liquid form (Calixto, 2000).

Since ages, people have developed belief in traditional system of medicine due to some known and obscure reasons. People have learned that the use of phytotherapeutic care through hit-and-trial method and herbal medicine are perceived to be free from side effects and the most effective medicines when compared with the modern style of treatment (Ekor, 2014; Patwardhan, 2016). Furthermore, herbal mixtures are believed as therapies, either as curative or preventive for all kinds of illnesses (Odiboh, Omojola, Okorie, & Ekanem, 2017).

Phytomedicine: A Treasure of Pharmacologically Active Products from Plants
https://doi.org/10.1016/B978-0-12-824109-7.00011-X

However, there is a lack of scientific knowledge to support all the medical philosophies of traditional medicines.

USPs as a marketing tool are specifically used for creation of distinct product image in market, magnetize attention, and modulate consumer's intention to purchase (Blythe, 2004). Marketing approach of phytomedicine advertisers has been reported as misleading. The purpose of marketing of drugs should have been informational rather creating illusion. But it has been witnessed that marketers do not follow the regulations of medicine marketing laid by the World Health Organization in global perspective or by authorities for respective geographies (Lewis & Strom, 2002). Misleading propositions and hyperbolic claims about phytomedicines are being frequently popularized by marketers, many of which are exaggerated, irrelevant without any substantial scientific proof (Tyler, 2000).

Misuse of marketing practices necessitates enforcement of strict regulations for careful screening of information disseminated through labeling and various other media platforms. Furthermore, lack of scientific knowledge to provide reliable clinical data and validate the beliefs about herbal formulations demands investment of extensive research and finance.

2.   Phytopharmaceutical market global scenario

The interest of people in healthcare products derived from plants has been witnessing a continuous rise in both industrially developed and developing nations (Blumenthal, 1999; Roberts & Tyler, 1998; WHO, 2004), with these herbal medicines being sold in various retail formats, besides drug stores. According to the World Health Organization (WHO), 80% of global population, which is around 4 billion people, use phytomedicines as primary healthcare (Nirmal et al., 2013; Pan et al., 2013; Singh et al., 2019; Srivastava et al., 2019). Around the world, 25% of modern drugs prescribed are derived from plants used in traditional medicine (Dettweiler et al., 2019; Raghavendra et al., 2009). The medical prescriptions in Germany account for 50% sale of phytomedicines, while in the United States around 25%–30% medical prescriptions dispensed include drugs obtained from plants (Nirmal et al., 2013). This has boosted international trade of phytomedicines enormously and has attracted the interest of multinationals and other big pharma companies. The world phytomedicine market was reported to be valued at USD 12.4 billion in 1994, growing to USD 24.1 billion in 2002 and USD 100 billion in 2011 (Saslis-Lagoudakis et al., 2015; Vasisht & Kumar, 2002). The industry has been projected to grow at AAGR (average annual growth rate) of 6%–8% worldwide to reach USD 111 billion by 2023 and USD 5 trillion by 2050 (Bhowmik, Kumar, Tripathi, & Chiranjib, 2009; Govt. of India, 2000; Neupane & Lamichhane, 2020; Seethapathy et al., 2018; Singh et al., 2019). Europe leads the international market of herbal medicines with 45% market share followed by Southeast Asia, Japan, and North America with 19%, 16%, and 11% market share, respectively (Sen & Chakraborty, 2015). From the impressive market value of USD 6 billion reported in 1994, Europe grew to USD 8.9 billion in 2002 and USD 44.48 billion in 2011 (Fig. 1) (Saslis-Lagoudakis et al., 2015). The demand for phytomedicines in European countries is dominated by Germany where around 50% of population is reported as traditional medicine consumers followed by France with 49% consumer base (Sen & Chakraborty, 2015). The United States is the fast-growing and largest market of phytomedicines in North American region. In the United States, 60 million consumers have been reported as taking phytomedicines for disease treatment and other healthcare purposes (American Botanical Council, 2011). In 1998 the US market was reported to be valued at USD 4.5 billion (Vasisht & Kumar, 2002). The demand for phytomedicines in the United States has witnessed a significant rise because people were seeking more preventive drugs for diseases against the drugs prescribed specifically for treatment after problem has been diagnosed.

Among Asian countries, China and India are the biggest producers and consumers of herbal medicines (Lange, 2006). In China, phytomedicines account for 30%–50% of total drug consumption (Nirmal et al., 2013). Phytotherapeutic care as primary healthcare system is used by around 40% of urban patients and around 90% of rural patients across China (Vasisht & Kumar, 2002). The sale of Chinese traditional medicine witnessed a tremendous growth from USD 2.5 in 1993 to USD 47.84 billion in 2010 (WHO, 2012). India is rich in plant diversity, and practice of traditional medicines in India is among the oldest traditional knowledge systems. It consists of Ayurveda, Yoga, Unani, Siddha, Naturopathy, and Homeopathy system (Ministry of Health and Family Welfare, 2015). Through AYUSH system, around 8000 phytomedicinal remedies are made available to public along with conventional medical care (National Health Portal, 2016). In India the herbal market is valued around USD 1 billion with

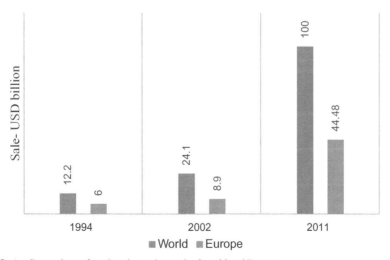

FIG. 1 Comparison of market size and growth of world and Europe.

USD 80 million export of crude herbal extracts (Gunjan et al., 2015; Pandey, Debnath, Gupta, & Chikara, 2011). The Indian phytomedicine market is project to reach USD 8 billion by 2022 (Sachitanand, 2017). The Japanese traditional medicine method is popularly known as Kampo medicine. The sale of Kampo medicine or phytomedicines witnessed around 30% growth from 1999 to 2002. The Kampo market was valued at USD 2.9 billion in 2002 from USD 2.2 billion in 1999 (Vasisht & Kumar, 2002). The most popular Japanese herbal formulations are Sho-saiko and Hochu-ekki, which accounted for around USD 163 million sales in 2001 (Stephen, 2001).

Out of total healthcare services delivered over past decade, the percentage pf population consuming medicines derived from plants in African countries and Latin America accounted for 80% of total population, followed by Chile and India with 71% and 65% of population dependent on herbal medicines, respectively. The percentage of population in different countries using herbal medicines is shown in Fig. 2 (Ekor, 2014; Vasisht & Kumar, 2002; WHO, 2002). This tremendous market growth of phytopharmaceutical industry across globe is attributed to reported effectiveness or efficacy of phytotherapeutic treatment, health benefits of phytomedicines, rising awareness of side effects of synthetic medicines, self-medication and increasing preference for preventive drugs, increase in R&D activities, rise in population, and inaccessibility of large portion of population around the world to conventional or synthetic drugs because of high cost (Ekor, 2014; National Health Portal, 2016; Srivastava et al., 2019). Moreover the rise in chronic diseases has forced the rise in consumption of bioactive ingredients from plants as a general perception of "free from side effects." WHO in 2001 has reported that around 60% of 56.5 million deaths occurring around the world are due to chronic diseases. In 2012, as per National Center for Chronic Disease Prevention and Health Promotion, in the United States, one among four adults suffer from more than two chronic diseases, and a total of 117 million people were reported having one or more chronic diseases in 2012 (The Vegan Business Magazine, 2019). The remarkable growth and promising potential of markets across globe have led to various developments in herbal medicine industry ranging from acquisitions, mergers, collaborations, and facility extensions by big market players in pharmaceutical industry (Table 1).

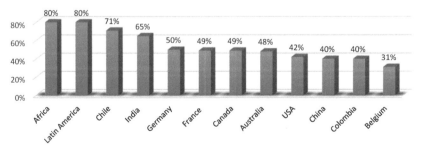

FIG. 2 Percentage of population in different countries using herbal medicines.

TABLE 1 Recent market developments in herbal medicine industry.

Company/country of origin	Acquisitions and facility development	Source
Bayer, Germany	Acquired Dihon Pharmaceutical, a China-based manufacturer of traditional Chinese medicine and OTC dermatology products for women	(BioSpectrum, 2015) www.biospectrumasia.com
Martin Bauer Group, Germany	Acquisition of BI Nutraceuticals–United States to strengthen its product portfolio of herbal ingredients	(Martin Bauer Group, 2019) www.martin-bauer-group.com
Blackmores, Australia	Acquired Global Therapeutics, a leading herbal medicine company of Australia	(Blackmores, 2016) www.blackmores.com
Arkopharma Laboratories, France	Acquired Distrifa, a Portuguese distributor to grow its market across Iberian Peninsula	(ARKOPHARMA, 2019) www.arkopharma.com
Givaudan, Switzerland	Acquired Golden Frog (Vietnam) to enhance its product portfolio of natural extracts and essential oils	(Givaudan, 2019) www.givaudan.coma
Dabur, India	Collaborated with Amazon to launch of Ayurveda e-marketplace	(Business World, 2017) www.businessworld.in
CANVAS, Australia	Partnership with Maiko Pte Ltd. to manage its sales, marketing, and branding in Singapore	(Business Insider, 2020) www.businessinsider.sg

3.   USP: A USP is what makes you different from your competitor

With the aid of advertising and marketing pioneers Claude C. Hopkins and John E. Kennedy, the advertising legend Albert Lasker founded modern advertising. The powerful and creative application of "reason-why" advertising by Lasker had a significant effect on modern advertising. They are viewed as the forerunners of the consumer-centric unique selling proposition (USP) approach (Basal, 2019; Cruikshank & Schultz, 2010). The unique factor of product advertising later became USP.

In 1961 "Rosser Reeves," a television advertising pioneer of Ted Bates & Company, first coined the term unique selling proposition (USP) and used it in advertising (Basal, 2019; Reeves, 1961). He argued that by using USP approach advertisement would be successful. Unique selling proposition is a marketing strategy that refers to some specific functional benefit or characteristics articulated through a proposition (e.g., antidandruff claim by Head & Shoulders and Himalaya Neem Face Wash that fights pimples and prevents marks) that differentiates the company, product, service, or brand uniquely from its competitors (Laskey, Day, & Crask, 1989; Clark, 2011).

As defined by Dr. James Blythe, USP "contains the one feature of the product that most stands out as different from the competition, and is usually a feature that conveys unique benefits to the consumer" (Blythe, 2004).

USP requires companies to make propositions to its customers that are:

- *Assertive:* The advertisement must convey to audience: "Buy the product and get the specific advantage."
- *Unique:* The benefit should be such that the competition cannot imitate or does not offer.
- *Persuasive:* Proposition must be able to convince new potential customers to embrace your brand.

The key elements behind the well-done USP are functional differentiation, ceaseless advertising, and mass appeal. The advertising should be focused on communicating one message to consumers repeatedly, because consumers retain only one strong claim made by the advertiser (Naik, 2007). The functional differentiation will not be able to serve the purpose for long because of possible imitation by competitors. It is the advertised benefit that moves millions of customers away from their existing brand preference (Clark, 2011).

4. Popular USPs in phytomedicine marketing

Every person is unique, so is every brand in terms of image, offering, and the advertising message. The propositions with which the phytomedicines or herbal drugs are promoted across globe are generally focused on efficacy, free from side effects and treatment of incurable and noncommunicable diseases (NCDs) by medication (Ahmad, Patel, Parimalakrishnan, Mohanta, & Nagappa, 2015; Patwardhan, 2016; Thanisorn & Bunchapattanasakda, 2011). But handsome efforts are invested to promote production-oriented propositions like being 100% organic and "handmade" (Rowey & Spiezia, 2006), where the attempt is made to build the high-quality product perception in consumers mind. Perception of brand has a significant relationship with consumer trust, consumer preference of brand, loyalty, and brand recognition (Thanisorn & Bunchapattanasakda, 2011). It reinforces attitude and describes consumer's positive or negative disposition toward brand (Guthrie & Kim, 2009). Furthermore, purchase decisions of consumers are based on perception about brand, consumer expectation from herbal medicines, and personal experience more than real statistical estimates (Fig. 3) (Patwardhan, 2016; Schiffman & Kanuk, 2004).

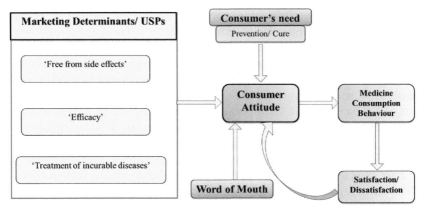

FIG. 3   Factors affecting consumer attitude toward herbal medicine consumption.

4.1   Free from side effects

The slogan "Free from side effects" has been widely associated with herbal drugs through its repetitive use in promotion of phytomedicines. This continuous reinforcement of message circulated via different media platforms, endorsed by eminent public personalities along with claims made by credible organizations like Central Council for Research in Ayurvedic Sciences—New Delhi (CCRAS), has translated it into a common belief among masses (Patwardhan, 2016). With this general belief about phytomedicines of being safe or free from side effects, people are displaying inclined tendency toward their use at any cost. Many researchers have advocated in their studies that high rating is being attributed to safety perception of herbal medicines among factors affecting purchase decisions (Chen, 2010; Suganya & Hamsalakshmi, 2017). In a recent study, Adlakha and Sharma (2019) have found that around 80% of people believe that herbal drugs are free from side effects with a mean score of 4.70. Another study conducted on medicine advertisements revealed that 95% of Ayurvedic products being promoted claimed on their labels or in advertisements as 100% safe (Ahmad et al., 2015). Some examples of the taglines, which infer safety of herbal medicines are, 100% natural, pure natural ingredients, organic, etc. However, the point of discussion remains whether all herbal medicines or phytotherapeutic drugs are actually free from adverse drug reactions (ADRs).

4.2   Efficacy

The quality of the raw material or plants in production of phytomedicines determines efficacy or effectiveness of drugs to a greater extent. The standard definition of quality given by Joseph M. Juran is "Fitness for purpose," whereas Philip B. Crosby has defined it as "Conformance to requirements"

(Malik & Choudhary, 2008, p. 6). The advertised effectiveness of herbal medicine to serve the intended purpose contributes to building perception of good quality in consumer's mind. Odiboh et al. (2017) while studying herbal marketing in Nigeria found that six out of ten focus groups featured instant efficacy of herbal medicines. Many studies have advocated that efficacy of phytomedicines is weighted as the highest influencer in buying herbal products (Adlakha & Sharma, 2019; Vos & Brennan, 2010; Wang, Keh, & Bolton, 2008). The effectiveness of the herbal drugs is greatly affected by the environmental conditions, harvesting time, purification and extraction methods applied, and storage conditions (Jun-ling et al., 2020; Meng, Xiaoliang, Xiumei, Vincieri, & Bilia, 2009; Wang et al., 2015). These factors play a vital role in regulating chemical constituents and pharmacological activities of the final extracts (Bao et al., 2017). The efficacy claims made by several herbal products are like 100% diabetes free, provides quick relief within 24 h by Proctopiles (Qatar.desertcart.com, 2020), Garcinia Cambogia 60% HCA weight loss by Bioveda Naturals, 100% pure plant herbal medicinal oil by Plant Gift (Aliexpress.com, 2020), FDA-approved powerful effective ingredients, enlargement of male genitals with 30-day course, and treatment of erectile dysfunction with single dose (Ahmad et al., 2015).

4.3 Treatment of incurable and NCDs by medication

Incurable is a category of diseases for which no drug may claim to cure or prevent or intend to convey to a potential consumer any treatment for any one or more ailments specified under the category. It includes diseases like diabetes, asthma, cancer, obesity, AIDS, atherosclerosis, cardiovascular disease, baldness, and Alzheimer's disease (Ahmad et al., 2015; Karimi, Majlesi, & Rafieian-Kopaei, 2015; Patwardhan, 2016). Frequently used adjectives for promoting scientifically unsubstantiated claims of phytomedicine product advertisers are "stop diabetes, reverse diabetes, 100% herbal treatment for obesity, tackle obesity, happy heart, all natural cardio cure, anticancer herbal medicine, destroy cancer cells, etc." These types of claims are popular in Asian region, particularly in India, made by manufacturers of Ayurvedic products (Patwardhan, 2013). The reason behind this is the large number of patients suffering from one or the other ailment and their high mortality rate (Patwardhan, 2016). Use of these slogans as unique selling propositions of difficult-to-treat ailment medicines is an attempt to capitalize on the helplessness and cold emotional state of patients. Helplessness is passiveness often experienced by patients suffering from diseases that cannot be controlled or cured (Konnikova, 2015) and is more often witnessed in old age people than young ones (Robnett & Chop, 2013). Furthermore, to fully leverage the benefit of the unsubstantiated claims, testimonial marketing is applied as a tool for connecting, interacting, and building trust among consumers (Gould, 2013).

5. USPs can be misleading

Advertising is supposed to be informative and reliable and not misleading. The frequent users of phytotherapeutic drugs and other herbal supplements are likely to accredit that claims made are true (Tayade & Kulkarni, 2011). The advertisements and the propositions generally made are persuasive than informational. The advertisers of some herbal medicine and supplements have tendency to make hyperbolic or superlative claims without scientific evidence (Ahmad et al., 2015; Tyler, 2000). The advertisement of product assertions is usually contentious due to difference in quality of information delivered to consumer and trueness of health benefits claimed (Crawford & Leventis, 2005). The reason behind this is herbal preparations barely meet the quality standards (Cardellina, 2002; Gunjan et al., 2015).

Herbal entities can please and displease, cure and harm, and give comfort and pain and can even cause death (Odiboh et al., 2017). The traditional medicines or phytotherapies are oldest unsubstantiated claimant of cure for all kinds of diseases. The common understanding that phytomedicines are free from side effects is misleading besides being untrue (Ekor, 2014). Plants contain hundreds of components such as ephedrine, phorbol esters, and pyrrolizidine alkaloids with toxic chemical properties (Calixto, 2000; Jordan & Haywood, 2007; Karimi et al., 2015). Herbal drugs are also capable of producing toxic effects due to incorrect dosage or overmedication for longer duration, contamination, adulteration, and reaction with conventional medicines (Drew & Myers, 1997; Gunjan et al., 2015; Kim et al., 2013). Sometimes mislabeling and misidentification of medicines derived from herbs also become potential causes or adverse drug interactions (Chee, Berlin, & Schatz, 2011; Radhika, 2017). Therefore continuous examination of phytomedicines for presence of possible adulterants becomes imperative (Ekor, 2014). Some examples of adverse drug reactions include hepatotoxicity and neurotoxicity caused by phytomedicines derived from plants like *Ephedra* (Schoepfer et al., 2007; Varlibas et al., 2009), *Tussilago farfara* potential cause of cirrhosis (Edgar, Colegate, Boppré, & Molyneux, 2011), gastrointestinal tract burning sensation by *Allium sativum* (garlic) extract (Rose, Croissant, Parliament, & Levin, 1990), *Ginseng* can cause high blood pressure and vaginal bleeding (Dunnick & Nyska, 2013) and overconsumption of *Piper methysticum* (Kava) that may result in diplopia and photophobia (Boullata & Nace, 2000). Several studies have also advocated the presence of heavy metals like mercury, arsenic, and lead in herbal extracts (Koch et al., 2011; Martena et al., 2010; Saper et al., 2008).

Furthermore, studies have found that there is significant increase in phytomedicine advertisements with embellished claims on efficacy and treatment of diseases classified as incurable by medication (Ahmad et al., 2015; Roush, 2016). For the former, scientists and medical experts have publicly denounced the availability of scientific justification for efficacy claims of herbal medicines (Hampton, 2005; Jordan & Haywood, 2007; Morris & Avorn, 2003), and for the latter, it is potentially not possible by medicines (Patwardhan, 2016). An apparent trend of adding analogue medicines and drugs to phytomedicines has been witnessed to improve their efficacy. Adulteration with drugs like silde-nafil, indomethacin, and lovastatin and other analogue medicines is trending globally (Dastjerdi, Akhgari, Kamali, & Mousavi, 2018; Patrick-Iwuanyanwu & Emerue, 2014; Rivera, Loya, & Ceballos, 2013). Consumption of adulterated products can cause acute, chronic, and sometimes disastrous health effects. The highest proportion of adulterated herbal products were reported to be sold in markets of Brazil with 68% of products being reported as adulterated followed by Taiwan and India with 32% and 31%, respectively, whereas the lowest per-centage was reported from China with only 19% of adulterated herbal product presence in market (Fig. 2).

However, treatments involving the use of phytotherapies have exhibited promising efficacy and are witnessed to be more advantageous than harmful (Ekor, 2014). Around one lakh deaths are reported every year in the United States due to synthetic drug toxicities, while as very few deaths or adverse drug reactions have been reported due to herbal medicines (Karimi et al., 2015). It indicates that adverse reactions of phytomedicines are less frequent than con-ventional drugs when used suitably.

Consumers believe that medicines promoted as "100% natural or organic" are safe inherently (Jordan & Haywood, 2007). Even though the Food and Drug Administration has strictly prohibited marketing of unsubstantiated claims, the regulations are dishonored repeatedly (Lewis & Strom, 2002). Several studies argue the potential failure of manufacturers and marketers to reveal information related to toxicities, contraindications, and adverse effects of phytomedicines on different media platforms (Ashar, Miller, Getz, & Pichard, 2003; Marcinkow, Parkhomchik, Schmode, & Yuksel, 2019; Schmidt, Sharma, Schifano, & Feinmann, 2011). This gross violation of law and information concealment ex-poses people to two kinds of risks. The first risk is to cause injury to user's health, and the second risk is to hurt consumers financially (Ahmad et al., 2015). Therefore there is a need of conducting more research into blatant delinquency of herbal abuse with a motive to provide suitable models of communication to ensure harmony between healthcare system and phytomedicine marketing.

6. Conclusion

Since ancient times, plants have been used as a potential source of medicine. The extracts obtained from herbs have played a significant role in healthcare. It has been realized that several modern medicines and tools of treatment are expensive and cumbersome and have severe side effects. There is a need for an alternative that is economical, effective, and safe. People in billions across the globe, from both developed and developing nations, depend on traditional system of health-care. The trend of phytomedicine consumption with a motive to treat several ailments and to boost immune system has been witnessing exponential rise. With this, issues relating to the misleading marketing practices and concerns about consumer's health are becoming more prevalent. There is an increased tendency of self-medication reinforced by the belief that products labeled as "natural" or extracted from natural source are nontoxic, harmless, and efficient. Efficacy and safety of herbal drugs besides being attributed as most influential predictors of consumer consumption decision are also perceived as discriminating factors of their growing preference over orthodox medicine. As a consequence, most of the selling propositions made by phytomedicine advertisers are found to be directed at communicating high safety- and efficacy-related attributes of drugs.

However, the marketing claims made by manufacturers with profiteering intentions has been tagged as unsubstantial and untrue by several studies. The unreliable assertions of magical health benefits backed by promoters' marketing validation through advertising and labeling are objectively made to drive sales while ignoring public health risks. Most of the propositions made as factors of competitive advantage are exaggerated claims of efficacy and unjustified generalizations without having any sufficient scientific proof behind them. This intensifies need for installation of stringent regulatory forces to curtail the spread of anecdotal information related to phytomedicines through misleading advertisement campaigns. Also, promotion of investment and commitment in research and development of phytopharmaceuticals for protection of public health become imperative.

References

Adlakha, K., & Sharma, S. (2019). Brand positioning using multidimensional scaling technique: An application to herbal healthcare brands in Indian market. *Vision, 24*(3), 345–355.

Ahmad, A., Patel, I., Parimalakrishnan, S., Mohanta, G. P., & Nagappa, A. N. (2015). Advertisement on medicines/treatment in newspapers violating Indian laws? *International Journal of Current Pharmaceutical, Review and Research, 6*(1), 49–58.

Aliexpress.com. (2020). Available from: https://www.aliexpress.com/item/32606056983.html. (Accessed 25 April 2020).

American Botanical Council. (2011). *Herb sales Up 3.3% in United States in 2010, American Botanical Council.* Available from: http://cms.herbalgram.org/press/2011/HerbMarketReport2010.html. (Accessed 22 April 2020).

ARKOPHARMA. (2019). *Ley dates and figures.* Available from: https://www.arkopharma.com/en-GB/key-dates-and-figures. (Accessed 2 May 2020).

Ashar, B. H., Miller, R. G., Getz, K. J., & Pichard, C. P. (2003). A critical evaluation of internet marketing of products that contain ephedra. *Mayo Clinic Proceedings-Elsevier, 78*(8), 944–946.

Bao, J., Ding, R. B., Liang, Y., Liu, F., Wang, K., Jia, X., et al. (2017). Differences in chemical component and anticancer activity of green and ripe forsythia-fructus. *The American Journal of Chinese Medicine, 45*(07), 1513–1536.

Basal, B. (2019). Historical transformation of unique selling proposition (USP) in advertising narration. In R. Yilmaz (Ed.), *Handbook of research on narrative advertising* (pp. 141–150). Pennsylvania: IGI Global.

Bhowmik, D., Kumar, K. P., Tripathi, P., & Chiranjib, B. (2009). Traditional herbal medicines: An overview. *Archives of Applied Science Research, 1*(2), 165–177.

BioSpectrum. (2015). *Industry records all time high mergers and acquisitions*. Available from: https://www.biospectrumasia.com/analysis/52/4411/industry-records-all-time-high-mergers-and-acquisitions.html. (Accessed 2 May 2020).

Blackmores. (2016). *Blackmores acquisition of global therapeutics*. Available from: https://www.blackmores.com.au/about-us/investor-centre/investor-webcasts/blackmores-acquisition-of-global-therapeutics. (Accessed 2 May 2020).

Blumenthal, M. (1999). Herb industry sees mergers, acquisitions, and entry by pharmaceutical giants in 1998. *Herbal Gram, 45*, 67–68.

Blythe, J. (2004). *Essentials of marketing* (3rd ed., p. 250). New York: Financial Times Prentice Hall, ISBN:9780273693581.

Boullata, J. I., & Nace, A. M. (2000). Safety issues with herbal medicine. *Pharmacotherapy: The Journal of Human Pharmacology and Drug Therapy, 20*(3), 257–269.

Business Insider. (2020). *Australian Beauty brand CANVAS set to strengthen presence in Singapore with New distributor*. Available from: https://www.businessinsider.sg/australian-beauty-brand-canvas-set-to-strengthen-presence-in-singapore-with-new-distributor. (Accessed 2 May 2020).

Business World. (2017). *Dabur India teams up with Amazon to launch Ayurveda E-marketplace*. Available from: http://www.businessworld.in/article/Dabur-India-Teams-Up-With-Amazon-To-Launch-Ayurveda-E-Marketplace/22-09-2017-126808. (Accessed 2 May 2020).

Calixto, J. B. (2000). Efficacy, safety, quality control, marketing and regulatory guidelines for herbal medicines (phytotherapeutic agents). *Brazilian Journal of Medical and Biological Research, 33*(2), 179–189.

Cardellina, J. H. (2002). Challenges and opportunities confronting the botanical dietary supplement industry. *Journal of Natural Products, 65*(7), 1073–1084.

Chee, B. W., Berlin, R., & Schatz, B. (2011). Predicting adverse drug events from personal health messages. *American Medical Informatics Association Annual Symposium Proceedings*, 217.

Chen, Y. S. (2010). The drivers of green brand equity: Green brand image, green satisfaction, and green trust. *Journal of Business Ethics, 93*(2), 307–319.

Clark, D. (2011). Reinventing your personal brand. *Harvard Business Review*. Available from: https://hbr.org/2011/03/reinventing-your-personal-brand. (Accessed 23 April 2020).

Crawford, S. Y., & Leventis, C. (2005). Herbal product claims: Boundaries of marketing and science. *Journal of Consumer Marketing, 22*(7), 432–436.

Cruikshank, J. L., & Schultz, A. W. (2010). *The man who sold America: The amazing (but true!) story of Albert D. Lasker and the creation of the advertising century*. Harvard Business Press.

Dastjerdi, A. G., Akhgari, M., Kamali, A., & Mousavi, Z. (2018). Principal component analysis of synthetic adulterants in herbal supplements advertised as weight loss drugs. *Complementary Therapies in Clinical Practice, 31*, 236–241.

Dettweiler, M., Lyles, J. T., Nelson, K., Dale, B., Reddinger, R. M., Zurawski, D. V., et al. (2019). American civil war plant medicines inhibit growth, biofilm formation, and quorum sensing by multidrug-resistant bacteria. *Scientific Reports, 9*(1), 1–12.

Drew, A., & Myers, S. P. (1997). Safety issues in herbal medicine: Implications for the health professions. *Medical Journal of Australia, 166*, 538–541.

Dunnick, J. K., & Nyska, A. (2013). The toxicity and pathology of selected dietary herbal medicines. *Toxicologic Pathology, 41*(2), 374–386.

Edgar, J. A., Colegate, S. M., Boppré, M., & Molyneux, R. J. (2011). Pyrrolizidine alkaloids in food: A spectrum of potential health consequences. *Food Additives & Contaminants: Part A, 28*(3), 308–324.

Ekor, M. (2014). The growing use of herbal medicines: Issues relating to adverse reactions and challenges in monitoring safety. *Frontiers in Pharmacology: Ethno Pharmacology, 4*(177), 1–10.

Givaudan. (2019a). *Givaudan completes acquisition of vietnamese flavour company Golden frog*. Available from: https://www.givaudan.com/media/media-releases/2019/acquisition-golden-frog-completed. (Accessed 2 May 2020).

Gould, L. (2013). *The power of testimonial marketing*. Available from: https://www.expertvoice.com/the-power-of-testimonial-marketing/. (Accessed 26 April 2020).

Govt. of India., Medicinal Plants Introduction, Indian System of Medicine and Homoeopathy (ISMH). (2000). *Department of ISMH, ministry of health and family welfare*. Available from: http://indianmedicine.nic.in/html/plants/mimain.htm. (Accessed 20 April 2020).

Gunjan, M., Naing, T. W., Saini, R. S., Ahmad, A., Naidu, J. R., & Kumar, I. (2015). Marketing trends & future prospects of herbal medicine in the treatment of various disease. *World Journal of Pharmaceutical Research, 4*(9), 132–155.

Guthrie, M., & Kim, H. (2009). The relationship between consumer involvement and brand perceptions of female cosmetic consumers. *Journal of Brand Management, 17*, 114–133.

Hampton, T. (2005). More scrutiny for dietary supplements? *Journal of the American Medical Association, 293*, 27–28.

Jordan, M. A., & Haywood, T. (2007). Evaluation of internet websites marketing herbal weight-loss supplements to consumers. *The Journal of Alternative and Complementary Medicine, 13*(9), 1035–1043.

Jun-ling, R., Ai-Hua, Z., Ling, K., Ying, H., Guang-Li, Y., Hui, S., et al. (2020). Analytical strategies for the discovery and validation of quality-markers of traditional Chinese medicine. *Phytomedicine, 67*.

Karimi, A., Majlesi, M., & Rafieian-Kopaei, M. (2015). Herbal versus synthetic drugs; beliefs and facts. *Journal of Nephropharmacology, 4*(1), 27–30.

Kim, E. J., Chen, Y., Huang, J. Q., Li, K. M., Razmovski-Naumovski, V., Poon, J., et al. (2013). Evidence-based toxicity evaluation and scheduling of Chinese herbal medicines. *Journal of Ethnopharmacology, 146*(1), 40–61.

Koch, I., Moriarty, M., House, K., Sui, J., Cullen, W. R., Saper, R. B., et al. (2011). Bio accessibility of lead and arsenic in traditional Indian medicines. *Science of the Total Environment, 409*(21), 4545–4552.

Konnikova, M. (2015). Trying to cure depression, but inspiring torture. In *The New Yorker*. Available from: https://www.newyorker.com/science/maria-konnikova/theory-psychology-justified-torture. (Accessed 26 April 2020).

Lange, D. (2006). In R. J. Bogers, L. E. Craker, & D. Lange (Eds.), *International trade in medicinal and aromatic plants: actors, volumes and commodities* (pp. 155–170). Netherlands: Frontis-Springer.

Laskey, H. A., Day, E., & Crask, M. R. (1989). Typology of main message strategies for television commercials. *Journal of Advertising, 18*(1), 36–41.

Lewis, J. D., & Strom, B. L. (2002). Balancing safety of dietary supplements with the free market. *Annals of Internal Medicine, 136*, 616–618.

Malik, K., & Choudhary, P. (2008). *Software quality: A practitioner's approach* (1st ed.). New Delhi: Tata McGraw-Hill Education.

Marcinkow, A., Parkhomchik, P., Schmode, A., & Yuksel, N. (2019). The quality of information on combined oral contraceptives available on the internet. *Journal of Obstetrics and Gynaecology Canada, 41*(11), 1599–1607.

Martena, M. J., Van Der Wielen, J. C., Rietjens, I. M., Klerx, W. N., De Groot, H. N., & Konings, E. J. (2010). Monitoring of mercury, arsenic, and lead in traditional Asian herbal preparations on the Dutch market and estimation of associated risks. *Food Additives and Contaminants, 27*(2), 190–205.

Martin Bauer Group. (2019). *Martin Bauer Group announces the acquisition of BI nutraceuticals.* Available from: https://www.martin-bauer-group.com/en/news/martin-bauer-group-announces-the-acquisition-of-bi-nutraceuticals/. (Accessed 2 May 2020).

Meng, W., Xiaoliang, R., Xiumei, G., Vincieri, F. F., & Bilia, A. R. (2009). Stability of active ingredients of traditional Chinese medicine (TCM). *Natural Product Communications*, *4*(12), 1761–1776.

Ministry of Health and Family Welfare. (2015). *Gazette Notification G.S.R. 918 (E).* Available from: http://www.cdsco.nic.in/writereaddata/GSR%20918. (Accessed 22 April 2020).

Morris, C. A., & Avorn, J. (2003). Internet marketing of herbal products. *Journal of the American Medical Association*, *290*, 1505–1509.

Naik, P. A. (2007). Integrated marketing communications: Provenance, practice and principles. In *The SAGE handbook of advertising* (pp. 34–53).

National Health Portal. (2016). *Introduction and importance of medicinal plants and herbs.* Available from: https://www.nhp.gov.in/introduction-and-importance-of-medicinal-plants-and-herbs_mtl. (Accessed 22 April 2020).

Neupane, P., & Lamichhane, J. (2020). Overview of Himalayan medicinal plants and phytomedicine. *International Journal of Innovative Science and Research Technology*, *5*(2), 857–866.

Nirmal, S. A., Pal, S. C., Otimenyin, S. O., Aye, T., Elachouri, M., Kundu, S. K., et al. (2013). Contribution of herbal products in global market. *The Pharma Review*, 95–104.

Odiboh, O., Omojola, O., Okorie, N., & Ekanem, T. (2017). Sobotone, ponkiriyon, herbal marketing communication and Nigeria's healthcare system. In *Proceedings of SOCIOINT 2017- 4th international conference on education, social sciences and humanities, 10–12 July 2017- Dubai* (pp. 1395–1401).

Pan, S. Y., Zhou, S. F., Gao, S. H., Yu, Z. L., Zhang, S. F., Tang, M. K., et al. (2013). New perspectives on how to discover drugs from herbal medicines: CAM's outstanding contribution to modern therapeutics. *Evidence-based Complementary and Alternative Medicine*, 1–25.

Pandey, M., Debnath, M., Gupta, S., & Chikara, S. K. (2011). Phytomedicine: An ancient approach turning into future potential source of therapeutics. *Journal of Pharmacognosy and Phytotherapy*, *3*(1), 113–117.

Patrick-Iwuanyanwu, K. C., & Emerue, J. A. (2014). Evaluation of acute and sub-chronic toxicities of ulcer fast®: A bi-herbal formula in male Wistar albino rats. *International Journal of Basic & Clinical Pharmacology*, *3*(6), 970–977.

Patwardhan, B. (2013). Time for evidence-based ayurveda: A clarion call for action. *Journal of Ayurveda and integrative medicine*, *4*(2), 63–66.

Patwardhan, B. (2016). Ayurvedic drugs in case: Claims, evidence, regulations and ethics. *Journal of Ayurveda and Integrative Medicine*, *7*, 135–137.

Qatar.desertcart.com, Vrinda ayurvedic piles capsules—Fast relieve In Bleeding, burning & pain with proctopiles cream, Available from: https://qatar.desertcart.com/products/178573816-vrinda-ayurvedic-piles-capsules-fast-relieve-in-bleeding-burning-pain-with-proctopiles-cream-absolutely-free-pack-of-2 (Accessed 25 April 2020), 2020.

Radhika, S. (2017). Herbs-are they safe? *Indian Journal of Cardiovascular Disease in Women WINCARS*, *2*(2), 3–4.

Raghavendra, H. L., Yogesh, H. S., Gopalakrishna, B., Chandrashekhar, V. M., Kumar, S., & Kumar, V. (2009). An overview of herbal medicine. *International Journal of Pharmaceutical Sciences*, *1*(1), 1–20.

Reeves, R. (1961). *Reality in advertising*. New York, NY: Alfred Knopf.

Rivera, J. O., Loya, A. M., & Ceballos, R. (2013). Use of herbal medicines and implications for conventional drug therapy medical sciences. *Alternative and Integrative Medicine*, *2*(6), 1–6.

Roberts, J. E., & Tyler, V. E. (1998). *Tyler's herbs of choice. The therapeutic use of phytomedicinals.* New York: The Haworth Press, Inc.

Robnett, R. H., & Chop, W. C. (2013). *Gerontology for the health care professional.* Jones & Bartlett Publishers.

Rose, K. D., Croissant, P. D., Parliament, C. F., & Levin, M. B. (1990). Spontaneous spinal epidural hematoma with associated platelet dysfunction from excessive garlic ingestion: A case report. *Neurosurgery, 26*(5), 880–882.

Roush, R. A. (2016). *Complementary and alternative medicine.* New York: Routledge: Taylor & Francis Group.

Rowey, J., & Spiezia, M. (2006). Spiezia organics an SME marketing. *The Marketing Review, 6,* 253–264.

Sachitanand, R. (2017). Why Indian companies like HUL, Patanjali, Dabur are taking crack at the market for Ayurvedic and herbal products. *The Economic Times.* Available from: http://economictimes.indiatimes.com. (Accessed 22 April 2020).

Saper, R. B., Phillips, R. S., Sehgal, A., Khouri, N., Davis, R. B., Paquin, J., et al. (2008). Lead, mercury, and arsenic in US-and Indian-manufactured Ayurvedic medicines sold via the internet. *Journal of the American Medical Association, 300*(8), 915–923.

Saslis-Lagoudakis, C. H., Bruun-Lund, S., Iwanycki, N. E., Seberg, O., Petersen, G., Jäger, A. K., et al. (2015). Identification of common horsetail (*Equisetum arvense* L.; Equisetaceae) using thin layer chromatography versus DNA barcoding. *Scientific Reports, 5,* 11942.

Schiffman, L. G., & Kanuk, L. L. (2004). *Consumer behaviour* (7th ed.). London: Prentice Hall.

Schmidt, M. M., Sharma, A., Schifano, F., & Feinmann, C. (2011). "Legal highs" on the net— Evaluation of UK-based websites, products and product information. *Forensic Science International, 206*(1–3), 92–97.

Schoepfer, A. M., Engel, A., Fattinger, K., Marbet, U. A., Criblez, D., Reichen, J., et al. (2007). Herbal does not mean innocuous: Ten cases of severe hepatotoxicity associated with dietary supplements from Herbalife® products. *Journal of Hepatology, 47*(4), 521–526.

Seethapathy, G. S., Tadesse, M., Urumarudappa, S. K. J., Gunaga, S. V., Vasudeva, R., Malterud, K. E., et al. (2018). Authentication of Garcinia fruits and food supplements using DNA barcoding and NMR spectroscopy. *Scientific Reports, 8*(1), 10561.

Sen, S., & Chakraborty, R. (2015). Towards the integration and advancement of herbal medicine: a focus on traditional Indian medicine. *Botanics: Targets and Therapy, 5,* 33–44.

Singh, A., Kalaivani, M., Chaudhary, P., Srivastava, S., Goyal, R. K., & Gupta, S. K. (2019). Opportunities and challenges in development of phytopharmaceutical drug in India-a SWOT analysis. *Journal of Young Pharmacists, 11*(3), 322–327.

Srivastava, A., Srivastava, P., Pandey, A., Khanna, V. K., & Pant, A. B. (2019). *Phytomedicine: A potential alternative medicine in controlling neurological disorders* (pp. 625–655). New Look to Phytomedicine-Academic Press.

Stephen, M. (2001). Hong Kong pivotal for Chinese medicines in Japan. *International Market News.* 21 June, Available from: http://www.tdctrade.com/imn/imn190/medicines01.htm. (Accessed 20 April 2020).

Suganya, R., & Hamsalakshmi, R. (2017). A study on customer buying behaviour of selected ayurvedic healthcare products. *International Journal of Advanced Research and Development, 2*(2), 13–18.

Tayade, M., & Kulkarni, N. B. (2011). Accuracy of the drug advertisements in medical journals in India. *Journal of Clinical and Diagnostic Research, 5*(3), 583–585.

Thanisorn, R., & Bunchapattanasakda. (2011). Marketing strategies of imported herbal cosmetic products in Thailand. *Information Management and Business Review, 3*(4), 217–221.

The Vegan Business Magazine. (2019). *Plant extracts market: Global scenario & market highlights.* Available from: https://vegconomist.com/studies-and-numbers/plant-extracts-market-global-scenario-market-highlights/. (Accessed 18 April 2020).

Tyler, V. E. (2000). Herbal medicine: From the past to the future. *Public Health Nutrition, 3*(4a), 447–452.

Varlibas, F., Delipoyraz, I., Yuksel, G., Filiz, G., Tireli, H., & Gecim, N. O. (2009). Neurotoxicity following chronic intravenous use of "Russian cocktail". *Clinical Toxicology, 47*(2), 157–160.

Vasisht, K., & Kumar, V. (2002). Trade and production of herbal medicines and natural health products. *ICS-UNIDO-Trieste, 3.*

Vos, L., & Brennan, R. (2010). Complementary and alternative medicine: Shaping a marketing research agenda. *Marketing Intelligence & Planning, 28*(3), 349–364.

Wang, W., Keh, H. T., & Bolton, L. E. (2008). Consumer perceptions of traditional Chinese versus Western medicine in China. *Advances in Consumer Research, 35*, 39–43.

Wang, Z. Y., Wang, H. L., Zhou, J., Ma, H. Y., Gong, Y., Yan, W. L., et al. (2015). Comparison of chemical composition between fresh and processed Bufonis Venenum by UPLC-TQ-MS. *China Journal of Chinese Materiamedica, 40*(20), 3967–3973.

WHO. (2002). *Traditional medicine strategy (2002–2005).* WHO/EDM/TRM/2002.1 Geneva, Switzerland: World Health Organization.

WHO. (2004). *WHO guidelines on safety monitoring of herbal medicines in pharmacovigilance systems.* Geneva, Switzerland: World Health Organization.

WHO. (2012). *Regional progress in traditional medicine 2011–2010.* Traditional Medicine. Available from: http://www.wpro.who.int. (Accessed 28 April 2020).

Index

Note: Page numbers followed by *f* indicate figures and *t* indicate tables.